Undergraduate Texts in Mathematics

Undergraduate Texts in Mathematics

Undergraduate Texts in Mathematics are generally aimed at third- and fourth-year undergraduate mathematics students at North American universities. These texts strive to provide students and teachers with new perspectives and novel approaches. The books include motivation that guides the reader to an appreciation of interrelations among different aspects of the subject. They feature examples that illustrate key concepts as well as exercises that strengthen understanding.

For further volumes:
http://www.springer.com/series/666

William A. Adkins • Mark G. Davidson

Ordinary Differential Equations

 Springer

William A. Adkins
Department of Mathematics
Louisiana State University
Baton Rouge, LA
USA

Mark G. Davidson
Department of Mathematics
Louisiana State University
Baton Rouge, LA
USA

ISSN 0172-6056
ISBN 978-1-4899-8767-9   ISBN 978-1-4614-3618-8 (eBook)
DOI 10.1007/978-1-4614-3618-8
Springer New York Heidelberg Dordrecht London

Mathematics Subject Classification (2010): 34-01
© Springer Science+Business Media New York 2012
Softcover re-print of the Hardcover 1st edition 2012

Printed on acid-free paper

Springer is part of Springer Science+Business Media (www.springer.com)

Preface

This text is intended for the introductory three- or four-hour one-semester sopho-more level differential equations course traditionally taken by students majoring in science or engineering. The prerequisite is the standard course in elementary calculus.

Engineering students frequently take a course on and use the Laplace transform as an essential tool in their studies. In most differential equations texts, the Laplace transform is presented, usually toward the end of the text, as an alternative method for the solution of constant coefficient linear differential equations, with particular emphasis on discontinuous or impulsive forcing functions. Because of its placement at the end of the course, this important concept is not as fully assimilated as one might hope for continued applications in the engineering curriculum. *Thus, a goal of the present text is to present the Laplace transform early in the text, and use it as a tool for motivating and developing much of the remaining differential equation concepts for which it is particularly well suited.*

There are several rewards for investing in an early development of the Laplace transform. The standard solution methods for constant coefficient linear differential equations are immediate and simplified. We are able to provide a proof of the existence and uniqueness theorems which are not usually given in introductory texts. The solution method for constant coefficient linear systems is streamlined, and we avoid having to introduce the notion of a defective or nondefective matrix or develop generalized eigenvectors. Even the Cayley–Hamilton theorem, used in Sect. 9.6, is a simple consequence of the Laplace transform. In short, the Laplace transform is an effective tool with surprisingly diverse applications.

Mathematicians are well aware of the importance of transform methods to simplify mathematical problems. For example, the Fourier transform is extremely important and has extensive use in more advanced mathematics courses. The wavelet transform has received much attention from both engineers and mathe-maticians recently. It has been applied to problems in signal analysis, storage and transmission of data, and data compression. We believe that students should be introduced to transform methods early on in their studies and to that end, the Laplace transform is particularly well suited for a sophomore level course in differential equations. It has been our experience that by introducing the Laplace transform near the beginning of the text, students become proficient in its use and comfortable with this important concept, while at the same time learning the standard topics in differential equations.

Chapter 1 is a conventional introductory chapter that includes solution techniques for the most commonly used first order differential equations, namely, separable and linear equations, and some substitutions that reduce other equations to one of these. There are also the Picard approximation algorithm and a description, without proof, of an existence and uniqueness theorem for first order equations.

Chapter 2 starts immediately with the introduction of the Laplace transform as an integral operator that turns a differential equation in t into an algebraic equation in another variable s. A few basic calculations then allow one to start solving some differential equations of order greater than one. The rest of this chapter develops the necessary theory to be able to efficiently use the Laplace transform. Some proofs, such as the injectivity of the Laplace transform, are delegated to an appendix. Sections 2.6 and 2.7 introduce the basic function spaces that are used to describe the solution spaces of constant coefficient linear homogeneous differential equations.

With the Laplace transform in hand, Chap. 3 efficiently develops the basic theory for constant coefficient linear differential equations of order 2. For example, the homogeneous equation $q(\boldsymbol{D})y = 0$ has the solution space $\mathcal{E}_q$ that has already been described in Sect. 2.6. The Laplace transform immediately gives a very easy procedure for finding the test function when teaching the method of undetermined coefficients. Thus, it is unnecessary to develop a rule-based procedure or the annihilator method that is common in many texts.

Chapter 4 extends the basic theory developed in Chap. 3 to higher order equations. All of the basic concepts and procedures naturally extend. If desired, one can simultaneously introduce the higher order equations as Chap. 3 is developed or very briefly mention the differences following Chap. 3.

Chapter 5 introduces some of the theory for second order linear differential equations that are not constant coefficient. Reduction of order and variation of parameters are topics that are included here, while Sect. 5.4 uses the Laplace transform to transform certain second order nonconstant coefficient linear differential equations into first order linear differential equations that can then be solved by the techniques described in Chap. 1.

We have broken up the main theory of the Laplace transform into two parts for simplicity. Thus, the material in Chap. 2 only uses continuous input functions, while in Chap. 6 we return to develop the theory of the Laplace transform for discontinuous functions, most notably, the step functions and functions with jump discontinuities that can be expressed in terms of step functions in a natural way. The Dirac delta function and differential equations that use the delta function are also developed here. The Laplace transform works very well as a tool for solving such differential equations. Sections 6.6–6.8 are a rather extensive treatment of periodic functions, their Laplace transform theory, and constant coefficient linear differential equations with periodic input function. These sections make for a good supplemental project for a motivated student.

Chapter 7 is an introduction to power series methods for linear differential equations. As a nice application of the Frobenius method, explicit Laplace inversion formulas involving rational functions with denominators that are powers of an irreducible quadratic are derived.

Chapter 8 is primarily included for completeness. It is a standard introduction to some matrix algebra that is needed for systems of linear differential equations. For those who have already had exposure to this basic algebra, it can be safely skipped or given as supplemental reading.

Chapter 9 is concerned with solving systems of linear differential equations. By the use of the Laplace transform, it is possible to give an explicit formula for the matrix exponential $e^{At} = \mathcal{L}^{-1}\{(sI - A)^{-1}\}$ that does not involve the use of eigenvectors or generalized eigenvectors. Moreover, we are then able to develop an efficient method for computing e^{At} known as Fulmer's method. Another thing which is somewhat unique is that we use the matrix exponential in order to solve a constant coefficient system $y' = Ay + f(t)$, $y(t_0) = y_0$ by means of an integrating factor. An immediate consequence of this is the existence and uniqueness theorem for higher order constant coefficient linear differential equations, a fact that is not commonly proved in texts at this level.

The text has numerous exercises, with answers to most odd-numbered exercises in the appendix. Additionally, a student solutions manual is available with solutions to most odd-numbered problems, and an instructors solution manual includes solutions to most exercises.

Chapter Dependence

The following diagram illustrates interdependence among the chapters.

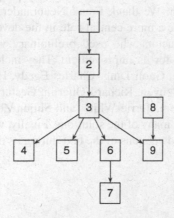

Suggested Syllabi

The following table suggests two possible syllabi for one semester courses.

3-Hour Course	4-Hour Course	Further Reading
Sections 1.1–1.6	Sections 1.1–1.7	
Sections 2.1–2.8	Sections 2.1–2.8	
Sections 3.1–3.6	Sections 3.1–3.7	
Sections 4.1–4.3	Sections 4.1–4.4	Section 4.5
Sections 5.1–5.3, 5.6	Sections 5.1–5.6	
Sections 6.1–6.5	Sections 6.1–6.5	Sections 6.6–6.8
	Sections 7.1–7.3	Section 7.4
Sections 9.1–9.5	Sections 9.1–9.5, 9.7	Section 9.6
		Sections A.1, A.5

Chapter 8 is on matrix operations. It is not included in the syllabi given above since some of this material is sometimes covered by courses that precede differential equations. Instructors should decide what material needs to be covered for their students. The sections in the Further Reading column are written at a more advanced level. They may be used to challenge exceptional students.

We routinely provide a basic table of Laplace transforms, such as Tables 2.6 and 2.7, for use by students during exams.

Acknowledgments

We would like to express our gratitude to the many people who have helped to bring this text to its finish. We thank Frank Neubrander who suggested making the Laplace transform have a more central role in the development of the subject. We thank the many instructors who used preliminary versions of the text and gave valuable suggestions for its improvement. They include Yuri Antipov, Scott Baldridge, Blaise Bourdin, Guoli Ding, Charles Egedy, Hui Kuo, Robert Lipton, Michael Malisoff, Phuc Nguyen, Richard Oberlin, Gestur Olafsson, Boris Rubin, Li-Yeng Sung, Michael Tom, Terrie White, and Shijun Zheng. We thank Thomas Davidson for proofreading many of the solutions. Finally, we thank the many many students who patiently used versions of the text during its development.

Baton Rouge, Louisiana   William A. Adkins
                                                                               Mark G. Davidson

Contents

List of Tables

Chapter 1
First Order Differential Equations

1.1   An Introduction to Differential Equations

Many problems of science and engineering require the description of some measurable quantity (position, temperature, population, concentration, electric current, etc.) as a function of time. Frequently, the scientific laws governing such quantities are best expressed as equations that involve the rate at which that quantity changes over time. Such laws give rise to differential equations. Consider the following three examples:

Example 1 (Newton's Law of Heating and Cooling). Suppose we are interested in the temperature of an object (e.g., a cup of hot coffee) that sits in an environment (e.g., a room) or space (called, ambient space) that is maintained at a constant temperature T_a. *Newton's law of heating and cooling* states that the *rate* at which the temperature $T(t)$ of the object changes is *proportional* to the *temperature difference between* the object and ambient space. Since rate of change of $T(t)$ is expressed mathematically as the derivative, $T'(t)$,[1] Newton's law of heating and cooling is formulated as the mathematical expression

$$T'(t) = r(T(t) - T_a),$$

where r is the constant of proportionality. Notice that this is an equation that relates the first derivative $T'(t)$ and the function $T(t)$ itself. It is an example of a differential equation. We will study this example in detail in Sect. 1.3.

Example 2 (Radioactive decay). Radioactivity results from the instability of the nucleus of certain atoms from which various particles are emitted. The atoms then

---

[1] In this text, we will generally use the prime notation, that is, y', y'', y''' (and $y^{(n)}$ for derivatives of order greater than 3) to denote derivatives, but the Leibnitz notation $\frac{dy}{dt}$, $\frac{d^2y}{dt^2}$, etc. will also be used when convenient.

W.A. Adkins and M.G. Davidson, *Ordinary Differential Equations*,
Undergraduate Texts in Mathematics, DOI 10.1007/978-1-4614-3618-8_1,
© Springer Science+Business Media New York 2012

decay into other isotopes or even other atoms. *The **law of radioactive decay** states that the **rate** at which the radioactive atoms disintegrate is **proportional** to the total number of radioactive atoms present.* If $N(t)$ represents the number of radioactive atoms at time t, then the rate of change of $N(t)$ is expressed as the derivative $N'(t)$. Thus, the law of radioactive decay is expressed as the equation

$$N'(t) = -\lambda N(t).$$

As in the previous example, this is an equation that relates the first derivative $N'(t)$ and the function $N(t)$ itself, and hence is a differential equation. We will consider it further in Sect. 1.3.

As a third example, consider the following:

Example 3 (Newton's Laws of Motion). Suppose $s(t)$ is a position function of some body with mass m as measured from some fixed origin. We assume that as time passes, forces are applied to the body so that it moves along some line. Its velocity is given by the first derivative, $s'(t)$, and its acceleration is given by the second derivative, $s''(t)$. *Newton's **second law of motion** states that the **net force** acting on the body is the **product of its mass and acceleration**.* Thus,

$$ms''(t) = F_{\text{net}}(t).$$

Now in many circumstances, the net force acting on the body depends on time, the object's position, and its velocity. Thus, $F_{\text{net}}(t) = F(t, s(t), s'(t))$, and this leads to the equation

$$ms''(t) = F(t, s(t), s'(t)).$$

A precise formula for F depends on the circumstances of the given problem. For example, the motion of a body in a spring-body-dashpot system is given by $ms''(t) + \mu s'(t) + k s(t) = f(t)$, where μ and k are constants related to the spring and dashpot and $f(t)$ is some applied external (possibly) time-dependent force. We will study this example in Sect. 3.6. For now though, we just note that this equation relates the second derivative to the function, its derivative, and time. It too is an example of a differential equation.

Each of these examples illustrates two important points:

- Scientific laws regarding physical quantities are frequently expressed and best understood in terms of how that quantity changes.
- The mathematical model that expresses those changes gives rise to equations that involve derivatives of the quantity, that is, differential equations.

We now give a more formal definition of the types of equations we will be studying. An ***ordinary differential equation*** is an equation relating an unknown function $y(t)$, some of the derivatives of $y(t)$, and the variable t, which in many applied problems will represent time. The domain of the unknown function is some interval of the real line, which we will frequently denote by the symbol I.[2] The **order** of a differential equation is the order of the highest derivative that appears in the differential equation. Thus, the order of the differential equations given in the above examples is summarized in the following table:

Differential equation	Order
$T'(t) = r(T(t) - T_a)$	1
$N'(t) = -\lambda N(t)$	1
$ms''(t) = F(t, s(t), s'(t))$	2

Note that $y(t)$ is our generic name for an unknown function, but in concrete cases, the unknown function may have a different name, such as $T(t)$, $N(t)$, or $s(t)$ in the examples above. The **standard form** for an ordinary differential equation is obtained by solving for the highest order derivative as a function of the unknown function $y = y(t)$, its lower order derivatives, and the independent variable t. Thus, a first order ordinary differential equation is expressed in standard form as

$$y'(t) = F(t, y(t)), \tag{1}$$

a second order ordinary differential equation in standard form is written

$$y''(t) = F(t, y(t), y'(t)), \tag{2}$$

and an nth order differential equation is expressed in standard form as

$$y^{(n)}(t) = F(t, y(t), \ldots, y^{(n-1)}(t)). \tag{3}$$

The standard form is simply a convenient way to be able to talk about various hypotheses to put on an equation to insure a particular conclusion, such as *existence and uniqueness of solutions* (discussed in Sect. 1.7) and to classify various types of equations (as we do in this chapter, for example) so that you will know which algorithm to apply to arrive at a solution. In the examples given above, the equations

$$T'(t) = r(T(t) - T_a),$$
$$N'(t) = -\lambda N(t)$$

are in standard form while the equation in Example 3 is not. However, simply dividing by m gives

$$s''(t) = \frac{1}{m} F(t, s(t), s'(t)),$$

a second order differential equation in standard form.

---

[2] Recall that the standard notations from calculus used to describe an interval I are (a, b), $[a, b)$, $(a, b]$, and $[a, b]$ where $a < b$ are real numbers. There are also the infinite length intervals $(-\infty, a)$ and (a, ∞) where a is a real number or $\pm\infty$.

In differential equations involving the unknown function $y(t)$, the variable t is frequently referred to as the *independent variable*, while y is referred to as the *dependent variable*, indicating that y has a functional dependence on t. In writing ordinary differential equations, it is conventional to suppress the implicit functional evaluations $y(t)$, $y'(t)$, etc. and write y, y', etc. Thus the differential equations in our examples above would be written

$$T' = r(T - T_a),$$

$$N' = -\lambda N,$$

$$\text{and} \quad s'' = \frac{1}{m} F(t, s, s'),$$

where the dependent variables are respectively, T, N, and s.

Sometimes we must deal with functions $u = u(t_1, t_2, \ldots, t_n)$ of two or more variables. In this case, a *partial differential equation* is an equation relating u, some of the partial derivatives of u with respect to the variables $t_1, \ldots, t_n$, and possibly the variables themselves. While there may be a time or two where we need to consider a partial differential equation, the focus of this text is on the study of ordinary differential equations. Thus, when we use the term differential equation without a qualifying adjective, you should assume that we mean *ordinary* differential equation.

Example 4. Consider the following differential equations. Determine their order, whether ordinary or partial, and the standard form where appropriate:

1. $y' = 2y$
2. $y' - y = t$
3. $y'' + \sin y = 0$
4. $y^{(4)} - y'' = y$
5. $ay'' + by' + cy = A \cos \omega t \quad (a \neq 0)$
6. $\dfrac{\partial^2 u}{\partial x^2} + \dfrac{\partial^2 u}{\partial y^2} = 0$

▶ **Solution.** Equations (1)–(5) are ordinary differential equations while (6) is a partial differential equation. Equations (1) and (2) are first order, (3) and (5) are second order, and (4) is fourth order. Equation (1) is in standard form. The standard forms for (2)–(5) are as follows:

2. $y' = y + t$
3. $y'' = -\sin y$
4. $y^{(4)} = y'' + y$
5. $y'' = -\dfrac{b}{a} y' - \dfrac{c}{a} y + \dfrac{A}{a} \cos \omega t$ ◀

Solutions

In contrast to algebraic equations, where the given and unknown objects are numbers, differential equations belong to the much wider class of *functional equations* in which the given and unknown objects are functions (scalar functions or vector functions) defined on some interval. A *solution of an ordinary differential equation* is a function $y(t)$ defined on some specific interval $I \subseteq \mathbb{R}$ such that substituting $y(t)$ for y and substituting $y'(t)$ for y', $y''(t)$ for y'', etc. in the equation gives a *functional identity*. That is, an identity which is satisfied for *all* $t \in I$. For example, if a first order differential equation is given in standard form as $y' = F(t, y)$, then a function $y(t)$ defined on an interval I is a solution if

$$y'(t) = F(t, y(t)) \quad \text{for all } t \in I.$$

More generally, $y(t)$, defined on an interval I, is a solution of an nth order differential equation expressed in standard form by $y^{(n)} = F(t, y, y', \ldots, y^{(n-1)})$ provided

$$y^{(n)}(t) = F(t, y(t), \ldots, y^{(n-1)}(t)) \quad \text{for all } t \in I.$$

It should be noted that it is not necessary to express the given differential equation in standard form in order to check that a function is a solution. Simply substitute $y(t)$ and the derivatives of $y(t)$ into the differential equation as it is given. The *general solution* of a differential equation is the set of all solutions. As the following examples will show, writing down the general solution to a differential equation can range from easy to difficult.

Example 5. Consider the differential equation

$$y' = y - t. \tag{4}$$

Determine which of the following functions defined on the interval $(-\infty, \infty)$ are solutions:

1. $y_1(t) = t + 1$
2. $y_2(t) = e^t$
3. $y_3(t) = t + 1 - 7e^t$
4. $y_4(t) = t + 1 + ce^t$ where c is an arbitrary scalar.

▶ **Solution.** In each case, we calculate the derivative and substitute the results in (4). The following table summarizes the needed calculations:

Function	$y'(t)$	$y(t) - t$
$y_1(t) = t + 1$	$y_1'(t) = 1$	$y_1(t) - t = t + 1 - t = 1$
$y_2(t) = e^t$	$y_2'(t) = e^t$	$y_2(t) - t = e^t - t$
$y_3(t) = t + 1 - 7e^t$	$y_3'(t) = 1 - 7e^t$	$y_3(t) - t = t + 1 - 7e^t - t = 1 - 7e^t$
$y_4(t) = t + 1 + ce^t$	$y_4'(t) = 1 + ce^t$	$y_4(t) - t = t + 1 + ce^t - t = 1 + ce^t$

For $y_i(t)$ to be a solution of (4), the second and third entries in the row for $y_i(t)$ must be the same. Thus, $y_1(t)$, $y_3(t)$, and $y_4(t)$ are solutions while $y_2(t)$ is not a

Fig. 1.1 The solutions
$y_g(t) = t + 1 + ce^t$ of
$y' = y - t$ for various c

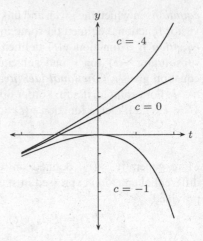

solution. Notice that $y_1(t) = y_4(t)$ when $c = 0$ and $y_3(t) = y_4(t)$ when $c = -7$. Thus, $y_4(t)$ actually already contains $y_1(t)$ and $y_3(t)$ by appropriate choices of the constant $c \in \mathbb{R}$, the real numbers. ◀

The differential equation given by (4) is an example of a first order *linear* differential equation. The theory of such equations will be discussed in Sect. 1.4, where we will show that *all* solutions to (4) are included in the function

$$y_4(t) = t + 1 + ce^t, \qquad t \in (-\infty, \infty)$$

of the above example by appropriate choice of the constant c. We call this the general solution of (4) and denote it by $y_g(t)$. Figure 1.1 is the graph of $y_g(t)$ for various choices of the constant c.

Observe that the general solution is parameterized by the constant c, so that there is a solution for each value of c and hence there are infinitely many solutions of (4). This is characteristic of many differential equations. Moreover, the domain is the same for each of the solutions, namely, the entire real line. With the following example, there is a completely different behavior with regard to the domain of the solutions. Specifically, the domain of each solution varies with the parameter c and is not the same interval for all solutions.

Example 6. Consider the differential equation

$$y' = -2t(1 + y)^2. \tag{5}$$

Show that the following functions are solutions:

1. $y_1(t) = -1$
2. $y_2(t) = -1 + (t^2 - c)^{-1}$, for any constant c

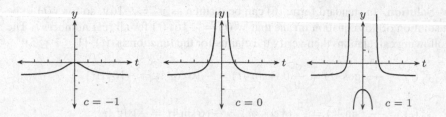

Fig. 1.2 The solutions $y_2(t) = -1 + (t^2 - c)^{-1}$ of $y' = -2t(1 + y)^2$ for various c

▶ **Solution.** Let $y_1(t) = -1$. Then $y_1'(t) = 0$ and $-2t(1 + y_1(t))^2 = -2t(0) = 0$, which is valid for all $t \in (-\infty, \infty)$. Hence, $y_1(t) = -1$ is a solution.

Now let $y_2(t) = -1 + (t^2 - c)^{-1}$. Straightforward calculations give

$$y_2'(t) = -2t(t^2 - c)^{-2}, \text{ and}$$

$$-2t(1 + y_2(t))^2 = -2t(1 + (-1 + (t^2 - c)^{-1}))^2 = -2t(t^2 - c)^{-2}.$$

Thus, $y_2'(t) = -2t(1 + y_2(t))^2$ so that $y_2(t)$ is a solution for any choice of the constant c. ◀

Equation (5) is an example of a *separable* differential equation. The theory of separable equations will be discussed in Sect. 1.3. It turns out that there are no solutions to (5) other than $y_1(t)$ and $y_2(t)$, so that these two sets of functions constitute the general solution $y_g(t)$. Notice that the intervals on which $y_2(t)$ is defined depend on the constant c. For example, if $c < 0$, then $y_2(t) = -1 + (t^2 - c)^{-1}$ is defined for all $t \in (-\infty, \infty)$. If $c = 0$, then $y_2(t) = -1 + t^{-2}$ is defined on two intervals: $t \in (-\infty, 0)$ or $t \in (0, \infty)$. Finally, if $c > 0$, then $y_2(t)$ is defined on three intervals: $(-\infty, -\sqrt{c})$, $(-\sqrt{c}, \sqrt{c})$, or $(\sqrt{c}, \infty)$. Figure 1.2 gives the graph of $y_2(t)$ for various choices of the constant c.

Note that the interval on which the solution $y(t)$ is defined is not at all apparent from looking at the differential equation (5).

Example 7. Consider the differential equation

$$y'' + 16y = 0. \tag{6}$$

Show that the following functions are solutions on the entire real line:

1. $y_1(t) = \cos 4t$
2. $y_2(t) = \sin 4t$
3. $y_3(t) = c_1 \cos 4t + c_2 \sin 4t$, where c_1 and c_2 are constants.

Show that the following functions are not solutions:

4. $y_4(t) = e^{4t}$
5. $y_5(t) = \sin t$.

▶ **Solution.** In standard form, (6) can be written as $y'' = -16y$, so for $y(t)$ to be a solution of this equation means that $y''(t) = -16y(t)$ for all real numbers t. The following calculations then verify the claims for the functions $y_i(t)$, $(1 \le i \le 5)$:

1. $y_1''(t) = \dfrac{d^2}{dt^2}(\cos 4t) = \dfrac{d}{dt}(-4 \sin 4t) = -16 \cos 4t = -16y_1(t)$

2. $y_2''(t) = \dfrac{d^2}{dt^2}(\sin 4t) = \dfrac{d}{dt}(4 \cos 4t) = -16 \sin 4t = -16y_2(t)$

3. $y_3''(t) = \dfrac{d^2}{dt^2}(c_1 \cos 4t + c_2 \sin 4t) = \dfrac{d}{dt}(-4c_1 \sin 4t + 4c_2 \cos 4t)$

   $= -16c_1 \cos 4t - 16c_2 \sin 4t = -16y_3(t)$

4. $y_4''(t) = \dfrac{d^2}{dt^2}(e^{4t}) = \dfrac{d}{dt}(4e^{4t}) = 16e^{4t} \neq -16y_4(t)$

5. $y_5''(t) = \dfrac{d^2}{dt^2}(\sin t) = \dfrac{d}{dt}(\cos t) = -\sin t \neq -16y_5(t)$ ◀

It is true, but not obvious, that letting c_1 and c_2 vary over all real numbers in $y_3(t) = c_1 \cos 4t + c_2 \sin 4t$ produces all solutions to $y'' + 16y = 0$, so that $y_3(t)$ is the general solution of (6). This differential equation is an example of a second order *constant coefficient linear* differential equation. These equations will be studied in Chap. 3.

The Arbitrary Constants

In Examples 5 and 6, we saw that the solution set of the given first order equation was parameterized by an arbitrary constant c (although (5) also had an extra solution $y_1(t) = -1$), and in Example 7, the solution set of the second order equation was parameterized by two constants c_1 and c_2. To understand why these results are not surprising, consider what is arguably the simplest of all first order differential equations:

$$y' = f(t),$$

where $f(t)$ is some continuous function on some interval I. Integration of both sides produces a solution

$$y(t) = \int f(t)\, dt + c, \tag{7}$$

where c is a constant of integration and $\int f(t)\, dt$ is any fixed antiderivative of $f(t)$. The fundamental theorem of calculus implies that all antiderivatives are of this form so (7) is the general solution of $y' = f(t)$. Generally speaking, solving any first order differential equation will implicitly involve integration. A similar calculation for the differential equation

$$y'' = f(t)$$

gives $y'(t) = \int f(t) \, dt + c_1$ so that a second integration gives

$$y(t) = \int y'(t) \, dt + c_2 = \int \left(\int f(t) \, dt + c_1 \right) dt + c_2$$

$$= \int \left(\int f(t) \, dt \right) dt + c_1 t + c_2,$$

where c_1 and c_2 are arbitrary scalars. The fact that we needed to integrate twice explains why there are two scalars. It is generally true that *the number of parameters (arbitrary constants) needed to describe the solution set of an ordinary differential equation is the same as the order of the equation.*

Initial Value Problems

As we have seen in the examples of differential equations and their solutions presented in this section, differential equations generally have infinitely many solutions. So to specify a particular solution of interest, it is necessary to specify additional data. What is usually convenient to specify for a first order equation is an initial value t_0 of the independent variable and an initial value $y(t_0)$ for the dependent variable evaluated at t_0. For a second order equation, one would specify an initial value t_0 for the independent variable, together with an initial value $y(t_0)$ and an initial derivative $y'(t_0)$ at t_0. There is an obvious extension to higher order equations. When the differential equation and initial values are specified, one obtains what is known as an *initial value problem*. Thus, a first order initial value problem in standard form is

$$y' = F(t, y), \quad y(t_0) = y_0, \tag{8}$$

while a second order equation in standard form is written

$$y'' = F(t, y, y'), \quad y(t_0) = y_0, \quad y'(t_0) = y_1. \tag{9}$$

Example 8. Determine a solution to each of the following initial value problems:

1. $y' = y - t$, $y(0) = -3$
2. $y'' = 2 - 6t$, $y(0) = -1$, $y'(0) = 2$

▶ **Solution.**
1. Recall from Example 5 that for each $c \in \mathbb{R}$, the function $y(t) = t + 1 + ce^t$ is a solution for $y' = y - t$. This is the function $y_4(t)$ from Example 5. Thus, our strategy is just to try to match one of the constants c with the required initial condition $y(0) = -3$. Thus,

$$-3 = y(0) = 1 + ce^0 = 1 + c$$

requires that we take $c = -4$. Hence,

$$y(t) = t + 1 - 4e^t$$

is a solution of the initial value problem.
2. The second equation is asking for a function $y(t)$ whose second derivative is the given function $2 - 6t$. But this is precisely the type of problem we discussed earlier and that you learned to solve in calculus using integration. Integration of y'' gives

$$y'(t) = \int y''(t) \, dt + c_1 = \int (2 - 6t) \, dt + c_1 = 2t - 3t^2 + c_1,$$

and evaluating at $t = 0$ gives the equation

$$2 = y'(0) = (2t - 3t^2 + c_1)\big|_{t=0} = c_1.$$

Thus, $c_1 = 2$ and $y'(t) = 2t - 3t^2 + 2$. Now integrate again to get

$$y(t) = \int y'(t) \, dt = \int (2 + 2t - 3t^2) \, dt = 2t + t^2 - t^3 + c_0,$$

and evaluating at $t = 0$ gives the equation

$$-1 = y(0) = (2t + t^2 - t^3 + c_0)\big|_{t=0} = c_0.$$

Hence, $c_0 = -1$ and we get $y(t) = -1 + 2t + t^2 - t^3$ as the solution of our second order initial value problem.   ◀

Some Concluding Comments

Because of the simplicity of the second order differential equation in the previous example, we indicated a rather simple technique for solving it, namely, integration repeated twice. This was not possible for the other examples, even of first order equations, due to the functional dependencies between y and its derivatives. In general, there is not a single technique that can be used to solve all differential equations, where by solve we mean to find an explicit functional description of the general solution $y_g(t)$ as an explicit function of t, possibly depending on some arbitrary constants. Such a $y_g(t)$ is sometimes referred to as a *closed form solution*. There are, however, solution techniques for certain types or categories of differential equations. In this chapter, we will study categories of first order differential equations such as:

- Separable
- Linear
- Homogeneous
- Bernoulli
- Exact

Each category will have its own distinctive solution technique. For higher order differential equations and systems of first order differential equations, the concept of *linearity* will play a very central role for it allows us to write the general solution in a concise way, and in the *constant coefficient* case, it will allow us to give a precise prescription for obtaining the solution set. This prescription and the role of the Laplace transform will occupy the two main important themes of the text. The role of the Laplace transform will be discussed in Chap. 2. In this chapter, however, we stick to a rather classical approach to first order differential equations and, in particular, we will discuss in the next section *direction fields* which allow us to give a pictorial explanation of solutions.

Exercises

1–3. In each of these problems, you are asked to model a scientific law by means of a differential equation.

1. *Malthusian Growth Law.* Scientists who study populations (whether populations of people or cells in a Petri dish) observe that over small periods of time, the **rate of growth** of the population is **proportional** to the **population present**. This law is called the **Malthusian growth law**. Let $P(t)$ represent the number of individuals in the population at time t. Assuming the Malthusian growth law, write a differential equation for which $P(t)$ is the solution.

2. *The Logistic Growth Law.* The Malthusian growth law does not account for many factors affecting the growth of a population. For example, disease, overcrowding, and competition for food are not reflected in the Malthusian model. The goal in this exercise is to modify the Malthusian model to take into account the birth rate and death rate of the population. Let $P(t)$ denote the population at time t. Let $b(t)$ denote the birth rate and $d(t)$ the death rate at time t.

   (a) Suppose the birth rate is proportional to the population. Model this statement in terms of $b(t)$ and $P(t)$.

   (b) Suppose the death rate is proportional to the square of the population. Model this statement in terms of $d(t)$ and $P(t)$.

   (c) The logistic growth law states that the overall growth rate is the difference of the birth rate and death rate, as given in parts (a) and (b). Model this law as a differential equation in $P(t)$.

3. *Torricelli's Law.* Suppose a cylindrical container containing a fluid has a drain on the side. Torricelli's law states that the change in the height of the fluid above the middle of the drain is proportional to the square root of the height. Let $h(t)$ denote the height of the fluid above the middle of the drain. Determine a differential equation in $h(t)$ that models Torricelli's law.

4–11. Determine the order of each of the following differential equations. Write the equation in standard form.

4. $y^2 y' = t^3$
5. $y' y'' = t^3$
6. $t^2 y' + ty = e^t$
7. $t^2 y'' + ty' + 3y = 0$
8. $3y' + 2y + y'' = t^2$
9. $t(y^{(4)})^3 + (y''')^4 = 1$
10. $y' + t^2 y = ty^4$
11. $y''' - 2y'' + 3y' - y = 0$

12–18. Following each differential equation are four functions $y_1, \ldots, y_4$. Determine which are solutions to the given differential equation.

12. $y' = 2y$

   (a) $y_1(t) = 0$
   (b) $y_2(t) = t^2$
   (c) $y_3(t) = 3e^{2t}$
   (d) $y_4(t) = 2e^{3t}$

13. $ty' = y$

   (a) $y_1(t) = 0$
   (b) $y_2(t) = 3t$
   (c) $y_3(t) = -5t$
   (d) $y_4(t) = t^3$

14. $y'' + 4y = 0$

   (a) $y_1(t) = e^{2t}$
   (b) $y_2(t) = \sin 2t$
   (c) $y_3(t) = \cos(2t - 1)$
   (d) $y_4(t) = t^2$

15. $y' = 2y(y - 1)$

   (a) $y_1(t) = 0$
   (b) $y_2(t) = 1$
   (c) $y_3(t) = 2$
   (d) $y_4(t) = \frac{1}{1-e^{2t}}$

16. $2yy' = 1$

   (a) $y_1(t) = 1$
   (b) $y_2(t) = t$
   (c) $y_3(t) = \ln t$
   (d) $y_4(t) = \sqrt{t - 4}$

17. $2yy' = y^2 + t - 1$

   (a) $y_1(t) = \sqrt{-t}$
   (b) $y_2(t) = -\sqrt{e^t - t}$
   (c) $y_3(t) = \sqrt{t}$
   (d) $y_4(t) = -\sqrt{-t}$

18. $y' = \dfrac{y^2 - 4yt + 6t^2}{t^2}$

   (a) $y_1(t) = t$
   (b) $y_2(t) = 2t$
   (c) $y_3(t) = 3t$
   (d) $y_4(t) = \dfrac{3t + 2t^2}{1 + t}$

19–25. Verify that each of the given functions $y(t)$ is a solution of the given differential equation on the given interval I. Note that all of the functions depend on an arbitrary constant $c \in \mathbb{R}$.

19. $y' = 3y + 12$; $y(t) = ce^{3t} - 4$, $I = (-\infty, \infty)$
20. $y' = -y + 3t$; $y(t) = ce^{-t} + 3t - 3$ $I = (-\infty, \infty)$
21. $y' = y^2 - y$; $y(t) = 1/(1 - ce^t)$ $I = (-\infty, \infty)$ if $c < 0, I = (-\ln c, \infty)$
    if $c > 0$
22. $y' = 2ty$; $y(t) = ce^{t^2}$, $I = (-\infty, \infty)$
23. $y' = -e^y - 1$; $y(t) = -\ln(ce^t - 1)$ with $c > 0$, $I = (-\ln c, \infty)$
24. $(t + 1)y' + y = 0$; $y(t) = c(t + 1)^{-1}$, $I = (-1, \infty)$
25. $y' = y^2$; $y(t) = (c - t)^{-1}$, $I = (-\infty, c)$

26–31. Solve the following differential equations.

26. $y' = t + 3$
27. $y' = e^{2t} - 1$
28. $y' = te^{-t}$
29. $y' = \dfrac{t + 1}{t}$
30. $y'' = 2t + 1$
31. $y'' = 6 \sin 3t$

32–38. Find a solution to each of the following initial value problems. See Exercises 19–31 for the general solutions of these equations.

32. $y' = 3y + 12$, $y(0) = -2$
33. $y' = -y + 3t$, $y(0) = 0$
34. $y' = y^2 - y$, $y(0) = 1/2$
35. $(t + 1)y' + y = 0$, $y(1) = -9$
36. $y' = e^{2t} - 1$, $y(0) = 4$
37. $y' = te^{-t}$, $y(0) = -1$
38. $y'' = 6 \sin 3t$, $y(0) = 1, y'(0) = 2$

1.2   Direction Fields

Suppose

$$y' = F(t, y) \qquad (1)$$

is a first order differential equation (in standard form), where $F(t, y)$ is defined in some region of the (t, y)-plane. The geometric interpretation of the derivative of a function $y(t)$ at t_0 as the slope of the tangent line to the graph of $y(t)$ at $(t_0, y(t_0))$ provides us with an elementary and often very effective method for the visualization of the **solution curves** (:= graphs of solutions) to (1). The visualization process involves the construction of what is known as a **direction field** or **slope field** for the differential equation. For this construction, we proceed as follows.

Construction of Direction Fields

1. Solve the given first order differential equation for y' to put it in the standard form $y' = F(t, y)$.
2. Choose a grid of points in a rectangular region

$$\mathcal{R} = \{(t, y) : a \le t \le b; c \le y \le d\}$$

   in the (t, y)-plane where $F(t, y)$ is defined. This means imposing a graph-paper-like grid of vertical lines $t = t_i$ for $a = t_1 < t_2 < \cdots < t_N = b$ and horizontal lines $y = y_j$ for $c = y_1 < y_1 < \cdots < y_M = d$. The points (t_i, y_j) where the grid lines intersect are the **grid points**.
3. At each point (t, y), the number $F(t, y)$ represents the slope of a solution curve through this point. For example, if $y' = y^2 - t$ so that $F(t, y) = y^2 - t$, then at the point $(1, 1)$ the slope is $F(1, 1) = 1^2 - 1 = 0$, at the point $(2, 1)$ the slope is $F(2, 1) = 1^2 - 2 = -1$, and at the point $(1, -2)$ the slope is $F(1, -2) = 3$.
4. Through the grid point (t_i, y_j), draw a small line segment having the slope $F(t_i, y_j)$. Thus, for the equation $y' = y^2 - t$, we would draw a small line segment of slope 0 through $(1, 1)$, slope -1 through $(2, 1)$, and slope 3 through $(1, -2)$. With a graphing calculator, one of the computer mathematics programs Maple, Mathematica, or MATLAB, or with pencil, paper, and a lot of patience, you can draw line segments of the appropriate slope at all of the points of the chosen grid. The resulting picture is called a **direction field** for the differential equation $y' = F(t, y)$.
5. With some luck with respect to scaling and the selection of the (t, y)-rectangle $\mathcal{R}$, you will be able to visualize some of the line segments running together to make a graph of one of the solution curves.

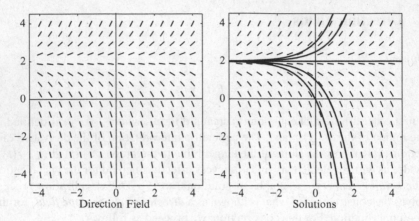

Fig. 1.3 Direction field and some solutions for $y' = y - 2$

6. *To sketch a solution curve of $y' = F(t, y)$ from a direction field, start with a point $P_0 = (t_0, y_0)$ on the grid, and sketch a short curve through P_0 with tangent slope $F(t_0, y_0)$. Follow this until you are at or close to another grid point $P_1 = (t_1, y_1)$. Now continue the curve segment by using the updated tangent slope $F(t_1, y_1)$. Continue this process until you are forced to leave your sample rectangle $\mathcal{R}$.* The resulting curve will be an approximate solution to the initial value problem $y' = F(t, y)$, $y(t_0) = y_0$. Generally speaking, more accurate approximations are obtained by taking finer grids. The solutions are sometimes called ***trajectories***.

Example 1. Draw the direction field for the differential equation $y' = y - 2$. Draw several solution curves on the direction field.

▶ **Solution.** We have chosen a rectangle $\mathcal{R} = \{(t, y) : -4 \leq t, y \leq 4\}$ for drawing the direction field, and we have chosen to use 16 sample points in each direction, which gives a total of 256 grid points where a slope line will be drawn. Naturally, this is being done by computer and not by hand. Figure 1.3 gives the completed direction field with five solution curves drawn. The solutions that are drawn in are the solutions of the initial value problems

$$y' = y - 2, \qquad y(0) = y_0,$$

where the initial value y_0 is 0, 1, 2, 2.5, and 3, reading from the bottom solution to the top.   ◀

You will note in this example that the line $y = 2$ is a solution. In general, any solution to (1) of the form $y(t) = y_0$, where y_0 is a constant, is called an ***equilibrium solution***. Its graph is called an ***equilibrium line***. Equilibrium solutions are those constant functions $y(t) = y_0$ determined by the constants y_0 for which $F(t, y_0) = 0$ for all t. For example, Newton's law of heating and cooling

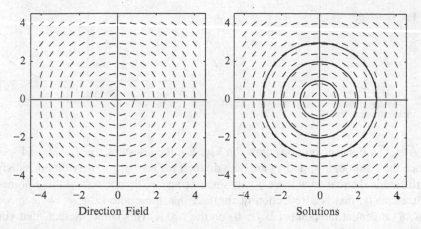

Direction Field   Solutions

Fig. 1.4 Direction field and some solutions for $y' = -t/y$

(Example 1 of Sect. 1.1) is modeled by the differential equation $T' = r(T - T_a)$ which has an equilibrium solution $T(t) = T_a$. This conforms with intuition since if the temperature of the object and the temperature of ambient space are the same, then no change in temperature takes place. The object's temperature is then said to be in *equilibrium*.

Example 2. Draw the direction field for the differential equation $yy' = -t$. Draw several solution curves on the direction field and deduce the family of solutions.

▶ **Solution.** Before we can draw the direction field, it is necessary to first put the differential equation $yy' = -t$ into standard form by solving for y'. Solving for y' gives the equation

$$y' = -\frac{t}{y}. \tag{2}$$

Notice that this equation is not defined for $y = 0$, even though the original equation is. Thus, we should be alert to potential problems arising from this defect. Again we have chosen a rectangle $\mathcal{R} = \{(t, y) : -4 \le t, y \le 4\}$ for drawing the direction field, and we have chosen to use 16 sample points in each direction. Figure 1.4 gives the completed direction field and some solutions. The solutions which are drawn in are the solutions of the initial value problems $yy' = -t$, $y(0) = \pm 1, \pm 2,$ ± 3. The solution curves appear to be circles centered at $(0, 0)$. In fact, the family of such circles is given by $t^2 + y^2 = c$, where $c > 0$. We can verify that functions determined implicitly by the family of circles $t^2 + y^2 = c$ are indeed solutions. For, by implicit differentiation of the equation $t^2 + y^2$ (with respect to the t variable), we get $2t + 2yy' = 0$ and solving for y' gives (2). Solving $t^2 + y^2 = c$ implicitly for y gives two families of continuous solutions, specifically, $y_1(t) = \sqrt{c - t^2}$ (upper semicircle) and $y_2(t) = -\sqrt{c - t^2}$ (lower semicircle). For both families of functions, c is a positive constant and the functions are defined on the interval

Fig. 1.5 Graph of
$f(t, y) = c$

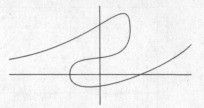

$(-\sqrt{c}, \sqrt{c})$. For the solutions drawn in Fig. 1.4, the constant c is 1, $\sqrt{2}$, and $\sqrt{3}$. Notice that, although y_1 and y_2 are both defined for $t = \pm\sqrt{c}$, they do not satisfy the differential equation at these points since y_1' and y_2' do not exist at these points. Geometrically, this is a reflection of the fact that the circle $t^2 + y^2 = c$ has a vertical tangent at the points $(\pm\sqrt{c}, 0)$ on the t-axis. This is the "defect" that you were warned could occur because the equation $yy' = -t$, when put in standard form $y' = -t/y$, is not defined for $y = 0$.   ◄

Note that in the examples given above, *the solution curves do not intersect*. This is no accident. We will see in Sect. 1.7 that under mild smoothness assumptions on the function $F(t, y)$, it is absolutely certain that the solution curves (trajectories) of an equation $y' = F(t, y)$ can never intersect.

Implicitly Defined Solutions

Example 2 is one of many examples where solutions are sometimes implicitly defined. Let us make a few general remarks when this occurs. Consider a relationship between the two variables t and y determined by the equation

$$f(t, y) = c. \tag{3}$$

We will say that a function $y(t)$ defined on an interval I is ***implicitly defined*** by (3) provided

$$f(t, y(t)) = c \quad \text{for all } t \in I. \tag{4}$$

This is a precise expression of what we mean by the statement:

Solve the equation $f(t, y) = c$ for y as a function of t.

To illustrate, we show in Fig. 1.5 a typical graph of the relation $f(t, y) = c$, for a particular c. We observe that there are three choices of solutions that are continuous functions. We have isolated these and call them $y_1(t)$, $y_2(t)$, and $y_3(t)$. The graphs of these are shown in Fig. 1.6. Observe that the maximal intervals of definition for y_1, y_2, and y_3 are not necessarily the same.

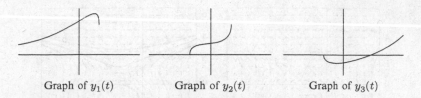

| Graph of $y_1(t)$ | Graph of $y_2(t)$ | Graph of $y_3(t)$ |

Fig. 1.6 Graphs of functions implicitly defined by $f(t, y) = c$

By differentiating[3] (4) with respect to t (using the chain rule from multiple variable calculus), we find

$$\frac{\partial f}{\partial t}(t, y(t)) + \frac{\partial f}{\partial y}(t, y(t))y'(t) = 0.$$

Since the constant c is not present in this equation, we conclude that *every* function implicitly defined by the equation $f(t, y) = c$, for any constant c, is a solution of the same first order differential equation

$$\frac{\partial f}{\partial t} + \frac{\partial f}{\partial y}y' = 0. \tag{5}$$

We shall refer to (5) as the **differential equation for the family of curves** $f(t, y) = c$. One valuable technique that we will encounter in Sect. 1.6 is that of solving a first order differential equation by recognizing it as the differential equation of a particular family of curves.

Example 3. Find the first order differential equation for the family of hyperbolas

$$ty = c$$

in the (t, y)-plane.

▶ **Solution.** Implicit differentiation of the equation $ty = c$ gives

$$y + ty' = 0$$

as the differential equation for this family. In standard form, this equation is $y' = -y/t$. Notice that this agrees with expectations, since for this simple family $ty = c$, we can solve explicitly to get $y = c/t$ (for $t \neq 0$) so that $y' = -c/t^2 = -y/t$.   ◀

It may happen that it is possible to express the solution for the differential equation $y' = F(t, y)$ as an explicit formula, but the formula is sufficiently complicated that it does not shed much light on the nature of the solution. In such

---

[3]In practice, this is just implicit differentiation.

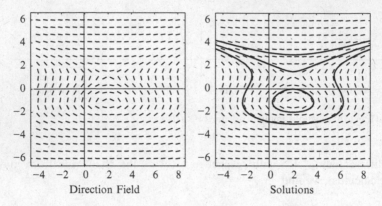

Fig. 1.7 Direction field and some solutions for $y' = \frac{2t-4}{3y^2-4}$

a situation, constructing a direction field and drawing the solution curves on the direction field can sometimes give useful insight concerning the solutions. The following example is a situation where the picture is more illuminating than the formula.

Example 4. Verify that

$$y^3 - 4y - t^2 + 4t = c \tag{6}$$

defines an implicit family of solutions to the differential equation

$$y' = \frac{2t - 4}{3y^2 - 4}.$$

▶ **Solution.** Implicit differentiation gives

$$3y^2 y' - 4y' - 2t + 4 = 0,$$

and solving for y', we get

$$y' = \frac{2t - 4}{3y^2 - 4}.$$

Solving (6) involves a messy cubic equation which does not necessarily shed great light upon the nature of the solutions as functions of t. However, if we compute the direction field of $y' = \frac{2t-4}{3y^2-4}$ and use it to draw some solution curves, we see information concerning the nature of the solutions that is not easily deduced from the implicit form given in (6). For example, Fig. 1.7 gives the direction field and some solutions. Some observations that can be deduced from the picture are:

- In the lower part of the picture, the curves seem to be deformed ovals centered about the point $P \approx (2, -1.5)$.
- Above the point $Q \approx (2, 1.5)$, the curves no longer are closed but appear to increase indefinitely in both directions. ◀

Exercises

1–3. For each of the following differential equations, use some computer math program to sketch a direction field on the rectangle $\mathcal{R} = \{(t, y) : -4 \leq t, y \leq 4\}$ with integer coordinates as grid points. That is, t and y are each chosen from the set $\{-4, -3, -2, -1, 0, 1, 2, 3, 4\}$.

1. $y' = t$
2. $y' = y^2$
3. $y' = y(y + t)$

4–9. A differential equation is given together with its direction field. One solution is already drawn in. Draw the solution curves through the points (t, y) as indicated. Keep in mind that the trajectories will not cross each other in these examples.

4. $y' = 1 - y^2$

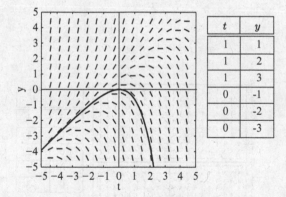

t	y
0	1
0	-1
0	2
-2	2
-2	-2
2	-2

5. $y' = y - t$

t	y
1	1
1	2
1	3
0	-1
0	-2
0	-3

6. $y' = -ty$

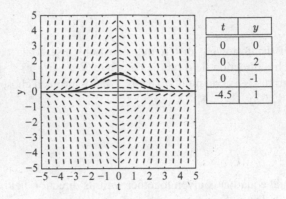

t	y
0	0
0	2
0	-1
-4.5	1

7. $y' = y - t^2$

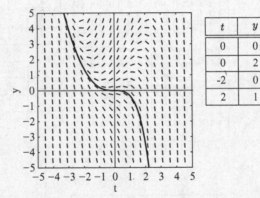

t	y
0	0
0	2
-2	0
2	1

8. $y' = ty^2$

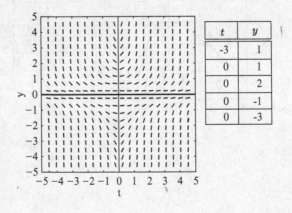

t	y
-3	1
0	1
0	2
0	-1
0	-3

9. $y' = \dfrac{ty}{1+y}$

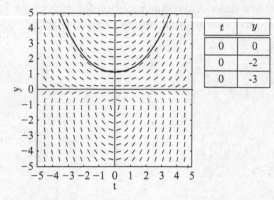

t	y
0	0
0	-2
0	-3

10–13. For the following differential equations, determine the equilibrium solutions, if they exist.

10. $y' = y^2$
11. $y' = y(y + t)$
12. $y' = y - t$
13. $y' = 1 - y^2$

14. The direction field given in Problem 5 for $y' = y - t$ suggests that there may be a linear solution. That is a solution of the form $y = at + b$. Find such a solution.

15. Below is the direction field and some trajectories for $y' = \cos(y + t)$. The trajectories suggest that there are linear solutions that act as asymptotes for the nonlinear trajectories. Find these linear solutions.

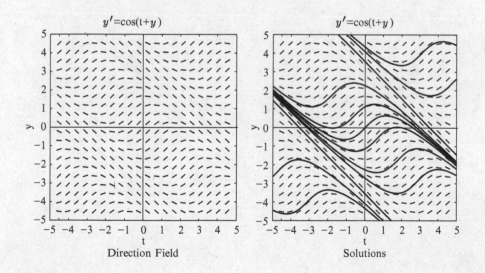

Direction Field   Solutions

16–19. Find the first order differential equation for each of the following families of curves. In each case, c denotes an arbitrary real constant.

16. $3t^2 + 4y^2 = c$
17. $y^2 - t^2 - t^3 = c$
18. $y = ce^{2t} + t$
19. $y = ct^3 + t^2$

1.3 Separable Differential Equations

In the next few sections, we will concentrate on solving particular categories of first order differential equations by means of explicit formulas and algorithms. These categories of equations are described by means of restrictions on the function $F(t, y)$ that appears on the right-hand side of a first order ordinary differential equation given in standard form

$$y' = F(t, y). \tag{1}$$

The first of the standard categories of first order equations to be studied is the class of equations with *separable variables*, that is, equations of the form

$$y' = h(t)g(y). \tag{2}$$

Such an equation is said to be a *separable differential equation* or just *separable*, for short. Thus, (1) *is separable if the right-hand side $F(t, y)$ can be written as a product of a function of t and a function of y.* Most functions of two variables cannot be written as such a product, so being separable is rather special. However, a number of important applied problems turn out to be modeled by separable differential equations. We will explore some of these at the end of this section and in the exercises.

Example 1. Identify the separable equations from among the following list of differential equations:

1. $y' = t^2 y^2$ 2. $y' = y - y^2$

3. $y' = \dfrac{t - y}{t + y}$ 4. $y' = \dfrac{t}{y}$

5. $(2t - 1)(y^2 - 1)y' + t - y - 1 + ty = 0$ 6. $y' = f(t)$

7. $y' = p(t)y$ 8. $y'' = ty$

▶ **Solution.** Equations (1), (2) and (4)–(7) are separable. For example, in (2), $h(t) = 1$ and $g(y) = y - y^2$; in (4), $h(t) = t$ and $g(y) = 1/y$; and in (6), $h(t) = f(t)$ and $g(y) = 1$. To see that (5) is separable, we bring all terms not containing y' to the other side of the equation, that is,

$$(2t - 1)(y^2 - 1)y' = -t + y + 1 - ty = -t(1 + y) + 1 + y = (1 + y)(1 - t).$$

Solving this equation for y' gives

$$y' = \frac{(1 - t)}{(2t - 1)} \cdot \frac{(1 + y)}{(y^2 - 1)},$$

which is separable with $h(t) = (1 - t)/(2t - 1)$ and $g(y) = (1 + y)/(y^2 - 1)$. Equation (3) is not separable because the right-hand side cannot be written as product of a function of t and a function of y. Equation (8) is not a separable equation, even though the right-hand side is $ty = h(t)g(y)$, since it is a *second order* equation and our definition of separable applies only to first order equations.

◄

Equation (2) in the previous example is worth emphasizing since it is typical of many commonly occurring separable differential equations. What is special is that it has the form

$$y' = g(y), \tag{3}$$

where the right-hand side depends only on the dependent variable y. That is, in (2), we have $h(t) = 1$. Such an equation is said to be **autonomous** or time independent. Some concrete examples of autonomous differential equations are the law of radioactive decay, $N' = -\lambda N$; Newton's law of heating and cooling, $T' = r(T - T_a)$; and the logistic growth model equation $P' = (a - bP)P$. These examples will be studied later in this section.

To motivate the general algorithm for solving separable differential equations, let us first consider a simple example.

Example 2. Solve

$$y' = -2t(1 + y)^2. \tag{4}$$

(See Example 6 of Sect. 1.1 where we considered this equation.)

► **Solution.** This is a separable differential equation with $h(t) = -2t$ and $g(y) = (1 + y)^2$. We first note that $y(t) = -1$ is an equilibrium solution. (See Sect. 1.2.) To proceed, assume $y \neq -1$. If we use the Leibniz form for the derivative, $y' = \frac{dy}{dt}$, then (4) can be rewritten as

$$\frac{dy}{dt} = -2t(1 + y)^2.$$

Dividing by $(1 + y)^2$ and multiplying by dt give

$$(1 + y)^{-2} dy = -2t \, dt. \tag{5}$$

Now integrate both sides to get

$$-(1 + y)^{-1} = -t^2 + c,$$

where c is the combination of the arbitrary constants of integration from both sides. To solve for y, multiply both sides by -1, take the reciprocal, and then add -1.

We then get $y = -1 + (t^2 - c)^{-1}$, where c is an arbitrary scalar. Remember that we have the equilibrium solution $y = -1$ so the solution set is

$$y = -1 + (t^2 - c)^{-1},$$
$$y = -1,$$

where $c \in \mathbb{R}$.   ◄

We note that the t and y variables in (5) have been *separated* by the equal sign, which is the origin of the name of this category of differential equations. The left-hand side is a function of y times $\mathrm{d}y$ and the right-hand side is a function of t times $\mathrm{d}t$. This process allows separate integration to give an implicit relationship between t and y. This can be done more generally as outlined in the following algorithm.

Algorithm 3. To solve a separable differential equation,

$$y' = h(t)g(y),$$

perform the following operations:

Solution Method for Separable Differential Equations

1. *Determine the equilibrium solutions.* These are all of the constant solutions $y = y_0$ and are determined by solving the equation $g(y) = 0$ for y_0.
2. *Separate the variables in a form convenient for integration.* That is, we formally write

$$\frac{1}{g(y)}\, \mathrm{d}y = h(t)\, \mathrm{d}t$$

   and refer to this equation as the **differential form** of the separable differential equation.
3. *Integrate both sides, the left-hand side with respect to y and the right-hand side with respect to t.*[4] This yields

$$\int \frac{1}{g(y)}\, \mathrm{d}y = \int h(t)\, \mathrm{d}t,$$

   which produces the implicit solution

$$Q(y) = H(t) + c,$$

   where $Q(y)$ is an antiderivative of $1/g(y)$ and $H(t)$ is an antiderivative of $h(t)$. Such antiderivatives differ by a constant c.
4. (*If possible, solve the implicit relation explicitly for y.*)   □

Note that Step 3 is valid as long as the antiderivatives exist on an interval. From calculus, we know that an antiderivative exists on an interval as long as the integrand is a continuous function on that interval. Thus, it is sufficient that $h(t)$ and $g(y)$ are continuous on appropriate intervals in t and y, respectively, and we will also need $g(y) \neq 0$ in order for $1/g(y)$ to be continuous.

In the following example, please note in the algebra how we carefully track the evolution of the constant of integration and how the equilibrium solution is folded into the general solution set.

Example 4. Solve

$$y' = 2ty. \tag{6}$$

▶ **Solution.** This is a separable differential equation: $h(t) = 2t$ and $g(y) = y$. Clearly, $y = 0$ is an equilibrium solution. Assume now that $y \neq 0$. Then (6) can be rewritten as $\frac{dy}{dt} = 2ty$. Separating the variables by dividing by y and multiplying by dt gives

$$\frac{1}{y}\, dy = 2t\, dt.$$

Integrating both sides gives

$$\ln|y| = t^2 + k_0, \tag{7}$$

where k_0 is an arbitrary constant. We thus obtain a family of implicitly defined solutions y. In this example, we will not be content to leave our answer in this implicit form but rather we will solve explicitly for y as a function of t. Carefully note the sequence of algebraic steps we give below. This same algebra is needed in several examples to follow. We first exponentiate both sides of (7) (remembering that $e^{\ln x} = x$ for all *positive* x, and $e^{a+b} = e^a e^b$ for all a and b) to get

$$|y| = e^{\ln|y|} = e^{t^2+k_0} = e^{k_0}e^{t^2} = k_1 e^{t^2}, \tag{8}$$

where $k_1 = e^{k_0}$ is a *positive* constant, since the exponential function is positive. Next we get rid of the absolute values to get

$$y = \pm|y| = \pm k_1 e^{t^2} = k_2 e^{t^2}, \tag{9}$$

---

[4]Technically, we are treating $y = y(t)$ as a function of t and both sides are integrated with respect to t, but the left-hand side becomes an integral with respect to y using the change of variables $y = y(t)$, $dy = y'dt$.

where $k_2 = \pm k_1$ is a *nonzero* real number. Now note that the equilibrium solution $y = 0$ can be absorbed into the family $y = k_2 e^{t^2}$ by allowing $k_2 = 0$. Thus, the solution set can be written

$$y = ce^{t^2}, \tag{10}$$

where c is an *arbitrary* constant.   ◄

Example 5. Find the solutions of the differential equation

$$y' = \frac{-t}{y}.$$

(This example was considered in Example 2 of Sect. 1.2 via direction fields.)

► **Solution.** We first rewrite the equation in the form $\frac{dy}{dt} = -t/y$ and separate the variables to get

$$y\, dy = -t\, dt.$$

Integration of both sides gives $\int y\, dy = -\int t\, dt$ or $\frac{1}{2}y^2 = -\frac{1}{2}t^2 + c$. Multiplying by 2 and adding t^2 to both sides, we get

$$y^2 + t^2 = c,$$

where we write c instead of $2c$ since twice an arbitrary constant c is still an arbitrary constant. This is the standard equation for a circle of radius $\sqrt{c}$ centered at the origin, for $c > 0$. Solving for y gives

$$y = \pm\sqrt{c - t^2},$$

the equations for the half circles we obtained in Example 2 of Sect. 1.2.   ◄

It may happen that a formula solution for the differential equation $y' = F(t, y)$ is possible, but the formula is sufficiently complicated that it does not shed much light on the nature of the solutions. In such a situation, it may happen that constructing a direction field and drawing the solution curves on the direction field gives useful insight concerning the solutions. The following example is such a situation.

Example 6. Find the solutions of the differential equation

$$y' = \frac{2t - 4}{3y^2 - 4}.$$

► **Solution.** Again we write y' as $\frac{dy}{dt}$ and separate the variables to get

$$(3y^2 - 4)\, dy = (2t - 4)\, dt.$$

Integration gives

$$y^3 - 4y = t^2 - 4t + c.$$

Solving this cubic equation explicitly for y is possible, but it is complicated and not very revealing so we shall leave our solution in implicit form.[5] The direction field for this example was given in Fig. 1.7. As discussed in Example 6 of Sect. 1.2, the direction field reveals much more about the solutions than the explicit formula derived from the implicit formula given above.   ◄

Example 7. Solve the initial value problem

$$y' = \frac{y^2 + 1}{t^2}, \qquad y(1) = 1/\sqrt{3}.$$

Determine the maximum interval on which this solution is defined.

► **Solution.** Since $y^2 + 1 \geq 1$, there are no equilibrium solutions. Separating the variables gives

$$\frac{dy}{y^2 + 1} = \frac{dt}{t^2},$$

and integration of both sides gives $\tan^{-1} y = -\frac{1}{t} + c$. In this case, it is a simple matter to solve for y by applying the tangent function to both sides of the equation. Since $\tan(\tan^{-1} y) = y$, we get

$$y(t) = \tan\left(-\frac{1}{t} + c\right).$$

To find c, observe that $1/\sqrt{3} = y(1) = \tan(-1 + c)$, which implies that $c - 1 = \pi/6$, so $c = 1 + \pi/6$. Hence,

$$y(t) = \tan\left(-\frac{1}{t} + 1 + \frac{\pi}{6}\right).$$

To determine the maximum domain on which this solution is defined, note that the tangent function is defined on the interval $(-\pi/2, \pi/2)$, so that $y(t)$ is defined for all t satisfying

$$-\frac{\pi}{2} < -\frac{1}{t} + 1 + \frac{\pi}{6} < \frac{\pi}{2}.$$

---

[5]The formula for solving a cubic equation is known as Cardano's formula after Girolamo Cardano (1501–1576), who was the first to publish it.

Since $-\frac{1}{t} + 1 + \frac{\pi}{6}$ is increasing and the limit as $t \to \infty$ is $1 + \frac{\pi}{6} < \frac{\pi}{2}$, the second of the above inequalities is valid for all $t > 0$. The first inequality is solved to give $t > 3/(3 + 2\pi)$. Thus, the maximum domain for the solution $y(t)$ is the interval $(3/(3 + 2\pi), \infty)$. ◀

Radioactive Decay

In Example 2 of Sect. 1.1, we discussed the law of radioactive decay, which states: *If $N(t)$ is the quantity[6] of radioactive isotopes at time t, then the rate of decay is proportional to $N(t)$.* Since rate of change is expressed as the derivative with respect to the time variable t, it follows that $N(t)$ is a solution to the differential equation

$$N' = -\lambda N,$$

where λ is a constant. You will recognize this as a separable differential equation, which can be written in differential form with variables separated as

$$\frac{dN}{N} = -\lambda \, dt.$$

Integrating the left side with respect to N and the right side with respect to t leads to $\ln |N| = -\lambda t + k_0$, where k_0 is an arbitrary constant. As in Example 4, we can solve for N as a function of t by applying the exponential function to both sides of $\ln |N| = -\lambda t + k_0$. This gives

$$|N| = e^{\ln|N|} = e^{-\lambda t + k_0} = e^{k_0} e^{-\lambda t} = k_1 e^{-\lambda t},$$

where $k_1 = e^{k_0}$ is a positive constant. Since $N = \pm |N| = \pm k_1 e^{-\lambda t}$, we conclude that $N(t) = c e^{-\lambda t}$, where c is an arbitrary constant. Notice that this family includes the equilibrium solution $N = 0$ when $c = 0$. Further, at time $t = 0$, we have $N(0) = c e^0 = c$. *The constant c therefore represents the quantity of radioactive isotopes at time $t = 0$ and is denoted N_0.* In summary, we find

$$N(t) = N_0 e^{-\lambda t}. \tag{11}$$

The constant λ is referred to as the ***decay constant***. However, most scientists prefer to specify the rate of decay by another constant known as the ***half-life*** of the radioactive isotope. The half-life is the time, τ, it takes for half the quantity to decay. Thus, τ is determined by the equation $N(\tau) = \frac{1}{2} N_0$ or $N_0 e^{-\lambda \tau} = \frac{1}{2} N_0$. Eliminating

---

[6] $N(t)$ could represent the number of atoms or the mass of radioactive material at time t.

N_0 gives $e^{-\lambda \tau} = \frac{1}{2}$, which can be solved for τ by applying the logarithm function ln to both sides of the equation to get $-\lambda \tau = \ln(1/2) = -\ln 2$ so

$$\tau = \frac{\ln 2}{\lambda}, \tag{12}$$

an inverse proportional relationship between the half-life and decay constant. Note, in particular, that the half-life does not depend on the initial amount N_0 present. Thus, it takes the same length of time to decay from 2 g to 1 g as it does to decay from 1 g to 1/2 g.

Carbon 14, ^{14}C, is a radioactive isotope of carbon that decays into the stable nonradioactive isotope nitrogen-14, ^{14}N, and has a half-life of 5730 years. Plants absorb atmospheric carbon during photosynthesis, so the ratio of ^{14}C to normal carbon in plants and animals when they die is approximately the same as that in the atmosphere. However, the amount of ^{14}C decreases after death from radioactive decay. Careful measurements allow the date of death to be estimated.[7]

Example 8. A sample of wood from an archeological site is found to contain 75% of ^{14}C (per unit mass) as a sample of wood found today. What is the age of the wood sample.

▶ **Solution.** At $t = 0$, the total amount, N_0, of carbon-14 in the wood sample begins to decay. Now the quantity is $0.75N_0$. This leads to the equation $0.75N_0 = N_0 e^{-\lambda t}$. Solving for t gives $t = -\frac{\ln 0.75}{\lambda}$. Since the half-life of ^{14}C is 5730 years, (12) gives the decay constant $\lambda = \frac{\ln 2}{5730}$. Thus, the age of the sample is

$$t = 5730 - \frac{\ln 0.75}{\ln 2} \approx 2378$$

years. If the wood sample can be tied to the site, then an archeologist might be able to conclude that the site is about 2378 years old.   ◀

Newton's Law of Heating and Cooling

Recall that Newton's law of heating and cooling (Example 1 of Sect. 1.1) states that *the rate of change of the temperature, $T(t)$, of a body is proportional to the*

---

[7]There are limitations, however. The ratio of ^{14}C to other forms of carbon in the atmosphere is not constant as originally supposed. This variation is due, among other things, to changes in the intensity of the cosmic radiation that creates ^{14}C. To compensate for this variation, dates obtained from radiocarbon laboratories are now corrected using standard calibration tables.

difference between that body and the temperature of ambient space. This law thus states that $T(t)$ is a solution to the differential equation

$$T' = r(T - T_a),$$

where T_a is the ambient temperature and r is the proportionality constant. This equation is separable and can be written in differential form as

$$\frac{dT}{T - T_a} = r\, dt.$$

Integrating both sides leads to $\ln |T - T_a| = rt + k_0$, where k_0 is an arbitrary constant. Solving for $T - T_a$ is very similar to the algebra we did in Example 4. We can solve for $T - T_a$ as a function of t by applying the exponential function to both sides of $\ln |T - T_a| = rt + k_0$. This gives

$$|T - T_a| = e^{\ln|T-T_a|} = e^{rt+k_0} = e^{k_0}e^{rt} = k_1 e^{rt},$$

where $k_1 = e^{k_0}$ is a positive constant. Since $T - T_a = \pm |T - T_a| = \pm k_1 e^{rt}$, we conclude that $T(t) - T_a = ce^{rt}$, where c is an arbitrary constant. Notice that this family includes the equilibrium solution $T = T_a$ when $c = 0$. In summary, we find

$$T(t) = T_a + ce^{rt}. \tag{13}$$

Example 9. A turkey, which has an initial temperature of $40°$ (Fahrenheit), is placed into a $350°$ oven. After one hour, the temperature of the turkey is $100°$. Use Newton's law of heating and cooling to find

1. The temperature of the turkey after 2 h
2. How many hours it takes for the temperature of the turkey to reach $170°$

▶ **Solution.** In this case, the oven is the surrounding medium and has a constant temperature of $T_a = 350°$ so (13) gives

$$T(t) = 350 + ce^{rt}.$$

The initial temperature is $T(0) = 40°$, and this implies $40 = 350 + c$ and hence $c = -310$. To determine r, note that we are given $T(1) = 100$. This implies $100 = T(1) = 350 - 310e^r$, and solving for r gives $r = \ln \frac{25}{31} \approx -0.21511$. To answer question (1), compute $T(2) = 350 - 310e^{2r} \approx 148.39°$. To answer question (2), we want to find t so that $T(t) = 170$, that is, solve $170 = T(t) = 350 - 310e^{rt}$. Solving this gives $rt = \ln \frac{18}{31}$ so $t \approx 2.53\,\text{h} \approx 2\,\text{h } 32$ min.   ◀

The Malthusian Growth Model

Let $P(t)$ represent the number of individuals in a population at time t. Two factors influence its growth: the birth rate and the death rate. Let $b(t)$ and $d(t)$ denote the birth and death rates, respectively. Then $P'(t) = b(t) - d(t)$. *In the Malthusian growth model, the birth and death rates are assumed to be proportional to the number in the population.* Thus

$$b(t) = \beta P(t) \quad \text{and} \quad d(t) = \delta P(t), \tag{14}$$

for some real constants β and δ. We are thus led to the differential equation

$$P'(t) = (\beta - \delta)P(t) = rP(t), \tag{15}$$

where $r = \beta - \delta$. Except for the notation, this equation is the same as the equation that modeled radioactive decay. The same calculation gives

$$P(t) = P_0 e^{rt}, \tag{16}$$

for the solution of this differential equation, where $P_0 = P(0)$ is the *initial population*. The constant r is referred to as the *Malthusian parameter*.

Example 10. Suppose 50 bacteria are placed in a Petri dish. After 30 min there are 120 bacteria. Assume the Malthusian growth model. How many bacteria will there be after 2 h? After 6 h?

▶ **Solution.** The initial population is $P_0 = 50$. After 30 min we have $120 = P(30) = P_0 e^{rt}|_{t=30} = 50e^{30r}$. This implies that the Malthusian parameter is $r = \frac{1}{30} \ln \frac{12}{5}$. In 120 min, we have

$$P(120) = P_0 e^{rt}\Big|_{t=120}$$

$$= 50e^{\frac{120}{30} \ln \frac{12}{5}} \approx 1,659.$$

In 6 h or 360 min, we have

$$P(360) = P_0 e^{rt}\Big|_{t=360}$$

$$= 50e^{\frac{360}{30} \ln \frac{12}{5}} \approx 1,826,017. \qquad ◀$$

While the Malthusian model may be a reasonable model for short periods of time, it does not take into account growth factors such as disease, overcrowding, and competition for food that come into play for large populations and are not seen in small populations. Thus, while the first calculation in the example above may be plausible, it is likely implausible for the second calculation.

The Logistic Growth Model

Here we generalize the assumptions made in the Malthusian model about the birth and death rates.[8] In a confined environment, resources are limited. Hence, it is reasonable to assume that as a population grows, the birth and death rates will decrease and increase, respectively. From (14), the per capita birth rate in the Malthusian model, $b(t)/P(t) = \beta$, is constant. Now assume it decreases linearly with the population, that is, $b(t)/P(t) = \beta - k_\beta P(t)$, for some positive constant k_β. Similarly, assume the per capita death rate increases linearly with the population, that is, $d(t)/P(t) = \delta + k_\delta P(t)$, for some positive constant k_δ. We are then led to the following birth and death rate models:

$$b(t) = (\beta - k_\beta P(t))P(t) \quad \text{and} \quad d(t) = (\delta + k_\delta P(t))P(t). \tag{17}$$

Since the rate of change of population is the difference between the birth rate and death rate, we conclude

$$
\begin{aligned}
P'(t) &= b(t) - d(t) \\
&= ((\beta - \delta) - (k_\beta + k_\delta)P(t))P(t) \\
&= (\beta - \delta)\left(1 - \frac{k_\beta + k_\delta}{\beta - \delta}P(t)\right)P(t) \\
&= r\left(1 - \frac{P(t)}{m}\right)P(t),
\end{aligned} \tag{18}
$$

where we set $r = \beta - \delta$ and $m = (\beta - \delta)/(k_\beta + k_\delta)$. Equation (18) shows that under the assumption (17), the population is a solution of the differential equation

$$P' = r\left(1 - \frac{P}{m}\right)P, \tag{19}$$

which is known as the *logistic differential equation* or the *Verhulst population differential equation* in its classical form. The parameter r is called the *Malthusian parameter*; it represents the rate at which the population would grow if it were unencumbered by environmental constraints. The number m is called the *carrying capacity* of the population, which, as explained below, represents the maximum population possible under the given model.

---

[8] A special case was discussed in Exercise 2 of Sect. 1.1.

To solve the logistic differential equation (19), first note that the equation is separable since the right-hand side of the equation depends only on the dependent variable P. Next observe there are two equilibrium solutions obtained by constant solutions of the equation

$$r\left(1 - \frac{P}{m}\right) P = 0.$$

These equilibrium (constant) solutions are $P(t) = 0$ and $P(t) = m$. Now proceed to the separation of variables algorithm. Separating the variables in (19) gives

$$\frac{1}{\left(1 - \frac{P}{m}\right) P} \, dP = r \, dt.$$

In order to integrate the left-hand side of this equation, it is first necessary to use partial fractions to get

$$\frac{1}{\left(1 - \frac{P}{m}\right) P} = \frac{m}{(m - P)P} = \frac{1}{m - P} + \frac{1}{P}.$$

Thus, the separated variables form of (19) suitable for integration is

$$\left(\frac{1}{m - P} + \frac{1}{P}\right) dP = r \, dt.$$

Integrating the left side with respect to P and the right side with respect to t gives

$$-\ln|m - P| + \ln|P| = \ln|P/(m - P)| = rt + k,$$

where k is the constant of integration. Using the same algebraic techniques employed in Example 4 and the earlier examples on radioactive decay and Newton's law of cooling, we get

$$\frac{P}{m - P} = ce^{rt},$$

for some real constant c. To solve for P, note that

$$P = ce^{rt}(m - P) = cme^{rt} - Pce^{rt} \implies P(1 + ce^{rt}) = cme^{rt}.$$

This gives

$$P(t) = \frac{cme^{rt}}{1 + ce^{rt}}$$

$$= \frac{cm}{e^{-rt} + c}, \tag{20}$$

where the second equation is obtained from the first by multiplying the numerator and denominator by e^{-rt}.

The equilibrium solution $P(t) = 0$ is obtained by setting $c = 0$. The equilibrium solution $P(t) = m$ does not occur for *any* choice of c, so this solution is an extra one. Also note that since $r = \beta - \delta > 0$ assuming that the birth rate exceeds the death rate, we have $\lim_{t\to\infty} e^{-rt} = 0$ so

$$\lim_{t\to\infty} P(t) = \lim_{t\to\infty} \frac{cm}{e^{-rt}+c} = \frac{cm}{c} = m,$$

independent of $c \neq 0$. What this means is that if we start with a positive population, then over time, the population will approach a maximum (sustainable) population m. This is the interpretation of the carrying capacity of the environment.

When $t = 0$, we get $P(0) = \frac{cm}{1+c}$ and solving for c gives $c = \frac{P(0)}{m-P(0)}$. Substituting c into (20) and simplifying give

$$P(t) = \frac{mP(0)}{P(0) + (m - P(0))e^{-rt}}. \tag{21}$$

Equation (21) is called the ***logistic equation***. That is, the logistic equation refers to the *solution* of the logistic differential equation. Below is its graph. You will note that the horizontal line $P = m$ is an asymptote. You can see from the graph that $P(t)$ approaches the limiting population m as t grows.

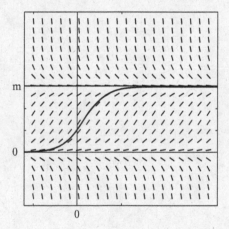

The graph of the logistic equation: $P(t) = \dfrac{mP(0)}{P(0) + (m - P(0))e^{-rt}}$

Example 11. Suppose biologists stock a lake with 200 fish and estimate the carrying capacity of the lake to be 10,000 fish. After two years, there are 2,500 fish. Assume the logistic growth model and estimate how long it will take for there to be 9,000 fish.

► **Solution.** The initial population is $P(0) = 200$ and the carrying capacity is $m = 10,000$. Thus, the logistic equation, (21), is

$$P(t) = \frac{mP(0)}{P(0) + (m - P(0))e^{-rt}} = \frac{2,000,000}{200 + 9,800e^{-rt}} = \frac{10,000}{1 + 49e^{-rt}}.$$

Now let $t = 2$ (assume that time is measured in years). Then we get $2,500 = 10,000/(1 + 49e^{-2r})$, and solving for r gives $r = \frac{1}{2}\ln\frac{49}{3}$. Now we want to know for what t is $P(t) = 9,000$. With r as above the equation $P(t) = 9,000$ can be solved for t to give

$$t = \frac{2\ln 441}{\ln\frac{49}{3}} \approx 4.36 \text{ years.}$$

The logistics growth model predicts that the population of fish will reach 9,000 in 4.36 years.   ◄

Exercises

1–9. In each of the following problems, determine whether or not the equation is separable. Do *not* solve the equations!

1. $y' = 2y(5 - y)$
2. $yy' = 1 - y$
3. $t^2 y' = 1 - 2ty$
4. $\dfrac{y'}{y} = y - t$
5. $ty' = y - 2ty$
6. $y' = ty^2 - y^2 + t - 1$
7. $(t^2 + 3y^2)y' = -2ty$
8. $y' = t^2 + y^2$
9. $e^t y' = y^3 - y$

10–30. Find the general solution of each of the following differential equations. If an initial condition is given, find the particular solution that satisfies this initial condition.

10. $yy' = t, \ y(2) = -1$
11. $(1 - y^2) - tyy' = 0$
12. $y^3 y' = t$
13. $y^4 y' = t + 2$
14. $y' = ty^2$
15. $y' + (\tan t)y = \tan t, \ -\frac{\pi}{2} < t < \frac{\pi}{2}$
16. $y' = t^m y^n$, where m and n are positive integers, $n \neq 1$.
17. $y' = 4y - y^2$
18. $yy' = y^2 + 1$
19. $y' = y^2 + 1$
20. $tyy' + t^2 + 1 = 0$
21. $y + 1 + (y - 1)(1 + t^2)y' = 0$
22. $2yy' = e^t$
23. $(1 - t)y' = y^2$
24. $\frac{dy}{dt} - y = y^2, \quad y(0) = 0.$
25. $y' = 4ty^2, \quad y(1) = 0$
26. $\frac{dy}{dx} = \frac{xy+2y}{x}, \quad y(1) = e$
27. $y' + 2yt = 0, \quad y(0) = 4$
28. $y' = \frac{\cot y}{t}, \quad y(1) = \frac{\pi}{4}$
29. $\frac{(u^2+1)}{y}\frac{dy}{du} = u, \quad y(0) = 2$
30. $ty - (t + 2)y' = 0$
31. Solve the initial value problem

$$y' = \frac{y^2 + 1}{t^2} \qquad y(1) = \sqrt{3}.$$

Determine the maximum interval (a, b) on which this solution is defined. Show that $\lim_{t \to b^-} y(t) = \infty$.

32–34. *Radioactive Decay*

32. A human bone is found in a melting glacier. The carbon-14 content of the bone is only one-third of the carbon-14 content in a similar bone today. Estimate the age of the bone? The half-life of C-14 is 5,730 years.

33. Potassium-40 is a radioactive isotope that decays into a single argon-40 atom and other particles with a half-life of 1.25 billion years. A rock sample was found that contained 8 times as many potassium-40 atoms as argon-40 atoms. Assume the argon-40 only comes from radioactive decay. Date the rock to the time it contained only potassium-40.

34. Cobalt 60 is a radioactive isotope used in medical radiotherapy. It has a half-life of 5.27 years. How long will it take for a sample of cobalt 60 to reduce to 30% of the original?

35–40. In the following problems, assume *Newton's law of heating and cooling.*

35. A bottle of your favorite beverage is at room temperature, 70°F, and it is then placed in a tub of ice water at time $t = 0$. After 30 min, the temperature is 55°. Assuming Newton's law of heating and cooling, when will the temperature drop to 45°.

36. A cup of coffee, brewed at 180° (Fahrenheit), is brought into a car with inside temperature 70°. After 3 min, the coffee cools to 140°. What is the temperature 2 min later?

37. The temperature outside a house is 90°, and inside it is kept at 65°. A thermometer is brought from the outside reading 90°, and after 2 min, it reads 85°. How long will it take to read 75°? What will the thermometer read after 20 min?

38. A cold can of soda is taken out of a refrigerator with a temperature of 40° and left to stand on the counter top where the temperature is 70°. After 2 h the temperature of the can is 60°. What was the temperature of the can 1 h after it was removed from the refrigerator?

39. A large cup of hot coffee is bought from a local drive through restaurant and placed in a cup holder in a vehicle. The inside temperature of the vehicle is 70° Fahrenheit. After 5 min the driver spills the coffee on himself and receives a severe burn. Doctors determine that to receive a burn of this severity, the temperature of the coffee must have been about 150°. If the temperature of the coffee was 142° six minutes after it was sold what was the temperature at which the restaurant served it.

40. A student wishes to have some friends over to watch a football game. She wants to have cold beer ready to drink when her friends arrive at 4 p.m. According to her taste, the beer can be served when its temperature is 50°. Her experience shows that when she places 80° beer in the refrigerator that is kept at a constant temperature of 40°, it cools to 60° in an hour. By what time should she put the beer in the refrigerator to ensure that it will be ready for her friends?

41–43. In the following problems, assume the *Malthusian growth model*.

41. The population of elk in a region of a national forest was 290 in 1980 and 370 in 1990. Forest rangers want to estimate the population of elk in 2010. They assume the Malthusian growth model. What is their estimate?

42. The time it takes for a culture of 40 bacteria to double its population is 3 h. How many bacteria are there at 30 h?

43. If it takes 5 years for a certain population to triple in size, how long does it take to double in size?

44–47. In the following problems, assume the *logistic growth model*.

44. The population of elk in a region of a national forest was 290 in 1980 and 370 in 1990. Experienced forest rangers estimate that the carrying capacity of the population of elk for the size and location of the region under consideration is 800. They want to estimate the population of elk in 2010. They assume the logistic growth model. What is their estimate?

45. A biologist is concerned about the rat population in a closed community in a large city. Initial readings 2 years ago gave a population of 2,000 rats, but now there are 3,000. The experienced biologist estimates that the community cannot support more than 5,000. The alarmed mayor tells the biologist not to reveal that to the general public. He wants to know how large the population will be at election time two years from now. What can the biologist tell him?

46. Assuming the logistics equation $P(t) = \frac{mP_0}{P_0 + (m - P_0)e^{-rt}}$, suppose $P(0) = P_0$, $P(t_0) = P_1$, and $P(2t_0) = P_2$. Show

$$m = \frac{P_1\left(P_1(P_0 + P_2) - 2P_0P_2\right)}{P_1^2 - P_0P_2},$$

$$r = \frac{1}{t_0}\ln\left(\frac{P_2(P_1 - P_0)}{P_0(P_2 - P_1)}\right).$$

47. Four hundred butterflies are introduced into a closed community of the rain-forest and subsequently studied. After 3 years, the population increases to 700, and 3 years thereafter, the population increases to 1,000. What is the carrying capacity?

1.4 Linear First Order Equations

A first order differential equation that can be written in the form

$$y' + p(t)y = f(t) \tag{1}$$

is called a *linear first order differential equation* or just *linear*, for short. We will say that (1) is the *standard form* for a linear differential equation.[9] We will assume that the *coefficient function*, p, and the *forcing function*, f, are continuous functions on an interval I. In later chapters, the continuity restrictions will be removed. Equation (1) is *homogeneous* if the forcing function is zero on I and *nonhomogeneous* if the forcing function f is not zero. Equation (1) is *constant coefficient* provided the coefficient function p is a constant function, that is, $p(t) = p_0 \in \mathbb{R}$ for all $t \in I$.

Example 1. Characterize the following list of first order differential equations:

1. $y' = y - t$
2. $y' + ty = 0$
3. $y' = \sec t$
4. $y' + y^2 = t$
5. $ty' + y = t^2$
6. $y' - \frac{3}{t}y = t^4$
7. $y' = 7y$
8. $y' = \tan(ty)$

▶ **Solution.** The presence of the term y^2 in (4) and the presence of the term $\tan(ty)$ in (8) prevent them from being linear. Equation (1), (5), and (7) can be written $y' - y = -t$, $y' + (1/t)y = t$, and $y' - 7y = 0$, respectively. Thus, all but (4) and (8) are linear. Equations (2) and (7) are homogeneous. Equations (1), (3), (5), and (6) are nonhomogeneous. Equations (1), (3), and (7) are constant coefficient. The interval of continuity for the forcing and coefficient functions is the real line for (1), (2), (7); an interval of the form $(-\frac{\pi}{2} + m\pi, \frac{\pi}{2} + m\pi)$, m an integer, for (3); and $(-\infty, 0)$ or $(0, \infty)$ for (5) and (6). ◀

Notice (2), (3), and (7) are also separable. Generally, this occurs when either the coefficient function, $p(t)$, is zero or the forcing function, $f(t)$, is zero. Thus, there is an overlap between the categories of separable and linear differential equations.

---

[9]This conflicts with the use of the term *standard form*, given in Sect. 1.1, where we meant a first order differential equation written in the form $y' = F(t, y)$. Nevertheless, in the context of first order linear differential equations, we will use the term *standard form* to mean an equation written as in (1).

A Solution Method for Linear Equations

Consider the following two linear differential equations:

$$ty' + 2y = 4, \tag{2}$$
$$t^2 y' + 2ty = 4t. \tag{3}$$

They both have the same standard form

$$y' + \frac{2}{t} y = \frac{4}{t} \tag{4}$$

so they both have the same solution set. Further, multiplying (4) by t and t^2 gives (2) and (3), respectively. Now (2) is simpler than (3) in that it does not have a needless extra factor of t. However, (3) has an important redeeming property that (2) does not have, namely, the left-hand side is a perfect derivative, by which we mean that $t^2 y' + 2ty = (t^2 y)'$. Just apply the product rule to check this. Thus, (3) can be rewritten in the form

$$(t^2 y)' = 4t, \tag{5}$$

and it is precisely this form that leads to the solution: Integrating both sides with respect to t gives $t^2 y = 2t^2 + c$ and dividing by t^2 gives

$$y = 2 + ct^{-2}, \tag{6}$$

where c is an arbitrary real number. Of course, we could have multiplied (4) by any function, but of all functions, it is only $\mu(t) = t^2$ (up to a multiplicative scalar) that simplifies the left-hand side as in (5).

The process just described generalizes to an arbitrary linear differential equation in *standard form*

$$y' + py = f. \tag{7}$$

That is, we can find a function μ so that when the left side of (7) is multiplied by μ, the result is a perfect derivative $(\mu y)'$ with respect to the t variable. This will require that

$$\mu y' + \mu p y = (\mu y)' = \mu y' + \mu' y. \tag{8}$$

Such a function μ is called an **integrating factor**. To find μ, note that simplifying (8) by canceling $\mu y'$ gives $\mu p y = \mu' y$, and we can cancel y (assuming that $y(t) \neq 0$) to get

$$\mu' = \mu p, \tag{9}$$

a separable differential equation for the unknown function μ. Separating variables gives $\mu'/\mu = p$ and integrating both sides gives $\ln|\mu| = P$, where $P = \int p\,dt$ is any antiderivative of p. Taking the exponential of both sides of this equation produces a formula for the integrating factor

$$\mu = e^P = e^{\int p\,dt}; \tag{10}$$

namely, μ *is the exponential of any antiderivative of the coefficient function.* For example, the integrating factor $\mu(t) = t^2$ found in the example above is derived as follows: the coefficient function in (4) is $p(t) = 2/t$. An antiderivative is $P(t) = 2\ln t = \ln t^2$ and hence $\mu(t) = e^{P(t)} = e^{\ln t^2} = t^2$.

Once an integrating factor is found, we can solve (7) in the same manner as the example above. First multiplying (7) by the integrating factor gives $\mu y' + \mu p y = \mu f$. By (8), $\mu y' + \mu p y = (\mu y)'$ so we get

$$(\mu y)' = \mu f. \tag{11}$$

Integrating both sides of this equation gives $\mu y = \int \mu f\,dt + c$, and hence,

$$y = \frac{1}{\mu}\int \mu f\,dt + \frac{c}{\mu},$$

where c is an arbitrary constant.

It is easy to check that the formula we have just derived is a solution. Further, we have just shown that any solution takes on this form. We summarize this discussion in the following theorem:

Theorem 2. *Let p and f be continuous functions on an interval I. A function y is a solution of the first order linear differential equation $y' + py = f$ on I if and only if*

Solution of First Order Linear Equations

$$y = \frac{1}{\mu}\int \mu f\,dt + \frac{c}{\mu},$$

---

where $c \in \mathbb{R}$, P is any antiderivative of p on the interval I, and $\mu = e^P$.

The steps needed to derive the solution when dealing with concrete examples are summarized in the following algorithm.

Algorithm 3. The following procedure is used to solve a first order linear differential equation.

Solution Method for First Order Linear Equations

1. *Put the given linear equation in standard form:* $y' + py = f$.
2. *Find an integrating factor,* μ: To do this, compute an antiderivative $P = \int p \, dt$ and set $\mu = e^P$.
3. *Multiply the equation (in standard form) by* μ: This yields

$$\mu y' + \mu' y = \mu f.$$

4. *Simplify the left-hand side:* Since $(\mu y)' = \mu y' + \mu' y$, we get

$$(\mu y)' = \mu f.$$

5. *Integrate both sides of the resulting equation:* This yields

$$\mu y = \int \mu f \, dt + c.$$

6. *Divide by* μ *to get the solution* y:

$$y = \frac{1}{\mu} \int \mu f \, dt + \frac{c}{\mu}. \tag{12}$$

Remark 4. You should *not* memorize formula (12). What you should remember instead is the sequence of steps in Algorithm 3, and apply these steps to each concretely presented linear first order differential equation.

Example 5. Find all solutions of the differential equation

$$t^2 y' + ty = 1$$

on the interval $(0, \infty)$.

▶ **Solution.** We shall follow Algorithm 3 closely. First, we put the given linear differential equation in standard form to get

$$y' + \frac{1}{t} y = \frac{1}{t^2}. \tag{13}$$

The coefficient function is $p(t) = 1/t$ and its antiderivative is $P(t) = \ln |t| = \ln t$ (since we are on the interval $(0, \infty)$, $t = |t|$). It follows that the integrating factor is $\mu(t) = e^{\ln t} = t$. Multiplying (13) by $\mu(t) = t$ gives

$$t y' + y = \frac{1}{t}.$$

Next observe that the left side of this equality is equal to $\frac{d}{dt}(ty)$. Thus,

$$\frac{d}{dt}(ty) = \frac{1}{t}.$$

Now take antiderivatives of both sides to get

$$ty = \ln t + c,$$

where $c \in \mathbb{R}$. Now dividing by the integrating factor $\mu(t) = t$ gives the solution

$$y(t) = \frac{\ln t}{t} + \frac{c}{t}. \qquad \blacktriangleleft$$

Example 6. Find all solutions of the differential equation

$$y' = (\tan t)y + \cos t$$

on the interval $I = (-\frac{\pi}{2}, \frac{\pi}{2})$.

▶ **Solution.** We first write the given equation in standard form to get

$$y' - (\tan t)y = \cos t.$$

The coefficient function is $p(t) = -\tan t$ (it is a common mistake to forget the minus sign). An antiderivative is $P(t) = -\ln \sec t = \ln \cos t$. It follows that $\mu(t) = e^{\ln \cos t} = \cos t$ is an integrating factor. We now multiply by μ to get $(\cos t)y' - (\sin t)y = \cos^2 t$, and hence,

$$((\cos t)y)' = \cos^2 t.$$

Integrating both sides, taking advantage of double angle formulas, gives

$$(\cos t)y = \int \cos^2 t \, dt = \int \frac{1}{2}(\cos 2t + 1) \, dt$$

$$= \frac{1}{2}t + \frac{1}{4}\sin 2t + c$$

$$= \frac{1}{2}(t + \sin t \cos t) + c.$$

Dividing by the integrating factor $\mu(t) = \cos t$ gives the solution

$$y = \frac{1}{2}(t \sec t + \sin t) + c \sec t.$$ ◄

Example 7. Find the solution of the differential equation

$$y' + y = t,$$

with initial condition $y(0) = 2$.

► **Solution.** The given differential equation is in standard form. The coefficient function is the constant function $p(t) = 1$. Thus, $\mu(t) = e^{\int 1\,dt} = e^t$. Multiplying by e^t gives $e^t y' + e^t y = te^t$ which is

$$(e^t y)' = te^t.$$

Integrating both sides gives $e^t y = \int te^t\,dt = te^t - e^t + c$, where $\int te^t\,dt$ is calculated using integration by parts. Dividing by e^t gives

$$y = t - 1 + ce^{-t}.$$

Now the initial condition implies $2 = y(0) = 0 - 1 + c$ and hence $c = 3$. Thus,

$$y = t - 1 + 3e^{-t}.$$ ◄

Initial Value Problems

In many practical problems, an initial value may be given. As in Example 7 above, the initial condition may be used to determine the arbitrary scalar in the general solution once the general solution has been found. It can be useful, however, to have a single formula that directly encodes the initial value in the solution. This is accomplished by the following corollary to Theorem 2, which also establishes the uniqueness of the solution.

Corollary 8. *Let p and f be continuous on an interval I, $t_0 \in I$, and $y_0 \in \mathbb{R}$. Then the unique solution of the initial value problem*

$$y' + py = f, \quad y(t_0) = y_0 \tag{14}$$

is given by

$$y(t) = e^{-P(t)} \int_{t_0}^{t} e^{P(u)} f(u)\, du + y_0 e^{-P(t)}, \tag{15}$$

where $P(t) = \int_{t_0}^{t} p(u)\, du$.

Proof. Let y be given by (15). Since P is an antiderivative of p, it follows that $\mu = e^{P}$ is an integrating factor. Replacing e^{P} and e^{-P} in the formula by μ and μ^{-1}, respectively, we see that y has the form given in Theorem 2. Hence, $y(t)$ is a solution of the linear first order equation $y' + p(t)y = f(t)$. Moreover, $P(t_0) = \int_{t_0}^{t_0} p(u)\, du = 0$, and

$$y(t_0) = y_0 e^{-P(t_0)} + e^{-P(t_0)} \int_{t_0}^{t_0} e^{P(u)} f(u)\, du = y_0,$$

so that $y(t)$ is a solution of the initial value problem given by (14). As to uniqueness, suppose that $y_1(t)$ is any other such solution. Set $y_2(t) = y(t) - y_1(t)$. Then

$$\begin{aligned}
y_2' + p y_2 &= y' - y_1' + p(y - y_1) \\
&= y' + py - (y_1' + p y_1) \\
&= f - f = 0.
\end{aligned}$$

Further, $y_2(t_0) = y(t_0) - y_1(t_0) = 0$. It follows from Theorem 2 that $y_2(t) = c e^{-\tilde{P}(t)}$ for some constant $c \in \mathbb{R}$ and an antiderivative $\tilde{P}(t)$ of $p(t)$. Since $y_2(t_0) = 0$ and $e^{-\tilde{P}(t_0)} \neq 0$, it follows that $c = 0$. Thus, $y(t) - y_1(t) = y_2(t) = 0$ for all $t \in I$. This shows that $y_1(t) = y(t)$ for all $t \in I$, and hence, $y(t)$ is the only solution of (14).   □

Example 9. Use Corollary 8 to find the solution of the differential equation

$$y' + y = t,$$

with initial condition $y(0) = 2$.

▶ **Solution.** The coefficient function is $p(t) = 1$. Thus, $P(t) = \int_0^t p(u)\, du = u\big|_0^t = t$, $e^{P(t)} = e^t$ and $e^{-P(t)} = e^{-t}$. Let y be given as in the corollary. Then

$$\begin{aligned}
y(t) &= e^{-P(t)} \int_{t_0}^{t} e^{P(u)} f(u)\, du + y_0 e^{-P(t)} \\
&= e^{-t} \int_0^t u e^u\, du + 2 e^{-t}
\end{aligned}$$

$$= e^{-t} \left(ue^u - e^u\right)\big|_0^t + 2e^{-t}$$
$$= e^{-t} \left(te^t - e^t - (-1)\right) + 2e^{-t}$$
$$= t - 1 + e^{-t} + 2e^{-t} = t - 1 + 3e^{-t}.$$

Rather than memorizing (15), it is generally easier to remember Algorithm 3 and solve such problems as we did in Example 7.   ◀

Example 10. Find the solution of the initial value problem

$$y' + y = \frac{1}{1-t},$$

with $y(0) = 0$ on the interval $(-\infty, 1)$.

▶ **Solution.** By Corollary 8, the solution is

$$y(t) = e^{-t} \int_0^t \frac{e^u}{1-u}\, du.$$

Since the function $\frac{e^u}{1-u}$ has no closed form antiderivative on the interval $(-\infty, 1)$, we might be tempted to stop at this point and say that we have solved the equation. While this is a legitimate statement, the present representation of the solution is of little practical use, and a further detailed study is necessary if you are "really" interested in the solution. Any further analysis (numerical calculations, qualitative analysis, etc.) would be based on what type of information you are attempting to ascertain about the solution.   ◀

Analysis of the General Solution Set

For a linear differential equation $y' + py = f$, where p and f are continuous functions on an interval I, Theorem 2 gives the general solution

$$y = \mu^{-1} \int \mu f\, dt + c\mu^{-1},$$

where $\mu = e^P$, $P = \int p\, dt$, and c is an arbitrary constant. We see that the solution is the sum of two parts.

The Particular Solution

When we set $c = 0$ in the general solution above, we get a single fixed solution which we denote by y_p. Specifically,

$$y_p = \mu^{-1} \int \mu f \, dt = e^{-P} \int e^P f \, dt.$$

This solution is called a *particular solution*. Keep in mind though that there are many particular solutions depending on the antiderivative chosen for $\int e^P f \, dt$. Once an antiderivative is fixed, then the particular solution is determined.

The Homogeneous Solutions

When we set $f = 0$ in $y + py = f$, we get what is called the *associated homogeneous equation*, $y' + py = 0$. Its solution is then obtained from the general solution. The integral reduces to 0 since $f = 0$ so all that is left is the second summand ce^{-P}. We set

$$y_h = c\mu^{-1} = ce^{-P}.$$

This solution is called the *homogeneous solution*. The notation is sometimes used a bit ambiguously. Mostly, we think of y_h as a family of functions parameterized by c. Yet, there are times when we will have a specific c in mind and then y_h will be an actual function. It will be clear from the context which is meant.

The General Solution

The relationship between the *general* solution y_g of $y' + py = f$, a *particular* solution y_p of this equation, and the a *homogeneous* solution y_h, is usually expressed as

$$y_g = y_p + y_h. \tag{16}$$

What this means is that *every* solution to $y' + py = f$ can be obtained by starting with a single particular solution y_p and adding to that the homogeneous solution y_h. The key observation is the following. Suppose that y_1 and y_2 are *any* two solutions of $y' + py = f$. Then

$$(y_2 - y_1)' + p(y_2 - y_1) = (y_2' + py_2) - (y_1' + py_1)$$
$$= f - f$$
$$= 0,$$

so that $y_2 - y_1$ is a homogeneous solution. Let $y_h = y_2 - y_1$. Then $y_2 = y_1 + y_h$. Therefore, given a solution $y_1(t)$ of $y' + p(t)y = f(t)$, any other solution y_2 is obtained from y_1 by adding a homogeneous solution y_h.

Mixing Problems

Example 11. Suppose a tank contains 10 L of a brine solution (salt dissolved in water). Assume the initial concentration of salt is 100 g/L. Another brine solution flows into the tank at a rate of 3 L/min with a concentration of 400 g/L. Suppose the mixture is well stirred and flows out of the tank at a rate of 3 L/min. Let $y(t)$ denote the amount of salt in the tank at time t. Find $y(t)$. The following diagram may be helpful to visualize this problem.

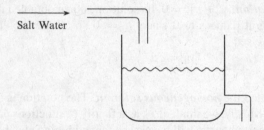

▶ **Solution.** Here again is a situation in which it is best to express how a quantity like $y(t)$ changes. If $y(t)$ represents the amount (in grams) of salt in the tank, then $y'(t)$ represents the total rate of change of salt in the tank with respect to time. It is given by the difference of two rates: the *input rate* of salt in the tank and the *output rate* of salt in the tank. Thus,

$$y' = \text{input rate} - \text{output rate.} \tag{17}$$

Both the input rate and output rate of salt are determined by product of the flow rate of the brine and the concentration of the brine solution. Specifically,

$$\frac{\text{amount of salt}}{\text{unit of time}} = \frac{\text{volume of brine}}{\text{unit of time}} \times \frac{\text{amount of salt}}{\text{volume of brine}}, \tag{18}$$

In this example, the units of measure are liters(L), grams(g), and minutes(min).

Input Rate of Salt   Since the inflow rate is 3 L/min and the concentration is fixed at 400 g/L we have

$$\text{input rate} = 3 \ \frac{L}{\min} \times 400 \ \frac{g}{L} = 1200 \ \frac{g}{\min}.$$

Output Rate of Salt   The outflow rate is likewise 3 L/min. However, the concentration of the solution that leaves the tank varies with time. Since the amount of salt in the tank is $y(t)$ g and the volume of fluid is always 10 L (the inflow and outflow rates are the same (3 L/min) so the volume never changes), the concentration of the fluid as it leaves the tank is $\frac{y(t)}{10} \ \frac{g}{L}$. We thus have

$$\text{output rate} = 3 \ \frac{L}{\min} \times \frac{y(t)}{10} \ \frac{g}{L} = \frac{3y(t)}{10} \ \frac{g}{\min}.$$

The initial amount of salt in the tank is $y(0) = 100 \ \frac{g}{L} \times 10 \ L = 1000$ g. Equation (17) thus leads us to the following linear differential equation

$$y' = 1,200 - \frac{3y}{10}, \tag{19}$$

with initial value $y(0) = 1,000$.

In standard form, (19) becomes $y' + \frac{3}{10}y = 1,200$. Multiplying by the integrating factor, $e^{3t/10}$, leads to $(e^{3t/10}y)' = 1,200e^{3t/10}$. Integrating and dividing by the integrating factor give

$$y(t) = 4,000 + ce^{-\frac{3}{10}t}.$$

Finally, the initial condition implies $1,000 = y(0) = 4,000 + c$ so $c = -3,000$. Hence,

$$y(t) = 4,000 - 3,000e^{-\frac{3}{10}t}. \tag{20}$$

Below we give the graph of (20), representing the amount of salt in the tank at time t.

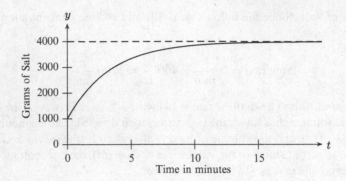

You will notice that $\lim_{t \to \infty}(4{,}000 - 3{,}000e^{-3t/10}) = 4{,}000$ as indicated by the horizontal asymptote in the graph. In the long term, the concentration of the brine in the tank approaches $\frac{4{,}000\,\text{g}}{10\,\text{L}} = 400$ g/L. This is expected since the brine coming into the tank has concentration 400 g/L.   ◀

Problems like Example 11 are called ***mixing problems*** and there are many variations. There is nothing special about the use of a salt solution. Any solvent with a solute will do. For example, one might want to study a pollutant flowing into a lake (the tank) with an outlet feeding a water supply for a town. In Example 11, the inflow and outflow rates were the same but we will consider examples where this is not the case. We will also consider situations where the concentration of salt in the inflow is not constant but varies as a function f of time. This function may even be discontinuous as will be the case in Chap. 6. In Chap. 9, we consider the case where two or more tanks are interconnected. Rather than memorize some general formulas, it is best to derive the appropriate differential equation on a case by case basis. Equations (17) and (18) are the basic principles to model such mixing problems.

Example 12. A large tank contains 100 gal of brine in which 50 lbs of salt are dissolved. Brine containing 2 lbs of salt per gallon flows into the tank at the rate of 6 gal/min. The mixture, which is kept uniform by stirring, flows out of the tank at the rate of 4 gal/min. Find the amount of salt in the tank at the end of t minutes. After 50 min, how much salt will be in the tank and what will be the volume of brine? (The units here are abbreviated: gallon(s)=gal, pound(s) =lbs, and minute(s) = min.)

▶ **Solution.** Let $y(t)$ denote the number of pounds of salt in the tank after t minutes. Note that in this problem, the difference between inflow and outflow rates is 2 gal/min. At time t, there will be $V(t) = 100 + 2t$ gal of brine, and so the concentration (in lbs/gal) will then be

$$\frac{y(t)}{V(t)} = \frac{y(t)}{100 + 2t}.$$

We use (18) to compute the input and output rates of salt:

Input Rate

$$\text{input rate} = 6 \frac{\text{gal}}{\text{min}} \times 2 \frac{\text{lbs}}{\text{gal}} = 12 \frac{\text{lbs}}{\text{min}}.$$

Output Rate

$$\text{output rate} = 4 \frac{\text{gal}}{\text{min}} \times \frac{y(t)}{100 + 2t} \frac{\text{lbs}}{\text{gal}} = \frac{4y(t)}{100 + 2t} \frac{\text{lbs}}{\text{min}}.$$

Applying (17) yields the initial value problem

$$y'(t) = 12 - \frac{4y(t)}{100 + 2t}, \quad y(0) = 50.$$

Simplifying and putting in standard form give $y' + \frac{2}{50+t}y = 12$. The coefficient function is $p(t) = \frac{2}{50+t}$, $P(t) = \int p(t)\, dt = 2\ln(50 + t) = \ln(50 + t)^2$, and the integrating factor is $\mu(t) = (50 + t)^2$. Thus, $((50 + t)^2 y)' = 12(50 + t)^2$. Integrating and simplifying give

$$y(t) = 4(50 + t) + \frac{c}{(50 + t)^2}.$$

The initial condition $y(0) = 50$ implies $c = -3(50)^3$ so

$$y = 4(50 + t) - \frac{3(50)^3}{(50 + t)^2}.$$

After $50\,\text{min}$, there will be $y(50) = 400 - \frac{3}{8}100 = 362.5\,\text{lbs}$ of salt in the tank and $200\,\text{gal}$ of brine. ◄

Exercises

1–25. Find the general solution of the given differential equation. If an initial condition is given, find the particular solution which satisfies this initial condition.

1. $y' + 3y = e^t$, $y(0) = -2$
2. $(\cos t)y' + (\sin t)y = 1$, $y(0) = 5$
3. $y' - 2y = e^{2t}$, $y(0) = 4$
4. $ty' + y = e^t$
5. $ty' + y = e^t$, $y(1) = 0$
6. $ty' + my = t \ln(t)$, where m is a constant
7. $y' = -\frac{y}{t} + \cos(t^2)$
8. $y' + 2y = \sin t$
9. $y' - 3y = 25 \cos 4t$
10. $t(t + 1)y' = 2 + y$
11. $z' = 2t(z - t^2)$
12. $y' + ay = b$, where a and b are constants
13. $y' + y \cos t = \cos t$, $y(0) = 0$
14. $y' - \dfrac{2}{t+1}y = (t + 1)^2$
15. $y' - \dfrac{2}{t}y = \dfrac{t+1}{t}$, $y(1) = -3$
16. $y' + ay = e^{-at}$, where a is a constant
17. $y' + ay = e^{bt}$, where a and b are constants and $b \neq -a$
18. $y' + ay = t^n e^{-at}$, where a is a constant
19. $y' = y \tan t + \sec t$
20. $ty' + 2y \ln t = 4 \ln t$
21. $y' - \dfrac{n}{t}y = e^t t^n$
22. $y' - y = te^{2t}$, $y(0) = a$
23. $ty' + 3y = t^2$, $y(-1) = 2$
24. $y' + 2ty = 1$, $y(0) = 1$
25. $t^2 y' + 2ty = 1$, $y(2) = a$

26–33. *Mixing Problems*

26. A tank contains 100 gal of brine made by dissolving 80 lb of salt in water. Pure water runs into the tank at the rate of 4 gal/min, and the mixture, which is kept uniform by stirring, runs out at the same rate. Find the amount of salt in the tank at any time t. Find the concentration of salt in the tank at any time t.

27. A tank contains 10 gal of brine in which 2 lb of salt is dissolved. Brine containing 1 lb of salt per gallon flows into the tank at the rate of 3 gal/min, and the stirred mixture is drained off the tank at the rate of 4 gal/min. Find the amount $y(t)$ of salt in the tank at any time t.

28. A tank holds 10 L of pure water. A brine solution is poured into the tank at a rate of 1 L/min and kept well stirred. The mixture leaves the tank at the same rate. If the brine solution has a concentration of 10 g of salt per liter, what will the concentration be in the tank after 10 min.

29. A 30-L container initially contains 10 L of pure water. A brine solution containing 20 g of salt per liter flows into the container at a rate of 4 L/min. The well-stirred mixture is pumped out of the container at a rate of 2 L/min.

   (a) How long does it take for the container to overflow?
   (b) How much salt is in the tank at the moment the tank begins to overflow?

30. A 100-gal tank initially contains 10 gal of fresh water. At time $t = 0$, a brine solution containing 0.5 lb of salt per gallon is poured into the tank at the rate of 4 gal/min while the well-stirred mixture leaves the tank at the rate of 2 gal/min.

   (a) Find the time T it takes for the tank to overflow.
   (b) Find the amount of salt in the tank at time T.
   (c) If $y(t)$ denotes the amount of salt present at time t, what is $\lim_{t \to \infty} y(t)$?

31. For this problem, our tank will be a lake and the brine solution will be polluted water entering the lake. Thus, assume that we have a lake with volume V which is fed by a polluted river. Assume that the rate of water flowing into the lake and the rate of water flowing out of the lake are equal. Call this rate r, let c be the concentration of pollutant in the river as it flows *into* the lake, and assume perfect mixing of the pollutant in the lake (this is, of course, a very unrealistic assumption).

   (a) Write down and solve a differential equation for the amount $P(t)$ of pollutant in the lake at time t and determine the limiting *concentration* of pollutant in the lake as $t \to \infty$.
   (b) At time $t = 0$, the river is cleaned up, so no more pollutant flows into the lake. Find expressions for how long it will take for the pollution in the lake to be reduced to (i) 1/2 and (ii) 1/10 of the value it had at the time of the cleanup.
   (c) Assuming that Lake Erie has a volume V of 460 km^3 and an inflow–outflow rate of $r = 175$ km^3/year, give numerical values for the times found in Part (b). Answer the same question for Lake Ontario, where it is assumed that $V = 1640$ km^3 and $r = 209$ km^3/year.

32. Two tanks, Tank 1 and Tank 2, are arranged so that the outflow of Tank 1 is the inflow of Tank 2. Each tank initially contains 10 L of pure water. A brine solution with concentration 100 g/L flows into Tank 1 at a rate of 4 L/min. The well-stirred mixture flows into Tank 2 at the same rate. In Tank 2, the mixture is

well stirred and flows out at a rate of 4 L/min. Find the amount of salt in Tank 2 at any time t. The following diagram may be helpful to visualize this problem.

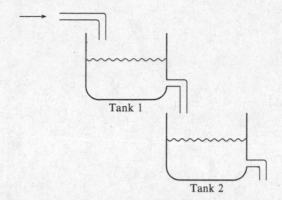

33. Two large tanks, Tank 1 and Tank 2, are arranged so that the outflow of Tank 1 is the inflow of Tank 2, as in the configuration of Exercise 32. Tank 1 initially contains 10 L of pure water while Tank 2 initially contains 5 L of pure water. A brine solution with concentration 10 g/L flows into Tank 1 at a rate of 4 L/min. The well-stirred mixture flows into Tank 2 at a rate of 2 L/min. In Tank 2, the mixture is well stirred and flows out at a rate of 1 liter per minute. Find the amount of salt in Tank 2 before either Tank 1 or Tank 2 overflows.

1.5 Substitutions

Just as a complicated integral may be simplified into a familiar integral form by a judicious substitution, so too may a differential equation. Generally speaking, substitutions take one of two forms. Either we replace y in $y' = F(t, y)$ by some function $y = \phi(t, v)$ to get a differential equation $v' = G(t, v)$ in v and t, or we set $z = \psi(t, y)$ for some expression $\psi(t, y)$ that appears in the formula for $F(t, y)$, resulting in a differential equation $z' = H(t, z)$ in z and t. Of course, the goal is that this new differential equation will fall into a category where there are known solution methods. Once the unknown function v (or z) is determined, then y can be determined by $y = \phi(t, v)$ (or implicitly by $z = \psi(t, y)$). In this section, we will illustrate this procedure by discussing the homogeneous and Bernoulli equations, which are simplified to separable and linear equations, respectively, by appropriate substitutions.

Homogeneous Differential Equations

A *homogeneous differential equation*[10] is a first order differential equation that can be written in the form

$$y' = f\left(\frac{y}{t}\right), \tag{1}$$

for some function f. To solve such a differential equation, we use the substitution $y = tv$, where $v = v(t)$ is some unknown function and rewrite (1) as a differential equation in terms of t and v. By the product rule, we have $y' = v + tv'$. Substituting this into (1) gives $v + tv' = f(tv/t) = f(v)$, a separable differential equation for which the variables t and v may be separated to give

$$\frac{dv}{f(v) - v} = \frac{dt}{t}. \tag{2}$$

Once v is found we may determine y by the substitution $y = tv$. *It is not useful to try to memorize* (2). *Rather, remember the substitution method that leads to it.* As an illustration, consider the following example.

Example 1. Solve the following differential equation:

$$y' = \frac{t + y}{t - y}.$$

---

[10]Unfortunately, the term "homogeneous" used here is the same term used in Sect. 1.4 to describe a linear differential equation in which the forcing function is zero. These meanings are different. Usually, context will determine the appropriate meaning.

▶ **Solution.** This differential equation is neither separable nor linear. However, it is homogeneous since we can write the right-hand side as

$$\frac{t+y}{t-y} = \frac{(1/t)\,t+y}{(1/t)\,t-y} = \frac{1+(y/t)}{1-(y/t)},$$

which is a function of y/t. Let $y = tv$, so that $y' = v + tv'$. The given differential equation can then be written as

$$v + tv' = \frac{1+v}{1-v}.$$

Now subtract v from both sides and simplify to get

$$tv' = \frac{1+v}{1-v} - v = \frac{1+v^2}{1-v}.$$

Separating the variables gives

$$\frac{dv}{1+v^2} - \frac{v\,dv}{1+v^2} = \frac{dt}{t}.$$

Integrating both sides gives us $\tan^{-1} v - (1/2)\ln(1+v^2) = \ln t + c$. Now substitute $v = y/t$ to get the implicit solution

$$\tan^{-1}(y/t) - \frac{1}{2}\ln(1+(y/t)^2) = \ln t + c.$$

This may be simplified slightly (multiply by 2 and combine the ln terms) to get

$$2\tan^{-1}(y/t) - \ln(t^2+y^2) = 2c. \qquad \blacktriangleleft$$

A differential equation $y' = F(t, y)$ is homogeneous if we can write $F(t, y) = f(y/t)$, for some function f. This is what we did in the example above. To help identify situations where the identification $F(t, y) = f(y/t)$ is possible, we introduce the concept of a homogeneous function. A function $p(t, y)$ is said to be **homogeneous of degree** n if $p(\alpha t, \alpha y) = \alpha^n p(t, y)$, for all $\alpha > 0$. To simplify the terminology, we will call a homogeneous function of degree zero just **homogeneous**. For example, the polynomials $t^3 + ty^2$ and y^3 are both homogeneous of degree 3. Their quotient $F(t, y) = (t^3 + ty^2)/y^3$ is a homogeneous function. Indeed, if $\alpha > 0$, then

$$F(\alpha t, \alpha y) = \frac{(\alpha t)^3 + (\alpha t)(\alpha y)^2}{(\alpha y)^3} = \frac{\alpha^3}{\alpha^3}\frac{t^3+ty^2}{y^3} = F(t, y).$$

In fact, it is easy to see that the quotient of any two homogeneous functions of the same degree is a homogeneous function. Suppose $t > 0$ and let $\alpha = 1/t$. For a homogeneous function F, we have

$$F(t,\, y) = F\left(\frac{1}{t}t,\, \frac{1}{t}y\right) = F\left(1,\, \frac{y}{t}\right).$$

Similarly, if $t < 0$ and $\alpha = -1/t$ then

$$F(t,\, y) = F\left(\frac{-1}{t}t,\, \frac{-1}{t}y\right) = F\left(-1,\, -\frac{y}{t}\right).$$

Therefore, if $F(t,\, y)$ is homogeneous, then $F(t,\, y) = f(t/y)$ where

$$f\left(\frac{y}{t}\right) = \begin{cases} F\left(1,\, \dfrac{y}{t}\right) & \text{if } t > 0, \\ F\left(-1,\, -\dfrac{y}{t}\right) & \text{if } t < 0 \end{cases}$$

so that we have the following useful criterion for identifying homogeneous functions:

Lemma 2. *If $F(t,\, y)$ is a homogeneous function, then it can be written in the form*

$$F(t,\, y) = f\left(\frac{y}{t}\right),$$

for some function f. Furthermore, the differential equation $y' = F(t,\, y)$ is a homogeneous differential equation.

Example 3. Solve the following differential equation

$$y' = \frac{t^2 + 2ty - y^2}{2t^2}.$$

▶ **Solution.** Since the numerator and denominator are homogeneous of degree 2, the quotient is homogeneous. Letting $y = tv$ and dividing the numerator and denominator of the right-hand side by t^2, we obtain

$$v + tv' = \frac{1 + 2v - v^2}{2}.$$

Subtracting v from both sides gives $tv' = \frac{1-v^2}{2}$. There are two equilibrium solutions: $v = \pm 1$. If $v \neq \pm 1$, separate the variables to get

$$\frac{2\,dv}{1 - v^2} = \frac{dt}{t},$$

and apply partial fractions to the left-hand side to conclude

$$\frac{dv}{1-v} + \frac{dv}{1+v} = \frac{dt}{t}.$$

Integrating gives $-\ln|1-v| + \ln|1+v| = \ln|t| + c$, exponentiating both sides gives $\frac{1+v}{1-v} = kt$ for $k \neq 0$, and simplifying leads to $v = \frac{kt-1}{kt+1}$, $k \neq 0$. Substituting $y = vt$ gives the solutions

$$y = \frac{kt^2 - t}{kt + 1}, \quad k \neq 0,$$

whereas the equilibrium solutions $v = \pm 1$ produce the solutions $y = \pm t$. Note that for $k = 0$, we get the solution $y = -t$, which is one of the equilibrium solutions, but the equilibrium solution $y = t$ does not correspond to any choice of k. ◀

Bernoulli Equations

A differential equation of the form

$$y' + p(t)y = f(t)y^n, \tag{3}$$

is called a ***Bernoulli equation***.[11] If $n = 0$, this equation is linear, while if $n = 1$, the equation is both separable and linear. Thus, we will assume $n \neq 0, 1$. Note, also, that if $n > 0$, then $y = 0$ is a solution. Start by dividing (3) by y^n to get

$$y^{-n}y' + p(t)y^{1-n} = f(t). \tag{4}$$

Use the coefficient of $p(t)$ as a new dependent variable. That is, use the substitution $z = y^{1-n}$. Thus, z is treated as a function of t and the chain rule gives $z' = (1 - n)y^{-n}y'$, which is the first term of (4) multiplied by $1 - n$. Therefore, substituting for z and multiplying by the constant $(1 - n)$ give,

$$z' + (1-n)p(t)z = (1-n)f(t), \tag{5}$$

which is a *linear* first order differential equation in the variables t and z. Equation (5) can then be solved by Algorithm 3 of Sect. 1.4, and the solution to (3) is obtained by solving $z = y^{1-n}$ for y (and including $y = 0$ in the case $n > 0$).

Example 4. Solve the Bernoulli equation $y' + y = 5(\sin 2t)y^2$.

---

[11]Named after Jakoub Bernoulli (1654–1705).

► **Solution.** Note first that $y = 0$ is a solution since $n = 2 > 0$. After dividing our equation by y^2, we get $y^{-2}y' + y^{-1} = 5 \sin 2t$. Let $z = y^{-1}$. Then $z' = -y^{-2}y'$ and substituting gives $-z' + z = 5 \sin 2t$. In the standard form for linear equations, this becomes

$$z' - z = -5 \sin 2t.$$

We can apply Algorithm 3 of Sect. 1.4 to this linear differential equation. The integrating factor will be e^{-t}. Multiplying by the integrating factor gives $(e^{-t}z)' = -5e^{-t} \sin 2t$. Now integrate both sides to get

$$e^{-t}z = \int -5e^{-t} \sin 2t \, dt = (\sin 2t + 2\cos 2t)e^{-t} + c,$$

where $\int -5e^{-t} \sin 2t \, dt$ is computed by using integration by parts twice. Hence,

$$z = \sin 2t + 2\cos 2t + ce^t.$$

Now go back to the original function y by solving $z = y^{-1}$ for y to get

$$y = z^{-1} = (\sin 2t + 2\cos 2t + ce^t)^{-1} = \frac{1}{\sin 2t + 2\cos 2t + ce^t}.$$

This function y, together with $y = 0$, constitutes the general solution of $y' + y = 5(\sin 2t)y^2$. ◄

Remark 5. We should note that in the z variable, the solution $z = \sin t - \cos t + ce^t$ is valid for all t in $\mathbb{R}$. This is to be expected. Since $z' - z = -5 \sin 2t$ is linear and the coefficient function, $p(t) = 1$, and the forcing function, $f(t) = -5 \sin 2t$, are continuous on $\mathbb{R}$, Theorem 2 of Sect. 1.2 implies that any solution will be defined on all of $\mathbb{R}$. However, in the y variable, things are very different. The solution $y = 1/(\sin t - \cos t + ce^t)$ to the given Bernoulli equation is valid only on intervals where the denominator is nonzero. This is precisely because of the inverse relationship between y and z: the location of the zeros of z is precisely where y has a vertical asymptote. In the figure below, we have graphed a solution in the z variable and the corresponding solution in the $y = 1/z$ variable. Note that in the y variable, the intervals of definition are determined by the zeros of z. Again, it is not at all evident from the differential equation, $y' - y = -2(\sin t)y^2$, that the solution curves would have "chopped up" intervals of definition. The linear case as described in Theorem 2 of Sect. 1.4 is thus rather special.

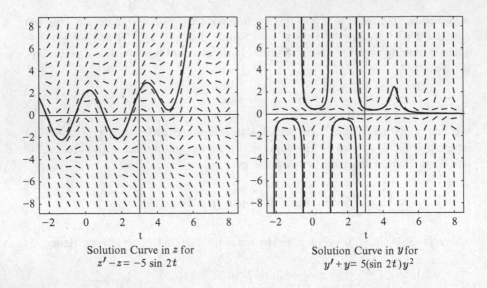

Solution Curve in z for
$z' - z = -5 \sin 2t$

Solution Curve in y for
$y' + y = 5(\sin 2t)y^2$

Linear Substitutions

Consider a differential equation of the form

$$y' = f(at + by + c), \tag{6}$$

where a, b, c are real constants. Let $z = at + by + c$. Then $z' = a + by'$ so
$y' = (z' - a)/b$. Substituting gives $(z' - a)/b = f(z)$ which in standard form is

$$z' = bf(z) + a,$$

a separable differential equation. Consider the following example.

Example 6. Solve the following differential equation:

$$y' = (t + y + 1)^2.$$

▶ **Solution.** Let $z = t + y + 1$. Then $z' = 1 + y'$ so $y' = z' - 1$. Substituting
gives $z' - 1 = z^2$, and in standard form, we get the separable differential equation
$z' = 1 + z^2$. Separating variables gives

$$\frac{dz}{1 + z^2} = dt.$$

Integrating gives $\tan^{-1} z = t + c$. Thus, $z = \tan(t + c)$. Now substitute $z = t + y + 1$ and simplify to get the solution

$$y = -t - 1 + \tan(t + c). \qquad \blacktriangleleft$$

Substitution methods are a general way to simplify complicated differential equations into familiar forms. These few that we have discussed are just some of many examples that can be solved by the technique of substitution. If you ever come across a differential equation you cannot solve, try to simplify it into a familiar form by finding the right substitution.

Exercises

1–8. Find the general solution of each of the following homogeneous equations. If an initial value is given, also solve the initial value problem.

1. $t^2 y' = y^2 + yt + t^2, \quad y(1) = 1$

2. $y' = \dfrac{4t - 3y}{t - y}$

3. $y' = \dfrac{y^2 - 4yt + 6t^2}{t^2}, \quad y(2) = 4$

4. $y' = \dfrac{y^2 + 2yt}{t^2 + yt}$

5. $y' = \dfrac{3y^2 - t^2}{2ty}$

6. $y' = \dfrac{t^2 + y^2}{ty}, \quad y(e) = 2e$

7. $ty' = y + \sqrt{t^2 - y^2}$

8. $t^2 y' = yt + y\sqrt{t^2 + y^2}$

9–17. Find the general solution of each of the following Bernoulli equations. If an initial value is given, also solve the initial value problem.

9. $y' - y = ty^2, \quad y(0) = 1$

10. $y' + y = y^2, \quad y(0) = 1$

11. $y' + ty = ty^3$

12. $y' + ty = t^3 y^3$

13. $(1 - t^2)y' - ty = 5ty^2$

14. $y' + \dfrac{y}{t} = y^{2/3}$

15. $yy' + ty^2 = t, \quad y(0) = -2$

16. $2yy' = y^2 + t - 1$

17. $y' + y = ty^3$

18. Write the logistic differential equation $P' = r(1 - \frac{P}{m})P$ ((19) of Sect. 1.3) as a Bernoulli equation and solve it using the technique described in this section.

19–22. Use an appropriate linear substitution to solve the following differential equations:

19. $y' = (2t - 2y + 1)^{-1}$

20. $y' = (t - y)^2$

21. $y' = \dfrac{1}{(t + y)^2}$

22. $y' = \sin(t - y)$

23–26. Use the indicated substitution to solve the given differential equation.

23. $2yy' = y^2 + t - 1, z = y^2 + t - 1$ (Compare with Exercise 16.)

24. $y' = \tan y + \dfrac{2\cos t}{\cos y}, z = \sin y$

25. $y' + y \ln y = ty, z = \ln y$

26. $y' = -e^y - 1, z = e^{-y}$

1.6 Exact Equations

When $y(t)$ is any function defined implicitly by an equation

$$V(t, y) = c,$$

that is,

$$V(t, y(t)) = c$$

for all t in an interval I, implicit differentiation of this equation with respect to t shows that $y(t)$ satisfies the equation

$$0 = \frac{d}{dt}c = \frac{d}{dt}V(t, y(t)) = \frac{\partial V}{\partial t}(t, y(t)) + \frac{\partial V}{\partial y}(t, y(t))y'(t),$$

and thus is a solution of the differential equation

$$\frac{\partial V}{\partial t} + \frac{\partial V}{\partial y}y' = 0.$$

For example, any solution $y(t)$ of the equation $y^2 + 2yt - y^2 = c$ is a solution of the differential equation

$$(2y - 2t) + (2y + 2t)y' = 0.$$

Conversely, suppose we are given a differential equation

$$M + Ny' = 0, \tag{1}$$

where M and N are functions of the two variables t and y. If there is a function $V(t, y)$ for which

$$M = \frac{\partial V}{\partial t} \quad \text{and} \quad N = \frac{\partial V}{\partial y}, \tag{2}$$

then we can work backward from the implicit differentiation in the previous paragraph to conclude that any implicitly defined solution of

$$V(t, y) = c, \tag{3}$$

for c an arbitrary constant, is a solution of the differential equation (1). To summarize, *if the given functions M and N in (1) are such that there is a function $V(t, y)$ for which equations (2) are satisfied, then the solution of the differential equation (1) is given by the implicitly defined solutions to (3)*. In this case, we will say that the differential equation $M + Ny' = 0$ is an ***exact differential equation***.

Suppose we are given a differential equation $M + Ny' = 0$, but we are not given a priori that $M = \partial V/\partial t$ and $N = \partial V/\partial t$. How can we determine if there is such a function $V(t, y)$, and if there is, how can we find it? That is, is there a *criterion* for determining if a given differential equation is exact, and if so, is there a *procedure* for producing the function $V(t, y)$ that determines the solution via $V(t, y) = c$. The answer to both questions is yes.

A criterion for exactness comes from the equality of mixed partial derivatives. Recall (from your calculus course) that all functions $V(t, y)$ whose second partial derivatives exist and are continuous satisfy[12]

$$\frac{\partial}{\partial y}\left(\frac{\partial V}{\partial t}\right) = \frac{\partial}{\partial t}\left(\frac{\partial V}{\partial y}\right). \tag{4}$$

If the equation $M + Ny' = 0$ is exact, then, by definition, there is a function $V(t, y)$ such that $\partial V/\partial t = M$ and $\partial V/\partial y = N$, so (4) gives

$$\frac{\partial M}{\partial y} = \frac{\partial}{\partial y}\left(\frac{\partial V}{\partial t}\right) = \frac{\partial}{\partial t}\left(\frac{\partial V}{\partial y}\right) = \frac{\partial N}{\partial t}.$$

Hence, a *necessary condition for exactness* of $M + Ny' = 0$ is

$$\frac{\partial M}{\partial y} = \frac{\partial N}{\partial t}. \tag{5}$$

We now ask if this condition is also sufficient for the differential equation $M + Ny' = 0$ to be exact. That is, if (5) is true, can we always find a function $V(t, y)$ such that $M = \partial V/\partial t$ and $N = \partial V/\partial y$? We can easily find $V(t, y)$ such that

$$\frac{\partial V}{\partial t} = M \tag{6}$$

by integrating M with respect to the t variable. After determining $V(t, y)$ so that (6) holds, can we also guarantee that $\partial V/\partial y = N$? Integrating (6) with respect to t, treating the y variable as a constant, gives

$$V = \int M\, dt + \phi(y). \tag{7}$$

The function $\phi(y)$ is an arbitrary function of y that appears as the "integration constant" since any function of y goes to 0 when differentiated with respect to t. The "integration constant" $\phi(y)$ can be determined by differentiating V in (7) with respect to y and equating this expression to N since $\partial V/\partial y = N$ if the differential

---

[12]This equation is known as Clairaut's theorem (after Alexis Clairaut (1713–1765)) on the equality of mixed partial derivatives.

equation is exact. Thus (since integration with respect to t and differentiation with respect to y commute)

$$\frac{\partial V}{\partial y} = \frac{\partial}{\partial y} \int M \, dt + \frac{d\phi}{dy} = \int \frac{\partial M}{\partial y} \, dt + \frac{d\phi}{dy} = N. \tag{8}$$

That is,

$$\frac{d\phi}{dy} = N - \int \frac{\partial M}{\partial y} \, dt. \tag{9}$$

The verification that the function on the right is really a function only of y (as it must be if it is to be the derivative $\frac{d\phi}{dy}$ of a function of y) is where condition (5) is needed. Indeed, using (5)

$$N - \int \frac{\partial M}{\partial y} \, dt = N - \int \frac{\partial N}{\partial t} \, dt = N - (N + \psi(y)) = -\psi(y). \tag{10}$$

Hence, (9) is a valid equation for determining $\phi(y)$ in (7).

We can summarize our conclusions in the following result.

Theorem 1 (Criterion for exactness). *A first order differential equation*

$$M(t, y) + N(t, y)y' = 0$$

is exact if and only if

$$\boxed{\frac{\partial M}{\partial y} = \frac{\partial N}{\partial t}.} \tag{11}$$

If this condition is satisfied, then the general solution of the differential equation is given by $V(t, y) = c$, where $V(t, y)$ is determined by (7) and (9).

The steps needed to derive the solution of an exact equation when dealing with concrete examples are summarized in the following algorithm.

Algorithm 2. The following procedure is used to solve an exact differential equation.

Solution Method for Exact Equations

1. *Check the equation $M + Ny' = 0$ for exactness:* To do this, check if

$$\frac{\partial M}{\partial y} = \frac{\partial N}{\partial t}.$$

2. *Integrate the equation $M = \partial V/\partial t$ with respect to t to get*

$$V(t, y) = \int M(t, y)\,dt + \phi(y), \tag{12}$$

   where the "integration constant" $\phi(y)$ is a function of y.

3. *Differentiate this expression for $V(t, y)$ with respect to y, and set it equal to N:*

$$\frac{\partial}{\partial y}\left(\int M(t, y)\,dt\right) + \frac{d\phi}{dy} = \frac{\partial V}{\partial y} = N(t, y).$$

4. *Solve this equation for $\frac{d\phi}{dy}$:*

$$\frac{d\phi}{dy} = N(t, y) - \frac{\partial}{\partial y}\left(\int M(t, y)\,dt\right). \tag{13}$$

5. *Integrate $\frac{d\phi}{dy}$ to get $\phi(y)$ and then substitute in (12) to find $V(t, y)$.*
6. *The solution of $M + Ny' = 0$ is then given by*

$$V(t, y) = c, \tag{14}$$

   where c is an arbitrary constant.

It is worthwhile to emphasize that *the solution to the exact equation $M + Ny' = 0$ is not the function $V(t, y)$ found by the above procedure but rather the functions $y(t)$ defined by the implicit relation $V(t, y) = c$.*

Example 3. Determine if the differential equation

$$(3t^2 + 4ty - 2) + (2t^2 + 6y^2)y' = 0$$

is exact. If it is exact, solve it.

▶ **Solution.** This equation is in the standard form $M + Ny'$ with $M(t, y) = 3t^2 + 4ty - 2$ and $N(t, y) = 2t^2 + 6y^2$. Then

$$\frac{\partial M}{\partial y} = 4t = \frac{\partial N}{\partial t},$$

so the exactness criterion is satisfied and the equation is exact. To solve the equation, follow the solution procedure by computing

$$V(t, y) = \int (3t^2 + 4ty - 2)\,dt + \phi(y) = t^3 + 2t^2y - 2t + \phi(y),$$

where $\phi(y)$ is a function only of y that is yet to be computed. Now differentiate this expression with respect to y to get

$$\frac{\partial V}{\partial y} = 2t^2 + \frac{d\phi}{dy}.$$

But since the differential equation is exact, we also have

$$\frac{\partial V}{\partial y} = N(t, y) = 2t^2 + 6y^2,$$

and combining these last two expressions, we conclude

$$\frac{d\phi}{dy} = 6y^2.$$

Integrating with respect to y gives $\phi(y) = 2y^3 + c_0$, so that

$$V(t, y) = t^3 + 2t^2y - 2t + 2y^3 + c_0,$$

where c_0 is a constant. The solutions of the differential equation are then given by the relation $V(t, y) = c$, that is,

$$t^3 + 2t^2y - 2t + 2y^3 = c,$$

where the constant c_0 has been incorporated in the constant c.   ◄

What happens if we try to solve the equation $M(t, y) + N(t, y)y' = 0$ by the procedure outlined above without *first* verifying that it is exact? If the equation is not exact, you will discover this fact when you get to (9), since $\frac{d\phi}{dy}$ will *not* be a function only of y, as the following example illustrates.

Example 4. Try to solve the equation $(t - 3y) + (2t + y)y' = 0$ by the solution procedure for exact equations.

► **Solution.** Note that $M(t, y) = t - 3y$ and $N(t, y) = 2t + y$. First apply (12) to get

$$V(t, y) = \int M(t, y),dt = \int (t - 3y)\,dt = \frac{t^2}{2} - 3ty + \phi(y),$$

and then determine $\phi(y)$ from (13):

$$\frac{d\phi}{dy} = N(t, y) - \frac{\partial}{\partial y} \int M(t, y)\, dt = (2t + y) - \frac{\partial}{\partial y}\left(\frac{t^2}{2} - 3ty + \phi(y)\right) = y - t.$$

But we see that there is a problem since $\frac{d\phi}{dy} = y - t$ involves both y and t. This is where it becomes obvious that you are not dealing with an exact equation, and you cannot proceed with this procedure. Indeed, $\partial M/\partial y = -3 \neq 2 = \partial N/\partial t$, so that this equation fails the exactness criterion (11).   ◀

Integrating Factors

The differential equation

$$3y^2 + 8t + 2tyy' = 0 \tag{15}$$

can be written in the form $M + Ny' = 0$ where

$$M = 3y^2 + 8t \qquad \text{and} \qquad N = 2ty.$$

Since

$$\frac{\partial M}{\partial y} = 6y, \qquad \frac{\partial N}{\partial t} = 2y,$$

the equation is not exact. However, if we multiply (15) by t^2, we arrive at an equivalent differential equation

$$3y^2t^2 + 8t^3 + 2t^3yy' = 0, \tag{16}$$

that is exact. Indeed, in the new equation, $M = 3y^2t^2 + 8t^3$ and $N = 2t^3y$ so that the exactness criterion is satisfied:

$$\frac{\partial M}{\partial y} = 6yt^2 = \frac{\partial N}{\partial t}.$$

Using the solution method for exact equations, we find

$$V(t, y) = \int \left(3y^2t^2 + 8t^3\right)\, dt = y^2t^3 + 2t^4 + \phi(y),$$

where

$$\frac{\partial V}{\partial y} = 2yt^3 + \frac{d\phi}{dy} = N = 2t^3y.$$

Thus, $\frac{d\phi}{dy} = 0$ so that

$$V(t, y) = y^2 t^3 + 2t^4 + c_0.$$

Therefore, incorporating the constant c_0 into the general constant c, we have

$$y^2 t^3 + 2t^4 = c,$$

as the solution of (16) and, hence, also of the equivalent equation (15). Thus, multiplication of the nonexact equation (15) by the function t^2 has produced an exact equation (16) that is easily solved.

This suggests the following question. Can we find a function $\mu(t, y)$ so that multiplication of a nonexact equation

$$M + Ny' = 0 \tag{17}$$

by μ produces an equation

$$\mu M + \mu N y' = 0 \tag{18}$$

that is exact? When there is such a μ, we call it an ***integrating factor*** for (17).

The exactness criterion for (18) is

$$\frac{\partial(\mu M)}{\partial y} = \frac{\partial(\mu N)}{\partial t}.$$

Written out, this becomes

$$\mu \frac{\partial M}{\partial y} + M \frac{\partial \mu}{\partial y} = \mu \frac{\partial N}{\partial t} + N \frac{\partial \mu}{\partial t},$$

which implies

$$\frac{\partial \mu}{\partial t} N - \frac{\partial \mu}{\partial y} M = \left(\frac{\partial M}{\partial y} - \frac{\partial N}{\partial t} \right) \mu. \tag{19}$$

Thus, it appears that the search for an integrating factor for the nonexact ordinary differential equation (17) involves the solution of (19), which is a partial differential equation for μ. In general, it is quite difficult to solve (19). However, there are some special situations in which it is possible to use (19) to find an integrating factor μ. One such example occurs if μ *is a function only of* t. In this case, $\partial \mu / \partial t = d\mu/dt$ and $\partial \mu / \partial y = 0$ so (19) can be written as

$$\frac{1}{\mu} \frac{d\mu}{dt} = \frac{\partial M/\partial y - \partial N/\partial t}{N}. \tag{20}$$

Since the left-hand side of this equation depends only on t, the same must be true of the right-hand side. We conclude that we can find an integrating factor $\mu = \mu(t)$

consisting of a function only of t provided the right-hand side of (20) is a function only of t. If this is true, then we put

$$\frac{\partial M/\partial y - \partial N/\partial t}{N} = p(t), \tag{21}$$

so that (20) becomes the linear differential equation $\mu' = p(t)\mu$, which has the solution

$$\mu(t) = e^{\int p(t)\,dt}. \tag{22}$$

After multiplication of (17) by this $\mu(t)$, the resulting equation is exact, and the solution is obtained by the solution method for exact equations.

Example 5. Find an integrating factor for the equation

$$3y^2 + 8t + 2tyy' = 0.$$

▶ **Solution.** This equation is (15). Since

$$\frac{\partial M/\partial y - \partial N/\partial t}{N} = \frac{6y - 2y}{2ty} = \frac{2}{t},$$

is a function of only t, we conclude that an integrating factor is

$$\mu(t) = e^{\int (2/t)\,dt} = e^{2\ln t} = t^2.$$

This agrees with what we already observed in (16).   ◀

Similar reasoning gives a criterion for (17) to have an integrating factor $\mu(y)$ that involves only the variable y. Namely, if the expression

$$\frac{\partial M/\partial y - \partial N/\partial t}{-M} \tag{23}$$

is a function of only y, say $q(y)$, then

$$\mu(y) = e^{\int q(y)\,dy} \tag{24}$$

is a function only of y that is an integrating factor for (17).

Remark 6. A linear differential equation $y' + p(t)y = f(t)$ can be rewritten in the form $M + Ny' = 0$ where $M = p(t)y - f(t)$ and $N = 1$. In this case, $\partial M/\partial y = p(t)$ and $\partial N/\partial t = 0$ so a linear equation is never exact unless $p(t) = 0$. However,

$$\frac{\partial M/\partial y - \partial N/\partial t}{N} = p(t)$$

so there is an integrating factor that depends only on t:

$$\mu(t) = e^{\int p(t)\,dt}.$$

This is exactly the same function that we called an integrating factor for the linear differential equation in Sect. 1.4. Thus, the integrating factor for a linear differential equation is a special case of the concept of an integrating factor to transform a general equation into an exact one.

Exercises

1–9. Determine if the equation is exact, and if it is exact, find the general solution.

1. $(y^2 + 2t) + 2tyy' = 0$
2. $y - t + (t + 2y)y' = 0$
3. $2t^2 - y + (t + y^2)y' = 0$
4. $y^2 + 2tyy' + 3t^2 = 0$
5. $(3y - 5t) + 2yy' - ty' = 0$
6. $2ty + (t^2 + 3y^2)y' = 0$, $y(1) = 1$
7. $2ty + 2t^3 + (t^2 - y)y' = 0$
8. $t^2 - y - ty' = 0$
9. $(y^3 - t)y' = y$
10. Find conditions on the constants a, b, c, d which guarantee that the differential equation $(at + by) = (ct + dy)y'$ is exact.

1.7 Existence and Uniqueness Theorems

Let us return to the general initial value problem

$$y' = F(t, y), \quad y(t_0) = y_0, \tag{1}$$

introduced in Sect. 1.1. We want to address the existence and uniqueness of solutions to (1). The main theorems we have in mind put certain relatively mild conditions on F to insure that a solution exists, is unique, and/or both. We say that a solution *exists* if there is a function $y = y(t)$ defined on an interval containing t_0 as an interior point and satisfying (1). We say that a solution is *unique* if there is only one such function $y = y(t)$. Why are the concepts of existence and uniqueness important? Differential equations frequently model real-world systems as a function of time. Knowing that a solution exists means that the system has predictable future states. Further, when that solution is also unique, then for each future time there is only *one* possible state, leaving no room for ambiguity.

To illustrate these points and some of the theoretical aspects of the existence and uniqueness theorems that follow, consider the following example.

Example 1. Show that the following two functions satisfy the initial value problem

$$y' = 3y^{2/3}, \quad y(0) = 0. \tag{2}$$

1. $y(t) = 0$
2. $y(t) = t^3$

▶ **Solution.** Clearly, the constant function $y(t) = 0$ for $t \in \mathbb{R}$ is a solution. For the second function $y(t) = t^3$, observe that $y' = 3t^2$ while $3y^{2/3} = 3(t^3)^{2/3} = 3t^2$. Further, $y(0) = 0$. It follows that both of the functions $y(t) = 0$ and $y(t) = t^3$ are solutions of (2). ◀

If this differential equation modeled a real-world system, we would have problems accurately predicting future states. Should we use $y(t) = 0$ or $y(t) = t^3$? It is even worse than this, for further analysis of this differential equation reveals that there are many other solutions from which to choose. (See Example 9.) What is it about (2) that allows multiple solutions? More precisely, what conditions could we impose on (1) to guarantee that a solution exists and is unique? These questions are addressed in Picard's existence and uniqueness theorem, Theorem 5, stated below.

Thus far, our method for proving the existence of a solution to an ordinary differential equation has been to explicitly find one. This has been a reasonable approach for the categories of differential equations we have introduced thus far. However, there are many differential equations that do not fall into any of these categories, and knowing that a solution exists is a fundamental piece of information in the analysis of any given initial value problem.

Suppose $F(t, y)$ is a *continuous* function of (t, y) in the rectangle

$$\mathcal{R} := \{(t, y) : a \le t \le b, c \le y \le d\}$$

and (t_0, y_0) is an interior point of $\mathcal{R}$. The key to the proof of existence and uniqueness is the fact that a continuously differentiable function $y(t)$ is a solution of (1) if and only if it is a solution of the integral equation

$$y(t) = y_0 + \int_{t_0}^{t} F(u, y(u)) \, du. \qquad (3)$$

To see the equivalence between (1) and (3), assume that $y(t)$ is a solution to (1), so that

$$y'(t) = F(t, y(t))$$

for all t in an interval containing t_0 as an interior point and $y(t_0) = y_0$. Replace t by u in this equation, integrate both sides from t_0 to t, and use the fundamental theorem of calculus to get

$$\int_{t_0}^{t} F(u, y(u)) \, du = \int_{t_0}^{t} y'(u) \, du = y(t) - y(t_0) = y(t) - y_0,$$

which implies that $y(t)$ is a solution of (3). Conversely, if $y(t)$ is a continuously differentiable solution of (3), it follows that

$$g(t) := F(t, y(t))$$

is a continuous function of t since $F(t, y)$ is a continuous function of t and y. Apply the fundamental theorem of calculus to get

$$y'(t) = \frac{d}{dt} y(t) = \frac{d}{dt} \left(y_0 + \int_{t_0}^{t} F(u, y(u)) \, du \right)$$

$$= \frac{d}{dt} \left(y_0 + \int_{t_0}^{t} g(u) \, du \right) = g(t) = F(t, y(t)),$$

which is what it means for $y(t)$ to be a solution of (1). Since

$$y(t_0) = y_0 + \int_{t_0}^{t_0} F(u, y(u)) \, du = y_0,$$

$y(t)$ also satisfies the initial value in (1).

We will refer to (3) as the *integral equation* corresponding to the initial value problem (1) and conversely, (1) is referred to as the *initial value problem* corresponding to the integral equation (3). What we have shown is that a solution to the initial value problem is a solution to the corresponding integral equation and vice versa.

Example 2. Find the integral equation corresponding to the initial value problem

$$y' = t + y, \quad y(0) = 1.$$

▶ **Solution.** In this case, $F(t, y) = t + y$, $t_0 = 0$ and $y_0 = 1$. Replace the independent variable t by u in $F(t, y(t))$ to get $F(u, y(u)) = u + y(u)$. Thus, the integral equation (3) corresponding to this initial value problem is

$$y(t) = 1 + \int_0^t (u + y(u)) \, du. \qquad \blacktriangleleft$$

For any continuous function y, define

$$\mathcal{T}y(t) = y_0 + \int_{t_0}^t F(u, y(u)) \, du.$$

That is, $\mathcal{T}y$ is the right-hand side of (3) for any continuous function y. Given a function y, $\mathcal{T}$ produces a new function $\mathcal{T}y$. If we can find a function y so that $\mathcal{T}y = y$, we say y is a **fixed point** of $\mathcal{T}$. A fixed point y of $\mathcal{T}$ is precisely a solution to (3) since if $y = \mathcal{T}y$, then

$$y = \mathcal{T}y = y_0 + \int_{t_0}^t F(u, y(u)) \, du,$$

which is what it means to be a solution to (3). To solve equations like the integral equation (3), mathematicians have developed a variety of so-called "fixed point theorems" for operators such as $\mathcal{T}$, each of which leads to an existence and/or uniqueness result for solutions to an integral equation. One of the oldest and most widely used of the existence and uniqueness theorems is due to Émile Picard (1856–1941). Assuming that the function $F(t, y)$ is sufficiently "nice," he first employed the **method of successive approximations**. This method is an iterative procedure which begins with a crude approximation of a solution and improves it using a step-by-step procedure that brings us as close as we please to an exact and unique solution of (3). The process should remind students of Newton's method where successive approximations are used to find numerical solutions to $f(t) = c$ for some function f and constant c. The algorithmic procedure follows.

Algorithm 3. Perform the following sequence of steps to produce an *approximate solution* of (3), and hence to the initial value problem, (1).

Picard Approximations

1. A rough initial approximation to a solution of (3) is given by the constant function

$$y_0(t) := y_0.$$

2. Insert this initial approximation into the right-hand side of (3) and obtain the first approximation

$$y_1(t) := y_0 + \int_{t_0}^{t} F(u, y_0(u))\, du.$$

3. The next step is to generate the second approximation in the same way, that is,

$$y_2(t) := y_0 + \int_{t_0}^{t} F(u, y_1(u))\, du.$$

4. At the nth stage of the process, we have

$$y_n(t) := y_0 + \int_{t_0}^{t} F(u, y_{n-1}(u))\, du,$$

which is defined by substituting the previous approximation $y_{n-1}(t)$ into the right-hand side of (3).

In terms of the operator $\mathcal{T}$ introduced above, we can write

$$y_1 = \mathcal{T} y_0$$
$$y_2 = \mathcal{T} y_1 = \mathcal{T}^2 y_0$$
$$y_3 = \mathcal{T} y_2 = \mathcal{T}^3 y_0$$
$$\vdots$$

The result is a sequence of functions $y_0(t), y_1(t), y_2(t), \ldots$, defined on an interval containing t_0. We will refer to y_n as the ***nth Picard approximation*** and the sequence $y_0(t), y_1(t), \ldots, y_n(t)$ as the ***first n Picard approximations***. Note that the first n Picard approximations actually consist of $n + 1$ functions, since the starting approximation $y_0(t) = y_0$ is included.

Example 4. Find the first three Picard approximations for the initial value problem

$$y' = t + y, \quad y(0) = 1.$$

▶ **Solution.** The corresponding integral equation was computed in Example 2:

$$y(t) = 1 + \int_0^t (u + y(u))\, du.$$

We have

$$y_0(t) = 1$$

$$y_1(t) = 1 + \int_0^t (u + y_0(u))\, du$$

$$= 1 + \int_0^t (u + 1)\, du$$

$$= 1 + \left(\frac{u^2}{2} + u\right)\Big|_0^t = 1 + \frac{t^2}{2} + t = 1 + t + \frac{t^2}{2}.$$

$$y_2(t) = 1 + \int_0^t \left(u + 1 + u + \frac{u^2}{2}\right) du = 1 + \int_0^t \left(1 + 2u + \frac{u^2}{2}\right) du$$

$$= 1 + \left(u + u^2 + \frac{u^3}{6}\right)\Big|_0^t = 1 + t + t^2 + \frac{t^3}{6}.$$

$$y_3(t) = 1 + \int_0^t \left(u + 1 + u + u^2 + \frac{u^3}{6}\right) du = 1 + \int_0^t \left(1 + 2u + u^2 + \frac{u^3}{6}\right) du$$

$$= 1 + \left(u + u^2 + \frac{u^3}{3} + \frac{u^4}{24}\right)\Big|_0^t = 1 + t + t^2 + \frac{t^3}{3} + \frac{t^4}{24}. \qquad ◀$$

It was one of Picard's great contributions to mathematics when he showed that the functions $y_n(t)$ converge to a unique, continuously differentiable solution $y(t)$ of the integral equation, (3), and thus of the initial value problem, (1), under the mild condition that the function $F(t, y)$ and its partial derivative $F_y(t, y) := \frac{\partial}{\partial y} F(t, y)$ are continuous functions of (t, y) on the rectangle $\mathcal{R}$.

Theorem 5 (Picard's Existence and Uniqueness Theorem).[13] *Let $F(t, y)$ and $F_y(t, y)$ be continuous functions of (t, y) on a rectangle*

$$\mathcal{R} = \{(t, y) : a \le t \le b, c \le y \le d\}.$$

If (t_0, y_0) is an interior point of $\mathcal{R}$, then there exists a unique solution $y(t)$ of

$$y' = F(t, y), \quad y(t_0) = y_0,$$

---

[13] A proof of this theorem can be found in G.F. Simmons' book *Differential Equations with Applications and Historical Notes*, 2nd edition McGraw-Hill, 1991.

on some interval $[a', b']$ with $t_0 \in [a', b'] \subset [a, b]$. Moreover, the sequence of approximations $y_0(t) := y_0$

$$y_n(t) := y_0 + \int_{t_0}^{t} F(u, y_{n-1}(u)) \, du,$$

computed by Algorithm 3 converges uniformly[14] to $y(t)$ on the interval $[a', b']$.

Example 6. Consider the initial value problem

$$y' = t + y \quad y(0) = 1.$$

For $n \geq 1$, find the nth Picard approximation and determine the limiting function $y = \lim_{n \to \infty} y_n$. Show that this function is a solution and, in fact, the only solution.

▶ **Solution.** In Example 4, we computed the first three Picard approximations:

$$y_1(t) = 1 + t + \frac{t^2}{2},$$

$$y_2(t) = 1 + t + t^2 + \frac{t^3}{3!},$$

$$y_3(t) = 1 + t + t^2 + \frac{t^3}{3} + \frac{t^4}{4!}.$$

It is not hard to verify that

$$y_4(t) = 1 + t + t^2 + \frac{t^3}{3} + \frac{t^4}{12} + \frac{t^5}{5!}$$

$$= 1 + t + 2\left(\frac{t^2}{2!} + \frac{t^3}{3!} + \frac{t^4}{4!}\right) + \frac{t^5}{5!}$$

and inductively,

$$y_n(t) = 1 + t + 2\left(\frac{t^2}{2!} + \frac{t^3}{3!} + \cdots + \frac{t^n}{n!}\right) + \frac{t^{n+1}}{(n+1)!}.$$

---

[14] *Uniform convergence* is defined as follows: for all $\epsilon > 0$, there exists n_0 such that the maximal distance between the graph of the functions $y_n(t)$ and the graph of $y(t)$ (for $t \in [a', b']$) is less than ϵ for all $n \geq n_0$. We will not explore in detail this kind of convergence, but we will note that it implies *pointwise convergence*. That is, for each $t \in [\alpha', \beta']$ $y_n(t) \to y(t)$.

Recall from calculus that $e^t = \sum_{n=0}^{\infty} \dfrac{t^n}{n!}$, so the part in parentheses in the expression for $y_n(t)$ is the first n terms of the expansion of e^t minus the first two:

$$\frac{t^2}{2!} + \frac{t^3}{3!} + \cdots + \frac{t^n}{n!} = \left(1 + t + \frac{t^2}{2!} + \frac{t^3}{3!} + \cdots + \frac{t^n}{n!}\right) - (1 + t).$$

Thus,

$$y(t) = \lim_{n \to \infty} y_n(t) = 1 + t + 2(e^t - (1 + t)) = -1 - t + 2e^t.$$

It is easy to verify by direct substitution that $y(t) = -1 - t + 2e^t$ is a solution to $y' = t + y$ with initial value $y(0) = 1$. Moreover, since the equation $y' = t + y$ is a first order linear differential equation, the techniques of Sect. 1.4 show that $y(t) = -1 - t + 2e^t$ is the *unique* solution since it is obtained by an explicit formula. Alternatively, Picard's theorem may be applied as follows. Consider any rectangle $\mathcal{R}$ about the point $(0, 1)$. Let $F(t, y) = t + y$. Then $F_y(t, y) = 1$. Both $F(t, y)$ and $F_y(t, y)$ are continuous functions on the whole (t, y)-plane and hence continuous on $\mathcal{R}$. Therefore, Picard's theorem implies that $y(t) = \lim_{n \to \infty} y_n(t)$ is the unique solution of the initial value problem

$$y' = t + y \quad y(0) = 1.$$

Hence, $y(t) = -1 - t + 2e^t$ is the only solution. ◄

Example 7. Consider the Riccati equation

$$y' = y^2 - t$$

with initial condition $y(0) = 0$. Determine whether Picard's theorem applies on a rectangle containing $(0, 0)$. What conclusions can be made? Determine the first three Picard approximations.

▶ **Solution.** Here, $F(t, y) = y^2 - t$ and $F_y(t, y) = 2y$ are continuous on all of $\mathbb{R}^2$ and hence on *any* rectangle that contains the origin. Thus, by Picard's Theorem, the initial value problem

$$y' = y^2 - t, \quad y(0) = 0$$

has a unique solution on some interval I containing 0. Picard's theorem does not tell us on what interval the solution is defined, only that there is *some* interval. The direction field for $y' = y^2 - t$ with the unique solution through the origin is given below and suggests that the maximal interval I_{max} on which the solution exists should be of the form $I_{max} = (a, \infty)$ for some $-\infty \leq a < -1$. However, without further analysis of the problem, we have no precise knowledge about the maximal domain of the solution.

Next we show how Picard's method of successive approximations works in this example. To use this method, we rewrite the initial value problem as an integral equation. In this example, $F(t, y) = y^2 - t$, $t_0 = 0$ and $y_0 = 0$. Thus, the corresponding integral equation is

$$y(t) = \int_0^t (y(u)^2 - u)\, du. \tag{4}$$

We start with our initial approximation $y_0(t) = 0$, plug it into (4), and obtain our first approximation

$$y_1(t) = \int_0^t (y_0(u)^2 - u)\, du = -\int_0^t u\, du = -\frac{1}{2}t^2.$$

The second iteration yields

$$y_2(t) = \int_0^t (y_1(u)^2 - u)\, du = \int_0^t \left(\frac{1}{4}u^4 - u\right) du = \frac{1}{4\cdot 5}t^5 - \frac{1}{2}t^2.$$

Since $y_2(0) = 0$ and

$$y_2(t)^2 - t = \frac{1}{4^2 \cdot 5^2}t^{10} - \frac{1}{4\cdot 5}t^7 + \frac{1}{4}t^4 - t = \frac{1}{4^2 \cdot 5^2}t^{10} - \frac{1}{4\cdot 5}t^7 + y_2'(t) \approx y_2'(t)$$

if t is close to 0, it follows that the second iterate $y_2(t)$ is already a "good" approximation of the exact solution for t close to 0. Since

$$y_2(t)^2 = \frac{1}{4^2 \cdot 5^2}t^{10} - \frac{1}{4\cdot 5}t^7 + \frac{1}{4}t^4,$$

it follows that

$$y_3(t) = \int_0^t \left(\frac{1}{4^2 \cdot 5^2} u^{10} - \frac{1}{4 \cdot 5} u^7 + \frac{1}{4} u^4 - u \right) du$$

$$= \frac{1}{11 \cdot 4^2 \cdot 5^2} t^{11} - \frac{1}{4 \cdot 5 \cdot 8} t^8 + \frac{1}{4 \cdot 5} t^5 - \frac{1}{2} t^2.$$

According to Picard's theorem, the successive approximations $y_n(t)$ converge toward the exact solution $y(t)$, so we expect that $y_3(t)$ is an even better approximation of $y(t)$ for t close enough to 0. The graphs of y_1, y_2, and y_3 are given below.

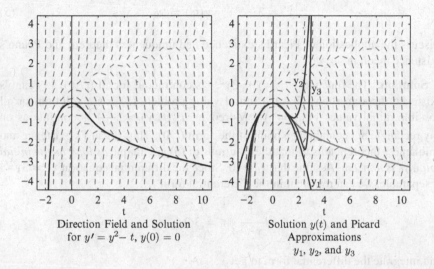

Direction Field and Solution
for $y' = y^2 - t$, $y(0) = 0$

Solution $y(t)$ and Picard
Approximations
y_1, y_2, and y_3

Although the Riccati equation looks rather simple in form, its solution cannot be obtained by methods developed in this chapter. In fact, the solution is not expressible in terms of elementary functions but requires special functions such as the Bessel functions. The calculation of the Picard approximations does not reveal a pattern by which we might guess what the nth term might be. This is rather typical. Only in special cases can we expect to find a such a general formula for y_n.   ◀

If one only assumes that the function $F(t, y)$ is continuous on the rectangle $\mathcal{R}$, but makes no assumptions about $F_y(t, y)$, then Guiseppe Peano (1858–1932) showed that the initial value problem (1) still has a solution on some interval I with $t_0 \in I \subset [a, b]$. However, in this case, the solutions are not necessarily unique.

Theorem 8 (Peano's Existence Theorem[15]). *Let $F(t, y)$ be a continuous functions of (t, y) on a rectangle*

---

[15]For a proof see, for example, A.N. Kolmogorov and S.V. Fomin, *Introductory Real Analysis*, Chap. 3, Sect. 11, Dover 1975.

$$\mathcal{R} = \{(t, y) : a \le t \le b, c \le y \le d\}.$$

If (t_0, y_0) is an interior point of $\mathcal{R}$, then there exists a solution $y(t)$ of

$$y' = F(t, y), \quad y(t_0) = y_0,$$

on some interval $[a', b']$ with $t_0 \in [a', b'] \subset [a, b]$.

Let us reconsider the differential equation introduced in Example 1.

Example 9. Consider the initial value problem

$$y' = 3y^{2/3}, \quad y(t_0) = y_0. \tag{5}$$

Discuss the application of Picard's existence and uniqueness theorem and Peano's existence theorem.

▶ **Solution.** The function $F(t, y) = 3y^{2/3}$ is continuous for all (t, y), so Peano's existence theorem shows that the initial value problem (5) has a solution for all possible initial values $y(t_0) = y_0$. Moreover, $F_y(t, y) = \frac{2}{y^{1/3}}$ is continuous on any rectangle not containing a point of the form $(t, 0)$. Thus, Picard's existence and uniqueness theorem tells us that the solutions of (5) are unique *as long as the initial value y_0 is nonzero*. Assume that $y_0 \ne 0$. Since the differential equation $y' = 3y^{2/3}$ is separable, we can rewrite it in the differential form

$$\frac{1}{y^{2/3}} \, dy = 3 dt,$$

and integrate the differential form to get

$$3y^{1/3} = 3t + c.$$

Thus, the functions $y(t) = (t + c)^3$ for $t \in \mathbb{R}$ are solutions for $y' = 3y^{2/3}$. Clearly, the equilibrium solution $y(t) = 0$ does not satisfy the initial condition. The constant c is determined by $y(t_0) = y_0$. We get $c = y_0^{1/3} - t_0$, and thus $y(t) = (t + y_0^{1/3} - t_0)^3$ is the unique solution of *if $y_0 \ne 0$. If $y_0 = 0$*, then (5) admits more than one solution. Two of them are given in Example 1. However, there are many more. In fact, the following functions are all solutions:

$$y(t) = \begin{cases} (t - \alpha)^3 & \text{if } t < \alpha \\ 0 & \text{if } \alpha \le t \le \beta \\ (t - \beta)^3 & \text{if } t > \beta, \end{cases} \tag{6}$$

where $t_0 \in [\alpha, \beta]$. The graph of one of these functions (where $\alpha = -1$, $\beta = 1$) is depicted below. What changes among the different functions is the length of the straight line segment joining α to β on the t-axis.

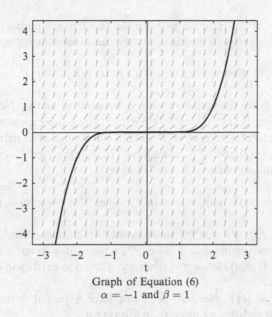

Graph of Equation (6)
$\alpha = -1$ and $\beta = 1$

◄

Picard's theorem, Theorem 5, is called a *local existence and uniqueness theorem* because it guarantees the existence of a unique solution in some subinterval $I \subset [a, b]$. In contrast, the following important variant of Picard's theorem yields a unique solution on the whole interval $[a, b]$.

Theorem 10. *Let $F(t, y)$ be a continuous function of (t, y) that satisfies a Lipschitz condition on a strip $S = \{(t, y) : a \leq t \leq b, -\infty < y < \infty\}$. That is, assume that*

$$|F(t, y_1) - F(t, y_2)| \leq K|y_1 - y_2|$$

for some constant $K > 0$ and for all (t, y_1) and (t, y_2) in S. If (t_0, y_0) is an interior point of S, then there exists a unique solution of

$$y' = F(t, y), \quad y(t_0) = y_0,$$

on the interval $[a, b]$.

Example 11. Show that the following differential equations have unique solutions on all of $\mathbb{R}$:

1. $y' = e^{\sin ty}$, $y(0) = 0$
2. $y' = |ty|$, $y(0) = 0$

▶ **Solution.** For each differential equation, we will show that Theorem 10 applies on the strip $\mathcal{S} = \{(t, y) : -a \leq t \leq a, -\infty < y < \infty\}$ and thus guarantees a unique solution on the interval $[-a, a]$. Since a is arbitrary, the solution exists on $\mathbb{R}$.

1. Let $F(t, y) = e^{\sin ty}$. Here we will use the fact that the partial derivative of F with respect to y exists so we can apply the mean value theorem:

$$F(t, y_1) - F(t, y_2) = F_y(t, y_0)(y_1 - y_2), \tag{7}$$

where y_1 and y_2 are real numbers with y_0 between y_1 and y_2. Now focus on the partial derivative $F_y(t, y) = e^{\sin ty} t \cos(ty)$. Since the exponential function is increasing, the largest value of $e^{\sin ty}$ occurs when sin is at its maximum value of 1. Since $|\cos ty| \leq 1$ and $|t| \leq a$, we have $|F_y(t, y)| = \left|e^{\sin ty} t \cos ty\right| \leq e^1 a = ea$. Now take the absolute value of (7) to get

$$|F(t, y_1) - F(t, y_2)| = \left|e^{\sin ty_1} - e^{\sin ty_2}\right| \leq ea\, |y_1 - y_2|.$$

It follows that $F(t, y) = e^{\sin ty}$ satisfies the Lipschitz condition with $K = ea$. Theorem 10 now implies that $y' = e^{\sin ty}$, with $y(0) = 0$ has a unique solution on the interval $[-a, a]$. Since a is arbitrary, a solution exists and is unique on all of $\mathbb{R}$.

2. Let $F(t, y) = |ty|$. Here F does not have a partial derivative at $(0, 0)$. Nevertheless, it satisfies a Lipschitz condition for

$$|F(t, y_1) - F(t, y_2)| = ||ty_1| - |ty_2|| \leq |t|\, |y_1 - y_2| \leq a\, |y_1 - y_2|,$$

since the maximum value of t on $[-a, a]$ is a. It follows that $F(t, y) = |ty|$ satisfies the Lipschitz condition with $K = a$. Theorem 10 now implies that

$$y' = |ty| \quad y(0) = 0,$$

has a unique solution on the interval $[-a, a]$. Since a is arbitrary, a solution exists and is unique on all of $\mathbb{R}$.   ◀

Remark 12.
1. When Picard's theorem is applied to the initial value problem $y' = e^{\sin ty}$, $y(0) = 0$, we can only conclude that there is a unique solution in an interval about the origin. Theorem 10 thus tells us much more, namely, that the solution is in fact defined on the entire real line.
2. In the case of $y' = |ty|$, $y(0) = 0$, Picard's theorem does not apply at all since the absolute value function is not differentiable at 0. Nevertheless, Theorem 10 tells us that a unique solution exists on all of $\mathbb{R}$. Now that you know this, can you guess what that unique solution is?

The Geometric Meaning of Uniqueness

The theorem on existence and uniqueness of solutions of differential equations, Theorem 5, has a particularly useful geometric interpretation. Suppose that $y' = F(t, y)$ is a first order differential equation for which Picard's theorem applies. If $y_1(t)$ and $y_2(t)$ denote two different solutions of $y' = F(t, y)$, then the graphs of $y_1(t)$ and $y_2(t)$ can *never* intersect. The reason for this is just that if (t_0, y_0) is a point of the plane which is common to both the graph of $y_1(t)$ and that of $y_2(t)$, then *both* of these functions will satisfy the initial value problem

$$y' = F(t, y), \quad y(t_0) = y_0.$$

But if $y_1(t)$ and $y_2(t)$ are different functions, this will violate the uniqueness provision of Picard's theorem.

To underscore this point, consider the following contrasting example: The differential equation

$$ty' = 3y \tag{8}$$

is linear (and separable). Thus, it is easy to see that $y(t) = ct^3$ is its general solution. In standard form, (8) is

$$y' = \frac{3y}{t} \tag{9}$$

and the right-hand side, $F(t, y) = 3y/t$, is continuous provided $t \neq 0$. Thus, assuming $t \neq 0$, Picard's theorem applies to give the conclusion that the initial value problem $y' = 3y/t$, $y(t_0) = y_0$ has a unique local solution, given by $y(t) = (y_0/t_0^3)t^3$. However, if $t_0 = 0$, Picard's theorem provides no information about the existence and uniqueness of solutions. Indeed, in its standard form, (9), it is not meaningful to talk about solutions of this equation at $t = 0$ since $F(t, y) = 3y/t$ is not even defined for $t = 0$. But in the originally designated form, (8), where the t appears as multiplication on the left side of the equation, then an initial value problem starting at $t = 0$ makes sense, and moreover, the initial value problem

$$ty' = 3y, \quad y(0) = 0$$

has infinitely many solutions of the form $y(t) = ct^3$ for *any* $c \in \mathbb{R}$, whereas the initial value problem

$$ty' = 3y, \quad y(0) = y_0$$

has no solution if $y_0 \neq 0$. See the figure below, where we have graphed the function $y(t) = ct^3$ for several values of c. Notice that all of them pass through the origin (i.e., $y(0) = 0$), but none pass through any other point on the y-axis.

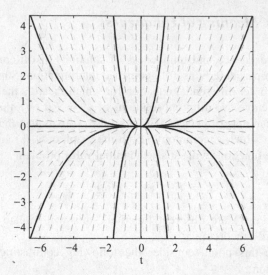

Thus, the situation depicted above where several solutions of the same differential equation go through the same point (in this case $(0, 0)$) can never occur for a differential equation which satisfies the hypotheses of Theorem 5.

The above remark can be exploited in the following way. The constant function $y_1(t) = 0$ is clearly a solution to the differential equation $y' = y^3 + y$. Since $F(t, y) = y^3 + y$ has continuous partial derivatives, Picard's theorem applies. Hence, if $y_2(t)$ is a solution of the equation for which $y_2(0) = 1$, the above observation takes the form of stating that $y_2(t) > 0$ for *all* t. This is because, in order for $y(t)$ to ever be negative, it must first cross the t-axis, which is the graph of $y_1(t)$, and we have observed that two solutions of the same differential equation can never cross.

Exercises

1–4. Write the corresponding integral equation for each of the following initial value problems.

1. $y' = ty$, $y(1) = 1$
2. $y' = y^2$ $y(0) = -1$
3. $y' = \dfrac{t - y}{t + y}$, $y(0) = 1$
4. $y' = 1 + t^2$, $y(0) = 0$

5–9. Find the first n Picard approximations for the following initial value problems.

5. $y' = ty$, $y(1) = 1$, $n = 3$
6. $y' = t - y$, $y(0) = 1$, $n = 4$
7. $y' = t + y^2$, $y(0) = 0$, $n = 3$
8. $y' = y^3 - y$, $y(0) = 1$, $n = 3$
9. $y' = 1 + (t - y)^2$, $y(0) = 0$, $n = 5$

10–14. Which of the following initial value problems are guaranteed a unique solution by Picard's theorem (Theorem 5)? Explain.

10. $y' = 1 + y^2$, $y(0) = 0$
11. $y' = \sqrt{y}$, $y(1) = 0$
12. $y' = \sqrt{y}$, $y(0) = 1$
13. $y' = \dfrac{t - y}{t + y}$, $y(0) = -1$
14. $y' = \dfrac{t - y}{t + y}$, $y(1) = -1$

15. Determine a formula for the nth Picard approximation for the initial value problem

$$y' = ay, \quad y(0) = 1,$$

where $a \in \mathbb{R}$. What is the limiting function $y(t) = \lim_{n \to \infty} y_n(t)$. Is it a solution? Are there other solutions that we may have missed?

16. (a) Find the exact solution of the initial value problem

$$y' = y^2, \quad y(0) = 1.$$

   (b) Calculate the first three Picard approximations $y_1(t)$, $y_2(t)$, and $y_3(t)$ and compare these results with the exact solution.

17. Determine whether the initial value problem

$$y' = \cos(t + y), \quad y(t_0) = y_0,$$

has a unique solution defined on all of $\mathbb{R}$.

18. Consider the linear differential equation $y' + p(t)y = f(t)$, with initial condition $y(t_0) = y_0$, where $p(t)$ and $f(t)$ are continuous on an interval

$I = [a, b]$ containing t_0 as an interior point. Use Theorem 10 to show that there is unique solution defined on $[a, b]$.

19. (a) Find the general solution of the differential equation

$$ty' = 2y - t.$$

Sketch several specific solutions from this general solution.
(b) Show that there is no solution satisfying the initial condition $y(0) = 2$. Why does this not contradict Theorem 5?

20. (a) Let t_0, y_0 be arbitrary and consider the initial value problem

$$y' = y^2, \quad y(t_0) = y_0.$$

Explain why Theorem 5 guarantees that this initial value problem has a solution on some interval $|t - t_0| \leq h$.
(b) Since $F(t, y) = y^2$ and $F_y(t, y) = 2y$ are continuous on all of the (t, y)-plane, one might hope that the solutions are defined for all real numbers t. Show that this is not the case by finding a solution of $y' = y^2$ which is defined for all $t \in \mathbb{R}$ and another solution which is *not* defined for all $t \in \mathbb{R}$. (Hint: Find the solutions with $(t_0, y_0) = (0, 0)$ and $(0, 1)$.)

21. Is it possible to find a function $F(t, y)$ that is continuous and has a continuous partial derivative $F_y(t, y)$ on a rectangle containing the origin such that the two functions $y_1(t) = t$ and $y_2(t) = t^2 - 2t$ are both solutions to $y' = F(t, y)$ on an interval containing 0?

22. Is it possible to find a function $F(t, y)$ that is continuous and has a continuous partial derivative $F_y(t, y)$ on a rectangle containing $(0, 1)$ such that the two functions $y_1(t) = (t + 1)^2$ and the constant function $y_2(t) = 1$ are both solutions to $y' = F(t, y)$ on an interval containing 0?

23. Show that the function

$$y_1(t) = \begin{cases} 0, & \text{for } t < 0 \\ t^3 & \text{for } t \geq 0 \end{cases}$$

is a solution of the initial value problem $ty' = 3y$, $y(0) = 0$. Show that $y_2(t) = 0$ for all t is a second solution. Explain why this does not contradict Theorem 5.

Chapter 2
The Laplace Transform

2.1 Laplace Transform Method: Introduction

The method for solving a first order linear differential equation $y' + p(t)y = f(t)$ (Algorithm 3 of Sect. 5) involves multiplying the equation by an integrating factor $\mu(t) = e^{\int p(t)\,dt}$ chosen so that the left-hand side of the resulting equation becomes a perfect derivative $(\mu(t)y)'$. Then the unknown function $y(t)$ can be retrieved by integration. When $p(t) = k$ is a constant, $\mu(t) = e^{kt}$ is an exponential function. Unfortunately, for higher order linear equations, there is not a corresponding type of integrating factor. There is, however, a useful method involving multiplication by an exponential function that can be used for solving an ***nth order constant coefficient linear differential equation***, that is, an equation

$$y^{(n)} + a_{n-1}y^{(n-1)} + \cdots + a_1 y' + a_0 y = f(t) \tag{1}$$

in which $a_0, a_1, \ldots, a_{n-1}$ are constants. The method proceeds by multiplying (1) by the exponential term e^{-st}, where s is another variable, and then integrating the resulting equation from 0 to ∞, to obtain the following equation involving the variable s:

$$\int_0^\infty e^{-st}\left(y^{(n)} + a_{n-1}y^{(n-1)} + \cdots + a_1 y' + a_0 y\right)dt = \int_0^\infty e^{-st} f(t)\,dt. \tag{2}$$

The integral $\int_0^\infty e^{-st} f(t)\,dt$ is called the ***Laplace transform*** of $f(t)$, and we will denote the Laplace transform of the function $f(t)$ by means of the corresponding capital letter $F(s)$ or the symbol $\mathcal{L}\{f(t)\}(s)$. Thus,

$$\mathcal{L}\{f(t)\}(s) = \int_0^\infty e^{-st} f(t)\,dt = F(s). \tag{3}$$

W.A. Adkins and M.G. Davidson, *Ordinary Differential Equations*,
Undergraduate Texts in Mathematics, DOI 10.1007/978-1-4614-3618-8_2,
© Springer Science+Business Media New York 2012

Note that $F(s)$ is a function of the new variable s, while the original function $f(t)$ is a function of the variable t. The integral involved is an *improper* integral since the domain of integration is of infinite length; at this point, we will simply assume that the integrals in question exist for all real s greater than some constant a.

Before investigating the left-hand side of (2), we will first calculate a couple of simple Laplace transforms to see what the functions $F(s)$ may look like.

Example 1. Find the Laplace transform of $f(t) = e^{at}$ and $g(t) = te^{at}$.

▶ **Solution.** First, we find the Laplace transform of $f(t) = e^{at}$.

$$
\begin{aligned}
F(s) = \mathcal{L}\{f(t)\}(s) &= \int_0^\infty e^{-st} e^{at}\, dt \\
&= \int_0^\infty e^{(a-s)t}\, dt = \lim_{r\to\infty} \int_0^r e^{(a-s)t}\, dt \\
&= \lim_{r\to\infty} \left.\frac{e^{(a-s)t}}{a-s}\right|_0^r = \lim_{r\to\infty}\left(\frac{e^{(a-s)r}}{a-s} - \frac{1}{a-s}\right) \\
&= \frac{1}{s-a} + \lim_{r\to\infty} \frac{e^{(a-s)r}}{a-s} \\
&= \frac{1}{s-a},
\end{aligned}
$$

provided $a - s < 0$, that is, $s > a$. In this situation, the limit of the exponential term is 0. Therefore,

$$\mathcal{L}\{e^{at}\}(s) = \frac{1}{s-a} \qquad \text{for } s > a. \tag{4}$$

A similar calculation gives the Laplace transform of $g(t) = te^{at}$, except that integration by parts will be needed in the calculation:

$$
\begin{aligned}
G(s) = \mathcal{L}\{te^{at}\}(s) &= \int_0^\infty e^{-st}(te^{at})\, dt = \int_0^\infty t e^{(a-s)t}\, dt \\
&= \lim_{r\to\infty} \left(\left.\frac{te^{(a-s)t}}{a-s}\right|_0^r - \int_0^r \frac{e^{(a-s)t}}{a-s}\, dt\right) \\
&= \lim_{r\to\infty} \left[\frac{te^{(a-s)t}}{a-s} - \frac{e^{(a-s)t}}{(a-s)^2}\right]_0^r \\
&= \lim_{r\to\infty} \left(\frac{re^{(a-s)r}}{a-s} - \frac{e^{(a-s)r}}{(a-s)^2} + \frac{1}{(a-s)^2}\right) \\
&= \frac{1}{(a-s)^2} \qquad \text{for } s > a.
\end{aligned}
$$

Therefore,

$$\mathcal{L}\{te^{at}\}(s) = \frac{1}{(s-a)^2} \qquad \text{for } s > a. \tag{5}$$

◀

In each of these formulas, the parameter a represents any real number. Thus, some specific examples of (4) and (5) are

$$\mathcal{L}\{1\} = \frac{1}{s} \qquad s > 0 \quad a = 0 \text{ in (4)},$$

$$\mathcal{L}\{e^{2t}\} = \frac{1}{s-2} \qquad s > 2 \quad a = 2 \text{ in (4)},$$

$$\mathcal{L}\{e^{-2t}\} = \frac{1}{s+2} \qquad s > -2 \quad a = -2 \text{ in (4)},$$

$$\mathcal{L}\{t\} = \frac{1}{s^2} \qquad s > 0 \quad a = 0 \text{ in (5)},$$

$$\mathcal{L}\{te^{2t}\} = \frac{1}{(s-2)^2} \quad s > 2 \quad a = 2 \text{ in (5)},$$

$$\mathcal{L}\{te^{-2t}\} = \frac{1}{(s+2)^2} \quad s > -2 \quad a = -2 \text{ in (5)}.$$

We now turn to the left-hand side of (2). Since the integral is additive, we can write the left-hand side as a sum of terms

$$a_j \int_0^\infty e^{-st} y^{(j)} \, dt = a_j \mathcal{L}\{y^{(j)}\}(s). \tag{6}$$

For now, we will not worry about whether the solution function $y(t)$ is such that the Laplace transform of $y^{(j)}$ exists. Ignoring the constant, the $j = 0$ term is the Laplace transform of $y(t)$, which we denote by $Y(s)$, and the $j = 1$ term is the Laplace transform of $y'(t)$, and this can also be expressed in terms of $Y(s) = \mathcal{L}\{y(t)\}(s)$ by use of integration by parts:

$$\int_0^\infty e^{-st} y'(t) \, dt = e^{-st} y(t) \big|_0^\infty + s \int_0^\infty e^{-st} y(t) \, dt$$

$$= \lim_{t \to \infty} e^{-st} y(t) - y(0) + s Y(s).$$

We will now further restrict the type of functions that we consider by requiring that

$$\lim_{t \to \infty} e^{-st} y(t) = 0.$$

We then conclude that

$$\mathcal{L}\{y'(t)\}(s) = \int_0^\infty e^{-st} y'(s) \, dt = s Y(s) - y(0). \tag{7}$$

Table 2.1 Basic Laplace
transform formulas

	$f(t)$	$\longleftrightarrow$	$F(s) = \mathcal{L}\{f(t)\}(s)$
1.	1	$\longleftrightarrow$	$\dfrac{1}{s}$
2.	t^n	$\longleftrightarrow$	$\dfrac{n!}{s^{n+1}}$
3.	e^{at}	$\longleftrightarrow$	$\dfrac{1}{s-a}$
4.	$t^n e^{at}$	$\longleftrightarrow$	$\dfrac{n!}{(s-a)^{n+1}}$
5.	$\cos bt$	$\longleftrightarrow$	$\dfrac{s}{s^2+b^2}$
6.	$\sin bt$	$\longleftrightarrow$	$\dfrac{b}{s^2+b^2}$
7.	$af(t)+bg(t)$	$\longleftrightarrow$	$aF(s)+bG(s)$
8.	$y'(t)$	$\longleftrightarrow$	$sY(s)-y(0)$
9.	$y''(t)$	$\longleftrightarrow$	$s^2Y(s)-sy(0)-y'(0)$

By use of repeated integration by parts, it is possible to express the Laplace transforms of all of the derivatives $y^{(j)}$ in terms of $Y(s)$ and values of $y^{(k)}(t)$ at $t = 0$. The formula is

$$\mathcal{L}\left\{y^{(n)}(t)\right\}(s) = \int_0^\infty e^{-st} y^{(n)} \, dt$$

$$= s^n Y(s) - (s^{n-1}y(0) + s^{n-2}y'(0) + \cdots + sy^{n-2}(0) + y^{n-1}(0)),$$

(8)

with the important special case for $n = 2$ being

$$\mathcal{L}\left\{y''(t)\right\}(s) = s^2Y(s) - sy(0) - y'(0). \tag{9}$$

The **_Laplace transform method_** for solving (1) is to use (8) to replace each Laplace transform of a derivative of $y(t)$ in (2) with an expression involving $Y(s)$ and initial values. This gives an _algebraic equation_ in $Y(s)$. Solve for $Y(s)$, and hopefully recognize $Y(s)$ as the Laplace transform of a known function $y(t)$. This latter recognition involves having a good knowledge of Laplace transforms of a wide variety of functions, which can be manifested by means of a table of Laplace transforms. A small table of Laplace transforms, Table 2.1, is included here for use in some examples. The table will be developed fully and substantially expanded, starting in the next section. For now, we will illustrate the Laplace transform method by solving some differential equations of orders 1 and 2. The examples of order 1 could also be solved by the methods of Chap. 1.

Example 2. Solve the initial value problem

$$y' + 2y = e^{-2t}, \qquad y(0) = 0, \tag{10}$$

by the Laplace transform method.

▶ **Solution.** As in (2), apply the Laplace transform to both sides of the given differential equation. Thus,

$$\mathcal{L}\left\{y' + 2y\right\}(s) = \mathcal{L}\left\{e^{-2t}\right\}(s). \tag{11}$$

The right-hand side of this equation is

$$\mathcal{L}\left\{e^{-2t}\right\}(s) = \frac{1}{s+2},$$

which follows from (4) with $a = -2$. The left-hand side of (11) is

$$\mathcal{L}\left\{y' + 2y\right\}(s) = \mathcal{L}\left\{y'\right\} + 2\mathcal{L}\left\{y\right\} = sY(s) - y(0) + 2Y(s) = (s+2)Y(s),$$

where $Y(s) = \mathcal{L}\left\{y(t)\right\}(s)$. Thus, (11) becomes

$$(s+2)Y(s) = \frac{1}{s+2},$$

which can be solved for $Y(s)$ to give

$$Y(s) = \frac{1}{(s+2)^2}.$$

From (5) with $a = -2$, we see that $Y(s) = \mathcal{L}\left\{te^{-2t}\right\}(s)$, which suggests that $y(t) = te^{-2t}$. By direct substitution, we can check that $y(t) = te^{-2t}$ is, in fact, the solution of the initial value problem (10). ◀

Example 3. Solve the second order initial value problem

$$y'' + 4y = 0, \qquad y(0) = 1, \quad y'(0) = 0, \tag{12}$$

by the Laplace transform method.

▶ **Solution.** As in the previous example, apply the Laplace transform to both sides of the given differential equation. Thus,

$$\mathcal{L}\left\{y'' + 4y\right\}(s) = \mathcal{L}\left\{0\right\} = 0. \tag{13}$$

The left-hand side is

$$\mathcal{L}\left\{y'' + 4y\right\}(s) = \mathcal{L}\left\{y''\right\} + 4\mathcal{L}\left\{y\right\} = s^2Y(s) - sy(0) - y'(0) + 4Y(s)$$
$$= -s + (s^2 + 4)Y(s),$$

where $Y(s) = \mathcal{L}\left\{y(t)\right\}(s)$. Thus, (13) becomes

$$(s^2 + 4)Y(s) - s = 0,$$

which can be solved for $Y(s)$ to give

$$Y(s) = \frac{s}{s^2 + 4}.$$

Item 5, with $b = 2$, in Table 2.1, shows that $Y(s) = \mathcal{L}\{\cos 2t\}(s)$, which suggests that the solution $y(t)$ of the differential equation is $y(t) = \cos 2t$. Straightforward substitution again shows that this function satisfies the initial value problem. ◄

Here is a slightly more complicated example.

Example 4. Use the Laplace transform method to solve

$$y'' + 4y' + 4y = 2te^{-2t}, \tag{14}$$

with initial conditions $y(0) = 1$ and $y'(0) = -3$.

► **Solution.** Let $Y(s) = \mathcal{L}\{y(t)\}$ where, as usual, $y(t)$ is the unknown solution. Applying the Laplace transform to both sides of (14) gives

$$\mathcal{L}\{y'' + 4y' + 4y\}(s) = \mathcal{L}\{2te^{-2t}\},$$

which, after applying items 7–9 from Table 2.1 to the left-hand side, and item 4 to the right-hand side, gives

$$s^2 Y(s) - sy(0) - y'(0) + 4(sY(s) - y(0)) + 4Y(s) = \frac{2}{(s+2)^2}.$$

Now, substituting the initial values gives the algebraic equation

$$s^2 Y(s) - s + 3 + 4(sY(s) - 1) + 4Y(s) = \frac{2}{(s+2)^2}.$$

Collecting terms, we get

$$(s^2 + 4s + 4)Y(s) = s + 1 + \frac{2}{(s+2)^2}$$

and solving for $Y(s)$, we get

$$Y(s) = \frac{s+1}{(s+2)^2} + \frac{2}{(s+2)^4}.$$

This $Y(s)$ is not immediately recognizable as a term in the table of Laplace transforms. However, a simple partial fraction decomposition, which will be studied in more detail later in this chapter, gives

$$Y(s) = \frac{s+1}{(s+2)^2} + \frac{2}{(s+2)^4}$$

$$= \frac{(s+2) - 1}{(s+2)^2} + \frac{2}{(s+2)^4}$$

$$= \frac{1}{s+2} - \frac{1}{(s+2)^2} + \frac{2}{(s+2)^4}.$$

Each of these terms can be identified as the Laplace transform of a function in Table 2.1. That is, item 4 with $a = -2$ and $n = 0, 1,$ and 3 gives

$$\mathcal{L}\left\{e^{-2t}\right\}(s) = \frac{1}{s+2}, \ \mathcal{L}\left\{te^{-2t}\right\}(s) = \frac{1}{(s+2)^2}, \text{ and } \mathcal{L}\left\{t^3e^{-2t}\right\} = \frac{3!}{(s+4)^4}.$$

Thus, we recognize

$$Y(s) = \mathcal{L}\left\{e^{-2t} - te^{-2t} + \frac{1}{3}t^3e^{-2t}\right\},$$

which suggests that

$$y(t) = e^{-2t} - te^{-2t} + \frac{1}{3}t^3e^{-2t},$$

is the solution to (14). As before, substitution shows that this function is in fact a solution to the initial value problem. ◄

The examples given illustrate how to use the Laplace transform to solve the nth order constant coefficient linear differential equation (1). The steps are summarized as the following algorithm.

Algorithm 5. Use the following sequence of steps to solve (1) by means of the Laplace transform.

Laplace Transform Method

1. Set the Laplace transform of the left-hand side of the equation equal to the Laplace transform of the function on the right-hand side.
2. Letting $Y(s) = \mathcal{L}\{y(t)\}(s)$, where $y(t)$ is the unknown solution to (1), use the derivative formulas for the Laplace transform to express the Laplace transform of the left-hand side of the equation as a function involving $Y(s)$, some powers of s, and the initial values $y(0)$, $y'(0)$, etç.
3. Now solve the resulting algebraic equation for $Y(s)$.
4. Identify $Y(s)$ as the Laplace transform of a known function $y(t)$. This may involve some algebraic manipulations of $Y(s)$, such as partial fractions, for example, so that you may identify individual parts of $Y(s)$ from a table of Laplace transforms, such as the short Table 2.1 reproduced above.

The Laplace transform is quite powerful for the types of differential equations to which it applies. However, steps 1 and 4 in the above summary will require a more extensive understanding of Laplace transforms than the brief introduction we have presented here. We will start to develop this understanding in the next section.

Exercises

1–14. Solve each of the following differential equations using the Laplace transform method. Determine both $Y(s) = \mathcal{L}\{y(t)\}$ and the solution $y(t)$.

1. $y' - 4y = 0, \quad y(0) = 2$
2. $y' - 4y = 1, \quad y(0) = 0$
3. $y' - 4y = e^{4t}, \quad y(0) = 0$
4. $y' + ay = e^{-at}, \quad y(0) = 1$
5. $y' + 2y = 3e^t, \quad y(0) = 2$
6. $y' + 2y = te^{-2t}, \quad y(0) = 0$
7. $y'' + 3y' + 2y = 0, \quad y(0) = 3, y'(0) = -6$
8. $y'' + 5y' + 6y = 0, \quad y(0) = 2, y'(0) = -6$
9. $y'' + 25y = 0, \quad y(0) = 1, y'(0) = -1$
10. $y'' + a^2 y = 0, \quad y(0) = y_0, y'(0) = y_1$
11. $y'' + 8y' + 16y = 0, \quad y(0) = 1, y'(0) = -4$
12. $y'' - 4y' + 4y = 4e^{2t}, \quad y(0) = -1, y'(0) = -4$
13. $y'' + 4y' + 4y = e^{-2t}, \quad y(0) = 0, y'(0) = 1$
14. $y'' + 4y = 8, \quad y(0) = 2, y'(0) = 1$

2.2 Definitions, Basic Formulas, and Principles

Suppose $f(t)$ is a continuous function defined for all $t \geq 0$. The **Laplace transform** of f is the function $F(s) = \mathcal{L}\{f(t)\}(s)$ defined by the improper integral equation

$$F(s) = \mathcal{L}\{f(t)\}(s) = \int_0^\infty e^{-st} f(t)\, dt := \lim_{r \to \infty} \int_0^r e^{-st} f(t)\, dt \qquad (1)$$

provided the limit exists at s. It can be shown that if the Laplace transform exists at $s = N$, then it exists for all $s \geq N$.[1] This means that there is a smallest number N, which will depend on the function f, so that the limit exists whenever $s > N$.

Let us consider this equation somewhat further. The function f with which we start will sometimes be called the **input function**. Generally, "t" will denote the variable for an input function f, while the Laplace transform of f, denoted $\mathcal{L}\{f(t)\}(s)$,[2] is a new function (the **output function** or **transform function**), whose variable will usually be "s". Thus, (1) is a formula for computing the value of the function $\mathcal{L}\{f\}$ at the particular point s, so that, for example, $F(2) = \mathcal{L}\{f\}(2) = \int_0^\infty e^{-2t} f(t)\, dt$ provided $s = 2$ is in the domain of $\mathcal{L}\{f(t)\}$.

When possible, we will use a lowercase letter to denote the input function and the corresponding uppercase letter to denote its Laplace transform. Thus, $F(s)$ is the Laplace transform of $f(t)$, $Y(s)$ is the Laplace transform of $y(t)$, etc. Hence, there are two distinct notations that we will be using for the Laplace transform of $f(t)$; if there is no confusion, we use $F(s)$, otherwise we will write $\mathcal{L}\{f(t)\}(s)$.

To avoid the notation becoming too heavy-handed, we will frequently write $\mathcal{L}\{f\}$ rather than $\mathcal{L}\{f\}(s)$. That is, the variable s may be suppressed when the meaning is clear. It is also worth emphasizing that, while the input function f must have a domain that includes $[0, \infty)$, the Laplace transform $\mathcal{L}\{f\}(s) = F(s)$ is only defined for all sufficiently large s, and the domain will depend on the particular input function f. In practice, this will not be an issue, and we will generally not emphasize the particular domain of $F(s)$.

Functions of Exponential Type

The fact that the Laplace transform is given by an improper integral imposes restrictions on the growth of the integrand in order to insure convergence. A function

---

[1] A nice proof of this fact can be found on page 442 of the text *Advanced Calculus* (second edition) by David Widder, published by Prentice Hall (1961).

[2] Technically, f is the function while $f(t)$ is the value of the function f at t. Thus, to be correct, the notation should be $\mathcal{L}\{f\}(s)$. However, there are times when the variable t needs to be emphasized or f is given by a formula such as in $\mathcal{L}\{e^{2t}\}(s)$. Thus, we will freely use both notations: $\mathcal{L}\{f(t)\}(s)$ and $\mathcal{L}\{f\}(s)$.

f on $[0, \infty)$ is said to be of ***exponential type with order*** a if there is a constant K such that

$$|f(t)| \leq Ke^{at}$$

for all $t \in [0, \infty)$. If the order is not important to the discussion, we will just say f is of ***exponential type***. The idea here is to limit the kind of growth that we allow f to have; it cannot grow faster than a multiple of an exponential function. The above inequality means

$$-Ke^{at} \leq f(t) \leq Ke^{at},$$

for all $t \in [0, \infty)$ as illustrated in the graph below. Specifically, the curve, $f(t)$, lies between the upper and lower exponential functions, $-Ke^{at}$ and Ke^{at}.

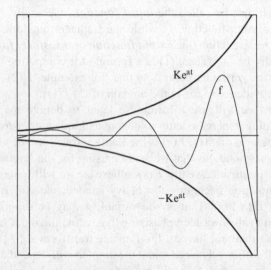

As we will see below, limiting growth in this way will assure us that f has a Laplace transform. If $f(t)$ is a bounded function, then there is a K so that $|f(t)| \leq K$ which implies that $f(t)$ is of exponential type of order $a = 0$. Hence, all bounded functions are of exponential type. For example, constant functions, $\cos bt$ and $\sin bt$ are of exponential type since they are bounded. Notice that if f is of exponential type of order a and $a < 0$, then $\lim_{t \to \infty} f(t) = 0$ and hence it is bounded. Since exponential type is a concept used to restrict the growth of a function, we will be interested only in exponential type of order $a \geq 0$.

The set of all functions of exponential type has the property that it is closed under addition and scalar multiplication. We will often see this property on sets of functions. A set $\mathcal{F}$ of functions (usually defined on some interval I) is a ***linear space*** (or ***vector space***) if it is closed under addition and scalar multiplication. More specifically, $\mathcal{F}$ is a linear space if

- $f_1 + f_2 \in \mathcal{F}$,
- $cf_1 \in \mathcal{F}$,

whenever $f_1, f_2 \in \mathcal{F}$ and c is a scalar. If the scalars in use are the real numbers, then $\mathcal{F}$ is referred to as a *real* linear space. If the scalars are the complex numbers, then $\mathcal{F}$ is a *complex* linear space. Unless otherwise stated, linear spaces will be real.

Proposition 1. *The set of functions of exponential type is a linear space.*

Proof. Suppose f_1 and f_2 are of exponential type and $c \in \mathbb{R}$ is a scalar. Then there are constants K_1, K_2, a_1, and a_2 so that $f_1(t) \leq K_1 e^{a_1 t}$ and $f_2(t) \leq K_2 e^{a_2 t}$. Now let $K = K_1 + K_2$ and let a be the larger of a_1 and a_2. Then

$$|f_1(t) + f_2(t)| \leq |f_2(t)| + |f_2(t)| \leq K_1 e^{a_1 t} + K_2 e^{a_2 t} \leq K_1 e^{at} + K_2 e^{at} = K e^{at}.$$

It follows that $f_1 + f_2$ is of exponential type. Further,

$$|cf_1(t)| \leq |c|\,|f_1(t)| \leq |c| K_1 e^{a_1 t}.$$

It follows that cf_1 is of exponential type. Thus, the set of all functions of exponential type is closed under addition and scalar multiplication, that is, is a linear space. $\square$

Lemma 2. *Suppose f is of exponential type of order a. Let $s > a$, then*

$$\lim_{t \to \infty} f(t)e^{-st} = 0.$$

Proof. Choose K so that $|f(t)| \leq K e^{at}$. Let $s > a$. Then

$$\left| f(t)e^{-st} \right| = \left| \frac{f(t)}{e^{at}} e^{-(s-a)t} \right| \leq K e^{-(s-a)t}.$$

Taking limits gives the result since $\lim_{t\to\infty} e^{-(s-a)t} = 0$ because $-(s-a) < 0$. $\square$

Proposition 3. *Let f be a continuous function of exponential type with order a. Then the Laplace transform $F(s) = \mathcal{L}\{f(t)\}(s)$ exists for all $s > a$ and, moreover, $\lim_{s\to\infty} F(s) = 0$.*

Proof. Let f be of exponential type of order a. Then $|f(t)| \leq K e^{at}$, for some K, so $|f(t)e^{-st}| \leq K e^{-(s-a)t}$ and

$$|F(s)| = \left| \int_0^\infty e^{-st} f(t)\, dt \right| \leq \int_0^\infty e^{-st} |f(t)|\, dt \leq K \int_0^\infty e^{-(s-a)t}\, dt = \frac{K}{s-a},$$

provided $s > a$. This shows that the integral converges absolutely, and hence the Laplace transform $F(s)$ exists for $s > a$, and in fact $|F(s)| \leq K/(s-a)$. Since $\lim_{s\to\infty} K/(s-a) = 0$, it follows that $\lim_{s\to\infty} F(s) = 0$. $\square$

It should be noted that many functions are not of exponential type. For example, in Exercises 39 and 40, you are asked to show that the function $y(t) = e^{t^2}$ is not of exponential type and does not have a Laplace transform. Proposition 3 should not be misunderstood. The restriction that f be of exponential type is a *sufficient* condition to guarantee that the Laplace transform exists. If a function is not of exponential type, it still may have a Laplace transform. See, for example, Exercise 41.

Lemma 4. *Suppose f is a continuous function defined on $[0, \infty)$ of exponential type of order $a \geq 0$. Then any antiderivative of f is also of exponential type and has order a if $a > 0$.*

Proof. Suppose $|f(t)| \leq K e^{at}$, for some K and a. Let $g(t) = \int_0^t f(u)\, du$. Suppose $a > 0$. Then

$$|g(t)| \leq \int_0^t |f(u)|\, du \leq \int_0^t K e^{au}\, du = \frac{K}{a}(e^{at} - 1) \leq \frac{K}{a} e^{at}.$$

It follows that g is of exponential type of order a. If $a = 0$, then $|f| \leq K$ for some K. The antiderivative g defined above satisfies $|g| \leq Kt$. Let $b > 0$. Then since $u \leq e^u$ (for u nonnegative), we have $bt \leq e^{bt}$. So $|g| \leq Kt \leq (K/b)e^{bt}$. It follows that g is of exponential type of order b for any positive b. Since any antiderivative of f has the form $g(t) + C$ for some constant C and constant functions are of exponential type, the lemma follows by Proposition 1.   ◄

We will restrict our attention to continuous input functions in this chapter. In Chap. 6, we ease this restriction and consider Laplace transforms of discontinuous functions.

Basic Principles and Formulas

A particularly useful property of the Laplace transform, both theoretically and computationally, is that of *linearity*. Specifically,

Theorem 5. *The Laplace transform is linear. In other words, if f and g are functions of exponential type and a and b are constants, then*

Linearity of the Laplace Transform

$$\mathcal{L}\{af + bg\} = a\mathcal{L}\{f\} + b\mathcal{L}\{g\}.$$

Proof. By Proposition 1, the function $af + bg$ is of exponential type and, by Proposition 3, has a Laplace transform. Since (improper) integration is linear, we have

$$\mathcal{L}\{af + bg\}(s) = \int_0^\infty e^{-st}(af(t) + bg(t))\, dt$$

$$= a\int_0^\infty e^{-st} f(t)\, dt + b\int_0^\infty e^{-st} g(t)\, dt$$

$$= a\mathcal{L}\{f\}(s) + b\mathcal{L}\{g\}(s). \qquad \square$$

The input derivative principles we derive below are the cornerstone principles of the Laplace transform. They are used to derive basic Laplace transform formulas and are the key to the Laplace transform method to solve differential equations.

Theorem 6. *Suppose $f(t)$ is a differentiable function on $[0, \infty)$ whose derivative $f'(t)$ is continuous and of exponential type of order $a \geq 0$. Then*

The Input Derivative Principle

The First Derivative

$$\mathcal{L}\{f'(t))\}(s) = s\mathcal{L}\{f(t)\}(s) - f(0), \quad s > a.$$

Proof. By Lemma 4, $f(t)$ is of exponential type. By Proposition 3, both $f(t)$ and $f'(t)$ have Laplace transforms. Using integration by parts (let $u = e^{-st}$, $dv = f'(t)\,dt$), we get

$$\mathcal{L}\{f'(t)\}(s) = \int_0^\infty e^{-st} f'(t)\,dt$$

$$= e^{-st} f(t)\big|_0^\infty - \int_0^\infty -se^{-st} f(t)\,dt$$

$$= -f(0) + s \int_0^\infty e^{-st} f(t)\,dt = s\mathcal{L}\{f(t)\}(s) - f(0).$$

For a function $g(t)$ defined of $[a, \infty)$, we use the notation $g(t)\big|_a^\infty$ to mean $\lim_{t \to \infty} (g(t) - g(a))$. ◄

Observe that if $f'(t)$ also satisfies the conditions of the input derivative principle, then we get

$$\mathcal{L}\{f''(t)\}(s) = s\mathcal{L}\{f'(t)\} - f'(0)$$

$$= s(s\mathcal{L}\{f(t)\} - f(0)) - f'(0)$$

$$= s^2 \mathcal{L}\{f(t)\} - sf(0) - f'(0).$$

We thus get the following corollary:

Corollary 7. *Suppose $f(t)$ is a differentiable function on $[0, \infty)$ with continuous second derivative of exponential type of order $a \geq 0$. Then*

Input Derivative Principle

The Second Derivative

$$\mathcal{L}\{f''(t)\}(s) = s^2 \mathcal{L}\{f(t)\}(s) - sf(0) - f'(0), \quad s > a.$$

Repeated applications of Theorem 6 give the following:

Corollary 8. *Suppose* $f(t)$ *is a differentiable function on* $[0, \infty)$ *with continuous nth derivative of exponential type of order* $a \geq 0$. *Then*

Input Derivative Principle
The nth Derivative

$$\mathcal{L}\left\{f^{(n)}(t)\right\} = s^n \mathcal{L}\left\{f(t)\right\} - s^{n-1} f(0) - \cdots - s f^{(n-2)}(0) - f^{(n-1)}(0),$$
$$\text{for } s > a.$$

We now compute the Laplace transform of some specific input functions that will be used frequently throughout the text.

Formula 9. Verify the Laplace transform formula:

Constant Functions

$$\mathcal{L}\left\{1\right\}(s) = \frac{1}{s}, \qquad s > 0.$$

▼ *Verification.* For the constant function 1, we have

$$\mathcal{L}\left\{1\right\}(s) = \int_0^{\infty} e^{-st} \cdot 1 \, dt = \lim_{r \to \infty} \left. \frac{e^{-ts}}{-s} \right|_0^r$$

$$= \lim_{r \to \infty} \frac{e^{-rs} - 1}{-s} = \frac{1}{s} \qquad \text{for } s > 0. \qquad\qquad ▲$$

For the limit above, we have used the basic fact that

$$\lim_{r \to \infty} e^{rc} = \begin{cases} 0 & \text{if } c < 0 \\ \infty & \text{if } c > 0. \end{cases}$$

Formula 10. Assume n is a nonnegative integer. Verify the Laplace transform formula:

Power Functions

$$\mathcal{L}\left\{t^n\right\}(s) = \frac{n!}{s^{n+1}}, \qquad s > 0.$$

▼ *Verification.* Let $b > 0$. Since $u \leq e^u$ for $u \geq 0$, it follows that $bt \leq e^{bt}$ for all $t \geq 0$. Thus, $t \leq e^{bt}/b$ and $t^n \leq e^{bnt}/b^n$. Since b is any positive number, it follows that t^n is of exponential type of order a for all $a > 0$ and thus has a Laplace transform for $s > 0$. Let $f(t) = t^n$ and observe that $f^{(k)}(0) = 0$ for $k = 1, \ldots, n - 1$ and $f^{(n)}(t) = n!$. Apply the nth order input derivative formula, Corollary 8, to get

$$\frac{n!}{s} = \mathcal{L}\{n!\}(s)$$

$$= \mathcal{L}\{f^{(n)}(t)\}(s)$$

$$= s^n \mathcal{L}\{t^n\} - s^{n-1} f(0) - \cdots - f^{(n-1)}(0)$$

$$= s^n \mathcal{L}\{t^n\}(s).$$

It follows that

$$\mathcal{L}\{t^n\}(s) = \frac{n!}{s^{n+1}}, \qquad s > 0. \qquad \blacktriangle$$

Example 11. Find the Laplace transform of

$$f(t) = 3 - 4t + 6t^3.$$

▶ **Solution.** Here we use the linearity and the basic Laplace transforms determined above:

$$\mathcal{L}\{3 - 4t + 6t^3\} = 3\mathcal{L}\{1\} - 4\mathcal{L}\{t\} + 6\mathcal{L}\{t^3\}$$

$$= 3\left(\frac{1}{s}\right) - 4\left(\frac{1}{s^2}\right) + 6\left(\frac{3!}{s^4}\right)$$

$$= \frac{3}{s} - \frac{4}{s^2} + \frac{36}{s^4}. \qquad \blacktriangleleft$$

The formula for the Laplace transform of t^n is actually valid even if the exponent is not an integer. Suppose that $\alpha \in \mathbb{R}$ is any real number. Then use the substitution $t = x/s$ in the Laplace transform integral to get

$$\mathcal{L}\{t^\alpha\} = \int_0^\infty t^\alpha e^{-st}\, dt = \int_0^\infty \left(\frac{x}{s}\right)^\alpha e^{-x}\, \frac{dx}{s} = \frac{1}{s^{\alpha+1}} \int_0^\infty x^\alpha e^{-x}\, dx \quad (s > 0).$$

The improper integral on the right converges as long as $\alpha > -1$, and it defines a function known as the **gamma function** or **generalized factorial function** evaluated at $\alpha + 1$. Thus, $\Gamma(\beta) = \int_0^\infty x^{\beta-1} e^{-x}\, dx$ is defined for $\beta > 0$, and the Laplace transform of the general power function is as follows:

Formula 12. If $\alpha > -1$, then

General Power Functions

$$\mathcal{L}\{t^{\alpha}\}(s) = \frac{\Gamma(\alpha+1)}{s^{\alpha+1}}, \qquad s > 0.$$

If n is a positive integer, then $\Gamma(n+1) = n!$ (see Exercise 42) so Formula 10 is a special case of Formula 12.

Formula 13. Assume $a \in \mathbb{R}$. Verify the Laplace transform formula:

Exponential Functions

$$\mathcal{L}\{e^{at}\}(s) = \frac{1}{s-a}, \qquad s > a.$$

▼ *Verification.* If $s > a$, then

$$\mathcal{L}\{e^{at}\}(s) = \int_0^{\infty} e^{-st} e^{at}\, dt = \int_0^{\infty} e^{-(s-a)t}\, dt = \left.\frac{e^{-(s-a)t}}{-(s-a)}\right|_0^{\infty} = \frac{1}{s-a}. \quad \blacktriangle$$

Formula 14. Let $b \in \mathbb{R}$. Verify the Laplace transform formulas:

Cosine Functions

$$\mathcal{L}\{\cos bt\}(s) = \frac{s}{s^2+b^2}, \qquad \text{for } s > 0.$$

and

Sine Functions

$$\mathcal{L}\{\sin bt\}(s) = \frac{b}{s^2+b^2}, \qquad \text{for } s > 0.$$

▼ *Verification.* Since both $\sin bt$ and $\cos bt$ are bounded continuous functions, they are of exponential type and hence have Laplace transforms. Let $f(t) = \cos bt$. Then $f'(t) = -b \sin bt$ and $f''(t) = -b^2 \cos bt$. The input derivative principle for the second derivative, Corollary 7, implies

$$-b^2 \mathcal{L}\{\cos bt\}(s) = \mathcal{L}\{f''(t)\}(s) = s^2 \mathcal{L}\{f(t)\} - sf(0) - f'(0)$$

$$= s^2 \mathcal{L}\{\cos bt\} - s(1) - (0)$$

$$= s^2 \mathcal{L}\{\cos bt\} - s.$$

Now subtract $s^2 \mathcal{L}\{\cos bt\}$ from both sides and combine terms to get

$$-(s^2 + b^2)\mathcal{L}\{\cos bt\} = -s.$$

Solving for $\mathcal{L}\{\cos bt\}$ gives

$$\mathcal{L}\{\cos bt\} = \frac{s}{s^2 + b^2}.$$

A similar calculation verifies the formula for $\mathcal{L}\{\sin bt\}$.   ▲

Example 15. Find the Laplace transform of

$$2e^{6t} + 3\cos 2t - 4\sin 3t.$$

▶ **Solution.** We use the linearity of the Laplace transform together with the formulas derived above to get

$$\mathcal{L}\{2e^{6t} + 3\cos 2t - 4\sin 3t\} = 2\mathcal{L}\{e^{6t}\} + 3\mathcal{L}\{\cos 2t\} - 4\mathcal{L}\{\sin 3t\}$$

$$= \frac{2}{s-6} + 3\left(\frac{s}{s^2 + 2^2}\right) - 4\left(\frac{3}{s^2 + 3^2}\right)$$

$$= \frac{2}{s-6} + \frac{3s}{s^2 + 4} - \frac{12}{s^2 + 9}.$$ ◀

Formula 16. Let n be a nonnegative integer and $a \in \mathbb{R}$. Verify the following Laplace transform formula:

Power-Exponential Functions

$$\mathcal{L}\{t^n e^{at}\}(s) = \frac{n!}{(s-a)^{n+1}}, \quad \text{for } s > a.$$

▼ *Verification.* Notice that

$$\mathcal{L}\{t^n e^{at}\}(s) = \int_0^\infty e^{-st} t^n e^{at}\, dt = \int_0^\infty e^{-(s-a)t} t^n\, dt = \mathcal{L}\{t^n\}(s-a).$$

What this formula says is that the Laplace transform of the function $t^n e^{at}$ evaluated at the point s is the same as the Laplace transform of the function t^n evaluated at the point $s - a$. Since $\mathcal{L}\{t^n\}(s) = n!/s^{n+1}$, we conclude

$$\mathcal{L}\{t^n e^{at}\}(s) = \frac{n!}{(s-a)^{n+1}}, \quad \text{for } s > a.$$ ▲

If the function t^n in Formula 16 is replaced by an arbitrary input function $f(t)$ with a Laplace transform $F(s)$, then we obtain the following:

Theorem 17. *Suppose f has Laplace transform $F(s)$. Then*

$$\boxed{\begin{array}{c} \textbf{\textit{First Translation Principle}} \\[4pt] \mathcal{L}\{e^{at} f(t)\}(s) = F(s-a) \end{array}}$$

Proof.

$$\mathcal{L}\{e^{at} f(t)\}(s) = \int_0^\infty e^{-st} e^{at} f(t)\, dt$$

$$= \int_0^\infty e^{-(s-a)} f(t)\, dt$$

$$= \mathcal{L}\{f(t)\}(s-a) = F(s-a). \qquad \square$$

In words, this formula says that to compute the Laplace transform of the product of $f(t)$ and e^{at}, it is only necessary to take the Laplace transform of $f(t)$ (namely, $F(s)$) and replace the variable s by $s-a$, where a is the coefficient of t in the exponential multiplier. It is convenient to use the following notation:

$$\mathcal{L}\{e^{at} f(t)\}(s) = F(s)|_{s \mapsto s-a}$$

to indicate this substitution.

Formula 18. Suppose $a, b \in \mathbb{R}$. Verify the Laplace transform formulas

$$\boxed{\mathcal{L}\{e^{at} \cos bt\}(s) = \frac{s-a}{(s-a)^2 + b^2}}$$

and

$$\boxed{\mathcal{L}\{e^{at} \sin bt\}(s) = \frac{b}{(s-a)^2 + b^2}.}$$

▼ *Verification.* From Example 14, we know that

$$\mathcal{L}\{\cos bt\}(s) = \frac{s}{s^2 + b^2} \quad \text{and} \quad \mathcal{L}\{\sin bt\}(s) = \frac{b}{s^2 + b^2}.$$

Replacing s by $s - a$ in each of these formulas gives the result. ▲

Example 19. Find the Laplace transform of

$$2e^{-t} \sin 3t \quad \text{and} \quad e^{3t} \cos \sqrt{2}t.$$

▶ **Solution.** Again we use linearity of the Laplace transform and the formulas derived above to get

$$\mathcal{L}\left\{2e^{-t}\sin 3t\right\} = 2\left.\frac{3}{s^2 + 3^2}\right|_{s \mapsto s+1} = \frac{6}{(s+1)^2 + 9} = \frac{6}{s^2 + 2s + 10},$$

$$\mathcal{L}\left\{e^{3t}\cos\sqrt{2}t\right\} = \left.\frac{s}{s^2 + \sqrt{2}^2}\right|_{s \mapsto s-3} = \frac{s-3}{(s-3)^2 + 2} = \frac{s-3}{s^2 - 6s + 11}. \quad ◀$$

We now introduce another useful principle that can be used to compute some Laplace transforms.

Theorem 20. *Suppose $f(t)$ is an input function and $F(s) = \mathcal{L}\{f(t)\}(s)$ is the transform function. Then*

Transform Derivative Principle

$$\mathcal{L}\{-tf(t)\}(s) = \frac{\mathrm{d}}{\mathrm{d}s}F(s).$$

Proof. By definition, $F(s) = \int_0^\infty e^{-st} f(t)\,\mathrm{d}t$, and thus,

$$F'(s) = \frac{\mathrm{d}}{\mathrm{d}s}\int_0^\infty e^{-st} f(t)\,\mathrm{d}t$$

$$= \int_0^\infty \frac{\mathrm{d}}{\mathrm{d}s}(e^{-st})f(t)\,\mathrm{d}t$$

$$= \int_0^\infty e^{-st}(-t)f(t)\,\mathrm{d}t = \mathcal{L}\{-tf(t)\}(s).$$

Interchanging the derivative and the integral can be justified. □

Repeated application of the transform derivative principle gives

---
Transform nth-Derivative Principle

$$(-1)^n \mathcal{L}\{t^n f(t)\}(s) = \frac{\mathrm{d}^n}{\mathrm{d}s^n}F(s).$$

Example 21. Use the transform derivative principle to compute

$$\mathcal{L}\{t\sin t\}(s).$$

▶ **Solution.** A direct application of the transform derivative principle gives

$$\mathcal{L}\{t\sin t\}(s) = -\mathcal{L}\{-t\sin t\}$$

$$= -\frac{d}{ds}\mathcal{L}\{\sin t\}(s)$$

$$= -\frac{d}{ds}\frac{1}{s^2+1}$$

$$= -\frac{-2s}{(s^2+1)^2} = \frac{2s}{(s^2+1)^2}.$$ ◀

Example 22. Compute the Laplace transform of t^2e^{2t} in two different ways: using the first translation principle and the transform derivative principle.

▶ **Solution.** Using the first translation principle, we get

$$\mathcal{L}\left\{t^2e^{2t}\right\}(s) = \mathcal{L}\left\{t^2\right\}(s)\big|_{s\mapsto s-2} = \frac{2}{(s-2)^3}.$$

Using the transform derivative principle, we get

$$\mathcal{L}\left\{t^2e^{2t}\right\} = \frac{d^2}{ds^2}\frac{1}{s-2} = \frac{d}{ds}\frac{-1}{(s-2)^2} = \frac{2}{(s-2)^3}.$$ ◀

Suppose $f(t)$ is a function and b is a positive real number. The function $g(t) = f(bt)$ is called the **dilation** of f by b. If the domain of f includes $[0, \infty)$, then so does any dilation of f since b is positive. The following theorem describes the Laplace transform of a dilation.

Theorem 23. *Suppose $f(t)$ is an input function and b is a positive real number. Then*

The Dilation Principle

$$\mathcal{L}\{f(bt))\}(s) = \frac{1}{b}\mathcal{L}\{f(t)\}(s/b).$$

Proof. This result follows from a change in variable in the definition of the Laplace transform:

$$\mathcal{L}\{f(bt)\}(s) = \int_0^\infty e^{-st}f(bt)\,dt$$

$$= \int_0^\infty e^{-(s/b)r}f(r)\frac{dr}{b}$$

$$= \frac{1}{b}\mathcal{L}\{f(t)\}(s/b).$$

Table 2.2 Basic Laplace transform formulas (We are assuming n is a nonnegative integer and a and b are real)

	$f(t)$	$\longleftrightarrow$	$F(s) = \mathcal{L}\{f(t)\}(s)$
1.	1	$\longleftrightarrow$	$\dfrac{1}{s}$
2.	t^n	$\longleftrightarrow$	$\dfrac{n!}{s^{n+1}}$
3.	e^{at}	$\longleftrightarrow$	$\dfrac{1}{s-a}$
4.	$t^n e^{at}$	$\longleftrightarrow$	$\dfrac{n!}{(s-a)^{n+1}}$
5.	$\cos bt$	$\longleftrightarrow$	$\dfrac{s}{s^2 + b^2}$
6.	$\sin bt$	$\longleftrightarrow$	$\dfrac{b}{s^2 + b^2}$
7.	$e^{at} \cos bt$	$\longleftrightarrow$	$\dfrac{s-a}{(s-a)^2 + b^2}$
8.	$e^{at} \sin bt$	$\longleftrightarrow$	$\dfrac{b}{(s-a)^2 + b^2}$

Table 2.3 Basic Laplace transform principles

Linearity	$\mathcal{L}\{af(t) + bg(t)\} = a\mathcal{L}\{f\} + b\mathcal{L}\{g\}$
Input derivative principles	$\mathcal{L}\{f'(t)\}(s) = s\mathcal{L}\{f(t)\} - f(0)$
	$\mathcal{L}\{f''(t)\}(s) = s^2\mathcal{L}\{f(t)\} - sf(0) - f'(0)$
First translation principle	$\mathcal{L}\{e^{at}f(t)\} = F(s-a)$
Transform derivative principle	$\mathcal{L}\{-tf(t)\}(s) = \dfrac{\mathrm{d}}{\mathrm{d}s}F(s)$
The dilation principle	$\mathcal{L}\{f(bt)\}(s) = \dfrac{1}{b}\mathcal{L}\{f(t)\}(s/b).$

To get the second line, we made the change of variable $t = r/b$. Since $b > 0$, the limits of integration remain unchanged.   □

Example 24. The formula $\mathcal{L}\left\{\frac{\sin t}{t}\right\}(s) = \cot^{-1}(s)$ will be derived later (in Sect. 5.4). Assuming this formula for now, determine $\mathcal{L}\left\{\frac{\sin bt}{t}\right\}(s)$.

▶ **Solution.** By linearity and the dilation principle, we have

$$\mathcal{L}\left\{\frac{\sin bt}{t}\right\}(s) = b\mathcal{L}\left\{\frac{\sin bt}{bt}\right\}(s)$$

$$= b\frac{1}{b}\,\mathcal{L}\left\{\frac{\sin t}{t}\right\}\bigg|_{s \mapsto s/b}$$

$$= \cot^{-1}(s)\,\big|_{s \mapsto s/b} = \cot^{-1}(s/b).$$ ◀

We now summarize in Table 2.2 the basic Laplace transform formulas and, in Table 2.3, the basic Laplace transform principles we have thus far derived. The student should learn these well as they will be used frequently throughout the text and exercises. With the use of these tables, we can find the Laplace transform of many functions. As we continue, several new formulas will be derived. Appendix C has a complete list of Laplace transform formulas and Laplace transform principles that we derive.

Exercises

1–4. Compute the Laplace transform of each function given below directly from the integral definition given in (1).

1. $3t + 1$
2. $5t - 9e^t$
3. $e^{2t} - 3e^{-t}$
4. te^{-3t}

5–18. Use Table 2.2 and linearity to find the Laplace transform of each given function.

5. $5e^{2t}$
6. $3e^{-7t} - 7t^3$
7. $t^2 - 5t + 4$
8. $t^3 + t^2 + t + 1$
9. $e^{-3t} + 7te^{-4t}$
10. $t^2 e^{4t}$
11. $\cos 2t + \sin 2t$
12. $e^t (t - \cos 4t)$
13. $(te^{-2t})^2$
14. $e^{-t/3} \cos \sqrt{6}t$
15. $(t + e^{2t})^2$
16. $5 \cos 3t - 3 \sin 3t + 4$
17. $\dfrac{t^4}{e^{4t}}$
18. $e^{5t} (8 \cos 2t + 11 \sin 2t)$

19–23. Use the transform derivative principle to compute the Laplace transform of the following functions.

19. te^{3t}
20. $t \cos 3t$
21. $t^2 \sin 2t$
22. $te^{-t} \cos t$
23. $tf(t)$ given that $F(s) = \mathcal{L}\{f\}(s) = \ln\left(\dfrac{s^2}{s^2 + 1}\right)$

24–25. Use the dilation principle to find the Laplace transform of each function. The given Laplace transforms will be established later.

24. $\dfrac{1-\cos 5t}{t}$; given $\mathcal{L}\left\{\dfrac{1-\cos t}{t}\right\} = \dfrac{1}{2} \ln\left(\dfrac{s^2}{s^2+1}\right)$
25. $\text{Ei}(6t)$; given $\mathcal{L}\{\text{Ei}(t)\} = \dfrac{\ln(s+1)}{s}$

26–31. Use trigonometric or hyperbolic identities to compute the Laplace transform of the following functions.

26. $\cos^2 bt$ (Hint: $\cos^2 \theta = \frac{1}{2}(1 + \cos 2\theta)$)
27. $\sin^2 bt$
28. $\sin bt \cos bt$
29. $\sin at \cos bt$
30. $\cosh bt$ (Recall that $\cosh bt = (e^{bt} + e^{-bt})/2$.)
31. $\sinh bt$ (Recall that $\sinh bt = (e^{bt} - e^{-bt})/2$.)

32–34. Use one of the input derivative formulas to compute the Laplace transform of the following functions.

32. e^{at}
33. $\sinh bt$
34. $\cosh bt$
35. Use the input derivative formula to derive the Laplace transform formula $\mathcal{L}\left\{\int_0^t f(u)\, du\right\} = F(s)/s$. Hint: Let $g(t) = \int_0^t f(u)\, du$ and note that $g'(t) = f(t)$. Now apply the input derivative formula to $g(t)$.

36–41. *Functions of Exponential Type:* Verify the following claims.

36. Suppose f is of exponential type of order a and $b > a$. Show f is of exponential type of order b.
37. Show that the product of two functions of exponential type is of exponential type.
38. Show that the definition given for a function of exponential type is equivalent to the following: A continuous function f on $[0, \infty)$ is of exponential type of order a if there are constants $K \geq 0$ and $N \geq 0$ such that $|f(t)| \leq Ke^{at}$ for all $t \geq N$ (i.e., we do not need to require that $N = 0$).
39. Show that the function $y(t) = e^{t^2}$ is not of exponential type.
40. Verify that the function $f(t) = e^{t^2}$ does not have a Laplace transform. That is, show that the improper integral that defines $F(s)$ does not converge for *any* value of s.
41. Let $y(t) = \sin(e^{t^2})$. Why is $y(t)$ of exponential type? Compute $y'(t)$ and show that it is not of exponential type. Nevertheless, show that $y'(t)$ has a Laplace transform. *The moral*: The derivative of a function of exponential type is not necessarily of exponential type, and there are functions that are not of exponential type that have a Laplace transform.
42. Recall from the discussion for Formula 12 that the *gamma function* is defined by the improper integral

$$\Gamma(\beta) = \int_0^\infty x^{\beta-1} e^{-x}\, dx, \qquad\qquad (\beta > 0).$$

(a) Show that $\Gamma(1) = 1$.

(b) Show that Γ satisfies the recursion formula $\Gamma(\beta + 1) = \beta\Gamma(\beta)$.
    (*Hint*: Integrate by parts.)

(c) Show that $\Gamma(n + 1) = n!$ when n is a nonnegative integer.

43. Show that $\int_0^\infty e^{-x^2} \, dx = \sqrt{\pi}/2$.
    (*Hint*: Let I be the integral and note that

$$I^2 = \left(\int_0^\infty e^{-x^2} \, dx \right) \left(\int_0^\infty e^{-y^2} \, dy \right) = \int_0^\infty \int_0^\infty e^{-(x^2+y^2)} \, dx \, dy.$$

Then evaluate the integral using polar coordinates.)

44. Use the integral from Exercise 43 to show that $\Gamma(\frac{1}{2}) = \sqrt{\pi}$. Then compute each of the following:

    (a) $\Gamma(\frac{3}{2})$ (b) $\Gamma(\frac{5}{2})$ (c) $\mathcal{L}\{\sqrt{t}\}$ (d) $\mathcal{L}\{t^{3/2}\}$.

2.3   Partial Fractions: A Recursive Algorithm for Linear Terms

A careful look at Table 2.2 reveals that the Laplace transform of each function we considered is a rational function. Laplace inversion, which is discussed in Sect. 2.5, will involve writing rational functions as sums of those simpler ones found in the table.

All students of calculus should be familiar with the technique of obtaining the partial fraction decomposition of a rational function. Briefly, a given proper rational function[3] $p(s)/q(s)$ is a sum of *partial fractions* of the form

$$\frac{A_j}{(s-r)^j} \quad \text{and} \quad \frac{B_k s + C_k}{(s^2 + cs + d)^k},$$

where A_j, B_k, and C_k are constants. The partial fractions are determined by the linear factors, $s - r$, and the irreducible quadratic factors, $s^2 + cs + d$, of the denominator $q(s)$, where the powers j and k occur up to the multiplicity of the factors. After finding a common denominator and equating the numerators, we obtain a system of linear equations to solve for the undetermined coefficients A_j, B_k, C_k. Notice that the degree of the denominator determines the number of coefficients that are involved in the form of the partial fraction decomposition. Even when the degree is relatively small, this process can be very tedious and prone to simple numerical mistakes.

Our purpose in this section and the next is to provide an alternate algorithm for obtaining the partial fraction decomposition of a rational function. This algorithm has the advantage that it is constructive (assuming the factorization of the denominator), recursive (meaning that only one coefficient at a time is determined), and self-checking. This recursive method for determining partial fractions should be well practiced by the student. It is the method we will use throughout the text and is an essential technique in solving nonhomogeneous differential equations discussed in Sect. 3.5. You may wish to review Appendix A.2 where notation and results about polynomials and rational functions are given.

In this section, we will discuss the algorithm in the linear case, that is, when the denominator has a linear term as a factor. In Sect. 2.4, we discuss the case where the denominator has an irreducible quadratic factor.

Theorem 1 (Linear Partial Fraction Recursion). *Suppose a proper rational function can be written in the form*

$$\frac{p_0(s)}{(s-r)^n q(s)}$$

---

[3] A *rational function* is the quotient of two polynomials. A rational function is *proper* if the degree of the numerator is less than the degree of the denominator.

and $q(r) \neq 0$. Then there is a unique number A_1 and a unique polynomial $p_1(s)$ such that

$$\frac{p_0(s)}{(s-r)^n q(s)} = \frac{A_1}{(s-r)^n} + \frac{p_1(s)}{(s-r)^{n-1} q(s)}. \tag{1}$$

The number A_1 and the polynomial $p_1(s)$ are given by

$$A_1 = \left. \frac{p_0(s)}{q(s)} \right|_{s=r} = \frac{p_0(r)}{q(r)} \quad and \quad p_1(s) = \frac{p_0(s) - A_1 q(s)}{s-r}. \tag{2}$$

Proof. After finding a common denominator in (1) and equating numerators, we get the polynomial equation $p_0(s) = A_1 q(s) + (s-r)p_1(s)$. Evaluating at $s = r$ gives $p_0(r) = A_1 q(r)$, and hence $A_1 = \frac{p_0(r)}{q(r)}$. Now that A_1 is determined, we have $p_0(s) - A_1 q(s) = (s-r)p_1(s)$, and hence $p_1(s) = \frac{p_0(s)-A_1 q(s)}{s-r}$. $\square$

Notice that in the calculation of p_1, it is necessary that $p_0(s) - A_1 q(s)$ have a factor of $s - r$. If such a factorization does not occur when working an example, then an error has been made. This is what is meant when we stated above that this recursive method is self-checking. In practice, we frequently factor $p_0(s) - A_1 q(s)$ and delete the $s - r$ factor. However, for large degree polynomials, it may be best to use the division algorithm for polynomials or synthetic division.

An application of Theorem 1 produces two items:

- The partial fraction of the form

$$\frac{A_1}{(s-r)^n}$$

- A remainder term of the form

$$\frac{p_1(s)}{(s-r)^{n-1} q(s)}$$

such that the original rational function $p_0(s)/(s-r)^n q(s)$ is the sum of these two pieces. We can now repeat the process on the new rational function $p_1(s)/((s-r)^{n-1} q(s))$, where the multiplicity of $(s-r)$ in the denominator has been reduced by 1, and continue in this manner until we have removed completely $(s-r)$ as a factor of the denominator. In this manner, we produce a sequence,

$$\frac{A_1}{(s-r)^n}, \ldots, \frac{A_n}{(s-r)},$$

which we will refer to as the $(s-r)$-**chain** for $p_0(s)/((s-r)^n q(s))$. The number of terms, n, is referred to as the **length** of the chain. The **chain table** below summarizes the data obtained.

The $(s - r)$-chain

$\dfrac{p_0(s)}{(s - r)^n q(s)}$	$\dfrac{A_1}{(s - r)^n}$
$\dfrac{p_1(s)}{(s - r)^{n-1}(s)q(s)}$	$\dfrac{A_2}{(s - r)^{n-1}}$
$\vdots$	$\vdots$
$\dfrac{p_{n-1}(s)}{(s - r)q(s)}$	$\dfrac{A_n}{(s - r)}$
$\dfrac{p_n(s)}{q(s)}$	

Notice that the partial fractions are placed in the second column while the remainder terms are placed in the first column under the previous remainder term. This form is conducive to the recursion algorithm. From the table, we get

$$\frac{p_0(s)}{(s - r)^n q(s)} = \frac{A_1}{(s - r)^n} + \cdots + \frac{A_n}{(s - r)} + \frac{p_n(s)}{q(s)}.$$

By factoring another linear term out of $q(s)$, the process can be repeated through all linear factors of $q(s)$. In the examples that follow, we will organize one step of the recursion process as follows:

**Partial Fraction Recursion Algorithm
by a Linear Term**

$$\frac{p_0(s)}{(s - r)^n q(s)} = \frac{A_1}{(s - r)^n} + \frac{p_1(s)}{(s - r)^{n-1}q(s)}$$

$$\text{where} \quad A_1 = \left.\frac{p_0(s)}{q(s)}\right|_{s=r} = \square$$

$$\text{and} \quad p_1(s) = \frac{1}{s - r}(p_0(s) - A_1 q(s)) = \square$$

The curved arrows indicate where the results of calculations are inserted. First, A_1 is calculated and inserted in two places: in the $(s-r)$-chain and in the calculation for $p_1(s)$. Afterward, $p_1(s)$ is calculated and the result inserted in the numerator of the remainder term. Now the process is repeated on $p_1(s)/((s-r)^{n-1}q(s))$ until the $(s-r)$-chain is completed.

Consider the following examples.

Example 2.   Find the partial fraction decomposition for

$$\frac{s-2}{(s-3)^2(s-4)}.$$

▶ **Solution.**   We will first compute the $(s-3)$ -chain. According to Theorem 1, we can write

$$\frac{s-2}{(s-3)^2(s-4)} = \frac{A_1}{(s-3)^2} + \frac{p_1(s)}{(s-3)(s-4)}$$

$$\text{where}\quad A_1 = \left.\frac{s-2}{s-4}\right|_{s=3} = -1$$

$$\text{and}\quad p_1(s) = \frac{1}{s-3}(s-2-(-1)(s-4)) = \frac{1}{s-3}(2s-6) = 2.$$

We thus get

$$\frac{s-2}{(s-3)^2(s-4)} = \frac{-1}{(s-3)^2} + \frac{2}{(s-3)(s-4)}.$$

We now repeat the recursion algorithm on the remainder term $\frac{2}{(s-3)(s-4)}$ to get

$$\frac{2}{(s-3)(s-4)} = \frac{A_2}{s-3} + \frac{p_2(s)}{s-4}$$

$$\text{where}\quad A_1 = \left.\frac{2}{s-4}\right|_{s=3} = -2$$

$$\text{and}\quad p_2(s) = \frac{1}{s-3}(2-(-2)(s-4)) = \frac{1}{s-3}(2s-6) = 2.$$

Putting these calculations together gives the $(s-3)$-chain

The $(s-3)$ -chain	
$\dfrac{s-2}{(s-3)^2(s-4)}$	$\dfrac{-1}{(s-3)^2}$
$\dfrac{2}{(s-3)(s-4)}$	$\dfrac{-2}{(s-3)}$
$\dfrac{2}{s-4}$	

The $(s-4)$-chain has length one and is given as the remainder entry in the $(s-3)$-chain; thus

$$\frac{s-2}{(s-3)^2(s-4)} = \frac{-1}{(s-3)^2} - \frac{2}{(s-3)} + \frac{2}{s-4}. \qquad \blacktriangleleft$$

A more substantial example is given next. The partial fraction recursion algorithm remains exactly the same so we will dispense with the curved arrows.

Example 3. Find the partial fraction decomposition for

$$\frac{16s}{(s+1)^3(s-1)^2}.$$

Remark 4. Before we begin with the solution, we remark that the traditional method for computing the partial fraction decomposition introduces the equation

$$\frac{16s}{(s+1)^3(s-1)^2} = \frac{A_1}{(s+1)^3} + \frac{A_2}{(s+1)^2} + \frac{A_3}{s+1} + \frac{A_4}{(s-1)^2} + \frac{A_5}{s-1}$$

and, after finding a common denominator, requires the simultaneous solution to a system of five equations in five unknowns, a doable task but one prone to simple algebraic errors.

▶ **Solution.** We will first compute the $(s+1)$ -chain. According to Theorem 1, we can write

$$\frac{16s}{(s+1)^3(s-1)^2} = \frac{A_1}{(s+1)^3} + \frac{p_1(s)}{(s+1)^2(s-1)^2},$$

$$\text{where} \quad A_1 = \frac{16s}{(s-1)^2}\bigg|_{s=-1} = -\frac{16}{4} = -4.$$

and $p_1(s) = \dfrac{1}{s+1}(16s - (-4)(s-1)^2)$

$\qquad\qquad = \dfrac{4}{s+1}(s^2 + 2s + 1) = 4(s+1).$

We now repeat the recursion step on the remainder term $\dfrac{4(s+1)}{(s+1)^2(s-1)^2}$ to get

$$\dfrac{4(s+1)}{(s+1)^2(s-1)^2} = \dfrac{A_2}{(s+1)^2} + \dfrac{p_2(s)}{(s+1)(s-1)^2},$$

where $A_2 = \dfrac{4(s+1)}{(s-1)^2}\Big|_{s=-1} = \dfrac{0}{4} = 0$

and $p_2(s) = \dfrac{1}{s+1}(4(s+1) - (0)(s-1)^2) = 4.$

Notice here that we could have canceled the $(s+1)$ term at the beginning and arrived immediately at $\frac{4}{(s+1)(s-1)^2}$. Then no partial fraction with $(s+1)^2$ in the denominator would occur. We chose though to continue the recursion step to show the process. The recursion process is now repeated on $\frac{4}{(s+1)(s-1)^2}$ to get

$$\dfrac{4}{(s+1)(s-1)^2} = \dfrac{A_3}{s+1} + \dfrac{p_3(s)}{(s-1)^2},$$

where $A_3 = \dfrac{4}{(s-1)^2}\Big|_{s=-1} = \dfrac{4}{4} = 1$

and $p_3(s) = \dfrac{1}{s+1}(4 - (1)(s-1)^2)$

$\qquad\qquad = \dfrac{-1}{s+1}(s^2 - 2s - 3) = \dfrac{-1}{s+1}(s+1)(s-3) = -(s-3).$

Putting these calculations together gives the $(s+1)$ -chain

The $(s+1)$ -chain	
$\dfrac{16}{(s+1)^3(s-1)^2}$	$\dfrac{-4}{(s+1)^3}$
$\dfrac{4(s+1)}{(s+1)^2(s-1)^2}$	$\dfrac{0}{(s+1)^2}$
$\dfrac{4}{(s+1)(s-1)^2}$	$\dfrac{1}{(s+1)}$
$\dfrac{-(s-3)}{(s-1)^2}$	

·We now compute the $(s-1)$ -chain for the remainder $\frac{-(s-3)}{(s-1)^2}$. It is implicit that $q(s) = 1$.

$$\frac{-(s-3)}{(s-1)^2} = \frac{A_4}{(s-1)^2} + \frac{p_4(s)}{s-1},$$

where $A_4 = -\frac{(s-3)}{1}\bigg|_{s=1} = 2$

and $p_4(s) = \frac{1}{s-1}(-(s-3)-(2)) = \frac{1}{s-1}(-s+1) = -1.$

The $(s-1)$ -chain is thus

The $(s-1)$ -chain	
$\dfrac{-(s-3)}{(s-1)^2}$	$\dfrac{2}{(s-1)^2}$
$\dfrac{-1}{(s-1)}$	

We now have

$$\frac{-(s-3)}{(s-1)^2} = \frac{2}{(s-1)^2} + \frac{-1}{s-1},$$

and putting this chain together with the $s+1$ -chain gives

$$\frac{16s}{(s+1)^3(s-1)^2} = \frac{-4}{(s+1)^3} + \frac{0}{(s+1)^2} + \frac{1}{s+1} + \frac{2}{(s-1)^2} + \frac{-1}{s-1}. \quad \blacktriangleleft$$

Product of Distinct Linear Factors

Let $p(s)/q(s)$ be a proper rational function. Suppose $q(s)$ is the product of distinct linear factors, that is,

$$q(s) = (s-r_1)\cdots(s-r_n),$$

where $r_1, \ldots r_n$ are distinct scalars. Then each chain has length one and the partial fraction decomposition has the form

$$\frac{p(s)}{q(s)} = \frac{A_1}{s - r_1} + \cdots + \frac{A_n}{(s - r_n)}.$$

The scalar A_i is the first and only entry in the $(s - r_i)$ -chain. Thus,

$$A_i = \left. \frac{p(s)}{q_i(s)} \right|_{s=r_i} = \frac{p(r_i)}{q_i(r_i)},$$

where $q_i(s) = q(s)/(s - r_i)$ is the polynomial obtained from $q(s)$ by factoring out $(s - r_i)$. If we do this for each $i = 1, \ldots, n$, it is unnecessary to calculate any remainder terms.

Example 5. Find the partial fraction decomposition of

$$\frac{-4s + 14}{(s - 1)(s + 4)(s - 2)}.$$

▶ **Solution.** The denominator $q(s) = (s - 1)(s + 4)(s - 2)$ is a product of distinct linear factors. Each partial fraction is determined as follows:

- For $\dfrac{A_1}{s - 1}$: $A_1 = \left. \dfrac{-4s + 14}{(s + 4)(s - 2)} \right|_{s=1} = \dfrac{10}{-5} = -2$

- For $\dfrac{A_2}{s + 4}$: $A_2 = \left. \dfrac{-4s + 14}{(s - 1)(s - 2)} \right|_{s=-4} = \dfrac{30}{30} = 1$

- For $\dfrac{A_3}{s - 2}$: $A_3 = \left. \dfrac{-4s + 14}{(s - 1)(s + 4)} \right|_{s=2} = \dfrac{6}{6} = 1$

The partial fraction decomposition is thus

$$\frac{-4s + 14}{(s - 1)(s + 4)(s - 2)} = \frac{-2}{s - 1} + \frac{1}{s + 4} + \frac{1}{s - 2}. \qquad ◀$$

Linear Partial Fractions and the Laplace Transform Method

In the example, notice how linear partial fraction recursion facilitates the Laplace transform method.

Example 6. Use the Laplace transform method to solve the following differential equation:

$$y'' + 3y' + 2y = e^{-t}, \tag{3}$$

with initial conditions $y(0) = 1$ and $y'(0) = 3$.

▶ **Solution.** We will use the Laplace transform to turn this differential equation in y into an algebraic equation in $Y(s) = \mathcal{L}\{y(t)\}$. Apply the Laplace transform to both sides. For the left-hand side, we get

$$\begin{aligned}
\mathcal{L}\{y'' + 3y' + 2y\} &= \mathcal{L}\{y''\} + 3\mathcal{L}\{y'\} + 2\mathcal{L}\{y\} \\
&= s^2 Y(s) - sy(0) - y'(0) + 3(sY(s) - y(0)) + 2Y(s) \\
&= (s^2 + 3s + 2)Y(s) - s - 6.
\end{aligned}$$

The first line uses the linearity of the Laplace transform, the second line uses the first and second input derivative principles, and the third line uses the given initial conditions and then simplifies the result. Since $\mathcal{L}\{e^{-t}\} = 1/(s+1)$, we get the algebraic equation

$$(s^2 + 3s + 2)Y(s) - s - 6 = \frac{1}{s+1},$$

from which it is easy to solve for $Y(s)$. Since $s^2 + 3s + 2 = (s+1)(s+2)$, we get

$$Y(s) = \frac{s+6}{(s+1)(s+2)} + \frac{1}{(s+1)^2(s+2)}.$$

We are now left with the task of finding an input function whose Laplace transform is $Y(s)$. To do this, we first compute the partial fraction decomposition of each term. The first term $\frac{s+6}{(s+1)(s+2)}$ has denominator which is a product of two distinct linear terms. Each partial fraction is determined as follows:

- For $\dfrac{A_1}{s+1}$: $\quad A_1 = \dfrac{s+6}{s+2}\bigg|_{s=-1} = \dfrac{5}{1} = 5$

- For $\dfrac{A_2}{s+2}$: $\quad A_2 = \dfrac{s+6}{s+1}\bigg|_{s=-2} = \dfrac{4}{-1} = -4$

The partial fraction decomposition is thus

$$\frac{s+6}{(s+1)(s+2)} = \frac{5}{s+1} - \frac{4}{s+2}.$$

For the second term $\frac{1}{(s+1)^2(s+2)}$, we compute the $(s+1)$-chain

The $(s+1)$ -chain	
$\dfrac{1}{(s+1)^2(s+2)}$	$\dfrac{1}{(s+1)^2}$
$\dfrac{-1}{(s+1)(s+2)}$	$\dfrac{-1}{s+1}$
$\dfrac{1}{s+2}$	

from which we get

$$\frac{1}{(s+1)^2(s+2)} = \frac{1}{(s+1)^2} - \frac{1}{s+1} + \frac{1}{s+2}.$$

It follows that

$$Y(s) = \frac{4}{s+1} - \frac{3}{s+2} + \frac{1}{(s+1)^2}.$$

Now we can determine the input function $y(t)$ directly from the basic Laplace transform table, Table 2.2. We get

$$y(t) = 4e^{-t} - 3e^{-2t} + te^{-t}.$$

This is the solution to (3).   ◄

Exercises

1–11. For each exercise below, compute the chain table through the indicated linear term.

1. $\dfrac{5s + 10}{(s - 1)(s + 4)}$; $(s - 1)$

2. $\dfrac{10s - 2}{(s + 1)(s - 2)}$; $(s - 2)$

3. $\dfrac{1}{(s + 2)(s - 5)}$; $(s - 5)$

4. $\dfrac{5s + 9}{(s - 1)(s + 3)}$; $(s + 3)$

5. $\dfrac{3s + 1}{(s - 1)(s^2 + 1)}$; $(s - 1)$

6. $\dfrac{3s^2 - s + 6}{(s + 1)(s^2 + 4)}$; $(s + 1)$

7. $\dfrac{s^2 + s - 3}{(s + 3)^3}$; $(s + 3)$

8. $\dfrac{5s^2 - 3s + 10}{(s + 1)(s + 2)^2}$; $(s + 2)$

9. $\dfrac{s}{(s + 2)^2(s + 1)^2}$; $(s + 1)$

10. $\dfrac{16s}{(s - 1)^3(s - 3)^2}$; $(s - 1)$

11. $\dfrac{1}{(s - 5)^5(s - 6)}$; $(s - 5)$

12–32. Find the partial fraction decomposition of each proper rational function.

12. $\dfrac{5s + 9}{(s - 1)(s + 3)}$

13. $\dfrac{8 + s}{s^2 - 2s - 15}$

14. $\dfrac{1}{s^2 - 3s + 2}$

15. $\dfrac{5s - 2}{s^2 + 2s - 35}$

16. $\dfrac{3s + 1}{s^2 + s}$

17. $\dfrac{2s + 11}{s^2 - 6s - 7}$

18. $\dfrac{2s^2 + 7}{(s - 1)(s - 2)(s - 3)}$

19. $\dfrac{s^2 + s + 1}{(s - 1)(s^2 + 3s - 10)}$

20. $\dfrac{s^2}{(s - 1)^3}$

21. $\dfrac{7}{(s + 4)^4}$

22. $\dfrac{s}{(s - 3)^3}$

23. $\dfrac{s^2 + s - 3}{(s + 3)^3}$

24. $\dfrac{5s^2 - 3s + 10}{(s + 1)(s + 2)^2}$

25. $\dfrac{s^2 - 6s + 7}{(s^2 - 4s - 5)^2}$

26. $\dfrac{81}{s^3(s + 9)}$

27. $\dfrac{s}{(s + 2)^2(s + 1)^2}$

28. $\dfrac{s^2}{(s + 2)^2(s + 1)^2}$

29. $\dfrac{8s}{(s - 1)(s - 2)(s - 3)^3}$

30. $\dfrac{25}{s^2(s - 5)(s + 1)}$

31. $\dfrac{s}{(s - 2)^2(s - 3)^2}$

32. $\dfrac{16s}{(s - 1)^3(s - 3)^2}$

33–38. Use the *Laplace transform method* to solve the following differential equations. (Give both $Y(s)$ and $y(t)$.)

33. $y'' + 2y' + y = 9e^{2t}, \quad y(0) = 0, y'(0) = 0$

34. $y'' + 3y' + 2y = 12e^{2t}, \quad y(0) = 1, y'(0) = -1$

35. $y'' - 4y' - 5y = 150t, \quad y(0) = -1, y'(0) = 1$

36. $y'' + 4y' + 4y = 4\cos 2t, \quad y(0) = 0, y'(0) = 1$

37. $y'' - 3y' + 2y = 4, \quad y(0) = 2, y'(0) = 3$

38. $y'' - 3y' + 2y = e^t, \quad y(0) = -3, y'(0) = 0$

2.4 Partial Fractions: A Recursive Algorithm for Irreducible Quadratics

We continue the discussion of Sect. 2.3. Here we consider the case where a real rational function has a denominator with irreducible quadratic factors.

Theorem 1 (Quadratic Partial Fraction Recursion). *Suppose a real proper rational function can be written in the form*

$$\frac{p_0(s)}{(s^2 + cs + d)^n q(s)},$$

where $s^2 + cs + d$ is an irreducible quadratic that is factored completely out of $q(s)$. Then there is a unique linear term $B_1 s + C_1$ and a unique polynomial $p_1(s)$ such that

$$\frac{p_0(s)}{(s^2 + cs + d)^n q(s)} = \frac{B_1 s + C_1}{(s^2 + cs + d)^n} + \frac{p_1(s)}{(s^2 + cs + d)^{n-1} q(s)}. \tag{1}$$

If $a + ib$ is a complex root of $s^2 + cs + d$, then $B_1 s + C_1$ and the polynomial $p_1(s)$ are given by

$$B_1 s + C_1 \big|_{s=a+bi} = \frac{p_0(s)}{q(s)} \bigg|_{s=a+bi} \quad \text{and} \quad p_1(s) = \frac{p_0(s) - (B_1 s + C_1) q(s)}{s^2 + cs + d}. \tag{2}$$

Proof. After finding a common denominator in (1) and equating numerators, we get the polynomial equation

$$p_0(s) = (B_1 s + C_1) q(s) + (s^2 + cs + d) p_1(s). \tag{3}$$

Evaluating at $s = a + ib$ gives $p_0(a + ib) = (B_1(a + ib) + C_1)(q(a + ib))$, and hence,

$$B_1(a + ib) + C_1 = \frac{p_0(a + ib)}{q(a + ib)}. \tag{4}$$

Equating the real and imaginary parts of both sides of (4) gives the equations

$$B_1 a + C_1 = \text{Re}\left(\frac{p_0(a + ib)}{q(a + ib)}\right),$$

$$B_1 b = \text{Im}\left(\frac{p_0(a + ib)}{q(a + ib)}\right).$$

Since $b \neq 0$ because the quadratic $s^2 + cs + d$ has no real roots, these equations can be solved for B_1 and C_1, so both B_1 and C_1 are determined by (4). Now solving for $p_1(s)$ in (3) gives

$$p_1(s) = \frac{p_0(s) - (B_1 s + C_1) q(s)}{s^2 + cs + d}.$$

$\square$

An application of Theorem 1 produces two items:

- The partial fraction of the form

$$\frac{B_1 s + C_1}{(s^2 + cs + d)^n}$$

- A remainder term of the form

$$\frac{p_1(s)}{(s^2 + cs + d)^{n-1} q(s)}$$

such that the original rational function $p_0(s)/(s^s + as + b)^n q(s)$ is the sum of these two pieces. We can now repeat the process in the same way as the linear case.

The result is called the $(s^2 + cs + d)$-***chain*** for the rational function $p_0(s)/(s^2 + cs + d)^n q(s)$. The table below summarizes the data obtained.

The $(s^2 + cs + d)$-chain	
$\dfrac{p_0(s)}{(s^2 + cs + d)^n q(s)}$	$\dfrac{B_1 s + C_1}{(s^2 + cs + d)^n}$
$\dfrac{p_1(s)}{(s^2 + cs + d)^{n-1} q(s)}$	$\dfrac{B_2 s + C_2}{(s^2 + cs + d)^{n-1}}$
$\vdots$	$\vdots$
$\dfrac{p_{n-1}(s)}{(s^2 + cs + d) q(s)}$	$\dfrac{B_n s + C_n}{(s^2 + cs + d)}$
$\dfrac{p_n(s)}{q(s)}$	

From this table, we can immediately read off the following decomposition:

$$\frac{p_0(s)}{(s^2 + cs + d)^n q(s)} = \frac{B_1 s + C_1}{(s^2 + cs + d)^n} + \cdots + \frac{B_n s + C_n}{(s^2 + cs + d)} + \frac{p_n(s)}{q(s)}.$$

In the examples that follow, we will organize one step of the recursion algorithm as follows:

**Partial Fraction Recursion Algorithm
by a Quadratic Term**

$$\frac{p_0(s)}{(s^2 + cs + d)^n q(s)} = \frac{B_1 s + C_1}{(s^2 + cs + d)^n} + \frac{p_1(s)}{(s^2 + cs + d)^{n-1} q(s)}$$

where $\quad B_1 s + C_1|_{s=a+bi} = \left. \dfrac{p_0(s)}{q(s)} \right|_{s=a+bi} \quad \Rightarrow B_1 = \square \text{ and } C_1 = \square$

and $\quad p_1(s) = \dfrac{1}{s^2 + cs + d}(p_0(s) - (B_1 s + C_1)q(s)) = \square$

As in the linear case, the curved arrows indicate where the results of calculations are inserted. First, B_1 and C_1 are calculated and inserted in two places: in the $(s^2 + cs + d)$-chain and in the calculation for $p_1(s)$. Afterward, $p_1(s)$ is calculated and the result inserted in the numerator of the remainder term. Now the process is repeated on $p_1(s)/((s^2 + cs + d)^{n-1} q(s))$ until the $(s^2 + cs + d)$-chain is completed.

Here are some examples of this process in action.

Example 2. Find the partial fraction decomposition for

$$\frac{5s}{(s^2 + 4)(s + 1)}.$$

▶ **Solution.** We have a choice of computing the linear chain through $s + 1$ or the quadratic chain through $s^2 + 4$. It is usually easier to compute linear chains. However, to illustrate the recursive algorithm for the quadratic case, we will compute the $(s^2 + 4)$-chain. The roots of $s^2 + 4$ are $s = \pm 2i$. We need only focus on one root and we will choose $s = 2i$.

According to Theorem 1, we can write

$$\frac{5s}{(s^2 + 4)(s + 1)} = \frac{B_1 s + C_1}{(s^2 + 4)} + \frac{p_1(s)}{s + 1},$$

where $\quad B_1(2i) + C_1 = \left. \dfrac{5s}{s+1} \right|_{s=2i} = \dfrac{10i}{2i + 1}$

$$= \frac{(10i)(-2i + 1)}{(2i + 1)(-2i + 1)} = \frac{20 + 10i}{5} = 4 + 2i$$

$$\Rightarrow B_1 = 1 \text{ and } C_1 = 4$$

$$\text{and}\quad p_1(s) = \frac{1}{s^2+4}(5s - (s+4)(s+1))$$

$$= \frac{1}{s^2+4}(-s^2-4) = -1.$$

It follows now that

$$\frac{5s}{(s^2+4)(s+1)} = \frac{s+4}{(s^2+4)} + \frac{-1}{s+1}.\qquad\blacktriangleleft$$

Note that in the above calculation, B_1 and C_1 are determined by comparing the real and imaginary parts of the complex numbers $2B_1 i + C_1 = 2i + 4$ so that the imaginary parts give $2B_1 = 2$ so $B_1 = 1$ and the real parts give $C_1 = 4$.

Example 3. Find the partial fraction decomposition for

$$\frac{30s+40}{(s^2+1)^2(s^2+2s+2)}.$$

Remark 4. We remark that since the degree of the denominator is 6, the traditional method of determining the partial fraction decomposition would involve solving a system of six equations in six unknowns.

▶ **Solution.** First observe that both factors in the denominator, $s^2 + 1$ and $s^2 + 2s + 2 = (s+1)^2 + 1$, are irreducible quadratics. We begin by determining the $s^2 + 1$-chain. Note that $s = i$ is a root of $s^2 + 1$.
Applying the recursive algorithm gives

$$\frac{30s+40}{(s^2+1)^2(s^2+2s+2)} = \frac{B_1 s + C_1}{(s^2+1)^2} + \frac{p_1(s)}{(s^2+1)(s^2+2s+2)},$$

$$\text{where}\quad B_1 i + C_1 = \frac{30s+40}{(s^2+2s+2)}\Big|_{s=i} = \frac{30i+40}{1+2i}$$

$$= \frac{(40+30i)(1-2i)}{(1+2i)(1-2i)} = \frac{100-50i}{5} = 20 - 10i$$

$$\Rightarrow B_1 = -10 \quad\text{and } C_1 = 20$$

$$\text{and}\quad p_1(s) = \frac{1}{s^2+1}(30s + 40 - (-10s+20)(s^2+2s+2))$$

$$= \frac{1}{s^2+1}(10s(s^2+1)) = 10s.$$

We now repeat the recursion algorithm on the remainder term

$$\frac{10s}{(s^2 + 1)(s^2 + 2s + 2)}.$$

$$\frac{10s}{(s^2 + 1)(s^2 + 2s + 2)} = \frac{B_2 s + C_2}{(s^2 + 1)} + \frac{p_2(s)}{(s^2 + 2s + 2)},$$

$$\text{where} \quad B_2 i + C_2 = \left. \frac{10s}{(s^2 + 2s + 2)} \right|_{s=i} = \frac{10i}{1 + 2i}$$

$$= \frac{(10i)(1 - 2i)}{(1 + 2i)(1 - 2i)} = \frac{20 + 10i}{5} = 4 + 2i$$

$$\Rightarrow B_2 = 2 \text{ and } C_2 = 4$$

$$\text{and} \quad p_2(s) = \frac{1}{s^2 + 1}(10s - (2s + 4)(s^2 + 2s + 2))$$

$$= \frac{1}{s^2 + 1}(-2(s + 4)(s^2 + 1)) = -2(s + 4).$$

We can now write down the $(s^2 + 1)$-chain.

The $(s^2 + 1)$-chain	
$\dfrac{30s + 40}{(s^2 + 1)^2(s^2 + 2s + 2)}$	$\dfrac{-10s + 20}{(s^2 + 1)^2}$
$\dfrac{10s}{(s^2 + 1)(s^2 + 2s + 2)}$	$\dfrac{2s + 4}{(s^2 + 1)}$
$\dfrac{-2(s + 4)}{(s^2 + 2s + 2)}$	

Since the last remainder term is already a partial fraction, we obtain

$$\frac{30s + 40}{(s^2 + 1)^2(s^2 + 2s + 2)} = \frac{-10s + 20}{(s^2 + 1)^2} + \frac{2s + 4}{s^2 + 1} + \frac{-2s - 8}{(s + 1)^2 + 1}. \qquad \blacktriangleleft$$

Quadratic Partial Fractions and the Laplace Transform Method

In the following example, notice how quadratic partial fraction recursion facilitates the Laplace transform method.

Example 5. Use the Laplace transform method to solve

$$y'' + 4y = \cos 3t, \tag{5}$$

with initial conditions $y(0) = 0$ and $y'(0) = 0$.

▶ **Solution.** Applying the Laplace transform to both sides of (5) and substituting the given initial conditions give

$$(s^2 + 4)Y(s) = \frac{s}{s^2 + 9}$$

and thus

$$Y(s) = \frac{s}{(s^2 + 4)(s^2 + 9)}.$$

Using quadratic partial fraction recursion, we obtain the $(s^2 + 4)$-chain

The $(s^2 + 4)$-chain	
$\dfrac{s}{(s^2 + 4)(s^2 + 9)}$	$\dfrac{s/5}{s^2 + 4}$
$\dfrac{-s/5}{s^2 + 9}$	

It follows from Table 2.2 that

$$y(t) = \frac{1}{5}(\cos 2t - \cos 3t).$$

This is the solution to (5).   ◀

Exercises

1–6. For each exercise below, compute the chain table through the indicated irreducible quadratic term.

1. $\dfrac{1}{(s^2 + 1)^2(s^2 + 2)}$; $(s^2 + 1)$

2. $\dfrac{s^3}{(s^2 + 2)^2(s^2 + 3)}$; $(s^2 + 2)$

3. $\dfrac{8s + 8s^2}{(s^2 + 3)^3(s^2 + 1)}$; $(s^2 + 3)$

4. $\dfrac{4s^4}{(s^2 + 4)^4(s^2 + 6)}$; $(s^2 + 4)$

5. $\dfrac{1}{(s^2 + 2s + 2)^2(s^2 + 2s + 3)^2}$; $(s^2 + 2s + 2)$

6. $\dfrac{5s - 5}{(s^2 + 2s + 2)^2(s^2 + 4s + 5)}$; $(s^2 + 2s + 2)$

7–16. Find the decomposition of the given rational function into partial fractions over $\mathbb{R}$.

7. $\dfrac{s}{(s^2 + 1)(s - 3)}$

8. $\dfrac{4s}{(s^2 + 1)^2(s + 1)}$

9. $\dfrac{9s^2}{(s^2 + 4)^2(s^2 + 1)}$

10. $\dfrac{9s}{(s^2 + 1)^2(s^2 + 4)}$

11. $\dfrac{2}{(s^2 - 6s + 10)(s - 3)}$

12. $\dfrac{30}{(s^2 - 4s + 13)(s - 1)}$

13. $\dfrac{25}{(s^2 - 4s + 8)^2(s - 1)}$

14. $\dfrac{s}{(s^2 + 6s + 10)^2(s + 3)^2}$

15. $\dfrac{s + 1}{(s^2 + 4s + 5)^2(s^2 + 4s + 6)^2}$

16. $\dfrac{s^2}{(s^2 + 5)^3 (s^2 + 6)^2}$

   Hint: Let $u = s^2$.

17–20. Use the *Laplace transform method* to solve the following differential equations. (Give both $Y(s)$ and $y(t)$.)

17. $y'' + 4y' + 4y = 4\cos 2t$, $\quad y(0) = 0,\ y'(0) = 1$

18. $y'' + 6y' + 9y = 50\sin t$, $\quad y(0) = 0,\ y'(0) = 2$

19. $y'' + 4y = \sin 3t$, $\quad y(0) = 0,\ y'(0) = 1$

20. $y'' + 2y' + 2y = 2\cos t + \sin t$, $\quad y(0) = 0,\ y'(0) = 0$

2.5 Laplace Inversion

In this section, we consider Laplace inversion and the kind of input functions that can arise when the transform function is a rational function. Given a transform function $F(s)$, we call an input function $f(t)$ the **inverse Laplace transform** of $F(s)$ if $\mathcal{L}\{f(t)\}(s) = F(s)$. We say *the* inverse Laplace transform because in most circumstances, it can be chosen uniquely. One such circumstance is when the input function is continuous. We state this fact as a theorem. For a proof of this result, see Appendix A.1

Theorem 1. *Suppose $f_1(t)$ and $f_2(t)$ are continuous functions defined on $[0, \infty)$ with the same Laplace transform. Then $f_1(t) = f_2(t)$.*

It follows from this theorem that if a transform function has a continuous input function, then it can have only one such input function. In Chap. 6, we will consider some important classes of discontinuous input functions, but for now, we will assume that all input functions are continuous and we write $\mathcal{L}^{-1}\{F(s)\}$ for the inverse Laplace transform of F. That is, $\mathcal{L}^{-1}\{F(s)\}$ *is the unique continuous function $f(t)$ that has $F(s)$ as its Laplace transform.* Symbolically,

> ### Defining Property of the Inverse Laplace Transform
>
> $$\mathcal{L}^{-1}\{F(s)\} = f(t) \iff \mathcal{L}\{f(t)\} = F(s).$$

We can thus view $\mathcal{L}^{-1}$ as an operation on transform functions $F(s)$ that produces input functions $f(t)$. Because of the defining property of the inverse Laplace transform, each formula for the Laplace transform has a corresponding formula for the inverse Laplace transform.

Example 2. List the corresponding inverse Laplace transform formula for each formula in Table 2.2.

▶ **Solution.** Each line of Table 2.4 corresponds to the same line in Table 2.2.   ◀

By identifying the parameters n, a, and b in specific functions $F(s)$, it is possible to read off $\mathcal{L}^{-1}\{F(s)\} = f(t)$ from Table 2.4 for some $F(s)$.

Example 3. Find the inverse Laplace transform of each of the following functions $F(s)$:

$$\text{1.} \quad \frac{6}{s^4} \qquad \text{2.} \quad \frac{2}{(s+3)^3} \qquad \text{3.} \quad \frac{5}{s^2+25} \qquad \text{4.} \quad \frac{s-1}{(s-1)^2+4}$$

▶ **Solution.**

1. $\mathcal{L}^{-1}\left\{\frac{6}{s^4}\right\} = t^3$ ($n = 3$ in Formula 2)

2. $\mathcal{L}^{-1}\left\{\frac{2}{(s+3)^3}\right\} = t^2 e^{-3t}$ ($n = 2$, $a = -3$ in Formula 4)

3. $\mathcal{L}^{-1}\left\{\frac{5}{s^2+25}\right\} = \sin 5t$ ($b = 5$ in Formula 6)

4. $\mathcal{L}^{-1}\left\{\frac{s-1}{(s-1)^2+4}\right\} = e^t \cos 2t$ ($a = 1$, $b = 2$ in Formula 7)   ◀

It is also true that each Laplace transform principle recorded in Table 2.3 results in a corresponding principle for the inverse Laplace transform. We will single out the linearity principle and the first translation principle at this time.

Theorem 4 (Linearity). *The inverse Laplace transform is linear. In other words, if $F(s)$ and $G(s)$ are transform functions with continuous inverse Laplace transforms and a and b are constants, then*

Linearity of the Inverse Laplace Transform

$$\mathcal{L}^{-1}\{aF(s) + bG(s)\} = a\mathcal{L}^{-1}\{F(s)\} + b\mathcal{L}^{-1}\{G(s)\}.$$

Proof. Let $f(t) = \mathcal{L}^{-1}\{F(s)\}$ and $g(t) = \mathcal{L}^{-1}\{G(s)\}$. Since the Laplace transform is linear by Theorem 5 of Sect. 2.2, we have

$$\mathcal{L}\{(af(t) + bg(t)\} = a\mathcal{L}\{f(t)\} + b\mathcal{L}\{g(t)\} = aF(s) + bG(s).$$

Since $af(t) + bg(t)$ is continuous, it follows that

$$\mathcal{L}^{-1}\{aF(s) + bG(s)\} = af(t) + bg(t) = a\mathcal{L}^{-1}\{F(s)\} + b\mathcal{L}^{-1}\{G(s)\}. \quad \square$$

Theorem 5. *If $F(s)$ has a continuous inverse Laplace transform, then*

Inverse First Translation Principle

$$\mathcal{L}^{-1}\{F(s-a)\} = e^{at}\mathcal{L}^{-1}\{F(s)\}$$

Proof. Let $f(t) = \mathcal{L}^{-1}\{F(s)\}$. Then the first translation principle (Theorem 17 of Sect. 2.2) gives

$$\mathcal{L}\{e^{at}f(t)\} = F(s-a),$$

and applying $\mathcal{L}^{-1}$ to both sides of this equation gives

$$\mathcal{L}^{-1}\{F(s-a)\} = e^{at}f(t) = e^{at}\mathcal{L}^{-1}\{F(s)\}. \qquad \square$$

Table 2.4 Basic inverse
Laplace transform formulas
(We are assuming n is a
nonnegative integer and a and
b are real)

1. $\mathcal{L}^{-1}\left\{\dfrac{1}{s}\right\} = 1$

2. $\mathcal{L}^{-1}\left\{\dfrac{n!}{s^{n+1}}\right\} = t^n$

3. $\mathcal{L}^{-1}\left\{\dfrac{1}{s-a}\right\} = e^{at}$

4. $\mathcal{L}^{-1}\left\{\dfrac{n!}{(s-a)^{n+1}}\right\} = t^n e^{at}$

5. $\mathcal{L}^{-1}\left\{\dfrac{s}{s^2+b^2}\right\} = \cos bt$

6. $\mathcal{L}^{-1}\left\{\dfrac{b}{s^2+b^2}\right\} = \sin bt$

7. $\mathcal{L}^{-1}\left\{\dfrac{s-a}{(s-a)^2+b^2}\right\} = e^{at}\cos bt$

8. $\mathcal{L}^{-1}\left\{\dfrac{b}{(s-a)^2+b^2}\right\} = e^{at}\sin bt$

Suppose $p(s)/q(s)$ is a proper rational function. Its partial fraction decomposition is a linear combination of the **simple (real) rational functions**, by which we mean rational functions of the form

$$\frac{1}{(s-a)^k}, \quad \frac{b}{((s-a)^2+b^2)^k}, \quad \text{and} \quad \frac{s-a}{((s-a)^2+b^2)^k}, \tag{1}$$

where a, b are real, $b > 0$, and k is a positive integer. The linearity of the inverse Laplace transform implies that Laplace inversion of rational functions reduces to finding the inverse Laplace transform of the three simple rational functions given above. The inverse Laplace transforms of the first of these simple rational functions can be read off directly from Table 2.4, while the last two can be determined from this table if $k = 1$. To illustrate, consider the following example.

Example 6. Suppose

$$F(s) = \frac{5s}{(s^2+4)(s+1)}.$$

Find $\mathcal{L}^{-1}\{F(s)\}$.

▶ **Solution.** Since $F(s)$ does not appear in Table 2.4 (or the equivalent Table 2.2), the inverse Laplace transform is not immediately evident. However, using the recursive partial fraction method we found in Example 2 of Sect. 2.4 that

$$\frac{5s}{(s^2+4)(s+1)} = \frac{s+4}{(s^2+4)} - \frac{1}{s+1}$$

$$= \frac{s}{s^2+2^2} + 2\frac{2}{s^2+2^2} - \frac{1}{s+1}.$$

By linearity of the inverse Laplace transform and perusal of Table 2.4, it is now evident that

$$\mathcal{L}^{-1}\left\{\frac{5s}{(s^2+4)(s+1)}\right\} = \cos 2t + 2\sin 2t - e^{-t}.\qquad\blacktriangleleft$$

When irreducible quadratics appear in the denominator, their inversion is best handled by completing the square and using the first translation principle as illustrated in the following example.

Example 7. Find the inverse Laplace transform of each rational function.

$$1.\ \frac{4s-8}{s^2+6s+25}\qquad 2.\ \frac{2}{s^2-4s+7}.$$

▶ **Solution.** In each case, the denominator is an irreducible quadratic. We will complete the square and use the translation principle.

1. Completing the square of the denominator gives

$$s^2 + 6s + 25 = s^2 + 6s + 9 + 25 - 9 = (s+3)^2 + 4^2.$$

In order to apply the first translation principle with $a = -3$, the numerator must also be rewritten with s translated. Thus, $4s - 8 = 4(s+3-3) - 8 = 4(s+3) - 20$. We now get

$$\mathcal{L}^{-1}\left\{\frac{4s-8}{s^2+6s+25}\right\} = \mathcal{L}^{-1}\left\{\frac{4(s+3)-20}{(s+3)^2+4^2}\right\}$$

$$= e^{-3t}\,\mathcal{L}^{-1}\left\{\frac{4s-20}{s^2+4^2}\right\}$$

$$= e^{-3t}\left(4\mathcal{L}\left\{\frac{s}{s^2+4^2}\right\} - 5\mathcal{L}\left\{\frac{4}{s^2+4^2}\right\}\right)$$

$$= e^{-3t}(4\cos 4t - 5\sin 4t).$$

Notice how linearity of Laplace inversion is used here.

2. Completing the square of the denominator gives

$$s^2 - 4s + 7 = s^2 - 4s + 4 + 3 = (s-2)^2 + \sqrt{3}^2.$$

We now get

$$\mathcal{L}^{-1}\left\{\frac{2}{s^2-4s+7}\right\} = \mathcal{L}^{-1}\left\{\frac{2}{(s-2)^2+\sqrt{3}^2}\right\}$$

$$= e^{2t}\,\mathcal{L}^{-1}\left\{\frac{2}{s^2+\sqrt{3}^2}\right\}$$

$$= e^{2t} \frac{2}{\sqrt{3}} \mathcal{L}^{-1} \left\{ \frac{\sqrt{3}}{s^2 + \sqrt{3}^2} \right\}$$

$$= \frac{2}{\sqrt{3}} e^{2t} \sin \sqrt{3}t. \qquad \blacktriangleleft$$

In the examples above, the first translation principle reduces the calculation of the inverse Laplace transform of a simple rational function involving irreducible quadratics with a translated s variable to one without a translation. More generally, the inverse first translation principle gives

$$\mathcal{L}^{-1} \left\{ \frac{b}{((s-a)^2 + b^2)^k} \right\} = e^{at} \mathcal{L}^{-1} \left\{ \frac{b}{(s^2 + b^2)^k} \right\},$$

$$\mathcal{L}^{-1} \left\{ \frac{s-a}{((s-a)^2 + b^2)^k} \right\} = e^{at} \mathcal{L}^{-1} \left\{ \frac{s}{(s^2 + b^2)^k} \right\}. \qquad (2)$$

Table 2.4 does not contain the inverse Laplace transforms of the functions on the right unless $k = 1$. Unfortunately, explicit formulas for these inverse Laplace transforms are not very simple for a general $k \geq 1$. There is however a recursive method for computing these inverse Laplace transform formulas which we now present. This method, which we call a *reduction of order formula*, may remind you of reduction formulas in calculus for integrating powers of trigonometric functions by expressing an integral of an nth power in terms of integrals of lower order powers.

Proposition 8 (Reduction of Order formulas). *If $b \neq 0$ is a real number and $k \geq 1$ is a positive integer, then*

$$\mathcal{L}^{-1} \left\{ \frac{1}{(s^2 + b^2)^{k+1}} \right\} = \frac{-t}{2kb^2} \mathcal{L}^{-1} \left\{ \frac{s}{(s^2 + b^2)^k} \right\} + \frac{2k-1}{2kb^2} \mathcal{L}^{-1} \left\{ \frac{1}{(s^2 + b^2)^k} \right\},$$

$$\mathcal{L}^{-1} \left\{ \frac{s}{(s^2 + b^2)^{k+1}} \right\} = \frac{t}{2k} \mathcal{L}^{-1} \left\{ \frac{1}{(s^2 + b^2)^k} \right\}.$$

Proof. Let $f(t) = \mathcal{L}^{-1} \left\{ \frac{1}{(s^2+b^2)^k} \right\}$. Then the transform derivative principle (Theorem 20 of Sect. 2.2) applies to give

$$\mathcal{L}\{tf(t)\} = -\frac{d}{ds} (\mathcal{L}\{f(t)\}) = -\frac{d}{ds} \left(\frac{1}{(s^2 + b^2)^k} \right)$$

$$= \frac{2ks}{(s^2 + b^2)^{k+1}}.$$

Now divide by $2k$ and take the inverse Laplace transform of both sides to get

$$\mathcal{L}^{-1}\left\{\frac{s}{(s^2+b^2)^{k+1}}\right\} = \frac{t}{2k}f(t) = \frac{t}{2k}\mathcal{L}^{-1}\left\{\frac{1}{(s^2+b^2)^k}\right\},$$

which is the second of the required formulas.

The first formula is done similarly. Let $g(t) = \mathcal{L}^{-1}\left\{\frac{s}{(s^2+b^2)^k}\right\}$. Then

$$\mathcal{L}\{tg(t)\} = -\frac{d}{ds}\left(\mathcal{L}\{g(t)\}\right) = -\frac{d}{ds}\left(\frac{s}{(s^2+b^2)^k}\right)$$

$$= -\frac{(s^2+b^2)^k - 2ks^2(s^2+b^2)^{k-1}}{(s^2+b^2)^{2k}}$$

$$= \frac{2s^2k - (s^2+b^2)}{(s^2+b^2)^{k+1}} = \frac{(2k-1)(s^2+b^2) - 2kb^2}{(s^2+b^2)^{k+1}}$$

$$= \frac{2k-1}{(s^2+b^2)^k} - \frac{2kb^2}{(s^2+b^2)^{k+1}}.$$

Divide by $2kb^2$, solve for the second term in the last line, and apply the inverse Laplace transform to get

$$\mathcal{L}^{-1}\left\{\frac{1}{(s^2+b^2)^{k+1}}\right\}$$

$$= \frac{-t}{2kb^2}g(t) + \frac{(2k-1)}{2kb^2}\mathcal{L}^{-1}\left\{\frac{1}{(s^2+b^2)^k}\right\}$$

$$= \frac{-t}{2kb^2}\mathcal{L}^{-1}\left\{\frac{s}{(s^2+b^2)^k}\right\} + \frac{(2k-1)}{2kb^2}\mathcal{L}^{-1}\left\{\frac{1}{(s^2+b^2)^k}\right\},$$

which is the first formula.   □

These equations are examples of one step recursion relations involving a pair of functions, both of which depend on a positive integer k. The kth formula for both families implies the $(k+1)^{\text{st}}$. Since we already know the formulas for

$$\mathcal{L}^{-1}\left\{\frac{1}{(s^2+b^2)^k}\right\} \quad \text{and} \quad \mathcal{L}^{-1}\left\{\frac{s}{(s^2+b^2)^k}\right\}, \tag{3}$$

when $k = 1$, the reduction formulas give the formulas for the case $k = 2$, which, in turn, allow one to calculate the formulas for $k = 3$, etc. With a little work, we can calculate these inverse Laplace transforms for any $k \geq 1$.

To see how to use these formulas, we will evaluate the inverse Laplace transforms in (3) for $k = 2$.

Formula 9. Use the formulas derived above to verify the following formulas:

$$\mathcal{L}^{-1}\left\{\frac{1}{(s^2+b^2)^2}\right\} = \frac{1}{2b^3}(-bt\cos bt + \sin bt),$$

$$\mathcal{L}^{-1}\left\{\frac{s}{(s^2+b^2)^2}\right\} = \frac{t}{2b}\sin bt.$$

▼ *Verification.* Here we use the reduction of order formulas for $k = 1$ to get

$$\mathcal{L}^{-1}\left\{\frac{1}{(s^2+b^2)^2}\right\} = \frac{-t}{2b^2}\mathcal{L}^{-1}\left\{\frac{s}{s^2+b^2}\right\} + \frac{1}{2b^2}\mathcal{L}^{-1}\left\{\frac{1}{s^2+b^2}\right\}$$

$$= \frac{-t}{2b^2}\cos bt + \frac{1}{2b^3}\sin bt$$

$$= \frac{1}{2b^3}(-bt\cos bt + \sin bt),$$

and $\quad \mathcal{L}^{-1}\left\{\dfrac{s}{(s^2+b^2)^2}\right\} = \dfrac{t}{2}\mathcal{L}^{-1}\left\{\dfrac{1}{s^2+b^2}\right\} = \dfrac{t}{2b}\sin bt.$ ▲

Using the calculations just done and applying the reduction formulas for $k = 2$ will then give the inverse Laplace transforms of (3) for $k = 3$. The process can then be continued to get formulas for higher values of k. In Table 2.5, we provide the inverse Laplace transform for the powers $k = 1, \ldots, 4$. You will be asked to verify them in the exercises. In Chap. 7, we will derive a closed formula for each value k.

By writing

$$\frac{cs+d}{(s^2+b^2)^{k+1}} = c\frac{s}{(s^2+b^2)^{k+1}} + d\frac{1}{(s^2+b^2)^{k+1}},$$

the two formulas in Proposition 8 can be combined using linearity to give a single formula:

Corollary 10. *Let b, c, and d be real numbers and assume $b \neq 0$. If $k \geq 1$ is a positive integer, then*

$$\mathcal{L}^{-1}\left\{\frac{cs+d}{(s^2+b^2)^{k+1}}\right\} = \frac{t}{2kb^2}\mathcal{L}^{-1}\left\{\frac{-ds+cb^2}{(s^2+b^2)^k}\right\} + \frac{2k-1}{2kb^2}\mathcal{L}^{-1}\left\{\frac{d}{(s^2+b^2)^k}\right\}.$$

As an application of this formula, we note the following result that expresses the *form* of $\mathcal{L}^{-1}\left\{\frac{cs+d}{(s^2+b^2)^k}\right\}$ in terms of polynomials, sines, and cosines.

Table 2.5 Inversion formulas involving irreducible quadratics

$\mathcal{L}^{-1}\left\{\dfrac{b}{(s^2+b^2)^k}\right\}$	$\longleftrightarrow \dfrac{b}{(s^2+b^2)^k}$
$\sin bt$	$\longleftrightarrow \dfrac{b}{(s^2+b^2)}$
$\frac{1}{2b^2}(\sin bt - bt\cos bt)$	$\longleftrightarrow \dfrac{b}{(s^2+b^2)^2}$
$\frac{1}{8b^4}\left((3-(bt)^2)\sin bt - 3bt\cos bt\right)$	$\longleftrightarrow \dfrac{b}{(s^2+b^2)^3}$
$\frac{1}{48b^6}\left((15-6(bt)^2)\sin bt - (15bt-(bt)^3)\cos bt\right)$	$\longleftrightarrow \dfrac{b}{(s^2+b^2)^4}$
$\mathcal{L}^{-1}\left\{\dfrac{s}{(s^2+b^2)^k}\right\}$	$\longleftrightarrow \dfrac{s}{(s^2+b^2)^k}$
$\cos bt$	$\longleftrightarrow \dfrac{s}{(s^2+b^2)}$
$\frac{1}{2b^2}bt\sin bt$	$\longleftrightarrow \dfrac{s}{(s^2+b^2)^2}$
$\frac{1}{8b^4}\left(bt\sin bt - (bt)^2\cos bt\right)$	$\longleftrightarrow \dfrac{s}{(s^2+b^2)^3}$
$\frac{1}{48b^6}\left((3bt-(bt)^3)\sin bt - 3(bt)^2\cos bt\right)$	$\longleftrightarrow \dfrac{s}{(s^2+b^2)^4}$

Corollary 11. *Let b, c, and d be real numbers and assume $b \neq 0$. If $k \geq 1$ is a positive integer, then there are polynomials $p_1(t)$ and $p_2(t)$ of degree at most $k - 1$ such that*

$$\mathcal{L}^{-1}\left\{\frac{cs+d}{(s^2+b^2)^k}\right\} = p_1(t)\sin bt + p_2(t)\cos bt. \tag{4}$$

Proof. We prove this by induction. If $k = 1$, this is certainly true since

$$\mathcal{L}^{-1}\left\{\frac{cs+d}{s^2+b^2}\right\} = \frac{d}{b}\sin bt + c\cos bt,$$

and thus, $p_1(t) = d/b$ and $p_2(t) = c$ are constants and hence polynomials of degree $0 = 1 - 1$. Now suppose for our induction hypothesis that $k \geq 1$ and that (4) is true for k. We need to show that this implies that it is also true for $k + 1$. By this assumption, we can find polynomials $p_1(t)$, $p_2(t)$, $q_1(t)$, and $q_2(t)$ of degree at most $k - 1$ so that

$$\mathcal{L}^{-1}\left\{\frac{-ds+cb^2}{(s^2+b^2)^k}\right\} = p_1(t)\sin bt + p_2(t)\cos bt$$

and $\quad \mathcal{L}^{-1}\left\{\dfrac{d}{(s^2+b^2)^k}\right\} = q_1(t)\sin bt + q_2(t)\cos bt.$

By Corollary 10,

$$\mathcal{L}^{-1}\left\{\frac{cs+d}{(s^2+b^2)^{k+1}}\right\} = \frac{t}{2kb^2}\mathcal{L}^{-1}\left\{\frac{-ds+cb^2}{(s^2+b^2)^k}\right\} + \frac{2k-1}{2kb^2}\mathcal{L}^{-1}\left\{\frac{d}{(s^2+b^2)^k}\right\}$$

$$= \frac{t}{2kb^2}(p_1(t)\sin bt + p_2(t)\cos bt)$$

$$+ \frac{2k-1}{2kb^2}(q_1(t)\sin bt + q_2(t)\cos bt)$$

$$= P_1(t)\sin bt + P_2(t)\cos bt,$$

where

$$P_1(t) = \frac{t}{2kb^2}p_1(t) + \frac{2k-1}{2kb^2}q_1(t) \quad \text{and} \quad P_2(t) = \frac{t}{2kb^2}p_2(t) + \frac{2k-1}{2kb^2}q_2(t).$$

Observe that $P_1(t)$ and $P_2(t)$ are obtained from the polynomials $p_i(t)$, $q_i(t)$ by multiplying by a term of degree at most 1. Hence, these are polynomials whose degrees are at most k, since the $p_i(t)$, $q_i(t)$ have degree at most $k-1$. This completes the induction argument. $\quad\square$

It follows from the discussion thus far that the inverse Laplace transform of *any* rational function can be computed by means of a partial fraction expansion followed by use of the formulas from Table 2.4. For partial fractions with irreducible quadratic denominator, the recursion formulas, as collected in Table 2.5, may be needed. Here is an example.

Example 12. Find the inverse Laplace transform of

$$F(s) = \frac{6s+6}{(s^2-4s+13)^3}.$$

▶ **Solution.** We first complete the square: $s^2 - 4s + 13 = s^2 - 4s + 4 + 9 = (s-2)^2 + 3^2$. Then

$$\mathcal{L}^{-1}\{F(s)\} = \mathcal{L}^{-1}\left\{\frac{6s+6}{((s-2)^2+3^2)^3}\right\} = \mathcal{L}^{-1}\left\{\frac{6(s-2)+18}{((s-2)^2+3^2)^3}\right\}$$

$$= e^{2t}\left(6\mathcal{L}^{-1}\left\{\frac{s}{(s^2+3^2)^3}\right\} + 6\mathcal{L}^{-1}\left\{\frac{3}{(s^2+3^2)^3}\right\}\right)$$

$$= \frac{6e^{2t}}{8 \cdot 3^4} \left(3t \sin 3t - (3t)^2 \cos 3t + (3 - (3t)^2) \sin 3t - 3(3t) \cos 3t \right)$$

$$= \frac{e^{2t}}{36} \left((1 + t - 3t^2) \sin 3t - (3t + 3t^2) \cos 3t\right).$$

The third line is obtained from Table 2.5 with $b = 3$ and $k = 3$.   ◄

Irreducible Quadratics and the Laplace Transform Method

We conclude with an example that uses the Laplace transform method, quadratic partial fraction recursion, and Table 2.5 to solve a second order differential equation.

Example 13. Use the Laplace transform method to solve the following differential equation:

$$y'' + 4y = 9t \sin t,$$

with initial conditions $y(0) = 0$, and $y'(0) = 0$.

► **Solution.** Table 2.5 gives $\mathcal{L}\{9t \sin t\} = 18s/(s^2 + 1)^2$. Now apply the Laplace transform to the differential equation to get

$$(s^2 + 4)Y(s) = \frac{18s}{(s^2 + 1)^2}$$

and hence

$$Y(s) = \frac{18s}{(s^2 + 1)^2(s^2 + 4)}.$$

Using quadratic partial fraction recursion, we obtain the $(s^2 + 1)$-chain

The $(s^2 + 1)$-chain	
$\dfrac{18s}{(s^2 + 1)^2(s^2 + 4)}$	$\dfrac{6s}{(s^2 + 1)^2}$
$\dfrac{-6s}{(s^2 + 1)(s^2 + 4)}$	$\dfrac{-2s}{s^2 + 1}$
$\dfrac{2s}{s^2 + 4}$	

Thus,

$$Y(s) = \frac{6s}{(s^2 + 1)^2} + \frac{-2s}{s^2 + 1} + \frac{2s}{s^2 + 4}.$$

Laplace inversion with Table 2.5 gives

$$y(t) = 3t \sin t - 2 \cos t + 2 \cos 2t. \qquad \blacktriangleleft$$

Exercises

1–18. Compute $\mathcal{L}^{-1}\{F(s)\}(t)$ for the given proper rational function $F(s)$.

1. $\dfrac{-5}{s}$

2. $\dfrac{3}{s-4}$

3. $\dfrac{3}{s^2}-\dfrac{4}{s^3}$

4. $\dfrac{4}{2s+3}$

5. $\dfrac{3s}{s^2+4}$

6. $\dfrac{2}{s^2+3}$

7. $\dfrac{2s-5}{s^2+6s+9}$

8. $\dfrac{2s-5}{(s+3)^3}$

9. $\dfrac{6}{s^2+2s-8}$

10. $\dfrac{s}{s^2-5s+6}$

11. $\dfrac{2s^2-5s+1}{(s-2)^4}$

12. $\dfrac{2s+6}{s^2-6s+5}$

13. $\dfrac{4s^2}{(s-1)^2(s+1)^2}$

14. $\dfrac{27}{s^3(s+3)}$

15. $\dfrac{8s+16}{(s^2+4)(s-2)^2}$

16. $\dfrac{5s+15}{(s^2+9)(s-1)}$

17. $\dfrac{12}{s^2(s+1)(s-2)}$

18. $\dfrac{2s}{(s-3)^3(s-4)^2}$

19–24. Use the first translation principle and Table 2.2 (or Table 2.4) to find the inverse Laplace transform.

19. $\dfrac{2s}{s^2 + 2s + 5}$

20. $\dfrac{1}{s^2 + 6s + 10}$

21. $\dfrac{s - 1}{s^2 - 8s + 17}$

22. $\dfrac{2s + 4}{s^2 - 4s + 12}$

23. $\dfrac{s - 1}{s^2 - 2s + 10}$

24. $\dfrac{s - 5}{s^2 - 6s + 13}$

25–34. Find the inverse Laplace transform of each rational function. Either the reduction formulas, Proposition 8, or the formulas in Table 2.5 can be used.

25. $\dfrac{8s}{(s^2 + 4)^2}$

26. $\dfrac{9}{(s^2 + 9)^2}$

27. $\dfrac{2s}{(s^2 + 4s + 5)^2}$

28. $\dfrac{2s + 2}{(s^2 - 6s + 10)^2}$

29. $\dfrac{2s}{(s^2 + 8s + 17)^2}$

30. $\dfrac{s + 1}{(s^2 + 2s + 2)^3}$

31. $\dfrac{1}{(s^2 - 2s + 5)^3}$

32. $\dfrac{8s}{(s^2 - 6s + 10)^3}$

33. $\dfrac{s - 4}{(s^2 - 8s + 17)^4}$

34. $\dfrac{2}{(s^2 + 4s + 8)^3}$

35–38. Use the *Laplace transform method* to solve the following differential equations. (Give both $Y(s)$ and $y(t)$.)

35. $y'' + y = 4 \sin t, \quad y(0) = 1, y'(0) = -1$

36. $y'' + 9y = 36t \sin 3t, \quad y(0) = 0, y'(0) = 3$

37. $y'' - 3y = 4t^2 \cos t, \quad y(0) = 0, y'(0) = 0$

38. $y'' + 4y = 32t \cos 2t, \quad y(0) = 0, y'(0) = 2$

39–44. Verify the following assertions. In each assertion, assume a, b, c are distinct. These are referred to as Heaviside expansion formulas of the first kind.

39. $\mathcal{L}^{-1}\left\{\dfrac{1}{(s-a)(s-b)}\right\} = \dfrac{e^{at}}{a-b} + \dfrac{e^{bt}}{b-a}$

40. $\mathcal{L}^{-1}\left\{\dfrac{s}{(s-a)(s-b)}\right\} = \dfrac{ae^{at}}{a-b} + \dfrac{be^{bt}}{b-a}$

41. $\mathcal{L}^{-1}\left\{\dfrac{1}{(s-a)(s-b)(s-c)}\right\} = \dfrac{e^{at}}{(a-b)(a-c)} + \dfrac{e^{bt}}{(b-a)(b-c)} +$
$\dfrac{e^{ct}}{(c-a)(c-b)}$

42. $\mathcal{L}^{-1}\left\{\dfrac{s}{(s-a)(s-b)(s-c)}\right\} = \dfrac{ae^{at}}{(a-b)(a-c)} + \dfrac{be^{bt}}{(b-a)(b-c)} +$
$\dfrac{ce^{ct}}{(c-a)(c-b)}$

43. $\mathcal{L}^{-1}\left\{\dfrac{s^2}{(s-a)(s-b)(s-c)}\right\} = \dfrac{a^2e^{at}}{(a-b)(a-c)} + \dfrac{b^2e^{bt}}{(b-a)(b-c)} +$
$\dfrac{c^2e^{ct}}{(c-a)(c-b)}$

44. $\mathcal{L}^{-1}\left\{\dfrac{s^k}{(s-r_1)\cdots(s-r_n)}\right\} = \dfrac{r_1^k e^{r_1 t}}{q'(r_1)} + \cdots + \dfrac{r_n^k e^{r_n t}}{q'(r_n)}$, where
$q(s) = (s-r_1)\cdots(s-r_k)$. Assume $r_1, \ldots, r_n$ are distinct.

45–50. Verify the following assertions. These are referred to as Heaviside expansion formulas of the second kind.

45. $\mathcal{L}^{-1}\left\{\dfrac{1}{(s-a)^2}\right\} = te^{at}$

46. $\mathcal{L}^{-1}\left\{\dfrac{s}{(s-a)^2}\right\} = (1 + at)e^{at}$

47. $\mathcal{L}^{-1}\left\{\dfrac{1}{(s-a)^3}\right\} = \dfrac{t^2}{2}e^{at}$

48. $\mathcal{L}^{-1}\left\{\dfrac{s}{(s-a)^3}\right\} = \left(t + \dfrac{at^2}{2}\right)e^{at}$

49. $\mathcal{L}^{-1}\left\{\dfrac{s^2}{(s-a)^3}\right\} = \left(1 + 2at + \dfrac{a^2t^2}{2}\right)e^{at}$

50. $\mathcal{L}^{-1}\left\{\dfrac{s^k}{(s-a)^n}\right\} = \left(\sum_{l=0}^{k}\binom{k}{l}a^{k-l}\dfrac{t^{n-l-1}}{(n-l-1)!}\right)e^{at}$

2.6   The Linear Spaces $\mathcal{E}_q$: Special Cases

Let $q(s)$ be any fixed polynomial. It is the purpose of this section and the next to efficiently determine all input functions having Laplace transforms that are proper rational functions with $q(s)$ in the denominator. In other words, we want to describe all the input functions $y(t)$ such that

$$\mathcal{L}\{y(t)\}(s) = \frac{p(s)}{q(s)},$$

where $p(s)$ may be any polynomial whose degree is less than that of $q(s)$. It turns out that our description will involve in a simple way the roots of $q(s)$ and their multiplicities and will involve the notion of linear combinations and spanning sets, which we introduce below. To get an idea why we seek such a description, consider the following example of a second order linear differential equation. The more general theory of such differential equations will be discussed in Chaps. 3 and 4.

Example 1.   Use the Laplace transform to find the solution set for the second order linear differential equation

$$y'' - 3y' - 4y = 0. \tag{1}$$

▶ **Solution.** Notice in this example that we are not specifying the initial conditions $y(0)$ and $y'(0)$. We may consider them arbitrary. Our solution set will be a family of solutions parameterized by two arbitrary constants (cf. the discussion in Sect. 1.1 under the subheading *The Arbitrary Constants*). We apply the Laplace transform to both sides of (1) and use linearity to get

$$\mathcal{L}\{y''\} - 3\mathcal{L}\{y'\} - 4\mathcal{L}\{y\} = 0.$$

Next use the input derivative principles:

$$\mathcal{L}\{y'(t)\}(s) = s\mathcal{L}\{y(t)\}(s) - y(0),$$

$$\mathcal{L}\{y''(t)\}(s) = s^2\mathcal{L}\{y(t)\}(s) - sy(0) - y'(0)$$

to get

$$s^2Y(s) - sy(0) - y'(0) - 3(sY(s) - y(0)) - 4Y(s) = 0,$$

where as usual we set $Y(s) = \mathcal{L}\{y(t)\}(s)$. Collect together terms involving $Y(s)$ and simplify to get

$$(s^2 - 3s - 4)Y(s) = sy(0) + y'(0) - 3y(0). \tag{2}$$

The polynomial coefficient of $Y(s)$ is $s^2 - 3s - 4$ and is called the ***characteristic polynomial*** of (1). To simplify the notation, let $q(s) = s^2 - 3s - 4$. Solving for $Y(s)$ gives

$$Y(s) = \frac{sy(0) + y'(0) - 3y(0)}{s^2 - 3s - 4} = \frac{p(s)}{q(s)}, \tag{3}$$

where $p(s) = sy(0) + y'(0) - 3y(0)$. Observe that $p(s)$ can be any polynomial of degree 1 since the initial values are unspecified and arbitrary. Our next step then is to find all the input functions whose Laplace transform has $q(s)$ in the denominator. Since $q(s) = s^2 - 3s - 4 = (s - 4)(s + 1)$, we see that the form of the partial fraction decomposition for $\frac{p(s)}{q(s)}$ is

$$Y(s) = \frac{p(s)}{q(s)} = c_1 \frac{1}{s - 4} + c_2 \frac{1}{s + 1}.$$

Laplace inversion gives

$$y(t) = c_1 e^{4t} + c_2 e^{-t}, \tag{4}$$

where c_1 and c_2 are arbitrary real numbers, that depend on the initial conditions $y(0)$ and $y'(0)$. We encourage the student to verify by substitution that $y(t)$ is indeed a solution to (1). We will later show that all such solutions are of this form. We may now write the solution set as

$$\left\{ c_1 e^{4t} + c_2 e^{-t} : c_1, c_2 \in \mathbb{R} \right\}. \tag{5}$$

◄

Let us make a few observations about this example. First observe that the characteristic polynomial $q(s) = s^2 - 3s - 4$, which is the coefficient of $Y(s)$ in (2), is easy to read off directly from the left side of the differential equation $y'' - 3y' - 4y = 0$; the coefficient of each power of s in $q(s)$ is exactly the coefficient of the corresponding order of the derivative in the differential equation. Second, with the characteristic polynomial in hand, we can jump to (3) to get the *form* of $Y(s)$, namely, $Y(s)$ is a proper rational function with the characteristic polynomial in the denominator. The third matter to deal with in this example is to compute $y(t)$ knowing that its Laplace transform $Y(s)$ has the special form $\frac{p(s)}{q(s)}$. It is this third matter that we address here. In particular, we find an efficient method to write down the solution set as given by (5) directly from any characteristic polynomial $q(s)$. The roots of $q(s)$ and their multiplicity play a decisive role in the description we give.

For any polynomial $q(s)$, we let $\mathcal{R}_q$ denote all the proper rational functions that *may* be written as $\frac{p(s)}{q(s)}$ for some polynomial $p(s)$. We let $\mathcal{E}_q$ denote the set of all input functions whose Laplace transform is in $\mathcal{R}_q$. In Example 1, we found the $\mathcal{E}_q = \{ c_1 e^{4t} + c_2 e^{-t} : c_1, c_2 \in \mathbb{R} \}$, where $q(s) = s^2 - 3s - 4$. If $c_1 = 1$ and $c_2 = 0$ then the function $e^{4t} = 1e^{4t} + 0e^{-t} \in \mathcal{R}_q$. Observe though that $\mathcal{L} \{ e^{4t} \} = \frac{1}{s-4}$. At first glance, it appears that $\mathcal{L} \{ e^{4t} \}$ is not in $\mathcal{E}_q$. However, we *may* write

$$\mathcal{L} \{ e^{4t} \} (s) = \frac{1}{s - 4} = \frac{1 (s + 1)}{(s - 4)(s + 1)} = \frac{s + 1}{q(s)}.$$

Thus, $\mathcal{L} \{ e^{4t} \} (s) \in \mathcal{R}_q$ and indeed e^{4t} is in $\mathcal{E}_q$. In a similar way, $e^{-t} \in \mathcal{E}_q$.

Recall from Sect. 2.2 the notion of a linear space of functions, namely, closure under addition and scalar multiplication.

Proposition 2.   *Both $\mathcal{E}_q$ and $\mathcal{R}_q$ are linear spaces.*

Proof. Suppose $\frac{p_1(s)}{q(s)}$ and $\frac{p_2(s)}{q(s)}$ are in $\mathcal{R}_q$ and $c \in \mathbb{R}$. Then $\deg p_1(s)$ and $\deg p_2(s)$ are less than $\deg q(s)$. Further,

- $\frac{p_1(s)}{q(s)} + \frac{p_2(s)}{q(s)} = \frac{p_1(s)+p_2(s)}{q(s)}$. Since addition of polynomials does not increase the degree, we have $\deg(p_1(s) + p_2(s)) < \deg q(s)$. It follows that $\frac{p_1(s)+p_2(s)}{q(s)}$ is in $\mathcal{R}_q$.

- $c\dfrac{p_1(s)}{q(s)} = \frac{cp_1(s)}{q(s)}$ is proper and has denominator $q(s)$; hence, it is in $\mathcal{R}_q$.

It follows that $\mathcal{R}_q$ is closed under addition and scalar multiplication, and hence, $\mathcal{R}_q$ is a linear space. Now suppose f_1 and f_2 are in $\mathcal{E}_q$ and $c \in \mathbb{R}$. Then $\mathcal{L}\{f_1\} \in \mathcal{R}_q$ and $\mathcal{L}\{f_2\} \in \mathcal{R}_q$. Further,

- $\mathcal{L}\{f_1 + f_2\} = \mathcal{L}\{f_1\} + \mathcal{L}\{f_2\} \in \mathcal{R}_q$. From this, it follows that $f_1 + f_2 \in \mathcal{E}_q$.
- $\mathcal{L}\{cf_1\} = c\mathcal{L}\{f_1\} \in \mathcal{R}_q$. From this it follows that $cf_1 \in \mathcal{E}_q$.

It follows that $\mathcal{E}_q$ is closed under addition and scalar multiplication, and hence, $\mathcal{E}_q$ is a linear space.   □

Description of $\mathcal{E}_q$ for $q(s)$ of Degree 2

The roots of a real polynomial of degree 2 occur in one of three ways:

1. Two distinct real roots as in Example 1
2. A real root with multiplicity two
3. Two complex roots

Let us consider an example of each type.

Example 3.   Find $\mathcal{E}_q$ for each of the following polynomials:

1. $q(s) = s^2 - 3s + 2$
2. $q(s) = s^2 - 2s + 1$
3. $q(s) = s^2 + 2s + 2$

▶ **Solution.**   In each case, $\deg q(s) = 2$; thus

$$\mathcal{R}_q = \left\{ \frac{p(s)}{q(s)} : \deg p(s) \le 1 \right\}.$$

1. Suppose $f(t) \in \mathcal{E}_q$. Since $q(s) = s^2 - 3s + 2 = (s-1)(s-2)$, a partial fraction decomposition of $\mathcal{L}\{f(t)\}(s) = \frac{p(s)}{q(s)}$ has the form $\frac{p(s)}{(s-1)(s-2)} = \frac{c_1}{s-1} + \frac{c_2}{s-2}$. Laplace inversion then gives $f(t) = \mathcal{L}^{-1}\left\{\frac{p(s)}{q(s)}\right\} = c_1 e^t + c_2 e^{2t}$. On the other hand, we have $e^t \in \mathcal{E}_q$ since $\mathcal{L}\{e^t\} = \frac{1}{s-1} = \frac{s-2}{(s-1)(s-2)} = \frac{s-2}{q(s)} \in \mathcal{R}_q$. Similarly, $e^{2t} \in \mathcal{E}_q$. Since $\mathcal{E}_q$ is a linear space, it follows that all functions of the form $c_1 e^t + c_2 e^{2t}$ are in $\mathcal{E}_q$. From these calculations, it follows that

$$\mathcal{E}_q = \left\{c_1 e^t + c_2 e^{2t} : c_1, c_2 \in \mathbb{R}\right\}.$$

2. Suppose $f(t) \in \mathcal{E}_q$. Since $q(s) = s^2 - 2s + 1 = (s-1)^2$, a partial fraction decomposition of $\mathcal{L}\{f(t)\}(s) = \frac{p(s)}{q(s)}$ has the form $\frac{p(s)}{(s-1)^2} = \frac{c_1}{s-1} + \frac{c_2}{(s-1)^2}$. Laplace inversion then gives $f(t) = \mathcal{L}^{-1}\left\{\frac{p(s)}{q(s)}\right\} = c_1 e^t + c_2 t e^t$. On the other hand, we have $e^t \in \mathcal{E}_q$ since $\mathcal{L}\{e^t\} = \frac{1}{s-1} = \frac{s-1}{(s-1)^2} = \frac{s-1}{q(s)} \in \mathcal{R}_q$. Similarly $\mathcal{L}\{t e^t\} = \frac{1}{(s-1)^2} \in \mathcal{R}_q$ so $t e^t \in \mathcal{E}_q$. Since $\mathcal{E}_q$ is a linear space, it follows that all functions of the form $c_1 e^t + c_2 t e^t$ are in $\mathcal{E}_q$. From these calculations, it follows that

$$\mathcal{E}_q = \left\{c_1 e^t + c_2 t e^t : c_1, c_2 \in \mathbb{R}\right\}.$$

3. We complete the square in $q(s)$ to get $q(s) = (s+1)^2 + 1$, an irreducible quadratic. Suppose $f(t) \in \mathcal{E}_q$. A partial fraction decomposition of $\mathcal{L}\{f(t)\} = \frac{p(s)}{q(s)}$ has the form

$$
\begin{aligned}
\frac{p(s)}{(s+1)^2 + 1} &= \frac{as+b}{(s+1)^2 + 1} \\
&= \frac{a(s+1) + b - a}{(s+1)^2 + 1} \\
&= c_1 \frac{s+1}{(s+1)^2 + 1} + c_2 \frac{1}{(s+1)^2 + 1},
\end{aligned}
$$

where $c_1 = a$ and $c_2 = b - a$. Laplace inversion then gives

$$f(t) = \mathcal{L}^{-1}\left\{\frac{p(s)}{q(s)}\right\} = c_1 e^{-t} \sin t + c_2 e^{-t} \cos t.$$

On the other hand, we have $e^{-t} \cos t \in \mathcal{E}_q$ since $\mathcal{L}\{e^{-t} \cos t\} = \frac{s+1}{(s+1)^2 + 1} \in \mathcal{R}_q$. Similarly, we have $e^{-t} \sin t \in \mathcal{E}_q$. Since $\mathcal{E}_q$ is a linear space, it follows that all functions of the form $c_1 e^{-t} \cos t + c_2 e^{-t} \sin t$ are in $\mathcal{E}_q$. It follows that

$$\mathcal{E}_q = \left\{c_1 e^{-t} \sin t + c_2 e^{-t} \cos t : c_1, c_2 \in \mathbb{R}\right\}. \qquad \blacktriangleleft$$

In Example 3, we observe that $\mathcal{E}_q$ takes the form

$$\mathcal{E}_q = \{c_1\phi_1 + c_2\phi_2 : c_1, c_2 \in \mathbb{R}\},$$

where

in case (1) $\phi_1(t) = e^t$ and $\phi_2(t) = e^{2t}$
in case (2) $\phi_1(t) = e^t$ and $\phi_2(t) = te^t$
in case (3) $\phi_1(t) = e^{-t}\sin t$ and $\phi_2(t) = e^{-t}\cos t$

We introduce the following useful concepts and notation that will allow us to rephrase the results of Example 3 in a more convenient way and, as we will see, will generalize to arbitrary polynomials $q(s)$. Suppose $\mathcal{F}$ is a linear space of functions and $\mathcal{S} = \{\phi_1, \ldots, \phi_n\} \subset \mathcal{F}$ a subset. A **linear combination** of $\mathcal{S}$ is a sum of the following form:

$$c_1\phi_1 + \cdots + c_n\phi_n,$$

where $c_1, \ldots, c_n$ are scalars in $\mathbb{R}$. Since $\mathcal{F}$ is a linear space (closed under addition and scalar multiplication), all such linear combinations are back in $\mathcal{F}$. The **span** of $\mathcal{S}$, denoted Span $\mathcal{S}$, is the set of all such linear combinations. Symbolically, we write

$$\text{Span } \mathcal{S} = \{c_1\phi_1 + \cdots + c_n\phi_n : c_1, \ldots, c_n \in \mathbb{R}\}.$$

If every function in $\mathcal{F}$ can be written as a linear combination of $\mathcal{S}$, then we say $\mathcal{S}$ **spans** $\mathcal{F}$. Alternately, $\mathcal{S}$ is referred to as a **spanning set** for $\mathcal{F}$. Thus, there are two things that need to be checked to determine whether $\mathcal{S}$ is a spanning set for $\mathcal{F}$:

- $\mathcal{S} \subset \mathcal{F}$.
- Each function in $\mathcal{F}$ is a linear combination of functions in $\mathcal{S}$.

Returning to Example 3, we can rephrase our results in the following concise way. For each $q(s)$, define $\mathcal{B}_q$ as given below:

1. $q(s) = s^2 - 3s + 2 = (s-1)(s-2)$ $\mathcal{B}_q = \{e^t, e^{2t}\}$

2. $q(s) = s^2 - 2s + 1 = (s-1)^2$ $\mathcal{B}_q = \{e^t, te^t\}$

3. $q(s) = (s+1)^2 + 1$ $\mathcal{B}_q = \{e^{-t}\cos t, e^{-t}\sin t\}$

Then, in each case,

$$\mathcal{E}_q = \text{Span } \mathcal{B}_q.$$

Notice how efficient this description is. In each case, we found two functions that make up $\mathcal{B}_q$. Once they are determined, then $\mathcal{E}_q = \text{Span } \mathcal{B}_q$ is the set of all linear combinations of the two functions in $\mathcal{B}_q$ and effectively gives all those functions whose Laplace transforms are rational functions with $q(s)$ in the denominator. The set $\mathcal{B}_q$ is called the **standard basis** of $\mathcal{E}_q$.

Example 3 generalizes in the following way for arbitrary polynomials of degree 2.

Theorem 4. *Suppose $q(s)$ is a polynomial of degree two. Define the standard basis $\mathcal{B}_q$ according to the way $q(s)$ factors as follows:*

1. $q(s) = (s - r_1)(s - r_2)$ $\mathcal{B}_q = \{e^{r_1 t},\ e^{r_2 t}\}$
2. $q(s) = (s - r)^2$ $\mathcal{B}_q = \{e^{rt},\ te^{rt}\}$
3. $q(s) = ((s - a)^2 + b^2)$ $\mathcal{B}_q = \{e^{at} \cos bt,\ e^{at} \sin bt\}$

We assume r_1, r_2, r, a, and b are real, $r_1 \neq r_2$, and $b > 0$. Let $\mathcal{E}_q$ be the set of input functions whose Laplace transform is a rational function with $q(s)$ in the denominator. Then

$$\mathcal{E}_q = \text{Span } \mathcal{B}_q.$$

Remark 5. Observe that these three cases may be summarized in terms of the roots of $q(s)$ as follows:

1. If $q(s)$ has distinct real roots r_1 and r_2, then $\mathcal{B}_q = \{e^{r_1 t},\ e^{r_2 t}\}$.
2. If $q(s)$ has one real root r with multiplicity 2, then $\mathcal{B}_q = \{e^{rt},\ te^{rt}\}$.
3. If $q(s)$ has complex roots $a \pm bi$, then $\mathcal{B}_q = \{e^{at} \cos bt,\ e^{at} \sin bt\}$. Since $\sin(-bt) = -\sin bt$ and $\cos(-bt) = \cos bt$, we may assume $b > 0$.

Proof. The proof follows the pattern set forth in Example 3.

1. Suppose $f(t) \in \mathcal{E}_q$. Since $q(s) = (s-r_1)(s-r_2)$, a partial fraction decomposition of $\mathcal{L}\{f(t)\}(s) = \frac{p(s)}{q(s)}$ has the form $\frac{p(s)}{(s-1)(s-2)} = \frac{c_1}{s-r_1} + \frac{c_2}{s-r_2}$. Laplace inversion then gives $f(t) = \mathcal{L}^{-1}\left\{\frac{p(s)}{q(s)}\right\} = c_1 e^{r_1 t} + c_2 e^{r_2 t}$. On the other hand, we have $e^{r_1 t} \in \mathcal{E}_q$ since $\mathcal{L}\{e^{r_1 t}\} = \frac{1}{s-r_1} = \frac{s-r_2}{(s-r_1)(s-r_2)} = \frac{s-r_2}{q(s)} \in \mathcal{R}_q$. Similarly, $e^{r_2 t} \in \mathcal{E}_q$. It now follows that

$$\mathcal{E}_q = \{c_1 e^{r_1 t} + c_2 e^{r_2 t} : c_1, c_2 \in \mathbb{R}\}.$$

2. Suppose $f(t) \in \mathcal{E}_q$. Since $q(s) = (s - r)^2$, a partial fraction decomposition of $\mathcal{L}\{f(t)\}(s) = \frac{p(s)}{q(s)}$ has the form $\frac{p(s)}{(s-r)^2} = \frac{c_1}{s-r} + \frac{c_2}{(s-r)^2}$. Laplace inversion then gives $f(t) = \mathcal{L}^{-1}\left\{\frac{p(s)}{q(s)}\right\} = c_1 e^{rt} + c_2 te^{rt}$. On the other hand, we have $e^{rt} \in \mathcal{E}_q$ since $\mathcal{L}\{e^{rt}\} = \frac{1}{s-r} = \frac{s-r}{(s-r)^2} = \frac{s-r}{q(s)} \in \mathcal{R}_q$. Similarly, $\mathcal{L}\{te^{rt}\}(s) = \frac{1}{(s-r)^2} \in \mathcal{R}_q$ so $te^{rt} \in \mathcal{E}_q$. It now follows that

$$\mathcal{E}_q = \{c_1 e^{rt} + c_2 te^{rt} : c_1, c_2 \in \mathbb{R}\}.$$

3. Suppose $f(t) \in \mathcal{E}_q$. A partial fraction decomposition of $\mathcal{L}\{f(t)\} = \frac{p(s)}{q(s)}$ has the form

$$\frac{p(s)}{((s-a)^2 + b^2)} = \frac{cs + d}{((s-a)^2 + b^2)}$$

$$= \frac{c(s-a) + d + ca}{((s-a)^2 + b^2)}$$

$$= c_1 \frac{s-a}{(s-a)^2 + b^2} + c_2 \frac{b}{(s-a)^2 + b^2},$$

where $c_1 = c$ and $c_2 = \frac{d+ca}{b}$. Laplace inversion then gives

$$f(t) = \mathcal{L}^{-1}\left\{\frac{p(s)}{q(s)}\right\} = c_1 e^{at} \cos bt + c_2 e^{at} \sin bt.$$

On the other hand, since $\mathcal{L}\{e^{at} \cos bt\} = \frac{s-a}{(s-a)^2+b^2} \in \mathcal{R}_q$, we have $e^{at} \cos bt \in \mathcal{E}_q$. Similarly, we have $e^{at} \sin bt \in \mathcal{E}_q$. It follows that

$$\mathcal{E}_q = \{c_1 e^{-t} \sin t + c_2 e^{-t} \cos t : c_1, c_2 \in \mathbb{R}\}.$$

In each case, $\mathcal{E}_q = \operatorname{Span} \mathcal{B}_q$, where $\mathcal{B}_q$ is prescribed as above.   □

This theorem makes it very simple to find $\mathcal{B}_q$ and thus $\mathcal{E}_q$ when $\deg q(s) = 2$. The prescription boils down to finding the roots and their multiplicities.

Example 6. Find the standard basis $\mathcal{B}_q$ of $\mathcal{E}_q$ for each of the following polynomials:

1. $q(s) = s^2 + 6s + 5$
2. $q(s) = s^2 + 4s + 4$
3. $q(s) = s^2 + 4s + 13$

▶ **Solution.** 1. Observe that $q(s) = s^2 + 6s + 5 = (s+1)(s+5)$. The roots are $r_1 = -1$ and $r_2 = -5$. Thus, $\mathcal{B}_q = \{e^{-t}, e^{-5t}\}$ and $\mathcal{E}_q = \operatorname{Span} \mathcal{B}_q$.
2. Observe that $q(s) = s^2 + 4s + 4 = (s+2)^2$. The root is $r = -2$ with multiplicity 2. Thus, $\mathcal{B}_q = \{e^{-2t}, te^{-2t}\}$ and $\mathcal{E}_q = \operatorname{Span} \mathcal{B}_q$.
3. Observe that $q(s) = s^2 + 4s + 13 = (s+2)^2 + 3^2$ is an irreducible quadratic. Its roots are $-2 \pm 3i$. Thus, $\mathcal{B}_q = \{e^{-2t} \cos 3t, e^{-2t} \sin 3t\}$ and $\mathcal{E}_q = \operatorname{Span} \mathcal{B}_q$.   ◀

As we go forward, you will see that the spanning set $\mathcal{B}_q$ for $\mathcal{E}_q$, for any polynomial $q(s)$, will be determined precisely by the roots of $q(s)$ and their multiplicities. We next consider two examples of a more general nature: when $q(s)$ is (1) a power of a single linear term and (2) a power of a single irreducible quadratic.

Power of a Linear Term

Let us now consider a polynomial which is a power of a linear term.

Proposition 7. *For $r \in \mathbb{R}$, let*

$$q(s) = (s - r)^n.$$

Then

$$\mathcal{B}_q = \left\{ e^{rt}, te^{rt}, \dots, t^{n-1}e^{rt} \right\}$$

is a spanning set for $\mathcal{E}_q$.

Remark 8. Observe that $\mathcal{B}_q$ only depends on the single root r and its multiplicity n. Further, $\mathcal{B}_q$ has exactly n functions which is the same as the degree of $q(s)$.

Proof. Suppose $f(t) \in \mathcal{E}_q$. Then a partial fraction decomposition of $\mathcal{L}\{f(t)\}(s) = \frac{p(s)}{q(s)}$ has the form

$$\frac{p(s)}{(s - r)^n} = a_1 \frac{1}{s - r} + a_2 \frac{1}{(s - r)^2} + \cdots + a_n \frac{1}{(s - r)^n}.$$

Laplace inversion then gives

$$f(t) = \mathcal{L}^{-1}\left\{ \frac{p(s)}{q(s)} \right\} = a_1 e^{rt} + a_2 t e^{rt} + a_3 \frac{t^2}{2!} e^{rt} + \cdots + a_n \frac{t^{n-1}}{(n-1)!} e^{rt}$$

$$= c_1 e^{rt} + c_2 t e^{rt} + c_3 t^2 e^{rt} + \cdots + c_n t^{n-1} e^{rt},$$

where, in the second line, we have relabeled the constants $\frac{a_k}{k!} = c_k$. Observe that

$$\mathcal{L}\{t^k e^{rt}\} = \frac{k!}{(s - r)^{k+1}} = \frac{k!(s - r)^{n-k-1}}{(s - r)^n} \in \mathcal{E}_q.$$

If

$$\mathcal{B}_q = \left\{ e^{rt}, te^{rt}, \dots, t^{n-1}e^{rt} \right\}$$

then it follows that

$$\mathcal{E}_q = \operatorname{Span} \mathcal{B}_q. \qquad \square$$

Example 9. Let $q(s) = (s - 5)^4$. Find all functions f so that $\mathcal{L}\{f\}(s)$ has $q(s)$ in the denominator. In other words, find $\mathcal{E}_q$.

▶ **Solution.** We simply observe from Proposition 7 that

$$\mathcal{B}_q = \left\{ e^{5t}, te^{5t}, t^2 e^{5t}, t^3 e^{5t} \right\}$$

and hence $\mathcal{E}_q = \operatorname{Span} \mathcal{B}_q.$ ◀

Power of an Irreducible Quadratic Term

Let us now consider a polynomial which is a power of an irreducible quadratic term. Recall that any irreducible quadratic $s^2 + cs + d$ may be written in the form $(s-a)^2 + b^2$, where $a \pm b$ are the complex roots.

Lemma 10. *Let n be a nonnegative integer; a, b real numbers; and $b > 0$. Let $q(s) = ((s-a)^2 + b^2)^n$. Then*

$$t^k e^{at} \cos bt \in \mathcal{E}_q \quad \text{and} \quad t^k e^{at} \sin bt \in \mathcal{E}_q,$$

for all $k = 0, \ldots, n-1$.

Proof. We use the translation principle and the transform derivative principle to get

$$\mathcal{L}\left\{t^k e^{at} \cos bt\right\}(s) = \mathcal{L}\left\{t^k \cos bt\right\}(s-a)$$

$$= (-1)^k \mathcal{L}\left\{\cos bt\right\}^{(k)}\Big|_{s \mapsto (s-a)}$$

$$= (-1)^k \left(\frac{s}{s^2 + b^2}\right)^{(k)}\Big|_{s \mapsto (s-a)}.$$

An induction argument which we leave as an exercise gives that $\left(\frac{s}{s^2+b^2}\right)^{(k)}$ is a proper rational function with denominator $(s^2 + b^2)^{k+1}$. Replacing s by $s-a$ gives

$$\mathcal{L}\left\{t^k e^{at} \cos bt\right\}(s) = \frac{p(s)}{((s-a)^2 + b^2)^{k+1}},$$

for some polynomial $p(s)$ with $\deg p(s) < 2(k+1)$. Now multiply the numerator and denominator by $((s-a)^2 + b^2)^{n-k-1}$ to get

$$\mathcal{L}\left\{t^k e^{at} \cos bt\right\}(s) = \frac{p(s)((s-a)^2 + b^2)^{n-k-1}}{((s-a^2)^2 + b^2)^n},$$

for $k = 0, \ldots, n-1$. Since the degree of the numerator is less that $2(k+1) + 2(n-k-1) = 2n$, it follows that $\mathcal{L}\left\{t^k e^{at} \cos bt\right\} \in \mathcal{E}_q$. A similar calculation gives $\mathcal{L}\left\{t^k e^{at} \sin bt\right\} \in \mathcal{E}_q$.   □

Proposition 11. *Let*

$$q(s) = ((s-a)^2 + b^2)^n$$

and assume $b > 0$. Then

$$\mathcal{B}_q = \left\{e^{at} \cos bt, e^{at} \sin bt, t e^{at} \cos bt, t e^{at} \sin bt, \right.$$

$$\left. \ldots, t^{n-1} e^{at} \cos bt, t^{n-1} e^{at} \sin bt\right\}$$

is a spanning set for $\mathcal{E}_q$.

Remark 12. Since $\cos(-bt) = \cos bt$ and $\sin(-bt) = -\sin bt$, we may assume that $b > 0$. Observe then that $\mathcal{B}_q$ depends only on the root $a + ib$, where $b > 0$, of $q(s)$ and the multiplicity n. Also $\mathcal{B}_q$ has precisely $2n$ functions which is the degree of $q(s)$.

Proof. By Lemma 10, each term in $\mathcal{B}_q$ is in $\mathcal{E}_q$. Suppose $f(t) \in \mathcal{E}_q$ and $\mathcal{L}\{f(t)\}(s) = \frac{p(s)}{q(s)}$ for some polynomial $p(s)$. A partial fraction decomposition of $\frac{p(s)}{q(s)}$ has the form

$$\frac{p(s)}{q(s)} = \frac{a_1 s + b_1}{((s-a)^2 + b^2)} + \frac{a_2 s + b_2}{((s-a)^2 + b^2)^2} + \cdots + \frac{a_n s + b_n}{((s-a)^2 + b^2)^n}. \qquad (6)$$

By Corollary 11 of Sect. 2.5 and the first translation principle, (2) of Sect. 2.5, the inverse Laplace transform of a term $\frac{a_k s + b_k}{((s-a)^2 + b^2)^k}$ has the form

$$\mathcal{L}^{-1}\left\{ \frac{a_k s + b_k}{((s-a)^2 + b^2)^k} \right\} = p_k(t)e^{at}\cos bt + q_k(t)e^{at}\sin bt,$$

where $p_k(t)$ and $q_k(t)$ are polynomials of degree at most $k-1$. We apply the inverse Laplace transform to each term in (6) and add to get

$$\mathcal{L}^{-1}\left\{ \frac{p(s)}{q(s)} \right\} = p(t)e^{at}\cos bt + q(t)e^{at}\sin bt,$$

where $p(t)$ and $q(t)$ are polynomials of degree at most $n - 1$. This means that $f(t) = \mathcal{L}^{-1}\left\{ \frac{p(s)}{q(s)} \right\}$ is a linear combination of functions from $\mathcal{B}_q$ as defined above. It follows now that

$$\mathcal{E}_q = \text{Span } \mathcal{B}_q. \qquad \qquad \square$$

Example 13. Let $q(s) = ((s-3)^2 + 2^2)^3$. Find all functions f so that $\mathcal{L}\{f\}(s)$ has $q(s)$ in the denominator. In other words, find $\mathcal{E}_q$.

▶ **Solution.** We simply observe from Proposition 11 that

$$\mathcal{B}_q = \{e^{3t}\cos 2t, \ e^{3t}\sin 2t, \ te^{3t}\cos 2t, \ te^{3t}\sin 2t, \ t^2 e^{3t}\cos 2t, \ t^2 e^{3t}\sin 2t, \}$$

and hence $\mathcal{E}_q = \text{Span } \mathcal{B}_q.$ ◀

Exercises

1–25. Find the standard basis $\mathcal{B}_q$ of $\mathcal{E}_q$ for each polynomial $q(s)$.

1. $q(s) = s - 4$
2. $q(s) = s + 6$
3. $q(s) = s^2 + 5s$
4. $q(s) = s^2 - 3s - 4$
5. $q(s) = s^2 - 6s + 9$
6. $q(s) = s^2 - 9s + 14$
7. $q(s) = s^2 - s - 6$
8. $q(s) = s^2 + 9s + 18$
9. $q(s) = 6s^2 - 11s + 4$
10. $q(s) = s^2 + 2s - 1$
11. $q(s) = s^2 - 4s + 1$
12. $q(s) = s^2 - 10s + 25$
13. $q(s) = 4s^2 + 12s + 9$
14. $q(s) = s^2 + 9$
15. $q(s) = 4s^2 + 25$
16. $q(s) = s^2 + 4s + 13$
17. $q(s) = s^2 - 2s + 5$
18. $q(s) = s^2 - s + 1$
19. $q(s) = (s + 3)^4$
20. $q(s) = (s - 2)^5$
21. $q(s) = s^3 - 3s^2 + 3s - 1$
22. $q(s) = (s + 1)^6$
23. $q(s) = (s^2 + 4s + 5)^2$
24. $q(s) = (s^2 - 8s + 20)^3$
25. $q(s) = (s^2 + 1)^4$

2.7   The Linear Spaces $\mathcal{E}_q$: The General Case

We continue the discussion initiated in the previous section. Let $q(s)$ be a fixed polynomial. We want to describe the linear space $\mathcal{E}_q$ of continuous functions that have Laplace transforms that are rational functions and have $q(s)$ in the denominator. In the previous section, we gave a description of $\mathcal{E}_q$ in terms of a spanning set $\mathcal{B}_q$ for polynomials of degree 2, a power of a linear term, and a power of an irreducible quadratic. We take up the general case here.

Exponential Polynomials

Let n be a nonnegative integer, and $a, b \in \mathbb{R}$, and assume $b \geq 0$. We will refer to functions of the form

$$t^n e^{at} \cos bt \quad \text{and} \quad t^n e^{at} \sin bt,$$

defined on $\mathbb{R}$, as *simple exponential polynomials*. We introduced these functions in Lemma 10 of Sect. 2.6. Note that if $b = 0$, then $t^n e^{at} \cos bt = t^n e^{at}$, and if both $a = 0$ and $b = 0$, then $t^n e^{at} \cos bt = t^n$. Thus, the terms

$$t^n e^{at} \quad \text{and} \quad t^n$$

are simple exponential polynomials for all nonnegative integers n and real numbers a. If $n = 0$ and $a = 0$, then $t^n e^{at} \cos bt = \cos bt$ and $t^n e^{at} \sin bt = \sin bt$. Thus, the basic trigonometric functions

$$\cos bt \quad \text{and} \quad \sin bt$$

are simple exponential polynomials. For example, all of the following functions are simple exponential polynomials:

   1. t^3 2. $t^5 \cos 2t$ 3. e^{-3t} 4. $t^2 e^{2t}$ 5. $t^4 e^{-8t} \sin 3t$

while *none* of the following are simple exponential polynomials:

   6. $t^{\frac{1}{2}}$ 7. $\dfrac{\sin 2t}{\cos 2t}$ 8. $t e^{t^2}$ 9. $\sin(e^t)$ 10. $\dfrac{e^t}{t}$

We refer to any linear combination of simple exponential polynomials as an *exponential polynomial*. In other words, an exponential polynomial is a function in the span of the simple exponential polynomials. We denote the set of all exponential polynomials by $\mathcal{E}$. All of the following are examples of exponential polynomials and are thus in $\mathcal{E}$:

1. $e^t + 2e^{2t} + 3e^{3t}$ 2. $t^2 e^t \sin 3t + 2t e^{7t} \cos 5t$ 3. $1 - t + t^2 - t^3 + t^4$

4. $t - 2t \cos 3t$ 5. $3e^{2t} + 4t e^{2t}$ 6. $2 \cos 4t - 3 \sin 4t$

Definition of $\mathcal{B}_q$

Recall that we defined $\mathcal{E}_q$ to be the set of input functions whose Laplace transform is in $\mathcal{R}_q$. We refine slightly our definition. We define $\mathcal{E}_q$ to be the set of *exponential polynomials* whose Laplace transform is in $\mathcal{R}_q$. That is,

$$\mathcal{E}_q = \{ f \in \mathcal{E} : \mathcal{L}\{f\} \in \mathcal{R}_q \}.$$

Thus, each function in $\mathcal{E}_q$ is defined on the real line even though the Laplace transform only uses the restriction to $[0, \infty)$.[4] We now turn our attention to describing $\mathcal{E}_q$ in terms of a spanning set $\mathcal{B}_q$. In each of the cases we considered in the previous section, $\mathcal{B}_q$ was made up of simple exponential polynomials. This will persist for the general case as well. Consider the following example.

Example 1. Let $q(s) = (s - 1)^3(s^2 + 1)^2$. Find a set $\mathcal{B}_q$ of simple exponential polynomials that spans $\mathcal{E}_q$.

▶ **Solution.** Recall that $\mathcal{E}_q$ consists of those input functions $f(t)$ such that $\mathcal{L}\{f(t)\}$ is in $\mathcal{R}_q$. In other words, $\mathcal{L}\{f(t)\}(s) = \frac{p(s)}{q(s)}$, for some polynomial $p(s)$ with degree less than that of $q(s)$. A partial fraction decomposition gives the following form:

$$\frac{p(s)}{(s-1)^3(s^2+1)^2} = \frac{a_1}{s-1} + \frac{a_2}{(s-1)^2} + \frac{a_3}{(s-1)^3} + \frac{a_4 s + a_5}{s^2 + 1} + \frac{a_6 s + a_7}{(s^2+1)^2}$$

$$= \frac{p_1(s)}{(s-1)^3} + \frac{p_2(s)}{(s^2+1)^2},$$

where $p_1(s)$ is a polynomial of degree at most 2 and $p_2(s)$ is a polynomial of degree at most 3. This decomposition allows us to treat Laplace inversion of $\frac{p(s)}{(s-1)^3(s^2+1)^2}$ in terms of the two pieces: $\frac{p_1(s)}{(s-1)^3}$ and $\frac{p_2(s)}{(s^2+1)^2}$. In the first case, the denominator is a power of a linear term, and in the second case, the denominator is a power of an irreducible quadratic. From Propositions 7 and 11 of Sect. 2.6, we get

$$\mathcal{L}^{-1}\left\{ \frac{p_1(s)}{q_1(s)} \right\} = c_1 e^t + c_2 t e^t + c_3 t^2 e^t,$$

---

[4]In fact, any function which has a power series with infinite radius of convergence, such as an exponential polynomial, is completely determined by it values on $[0, \infty)$. This is so since $f(t) = \sum_{n=0}^{\infty} \frac{f^{(n)}(0)}{n!} t^n$ and $f^{(n)}(0)$ are computed from $f(t)$ on $[0, \infty)$.

$$\mathcal{L}^{-1}\left\{\frac{p_2(s)}{q_2(s)}\right\} = c_4\cos t + c_5\sin t + c_6 t\cos t + c_7 t\sin t,$$

where $c_1,\ldots,c_7$ are scalars, $q_1(s) = (s-1)^3$, and $q_2(s) = (s^2+1)^2$. It follows now by linearity of the inverse Laplace transform that

$$f(t) = \mathcal{L}^{-1}\left\{\frac{p(s)}{(s-1)^3(s^2+1)^2}\right\}$$

$$= c_1 e^t + c_2 t e^t + c_3 t^2 e^t + c_4\cos t + c_5\sin t + c_6 t\cos t + c_7 t\sin t.$$

Thus, if

$$\mathcal{B}_q = \left\{e^t, te^t, t^2 e^t, \cos t, \sin t, t\cos t, t\sin t\right\}$$

then the above calculation gives $\mathcal{E}_q = \operatorname{Span}\mathcal{B}_q$. Also observe that we have shown that $\mathcal{B}_q = \mathcal{B}_{q_1} \cup \mathcal{B}_{q_2}$. Further, the order of $\mathcal{B}_q$, which is 7, matches the degree of $q(s)$. ◀

From this example, we see that $\mathcal{B}_q$ is the collection of simple exponential polynomials obtained from both $\mathcal{B}_{q_1}$ and $\mathcal{B}_{q_2}$, where $q_1(s) = (s-1)^3$ and $q_2(s) = (s^2+1)^2$ are the factors of $q(s)$. More generally, suppose that $q(s) = q_1(s)\cdots q_R(s)$, where $q_i(s)$ is a power of a linear term or a power of an irreducible quadratic term. Further assume that there is no repetition among the linear or quadratic terms. Then a partial fraction decomposition can be written in the form

$$\frac{p(s)}{q(s)} = \frac{p_1(s)}{q_1(s)} + \cdots + \frac{p_R(s)}{q_R(s)}.$$

We argue as in the example above and see that $\mathcal{L}^{-1}\left\{\frac{p(s)}{q(s)}\right\}$ is a linear combination of those simple exponential polynomial gotten from $\mathcal{B}_{q_1},\ldots,\mathcal{B}_{q_R}$. If we define

$$\mathcal{B}_q = \mathcal{B}_{q_1} \cup \cdots \cup \mathcal{B}_{q_R}$$

then we get the following theorem:

Theorem 2. *Let $q(s)$ be a fixed polynomial of degree n. Suppose $q(s) = q_1(s)\cdots q_R(s)$ where $q_i(s)$ is a power of a linear or an irreducible term and there is no repetition among the terms. Define $\mathcal{B}_q = \mathcal{B}_{q_1} \cup \cdots \cup \mathcal{B}_{q_R}$. Then*

$$\operatorname{Span}\mathcal{B}_q = \mathcal{E}_q.$$

Further, the degree of $p(s)$ is the same as the order of $\mathcal{B}_q$.

Proof. The essence of the proof is given in the argument in the previous paragraph. More details can be found in Appendix A.3. □

It is convenient to also express $\mathcal{B}_q$ in terms of the roots of $q(s)$ and their multiplicities. We put this forth in the following algorithm.

Algorithm 3. Let $q(s)$ be a polynomial. The following procedure is used to construct $\mathcal{B}_q$, a spanning set of $\mathcal{E}_q$.

Description of $\mathcal{B}_q$

Given a polynomial $q(s)$:

1. *Factor $q(s)$ and determine the roots and their multiplicities.*

2. *For each real root r with multiplicity m, the spanning set $\mathcal{B}_q$ will contain the simple exponential functions:*

$$e^{rt},\ te^{rt},\ldots,t^{m-1}e^{rt}.$$

3. *For each complex root $a \pm ib$ $(b > 0)$ with multiplicity m, the spanning set $\mathcal{B}_q$ will contain the simple exponential functions:*

$$e^{at}\cos bt,\ e^{at}\sin bt,\ \ldots,t^{m-1}e^{at}\cos bt,\ t^{m-1}e^{at}\sin bt.$$

Example 4. Find $\mathcal{B}_q$ if

1. $q(s) = 4(s-3)^2(s-6)$
2. $q(s) = (s+1)(s^2+1)$
3. $q(s) = 7(s-1)^3(s-2)^2((s-3)^2+5^2)^2$

▶ **Solution.**

1. The roots are $r_1 = 3$ with multiplicity 2 and $r_2 = 6$ with multiplicity 1. Thus,

$$\mathcal{B}_q = \left\{ e^{3t},\ te^{3t},\ e^{6t} \right\}.$$

2. The roots are $r = -1$ with multiplicity 1 and $a \pm ib = 0 \pm i$. Thus, $a = 0$ and $b = 1$. We now get

$$\mathcal{B}_q = \left\{ e^{-t},\ \cos t,\ \sin t \right\}.$$

3. The roots are $r_1 = 1$ with multiplicity 3, $r_2 = 2$ with multiplicity 2, and $a \pm ib = 3 \pm 5i$ with multiplicity 2. We thus get

$$\mathcal{B}_q = \left\{ e^t,\ te^t,\ t^2e^t,\ e^{2t},\ te^{2t},\ e^{3t}\cos 5t,\ e^{3t}\sin 5t, te^{3t}\cos 5t, te^{3t}\sin 5t \right\}. \ ◀$$

Laplace Transform Correspondences

We conclude this section with two theorems. The first relates $\mathcal{E}$ and $\mathcal{R}$ by way of the Laplace transform. The second relates $\mathcal{E}_q$ and $\mathcal{R}_q$. The notion of linearity is central so we establish that $\mathcal{E}$ and $\mathcal{R}$ are linear spaces. First note the following lemma.

Lemma 5. *Suppose S is a set of functions on an interval I. Let $\mathcal{F} =$ Span S. Then $\mathcal{F}$ is a linear space.*

Proof. If f and g are in Span S, then there are scalars $a_1, \ldots, a_n$ and $b_1, \ldots, b_m$ so that
$$f = a_1 f_1 + \cdots + a_n f_n \quad \text{and} \quad g = b_1 g_1 + \cdots + b_m g_m,$$
where $f_1, \ldots, f_n$ and $g_1, \ldots, g_m$ are in S. The sum
$$f + g = a_1 f_1 + \cdots + a_n f_n + b_1 g_1 + \cdots + b_m g_m,$$

is again a linear combination of function in S, and hence $f + g \in$ Span S. In a similar way, if c is a scalar and $f = a_1 f_1 + \cdots + a_n f_n$ is in Span S, then $cf = ca_1 f_1 + \cdots + ca_n f_n$ is a linear combinations of functions in S, and hence $cf \in$ Span S. It follows that Span S is closed under addition and scalar multiplication and hence is a linear space.   □

   Recall that we defined the set $\mathcal{E}$ of exponential polynomials as the span of the set of all simple exponential polynomials. Lemma 5 gives the following result.

Proposition 6. *The set of exponential polynomials $\mathcal{E}$ is a linear space.*

Proposition 7. *The set $\mathcal{R}$ of proper rational functions is a linear space.*

Proof. Suppose $\frac{p_1(s)}{q_1(s)}$ and $\frac{p_2(s)}{q_2(s)}$ are in $\mathcal{R}$ and $c \in \mathbb{R}$. Then

• $\frac{p_1(s)}{q_1(s)} + \frac{p_2(s)}{q_2(s)} = \frac{p_1(s)q_2(s)+p_2(s)q_1(s)}{q_1(s)q_2(s)}$ is again a proper rational function and hence in $\mathcal{R}$.
• $c\frac{p_1(s)}{q_1(s)} = \frac{cp_1(s)}{q_1(s)}$ is a again a proper rational function and hence in $\mathcal{R}$.

It follows that $\mathcal{R}$ is closed under addition and scalar multiplication, and hence $\mathcal{R}$ is a linear space.   □

Theorem 8. *The Laplace transform*
$$\mathcal{L} : \mathcal{E} \to \mathcal{R}$$

establishes a linear one-to-one correspondence between the linear space of exponential polynomials, $\mathcal{E}$, and the linear space of proper rational functions, $\mathcal{R}$.

Remark 9. This theorem means the following:

1. The Laplace transform is linear, which we have already established.
2. The Laplace transform of each $f \in \mathcal{E}$ is a rational function.
3. For each proper rational function $r \in \mathcal{R}$, there is a unique exponential polynomial $f \in \mathcal{E}$ so that $\mathcal{L}\{f\} = r$.

Proof. By Lemma 10 of Sect. 2.6, the Laplace transforms of the simple exponential polynomials $t^n e^{at} \cos bt$ and $t^n e^{at} \sin bt$ are in $\mathcal{R}$. Let $\phi \in \mathcal{E}$. There are simple exponential polynomials $\phi_1, \ldots, \phi_m$ such that $\phi = c_1 \phi_1 + \cdots + c_m \phi_m$, where $c_1, \ldots, c_m \in \mathbb{R}$. Since the Laplace transform is linear, we have $\mathcal{L}\{\phi\} = c_1 \mathcal{L}\{f_1\} + \cdots + c_m \mathcal{L}\{\phi_m\}$. Now each term $\mathcal{L}\{\phi_i\} \in \mathcal{R}$. Since $\mathcal{R}$ is a linear space, we $\mathcal{L}\{\phi\} \in \mathcal{R}$. It follows that the Laplace transform of any exponential polynomial is a rational function.

On the other hand, a proper rational function is a linear combination of the simple rational functions given in (1) in Sect. 2.5. Observe that

$$\mathcal{L}^{-1}\{1/(s-a)^n\}(s) = \frac{t^{n-1}}{(n-1)!}e^{at}$$

is a scalar multiple of a simple exponential polynomial. Also, Corollary 11 of Sect. 2.5 and the first translation principle establish that both $1/((s-a)^2 + b^2)^k$ and $s/((s-a)^2 + b^2)^k$ have inverse Laplace transforms that are linear combinations of $t^n e^{at} \sin bt$ and $t^n e^{at} \cos bt$ for $0 \le k < n$. It now follows that the inverse Laplace transform of any rational function is an exponential polynomial. Since the Laplace transform is one-to-one by Theorem 1 of Sect. 2.5, it follows that the Laplace transform establishes a one-to-one correspondence between $\mathcal{E}$ and $\mathcal{R}$.   □

We obtain by restricting the Laplace transform the following fundamental theorem.

Theorem 10. *The Laplace transform establishes a linear one-to-one correspondence between $\mathcal{E}_q$ and $\mathcal{R}_q$. In other words,*

$$\mathcal{L} : \mathcal{E}_q \to \mathcal{R}_q$$

is one-to-one and onto.

Exercises

1–11. Determine which of the following functions are in the linear space $\mathcal{E}$ of exponential polynomials.

1. $t^2 e^{-2t}$
2. $t^{-2} e^{2t}$
3. t/e^t
4. e^t/t
5. $t \sin \left(4t - \dfrac{\pi}{4} \right)$
6. $(t + e^t)^2$
7. $(t + e^t)^{-2}$
8. $t e^{t/2}$
9. $t^{1/2} e^t$
10. $\sin 2t / e^{2t}$
11. $e^{2t} / \sin 2t$

12–28. Find the standard basis $\mathcal{B}_q$ of $\mathcal{E}_q$ for each polynomial $q(s)$.

12. $q(s) = s^3 + s$
13. $q(s) = s^4 - 1$
14. $q(s) = s^3(s + 1)^2$
15. $q(s) = (s - 1)^3(s + 7)^2$
16. $q(s) = (s + 8)^2(s^2 + 9)^3$
17. $q(s) = (s + 2)^3(s^2 + 4)^2$
18. $q(s) = (s + 5)^2(s - 4)^2(s + 3)^2$
19. $q(s) = (s - 2)^2(s + 3)^2(s + 3)$
20. $q(s) = (s - 1)(s - 2)^2(s - 3)^3$
21. $q(s) = (s + 4)^2(s^2 + 6s + 13)^2$
22. $q(s) = (s + 5)(s^2 + 4s + 5)^2$
23. $q(s) = (s - 3)^3(s^2 + 2s + 10)^2$
24. $q(s) = s^3 + 8$
25. $q(s) = 2s^3 - 5s^2 + 4s - 1$
26. $q(s) = s^3 + 2s^2 - 9s - 18$
27. $q(s) = s^4 + 5s^2 + 6$
28. $q(s) = s^4 - 8s^2 + 16$

29–33. Verify the following *closure properties of the linear space of proper rational functions*.

29. *Multiplication*. Show that if $r_1(s)$ and $r_2(s)$ are in $\mathcal{R}$, then so is $r_1(s)r_2(s)$.
30. *Translation*. Show that if $r(s)$ is in $\mathcal{R}$, so is any translation of $r(s)$, that is, $r(s-a) \in \mathcal{R}$ for any a.
31. *Differentiation*. Show that if $r(s)$ is in $\mathcal{R}$, then so is the derivative $r'(s)$.
32. Let $q(s) = (s-a)^2 + b^2$. Suppose $r(s) \in \mathcal{R}_{q^n}$ but $r(s) \notin \mathcal{R}_{q^{n-1}}$. Then $r'(s) \in \mathcal{R}_{q^{n+1}}$ but $r'(s) \notin \mathcal{R}_{q^n}$.
33. Let $q(s) = (s-a)^2 + b^2$. Let $r \in \mathcal{R}_q$. Then $r^{(n)} \in \mathcal{R}_{q^{n+1}}$ but not in $\mathcal{R}_{q^n}$.

34–38. Verify the following *closure properties of the linear space of exponential polynomials*.

34. *Multiplication*. Show that if f and g are in $\mathcal{E}$, then so is fg.
35. *Translation*. Show that if f is in $\mathcal{E}$, so is any translation of f, i.e. $f(t - t_0) \in \mathcal{E}$, for any t_0.
36. *Differentiation*. Show that if f is in $\mathcal{E}$, then so is the derivative f'.
37. *Integration*. Show that if f is in $\mathcal{E}$, then so is $\int f(t)\, dt$. That is, any antiderivative of f is in $\mathcal{E}$.
38. Show that $\mathcal{E}$ is not closed under inversion. That is, find a function f so that $1/f$ is not in $\mathcal{E}$.

39–41. Let $q(s)$ be a fixed polynomial. Verify the following *closure properties of the linear space $\mathcal{E}_q$*.

39. *Differentiation*. Show that if f is in $\mathcal{E}_q$, then f' is in $\mathcal{E}_q$.
40. Show that if f is in $\mathcal{E}_q$, then the nth derivative of f, $f^{(n)}$, is in $\mathcal{E}_q$.
41. Show that if $f \in \mathcal{E}_q$, then any translate is in $\mathcal{E}_q$. That is, if $t_0 \in \mathbb{R}$, then $f(t - t_0) \in \mathcal{E}_q$.

2.8 Convolution

Table 2.3 shows many examples of operations defined on the input space that induce via the Laplace transform a corresponding operation on transform space, and vice versa. For example, multiplication by $-t$ in input space corresponds to differentiation in transform space. If $F(s)$ is the Laplace transform of $f(t)$, then this correspondence can be indicated as follows:

$$-tf(t) \longleftrightarrow \frac{\mathrm{d}}{\mathrm{d}s} F(s).$$

Our goal in this section is to study another such operational identity. Specifically, we will be concentrating on the question of what is the operation in the input space that corresponds to ordinary multiplication of functions in the transform space. Put more succinctly, suppose $f(t)$ and $g(t)$ are input functions with Laplace transforms $F(s)$ and $G(s)$, respectively. What input function $h(t)$ corresponds to the product $H(s) = F(s)G(s)$ under the Laplace transform? In other words, how do we fill in the following question mark in terms of $f(t)$ and $g(t)$?

$$\boxed{h(t) = ?} \longleftrightarrow F(s)G(s).$$

You might guess that $h(t) = f(t)g(t)$. That is, you would be guessing that multiplication in the input space corresponds to multiplication in the transform space. This guess is *wrong* as you can quickly see by looking at almost any example. For a concrete example, let

$$F(s) = 1/s \quad \text{and} \quad G(s) = 1/s^2.$$

Then $H(s) = F(s)G(s) = 1/s^3$ and $h(t) = t^2/2$. However, $f(t) = 1$, $g(t) = t$, and, hence, $f(t)g(t) = t$. Thus $h(t) \neq f(t)g(t)$.

Suppose f and g are continuous functions on $[0, \infty)$. We define the **convolution (product)**, $(f * g)(t)$, of f and g by the following integral:

$$(f * g)(t) = \int_0^t f(u)g(t-u)\,\mathrm{d}u. \tag{1}$$

The variable of integration we chose is u but any variable other than t can be used. Admittedly, convolution is an unusual product. It is not at all like the usual product of functions where the value (or state) at time t is determined by knowing just the value of each factor at time t. Rather, (1) tells us that the value at time t depends on knowing the values of the input function f and g for all u between 0 and t. They are then "meshed" together to give the value at t.

The following theorem, the convolution theorem, explains why convolution is so important. It is the operation of convolution in input space that corresponds under the Laplace transform to ordinary multiplication of transform functions.

Theorem 1 (The Convolution Theorem). *Let $f(t)$ and $g(t)$ be continuous functions of exponential type. Then $f * g$ is of exponential type. Further, if $F(s) = \mathcal{L}\{f(t)\}(s)$ and $G(s) = \mathcal{L}\{g(t)\}(s)$, then*

The Convolution Principle

$$\mathcal{L}\{(f * g)(t)\}(s) = F(s)G(s)$$

$$or \quad (f * g)(t) = \mathcal{L}^{-1}\{F(s) \cdot G(s))\}(t).$$

The second formula is just the Laplace inversion of the first formula. The proof of the convolution principle will be postponed until Chap. 6, where it is proved for a broader class of functions.

Let us consider a few examples that confirm the convolution principle.

Example 2. Let n be a positive integer, $f(t) = t^n$, and $g(t) = 1$. Compute the convolution $f * g$ and verify the convolution principle.

▶ **Solution.** Observe that $\mathcal{L}\{t^n\}(s) = n!/s^{n+1}$, $\mathcal{L}\{1\}(s) = 1/s$, and

$$f * g(t) = \int_0^t f(u)g(t-u)\,du = \int_0^t u^n \cdot 1\,du = \frac{u^{n+1}}{n+1}\Big|_0^t = \frac{t^{n+1}}{n+1}.$$

Further,

$$\mathcal{L}\{f * g\}(s) = \mathcal{L}\left\{\frac{t^{n+1}}{n+1}\right\}(s) = \frac{1}{n+1}\frac{(n+1)!}{s^{n+2}} = \frac{n!}{s^{n+1}} \cdot \frac{1}{s}$$

$$= \mathcal{L}\{f\} \cdot \mathcal{L}\{g\}$$

thus verifying the convolution principle.   ◀

Example 3. Compute $t^2 * t^3$ and verify the convolution principle.

▶ **Solution.** Here we let $f(t) = t^3$ and $g(t) = t^2$ in (1) to get

$$t^3 * t^2 = \int_0^t f(u)g(t-u)\,du = \int_0^t u^3(t-u)^2\,du$$

$$= \int_0^t t^2 u^3 - 2tu^4 + u^5\,du = \frac{t^6}{4} - 2\frac{t^6}{5} + \frac{t^6}{6} = \frac{t^6}{60}.$$

Additionally,

$$\mathcal{L}\{f * g\}(s) = \mathcal{L}\left\{\frac{t^6}{60}\right\}(s) = \frac{1}{60}\frac{6!}{s^7} = \frac{3!}{s^4} \cdot \frac{2}{s^3}$$

$$= \mathcal{L}\{f\} \cdot \mathcal{L}\{g\}. \qquad \blacktriangleleft$$

Example 4. Let $f(t) = \sin t$ and $g(t) = 1$. Compute $(f * g)(t)$ and verify the convolution principle.

▶ **Solution.** Observe that $F(s) = \mathcal{L}\{\sin t\} = 1/(s^2 + 1)$, $G(s) = \mathcal{L}\{1\} = 1/s$, and

$$(f * g)(t) = \int_0^t f(u)g(t-u)\,\mathrm{d}u = \int_0^t \sin u\,\mathrm{d}u = -\cos u\big|_0^t = 1 - \cos t.$$

Further,

$$\mathcal{L}\{f * g\}(s) = \mathcal{L}\{1 - \cos t\}(s) = \frac{1}{s} - \frac{s}{s^2 + 1}$$

$$= \frac{s^2 + 1 - s^2}{s(s^2 + 1)} = \frac{1}{s(s^2 + 1)} = \frac{1}{s^2 + 1} \cdot \frac{1}{s}$$

$$= \mathcal{L}\{f\} \cdot \mathcal{L}\{g\}. \qquad \blacktriangleleft$$

Properties of the Convolution Product

Convolution is sometimes called the convolution product because it behaves in many ways like an ordinary product. In fact, below are some of its properties:

Commutative property: $f * g = g * f$

Associative property: $(f * g) * h = f * (g * h)$

Distributive property: $f * (g + h) = f * g + f * h$

$$f * 0 = 0 * f = 0$$

Indeed, these properties of convolution are easily verified from the definition given in (1). For example, the commutative property is verified by a change of variables:

$$f * g(t) = \int_0^t f(u)g(t-u)\,\mathrm{d}u$$

Let $x = t - u$ then $\mathrm{d}x = -\mathrm{d}t$ and we get

$$= \int_{t}^{0} f(t-x)g(x)\,(-1)\mathrm{d}x$$

$$= \int_{0}^{t} g(x)f(t-x)\,\mathrm{d}x$$

$$= g * f.$$

It follows, for example, that

$$t^n * 1(t) = \int_{0}^{t} u^n \, \mathrm{d}u = \int_{0}^{t} (u-t)^n \, \mathrm{d}u = t^n * 1.$$

Both integrals are equal. The decision about which one to use depends on which you regard as the easiest to compute. You should verify the other properties listed.

There is one significant difference that convolution has from the ordinary product of functions, however. Examples 2 and 4 imply that the constant function 1 does not behave like a multiplicative identity. In fact, no such "function" exists.[5] Nevertheless, convolution by $f(t) = 1$ is worth singling out as a special case of the convolution principle.

Theorem 5. *Let $g(t)$ be a continuous function of exponential type and $G(s)$ its Laplace transform. Then $(1 * g)(t) = \int_{0}^{t} g(u)\,\mathrm{d}u$ and*

> ### Input Integral Principle
> $$\mathcal{L}\left\{\int_{0}^{t} g(u)\,\mathrm{d}u\right\} = \frac{G(s)}{s}.$$

Proof. Since $(1 * g)(t) = \int_{0}^{t} g(u)\,\mathrm{d}u$, the theorem follows directly from the convolution principle. However, it is noteworthy that the input integral principle follows from the input derivative principle. Here is the argument. Since g is of exponential type, so is any antiderivative by Lemma 4 of Sect. 2.2. Suppose $h(t) = \int_{0}^{t} g(u)\,\mathrm{d}u$. Then $h'(t) = g(t)$, and $h(0) = 0$ so the input derivative principle gives

$$G(s) = \mathcal{L}\{g(t)\} = \mathcal{L}\{h'(t)\} = sH(s) - h(0) = sH(s).$$

Hence, $H(s) = (1/s)G(s)$, and thus,

$$\mathcal{L}\left\{\int_{0}^{t} g(u)\,\mathrm{d}u\right\} = \mathcal{L}\{h(t)\}(s) = \frac{1}{s}G(s). \qquad \square$$

---

[5] In Chap. 6, we will discuss a so-called "generalized function" that will act as a multiplicative identity for convolution.

Remark 6. The requirement that g be of exponential type can be relaxed. It can be shown that if g is continuous on $[0, \infty)$ and has a Laplace transform so does any antiderivative, and the input integral principle remains valid.[6]

The input integral and convolution principles can also be used to compute the inverse Laplace transform of rational functions. Consider the following two examples

Example 7. Find the inverse Laplace transform of

$$\frac{1}{s(s^2+1)}.$$

▶ **Solution.** Instead of using partial fractions, we will use the input integral principle. Since $\mathcal{L}^{-1}\left\{\frac{1}{s^2+1}\right\} = \sin t$, we have

$$\mathcal{L}^{-1}\left\{\frac{1}{s(s^2+1)}\right\} = \int_0^t \sin u \, du$$

$$= -\cos u|_0^t = 1 - \cos t. \qquad ◀$$

Example 8. Compute the inverse Laplace transform of $\frac{s}{(s-1)(s^2+9)}$.

▶ **Solution.** The inverse Laplace transforms of $\frac{s}{s^2+9}$ and $\frac{1}{s-1}$ are $\cos 3t$ and e^t, respectively. The convolution theorem now gives

$$\mathcal{L}^{-1}\left\{\frac{s}{(s-1)(s^2+9)}\right\} = \cos 3t * e^t$$

$$= \int_0^t \cos 3u \, e^{t-u} \, du$$

$$= e^t \int_0^t e^{-u} \cos 3u \, du$$

$$= \frac{e^t}{10} (-e^{-u} \cos 3u + 3e^{-u} \sin 3u)|_0^t$$

$$= \frac{1}{10}(-\cos 3t + 3\sin 3t + e^t).$$

The computation of the integral involves integration by parts. We leave it to the student to verify this calculation. Of course, this calculation agrees with Laplace inversion using the method of partial fractions. ◀

---

[6] For a proof, see Theorem 6 and the remark that follows on page 450 of the text *Advanced Calculus* (second edition) by David Widder, published by Prentice Hall (1961).

Table 2.11 contains several general convolution formulas. The next few formulas verify some of the entries.

Formula 9. Verify the convolution product

$$e^{at} * e^{bt} = \frac{e^{at} - e^{bt}}{a - b},$$

(2)

where $a \neq b$, and verify the convolution principle.

▼ *Verification.* Use the defining equation (1) to get

$$e^{at} * e^{bt} = \int_0^t e^{au} e^{b(t-u)} \, du = e^{bt} \int_0^t e^{(a-b)u} \, du = \frac{e^{at} - e^{bt}}{a - b}.$$

Observe that

$$\mathcal{L}\left\{ \frac{e^{at} - e^{bt}}{a - b} \right\} = \frac{1}{a - b} \left(\frac{1}{s - a} - \frac{1}{s - b} \right)$$

$$= \frac{1}{(s-a)(s-b)} = \frac{1}{s - a} \cdot \frac{1}{s - b}$$

$$= \mathcal{L}\{e^{at}\} \cdot \mathcal{L}\{e^{bt}\},$$

so this calculation is in agreement with the convolution principle.   ▲

Formula 10. Verify the convolution product

$$e^{at} * e^{at} = t e^{at}$$

(3)

and verify the convolution principle.

▼ *Verification.* Computing from the definition:

$$e^{at} * e^{at} = \int_0^t e^{au} e^{a(t-u)} \, du = e^{at} \int_0^t du = t e^{at}.$$

As with the previous example, note that the calculation

$$\mathcal{L}\{t e^{at}\} = \frac{1}{(s-a)^2} = \mathcal{L}\{e^{at}\} \mathcal{L}\{e^{at}\},$$

which agrees with the convolution principle.   ▲

Remark 11. Since

$$\lim_{a \to b} \frac{e^{at} - e^{bt}}{a - b} = \frac{d}{da} e^{at} = t e^{at},$$

the previous two examples show that

$$\lim_{a \to b} e^{at} * e^{bt} = t e^{at} = e^{at} * e^{at},$$

so that the convolution product is, in some sense, a continuous operation.

Formula 12. Verify the following convolution product where $m, n \geq 0$:

$$t^m * t^n = \frac{m! \, n!}{(m + n + 1)!} t^{m+n+1}.$$

▼ *Verification.* Our method for this computation is to use the convolution theorem, Theorem 1. We get

$$\mathcal{L}\{t^m * t^n\} = \mathcal{L}\{t^m\} \mathcal{L}\{t^n\} = \frac{m!}{s^{m+1}} \frac{n!}{s^{n+1}} = \frac{m! \, n!}{s^{m+n+2}}.$$

Now take the inverse Laplace transform to conclude

$$t^m * t^n = \mathcal{L}^{-1}\left\{ \frac{m! \, n!}{s^{m+n+2}} \right\} = \frac{m! \, n!}{(m + n + 1)!} t^{m+n+1}. \qquad \blacktriangle$$

As special cases of this formula, note that

$$t^2 * t^3 = \frac{1}{60} t^6 \quad \text{and} \quad t * t^4 = \frac{1}{30} t^6.$$

The first was verified directly in Example 3.

In the next example, we revisit a simple rational function whose inverse Laplace transform can be computed by the techniques of Sect. 2.5.

Example 13. Compute the inverse Laplace transform of $\frac{1}{(s^2+1)^2}$.

▶ **Solution.** The inverse Laplace transform of $1/(s^2 + 1)$ is $\sin t$. By the convolution theorem, we have

$$\mathcal{L}^{-1}\left\{\frac{1}{(s^2+1)^2}\right\} = \sin t * \sin t.$$

$$= \int_0^t \sin u \sin(t-u)\,du$$

$$= \int_0^t \sin u(\sin t \cos u - \sin u \cos t)\,du$$

$$= \sin t \int_0^t \sin u \cos u\,du - \cos t \int_0^t \sin^2 u\,du$$

$$= \left(\sin t \frac{\sin^2 t}{2} - \cos t \frac{t - \sin t \cos t}{2}\right)$$

$$= \frac{\sin t - t \cos t}{2}.$$ ◀

Now, one should see how to handle $1/(s^2+1)^3$ and even higher powers: repeated applications of convolution. Let f^{*k} denote the convolution of f with itself k times. In other words,

$$f^{*k} = f * f * \cdots * f, \qquad \text{k times.}$$

Then it is easy to see that

$$\mathcal{L}^{-1}\left\{\frac{1}{(s^2+1)^{k+1}}\right\} = \sin^{*(k+1)} t$$

and $$\mathcal{L}^{-1}\left\{\frac{s}{(s^2+1)^{k+1}}\right\} = \cos t * \sin^{*k} t.$$

These rational functions with powers of irreducible quadratics in the denominator were introduced in Sect. 2.5 where recursion formulas were derived.

Computing convolution products can be tedious and time consuming. In Table 2.11, we provide a list of common convolution products. Students should familiarize themselves with this list so as to know when they can be used.

Exercises

1–4. Use the definition of the convolution to compute the following convolution products.

1. $t * t$
2. $t * t^3$
3. $3 * \sin t$
4. $(3t + 1) * e^{4t}$

5–9. Compute the following convolutions using the table or the convolution principle.

5. $\sin 2t * e^{3t}$
6. $(2t + 1) * \cos 2t$
7. $t^2 * e^{-6t}$
8. $\cos t * \cos 2t$
9. $e^{2t} * e^{-4t}$

10–15. Use the convolution principle to determine the following convolutions and thus verify the entries in the convolution table.

10. $t * t^n$
11. $e^{at} * \sin bt$
12. $e^{at} * \cos bt$
13. $\sin at * \sin bt$
14. $\sin at * \cos bt$
15. $\cos at * \cos bt$

16–21. Compute the Laplace transform of each of the following functions.

16. $f(t) = \int_0^t (t - x) \cos 2x \, dx$
17. $f(t) = \int_0^t (t - x)^2 \sin 2x \, dx$
18. $f(t) = \int_0^t (t - x)^3 e^{-3x} \, dx$
19. $f(t) = \int_0^t x^3 e^{-3(t-x)} \, dx$
20. $f(t) = \int_0^t \sin 2x \cos(t - x) \, dx$
21. $f(t) = \int_0^t \sin 2x \sin 2(t - x) \, dx$

22–31. In each of the following exercises, use the convolution theorem to compute the inverse Laplace transform of the given function.

22. $\dfrac{1}{(s-2)(s+4)}$

23. $\dfrac{1}{s^2-6s+5}$

24. $\dfrac{1}{(s^2+1)^2}$

25. $\dfrac{s}{(s^2+1)^2}$

26. $\dfrac{1}{(s+6)s^3}$

27. $\dfrac{2}{(s-3)(s^2+4)}$

28. $\dfrac{s}{(s-4)(s^2+1)}$

29. $\dfrac{1}{(s-a)(s-b)} \quad a \neq b$

30. $\dfrac{G(s)}{s+2}$

31. $G(s)\dfrac{s}{s^2+2}$

32. Let f be a function with Laplace transform $F(s)$. Show that

$$\mathcal{L}^{-1}\left\{\frac{F(s)}{s^2}\right\} = \int_0^t \int_0^{x_1} f(x_2)\, dx_2\, dx_1.$$

More generally, show that

$$\mathcal{L}^{-1}\left\{\frac{F(s)}{s^n}\right\} = \int_0^t \int_0^{x_1} \cdots \int_0^{x_{n-1}} f(x_n)\, dx_n \ldots dx_2\, dx_1.$$

33–38. Use the input integral principle or, more generally, the results of Problem 32 to compute the inverse Laplace transform of each function.

33. $\dfrac{1}{s^2(s^2+1)}$

34. $\dfrac{1}{s^2(s^2-4)}$

35. $\dfrac{1}{s^3(s+3)}$

36. $\dfrac{1}{s^2(s-2)^2}$

37. $\dfrac{1}{s(s^2+9)^2}$

38. $\dfrac{1}{s^3+s^2}$

2.9 Summary of Laplace Transforms and Convolutions

Laplace transforms and convolutions presented in Chap. 2 are summarized in Tables 2.6–2.11.

Table 2.6 Laplace transform rules

$f(t)$	$F(s)$	Page
Definition of the Laplace transform		
1. $f(t)$	$F(s) = \int_0^\infty e^{-st} f(t)\,dt$	111
Linearity		
2. $a_1 f_1(t) + a_2 f_2(t)$	$a_1 F_1(s) + a_2 F_2(s)$	114
Dilation principle		
3. $f(at)$	$\dfrac{1}{a} F\left(\dfrac{s}{a}\right)$	122
First Translation principle		
4. $e^{at} f(t)$	$F(s-a)$	120
Input derivative principle: first order		
5. $f'(t)$	$sF(s) - f(0)$	115
Input derivative principle: second order		
6. $f''(t)$	$s^2 F(s) - sf(0) - f'(0)$	115
Input derivative principle: nth order		
7. $f^{(n)}(t)$	$s^n F(s) - s^{n-1} f(0) - s^{n-2} f'(0) - \cdots - sf^{(n-2)}(0) - f^{(n-1)}(0)$	116
Transform derivative principle: first order		
8. $t f(t)$	$-F'(s)$	121
Transform derivative principle: second order		
9. $t^2 f(t)$	$F''(s)$	
Transform derivative principle: nth order		
10. $t^n f(t)$	$(-1)^n F^{(n)}(s)$	121
Convolution principle		
11. $(f * g)(t)$ $= \int_0^t f(\tau)g(t-\tau)\,d\tau$	$F(s)G(s)$	188
Input integral principle		
12. $\int_0^t f(v)dv$	$\dfrac{F(s)}{s}$	190

Table 2.7 Basic Laplace transforms

	$f(t)$	$F(s)$	Page
1.	1	$\dfrac{1}{s}$	116
2.	t	$\dfrac{1}{s^2}$	
3.	$t^n \quad (n = 0, 2, 3, \ldots)$	$\dfrac{n!}{s^{n+1}}$	116
4.	$t^\alpha \quad (\alpha > 0)$	$\dfrac{\Gamma(\alpha + 1)}{s^{\alpha+1}}$	118
5.	e^{at}	$\dfrac{1}{s - a}$	118
6.	$t e^{at}$	$\dfrac{1}{(s - a)^2}$	
7.	$t^n e^{at} \quad (n = 1, 2, 3, \ldots)$	$\dfrac{n!}{(s - a)^{n+1}}$	119
8.	$\sin bt$	$\dfrac{b}{s^2 + b^2}$	118
9.	$\cos bt$	$\dfrac{s}{s^2 + b^2}$	118
10.	$e^{at} \sin bt$	$\dfrac{b}{(s - a)^2 + b^2}$	120
11.	$e^{at} \cos bt$	$\dfrac{s - a}{(s - a)^2 + b^2}$	120

Table 2.8 Heaviside formulas

	$f(t)$	$F(s)$
1.	$\dfrac{r_1^k e^{r_1 t}}{q'(r_1)} + \cdots + \dfrac{r_n^k e^{r_n t}}{q'(r_n)},$ $q(s) = (s - r_1)\cdots(s - r_n)$	$\dfrac{s^k}{(s - r_1)\cdots(s - r_n)},$ $r_1, \ldots, r_n,$ distinct
2.	$\dfrac{e^{at}}{a - b} + \dfrac{e^{bt}}{b - a}$	$\dfrac{1}{(s - a)(s - b)}$
3.	$\dfrac{a e^{at}}{a - b} + \dfrac{b e^{bt}}{b - a}$	$\dfrac{s}{(s - a)(s - b)}$
4.	$\dfrac{e^{at}}{(a - b)(a - c)} + \dfrac{e^{bt}}{(b - a)(b - c)} + \dfrac{e^{ct}}{(c - a)(c - b)}$	$\dfrac{1}{(s - a)(s - b)(s - c)}$
5.	$\dfrac{a e^{at}}{(a - b)(a - c)} + \dfrac{b e^{bt}}{(b - a)(b - c)} + \dfrac{c e^{ct}}{(c - a)(c - b)}$	$\dfrac{s}{(s - a)(s - b)(s - c)}$

(continued)

Table 2.8   (continued)

6.	$\dfrac{a^2 e^{at}}{(a-b)(a-c)} + \dfrac{b^2 e^{bt}}{(b-a)(b-c)} + \dfrac{c^2 e^{ct}}{(c-a)(c-b)}$	$\dfrac{s^2}{(s-a)(s-b)(s-c)}$
7.	$\left(\sum_{l=0}^{k} \binom{k}{l} a^{k-l} \dfrac{t^{n-l-1}}{(n-l-1)!} \right) e^{at}$	$\dfrac{s^k}{(s-a)^n}$
8.	$t e^{at}$	$\dfrac{1}{(s-a)^2}$
9.	$(1 + at)e^{at}$	$\dfrac{s}{(s-a)^2}$
10.	$\dfrac{t^2}{2} e^{at}$	$\dfrac{1}{(s-a)^3}$
11.	$\left(t + \dfrac{at^2}{2} \right) e^{at}$	$\dfrac{s}{(s-a)^3}$
12.	$\left(1 + 2at + \dfrac{a^2 t^2}{2} \right) e^{at}$	$\dfrac{s^2}{(s-a)^3}$

In each case, a, b, and c are distinct. See Page 165.

Table 2.9   Laplace transforms involving irreducible quadratics

	$f(t)$	$F(s)$
1.	$\sin bt$	$\dfrac{b}{(s^2 + b^2)}$
2.	$\dfrac{1}{2b^2} (\sin bt - bt \cos bt)$	$\dfrac{b}{(s^2 + b^2)^2}$
3.	$\dfrac{1}{8b^4} \left((3 - (bt)^2) \sin bt - 3bt \cos bt \right)$	$\dfrac{b}{(s^2 + b^2)^3}$
4.	$\dfrac{1}{48b^6} \left((15 - 6(bt)^2) \sin bt - (15bt - (bt)^3) \cos bt \right)$	$\dfrac{b}{(s^2 + b^2)^4}$
5.	$\cos bt$	$\dfrac{s}{(s^2 + b^2)}$
6.	$\dfrac{1}{2b^2} bt \sin bt$	$\dfrac{s}{(s^2 + b^2)^2}$
7.	$\dfrac{1}{8b^4} (bt \sin bt - (bt)^2 \cos bt)$	$\dfrac{s}{(s^2 + b^2)^3}$
8.	$\dfrac{1}{48b^6} \left((3bt - (bt)^3) \sin bt - 3(bt)^2 \cos bt \right)$	$\dfrac{s}{(s^2 + b^2)^4}$

Table 2.10   Reduction of order formulas

$$\mathcal{L}^{-1} \left\{ \frac{1}{(s^2 + b^2)^{k+1}} \right\} = \frac{-t}{2kb^2} \mathcal{L}^{-1} \left\{ \frac{s}{(s^2 + b^2)^k} \right\} + \frac{2k-1}{2kb^2} \mathcal{L}^{-1} \left\{ \frac{1}{(s^2 + b^2)^k} \right\}$$

$$\mathcal{L}^{-1} \left\{ \frac{s}{(s^2 + b^2)^{k+1}} \right\} = \frac{t}{2k} \mathcal{L}^{-1} \left\{ \frac{1}{(s^2 + b^2)^k} \right\}$$

See Page 155.

Table 2.11 Basic convolutions

	$f(t)$	$g(t)$	$(f * g)(t)$	Page
1.	$f(t)$	$g(t)$	$f * g(t) = \int_0^t f(u)g(t-u)\,\mathrm{d}u$	187
2.	1	$g(t)$	$\int_0^t g(\tau)\,\mathrm{d}\tau$	190
3.	t^m	t^n	$\frac{m!n!}{(m+n+1)!}t^{m+n+1}$	193
4.	t	$\sin at$	$\dfrac{at - \sin at}{a^2}$	
5.	t^2	$\sin at$	$\dfrac{2}{a^3}\left(\cos at - (1 - \frac{a^2 t^2}{2})\right)$	
6.	t	$\cos at$	$\dfrac{1 - \cos at}{a^2}$	
7.	t^2	$\cos at$	$\dfrac{2}{a^3}(at - \sin at)$	
8.	t	e^{at}	$\dfrac{\mathrm{e}^{at} - (1 + at)}{a^2}$	
9.	t^2	e^{at}	$\dfrac{2}{a^3}\left(\mathrm{e}^{at} - \left(a + at + \frac{a^2 t^2}{2}\right)\right)$	
10.	e^{at}	e^{bt}	$\dfrac{1}{b-a}(\mathrm{e}^{bt} - \mathrm{e}^{at})\quad a \neq b$	192
11.	e^{at}	e^{at}	$t\mathrm{e}^{at}$	192
12.	e^{at}	$\sin bt$	$\dfrac{1}{a^2 + b^2}(b\mathrm{e}^{at} - b\cos bt - a\sin bt)$	195
13.	e^{at}	$\cos bt$	$\dfrac{1}{a^2 + b^2}(a\mathrm{e}^{at} - a\cos bt + b\sin bt)$	195
14.	$\sin at$	$\sin bt$	$\dfrac{1}{b^2 - a^2}(b\sin at - a\sin bt)\quad a \neq b$	195
15.	$\sin at$	$\sin at$	$\dfrac{1}{2a}(\sin at - at\cos at)$	195
16.	$\sin at$	$\cos bt$	$\dfrac{1}{b^2 - a^2}(a\cos at - a\cos bt)\quad a \neq b$	195
17.	$\sin at$	$\cos at$	$\dfrac{1}{2}t\sin at$	195
18.	$\cos at$	$\cos bt$	$\dfrac{1}{a^2 - b^2}(a\sin at - b\sin bt)\quad a \neq b$	195
19.	$\cos at$	$\cos at$	$\dfrac{1}{2a}(at\cos at + \sin at)$	195

Chapter 3
Second Order Constant Coefficient Linear Differential Equations

This chapter begins our study of second order linear differential equations, which are equations of the form

$$a(t)y'' + b(t)y' + c(t)y = f(t), \tag{1}$$

where $a(t)$, $b(t)$, $c(t)$, called the *coefficient functions*, and $f(t)$, known as the *forcing function*, are all defined on a common interval I. Equation (1) is frequently made into an *initial value problem* by imposing *initial conditions*: $y(t_0) = y_0$ and $y'(t_0) = y_1$, where $t_0 \in I$. Many problems in mathematics, engineering, and the sciences may be modeled by (1) so it is important to have techniques to solve these equations and to analyze the resulting solutions. This chapter will be devoted to the simplest version of (1), namely the case where the coefficient functions are constant. In Chap. 2, we introduced the Laplace transform method that codifies in a single procedure a solution method for (1), in the case where $f \in \mathcal{E}$ and initial values $y(0)$ and $y'(0)$ are given. However, it will be our approach going forward to first find the general solution to (1), without regard to initial conditions. When initial conditions are given, they determine a single function in the general solution set. The Laplace transform will still play a central role in most all that we do.

We wish to point out that our development of the Laplace transform thus far allows us to easily handle nth order constant coefficient linear differential equations for arbitrary n. Nevertheless, we will restrict our attention in this chapter to the second order case, which is the most important case for applications. Understanding this case well will provide an easy transition to the more general case to be studied in Chap. 4.

W.A. Adkins and M.G. Davidson, *Ordinary Differential Equations*,
Undergraduate Texts in Mathematics, DOI 10.1007/978-1-4614-3618-8_3,
© Springer Science+Business Media New York 2012

3.1 Notation, Definitions, and Some Basic Results

For the remainder of this chapter, we will assume that the coefficient functions in (1) are constant. Thus, our focus will be the equation

$$ay'' + by' + cy = f(t), \tag{1}$$

where a, b, and c are real numbers and the forcing function $f(t)$ is a continuous function on an interval I. We assume the **leading coefficient** a is nonzero, otherwise (1) is first order. Equation (1) is called a **second order constant coefficient linear differential equation**.

The left-hand side of (1) is made up of a combination of differentiations and multiplications by constants. To be specific and to introduce useful notation, let D denote the derivative operator: $D(y) = y'$. In a similar way, let D^2 denote the second derivative operator: $D^2(y) = D(Dy) = Dy' = y''$. If

$$L = aD^2 + bD + c, \tag{2}$$

where a, b, and c are the same constants given in (1), then

$$L(y) = ay'' + by' + cy.$$

We call L a **(second order) constant coefficient linear differential operator**. Another useful way to describe L is in terms of the polynomial $q(s) = as^2 + bs + c$: L is obtained from q by substituting D for s. We will write $L = q(D)$. For this reason, L is also called a **polynomial differential operator**. Equation (1) can now be rewritten

$$L(y) = f \quad \text{or} \quad q(D)y = f.$$

The polynomial q is referred to as the **characteristic polynomial** of L and will play a fundamental role in determining the solution set to (1).

The operator L can be thought of as taking a function y that has at least 2 continuous derivatives and producing a continuous function.

Example 1. Suppose $L = D^2 - 4D + 3$. Find

$$L(te^t), \qquad L(e^t), \qquad \text{and} \quad L(e^{3t}).$$

▶ **Solution.**

- $L(te^t) = D^2(te^t) - 4D(te^t) + 3(te^t)$
  $= D(e^t + te^t) - 4(e^t + te^t) + 3te^t$
  $= 2e^t + te^t - 4e^t - 4te^t + 3te^t$
  $= -2e^t$.

- $L(e^t) = D^2(e^t) - 4D(e^t) + 3(e^t)$
  $\quad\; = e^t - 4e^t + 3e^t$
  $\quad\; = 0.$
- $L(e^{3t}) = D^2(e^{3t}) - 4D(e^{3t}) + 3(e^{3t})$
  $\quad\;\; = 9e^{3t} - 12e^{3t} + 3e^{3t}$
  $\quad\;\; = 0.$ ◄

The adjective "linear" describes an important property that L satisfies. To explain this, let us start with the familiar derivative operator D. One learns early on in calculus the following two properties:

1. If y_1 and y_2 are continuously differentiable functions, then

$$D(y_1 + y_2) = D(y_1) + D(y_2).$$

2. If y is a continuously differentiable function and c is a scalar, then

$$D(cy) = cD(y).$$

Simply put, D preserves addition and scalar multiplication of functions. When an operation on functions satisfies these two properties, we call it **linear**. The second derivative operator D^2 is also linear:

$$D^2(y_1 + y_2) = D(Dy_1 + Dy_2) = D^2y_1 + D^2y_2$$
$$D^2(cy) = D(D(cy)) = DcDy = cD^2y.$$

It is easy to verify that sums and scalar products of linear operators are also linear operators, which means that any polynomial differential operator is linear. This is formalized in the following result.

Proposition 2. *The operator*

$$L = aD^2 + bD + c$$

is linear. Specifically,

1. If y_1 and y_2 have sufficiently many derivatives, then

$$L(y_1 + y_2) = L(y_1) + L(y_2).$$

2. If y has sufficiently many derivatives and c is a scalar, then

$$L(cy) = cL(y).$$

Proof. Suppose y, y_1, and y_2 are 2-times differentiable functions and c is a scalar. Then

$$
\begin{aligned}
L(y_1 + y_2) &= a\boldsymbol{D}^2(y_1 + y_2) + b\boldsymbol{D}(y_1 + y_2) + c(y_1 + y_2) \\
&= a(\boldsymbol{D}^2 y_1 + \boldsymbol{D}^2 y_2) + b(\boldsymbol{D} y_1 + \boldsymbol{D} y_2) + c(y_1 + y_2) \\
&= a\boldsymbol{D}^2 y_1 + b\boldsymbol{D} y_1 + c y_1 + a\boldsymbol{D}^2 y_2 + b\boldsymbol{D} y_2 + c y_2 \\
&= \boldsymbol{L} y_1 + \boldsymbol{L} y_2.
\end{aligned}
$$

Thus $\boldsymbol{L}$ preserves addition. In a similar way, $\boldsymbol{L}$ preserves scalar multiplication and hence is linear. $\qquad\square$

To illustrate the power of linearity, consider the following example.

Example 3. Let $\boldsymbol{L} = \boldsymbol{D}^2 - 4\boldsymbol{D} + 3$. Use linearity to determine

$$
\boldsymbol{L}(3e^t + 4te^t + 5e^{3t}).
$$

▶ **Solution.** Recall from Example 1 that

- $\boldsymbol{L}(e^t) = 0$.
- $\boldsymbol{L}(te^t) = -2e^t$.
- $\boldsymbol{L}(e^{3t}) = 0$.

Using linearity, we obtain

$$
\begin{aligned}
\boldsymbol{L}(3e^t + 4te^t + 5e^{3t}) &= 3\boldsymbol{L}(e^t) + 4\boldsymbol{L}(te^t) + 5\boldsymbol{L}(e^{3t}) \\
&= 3 \cdot 0 + 4 \cdot (-2e^t) + 5 \cdot 0 \\
&= -8e^t. \qquad\blacktriangleleft
\end{aligned}
$$

Solutions

An important consequence of linearity is that the set of all solutions to (1) has a particularly simple structure. We begin the description of that structure with the special case where the forcing function is identically zero. In this case, (1) becomes

$$
\boldsymbol{L}(y) = 0 \tag{3}
$$

and we refer to such an equation as **homogeneous**.

Proposition 4. *Suppose $\boldsymbol{L}$ is a linear differential operator. Then the solution set to $\boldsymbol{L} y = 0$ is a linear space. Specifically, suppose y, y_1, and y_2 are solutions to $\boldsymbol{L} y = 0$ and k is a scalar. Then $y_1 + y_2$ and ky are solutions to $\boldsymbol{L} y = 0$.*

Proof. By Proposition 2, we have

$$L(y_1 + y_2) = L(y_1) + L(y_2) = 0 + 0 = 0$$
$$L(ky) = kL(y) = k \cdot 0 = 0.$$

These equations show that the solution set is closed under addition and scalar multiplication and hence is a linear space.   □

Example 5. Use Example 1 to find as many solutions as possible to the homogeneous equation $Ly = 0$, where

$$L = D^2 - 4D + 3.$$

▶ **Solution.** In Example 1, we found that $L(e^t) = 0$ and $L(e^{3t}) = 0$. Now using Proposition 4, we have

$$c_1 e^t + c_2 e^{3t}$$

is a solution to $Ly = 0$, for all scalars c_1 and c_2. In other words,

$$L(c_1 e^t + c_2 e^{3t}) = 0,$$

for all scalars c_1 and c_2. We will later show that *all* of the solutions to $Ly = 0$ are of this form.   ◀

It is hard to overemphasize the importance of Proposition 4, since it indicates that once a few specific solutions to $L(y) = 0$ are known, then all linear combinations are likewise solutions. This gives a strategy for describing all solutions to $L(y) = 0$ provided we can find a few distinguished solutions. The linearity proposition will also allow for a useful way to describe all solutions to the general differential equation $L(y) = f(t)$ by reducing it to the homogeneous differential equation $L(y) = 0$, which we refer to as the *associated homogeneous differential equation*. The following theorem describes this relationship.

Theorem 6. *Suppose L is a linear differential operator and f is a continuous function. If y_p is a fixed particular solution to $L(y) = f$ and y_h is any solution to the associated homogeneous differential equation $L(y) = 0$, then*

$$y_p + y_h$$

is a solution to $L(y) = f$. Furthermore, any solution y to $L(y) = f$ has the form

$$y = y_p + y_h.$$

Proof. Suppose y_p satisfies $L(y_p) = f$ and y_h satisfies $L(y_h) = 0$. Then by linearity,

$$L(y_p + y_h) = L(y_p) + L(y_h) = f + 0 = f.$$

Thus $y_p + y_h$ is a solution to $L(y) = f$. On the other hand, suppose $y(t)$ is any solution to $L(y) = f$. Let $y_h = y - y_p$. Then, again by linearity,

$$L(y_h) = L(y - y_p) = L(y) - L(y_p) = f - f = 0.$$

Thus, $y_h(t)$ is a solution to $L(y) = 0$ and $y = y_p + y_h$.   □

This theorem actually provides an effective strategy for describing the solution set to a second order linear constant coefficient differential equation, which we formalize in the following algorithm. By an abuse of language, we will sometimes refer to solutions of the associated homogeneous equation $L(y) = 0$ as *homogeneous solutions*.

Algorithm 7. The general solution to a linear differential equation

$$L(y) = f(t)$$

can be found as follows:

Solution Method for Second Order Linear Equations

1. Find all the solutions y_h to the associated homogeneous differential equation $L y = 0$.
2. Find one particular solution y_p to $L(y) = f$.
3. Add the particular solution to the homogeneous solutions:

$$y_p + y_h.$$

As y_h varies over all homogeneous solutions, we obtain all solutions to $L(y) = f$

Example 8. Use Algorithm 7 to solve

$$y'' - 4y' + 3y = -2e^t.$$

▶ **Solution.** The left-hand side can be written $L(y)$, where L is the linear differential operator

$$L = D^2 - 4D + 3.$$

From Example 5, we found that

$$y_h(t) = c_1 e^t + c_2 e^{3t},$$

where c_1, $c_2 \in \mathbb{R}$, are all the solutions to the associated homogeneous equation $L(y) = 0$. By Example 1, a particular solution to $L(y) = -2e^t$ is $y_p(t) = te^t$. By Theorem 6, we have

$$y_p(t) + y_h(t) = te^t + c_1 e^t + c_2 e^{3t}$$

is a solution to $L(y) = 2e^t$, for all scalars c_1 and c_2.   ◀

The strategy outlined in Algorithm 7 is the strategy we will follow. Section 3.3 will be devoted to determining solutions to the associated homogeneous differential equation. Sections 3.4 and 3.5 will show effective methods for finding a particular solution when the forcing function $f(t) \in \mathcal{E}$ is an exponential polynomial. A more general method is found in Sect. 5.6.

Initial Value Problems

Suppose L is a constant coefficient linear differential operator, $f(t)$ is a function defined on an interval I, and $t_0 \in I$. To the equation

$$L(y) = f$$

we can associate *initial conditions* of the form

$$y(t_0) = y_0, \quad \text{and} \quad y'(t_0) = y_1.$$

The differential equation $L(y) = f$, together with the initial conditions, is called an *initial value problem*, just as in the case of first order differential equations. After finding the general solution to $L(y) = f$, the initial conditions are used to determine specific values for the arbitrary constants that parameterize the solution set. Here is an example.

Example 9. Use Example 8 to find the solution to the following initial value problem

$$L(y) = -2e^t, \qquad y(0) = 1, \ y'(0) = -2,$$

where $L = D^2 - 4D + 3$.

▶ **Solution.** In Example 8, we verified that

$$y(t) = te^t + c_1 e^t + c_2 e^{3t},$$

is a solution to $L(y) = -2e^t$ for every $c_1, c_2 \in \mathbb{R}$. Observe that $y'(t) = e^t + te^t + c_1 e^t + 3c_2 e^{3t}$. Setting $t = 0$ in both $y(t)$ and $y'(t)$ gives

$$1 = y(0) = c_1 + c_2$$
$$-2 = y'(0) = 1 + c_1 + 3c_2.$$

Solving these equations gives $c_1 = 3$, and $c_2 = -2$. Thus, $y(t) = te^t + 3e^t - 2e^{3t}$ is a solution to the given initial value problem. The existence and uniqueness theorem given below implies it is the only solution.   ◄

Initial Values Not Based at the Origin

You may have noticed that the initial conditions in the examples given in Chap. 2 and in Example 9 above are given at $t_0 = 0$. If the initial conditions are given elsewhere, then a simple translation can be used to shift the initial conditions back to the origin as follows. Suppose f is a function defined on an interval $[a, b]$, $t_0 \in [a, b]$, and the initial conditions are given by $y(t_0) = y_0$ and $y'(t_0) = y_1$. Let $g(t) = f(t + t_0)$ and $w(t) = y(t + t_0)$. Then $w'(t) = y'(t + t_0)$ and $w''(t) = y''(t + t_0)$. The initial value problem given by (1) in y becomes

$$aw'' + bw' + cw = g,$$

with initial conditions $w(0) = y(t_0) = y_0$ and $w'(0) = y'(t_0) = y_1$. We now solve for w; it has initial conditions at 0. The function $y(t) = w(t - t_0)$ will then be the solution to the original initial value problem on the interval $[a, b)$. Thus, it is not restrictive to give examples and base our results for initial values at $t_0 = 0$.

The Existence and Uniqueness Theorem

The existence and uniqueness theorem, as expressed by Corollary 8 of Sect. 1.4, for first order linear differential equations has an extension for second order linear differential equations. Its proof will be given in Chap. 9 in a much broader setting.

Theorem 10 (The Existence and Uniqueness Theorem). *Suppose $f(t)$ is a continuous real-valued function on an interval I. Let $t_0 \in I$. Then there is a unique real-valued function y defined on I satisfying*

$$ay'' + by' + cy = f(t), \tag{4}$$

with initial conditions $y(t_0) = y_0$ and $y'(t_0) = y_1$. If $f(t)$ is of exponential type, so are the solution $y(t)$ and its derivatives $y'(t)$ and $y''(t)$. Furthermore, if $f(t)$ is in $\mathcal{E}$, then $y(t)$ is also in $\mathcal{E}$.

You will notice that the kind of solution we obtain depends on the kind of forcing function. In particular, when the forcing function is an exponential polynomial, then so is the solution. This theorem thus provides the basis for applying the Laplace transform method. Specifically, when the Laplace transform is applied to both sides of (4), we *presumed* in previous examples that the solution y and its first and second derivative have Laplace transforms. The existence and uniqueness theorem thus justifies the Laplace transform method when the forcing function is of exponential type or, more specifically, an exponential polynomial.

Exercises

1–10. Determine which of the following are second order constant coefficient linear differential equations. In those cases where it is, write the equation in the form $L(y) = f(t)$, give the characteristic polynomial, and state whether the equation is homogeneous.

1. $y'' - yy' = 6$
2. $y'' - 3y' = e^t$
3. $y''' + y' + 4y = 0$
4. $y'' + \sin(y) = 0$
5. $ty' + y = \ln t$
6. $y'' + 2y' + 3y = e^{-t}$
7. $y'' - 7y' + 10y = 0$
8. $y' + 8y = t$
9. $y'' + 2 = \cos t$
10. $2y'' - 12y' + 18y = 0$

11–14. For the linear operator L, determine $L(y)$.

11. $L = D^2 + 3D + 2$.

    (a) $y = e^t$
    (b) $y = e^{-t}$
    (c) $y = \sin t$

12. $L = D^2 - 2D + 1$.

    (a) $y = 4e^t$
    (b) $y = \cos t$
    (c) $y = -e^{2t}$

13. $L = D^2 + 1$.

    (a) $y = -4 \sin t$
    (b) $y = 3 \cos t$
    (c) $y = 1$

14. $L = D^2 - 4D + 8$.

    (a) $y = e^{2t}$
    (b) $y = e^{2t} \sin 2t$
    (c) $y = e^{2t} \cos 2t$

15. Suppose L is a polynomial differential operator of order 2 and

    • $L(\cos 2t) = 10 \sin 2t$
    • $L(e^t) = 0$
    • $L(e^{4t}) = 0$.

Use this information to find other solutions to $L(y) = 10 \sin 2t$.

16. Suppose L is a polynomial differential operator of order 2 and

   - $L(te^{3t}) = 5e^{3t}$
   - $L(e^{3t}) = 0$
   - $L(e^{-2t}) = 0$.

   Use this information to find other solutions to $L(y) = 5e^{3t}$.

17. Let L be as in Exercise 15. Use the results there to solve the initial value problem
$$L(y) = 10 \sin 2t,$$
   where $y(0) = 1$ and $y'(0) = -3$.

18. Let L be as in Exercise 16. Use the results there to solve the initial value problem
$$L(y) = 5e^{3t},$$
   where $y(0) = -1$ and $y'(0) = 8$.

19. If $L = aD^2 + bD + c$ where a, b, c are real numbers, then show that $L(e^{rt}) = (ar^2 + br + c)e^{rt}$. That is, the effect of applying the operator L to the exponential function e^{rt} is to multiply e^{rt} by the number $ar^2 + br + c$.

20–21. Use the existence and uniqueness theorem to establish the following.

20. Suppose $\phi(t)$ is a solution to

$$y'' + ay' + by = 0,$$

   where a and b are real constants. Show that if the graph of ϕ is tangent to the t-axis, then $\phi = 0$.

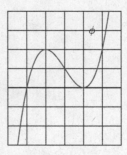

21. More generally, suppose ϕ_1 and ϕ_2 are solutions to

$$y'' + ay' + by = f,$$

where a and b are real constants and f is a continuous function on an interval I. Show that if the graphs of ϕ_1 and ϕ_2 are tangent at some point, then $\phi_1 = \phi_2$.

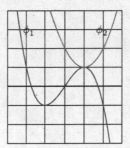

3.2 Linear Independence

In Sects. 2.6 and 2.7, we introduced $\mathcal{B}_q$, for any polynomial q, and referred to it as the standard basis of the linear space $\mathcal{E}_q$. For a linear space, $\mathcal{F}$, of functions defined on an interval I, a **basis** of $\mathcal{F}$ is a subset $\mathcal{B}$ that satisfies two properties:

1. Span $\mathcal{B} = \mathcal{F}$.
2. $\mathcal{B}$ is linearly independent.

The notion of a spanning set was developed in Sect. 2.6 where we showed that $\mathcal{B}_q$ spanned the linear space $\mathcal{E}_q$. It is the purpose of this section to explain the notion of linear independence and consider some of its consequences. We will then show that $\mathcal{B}_q$ is linearly independent, thus justifying that $\mathcal{B}_q$ is a basis of $\mathcal{E}_q$ in the precise sense given above.

A set of functions $\{f_1, \ldots, f_n\}$, defined on some interval I, is said to be **linearly independent** if the equation

$$a_1 f_1 + \cdots + a_n f_n = 0 \tag{1}$$

implies that all the coefficients $a_1, \ldots, a_n$ are zero. Otherwise, we say that $\{f_1, \ldots, f_n\}$ is **linearly dependent**.[1]

One must be careful about this definition. We do not try to solve equation (1) for the variable t. Rather, we are given that this equation is valid for all $t \in I$. With this information, the focus is on what this says about the coefficients $a_1, \ldots, a_n$: are they all necessarily zero or not.

We illustrate the definition with a very simple example.

Example 1. Show that the set $\{e^t, e^{-t}\}$ defined on $\mathbb{R}$ is linearly independent.

▶ **Solution.** Consider the equation

$$a_1 e^t + a_2 e^{-t} = 0, \tag{2}$$

for all $t \in \mathbb{R}$. In order to conclude linear independence, we need to show that a_1 and a_2 are zero. There are many ways this can be done. Below we show three approaches. Of course, only one is necessary.

Method 1: Evaluation at Specified Points

Let us evaluate (2) at two points: $t = 0$ and $t = 1$:

---

[1] A grammatical note: We say $f_1, \ldots, f_n$ *are* linearly independent (dependent) if the *set* $\{f_1, \ldots, f_n\}$ is linearly independent (dependent).

$$t = 0, \qquad a_1 + a_2 \quad = 0,$$
$$t = 1, \qquad a_1(e) + a_2(1/e) = 0.$$

Multiply the first equation by e and subtract the result from the second equation. We get $a_2((1/e) - e) = 0$. So $a_2 = 0$ and this in turn gives $a_1 = 0$. Thus, $\{e^t, e^{-t}\}$ is linearly independent.

Method 2: Differentiation

Take the derivative of (2) to get $a_1 e^t - a_2 e^{-t} = 0$. Evaluating equation (2) and the derivative at $t = 0$ gives

$$a_1 + a_2 = 0,$$
$$a_2 - a_2 = 0.$$

The only solution to these equations is $a_1 = 0$ and $a_2 = 0$. Hence, $\{e^t, e^{-t}\}$ is linearly independent. (This method will be discussed more generally when we introduce the Wronskian below.)

Method 3: The Laplace Transform

Here we take the Laplace transform of (2) to get

$$\frac{a_1}{s - 1} + \frac{a_2}{s + 1} = 0,$$

which is an equation valid for all $s > 1$ (since the Laplace transform of e^t is valid for $s > 1$). However, as an equation of rational functions, Corollary 7 of Appendix A.2 implies equality for all $s \neq 1, -1$. Now consider the limit as s approaches 1. If a_1 is not zero, then $\frac{a_1}{s-1}$ has an infinite limit while the second term $\frac{a_2}{s+1}$ has a finite limit. But this cannot be as the sum is 0. It must be that $a_1 = 0$ and therefore $\frac{a_2}{s+1} = 0$. This equation in turn implies $a_2 = 0$. Now it follows that $\{e^t, e^{-t}\}$ is linearly independent. (This method is a little more complicated than Methods 1 and 2 but will be the method we use to prove $\mathcal{B}_q$ is linearly independent.) ◀

Remark 2. Let us point out that when we are asked to determine whether a set of *two* functions $\{f_1, f_2\}$ is linearly dependent, it is enough to see that they are multiples of each other. For if $\{f_1, f_2\}$ is linearly dependent, then there are constants c_1 and c_2, not both zero, such that $c_1 f_1 + c_2 f_2 = 0$. By renumbering the functions if necessary, we may assume $c_1 \neq 0$. Then $f_1 = -\frac{c_2}{c_1} f_2$. Hence, f_1 and f_2 are multiples of each other. On the other hand, if one is a multiple of the other, that is, if $f_1 = mf_2$, then $f_1 - mf_2 = 0$ and this implies linear dependence. Thus, it is *immediate* that e^t and e^{-t} are linearly independent since they are not multiples of each other.

Example 3. Show that the set $\{e^t, \cos t, \sin t\}$ is linearly independent.

▶ **Solution.** We will use method 3 to show linear independence. Suppose $a_1 e^t + a_2 \cos t + a_3 \sin t = 0$. Take the Laplace transform to get

$$\frac{a_1}{s-1} + \frac{a_2 s + a_3}{s^2 + 1} = 0,$$

valid for all $s \neq 1, i, -i$, by Corollary 7 of Appendix A.2. Now consider the limit as s approaches 1. If $a_1 \neq 0$, then the first term becomes infinite while the second term is finite. Since the sum is 0, this is impossible so $a_1 = 0$. Thus $\frac{a_2 s + a_3}{s^2 + 1} = 0$. Now consider the limit as s approaches i. If either a_2 or a_3 is nonzero, then the quotient becomes infinite which cannot be. Thus we have $a_2 = a_3 = 0$. It now follows that $\{e^t, \cos t, \sin t\}$ is linearly independent. ◀

Let us consider an example of a set that is not linearly independent.

Example 4. Show that the set $\{e^t, e^{-t}, e^t + e^{-t}\}$ is linearly dependent.

▶ **Solution.** To show that a set is linearly dependent, we need only show that we can find a linear combination that adds up to 0 with coefficients not all zero. One such is

$$(1)e^t + (1)e^{-t} + (-1)(e^t + e^{-t}) = 0.$$

The coefficients, highlighted by the parentheses, are $1, 1$, and -1 and are not all zero. Thus, $\{e^t, e^{-t}, e^t + e^{-t}\}$ is linearly dependent. The dependency is clearly seen in that the third function is the sum of the first two. ◀

This example illustrates the following more general theorem.

Theorem 5. *A set of functions* $\{f_1, \ldots, f_n\}$ *is linearly dependent if and only if one of the functions is a linear combination of the others.*

Proof. Let us assume that one of the functions, f_1 say, is a linear combination of the other functions. Then we can write $f_1 = a_2 f_2 + \cdots + a_n f_n$, which is equivalent to

$$f_1 - a_2 f_2 - a_3 f_3 - \cdots - a_n f_n = 0.$$

Since not all of the coefficients are zero (the coefficient of f_1 is 1), it follows that $\{f_1, \ldots, f_n\}$ is linearly dependent. On the other hand, suppose $\{f_1, \ldots, f_n\}$ is linearly dependent. Then there are scalars, not all zero, such that $a_1 f_1 + \cdots + a_n f_n = 0$. By reordering if necessary, we may assume that $a_1 \neq 0$. Now we can solve for f_1 to get

$$f_1 = -\frac{a_2}{a_1} f_2 - \cdots - \frac{-a_n}{a_1} f_n,$$

Thus, one of the functions is a linear combination of the others. □

Solving Equations Involving Linear Combinations

Linear independence is precisely the requirement we need to be able to solve for coefficients involving a linear combination of functions. Consider the following example:

Example 6. Suppose we are given the following equation:

$$(c_1 + 4)e^t - (c_2 + 1)e^{-t} = 3e^t - 4c_1 e^{-t},$$

valid for all $t \in \mathbb{R}$. Determine c_1 and c_2.

▶ **Solution.** Subtracting the right side from both sides gives

$$(c_1 + 1)e^t + (4c_1 - c_2 - 1)e^{-t} = 0. \tag{3}$$

In Example 1, we showed that $\{e^t, e^{-t}\}$ is linear independent. Thus, we can now say that the coefficients in (3) are zero, giving us

$$
\begin{aligned}
c_1 + 1 &= 0 \\
4c_1 - c_2 - 1 &= 0
\end{aligned}
$$

Solving these equations simultaneously gives $c_1 = -1$ and $c_2 = -5$.   ◀

Notice that the equations obtained by setting the coefficients equal to zero in (3) are the same as equating corresponding coefficients in the original equations: $c_1 + 4 = 3$ and $-(c_2 + 1) = -4c_1$. More generally, we have the following theorem:

Theorem 7. *Suppose $\{f_1, \ldots, f_n\}$ is a linearly independent set. If*

$$a_1 f_1 + \cdots + a_n f_n = b_1 f_1 + \cdots + b_n f_n$$

then $a_1 = b_1, a_2 = b_2, \ldots, a_n = b_n$.

Proof. The given equation implies

$$(a_1 - b_1)f_1 + \cdots + (a_n - b_n)f_n = 0.$$

Linear independence implies that the coefficients are zero. Thus, $a_1 = b_1, a_2 = b_2$, ..., $a_n = b_n$.   □

The Linear Independence of $\mathcal{B}_q$

Let $q(s) = s^2 - 1 = (s-1)(s+1)$. Then $\mathcal{B}_q = \{e^t, e^{-t}\}$. In Example 2, we showed that $\mathcal{B}_q$ is linearly independent and used that fact in Example 6. In like manner, if $q(s) = (s-1)(s^2+1)$, then $\mathcal{B}_q = \{e^t, \cos t, \sin t\}$, and we showed in Example 3 that it too was linearly independent. The following theorem establishes the linear independence of $\mathcal{B}_q$ for any polynomial q. The method of proof is based on Method 3 in Example 2 and can be found in Appendix A.3.

Theorem 8. *Let q be a nonconstant polynomial. View $\mathcal{B}_q$ as a set of functions on $I = [0, \infty)$. Then $\mathcal{B}_q$ is linearly independent.*

One useful fact about linearly independent sets is that any subset is also linearly independent. Specifically,

Theorem 9. *Suppose S is a finite set of functions on an interval I which is linearly independent. Then any subset of S is also linearly independent.*

Proof. Suppose $S = \{f_1, \ldots, f_n\}$ is linearly independent and suppose $S_\circ$ is a subset of S. We may assume by reordering if necessary that $S_\circ = \{f_1, \ldots, f_k\}$, for some $k \leq n$. Suppose $c_1 f_1 + \cdots c_k f_k = 0$ for some constants $c_1, \ldots, c_k$. Let $c_{k+1} = \cdots = c_n = 0$. Then $c_1 f_1 + \cdots c_n f_n = 0$. Since S is linearly independent, $c_1, \ldots, c_n$ are all zero. It follows that $S_\circ$ is linearly independent. $\square$

Since $\mathcal{B}_q$ is linearly independent, it follows from Theorem 9 that any subset is also linearly independent.

Example 10. Show that the following sets are linearly independent:

1. $\{e^t, e^{-t}, te^{2t}\}$
2. $\{e^t, e^{-t}, te^{3t}\}$
3. $\{\cos t, \sin 2t\}$

▶ **Solution.**

1. Let $q(s) = (s-1)(s+1)(s-2)^2$. Then $\mathcal{B}_q = \{e^t, e^{-t}, e^{2t}, te^{2t}\}$ and $\{e^t, e^{-t}, te^{2t}\} \subset \mathcal{B}_q$. Theorem 9 implies linear independence.

2. Let $q(s) = (s-1)(s+1)(s-3)^2$. Then $\mathcal{B}_q = \{e^t, e^{-t}, e^{3t}, te^{3t}\}$ and $\{e^t, e^{-t}, te^{3t}\} \subset \mathcal{B}_q$. Linear independence follows from Theorem 9.

3. Let $q(s) = (s^2+1)(s^2+4)$. Then $\mathcal{B}_q = \{\cos t, \sin t, \cos 2t, \sin 2t\}$ and $\{\cos t, \sin 2t\} \subset \mathcal{B}_q$. Linear independence follows from Theorem 9.   ◀

Restrictions to Subintervals

It is important to keep in mind that the common interval of definition of a set of functions plays an implicit role in the definition of linear independence. Consider the following example.

Example 11. Let $f_1(t) = |t|$ and $f_2(t) = t$. Show that $\{f_1, f_2\}$ is linearly independent if the interval of definition is $I = \mathbb{R}$ and linearly dependent if $I = [0, \infty)$.

▶ **Solution.** Suppose the interval of definition is $I = \mathbb{R}$. Suppose

$$c_1 f_1(t) + c_2 f_2(t) = c_1 |t| + c_2 t = 0.$$

Evaluation at $t = 1$ and $t = -1$ gives

$$c_1 + c_2 = 0$$
$$c_1 - c_2 = 0$$

These equations reduce to $c_1 = 0$ and $c_2 = 0$. Thus, $\{f_1, f_2\}$ is linearly independent on $I = \mathbb{R}$. On the other hand, suppose $I = [0, \infty)$. Then $f_1(t) = t = f_2(t)$ (on I). Lemma 5 implies $\{f_1, f_2\}$ is linearly dependent.   ◄

Admittedly, the previous example is rather special. However, it does teach us that we cannot presume that restricting to a smaller interval will preserve linear independence. On the other hand, if a set of functions is defined on an interval I and linearly independent when restricted to a subset of I, then the set of functions is linearly independent on I. For example, Theorem 8 says that the set $\mathcal{B}_q$ is linearly independent on $[0, \infty)$ yet defined on all of $\mathbb{R}$. Thus, $\mathcal{B}_q$ is linearly independent as functions on $\mathbb{R}$.

The Wronskian

Suppose $f_1, \ldots, f_n$ are functions on an interval I with $n - 1$ derivatives. The **Wronskian** of $f_1, \ldots, f_n$ is given by

$$w(f_1, \ldots, f_n)(t) = \det \begin{bmatrix} f_1(t) & f_2(t) & \ldots & f_n(t) \\ f_1'(t) & f_2'(t) & \ldots & f_n'(t) \\ \vdots & & \ddots & \vdots \\ f_1^{(n-1)}(t) & f_2^{(n-1)}(t) & \ldots & f_n^{(n-1)(t)} \end{bmatrix}.$$

Clearly, the Wronskian is a function on I. We sometimes refer to the $n \times n$ matrix given above as the **Wronskian matrix** and denote it by $W(f_1, \ldots, f_n)(t)$.

Example 12. Find the Wronskian of the following sets of functions:

1. $\{t, 1/t\}$
2. $\{e^{2t}, e^{-t}\}$
3. $\{e^{t}, e^{-t}, e^{t} + e^{-t}\}$

▶ **Solution.**

1. $w(t, 1/t) = \det \begin{bmatrix} t & 1/t \\ 1 & -1/t^2 \end{bmatrix} = \frac{-1}{t} - \frac{1}{t} = \frac{-2}{t}.$

2. $w(e^{2t}, e^{-t}) = \det \begin{bmatrix} e^{2t} & e^{-t} \\ 2e^{2t} & -e^{-t} \end{bmatrix} = -e^{t} - 2e^{t} = -3e^{-t}.$

3.

$$w(e^{t}, e^{-t}, e^{t} + e^{-t}) = \det \begin{bmatrix} e^{t} & e^{-t} & e^{t} + e^{-t} \\ e^{t} & -e^{-t} & e^{t} - e^{-t} \\ e^{t} & e^{-t} & e^{t} + e^{-t} \end{bmatrix}$$

$$= e^{t} \det \begin{bmatrix} -e^{-t} & e^{t} - e^{-t} \\ e^{-t} & e^{t} + e^{-t} \end{bmatrix}$$

$$-e^{t} \det \begin{bmatrix} e^{-t} & e^{t} + e^{-t} \\ e^{-t} & e^{t} + e^{-t} \end{bmatrix}$$

$$+e^{t} \det \begin{bmatrix} e^{-t} & e^{t} + e^{-t} \\ -e^{-t} & e^{t} - e^{-t} \end{bmatrix}$$

$$= e^{t}(-2) - e^{t}(0) + e^{t}(2) = 0. \qquad \blacktriangleleft$$

Theorem 13. *Suppose $f_1, f_2, \ldots, f_n$ are functions on an interval I with $n - 1$ derivatives. Suppose the Wronskian $w(f_1, f_2, \ldots, f_n)$ is nonzero for some $t_0 \in I$. Then $\{f_1, f_2, \ldots, f_n\}$ is linearly independent.*

Proof. [2]Suppose $c_1, c_2, \ldots, c_n$ are scalars such that

$$c_1 f_1 + c_2 f_2 + \cdots + c_n f_n = 0 \qquad (4)$$

on I. We must show that the coefficients $c_1, \ldots, c_n$ are zero. Consider the $n - 1$ derivatives of (4):

---

[2]We assume in this proof some familiarity with matrices and determinants. See Chap. 8 for details.

$$
\begin{aligned}
c_1 f_1 \quad + \quad \cdots \quad + \quad c_2 f_2 \quad &= \quad 0 \\
c_1 f_1' \quad + \quad \cdots \quad + \quad c_2 f_2' \quad &= \quad 0 \\
\vdots \qquad\qquad \vdots \qquad\qquad \vdots \qquad\qquad &\vdots \\
c_1 f_1^{(n-1)} \quad + \quad \cdots \quad + \quad c_n f_n^{(n-1)} \quad &= \quad 0
\end{aligned}
$$

We can write this system in matrix form as

$$
W(f_1, \ldots, f_n)\boldsymbol{c} = \boldsymbol{0},
$$

where $W(f_1, \ldots, f_n)$ is the $n \times n$ Wronskian matrix,

$$
\boldsymbol{c} = \begin{bmatrix} c_1 \\ \vdots \\ c_n \end{bmatrix}, \quad \text{and} \quad \boldsymbol{0} = \begin{bmatrix} 0 \\ \vdots \\ 0 \end{bmatrix}.
$$

Since $w(f_1, \ldots, f_n)(t_0) \neq 0$, it follows that the Wronskian matrix at t_0, $W(f_1, \ldots, f_n)(t_0)$, is invertible. Thus,

$$
\boldsymbol{c} = W^{-1}(f_1, \ldots, f_n)(t_0)\boldsymbol{0} = \boldsymbol{0}.
$$

This means $c_1, \ldots, c_n$ are zero and $\{f_1, \ldots, f_n\}$ is linearly independent.   □

In Example 12, we saw that $w(t, 1/t) = -2/t^2$ and $w(e^{2t}, e^{-t}) = -3e^{-t}$, both nonzero functions. Hence, $\{t, 1/t\}$ and $\{e^{2t}, e^{-t}\}$ are linearly independent. A frequent mistake in the application of Theorem 13 is to assume the converse is true. Specifically, if the Wronskian, $w(f_1, \ldots, f_n)$, is zero, we may not conclude that $f_1, \ldots, f_n$ are linearly dependent. In Example 12, we saw that $w(e^t, e^{-t}, e^t + e^{-t}) = 0$. We cannot conclude from Theorem 13 that $\{e^t, e^{-t}, e^t + e^{-t}\}$ is linearly dependent. Nevertheless, linear dependence was shown in Example 4. However, Exercises 26 and 27 give a simple example of two linearly independent functions with zero Wronskian. If we add an additional assumption to the hypothesis of Theorem 13, then the converse will hold. This is the content of Theorem 8 of Sect. 3.3 given in the next section for the case $n = 2$ and, generally, in Theorem 6 of Sect. 4.2.

We conclude this section with a summary of techniques that can be used to show linear independence or dependence. Suppose $\mathcal{S} = \{f_1, \ldots, f_n\}$ is a set of functions defined on an interval I.

To Show that S is Linearly Independent

1. Evaluate a linear combination of S at selected points in I to get a linear system. If the solution is trivial, that is, all coefficients are zero, then S is linearly independent.
2. Compute the Wronskian, $w(f_1, \ldots, f_n)$. If it is nonzero, then S is linearly independent.
3. If $S \subset \mathcal{B}_q$ for some q, then S is linearly independent.

To Show that S is Linearly Dependent

1. Show there is a linear relation among the functions $f_1, \ldots, f_n$. That is, show that one of the functions is a linear combination of the others.
2. *Warning:* If the Wronskian, $w(f_1, \ldots, f_n)$, is zero, you *cannot* conclude that S is linearly dependent.

Exercises

1–13. Determine whether the given set of functions is linearly independent or linearly dependent. Unless otherwise indicated, assume the interval of definition is $I = \mathbb{R}$.

1. $\{t, t^2\}$
2. $\{e^t, e^{2t}\}$
3. $\{e^t, e^{t+2}\}$
4. $\{\ln 2t, \ln 5t\}$, on $I = (0, \infty)$
5. $\{\ln t^2, \ln t^5\}$
6. $\{\cos(t + \pi), \cos(t - \pi)\}$
7. $\{t, 1/t\}$, on $I = [0, \infty)$
8. $\{1, t, t^2\}$
9. $\{1, 1/t, 1/t^2\}$ on $I = (0, \infty)$
10. $\{\cos^2 t, \sin^2 t, 1\}$
11. $\{e^t, 1, e^{-t}\}$
12. $\{e^t, e^t \sin 2t\}$
13. $\{t^2 e^t, t^3 e^t, t^4 e^t\}$

14–21. Compute the Wronskian of the following set of functions.

14. $\{e^{3t}, e^{5t}\}$
15. $\{t, t \ln t\}$, $I = (0, \infty)$
16. $\{t \cos(3 \ln t), t \sin(3 \ln t)\}$, $I = (0, \infty)$
17. $\{t^{10}, t^{20}\}$
18. $\{e^{2t}, e^{3t}, e^{4t}\}$
19. $\{e^{r_1 t}, e^{r_2 t}, e^{r_3 t}\}$
20. $\{1, t, t^2\}$
21. $\{1, t, t^2, t^3\}$

22–25. Solve the following equations for the unknown coefficients.

22. $(a + b) \cos 2t - 3 \sin 2t = 2 \cos 2t + (a - b) \sin 2t$
23. $(25c_1 + 10c_2)e^{2t} + 25c_2 t e^{2t} = 25t e^{2t}$
24. $3a_1 t - a_2 t \ln t^3 = (a_2 + 1)t + (a_1 - a_2)t \ln t^5$
25. $a_1 + 3t - a_2 t^2 = a_2 + a_1 t - 3t^2$

26–27. In these two problems, we see an example of two linearly independent functions with zero Wronskian.

26. Verify that $y(t) = t^3$ and $y_2(t) = |t^3|$ are linearly independent on $(-\infty, \infty)$.
27. Show that the Wronskian, $w(y_1, y_2)(t) = 0$ for all $t \in \mathbb{R}$.

3.3 Linear Homogeneous Differential Equations

In this section, we focus on determining the solution set to a homogeneous second order constant coefficient linear differential equation. Recall that if $q(s)$ is a polynomial, we have defined the linear space $\mathcal{E}_q$ to be the set of all exponential polynomials whose Laplace transform is in $\mathcal{R}_q$, that is, can be written with denominator $q(s)$. (See Sects. 2.6 and 2.7.) Moreover, we have developed a very specific description of the space $\mathcal{E}_q$ by giving what we have called a standard basis $\mathcal{B}_q$ of $\mathcal{E}_q$ so that

$$\text{Span } \mathcal{B}_q = \mathcal{E}_q.$$

Lemma 1. *Let $q(s) = as^2 + bs + c$. If y is a function whose second derivative is of exponential type, then*

$$\mathcal{L}\{q(\boldsymbol{D})y\} = q(s)\mathcal{L}\{y\}(s) - p(s),$$

where $p(s) = ay_0 s + (ay_1 + by_0)$ is a polynomial of degree 1.

Proof. If $y''(t)$ is of exponential type, then so are $y(t)$ and $y'(t)$ by Lemma 4 of Sect. 2.2. We let $\mathcal{L}\{y(t)\} = Y(s)$ and apply linearity and the input derivative principles to get

$$
\begin{aligned}
\mathcal{L}\{q(\boldsymbol{D})y\} &= \mathcal{L}\{ay'' + by' + cy\} \\
&= a\mathcal{L}\{y''(t)\} + b\mathcal{L}\{y'(t)\} + c\mathcal{L}\{y(t)\} \\
&= as^2 Y(s) - asy_0 - ay_1 + bsY(s) - by_0 + cY(s) \\
&= (as^2 + bs + c)Y(s) - ay_0 s - ay_1 - by_0 \\
&= q(s)Y(s) - p(s),
\end{aligned}
$$

where $p(s) = ay_0 s + (ay_1 + by_0)$. $\qquad\square$

Theorem 2. *Let $q(s)$ be a polynomial of degree 2. Then the solution set to*

$$q(\boldsymbol{D})y = 0$$

is $\mathcal{E}_q$.

Proof. The forcing function $f(t) = 0$ is in $\mathcal{E}$. Thus, by Theorem 10 of Sect. 3.1, any solution to $q(\boldsymbol{D})y = 0$ is in $\mathcal{E}$. Suppose y is a solution. By Lemma 1 we have $\mathcal{L}\{q(\boldsymbol{D})y\}(s) = q(s)\mathcal{L}\{y\}(s) - p(s) = 0$ where p(s) is a polynomial of degree at most 1 depending on the initial values $y(0)$ and $y'(0)$. Solving for $\mathcal{L}\{y\}$ gives

$$\mathcal{L}\{y\}(s) = \frac{p(s)}{q(s)} \in \mathcal{R}_q.$$

This implies $y \in \mathcal{E}_q$. On the other hand, suppose $y \in \mathcal{E}_q$. Then $\mathcal{L}\{y\}(s) = \frac{p(s)}{q(s)} \in \mathcal{R}_q$, and by Lemma 1, we have

$$\mathcal{L}\{q(\boldsymbol{D})y\}(s) = q(s)\frac{p(s)}{q(s)} - p_1(s) = p(s) - p_1(s),$$

where $p_1(s)$ is a polynomial that depends on the initial conditions. Note, however, that $p(s) - p_1(s)$ is a polynomial in $\mathcal{R}$, the set of *proper* rational functions, and therefore must be identically 0. Thus, $\mathcal{L}\{q(\boldsymbol{D})y\} = 0$ and this implies $q(\boldsymbol{D})y = 0$ on $[0, \infty)$. Since $q(\boldsymbol{D})y$ and 0 are exponential polynomials and equal of $[0, \infty)$, they are equal on all of $\mathbb{R}$. It follows that the solution set to $q(\boldsymbol{D})y = 0$ is $\mathcal{E}_q$. ☐

Combining Theorem 2 and the prescription for the standard basis $\mathcal{B}_q$ given in Theorem 4 of Sect. 2.6, we get the following corollary.

Corollary 3. *Suppose $q(s) = c_2 s^2 + c_1 s + c_0$ is a polynomial of degree 2. If*

1. $q(s) = c_2(s - r_1)(s - r_2)$, where $r_1 \neq r_2$ are real, then $\mathcal{B}_q = \{e^{r_1 t}, e^{r_2 t}\}$.
2. $q(s) = c_2(s - r)^2$ then $\mathcal{B}_q = \{e^{rt}, te^{rt}\}$.
3. $q(s) = c_2((s - a)^2 + b^2)$, $b > 0$, then $\mathcal{B}_q = \{e^{at}\cos bt, e^{at}\sin bt\}$.

In each case, the solution set to

$$q(\boldsymbol{D})y = 0$$

is given by $\mathcal{E}_q = \mathrm{Span}\,\mathcal{B}_q$. That is, if $\mathcal{B}_q = \{y_1(t), y_2(t)\}$, then

$$\mathcal{E}_q = \{c_1 y_1(t) + c_2 y_2(t) : c_1, c_2 \in \mathbb{R}\}.$$

Remark 4. Observe that the solutions to $q(\boldsymbol{D})y = 0$ in these three cases may be summarized in terms of the roots of $q(s)$ as follows:

1. If $q(s)$ has distinct real roots r_1 and r_2, then all solutions are given by

$$y(t) = c_1 e^{r_1 t} + c_2 e^{r_2 t} : c_1, c_2 \in \mathbb{R}.$$

2. If $q(s)$ has one real root r with multiplicity 2, then all solutions are given by

$$y(t) = c_1 e^{rt} + c_2 t e^{rt} : c_1, c_2 \in \mathbb{R}.$$

3. If $q(s)$ has complex roots $a \pm bi$, then all solutions are given by

$$y(t) = c_1 e^{at}\cos bt + c_2 e^{at}\sin bt : c_1, c_2 \in \mathbb{R}.$$

Example 5. Find the general solution to the following differential equations:

1. $y'' + 3y' + 2y = 0$

2. $y'' + 2y' + y = 0$

3. $y'' - 6y' + 10y = 0$

▶ **Solution.** 1. The characteristic polynomial for $y'' + 3y' + 2y = 0$ is

$$q(s) = s^2 + 3s + 2 = (s + 1)(s + 2).$$

The roots are -1 and -2. The standard basis for $\mathcal{E}_q$ is $\{e^{-t}, e^{-2t}\}$. Thus, the solutions are

$$y(t) = c_1 e^{-t} + c_2 e^{-2t} : c_1, c_2 \in \mathbb{R}.$$

2. The characteristic polynomial of $y'' + 2y' + y = 0$ is

$$q(s) = s^2 + 2s + 1 = (s + 1)^2,$$

which has root $s = -1$ with multiplicity 2. The standard basis for $\mathcal{E}_q$ is $\{e^{-t}, te^{-t}\}$. The solutions are

$$y(t) = c_1 e^{-t} + c_2 t e^{-t} : c_1, c_2 \in \mathbb{R}.$$

3. The characteristic polynomial for $y'' - 6y' + 10y = 0$ is

$$q(s) = s^2 - 6s + 10 = (s - 3)^2 + 1.$$

From this, we see that the roots of q are $3 + i$ and $3 - i$. The standard basis for $\mathcal{E}_q$ is $\{e^{3t} \cos t, e^{3t} \sin t\}$. Thus, the solutions are

$$y(t) = c_1 e^{3t} \cos t + c_2 e^{3t} \sin t : c_1, c_2 \in \mathbb{R}. \qquad ◀$$

These examples show that it is a relatively easy process to write down the solution set to a homogeneous constant coefficient linear differential equation once the characteristic polynomial has been factored. We codify the process in the following algorithm.

Algorithm 6. Given a second order constant coefficient linear differential equation

$$q(\boldsymbol{D})y = 0,$$

the solution set is determined as follows:

Solution Method for
Second Order Homogeneous Linear Equations

1. Determine the characteristic polynomial, $q(s)$.
2. Factor $q(s)$ according to the three possibilities: distinct real roots, a double root, complex roots.
3. Construct $\mathcal{B}_q = \{y_1(t), y_2(t)\}$ as given in Sect. 2.6.
4. The solutions $y(t)$ are all linear combinations of the functions in the standard basis $\mathcal{B}_q$. In other words,

$$y(t) = c_1 y_1(t) + c_2 y_2(t),$$

for c_1, $c_2 \in \mathbb{R}$.

Initial Value Problems

Now suppose initial conditions, $y(0) = y_0$ and $y'(0) = y_1$, are associated to a differential equation $q(D)y = 0$. To determine the unique solution guaranteed by Theorem 10 of Sect. 3.1, we first find the general solution in terms of the standard basis, $\mathcal{B}_q$. Then the undetermined scalars given in the solution can be determined by substituting the initial values into $y(t)$ and $y'(t)$. This gives an alternate approach to using the Laplace transform method which incorporates the initial conditions from the beginning.

Example 7. Solve the initial value problem

$$y'' + 2y' + y = 0,$$

with initial conditions $y(0) = 2$ and $y'(0) = -3$.

▶ **Solution.** We first find the general solution. Observe that $q(s) = s^2 + 2s + 1 = (s + 1)^2$ is the characteristic polynomial for $y'' + 2y' + y = 0$. Thus, $\mathcal{B}_q = \{e^{-t}, te^{-t}\}$ and any solution is of the form $y(t) = c_1 e^{-t} + c_2 te^{-t}$. Observe that $y'(t) = -c_1 e^{-t} + c_2 (e^{-t} - te^{-t})$. Now evaluate both equations at $t = 0$ to get

$$\begin{aligned} c_1 \quad\quad &= 2 \\ -c_1 + c_2 &= -3, \end{aligned}$$

which implies that $c_1 = 2$ and $c_2 = -1$. Thus, the unique solution is

$$y(t) = 2e^{-t} - te^{-t}.$$

◀

Abel's Formula

We conclude this section with a converse to Theorem 13 of Sect. 3.2 in the second order case.

Theorem 8 (Abel's Formula). *Let $q(s) = s^2 + bs + c$ and suppose f_1 and f_2 are solutions to $q(D)y = 0$. Then the Wronskian satisfies*

$$w(f_1, f_2) = Ke^{-bt}, \tag{1}$$

for some constant K and $\{f_1, f_2\}$ is linearly independent if and only if $w(f_1, f_2)$ is nonzero.

Proof. First observe that since f_1 is a solution to $q(D)y = 0$ we have $f_1'' = -bf_1' - cf_1$ and similarly for f_2. To simplify the notation let $w = w(f_1, f_2) = f_1 f_2' - f_2 f_1'$. Then the product rule gives

$$\begin{aligned}
w' &= f_1' f_2' + f_1 f_2'' - (f_2' f_1' + f_2 f_1'') \\
&= f_1 f_2'' - f_2 f_1'' \\
&= f_1(-bf_2' - cf_2) - f_2(-bf_1' - cf_1) \\
&= -b(f_1 f_2' - f_2 f_1') \\
&= -bw.
\end{aligned}$$

Therefore, w satisfies the differential equation $w' + bw = 0$. By Theorem 2 of Sect. 1.4, there is a constant K so that

$$w(t) = Ke^{-bt}.$$

This gives (1). If $K \neq 0$, then $w \neq 0$ and it follows from Theorem 13 of Sect. 3.2 that $\{f_1, f_2\}$ is a linearly independent set.

Now suppose that $w = 0$. We may assume f_1 and f_2 are nonzero functions for otherwise it is automatic that $\{f_1, f_2\}$ is linearly dependent. Let t_0 be a real number where either $f_1(t_0)$ or $f_2(t_0)$ is nonzero and define $z(t) = f_2(t_0)f_1(t) - f_1(t_0)f_2(t)$. Then z is a solution to $q(D)y = 0$ and

$$\begin{aligned}
z(t_0) &= f_2(t_0)f_1(t_0) - f_1(t_0)f_2(t_0) = 0 \\
z'(t_0) &= f_2(t_0)f_1'(t_0) - f_1(t_0)f_2'(t_0) = 0.
\end{aligned}$$

The second line is obtained because $w = 0$. By the uniqueness and existence theorem, Theorem 10 of Sect. 3.1, it follows that $z(t) = 0$. This implies that $\{f_1, f_2\}$ is linearly dependent. $\square$

Remark 9. Notice that we require the leading coefficient of $q(s)$ be one. If $q(s) = as^2 + bs + c$, then the Wronskian is a multiple of $e^{\frac{-bt}{a}}$. Equation (1) is known as **Abel's formula**. Notice that the Wronskian is never zero or identically zero, depending on K. This will persist in the generalizations that you will see later.

Exercises

1–12. Determine the solution set to the following homogeneous differential equations. Write your answer as a linear combination of functions from the standard basis.

1. $y'' - y' - 2y = 0$
2. $y'' + y' - 12y = 0$
3. $y'' + 10y' + 24y = 0$
4. $y'' - 4y' - 12y = 0$
5. $y'' + 8y' + 16y = 0$
6. $y'' - 3y' - 10y = 0$
7. $y'' + 2y' + 5y = 0$
8. $2y'' - 12y' + 18y = 0$
9. $y'' + 13y' + 36y = 0$
10. $y'' + 8y' + 25y = 0$
11. $y'' + 10y' + 25y = 0$
12. $y'' - 4y' - 21y = 0$

13–16. Solve the following initial value problems.

13. $y'' - y = 0, \quad y(0) = 0, y'(0) = 1$
14. $y'' - 3y' - 10y = 0, \quad y(0) = 5, y'(0) = 4$
15. $y'' - 10y' + 25y = 0, \quad y(0) = 0, y'(0) = 1$
16. $y'' + 4y' + 13y = 0, \quad y(0) = 1, y'(0) = -5$

17–22. Determine a polynomial q so that $\mathcal{B}_q$ is the given set of functions. Compute the Wronskian and determine the constant K in Abel's formula.

17. $\{e^{3t}, e^{-7t}\}$
18. $\{e^{r_1 t}, e^{r_2 t}\}$
19. $\{e^{3t}, te^{3t}\}$
20. $\{e^{rt}, te^{rt}\}$
21. $\{e^t \cos 2t, e^t \sin 2t\}$
22. $\{e^{at} \cos bt, e^{at} \sin bt\}$

3.4 The Method of Undetermined Coefficients

In this section (and the next), we address the second part of Algorithm 7 of Sect. 3.3, namely, finding a particular solution to

$$q(D)y = f(t). \tag{1}$$

We will assume the characteristic polynomial $q(s)$ has degree 2. If we assume, $f(t) \in \mathcal{E}$ then the existence and uniqueness theorem, Theorem 10 of Sect. 3.1, implies that a solution $y(t) = y_p(t)$ to (1) is in $\mathcal{E}$. Therefore, the form of $y_p(t)$ is

$$y_p(t) = a_1 y_1(t) + \cdots + a_n y_n(t),$$

where each $y_i(t)$, for $i = 1, \ldots n$, is a *simple* exponential polynomial and the coefficients $a_1, \ldots, a_n$ are to be determined. We call y_p a *test function*. The *method of undetermined coefficients* can be broken into two parts. First, determine which simple exponential polynomials, $y_1(t), \ldots, y_n(t)$, will arise in a test function. Second, determine the coefficients, $a_1, \ldots, a_n$.

Before giving the general procedure let us consider the essence of the method in a simple example.

Example 1. Find the general solution to

$$y'' - y' - 6y = e^{-t}. \tag{2}$$

▶ **Solution.** Let us begin by finding the solution set to the associated homogeneous equation

$$y'' - y' - 6y = 0. \tag{3}$$

Observe that the characteristic polynomial is $q(s) = s^2 - s - 6 = (s-3)(s+2)$. Thus, $\mathcal{B}_q = \{e^{3t}, e^{-2t}\}$ and a homogeneous solution is of the form

$$y_h = c_1 e^{3t} + c_2 e^{-2t}. \tag{4}$$

Let us now find a particular solution to $y'' - y' - 6y = e^{-t}$. Since any particular solution will do, consider the case where $y(0) = 0$ and $y'(0) = 0$. Since $f(t) = e^{-t} \in \mathcal{E}$, we conclude by the uniqueness and existence theorem, Theorem 10 of Sect. 3.1, that the solution $y(t)$ is in $\mathcal{E}$. We apply the Laplace transform to both sides of (2) and use Lemma 1 of Sect. 3.3 to get

$$q(s)\mathcal{L}\{y(t)\}(s) = \frac{1}{s+1},$$

Solving for $\mathcal{L}\{y(t)\}$ gives

$$\mathcal{L}\{y(t)\}(s) = \frac{1}{(s+1)q(s)} = \frac{1}{(s+1)(s-3)(s+2)}.$$

It follows that $y(t) \in \mathcal{E}_{(s+1)(s-3)(s+2)}$. It is easy to see that $\mathcal{B}_{(s+1)(s-3)(s+2)} = \{e^{-t}, e^{3t}, e^{-2t}\}$. Thus,

$$y(t) = a_1 e^{-t} + a_2 e^{3t} + a_3 e^{-2t},$$

for some a_1, a_2, and a_3. Observe now that $a_2 e^{3t} + a_3 e^{-2t}$ is a homogeneous solution. This means then that the leftover piece

$$y_p(t) = a_1 e^{-t}$$

is a particular solution for some $a_1 \in \mathbb{R}$. This is the test function. Let us now determine a_1 by plugging $y_p(t)$ into the differential equation. First, observe that $y_p'(t) = -a_1 e^{-t}$ and $y_p''(t) = a_1 e^{-t}$. Thus, (3) gives

$$\begin{aligned} e^{-t} &= y_p'' - y_p' - 6y_p \\ &= a_1 e^{-t} + a_1 e^{-t} - 6a_1 e^{-t} \\ &= -4a_1 e^{-t}. \end{aligned}$$

From this we conclude $1 = -4a_1$ or $a_1 = -1/4$. Therefore,

$$y_p(t) = \frac{-1}{4} e^{-t}$$

is a particular solution and the general solution is obtained by adding the homogeneous solution to it. Thus, the functions

$$y(t) = y_p(t) + y_h(t) = \frac{-1}{4} e^{-t} + c_1 e^{3t} + c_2 e^{-2t},$$

where c_1 and c_2 are real numbers, make up the set of all solutions.   ◄

Remark 2. Let $v(s) = (s+1)$ be the denominator of $\mathcal{L}\{e^{-t}\}$. Then $v(s)q(s) = (s+1)(s-3)(s+2)$, and the standard basis for $\mathcal{E}_{vq}$ is $\{e^{-t}, e^{3t}, e^{-2t}\}$. The standard basis for $\mathcal{E}_q$ is $\{e^{3t}, e^{-2t}\}$. Observe that the test function $y_p(t)$ is made up of functions from the standard basis of $\mathcal{E}_{vq}$ that are not in the standard basis of $\mathcal{E}_q$. This will always happen. The general argument is in the proof of the following theorem.

Theorem 3. *Suppose $L = q(D)$ is a polynomial differential operator and $f \in \mathcal{E}$. If $\mathcal{L}\{f\} = u/v$, and $\mathcal{B}_q$ is the standard basis for $\mathcal{E}_q$, then there is a particular solution $y_p(t)$ to*

$$L(y) = f(t)$$

which is a linear combination of terms that are in $\mathcal{B}_{qv}$ but not in $\mathcal{B}_q$.

Proof. By Theorem 10 of Sect. 3.1, any solution to $L(y) = f(t)$ is in $\mathcal{E}$, and hence, it and its derivatives have Laplace transforms. Thus,

$$\mathcal{L}\{q(\boldsymbol{D})y\} = \mathcal{L}\{f\} = \frac{u(s)}{v(s)}$$

$$\implies q(s)\mathcal{L}\{y\} - p(s) = \frac{u(s)}{v(s)} \quad \text{by Lemma 1 of Sect. 3.3}$$

$$\implies \mathcal{L}\{y(t)\} = \frac{p(s)}{q(s)} + \frac{u(s)}{q(s)v(s)} = \frac{u(s) + p(s)v(s)}{q(s)v(s)}.$$

It follows that $y(t)$ is in $\mathcal{E}_{qv}$ and hence a linear combination of terms in $\mathcal{B}_{qv}$. Since $\mathcal{B}_q \subset \mathcal{B}_{qv}$, we can write $y(t) = y_h(t) + y_p(t)$, where $y_h(t)$ is a linear combination of terms in $\mathcal{B}_q$ and $y_p(t)$ is a linear combination of terms in $\mathcal{B}_{qv}$ but not in $\mathcal{B}_q$. Since $y_h(t)$ is a homogeneous solution, it follows that $y_p(t) = y(t) - y_h(t)$ is a particular solution of $L(y) = f(t)$ of the required form. $\qquad\square$

Theorem 3 is the basis for the following algorithm.

Algorithm 4. A general solution to a second order constant coefficient differential equation

$$q(\boldsymbol{D})(y) = f(t)$$

can be found by the following method.

The Method of Undetermined Coefficients

1. Compute the standard basis, $\mathcal{B}_q$, for $\mathcal{E}_q$.
2. Determine the denominator v so that $\mathcal{L}\{f\} = u/v$. This means that $f(t) \in \mathcal{E}_v$.
3. Compute the standard basis, $\mathcal{B}_{vq}$, for $\mathcal{E}_{vq}$.
4. The test function, $y_p(t)$, is the linear combination with arbitrary coefficients of functions in $\mathcal{B}_{vq}$ that are not in $\mathcal{B}_q$.
5. The coefficients in $y_p(t)$ are determined by plugging $y_p(t)$ into the differential equation $q(\boldsymbol{D})(y) = f(t)$.
6. The general solution is given by

$$y(t) = y_p(t) + y_h(t),$$

where $y_h(t)$ is an arbitrary function in $\mathcal{E}_q$.

Example 5. Find the general solution to $y'' - 5y' + 6y = 4e^{2t}$.

▶ **Solution.** The characteristic polynomial is

$$q(s) = s^2 - 5s + 6 = (s - 2)(s - 3)$$

and the standard basis for $\mathcal{E}_q$ is $\mathcal{B}_q = \{e^{2t}, e^{3t}\}$. Since $\mathcal{L}\{4e^{2t}\} = 4/(s - 2)$, we have $v(s) = s - 2$ so $v(s)q(s) = (s - 2)^2(s - 3)$ and the standard basis for $\mathcal{E}_{vq}$ is $\mathcal{B}_{vq} = \{e^{2t}, te^{2t}, e^{3t}\}$. The only function in $\mathcal{B}_{vq}$ that is not in $\mathcal{B}_q$ is te^{2t}. Therefore, our test function is $y_p(t) = a_1 te^{2t}$. A simple calculation gives

$$y_p = a_1 te^{2t}$$
$$y_p' = a_1 e^{2t} + 2a_1 te^{2t}$$
$$y_p'' = 4a_1 e^{2t} + 4a_1 te^{2t}.$$

Substitution into $y'' - 5y' + 6y = 4e^{2t}$ gives

$$\begin{aligned} 4e^{2t} &= y_p'' - 5y_p' + 6y_p \\ &= (4a_1 e^{2t} + 4a_1 te^{2t}) - 5(a_1 e^{2t} + 2a_1 te^{2t}) + 6(a_1 te^{2t}) \\ &= -a_1 e^{2t}. \end{aligned}$$

From this, it follows that $a_1 = -4$ and $y_p = -4te^{2t}$. The general solution is thus given by

$$y(t) = y_p(t) + y_h(t) = -4te^{2t} + c_1 e^{2t} + c_2 e^{3t},$$

where c_1 and c_2 are arbitrary real constants. ◀

Remark 6. Based on Example 1, one might have expected that the test function in Example 5 would be $y_p = c_1 e^{2t}$. But this cannot be since e^{2t} is a homogeneous solution, that is, $L(e^{2t}) = 0$, so it cannot possibly be true that $L(a_1 e^{2t}) = 4e^{2t}$. Observe that $v(s) = s - 2$ and $q(s) = (s - 2)(s + 3)$ share a common root, namely, $s = 2$, so that the product has root $s = 2$ with multiplicity 2. This produces te^{2t} in the standard basis for $\mathcal{E}_{vq}$ that does not appear in the standard basis for $\mathcal{E}_q$. In Example 1, all the roots of vq are distinct so this phenomenon does not occur. There is thus a qualitative difference between the cases when $v(s)$ and $q(s)$ have common roots and the cases where they do not. However, Algorithm 4 does not distinguish this difference. It will always produce a test function that leads to a particular solution.

Example 7. Find the general solution to

$$y'' - 3y' + 2y = 2te^t.$$

▶ **Solution.** The characteristic polynomial is

$$q(s) = s^2 - 3s + 2 = (s - 1)(s - 2).$$

Hence, the standard basis for $\mathcal{E}_q$ is $\mathcal{B}_q = \{e^t, e^{2t}\}$. The Laplace transform of $2te^t$ is $2/(s-1)^2$ so $v(s) = (s-1)^2$, and thus, $v(s)q(s) = (s-1)^3(s-2)$. Therefore, $y(t) \in \mathcal{E}_{vq}$, which has standard basis $\mathcal{B}_{vq} = \{e^t, te^t, t^2e^t, e^{2t}\}$. Since te^t and t^2e^t are the only functions in $\mathcal{B}_{vq}$ but not in $\mathcal{B}_q$, it follows that our test function has the form $y_p(t) = a_1te^t + a_2t^2e^t$ for unknown constants a_1 and a_2. We determine a_1 and a_2 by plugging $y_p(t)$ into the differential equation. A calculation of derivatives gives

$$y_p = a_1te^t + a_2t^2e^t$$
$$y_p' = a_1e^t + (a_1 + 2a_2)te^t + a_2t^2e^t$$
$$y_p'' = (2a_1 + 2a_2)e^t + (a_1 + 4a_2)te^t + a_2t^2e^t.$$

Substitution into $y'' - 3y' + 2y = 2te^t$ gives

$$2te^t = y_p'' - 3y_p' + 2y_p$$
$$= (-a_1 + 2a_2)e^t - 2a_2te^t.$$

Here is an example where we invoke the linear independence of $\mathcal{B}_{vq}$. By Theorems 7 and 8 of Sect. 3.2, we have that the coefficients a_1 and a_2 satisfy

$$-a_1 + 2a_2 = 0$$
$$-2a_2 = 2.$$

From this, we find $a_2 = -1$ and $a_1 = 2a_2 = -2$. Hence, a particular solution is

$$y_p(t) = -2te^{-t} - t^2e^{-t}$$

and the general solution is

$$y(t) = y_p(t) + y_h(t) = -2te^{-t} - t^2e^{-t} + c_1e^t + c_2e^{2t},$$

where c_1 and c_2 are arbitrary real constants.   ◄

Linearity is particularly useful when the forcing function $f(t)$ is a sum of terms. Here is the general principle, sometimes referred to as the ***superposition principle***.

Theorem 8 (Superposition Principle). *Suppose y_{p_1} is a solution to $L(y) = f_1$ and y_{p_2} is a solution to $L(y) = f_2$, where L is a linear differential operator. Then $a_1y_{p_1} + a_2y_{p_2}$ is a solution to $Ly = a_1f_1 + a_2f_2$.*

Proof. By linearity,

$$L(a_1y_{p_1} + a_2y_{p_2}) = a_1L(y_{p_1}) + a_2L(y_{p_2}) = a_1f_1 + a_2f_2.$$ □

Example 9. Find the general solution to

$$y'' - 5y' + 6y = 12 + 4e^{2t}. \tag{5}$$

► **Solution.** Theorem 8 allows us to find a particular solution by adding together the particular solutions to

$$.y'' - 5y' + 6y = 12 \tag{6}$$

and

$$y'' - 5y' + 6y = 4e^{2t}. \tag{7}$$

In both cases, the characteristic polynomial is $q(s) = s^2 - 5s + 6 = (s-2)(s-3)$. In (6), the Laplace transform of 12 is $12/s$. Thus, $v(s) = s$, $q(s)v(s) = s(s-2)(s-3)$, and $\mathcal{B}_{qv} = \{1, e^{2t}, e^{3t}\}$. The test function is $y_{p_1} = a$. It is easy to see that $y_{p_1} = 2$. In Example 5, we found that $y_{p_2} = -4te^{2t}$ is a particular solution to (7). By Theorem 8, $y_p = 2 - 4te^{2t}$ is a particular solution to (5). Thus,

$$y = 2 - 4te^{2t} + c_1 e^{2t} + c_2 e^{3t}$$

is the general solution.   ◄

Exercises

1–9. Given $q(s)$ and $v(s)$ below, determine the test function y_p for the differential equation $q(\boldsymbol{D})y = f$, where $\mathcal{L}f = u/v$.

1. $q(s) = s^2 - s - 2 \quad v(s) = s - 3$
2. $q(s) = s^2 + 6s + 8, \quad v(s) = s + 3$
3. $q(s) = s^2 - 5s + 6, \quad v(s) = s - 2$
4. $q(s) = s^2 - 7s + 12, \quad v(s) = (s - 4)^2$
5. $q(s) = (s - 5)^2, \quad v(s) = s^2 + 25$
6. $q(s) = s^2 + 1, \quad v(s) = s^2 + 4$
7. $q(s) = s^2 + 4, \quad v(s) = s^2 + 4$
8. $q(s) = s^2 + 4s + 5, \quad v(s) = (s - 1)^2$
9. $q(s) = (s - 1)^2, \quad v(s) = s^2 + 4s + 5$

10–24. Find the general solution for each of the differential equations given below.

10. $y'' + 3y' - 4y = e^{2t}$
11. $y'' - 3y' - 10y = 7e^{-2t}$
12. $y'' + 2y' + y = e^t$
13. $y'' + 2y' + y = e^{-t}$
14. $y'' + 3y' + 2y = 4$
15. $y'' + 4y' + 5y = e^{-3t}$
16. $y'' + 4y = 1 + e^t$
17. $y'' - y = t^2$
18. $y'' - 4y' + 4y = e^t$
19. $y'' - 4y' + 4y = e^{2t}$
20. $y'' + y = 2\sin t$
21. $y'' + 6y' + 9y = 25te^{2t}$
22. $y'' + 6y' + 9y = 25te^{-3t}$
23. $y'' + 6y' + 13y = e^{-3t}\cos 2t$
24. $y'' - 8y' + 25y = 104\sin 3t$

25–28. Solve each of the following initial value problems.

25. $y'' - 5y' - 6y = e^{3t}, \quad y(0) = 2, y'(0) = 1$
26. $y'' + 2y' + 5y = 8e^{-t}, \quad y(0) = 0, y'(0) = 8$
27. $y'' + y = 10e^{2t}, \quad y(0) = 0, y'(0) = 0$
28. $y'' - 4y = 2 - 8t, \quad y(0) = 0, y'(0) = 5$

3.5 The Incomplete Partial Fraction Method

In this section, we provide an alternate method for finding the solution set to the nonhomogeneous differential equation

$$q(\mathbf{D})(y) = f(t), \tag{1}$$

where $f \in \mathcal{E}$. This alternate method begins with the Laplace transform method and exploits the efficiency of the partial fraction decomposition algorithm developed in Sects. 2.3 and 2.4. However, the partial fraction decomposition applied to $Y(s) = \mathcal{L}\{y\}(s)$ is not needed to its completion. We therefore refer to this method as the *incomplete partial fraction method*. For purposes of illustration and comparison to the method of undetermined coefficients let us reconsider Example 5 of Sect. 3.4.

Example 1. Find the general solution to

$$y'' - 5y' + 6y = 4e^{2t}.$$

▶ **Solution.** Our goal is to find a particular solution $y_p(t)$ to which we will add the homogeneous solutions $y_h(t)$. Since any particular solution will do, we begin by applying the Laplace transform with initial conditions $y(0) = 0$ and $y'(0) = 0$. The characteristic polynomial is $q(s) = s^2 - 5s + 6 = (s-2)(s-3)$ and $\mathcal{L}\{4e^{2t}\} = 4/(s-2)$. If, as usual, we let $Y(s) = \mathcal{L}\{y\}(s)$ and take the Laplace transform of the differential equation with the given initial conditions, then $q(s)Y(s) = 4/(s-2)$ so that

$$Y(s) = \frac{4}{(s-2)q(s)} = \frac{4}{(s-2)^2(s-3)}.$$

In the table below, we compute the $(s-2)$-chain for $Y(s)$ but stop when the denominator reduces to $q(s) = (s-2)(s+3)$.

Incomplete $(s-2)$-chain	
$\dfrac{4}{(s-2)^2(s-3)}$	$\dfrac{-4}{(s-2)^2}$
$\dfrac{p(s)}{(s-2)(s-3)}$	

The table tells us that

$$\frac{4}{(s-2)^2(s-3)} = \frac{-4}{(s-2)^2} + \frac{p(s)}{(s-2)(s-3)}.$$

There is no need to compute $p(s)$ and finish out the table since the inverse Laplace transform of $\frac{p(s)}{(s-2)(s-3)} = \frac{p(s)}{q(s)}$ is in $\mathcal{E}_q$ and hence a solution of the associated homogeneous equation. If $Y_p(s) = \frac{-4}{(s-2)^2}$, then $y_p(t) = \mathcal{L}^{-1}\{Y_p\}(t) = -4te^{2t}$ is a particular solution. By linearity, we get the general solution:

$$y(t) = -4te^{2t} + c_1e^{2t} + c_2e^{3t}. \qquad \blacktriangleleft$$

Observe that the particular solution $y_p(t)$ we have obtained here is exactly what we derived using the method of undetermined coefficients in Example 5 of Sect. 3.4 and is obtained by one iteration of the partial fraction decomposition algorithm.

The Incomplete Partial Fraction Method

We now proceed to describe the procedure generally. Consider the differential equation

$$q(D)(y) = f(t),$$

where $q(s)$ is a polynomial of degree 2 and $f \in \mathcal{E}$. Suppose $\mathcal{L}\{f\} = u(s)/v(s)$. Since we are only interested in finding a particular solution, we can choose initial conditions that are convenient for us. Thus, we will assume that $y(0) = 0$ and $y'(0) = 0$, and as usual, let $Y(s) = \mathcal{L}\{y\}(s)$. Then applying the Laplace transform to the differential equation $q(D)(y) = f(t)$ and solving for $Y(s)$ give

$$Y(s) = \frac{u(s)}{v(s)q(s)}.$$

Let us consider the linear case where $v(s) = (s - \gamma)^m$ and $u(s)$ has no factors of $s - \gamma$. (The quadratic case, $v(s) = (s^2 + cs + d)^m$ is handled similarly.) This means $f(t) = p(t)e^{\gamma t}$, where the degree of p is $m - 1$. It could be the case that γ is a root of $q(s)$ with multiplicity j, in which case we can write

$$q(s) = (s - \gamma)^j q_\gamma(s),$$

where $q_\gamma(\gamma) \neq 0$. Thus,

$$Y(s) = \frac{u(s)}{(s - \gamma)^{m+j} q_\gamma(s)}.$$

For convenience, let $p_0(s) = u(s)$. We now iterate the partial fraction decomposition algorithm until the denominator is $q(s)$. This occurs after m iterations. The incomplete $(s - \gamma)$-chain is given by the table below.

Incomplete $(s - \gamma)$-chain	
$\dfrac{p_0(s)}{(s - \gamma)^{m+j} q_\gamma(s)}$	$\dfrac{A_1}{(s - \gamma)^{m+j}}$
$\dfrac{p_1(s)}{(s - \gamma)^{m+j-1} q_\gamma(s)}$	$\dfrac{A_2}{(s - \gamma)^{m+j-1}}$
$\vdots$	$\vdots$
$\dfrac{p_{m-1}(s)}{(s - \gamma)^{j+1} q_\gamma(s)}$	$\dfrac{A_m}{(s - \gamma)^{j+1}}$
$\dfrac{p_m(x)}{(s - \gamma)^{j} q_\gamma(s)}$	

The last entry in the first column has denominator $q(s) = (s - \gamma)^j q_\gamma(s)$, and hence, its inverse Laplace transform is a solution of the associated homogeneous equation, It follows that if $Y_p(s) = \frac{A_1}{(s-\gamma)^{m+j}} + \cdots + \frac{A_m}{(s-\gamma)^{j+1}}$, then

$$y_p(t) = \frac{A_1}{(m + j - 1)!} t^{m+j-1} e^{\gamma t} + \cdots + \frac{A_m}{(j)!} t^{j-1} e^{\gamma t}$$

is a particular solution.

To illustrate this general procedure, let us consider two further examples.

Example 2. Find the general solution to

$$y'' + 4y' + 4y = 2te^{-2t}.$$

▶ **Solution.** The characteristic polynomial is $q(s) = s^2 + 4s + 4 = (s + 2)^2$ and $\mathcal{L}\{2te^{-2t}\} = 2/(s + 2)^2$. Again assume $y(0) = 0$ and $y'(0) = 0$ then

$$Y(s) = \frac{2}{(s + 2)^4}.$$

But this term is a partial fraction. In other words, the incomplete $(s + 2)$-chain for $Y(s)$ degenerates:

Incomplete $(s+2)$-chain	
$\dfrac{2}{(s+2)^4}$	$\dfrac{2}{(s+2)^4}$
0	

Let $Y_p(s) = 2/(s+2)^4$. Then a particular solution is

$$y_p(t) = \mathcal{L}^{-1}\{Y_p\}(t) = \frac{2}{3!}t^3 e^{-2t} = \frac{t^3}{3}e^{-2t}.$$

The homogeneous solution can be read off from the roots of $q(s) = (s+2)^2$. Thus, the general solution is

$$y = \frac{1}{3}t^3 e^{-2t} + c_1 e^{-2t} + c_2 t e^{-2t}. \qquad \blacktriangleleft$$

Example 3. Find the general solution to

$$y'' + 4y = 16t e^{2t}$$

▶ **Solution.** The characteristic polynomial is $q(s) = s^2 + 4$ and

$$\mathcal{L}\left\{16t e^{2t}\right\}(s) = \frac{16}{(s-2)^2}.$$

Again assume $y(0) = 0$ and $y'(0) = 0$. Then

$$Y(s) = \frac{16}{(s^2 + 4)(s-2)^2}.$$

The incomplete $s - 2$-chain for $Y(s)$ is:

Incomplete $s - 2$ -chain	
$\dfrac{16}{(s^2 + 4)(s-2)^2}$	$\dfrac{2}{(s-2)^2}$
$\dfrac{-2(s+2)}{(s^2 + 4)(s-2)}$	$\dfrac{-1}{s-2}$
$\dfrac{p(s)}{s^2 + 4}$	

Let $Y_p(s) = \frac{2}{(s-2)^2} - \frac{1}{s-2}$. Then $y_p(t) = 2te^{2t} - e^{2t}$. The homogeneous solutions are $y_h = c_1 \cos 2t + c_2 \sin 2t$. Thus, the general solution is

$$y(t) = y_p(t) + y_h(t) = 2te^{2t} - e^{2t} + c_1 \cos 2t + c_2 \sin 2t. \qquad \blacktriangleleft$$

Exercises

1–12. Use the incomplete partial fraction method to solve the following differential equations.

1. $y'' - 4y = e^{-6t}$
2. $y'' + 2y' - 15y = 16e^{t}$
3. $y'' + 5y' + 6y = e^{-2t}$
4. $y'' + 3y' + 2y = 4$
5. $y'' + 2y' - 8y = 6e^{-4t}$
6. $y'' + 3y' - 10y = \sin t$
7. $y'' + 6y' + 9y = 25te^{2t}$
8. $y'' - 5y' - 6y = 10te^{4t}$
9. $y'' - 8y' + 25y = 36te^{4t}\sin(3t)$
10. $y'' - 4y' + 4y = te^{2t}$
11. $y'' + 2y' + y = \cos t$
12. $y'' + 2y' + 2y = e^{t}\cos t$

3.6 Spring Systems

In this section, we illustrate how a second order constant coefficient differential equation arises from modeling a spring-body-dashpot system. This model may arise in a simplified version of a suspension system on a vehicle or a washing machine. Consider the three main objects in the diagram below: the spring, the body, and the dashpot (shock absorber).

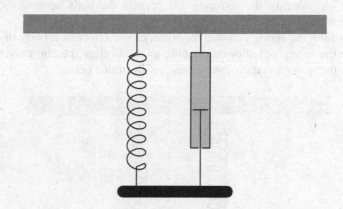

We assume the body only moves vertically without any twisting. Our goal is to determine the motion of the body in such a system. Various forces come into play. These include the force of gravity, the restoring force of the spring, the damping force of the dashpot, and perhaps an external force. Let us examine each of these forces and how they contribute to the overall motion of the body.

Force of Gravity

First, assume that the body has mass m. The force of gravity, F_G, acts on the body by the familiar formula

$$F_G = mg, \tag{1}$$

where g is the acceleration due to gravity. Our measurements will be positive in the downward direction so F_G is positive.

Restoring Force

When a spring is suspended with no mass attached, the end of the spring will lie at a reference point ($u = 0$). When the spring is stretched or compressed, we will denote the *displacement* by u. The force exerted by the spring that acts in the opposite direction to a force that stretches or compresses a spring is called the ***restoring force***. It depends on the displacement and is denoted by $F_R(u)$. Hooke's law says that the restoring force of many springs is proportional to the displacement, as long as the displacement is not too large. We will assume this. Thus, if u is the displacement we have

$$F_R(u) = -ku, \tag{2}$$

where k is a positive constant, called the **spring constant**. When the displacement is positive (downward), the restoring force pulls the body upward, hence the negative sign.

   To determine the spring constant k, consider the effect of a body of mass m attached to the spring and allowed to come to equilibrium (i.e., no movement). It will stretch the spring a certain distance, u_0, as illustrated below:

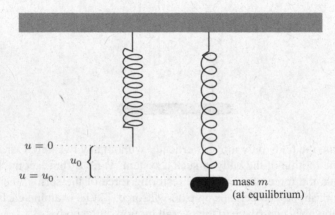

At equilibrium, the restoring force of the spring will cancel the gravitational force on the mass m. Thus, we get

$$F_R(u_0) + F_G = 0. \tag{3}$$

Combining equations (1), (2), and (3) gives us $mg - ku_0 = 0$, and hence,

$$k = \frac{mg}{u_0}. \tag{4}$$

Damping Force

In any practical situation, there will be some kind of resistance to the motion of the body. In our system, this resistance is represented by a dashpot, which in many situations is a shock absorber. The force exerted by the dashpot is called the *damping force*, F_D. It depends on a lot of factors, but an important factor is the velocity of the body. To see that this is reasonable, imagine the difference in the forces against your body when you dive into a swimming pool off a 3-meter board and when you dive from the side of the pool. The greater the velocity when you enter the pool the greater the force that decelerates your body. We will assume that the damping force is proportional to the velocity. We thus have

$$F_D = -\mu v = -\mu u',$$

where $v = u'$ is velocity and μ is a positive constant known as the *damping constant*. The damping force acts in a direction opposite the velocity, hence the negative sign.

External Forces and Newton's Law of Motion

We will let $f(t)$ denote an external force acting on the body. For example, this could be the varying vertical forces acting on a suspension system of a vehicle due to driving over a bumpy road. If $a = u''$ is acceleration, then Newton's second law of motion says that the total force of a body, given by mass times acceleration, is the sum of the forces acting on that body. We thus have

$$\text{Total Force} = F_G + F_R + F_D + \text{External Force},$$

which implies the equation

$$mu'' = mg - ku - \mu u' + f(t).$$

Equation (4) implies $mg = -ku_0$. Substituting and combining terms give

$$mu'' + \mu u' + k(u - u_0) = f(t).$$

If $y = u - u_0$, then y measures the displacement of the body from the equilibrium point, u_0. In this new variable, we obtain

$$my'' + \mu y' + ky = f(t). \tag{5}$$

This second order constant coefficient differential equation is a mathematical model for the spring-body-dashpot system. The solutions that can be obtained vary dramatically depending on the constants m, μ, and k, and, of course, the external force, $f(t)$. The initial conditions,

$$y(0) = y_0 \quad \text{and} \quad y'(0) = v_0$$

represent the initial position, $y(0) = y_0$, and the initial velocity, $y'(0) = v_0$, of the given body. Once the constants, external force, and initial conditions are determined, the uniqueness and existence theorem, Theorem 10 of Sect. 3.1, guarantees a unique

Table 3.1 Units of measure in metric and English systems

System	Time	Distance	Mass	Force
Metric	seconds (s)	meters (m)	kilograms (kg)	newtons (N)
English	seconds (s)	feet (ft)	slugs (sl)	pounds (lbs)

Table 3.2 Derived quantities

Quantity	Formula
velocity (v)	distance / time
acceleration (a)	velocity /time
force (F)	mass · acceleration
spring constant (k)	force / distance
damping constant (μ)	force /velocity

solution. Of course, we should always keep in mind that (5) is a mathematical model of a real phenomenon and its solution is an approximation to what really happens. However, as long as our assumptions about the spring and damping constants are in effect, which usually require that $y(t)$ and $y'(t)$ be relatively small in magnitude, and the mass of the spring is negligible compared to the mass of the body, the solution will be a reasonably good approximation.

Units of Measurement

Before we consider specific examples, we summarize the two commonly used units of measurement: The English and metric systems. Table 3.1 summarizes the units.

The main units of the metric system are kilograms and meters while in the English system they are pounds and feet. The time unit is common to both. Table 3.2 summarizes quantities derived from these units.

In the metric system, one Newton of force (N) will accelerate a 1 kilogram mass (kg) 1 m/s^2. In the English system, a 1 pound force (lb) will accelerate a 1 slug mass (sl) 1 ft/s^2. To compute the mass of a body in the English system, one must divide the weight by the acceleration due to gravity, which is $g = 32$ ft/s^2 near the surface of the earth. Thus, a body weighing 64 lbs has a mass of 2 slugs. To compute the force a body exerts due to gravity in the metric system, one must multiply the mass by the acceleration due to gravity, which is $g = 9.8$ m/s^2. Thus, a 5 kg mass exerts a gravitational force of 49 N.

The units for the spring constant k and damping constant μ are given according to the following table.

	k	μ
Metric	N/m	N s/m
English	lbs/ft	lbs s/ft

Example 1. 1. A dashpot exerts a damping force of 10 pounds when the velocity of the mass is 2 feet per second. Find the damping constant.
2. A dashpot exerts a damping force of 6 Newtons when the velocity is 40 centimeters per second. Find the damping constant.
3. A body weighing 4 pounds stretches a spring 2 inches. Find the spring constant.
4. A mass of 8 kilograms stretches a spring 20 centimeters. Find the spring constant.

▶ **Solution.** 1. The force is 10 pounds and the velocity is 2 feet per second. The damping constant is given by $\mu = $ force/velocity $= 10/2 = 5$ lbs s/ft.
2. The force is 6 Newtons and the velocity is .4 meters per second The damping constant is given by $\mu = $ force/velocity $= 6/.4 = 15$ N s/m.
3. The force is 4 pounds. A length of 2 inches is $1/6$ foot. The spring constant is $k = $ force/distance $= 4/(1/6) = 24$ lbs/ft.
4. The force exerted by a mass of 8 kilograms is $8 \cdot 9.8 = 78.4$ Newtons. A length of 20 centimeters is .2 meters. The spring constant is given by $k = $ force/distance $= 78.4/.2 = 392$ N/m.   ◀

Example 2. A spring is stretched 20 centimeters by a force of 5 Newtons. A body of mass 4 kilogram is attached to such a spring with an accompanying dashpot. At $t = 0$, the mass is pulled down from its equilibrium position a distance of 50 centimeters and released with a downward velocity of 1 meter per second. Suppose the damping force is 5 Newtons when the velocity of the body is .5 meter per second. Find a mathematical model that represents the motion of the body.

▶ **Solution.** We will model the motion by (5). Units are converted to the kilogram-meter units of the metric system. The mass is $m = 4$. The spring constant k is given by $k = 5/(.2) = 25$. The damping constant is given by $\mu = 5/.5 = 10$. Since no external force is mentioned, we may assume it is zero. The initial conditions are $y(0) = .5$ and $y'(0) = 1$. The following equation

$$4y''(t) + 10y'(t) + 25y = 0, \qquad y(0) = .5, \quad y'(0) = 1$$

represents the model for the motion of the body.   ◀

Example 3. A body weighing 4 pounds will stretch a spring 3 in. This same body is attached to such a spring with an accompanying dashpot. At $t = 0$, the mass is pulled down from its equilibrium position a distance of 1 foot and released. Suppose the damping force is 8 pounds when the velocity of the body is 2 feet per second. Find a mathematical model that represents the motion of the body.

▶ **Solution.** Units are converted to the pound-foot units in the English system. The mass is $m = 4/32 = 1/8$ slugs. The spring constant k is given by $k = 4/(3/12) = 16$. The damping constant is given by $\mu = 8/2 = 4$. Since no external force is mentioned, we may assume it is zero. The initial conditions are $y(0) = 1$ and $y'(0) = 0$. The following equation

$$\frac{1}{8}y''(t) + 4y'(t) + 16y = 0, \qquad y(0) = 1, \quad y'(0) = 0$$

models the motion of the body.   ◄

Let us now turn our attention to an analysis of (5). The zero-input response models the motion of the body with no external forces. We refer to this motion as *free motion* ($f(t) \equiv 0$). Otherwise, we refer to the motion as *forced motion* ($f(t) \neq 0$). In turn, each of these are divided into *undamped* ($\mu = 0$) and *damped* ($\mu \neq 0$).

Undamped Free Motion

When the damping constant is zero and there is no externally applied force, the resulting motion of the object is called *undamped free motion* or *simple harmonic motion*. This is an idealized situation, for seldom, if ever, will a system be free of any damping effects. Nevertheless, (5) becomes

$$my'' + ky = 0, \tag{6}$$

with $m > 0$ and $k > 0$. The characteristic polynomial of this equation is $q(s) = ms^2 + k = m(s^2 + k/m) = m(s^2 + \beta^2)$, where $\beta = \sqrt{k/m}$. Further, we have $\mathcal{B}_q = \{\sin \beta t, \cos \beta t\}$. Hence, (6) has the general solution

$$y = c_1 \cos \beta t + c_2 \sin \beta t. \tag{7}$$

Using the trigonometric identity $\cos(\beta t - \delta) = \cos(\beta t) \cos \delta + \sin(\beta t) \sin \delta$, (7) can be rewritten as

$$y = A \cos(\beta t - \delta), \tag{8}$$

where $c_1 = A \cos \delta$ and $c_2 = A \sin \delta$. Solving these equations for A and δ gives $A = \sqrt{c_1^2 + c_2^2}$ and $\tan \delta = c_2/c_1$. Therefore, the graph of $y(t)$ satisfying (6) is a pure cosine function with *frequency* β and with *period*

$$T = \frac{2\pi}{\beta} = 2\pi \sqrt{\frac{m}{k}}.$$

The numbers A and δ are commonly referred to as the *amplitude* and *phase angle* of the system. Equation (8) is called the *phase-amplitude form* of the solution to (6). From (8), we see that A is the maximum possible value of the function $y(t)$, and hence the maximum displacement from equilibrium, and that $|y(t)| = A$ precisely when $t = (\delta + n\pi)/\beta$, where $n \in \mathbb{Z}$.

The graph of (8) is given below and well represents the oscillating motion of the body. This idealized kind of motion occurs ubiquitously in the sciences.

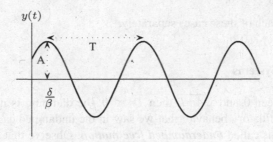

Example 4. Find the amplitude, phase angle, and frequency of the damped free motion of a spring-body-dashpot system with unit mass and spring constant 3. Assume the initial conditions are $y(0) = -3$ and $y'(0) = 3$.

▶ **Solution.** The initial value problem that models such a system is

$$y'' + 3y = 0, \qquad y(0) = -3, \ y'(0) = 3.$$

An easy calculation gives $y = -3\cos\sqrt{3}t + \sqrt{3}\sin\sqrt{3}t$. The amplitude is given by $A = ((-3)^2 + (\sqrt{3})^2)^{\frac{1}{2}} = 2\sqrt{3}$ and the phase angle is given implicitly by $\tan\delta = -\sqrt{3}/3$ hence $\delta = -\pi/6$. Thus,

$$y = 2\sqrt{3}\cos\left(\sqrt{3}t + \frac{\pi}{6}\right). \qquad ◀$$

Damped Free Motion

In this case, we include the damping term $\mu y'$ with $\mu > 0$. In applications, the coefficient μ represents the presence of friction or resistance, which can never be completely eliminated. Thus, we want solutions to the differential equation

$$my'' + \mu y' + ky = 0. \tag{9}$$

The characteristic polynomial $q(s) = ms^2 + \mu s + k$ has roots r_1 and r_2 given by the quadratic formula

$$r_1, r_2 = \frac{-\mu \pm \sqrt{\mu^2 - 4mk}}{2m}. \tag{10}$$

The nature of the solutions of (9) are determined by whether the discriminant $D = \mu^2 - 4mk$ is negative (complex roots), zero (double root), or positive (distinct roots). We say that the system is

- **Underdamped** if $D < 0$.
- **Critically damped** if $D = 0$.
- **Overdamped** if $D > 0$.

Let us consider each of these cases separately.

Underdamped Systems

When μ is between 0 and $\sqrt{4mk}$ then $D < 0$, the damping is not sufficient to overcome the oscillatory behavior that we saw in the undamped case, $\mu = 0$. The resulting motion is called **underdamped free motion**. Observe that in this case we can write

$$q(s) = m \left(s^2 + \frac{\mu}{m} s + \frac{k}{m} \right) = m((s - \alpha)^2 + \beta^2),$$

where $\alpha = -\frac{\mu}{2m}$ and $\beta = \frac{\sqrt{4mk - \mu^2}}{2m}$. Since $\mathcal{B}_q = \{e^{\alpha t} \cos \beta t, e^{\alpha t} \sin \beta t\}$, the solution to (9) is

$$y(t) = e^{\alpha t} \left(c_1 \cos \beta t + c_2 \sin \beta t \right),$$

which can be rewritten as

$$y(t) = A e^{\alpha t} \cos(\beta t - \delta), \tag{11}$$

where $A = \sqrt{c_1^2 + c_2^2}$ and $\tan \delta = c_2/c_1$, as we did earlier for the undamped case. Again we refer to (11) as the **phase-amplitude** form of the solution. A typical graph of (11) is given below.

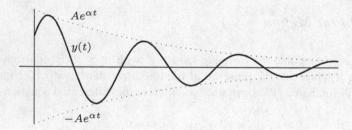

Notice that y appears to be a cosine curve in which the amplitude oscillates between $A e^{\alpha t}$ and $-A e^{\alpha t}$. The motion of the body passes through equilibrium at regular intervals. Since $\alpha < 0$, the amplitude decreases with time. One may imagine the suspension on an automobile with rather weak shock absorbers. A push on the fender will send the vehicle oscillating as in the graph above.

Critically Damped Systems

If $\mu = \sqrt{4mk}$, then the discriminant D is zero. At this critical point, the damping is just large enough to overcome oscillatory behavior. The resulting motion is called *critically damped free motion*. Observe that the characteristic polynomial can be written

$$q(s) = m(s - r)^2,$$

where $r = -\mu/2m < 0$. Since $\mathcal{B}_q = \{e^{rt}, te^r t\}$ the general solution of (9) is

$$y(t) = c_1 e^{rt} + c_2 t e^{rt} = (c_1 + c_2 t)e^{rt}. \tag{12}$$

In this case, there is no oscillatory behavior. In fact, the system will pass through equilibrium only if $t = -c_1/c_2$, and since $t > 0$, this only occurs if c_1 and c_2 have opposite signs. The following graph represents the two possibilities.

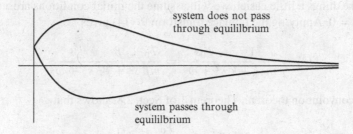

system does not pass
through equililbrium

system passes through
equililbrium

Overdamped Systems

When $\mu > \sqrt{4mk}$, then the discriminant D is positive. The resulting motion is called *overdamped free motion*. The characteristic polynomial $q(s)$ has two distinct real roots:

$$r_1 = \frac{-\mu + \sqrt{\mu^2 - 4mk}}{2m} \quad \text{and} \quad r_2 = \frac{-\mu - \sqrt{\mu^2 - 4mk}}{2m}.$$

Both roots are negative. Since $\mathcal{B}_q = \{e^{r_1 t}, e^{r_2 t}\}$, the general solution of (9) is

$$y(t) = c_1 e^{r_1 t} + c_2 e^{r_2 t}. \tag{13}$$

The graphs shown for the critically damped case are representative of the possible graphs for the present case as well.

Notice that in all three cases

$$\lim_{t \to \infty} y(t) = 0$$

and thus, the motion of $y(t)$ dies out as t increases.

Undamped Forced Motion

Undamped forced motion refers to the motion of a body governed by a differential equation

$$my'' + ky = f(t),$$

where $f(t)$ is a nonzero forcing function. We will only consider the special case where the forcing function is given by $f(t) = F_0 \cos \omega t$ where F_0 is a nonzero constant. Thus, we are interested in describing the solutions of the differential equation

$$my'' + ky = F_0 \cos \omega t, \tag{14}$$

where, as usual, $m > 0$ and $k > 0$. Imagine an engine embedded within a spring-body-dashpot system. The spring system has a characteristic frequency $\beta = \sqrt{k/m}$ while the engine exerts a cyclic force with frequency ω.

To make things a little easier, we will assume the initial conditions are $y(0) = 0$ and $y'(0) = 0$. Applying the Laplace transform to (14) gives

$$Y(s) = \frac{1}{ms^2 + k} \frac{F_0 s}{s^2 + \omega^2} = \frac{F_0}{m\beta} \frac{\beta}{s^2 + \beta^2} \frac{s}{s^2 + \omega^2}. \tag{15}$$

Then the convolution theorem, Theorem 1 of Sect. 2.8, shows that

$$y(t) = \mathcal{L}^{-1}(Y(s)) = \frac{F_0}{m\beta} \sin \beta t * \cos \omega t. \tag{16}$$

The following convolution formula comes from Table 2.11:

$$\sin \beta t * \cos \omega t = \begin{cases} \dfrac{\beta}{\beta^2 - \omega^2} (\cos \omega t - \cos \beta t) & \text{if } \beta \neq \omega \\ \dfrac{1}{2} t \sin \omega t & \text{if } \beta = \omega. \end{cases} \tag{17}$$

Combining equations (16) and (17) gives

$$y(t) = \begin{cases} \dfrac{F_0}{m(\beta^2 - \omega^2)} (\cos \omega t - \cos \beta t) & \text{if } \beta \neq \omega \\ \dfrac{F_0}{2m\omega} t \sin \omega t & \text{if } \beta = \omega. \end{cases} \tag{18}$$

We will first consider the case $\beta \neq \omega$ in (18). Notice that, in this case, the solution $y(t)$ is the sum of two cosine functions with equal amplitude $\left(= F_0/m(\beta^2 - \omega^2)\right)$, but different frequencies β and ω. Recall the trigonometric identity

$$\cos(\theta - \phi) - \cos(\theta + \phi) = 2 \sin \theta \sin \phi.$$

If we set $\theta - \phi = \omega t$ and $\theta + \phi = \beta t$ and solve for $\theta = (\beta + \omega)t/2$ and $\phi = (\beta - \omega)t/2$, we see that we can rewrite the first part of (18) in the form

$$y(t) = \frac{2F_0}{a(\beta^2 - \omega^2)} \sin \frac{(\beta - \omega)t}{2} \sin \frac{(\beta + \omega)t}{2}. \tag{19}$$

One may think of the function $y(t)$ as a sine function, namely, $\sin((\beta + \omega)t/2)$ (with frequency $(\beta + \omega)/2$), which is multiplied by another function, namely,

$$\frac{2F_0}{a(\beta^2 - \omega^2)} \sin \frac{(\beta - \omega)t}{2},$$

which functions as a time varying amplitude function.

An interesting case is when β is close to ω so that $\beta + \omega$ is close to 2ω and $\beta - \omega$ is close to 0. In this situation, one sine function changes very rapidly, while the other, which represents the change in amplitude, changes very slowly as is illustrated below. (In order for the solution to be periodic, we must also require that β/ω be a rational number. Periodicity of the solution is further discussed in Sect. 6.8.)

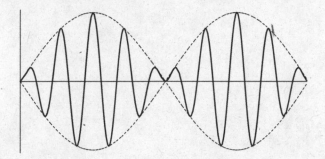

One might observe this type of motion in an unbalanced washing machine. The spinning action exerts a cyclic force, $F_0 \cos \omega t$, on the spring system which operates at a characteristic frequency β which is close to ω. The chaotic motion that results settles down momentarily only to repeat itself again. In music, this type of phenomenon, known as **beats**, can be heard when one tries to tune a piano. When the frequency of vibration of the string is close to that of the tuning fork, one hears a pulsating beat which disappears when the two frequencies coincide. The piano is slightly out of tune.

In the case where the input frequency and characteristic frequency are equal, $\beta = \omega$, in (18), the solution

$$y(t) = \frac{F_0}{2a\omega} t \sin \omega t$$

is unbounded as $t \to \infty$ as illustrated below.

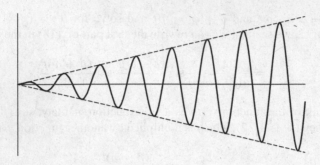

The resulting amplification of vibration eventually becomes large enough to destroy the mechanical system. This is a manifestation of **resonance** discussed further in Sects. 4.5 and 6.8.

Exercises

Assume forces are in pounds or newtons and lengths are in feet or meters.

1–4. Assume Hooke's law.

1. A body weighing 16 lbs stretches a spring 6 in. Find the spring constant.
2. The spring constant of a certain spring is $k = 20$ lbs/ft. If a body stretches the spring 9 inches, how much does it weigh?
3. A mass of 40 kilograms stretches a spring 80 centimeters. Find the spring constant.
4. The spring constant of a certain spring is $k = 784$ N/m. How far will a mass of 20 kilograms stretch the spring?

5–8. Assume the damping force of a dashpot is proportional to the velocity of the body.

5. A dashpot exerts a damping force of 4 lbs when the velocity of the mass is 6 inches per second. Find the damping constant.
6. A dashpot exerts a damping force of 40 Newtons when the velocity is 30 centimeters per second. Find the damping constant.
7. A dashpot has a damping constant $\mu = 100$ lbs s/ft and decelerates a body by 4 ft per second. What was the force exerted by the body?
8. A force of 40 N is applied on a body connected to a dashpot having a damping constant $\mu = 200$ N s/m. By how much will the dashpot decelerates the body?

9–14. For each exercise, investigate the motion of the mass. For undamped or underdamped motion, express the solution in the amplitude-phase form: that is, $y = Ae^{\alpha t} \cos(\beta t + \phi)$.

9. A spring is stretched 10 centimeters by a force of 2 Newtons. A body of mass 6 kilogram is attached to such a spring with no dashpot. At $t = 0$, the mass is pulled down from its equilibrium position a distance of 10 centimeters and released. Find a mathematical model that represents the motion of the body and solve. Determine the resulting motion. What is the amplitude, frequency, and phase shift?
10. A body of mass 4 kg will stretch a spring 80 centimeters. This same body is attached to such a spring with an accompanying dashpot. Suppose the damping force is 98 N when the velocity of the body is 2 m/s. At $t = 0$, the mass is given an initial upward velocity of 50 centimeters per second from its equilibrium position. Find a mathematical model that represents the motion of the body and solve. Determine the resulting motion. After release, does the mass every cross equilibrium? If so, when does it first cross equilibrium?
11. A body weighing 16 pounds will stretch a spring 6 inches. This same body is attached to such a spring with an accompanying dashpot. Suppose the damping force is 4 pounds when the velocity of the body is 2 feet per second. At $t = 0$,

the mass is pulled down from its equilibrium position a distance of 1 foot and released with a downward velocity of 1 foot per second. Find a mathematical model that represents the motion of the body and solve.

12. A body weighing 32 pounds will stretch a spring 2 feet. This same body is attached to such a spring with an accompanying dashpot. Suppose the damping constant is 8 lbs s/ft. At $t = 0$, the mass is pulled up from its equilibrium position a distance of 1 foot and released. Find a mathematical model that represents the motion of the body and solve. Determine the resulting motion. After release, does the mass every cross equilibrium?

13. A body weighing 2 pounds will stretch a spring 4 inches. This same body is attached to such a spring with no accompanying dashpot. At $t = 0$, the body is pushed downward from equilibrium with a velocity of 8 inches per second. Find a mathematical model that represents the motion of the body and solve. Determine the resulting motion. After release does the mass every cross equilibrium?

14. A spring is stretched 1 m by a force of 5 N. A body of mass 2 kg is attached to the spring with, accompanying dashpot. Suppose the damping force of the dashpot is 6 N when the velocity of the body is 1 m/s. At $t = 0$, the mass is pulled down from its equilibrium position a distance of 10 centimeters and given an initial downward velocity of 10 centimeters per second. Find a mathematical model that represents the motion of the body and solve. Determine the resulting motion. After release, does the mass every cross equilibrium?

15. Suppose m, μ, and k are positive. Show that the roots of the polynomial $q(s) = ms^2 + \mu s + k$

   1. Are negative if the roots are real.
   2. Have negative real parts if the roots are complex.

   Conclude that a solution to $my'' + \mu y' + ky = 0$ satisfies

   $$\lim_{t \to \infty} y(t) = 0.$$

16. Prove that a solution to an overdamped or critically damped system

   $$my'' + \mu y' + ky = 0$$

   crosses equilibrium at most once regardless of the initial conditions.

3.7 RCL Circuits

In this section, we consider RCL series circuits. These are simple electrical circuits with a resistor, capacitor, and inductor connected to a power source in series. The diagram below gives the basic components which we discuss below.

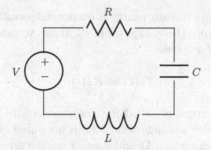

Charge

A charge is a fundamental property of matter that exhibits electrostatic attraction or repulsion. An electron is said to have a charge of -1 and a proton has a charge of $+1$. A *coulomb*, abbreviated C, the basic unit of measuring charge, is equivalent to the charge of about 6.242×10^{18} protons.

Current

The rate at which charged particles flow through a conductor is called *current*. Thus, if $q(t)$ represents the charge at a cross section in the circuit at time t, then current, $I(t)$, is given by

$$I(t) = q'(t).$$

Current is measured in *amperes*, abbreviated amp, which is coulombs per second, coulomb/s. In *conventional flow*, $I(t)$ is positive when charge flows from the positive terminal of the supply V.

Voltage

As electrons move through a circuit, they exchange energy with its components. The standard unit of energy is the *joule*, abbreviated J. *Voltage* is defined to be the quotient between energy and charge and is measured in joules per coulomb, J/C. We let $E(t)$ be the source voltage, V, in the diagram.

The charge, current, and voltage all obey basic fundamental laws with respect to the components in a RCL circuit. We discuss these laws for each component below.

Resistor

As current flows through a resistor, energy is exchanged (usually in the form of heat) resulting in a voltage drop. The voltage drop, V_R, from one side of the resistor to the other is governed by *Ohm's law*:

$$V_{\text{R}}(t) = RI(t), \tag{1}$$

where R is a positive constant called the **resistance** of the resistor. The unit of resistance, called the **ohm** and abbreviated Ω, is measured in voltage per ampere, V/A. For example, a resistor of 1 Ω will cause a voltage drop of 1 V if the current is 1 A.

Capacitor

A **capacitor** consists of two parallel conducting plates separated by a thin insulator. Current flows into a plate increasing the positive charge on one side and the negative charge on the other. The result is an electric field in the insulator between the plates that stores energy. That energy comes from a voltage drop across the capacitor and is governed by the **capacitor law**:

$$V_{\text{C}}(t) = \frac{1}{C}q(t),$$

where $q(t)$ is the charge on the capacitor and C is a positive constant called the **capacitance** of the capacitor. The unit of capacitance, called a *farad* and abbreviated F, is measured in charge per voltage, C/V. For example, a 1-farad capacitor will hold a charge of 1 C at a voltage drop of 1 volt. Since a coulomb is so large, capacitance is sometimes measured in millifarad (1 mF = 10^{-3} F) or microfarad (1 μF = 10^{-6}F).

Inductor

An **inductor** is typically built by wrapping a conductor such as copper wire in the shape of a coil around a core of ferromagnetic material.[3] As a current flows, a

---

[3]Sometimes other materials are present.

Table 3.3 Standard units of measurement for RCL circuits

Quantity	Symbol	Unit of measurement	unit symbol	relation to other units
Energy		Joule	J	
Time	t	Second	s	
Charge	q	Coulomb	C	
Voltage	E or V	Volt	V	J/C
Current	I	Ampere	A	C/s
Resistance	R	Ohm	Ω	V/A
Capacitance	C	Farad	F	C/V
Inductance	L	Henry	H	V/(A/s)

magnetic field about the inductor forms which stores energy and resists any change in current. The resulting voltage drop is governed by *Faraday's law*:

$$V_L = LI'(t), \tag{2}$$

where L is a positive constant called the **inductance** of the inductor. The unit of inductance, called the **henry** and abbreviated H, is measured in voltage per change in ampere, V/(A/s). For example, an inductor with an inductance of 1 H produces a voltage drop of 1 V when the current through the inductor changes at a rate of 1 A/s.

Table 3.3 summarizes the standard units of measurement for RCL circuits.

Kirchoff's Laws

There are two laws that govern the behavior of the current and voltage drops in a closed circuit due to Gustaf Kirchoff.

Kirchoff's Current Law

The sum of the currents flowing into and out of a point on a closed circuit is zero.

Kirchoff's Voltage Law

The sum of the voltage drops around a closed circuit is zero.

The voltage drop across the voltage source is $-E(t)$ in conventional flow. Thus, Kirchoff's voltage law implies $V_R + V_C + V_L - E = 0$. Now using Ohm's law, the capacitor law, and Faraday's law, we get

$$RI(t) + \frac{1}{C}q(t) + LI'(t) = E(t). \tag{3}$$

Table 3.4   Spring-body-mass and RCL circuit correspondence

Spring system		RCL circuit	
$my'' + \mu y' + ky = f(t)$		$Lq'' + Rq' + (1/C)q = E(t)$	
Displacement	y	Charge	q
Velocity	y'	Current	$q' = I$
Mass	m	Inductance	L
Damping constant	μ	Resistance	R
Spring constant	k	(Capacitance)$^{-1}$	$1/C$
Forcing function	$f(t)$	Applied voltage	$E(t)$

Now using the fact that $q'(t) = I(t)$, we can rewrite (3) to get

$$Lq''(t) + Rq'(t) + \frac{1}{C}q(t) = E(t). \tag{4}$$

In many applications, we are interested in the current $I(t)$. If we differentiate (4) and use $I(t) = q'(t)$, we get

$$LI''(t) + RI' + \frac{1}{C}I = E'(t). \tag{5}$$

Frequently, we are given the initial charge on the capacitor $q(0)$ and the initial current $I(0)$. By evaluating (3) at $t = 0$, we obtain $I'(0)$.

    The equations that model RCL circuits, (4) and (5), and the equation that models the spring-body-dashpot system, (5) of Sect. 3.6, are essentially the same. Both are second order constant coefficient linear differential equations. Table 3.4 gives the correspondence between the coefficients. The formulas in the coefficients m, μ, and k that describe concepts such as harmonic motion; underdamped, critically damped, and overdamped motion; resonance; etc. for spring systems, have a correspondence in the coefficients R, C, and L for RCL simple circuits. For example, simple harmonic motion with frequency $\beta = \sqrt{k/m}$ in a spring system occurs when there is no dashpot. Likewise, simple harmonic motion with frequency $\beta = 1/\sqrt{LC}$ in an RCL circuit occurs when there is no resistor.

Example 1. A circuit consists of a capacitor and inductor joined in series as illustrated.

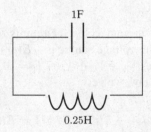

1F

0.25H

There is no voltage supply. Suppose the capacitor has a capacitance of 1 F and initial charge of $q(0) = 0.2$ C. Suppose the inductor has inductance 0.25 H. If there is no initial current find the current, $I(t)$, at time t. What is the system frequency, amplitude, and phase angle? What is the charge on the capacitor at time $t = \pi/4$.

▶ **Solution.** We use (5) to model the current and get

$$0.25I'' + I = 0. \tag{6}$$

We have $I(0) = 0$. To determined $I'(0)$, we evaluate (3) at $t = 0$ and use $I(0) = 0$, $q(0) = 0.2$, and $E(0) = 0$ to get $0.2 + 0.25I'(0) = 0$, and hence, $I'(0) = -0.2/0.25 = -0.8$. We now multiply (6) by 4 to get the following initial value problem:

$$I'' + 4I = 0, \quad I(0) = 0, \ I'(0) = -0.8.$$

A simple calculation gives

$$I(t) = -0.4 \sin 2t = 0.4 \cos \left(2t + \frac{\pi}{2}\right).$$

From the phase-amplitude form, it follows that the system frequency is 2, the amplitude is 0.4, and the phase angle is $-\pi/2$ (see the explanation after 3.6.(8) where these terms are defined). Since

$$q(t) - 0.2 = q(t) - q(0)$$

$$= \int_0^t I(\tau) \, d\tau$$

$$= (-0.4) \left.\frac{-\cos 2\tau}{2}\right|_0^t = 0.2(\cos 2t - 1)$$

it follows that $q(t) = 0.2 \cos 2t$. Hence, the charge on the capacitor at $t = \pi/4$ is $q(\pi/4) = 0$ coulombs.   ◀

Example 2. A resistor, capacitor, and inductor, are connected in series with a voltage supply of 14 V as illustrated below.

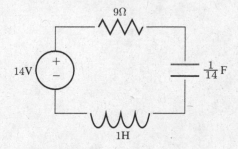

Find the charge on the capacitor at time t if the initial charge and initial current are 0. In the long term, what will be the charge on the capacitor and the current in the circuit.

▶ **Solution.** We have $q(0) = 0$ and $q'(0) = I(0) = 0$. Equation (4) gives

$$q'' + 9q' + 14q = 14, \quad q(0) = 0, \ q'(0) = 0.$$

The Laplace transform method gives

$$Q(s) = \frac{14}{s(s+2)(s+7)} = \frac{1}{s} + \frac{2}{5}\frac{1}{s+7} - \frac{7}{5}\frac{1}{s+2}.$$

Hence,

$$q(t) = 1 + \frac{2}{5}e^{-7t} - \frac{7}{5}e^{-2t}.$$

Observe that $\lim_{t\to\infty} q(t) = 1$. This means that in the long term the charge on the capacitor will be 1C. Since $I(t) = q'(t) = \frac{14}{5}\left(e^{-2t} - e^{-7t}\right)$ and $\lim_{t\to\infty} I(t) = 0$, there will be no current flowing in the long term. ◀

Exercises

1–4. Find the current and charge for each RCL series circuit from the data given below. Where appropriate, express the current in phase-amplitude form: that is, $I(t) = Ae^{-\alpha t} \cos(\beta t - \delta)$.

1. $R = 10\,\Omega, C = 5\,\text{mF}, L = 0.25\,\text{H}, V = 6\,\text{V}, q(0) = 0\,\text{C}, I(0) = 0\,\text{A}$.
2. $R = 5\,\Omega, C = .025\,\text{F}, L = 0.1\,\text{H}, V = 0\,\text{V}, q(0) = 0.01\,\text{C}, I(0) = 0\,\text{A}$.
3. $R = 4\,\Omega, C = .05\,\text{F}, L = 0.2\,\text{H}, V = 25\sin 5t\,\text{V}, q(0) = 0\,\text{C}, I(0) = 2\,\text{A}$.
4. $R = 11\,\Omega, C = 1/30\,\text{F}, L = 1\,\text{H}, V = 10e^{-5t}\,\text{V}, q(0) = 1\,\text{C}, I(0) = 2\,\text{A}$.

5. An RCL circuit consists of a 0.1-H inductor and a 0.1-F capacitor. The capacitor will fail if it reaches a charge greater than 2 C. Assume there is no initial charge on the capacitor and no initial current. A voltage supply is connected to the circuit with alternating current given by $V = (1/10)\cos 5t$. Determine the charge and whether the capacitor will fail.

6. An RCL circuit consists of a 0.1-H inductor and a 0.1-F capacitor. The capacitor will fail if it reaches a charge greater than 2 C. Assume there is no initial charge on the capacitor and no initial current. A voltage supply is connected to the circuit with alternating current given by $V = (1/10)\cos 10t$. Determine the charge and whether the capacitor will fail.

Chapter 4
Linear Constant Coefficient Differential Equations

Two springs systems (without dashpots) are coupled as illustrated in the diagram below:

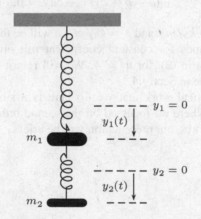

Let y_1 and y_2 denote the displacements of the bodies of mass m_1 and m_2 from their equilibrium positions, $y_1 = 0$ and $y_2 = 0$, respectively, where distances are measured in the downward direction. In these coordinates, $y_1(t)$ and $y_2(t) - y_1(t)$ represent the length the upper and lower springs are stretched at time t. There are two spring forces acting on the upper body. By Hooke's law, the force of the upper spring is $-k_1 y_1(t)$ while the force of the lower spring is given by $k_2(y_2(t) - y_1(t))$, where k_1 and k_2 are the respective spring constants. Newton's law of motion then implies

$$m_1 y_1''(t) = -k_1 y_1(t) + k_2(y_2(t) - y_1(t)).$$

W.A. Adkins and M.G. Davidson, *Ordinary Differential Equations*,
Undergraduate Texts in Mathematics, DOI 10.1007/978-1-4614-3618-8_4,
© Springer Science+Business Media New York 2012

The lower spring alone exerts a force of $-k_2(y_2(t) - y_1(t))$ on the lower body. Again Newton's law of motion implies

$$m_1 y_2''(t) = -k_2(y_2(t) - y_1(t)).$$

We are thus led to the following differential equations:

$$m_1 y_1''(t) + (k_1 + k_2)y_1(t) = k_2 y_2(t)$$
$$m_2 y_2''(t) + k_2 y_2(t) = k_2 y_1(t). \tag{1}$$

Equation (1) is an example of a *coupled system*[1] of differential equations. Notice the dependency of each equation on the other: the solution of one equation becomes (up to a multiplicative constant) the forcing function of the other equation. In Sect. 4.4, we will show that both y_1 and y_2 satisfy the fourth order differential equation

$$y^{(4)} + (a + b + c)y'' + cay = 0, \tag{2}$$

where $a = k_1/m_1$, $b = k_2/m_2$, and $c = k_2/m_2$. It will be the work of this chapter to show solution methods for constant coefficient nth order linear differential equations such as given in (2), for $n = 4$. We will return to the coupled spring problem presented above in Sect. 4.4.

Much of the theoretical work that we do here is a simple extension of the work done in Chap. 3 where we focussed on the second order constant coefficient differential equations. Thus, our presentation will be brief.

---

[1]In Chap. 9, we will study first order coupled systems in greater detail.

4.1 Notation, Definitions, and Basic Results

Suppose $a_0, \ldots, a_n$ are scalars, $a_n \neq 0$, and f is a function defined on some interval I. A differential equation of the form

$$a_n y^{(n)} + a_{n-1} y^{(n-1)} + \cdots + a_1 y' + a_0 y = f(t), \tag{1}$$

is called a ***constant coefficient nth order linear differential equation***. The constants $a_0, \ldots, a_n$ are called the ***coefficients***, and a_n is called the ***leading coefficient***. By dividing by a_n, if necessary, we can assume that the leading coefficient is 1. When $f = 0$, we call (1) ***homogeneous***, otherwise, it is called ***nonhomogeneous***.

The left-hand side of (1) is made up of a linear combination of differentiations and multiplications by constants. If D denotes the derivative operator as in Sect. 3.3, then $D^k(y) = y^{(k)}$. Let

$$L = a_n D^n + \cdots + a_1 D + a_0, \tag{2}$$

where $a_0, \ldots a_n$ are the same constants given in (1). Then $L(y) = a_n y^{(n)} + \cdots + a_1 y' + a_0 y$ and (1) can be rewritten

$$L(y) = f.$$

We call L in (2) a ***linear constant coefficient differential operator of order*** n (assuming, of course, that $a_n \neq 0$). Since L is a linear combination of powers of D, we will sometimes refer to it as a ***polynomial differential operator***. Specifically, if

$$q(s) = a_n s^n + a_{n-1} s^{n-1} + \cdots + a_1 s + a_0,$$

then q is a polynomial of order n. Since L is obtained from q by substituting D for s, and we will write $L = q(D)$. The polynomial q is referred as the ***characteristic polynomial*** of L. The operator L takes a function y that has at least n continuous derivatives and produces a continuous function.

Example 1. Suppose $L = D^3 + D$. Find

$$L(\sin t), \qquad L(\cos t), \qquad L(1), \qquad \text{and} \qquad L(e^t)$$

▶ **Solution.** • $L(\sin t) = D^3(\sin t) + D(\sin t)$
$$= -\cos t + \cos t$$
$$= 0.$$

• $L(\cos t) = D^3(\cos t) + D(\cos t)$
$$= \sin t - \sin t$$
$$= 0.$$

- $L(1) = D^3(1) + D(1)$
  $= 0.$
- $L(e^t) = D^3(e^t) + D(e^t)$
  $= (e^t) + (e^t)$
  $= 2e^t.$ ◄

We say a function y is a **homogeneous solution** of L if $Ly = 0$. Example 1 shows that $\sin t$, $\cos t$, and 1 are homogeneous solutions of $L = D^3 + D$.

Linearity and Consequences

It is easy to show by induction that $D^k(y_1 + y_2) = D^k y_1 + D^k y_2$ and $D^k cy = cD^k y$. The proofs of Propositions 2 and 4 of Sect. 3.3 and Theorem 6 of Sect. 3.3 extend to give what we put below in one theorem.

Theorem 2. *1. The operator*

$$L = a_n D^n + \ldots + a_1 D + a_0$$

given by (2) *is linear. Specifically,*

a. If y_1 and y_2 have n derivatives, then

$$L(y_1 + y_2) = L(y_1) + L(y_2).$$

b. If y has n derivatives and c is a scalar, then

$$L(cy) = cL(y).$$

2. If y_1 and y_2 are homogeneous solutions of L and c_1 and c_2 are scalars, then

$$c_1 y_1 + c_2 y_2$$

is a homogeneous solution.
3. Suppose f is a function. If y_p is a fixed particular solution to $Ly = f$ and y_h is any solution to the associated homogeneous differential equation $Ly = 0$, then

$$y_p + y_h$$

is a solution to $Ly = f$. Furthermore, any solution y to $Ly = f$ has the form

$$y = y_p + y_h.$$

Given a constant coefficient linear differential equation $Ly = f$, we call $Ly = 0$ the *associated homogeneous differential equation*. The following algorithm follows from Theorem 2 and outlines an effective strategy for finding the solution set to $Ly = f$.

Algorithm 3. The general solution to a linear differential equation

$$Ly = f$$

can be found as follows:

Solution Method for nth Order Linear Equations

1. Find all the solutions to the associated homogeneous differential equation
   $Ly = 0$
2. Find one particular solution y_p
3. Add the particular solution to the homogeneous solutions:

$$y_p + y_h.$$

As y_h varies over all homogeneous solutions, we obtain all solutions to
$Ly = f$.

You should notice that this is the same strategy as in the second order case discussed in Algorithm 7 of Sect. 3.3. This is the strategy we will follow. Section 4.2 will be devoted to determining solutions to the associated homogeneous differential equation. Section 4.3 will show how to find a particular solution.

Example 4. Use Algorithm 3 and Example 1 to find solutions to

$$y''' + y' = 2e^t.$$

▶ **Solution.** The left-hand side can be written Ly, where L is the differential operator
$$L = D^3 + D.$$

From Example 1, we found

- $L(e^t) = 2e^t$,
- $L(\sin t) = 0$,
- $L(\cos t) = 0$,
- $L(1) = 0$.

Notice that the first equation tells us that a particular solution is $y_p = e^t$. The second, third, and fourth equations give $\sin t$, $\cos t$, and 1 as solutions to the associated homogeneous differential equation $Ly = 0$. Thus, for each scalar c_1, c_2, and c_3, the function $y_h = c_1 \sin t + c_2 \cos t + c_3$ is a homogeneous solution. Now applying Algorithm 3, we have

$$y = y_p + y_h = e^t + c_1 \sin t + c_2 \cos t + c_3$$

is a solution to $Ly = 2e^t$, for all scalars c_1, c_2, and c_3. At this point, we cannot say that this is the solution set, but, in fact, it is.   ◄

Initial Value Problems

Suppose L is a constant coefficient linear differential operator of order n and f is a function defined on an interval I. Let $t_0 \in I$. To the equation

$$Ly = f$$

we can associate *initial conditions* of the form

$$y(t_0) = y_1, \ y'(t_0) = y_1, \ldots, y^{(n-1)}(t_0) = y_{n-1}.$$

We refer to the initial conditions and the differential equation $Ly = f$ as an *initial value problem*, just as in the second order case.

Theorem 5 (The Existence and Uniqueness Theorem). *Suppose p is an nth order real polynomial, $L = p(D)$, and f is a continuous real-valued function on an interval I. Let $t_0 \in I$. Then there is a unique real-valued function y defined on I satisfying*

$$Ly = f \quad y(t_0) = y_0, \ y'(t_0) = y_1, \ \ldots, \ y^{(n-1)}(t_0) = y_{n-1}, \tag{3}$$

where $y_0, \ y_1, \ \ldots, y_{n-1} \in \mathbb{R}$, If $I \supset [0, \infty)$ and f has a Laplace transform then so does the solution y. Furthermore, if f is in $\mathcal{E}$, then y is in $\mathcal{E}$.

When $n = 2$, we get the statement of the existence and uniqueness theorem, Theorem 10 of Sect. 3.1. Theorem 5 will be proved in a more general context in Sect. 9.5.

Example 6. Use Example 4 to find the solution to the following initial value problem:

$$Ly = 2e^t, \qquad y(0) = 2, \ y'(0) = 2, \ y''(0) = 3,$$

where $L = D^3 + D$.

▶ **Solution.** In Example 4, we verified that

$$y = e^t + c_1 \sin t + c_2 \cos t + c_3$$

is a solution to $Ly = 2e^t$ for all scalars c_1, c_2, and c_3. The initial conditions imply the following system:

$$
\begin{aligned}
1 \quad\quad + c_2 + c_3 &= y(0) = 2 \\
1 + c_1 \quad\quad\quad\quad &= y'(0) = 2 \\
1 \quad\quad - c_2 \quad\quad &= y''(0) = 3
\end{aligned}
$$

Solving this system gives $c_1 = 1$, $c_2 = -2$, and $c_3 = 3$. It follows that

$$y = e^t + \sin t - 2 \cos t + 3$$

and the existence and uniqueness theorem implies that y is the unique solution. ◀

Exercises

1–4. Determine which of the following are constant coefficient linear differential equations. In these cases, write the equation in the form $Ly = f$ and, determine the order, the characteristic polynomial, and whether they are homogeneous.

1. $y''' - 3y' = e^t$
2. $y^{(4)} + y' + 4y = 0$
3. $y^{(4)} + y^4 = 0$
4. $y^{(5)} + ty'' - 3y = 0$

5–8. For the linear operator L, determine $L(y)$.

5. $L = D^3 - 4D$

   (a) $y = e^{2t}$
   (b) $y = e^{-2t}$
   (c) $y = 2$

6. $L = D - 2$

   (a) $y = e^{-2t}$
   (b) $y = 3e^{2t}$
   (c) $y = \tan t$

7. $L = D^4 + 5D^2 + 4$

   (a) $y = e^{-t}$
   (b) $y = \cos t$
   (c) $y = \sin 2t$

8. $L = D^3 - D^2 + D - 1$

   (a) $y = e^t$
   (b) $y = te^t$
   (c) $y = \cos t$
   (d) $y = \sin t$

9. Suppose L is a polynomial differential operator and

   - $L(te^{2t}) = 8e^{2t}$
   - $L(e^{2t}) = 0$
   - $L(e^{-2t}) = 0$
   - $L(1) = 0$

   Use this information to find other solutions to $Ly = 8e^{2t}$.
10. Suppose L is a polynomial differential operator and

   - $L(\sin t) = -15 \sin t$
   - $L(e^{2t}) = 0$

- $L(e^{-2t}) = 0$
- $L(\sin 2t) = 0$
- $L(\cos 2t) = 0$

Use this information to find other solutions to $Ly = -15 \sin t$.

11. Let L be as in Exercise 9. Use the results there to solve the initial value problem

$$Ly = 8e^{2t},$$

where $y(0) = 2$, $y'(0) = -1$, and $y''(0) = 16$.

12. Let L be as in Exercise 10. Use the results there to solve the initial value problem

$$Ly = -15 \sin t,$$

where $y(0) = 0$, $y'(0) = 3$, $y''(0) = -16$, and $y'''(0) = -9$.

4.2   Linear Homogeneous Differential Equations

Suppose $L = q(D)$ where $q(s) = a_n s^n + \cdots + a_1 s + a_0$. As we will see in this section, the characteristic polynomial, q, plays a decisive role in determining the solution set to $Ly = 0$.

Lemma 1. *Let $q(s) = a_n s^n + \cdots + a_1 s + a_0$. If y is a function whose nth derivative is of exponential type, then*

$$\mathcal{L}\{q(D)y\} = q(s)\mathcal{L}\{y\} - p(s),$$

where $p(s)$ is a polynomial of degree at most $n - 1$ and depends on the coefficients of q and the initial conditions $y(0), y'(0), \ldots, y^{(n-1)}(0)$.

Proof. If $y^{(n)}$ is of exponential type, then so are all derivatives $y^{(k)}$, $k = 0, 1, \ldots$ $n - 1$ by Lemma 4 of Sect. 2.2. By the nth transform derivative principle, Corollary 8 of Sect. 2.2, we have

$$\mathcal{L}\left\{y^{(k)}\right\}(s) = s^k \mathcal{L}\{y\}(s) - p_k(s),$$

where $p_k(s)$ is a polynomial of order at most $k - 1$ depending on the initial conditions $y(0), y'(0), \ldots, y^{(k-1)}$. Let $\mathcal{L}\{y(t)\} = Y(s)$. Then

$$
\begin{aligned}
\mathcal{L}\{q(D)y\}(s) &= a_n \mathcal{L}\left\{y^{(n)}\right\} + \cdots + a_1 \mathcal{L}\{y'\} + a_0 \mathcal{L}\{y\} \\
&= a_n s^n Y(s) + \cdots + a_1 s Y(s) + a_0 Y(s) - p(s) \\
&= q(s)Y(s) - p(s),
\end{aligned}
$$

where $p(s) = a_n p_n + \cdots + a_0 p_0$ is a polynomial of degree at most $n - 1$ and depends on the initial values $y(0), y'(0), \ldots, y^{(n-1)}(0)$.   □

We now turn our attention to the solution set to

$$Ly = 0. \tag{1}$$

By Theorem 5 of Sect. 4.1, any solution to $L\{y\} = 0$ is in $\mathcal{E}$. Suppose y is such a solution. By Lemma 1, we have $\mathcal{L}\{Ly\} = q(s)\mathcal{L}\{y\} - p(s) = 0$. Solving for $\mathcal{L}\{y\}$ gives

$$\mathcal{L}\{y\}(s) = \frac{p(s)}{q(s)} \in \mathcal{R}_q.$$

It follows that $y \in \mathcal{E}_q$. Now suppose $y \in \mathcal{E}_q$. Then $\mathcal{L}\{y\}(s) = \frac{p(s)}{q(s)} \in \mathcal{R}_q$ and

$$\mathcal{L}\{Ly\}(s) = q(s)\frac{p(s)}{q(s)} - p_1(s) = p(s) - p_1(s),$$

where $p_1(s)$ is a polynomial that depends on the initial conditions. Note, however, that $p(s) - p_1(s)$ is a polynomial in $\mathcal{R}$ and therefore must be identically 0. Thus, $\mathcal{L}\{Ly\} = 0$ and this implies $Ly = 0$. The discussion given above implies the following theorem.

Theorem 2. *Let $q(s)$ be the characteristic polynomial of a linear differential operator $q(D)$ of order n. Then the solution set to*

$$q(D)y = 0$$

is $\mathcal{E}_q$. Thus, if $\mathcal{B}_q = \{y_1, y_2, \ldots, y_n\}$ is the standard basis of $\mathcal{E}_q$, then a solution to $q(D)y = 0$ is of the form

$$y = c_1 y_1 + c_2 y_2 + \cdots + c_n y_n,$$

where $c_1, c_2, \ldots, c_n$ are scalars.

The following algorithm codifies the procedure needed to find the solution set.

Algorithm 3. Given an nth order constant coefficient linear differential equation

$$q(D)y = 0$$

the solution set is determined as follows:

Solution Method for nth order Homogeneous Linear Differential Equations

1. Determine the characteristic polynomial, $q(s)$.
2. Factor $q(s)$ and construct $\mathcal{B}_q = \{y_1, y_2, \ldots, y_n\}$.
3. The solution set $\mathcal{E}_q$ is the set of all linear combinations of the functions in the standard basis $\mathcal{B}_q$. In other words,

$$\mathcal{E}_q = \text{Span } \mathcal{B}_q = \{c_1 y_1(t) + \cdots + c_n y_n(t) : c_1, \ldots, c_n \in \mathbb{R}\}.$$

Example 4. Find the general solution to the following differential equations:

1. $y''' + y' = 0$
2. $y^{(4)} - y = 0$
3. $y^{(5)} - 8y^{(3)} + 16y' = 0$

▶ **Solution.** 1. The characteristic polynomial for $y''' + y' = 0$ is

$$q(s) = s^3 + s = s(s^2 + 1).$$

The standard basis is $\mathcal{B}_q = \{1, \cos t, \sin t\}$. Thus the solution set is

$$\mathcal{E}_q = \{c_1 + c_2 \cos t + c_3 \sin t : c_1, c_2, c_3 \in \mathbb{R}\}.$$

2. The characteristic polynomial for $y^{(4)} - y = 0$ is

$$q(s) = s^4 - 1 = (s^2 + 1)(s - 1)(s + 1).$$

The standard basis is $\mathcal{B}_q = \{\cos t, \sin t, e^t, e^{-t}\}$. Thus, the solution set is

$$\mathcal{E}_q = \{c_1 \cos t + c_2 \sin t + c_3 e^t + c_4 e^{-t} : c_1, c_2, c_3, c_4 \in \mathbb{R}\}.$$

3. The characteristic polynomial for $y^{(5)} - 8y^{(3)} + 16y' = 0$ is

$$q(s) = s^5 - 8s^3 + 16s = s(s^4 + 8s^2 + 16) = s(s^2 - 4)^2 = s(s - 2)^2(s + 2)^2.$$

The standard basis is $\mathcal{B}_q = \{1, e^{2t}, te^{2t}, e^{-2t}, te^{-2t}\}$. Thus, the solution set is

$$\mathcal{E}_q = \{c_1 + c_2 e^{2t} + c_3 t e^{2t} + c_4 e^{-2t} + c_5 t e^{-2t} : c_1, \ldots, c_5 \in \mathbb{R}\}. \quad \blacktriangleleft$$

Initial Value Problems

Let $q(D)y = 0$ be a homogeneous constant coefficient differential equation, where $\deg p = n$. Suppose $y(t_0) = y_0, \ldots, y^{(n-1)}(t_0) = y_{n-1}$ are initial conditions. The existence and uniqueness Theorem 5 of Sect. 4.1 states there is a unique solution to the initial value problem. However, the general solution is a linear combination of the n functions in the standard basis $\mathcal{E}_q$. It follows then that the initial conditions uniquely determine the coefficients.

Example 5. Find the solution to

$$y''' - 4y'' + 5y' - 2y = 0; \quad y(0) = 1, \ y'(0) = 2, \ y''(0) = 0.$$

▶ **Solution.** The characteristic polynomial is $q(s) = s^3 - 4s^2 + 5s - 2$ and factors as

$$q(s) = (s - 1)^2(s - 2).$$

The basis for $\mathcal{E}_q$ is $\{e^t, te^t, e^{2t}\}$ and the general solution is

$$y = c_1 e^t + c_2 t e^t + c_3 e^{2t}.$$

We first calculate the derivatives and simplify to get

$$y(t) = c_1 e^t + c_2 t e^t + c_3 e^{2t}$$
$$y'(t) = (c_1 + c_2) e^t + c_2 t e^t + 2c_3 e^{2t}$$
$$y''(t) = (c_1 + 2c_2) e^t + c_2 t e^t + 4c_3 e^{2t}.$$

To determine the coefficients c_1, c_2, and c_3, we use the initial conditions. Evaluating at $t = 0$ gives

$$1 = y(0) = c_1 + c_3$$
$$2 = y'(0) = c_1 + c_2 + 2c_3$$
$$0 = y''(0) = c_1 + 2c_2 + 4c_3.$$

Solving these equations gives $c_1 = 4$, $c_2 = 4$, and $c_3 = -3$. The unique solution is thus

$$y(t) = 4e^t + 4t e^t - 3e^{2t}.$$ ◀

Abel's Formula

Theorem 6 (Abel's Formula). *Let $q(s) = s^n + a_{n-1} s^{n-1} + \cdots + a_1 s + a_0$ and suppose $f_1, f_2, \ldots, f_n$ are solutions to $q(\boldsymbol{D})y = 0$. Then the Wronskian satisfies*

$$w(f_1, f_2, \ldots, f_n) = K e^{-a_{n-1} t}, \tag{2}$$

for some constant K and $\{f_1, f_2, \ldots, f_n\}$ is linearly independent if and only if $w(f_1, f_2, \ldots, f_n)$ is nonzero.

Proof. The essential idea of the proof is the same as the proof of Abel's Formula for $n = 2$ (see Theorem 8 of Sect. 3.3). First observe that since f_1 is a solution to $q(\boldsymbol{D})y = 0$, we have $f_1^{(n)} = -a_{n-1} f_1^{(n-1)} - a_{n-2} f_1^{(n-2)} - \cdots - a_1 f_1' - a_0 f_1$ and similarly for $f_2, \ldots, f_n$. To simplify the notation, let $w = w(f_1, f_2, \ldots, f_n)$ and let $\boldsymbol{F} = f_1, f_2, \ldots, f_n$ be the first row of the Wronskian matrix $W(f_1, \ldots, f_n)$. Then $\boldsymbol{F}^{(k)} = f_1^{(k)}, \ldots, f_n^{(k)}$ is the $k + 1st$ row of the Wronskian matrix, for all $k = 1, 2, \ldots, n - 1$, and

$$\boldsymbol{F}^{(n)} = -a_{n-1} \boldsymbol{F}^{(n-1)} - a_{n-2} \boldsymbol{F}^{(n-2)} - \cdots - a_1 \boldsymbol{F}' - a_0 \boldsymbol{F}. \tag{3}$$

We can thus write

$$w = \det \begin{bmatrix} F \\ F' \\ F'' \\ \vdots \\ F^{(n-2)} \\ F^{(n-1)} \end{bmatrix}.$$

The derivative of w is the sum of the determinants of the matrices obtained by differentiating each row one at a time. Thus,

$$w' = \det \begin{bmatrix} F' \\ F' \\ F'' \\ \vdots \\ F^{(n-2)} \\ F^{(n-1)} \end{bmatrix} + \det \begin{bmatrix} F \\ F'' \\ F'' \\ \vdots \\ F^{(n-2)} \\ F^{(n-1)} \end{bmatrix} + \cdots + \det \begin{bmatrix} F \\ F' \\ F'' \\ \vdots \\ F^{(n-1)} \\ F^{(n-1)} \end{bmatrix} + \det \begin{bmatrix} F \\ F' \\ F'' \\ \vdots \\ F^{(n-2)} \\ F^{(n)} \end{bmatrix}$$

All determinants but the last are zero because the matrices have two equal rows. Using (3) and linearity of the determinant gives

$$w' = -a_{n-1} \det \begin{bmatrix} F \\ F' \\ \vdots \\ F^{(n-2)} \\ F^{(n-1)} \end{bmatrix} - a_{n-2} \det \begin{bmatrix} F \\ F' \\ \vdots \\ F^{(n-2)} \\ F^{(n-2)} \end{bmatrix} - \cdots - a_0 \det \begin{bmatrix} F \\ F' \\ \vdots \\ F^{(n-2)} \\ F \end{bmatrix}$$

$$= -a_{n-1} w.$$

The first determinant is w and the remaining determinants are zero, again because the matrices have two equal rows. Therefore, w satisfies the differential equation $w' + a_{n-1}w = 0$. By Theorem 2 of Sect. 1.5, there is a constant K so that

$$w(t) = K e^{-a_{n-1}t}.$$

If $w \neq 0$, then it follows from Theorem 13 of Sect. 3.2 that $\{f_1, f_2, \ldots, f_n\}$ is a linearly independent set.

If $w = 0$, then the Wronskian matrix at $t = 0$ is singular and there are scalars $c_1, \ldots, c_n$, not all zero, so that

$$
W(0) \begin{bmatrix} c_1 \\ c_2 \\ \vdots \\ c_n \end{bmatrix} = \begin{bmatrix} F(0) \\ F'(0) \\ F''(0) \\ \vdots \\ F^{(n-1)}(0) \end{bmatrix} \begin{bmatrix} c_1 \\ c_2 \\ \vdots \\ c_n \end{bmatrix} = \begin{bmatrix} 0 \\ 0 \\ \vdots \\ 0 \end{bmatrix}.
$$

Let $z(t) = c_1 f_1(t) + \ldots c_n f_n(t)$. Then z is a linear combination of $f_1, \ldots, f_n$ and hence is a solution to $q(D)y = 0$. Further, from the matrix product given above, $z(0) = 0, z'(0) = 0, \ldots, z^{(n-1)}(0) = 0$. By the uniqueness and existence theorem, Theorem 5 of Sect. 4.1, it follows that $z(t) = 0$ for all $t \in \mathbb{R}$. This implies that $\{f_1, f_2, \ldots, f_n\}$ is linearly dependent. $\square$

Exercises

1–9. Determine the solution set to the following homogeneous differential equations. Write your answer as a linear combination of functions from the standard basis.

1. $y''' - y = 0$
2. $y''' - 6y'' + 12y' - 8y = 0$
3. $y^{(4)} - y = 0$
4. $y''' + 2y'' + y' = 0$
5. $y^{(4)} - 5y'' + 4y = 0$
6. $(D - 2)(D^2 - 25)y = 0$
7. $(D + 2)(D^2 + 25)y = 0$
8. $(D^2 + 9)^3 y = 0$
9. $(D + 3)(D - 1)(D + 3)^2 y = 0$

10–11. Solve the following initial value problems.

10. $y''' + y'' - y' - y = 0$,
    $y(0) = 1, y'(0) = 4, y''(0) = -1$
11. $y^{(4)} - y = 0$,
    $y(0) = -1, y'(0) = 6, y''(0) = -3, y'''(0) = 2$

4.3 Nonhomogeneous Differential Equations

In this section, we are concerned with determining a particular solution to

$$q(\boldsymbol{D})y = f,$$

where $q(s)$ is a polynomial of degree n and $f \in \mathcal{E}$. In Chap. 3, we discussed two methods for the second order case: The method of undetermined coefficients and the incomplete partial fraction method. Both of these methods are not dependent on the degree of q so extend quite naturally.

The Method of Undetermined Coefficients

Theorem 1. *Suppose $\boldsymbol{L} = q(\boldsymbol{D})$ is a polynomial differential operator and $f \in \mathcal{E}$. Suppose $\mathcal{L}f = \frac{u}{v}$. Let $\mathcal{B}_q$ denote the standard basis for $\mathcal{E}_q$. Then there is a particular solution y_{p} to*

$$\boldsymbol{L}y = f$$

which is a linear combination of terms in $\mathcal{B}_{qv}$ but not in $\mathcal{B}_q$.

Proof. By Theorem 5 of Sect. 4.1, any solution to $\boldsymbol{L}y = f$ is in $\mathcal{E}$ and hence has a Laplace Transform. Thus,

$$\mathcal{L}\{q(\boldsymbol{D})\}\,y) = \mathcal{L}\{f\} = \frac{u(s)}{v(s)}$$

$$q(s)\mathcal{L}\{y\} - p(s) = \frac{u(s)}{v(s)} \quad \text{by Lemma 1 of Sect. 4.2}$$

$$\mathcal{L}\{y\} = = \frac{p(s)}{q(s)} + \frac{u(s)}{q(s)v(s)} = \frac{u(s) + p(s)v(s)}{q(s)v(s)}.$$

It follows that y is in $\mathcal{E}_{qv}$ and hence a linear combination of terms in $\mathcal{B}_{qv}$. Since $\mathcal{B}_q \subset \mathcal{B}_{qv}$, we can write $y = y_{\mathrm{h}} + y_{\mathrm{p}}$, where y_{h} is the linear combination of terms in $\mathcal{B}_q$ and y_{p} is a linear combination of terms in $\mathcal{B}_{qv}$ but not in $\mathcal{B}_q$. Since y_{h} is a homogeneous solution, it follows that $y_{\mathrm{p}} = y - y_{\mathrm{h}}$ is a particular solution of the required form. $\square$

If $\{\phi_1, \ldots, \phi_m\}$ are the functions in $\mathcal{B}_{qv}$ but not in $\mathcal{B}_q$, then a linear combination

$$a_1\phi_1 + \cdots + a_m\phi_m$$

will be referred to as a ***test function***.

Example 2. Find the general solution to

$$y^{(4)} - y = 4\cos t.$$

▶ **Solution.** The characteristic polynomial is $q(s) = s^4 - 1 = (s^2 + 1)(s^2 - 1) = (s^2 + 1)(s - 1)(s + 1)$. It follows that $\mathcal{B}_q = \{\cos t, \sin t, e^t, e^{-t}\}$, and hence, the homogeneous solutions take the form

$$y_h = c_1 \cos t + c_2 \sin t + c_3 e^t + c_4 e^{-t}.$$

For the particular solution, note that $\mathcal{L}\{\cos t\} = \frac{s}{s^2+1}$. Let $v(s) = s^2 + 1$ be the denominator. Then $q(s)v(s) = (s^2 + 1)^2(s - 1)(s + 1)$ and

$$\mathcal{B}_{qv} = \{\cos t, t\cos t, \sin t, t\sin t, e^t, e^{-t}\}.$$

The only functions in $\mathcal{B}_{qv}$ that are not in $\mathcal{B}_q$ are $t\cos t$ and $t\sin t$. It follows that a test function takes the form $y_p = a_1 t \cos t + a_2 t \sin t$. The coefficients are determined by substituting y_p into the given differential equation. To that end, observe that

$$y_p = a_1 t \cos t + a_2 t \sin t$$
$$y_p' = a_1(\cos t - t \sin t) + a_2(\sin t + t \cos t)$$
$$y_p'' = a_1(-2 \sin t - t \cos t) + a_2(2 \cos t - t \sin t)$$
$$y_p''' = a_1(-3 \cos t + t \sin t) + a_2(-3 \sin t - t \cos t)$$
$$y_p^{(4)} = a_1(4 \sin t + t \cos t) + a_2(-4 \cos t + t \sin t)$$

and thus,

$$y_p^{(4)} - y_p = 4a_1 \sin t + -4a_2 \cos t = 4\cos t.$$

The linear independence of $\{\cos t, \sin t\}$ and Theorem 7 of Sect. 3.2 imply

$$4a_1 = 0$$
$$-4a_2 = 4.$$

Hence, $a_1 = 0$ and $a_2 = -1$ from which we get

$$y_p = -t \sin t.$$

The general solution is thus

$$y = y_p + y_h = -t \sin t + c_1 \cos t + c_2 \sin t + c_3 e^t + c_4 e^{-t}. \qquad \blacktriangleleft$$

We can summarize Theorem 1 and the previous example in the following algorithm.

Algorithm 3. If

$$q(\boldsymbol{D})y = f,$$

where q is a polynomial of degree n and $f \in \mathcal{E}$, then the general solution can be obtained as follows:

The Method of Undetermined Coefficients

1. Compute the standard basis, $\mathcal{B}_q$, for $\mathcal{E}_q$.
2. Determine v so that $\mathcal{L}\{f\} = \frac{u}{v}$. That is, $f \in \mathcal{E}_v$.
3. Compute the standard basis, $\mathcal{B}_{vq}$, for $\mathcal{E}_{vq}$.
4. The test function, y_p is the linear combination with arbitrary coefficients of functions in $\mathcal{B}_{vq}$ that are not in $\mathcal{B}_q$.
5. The coefficients in y_p are determined by plugging y_p into the differential equation $q(\boldsymbol{D})y = f$.
6. The general solution is given by

$$y_p + y_h,$$

where $y_h \in \mathcal{E}_q$.

The Incomplete Partial Fraction Method

The description of the incomplete partial fraction method given in Sect. 3.5 was independent of the degree of q and so applies equally well here. To summarize the method, we record the following algorithm.

Algorithm 4. If

$$q(\boldsymbol{D})y = f,$$

where q is a polynomial of degree n and $f \in \mathcal{E}$, then the general solution can be obtained as follows:

The Incomplete Partial Fraction Method

1. Compute $\mathcal{L}\{f\} = \frac{u}{v}$.
2. With trivial initial conditions assumed, write $\mathcal{L}\{y\} = \frac{u}{qv}$.
3. Apply the partial fraction algorithm on $\frac{u}{qv}$ until the denominator of the remainder term is q.
4. A particular solution y_p is obtained by adding up the inverse Laplace transform of the resulting partial fractions.
5. The general solution is given by

$$y_p + y_h,$$

where $y_h \in \mathcal{E}_q$.

To illustrate this general procedure, let us consider two examples.

Example 5. Find the general solution to

$$y''' - 2y'' + y' = te^t.$$

▶ **Solution.** The characteristic polynomial is $q(s) = s^3 - 2s^2 + s = s(s-1)^2$ and $\mathcal{L}\{te^t\} = \frac{1}{(s-1)^2}$. Assuming trivial initial conditions, $y(0) = 0$, $y'(0) = 0$, and $y''(0) = 0$, we get

$$Y(s) = \mathcal{L}\{y\} = \frac{1}{(s-1)^4 s}.$$

The incomplete $(s-1)$-chain for $Y(s)$ is

Incomplete $(s-1)$-chain	
$\dfrac{1}{(s-1)^4 s}$	$\dfrac{1}{(s-1)^4}$
$\dfrac{-1}{(s-1)^3 s}$	$\dfrac{-1}{(s-1)^3}$
$\dfrac{p(s)}{(s-1)^2 s}$	

There is no need to compute $p(s)$ since the denominator is $q(s) = (s-1)^2 s$ and the resulting inverse Laplace transform is a homogeneous solution. Let $Y_p(s) = \frac{1}{(s-1)^4} - \frac{1}{(s-1)^3}$. Then a particular solution is

$$y_p = \mathcal{L}^{-1}\{Y_p\} = \frac{1}{3!}t^3 e^t - \frac{1}{2!}t^2 e^t.$$

Since $\mathcal{B}_q = \{1, e^t, e^{-t}\}$, we get the general solution

$$y = \frac{1}{6}t^3 e^t - \frac{1}{2}t^2 e^t + c_1 + c_2 e^t + c_2 t e^t. \qquad \blacktriangleleft$$

Example 6. Find the general solution to

$$y^{(4)} - y = 4\cos t.$$

▶ **Solution.** The characteristic polynomial is $q(s) = s^4 - 1 = (s^2 - 1)(s^2 + 1)$ and $\mathcal{L}\{4\cos t\} = \frac{4s}{s^2+1}$. Again assume $y(0) = 0$ and $y'(0) = 0$. Then

$$Y(s) = \frac{4s}{(s^2 - 1)(s^2 + 1)^2}.$$

Using the irreducible quadratic partial fraction method, we obtain the incomplete $(s^2 + 1)$-chain for $Y(s)$:

Incomplete $(s^2 + 1)$-chain	
$\dfrac{4s}{(s^2 - 1)(s^2 + 1)^2}$	$\dfrac{-2s}{(s^2 + 1)^2}$
$\dfrac{p(s)}{(s^2 - 1)(s^2 + 1)}$	

By Table 2.9, we have

$$y_p = \mathcal{L}^{-1}\left\{\frac{-2s}{(s^2 + 1)^2}\right\} = -t\sin t.$$

The homogeneous solution as $y_h = c_1 e^t + c_2 e^{-t} + c_3 \cos t + c_4 \sin t$, and thus, the general solution is

$$y = y_p + y_h = -t\sin t + c_1 e^t + c_2 e^{-t} + c_3 \cos t + c_4 \sin t. \qquad \blacktriangleleft$$

Exercises

1–4. Given q and v below, determine the test function y_p for the differential equation $q(D)y = f$, where $\mathcal{L}f = \frac{u}{v}$.

1. $q(s) = s^3 - s$ $v(s) = s + 1$
2. $q(s) = s^3 - s^2 - s + 1$ $v(s) = s - 1$
3. $q(s) = s^3 - s$ $v(s) = s - 2$
4. $q(s) = s^4 - 81$ $v(s) = s^2 + 9$

5–8. Use the method of undetermined coefficients to find the general solution for each of the differential equations given below.

5. $y''' - y' = e^t$
6. $y''' - y'' + y' - y = 4\cos t$
7. $y^{(4)} - 5y'' + 4y = e^{2t}$
8. $y^{(4)} - y = e^t + e^{-t}$

9–14. Use the incomplete partial fraction method to solve the following differential equations.

9. $y''' - y' = e^t$
10. $y''' - 4y'' + 4y' = 4te^{2t}$
11. $y''' + 4y' = t$
12. $y^{(4)} - 5y'' + 4y = e^{2t}$
13. $y''' - y'' + y' - y = 4\cos t$
14. $y^{(4)} - y = e^t + e^{-t}$

4.4   Coupled Systems of Differential Equations

Suppose $L_1 = q_1(D)$ and $L_2 = q_2(D)$ are polynomial differential operators of order m and n, respectively. A system of two differential equations of the form

$$L_1 y_1 = \lambda_1 y_2$$
$$L_2 y_2 = \lambda_2 y_1 \tag{1}$$

is an example of a **coupled system of differential equations**. We assume λ_1 and λ_2 are scalars. Typically, there are initial conditions

$$y_1(0) = a_0, \quad y_1'(0) = a_1, \quad \ldots, \quad y_1^{(m-1)}(0) = a_{m-1}$$
$$y_2(0) = b_0, \quad y_2'(0) = b_1, \quad \ldots, \quad y_2^{(n-1)}(0) = b_{n-1} \tag{2}$$

each given up to one less than the order of the corresponding differential operator. Notice how the solution of one equation becomes the input (up to a scalar multiple) of the other. The spring system given in the introduction provides an example where L_1 and L_2 are both of order 2. It is the goal of this section to show how such systems together with the initial conditions can be solved using higher order constant coefficient differential equations.[2] The result is a uniqueness and existence theorem. We will then return to the coupled spring problem considered in the introduction.

The solution method we describe here involves a basic fact about the algebra of polynomial differential operators.

The Commutativity of Polynomial Differential Operators

Just as polynomials can be multiplied and factored, so too polynomial differential operators. Consider a very simple example. Suppose $q_1(s) = s-1$ and $q_2(s) = s-2$. Then $q_1(D) = D - 1$ and $q_2(D) = D - 2$. Observe,

$$(D - 1)(D - 2)y = (D - 1)(y' - 2y)$$
$$= (D - 1)y' - (D - 1)(2y)$$
$$= y'' - y' - (2y' - 2y)$$

---

[2]There are other methods. For example, in the exercises, a nice Laplace transform approach will be developed. Theoretically, this is a much nicer approach. However, it is not necessarily computationally easier. In Chap. 9, we will consider systems of first order differential equations and show how (1) can fit in that context.

$$= y'' - 3y' + 2y$$
$$= (D^2 - 3D + 2)y = q(D)y,$$

where $q(s) = s^2 - 3s + 2 = (s - 1)(s - 2)$. Therefore, $q(D) = (D - 1)(D - 2)$, and since there was nothing special about the order, we can also write $q(D) = (D - 2)(D - 1)$. More generally, if $q(s) = q_1(s)q_2(s)$, then the corresponding differential operators multiply to give $q(D) = q_1(D)q_2(D)$. This discussion leads to the following commutative principle: If $L_1 = q_1(D)$ and $L_2 = q_2(D)$ are polynomial differential operators, then

$$L_1 L_2 = L_2 L_1.$$

Extending the Initial Conditions

Suppose y_1 and y_2 are solutions to (1) satisfying the initial values given in (2). It turns out then that these equations determine the initial conditions $y_1^{(k)}(0)$ and $y_2^{(k)}(0)$ for all $k = 1, 2, \ldots$. To see this, suppose $L_1 = q_1(D) = c_m D^m + \cdots c_1 D + c_0$. Now evaluate $L_1 y_1 = \lambda_1 y_2$ at $t = 0$ to get $L_1 y_1(0) = \lambda y_2(0)$ or

$$c_m y_1^{(m)}(0) + c_{m-1} y_1^{(m-1)}(0) + \ldots c_0 y_1(0) = \lambda_1 y_2(0).$$

Since all initial values except $y_1^{(m)}(0)$ are known, we can solve for $y_1^{(m)}(0)$. Hence, let $a_m = y_1^{(m)}(0)$ be the unique solution to $L_1 y_1(0) = \lambda_1 y_2(0)$. Now differentiate the equation $L_1 y_1 = \lambda_1 y_2$ and evaluate at $t = 0$ to get

$$c_m y^{(m+1)}(0) + c_{m-1} y^{(m)}(0) + \ldots c_0 y_1'(0) = \lambda_1 y_2'(0).$$

Now, all initial values except $y^{(m+1)}(0)$ are known. We define $a_{m+1} = y^{(m+1)}(0)$ to be the unique solution to $D L_1 y_1(0) = \lambda_1 D y_2(0)$. We can repeat this procedure recursively up to the $(n - 1)$st derivative of y_2 to get $a_m = y_1^{(m)}(0), \ldots, a_{n+m-1} = y_1^{(n+m-1)}(0)$ where a_{n+k} is the unique solution to

$$D^k L_1 y_1(0) = \lambda_1 D^k y_2(0).$$

In a similar way, we can recursively extend the initial values of y_2 to get $b_n = y_2^{(n)}(0), \ldots, b_{n+m-1} = y_2^{(n+m-1)}(0)$ where b_{m+k} is the unique solution to $D^k L_2 y_2(0) = \lambda_2 D^k y_1(0)$. It is important to notice that this procedure does not actually depend on explicitly knowing the functions y_1 and y_2: the values of a_{n+k} and b_{m+k} only depend on the recursive solutions to $D^k L_1 y_1(0) = \lambda_1 D y_2(0)$ and $D^k L_2 y_2(0) = \lambda_2 D y_1(0)$, respectively. It is also useful to observe that since the values of $a_k = y_1^{(k)}(0)$ and $b_k = y_2^{(k)}(0)$ are now known for $k = 0, \ldots, n + m - 1$, we can repeat the recursion process indefinitely to compute all higher order values of a_k and b_k. The key thing to note is that a_k and b_k are the (recursive) solutions to

$$D^k L_1 y_1(0) = \lambda_1 D^k y_2(0) \quad \text{and} \quad D^k L_1 y_1(0) = \lambda_1 D^k y_2(0), \tag{3}$$

respectively.

The following example illustrates the process outlined above.

Example 1. Consider the coupled system of differential equations

$$y_1'(t) + 3y_1(t) = 4y_2(t) \tag{4}$$

$$y_2''(t) + 3y_2(t) = 2y_1(t), \tag{5}$$

with initial conditions $y_1(0) = 3$, $y_2(0) = 1$, and $y_2'(0) = 3$. Determine the extended initial values $y_1^{(k)}(0)$ and $y_2^{(k)}(0)$, for $k = 0, \ldots, 4$.

▶ **Solution.** We evaluate (4) at $t = 0$ to get $y_1'(0) + 3y_1(0) = 4y_2(0)$. Since $y_1(0) = 3$ and $y_2(0) = 1$, it follows that $y_1'(0) = -5$. Apply D to (4) at $t = 0$ to get $y_1''(0) + 3y_1'(0) = 4y_2'(0)$. Since $y_1'(0) = -5$ and $y_2'(0) = 3$, we get $y_1''(0) = 27$. Now evaluate (5) at $t = 0$ to get $y_2''(0) + 3y_2(0) = 2y_1(0)$. It follows that $y_2''(0) = 3$. Apply D to (5) at $t = 0$ to get $y_2'''(0) + 3y_2'(0) = 2y_1'(0)$ which gives $y_2'''(0) = -19$. Apply D^2 to (5) at $t = 0$ to get $y_2^{(4)}(0) + 3y_2''(0) = 2y_1''(0)$ which implies $y_2^{(4)}(0) = 45$. Now apply D^2 to (4) at $t = 0$ to get $y_1'''(0) + 3y_1''(0) = 4y_2''(0)$ which implies $y_1'''(0) = -69$. Finally, apply D^3 to (4) at $t = 0$ to get $y_1^{(4)}(0) + 3y_1'''(0) = 4y_2'''(0)$ which implies $y_1^{(4)}(0) = 131$. The following table summarizes the data obtained.

$y_1(0) = 3$	$y_2(0) = 1$
$y_1'(0) = -5$	$y_2'(0) = 3$
$y_1''(0) = 27$	$y_2''(0) = 3$
$y_1'''(0) = -69$	$y_2'''(0) = -19$
$y_1^{(4)}(0) = 131$	$y_2^{(4)}(0) = 45.$ (6)

With patience, it is possible to compute any initial value for y_1 or y_2 at $t = 0$. ◀

We are now in a position to state the main theorem.

Theorem 2. *Suppose L_1 and L_2 are polynomial differential operators of order m and n, respectively. The unique solution (y_1, y_2) to the coupled system of differential equations*

$$L_1 y_1 = \lambda_1 y_2 \tag{7}$$

$$L_2 y_2 = \lambda_2 y_1 \tag{8}$$

with initial conditions

$$y_1(0) = a_0, \quad y_1'(0) = a_1, \quad \ldots, \quad y_1^{(m-1)}(0) = a_{m-1}$$

$$y_2(0) = b_0, \quad y_2'(0) = b_1, \quad \ldots, \quad y_2^{(n-1)}(0) = b_{n-1} \tag{9}$$

is given as follows: Extend the initial values a_k and b_k, for $k = 1, \ldots n + m - 1$, as determined by (3). Let $L = L_1 L_2 - \lambda_1 \lambda_2$. Then y_1 and y_2 are the homogeneous solutions to $L y = 0$ with initial conditions $y_1^{(k)}(0) = a_k$ and $y_2^{(k)}(0) = b_k$, $k = 0, \ldots, n + m - 1$, respectively.

Proof. First, let us suppose that (y_1, y_2) is a solution to the coupled system and satisfies the given initial conditions. Then y_1 and y_2 satisfy the initial conditions given by (3). Now apply L_2 to (7) and L_1 to (8) to get

$$L_2 L_1 y_1 = \lambda_1 L_2 y_2 = \lambda_1 \lambda_2 y_1$$

$$L_1 L_2 y_2 = \lambda_2 L_1 y_1 = \lambda_1 \lambda_2 y_2.$$

Both of these equations imply that y_1 and y_2 are solutions to $L y = 0$, where $L = L_1 L_2 - \lambda_1 \lambda_2$, satisfying the initial conditions $a_k = y_1^{(k)}(0)$ and $b_k = y_2^{(k)}(0)$, $k = 0, \ldots, n + m - 1$. By the uniqueness part of Theorem 5 of Sect. 4.1, y_1 and y_2 are uniquely determined. To show that solutions exist, suppose y_1 and y_2 are the solutions to $L y = 0$ with initial conditions (obtained recursively) $a_k = y_1^{(k)}(0)$ and $b_k = y_2^{(k)}(0)$, $k = 1, 2, \ldots, n + m - 1$. We will show that $L_1 y_1 = \lambda_1 y_2$. The argument that $L_2 y_2 = \lambda_1 y_1$ is similar. Let $z_1 = L_1 y_1$ and $z_2 = \lambda_1 y_2$. Then

$$L_1 L_2 z_1 = L_1 L_1 L_2 y_1 = \lambda_1 \lambda_2 L_1 y_1 = \lambda_1 \lambda_2 z_1$$

$$L_1 L_2 z_2 = \lambda_1 L_1 L_2 y_2 = \lambda_1 \lambda_1 \lambda_2 y_2 = \lambda_1 \lambda_2 z_2$$

It follows that z_1 and z_2 are homogeneous solutions to $L y = 0$. By (3), z_1 and z_2 satisfy the same initial values for all $k = 0, \ldots, n + m - 1$. By the existence and uniqueness theorem, Theorem 5 of Sect. 4.1, we have $z_1 = z_2$.   □

Algorithm 3. Suppose L_1 and L_2 are polynomial differential operators of order m and n, respectively. Then the solution to (1) with initial conditions given by (2) is given by the following algorithm:

Solution Method for Coupled Systems

1. Extend (recursively) the initial conditions $y_1^{(k)}(0)$ and $y_2^{(k)}(0)$, for $k = 1, 2, \ldots, n + m - 1$.
2. Let $L = L_1 L_2 - \lambda_1 \lambda_2$ and solve

$$L y = 0,$$

for each set of initial conditions.

Example 4. Solve the following coupled system:

$$y_1' + 3y_1 = 4y_2$$
$$y_2'' + 3y_2 = 2y_1,$$

$y_1(0) = 3$, $y_2(0) = 1$, and $y_2'(0) = 3$.

▶ **Solution.** In Example 1, we extended the initial values (see (6), where we will only need the first 3 rows). We let $L_1 = D+3$ and $L_2 = D^2+3$. Then y_1 and y_2 are homogeneous solutions to $L = L_1 L_2 - 8 = D^3 + 3D^2 + 3D + 1$. The characteristic polynomial is $q(s) = s^3 + 3s^2 + 3s + 1 = (s+1)^3$. Since $\mathcal{B}_q = \{e^{-t}, te^{-t}, t^2 e^{-t}\}$ it follows that the homogeneous solutions take the form

$$y = c_1 e^{-t} + c_2 t e^{-t} + c_3 t^2 e^{-t}.$$

To find y_1, we set $y_1 = c_1 e^{-t} + c^2 t e^{-t} + c_3 e^{-t}$. We compute y_1' and y_1'' and substitute the initial conditions $y_1(0) = 3$, $y_1'(0) = -5$, and $y_1''(0) = 27$ to get the following system:

$$
\begin{aligned}
c_1 & & &= 3 \\
-c_1 &+ c_2 & &= -5 \\
c_1 &- 2c_2 &+ 2c_3 &= 27.
\end{aligned}
$$

A short calculation gives $c_1 = 3$, $c_2 = -2$, and $c_3 = 10$. It follows that $y_1(t) = (3 - 2t + 10t^2)e^{-t}$. In a similar way, setting $y_2 = c_1 e^{-t} + c^2 t e^{-t} + c_3 e^{-t}$ leads to the system

$$
\begin{aligned}
c_1 & & &= 1 \\
-c_1 &+ c_2 & &= 3 \\
c_1 &- 2c_2 &+ 2c_3 &= 3.
\end{aligned}
$$

A short calculation gives $c_1 = 1$, $c_2 = 4$, and $c_3 = 5$, and hence, $y_2(t) = (1 + 4t + 5t^2)e^{-t}$. It follows that

$$y_1(t) = (3 - 2t + 10t^2)e^{-t} \quad \text{and} \quad y_2(t) = (1 + 4t + 5t^2)e^{-t}$$

are the solutions to the coupled system.   ◀

Coupled Spring Systems

We now return to the problem posed at the beginning of this chapter. Two springs systems (without dashpots) are coupled as illustrated in the diagram below:

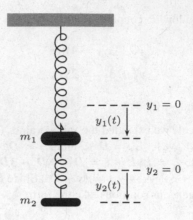

If y_1 and y_2 denote the displacements of the bodies of mass m_1 and m_2 from their equilibrium positions, $y_1 = 0$ and $y_2 = 0$, then we showed that y_1 and y_2 satisfy

$$m_1 y_1''(t) + (k_1 + k_2)y_1(t) = k_2 y_2(t)$$
$$m_2 y_2''(t) + k_2 y_2(t) = k_2 y_1(t). \tag{10}$$

If we divide each equation by the leading coefficient and let

$$a = \frac{k_1}{m_1}, \quad b = \frac{k_2}{m_1}, \quad \text{and} \quad c = \frac{k_2}{m_2}, \tag{11}$$

then we get the coupled system

$$L_1 y_1 = b y_2, \tag{12}$$
$$L_2 y_2 = c y_1, \tag{13}$$

where $L_1 = D^2 + a + b$ and $L_2 = D^2 + c$. We observe that

$$L = L_1 L_2 - bc$$
$$= D^4 + (a + b + c)D^2 + ac.$$

As to initial conditions, suppose we are given that $y_1(0) = A_0$, $y_1'(0) = A_1$, $y_2(0) = B_0$, and $y_2'(0) = B_1$. Then (12) and its derivative imply

$$y_1''(0) = -(a + b)y_1(0) + b y_2(0) = -(a + b)A_0 + b B_0$$
$$y_1'''(0) = -(a + b)y_1'(0) + b y_2'(0) = -(a + b)A_1 + b B_1.$$

In a similar way, (13) implies

$$y_2''(0) = -cy_2(0) + cy_1(0) = -cB_0 + cA_0$$
$$y_2'''(0) = -cy_2'(0) + cy_1'(0) = -cB_1 + cA_1.$$

Summarizing, we obtain that y_1 and y_2 both satisfy the same 4th order differential equation

$$y^{(4)} + (a + b + c)y'' + acy = 0 \tag{14}$$

with initial conditions

$$
\begin{array}{ll}
y_1(0) = A_0 & y_2(0) = B_0 \\
y_1'(0) = A_1 & y_2'(0) = B_1 \\
y_1''(0) = -(a+b)A_0 + bB_0 & y_2''(0) = -cB_0 + cA_0 \\
y_1'''(0) = -(a+b)A_1 + bB_1 & y_2'''(0) = -cB_1 + cA_1.
\end{array} \tag{15}
$$

Example 5. Consider the coupled spring system with $m_1 = 3$, $m_2 = 4$, $k_1 = 9$, and $k_2 = 12$. At $t = 0$, both masses are pulled downward a distance of 1m from equilibrium and released without imparting any momentum. Determine the motion of the system.

▶ **Solution.** From (11), we get

$$a = \frac{9}{3} = 3, \quad b = \frac{12}{3} = 4, \quad \text{and} \quad c = \frac{12}{4} = 3. \tag{16}$$

Thus, y_1 and y_2 satisfy the following coupled system of differential equations:

$$y_1'' + 7y_1 = 4y_2$$
$$y_2'' + 3y_2 = 3y_1.$$

By Theorem 2, y_1 and y_2 are homogeneous solutions to the 4th degree equation $q(\boldsymbol{D})y = 0$ where $q(s) = (s^2+7)(s^2+3) - 12 = s^4 + 10s^2 + 9 = (s^2+1)(s^2+9)$. It follows that y_1 and y_2 are linear combinations of $\mathcal{B}_q = \{\cos t, \sin t, \cos 3t, \sin 3t\}$. By (15), the initial conditions of y_1 and y_2 at $t = 0$ up to order 3 are

$$
\begin{array}{ll}
y_1(0) = 1 & y_2(0) = 1 \\
y_1'(0) = 0 & y_2'(0) = 0 \\
y_1''(0) = -3 & y_2''(0) = 0 \\
y_1'''(0) = 0 & y_2'''(0) = 0.
\end{array}
$$

Therefore, if $y_1 = c_1 \cos t + c_2 \sin t + c_3 \cos 3t + c_4 \sin 3t$, then the coefficients satisfy

$$
\begin{aligned}
c_1 + \quad c_3 \qquad\qquad &= 1 \\
c_2 \qquad + 3c_4 &= 0 \\
-c_1 + \quad -9c_3 \qquad\qquad &= -3 \\
-c_2 \qquad - 27c_4 &= 0
\end{aligned}
$$

from which we get $c_1 = \frac{3}{4}$, $c_2 = 0$, $c_3 = \frac{1}{4}$, and $c_4 = 0$. It follows that

$$
y_1(t) = \frac{3}{4}\cos t + \frac{1}{4}\cos 3t.
$$

On the other hand, if $y_2 = c_1\cos t + c_2\sin t + c_3\cos 3t + c_4\sin 3t$, then the coefficients satisfy

$$
\begin{aligned}
c_1 + \quad c_3 \qquad\qquad &= 1 \\
c_2 \qquad + 3c_4 &= 0 \\
-c_1 + \quad -9c_3 \qquad\qquad &= 0 \\
-c_2 \qquad - 27c_4 &= 0
\end{aligned}
$$

from which we get $c_1 = \frac{9}{8}$, $c_2 = 0$, $c_3 = \frac{-1}{8}$, and $c_4 = 0$. It follows that

$$
y_2(t) = \frac{9}{8}\cos t - \frac{1}{8}\cos 3t.
$$

Since y_1 and y_2 are periodic with period 2π, the motion of the masses (given below) are likewise periodic. Their graphs are simultaneously given below.

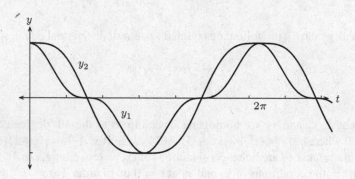

◀

Exercises

1–8. Solve the following coupled systems.

1.

$$y_1' - 6y_1 = -4y_2$$
$$y_2' = 2y_1$$

with initial conditions $y_1(0) = 2$ and $y_2(0) = -1$.

2.

$$y_1' - 3y_1 = -4y_2$$
$$y_2' + y_2 = y_1$$

with initial conditions $y_1(0) = 1$ and $y_2(0) = 1$.

3.

$$y_1' = 2y_2$$
$$y_2' = -2y_1$$

with initial conditions $y_1(0) = 1$ and $y_2(0) = -1$.

4.

$$y_1' - 2y_1 = 2y_2$$
$$y_2'' + 2y_2' + y_2 = -2y_1,$$

with initial conditions $y_1(0) = 3$, $y_2(0) = 0$, and $y_2'(0) = 3$.

5.

$$y_1' + 4y_1 = 10y_2$$
$$y_2'' - 6y_2' + 23y_2 = 9y_1,$$

with initial conditions $y_1(0) = 0$, $y_2(0) = 2$, and $y_2'(0) = 2$.

6.

$$y_1' - 2y_1 = -2y_2$$
$$y_2'' + y_2' + 6y_2 = 4y_1,$$

with initial conditions $y_1(0) = 1$, $y_2(0) = 5$, and $y_2'(0) = 4$.

7.

$$y_1'' + 2y_1' + 6y_1 = 5y_2$$
$$y_2'' - 2y_2' + 6y_2 = 9y_1,$$

with initial conditions $y_1(0) = 0$, $y_1'(0) = 0$, $y_2(0) = 6$, and $y_2'(0) = 6$.

8.

$$y_1'' + 2y_1 = -3y_2$$
$$y_2'' + 2y_2' - 9y_2 = 6y_1,$$

with initial conditions $y_1(0) = -1$, $y_1'(0) = -4$, $y_2(0) = 1$, and $y_2'(0) = 2$.

9–10. Solve the coupled spring systems for the given parameters m_1, m_2, k_1, and k_2 and initial conditions.

9. $m_1 = 2$, $m_2 = 1$, $k_1 = 4$, and $k_2 = 2$ with initial conditions $y_1(0) = 3$, $y_1'(0) = 3$, $y_2(0) = 0$, and $y_2'(0) = 0$

10. $m_1 = 4$, $m_3 = 1$, $k_1 = 8$, and $k_2 = 12$ with initial conditions $y_1(0) = 1$, $y_1'(0) = 0$, $y_2(0) = 6$, and $y_2'(0) = 0$

11. *The Laplace Transform Method for Coupled Systems*: In this exercise, we will see how the Laplace transform may be used to solve a coupled system:

$$L_1 y_1 = \lambda_1 y_2$$
$$L_2 y_2 = \lambda_2 y_1$$

with initial conditions

$$y_1(0) = a_0, \quad y_1'(0) = a_1, \quad \ldots, \quad y_1^{(m-1)}(0) = a_{m-1}$$
$$y_2(0) = b_0, \quad y_2'(0) = b_1, \quad \ldots, \quad y_2^{(n-1)}(0) = b_{n-1}.$$

Let $L_1 = q_1(D)$ and $L_2 = q_2(D)$ be polynomial differential operators. Let $Y_1 = \mathcal{L}\{y_1\}$ and $Y_2 = \mathcal{L}\{y_2\}$. Suppose $\mathcal{L}\{L_1 y_1\}(s) = q_1(s)Y_1(s) - p_1(s)$ and $\mathcal{L}\{L_2 y_2\}(s) = q_2(s)Y_2(s) - p_2(s)$, where $p_1(s)$ and $p_2(s)$ are polynomials determined by the initial conditions.

1. Show that Y_1 and Y_2 satisfy the following matrix relation:

$$\begin{pmatrix} q_1(s) & -\lambda_1 \\ -\lambda_2 & q_2(s) \end{pmatrix} \begin{pmatrix} Y_1(s) \\ Y_2(s) \end{pmatrix} = \begin{pmatrix} p_1(s) \\ p_2(s) \end{pmatrix}.$$

2. Show that this system has a solution given by

$$Y_1(s) = \frac{p_1(s)q_2(s) + \lambda_1 p_2(s)}{q_1(s)q_2(s) - \lambda_1\lambda_2},$$

$$Y_2(s) = \frac{p_2(s)q_1(s) + \lambda_2 p_1(s)}{q_1(s)q_2(s) - \lambda_1\lambda_2}.$$

12–16. Use the Laplace transform method developed in Exercise 11 to solve the following coupled systems.

12.

$$y_1' = -y_2$$
$$y_2' - 2y_2 = y_1$$

with initial conditions $y_1(0) = 1$ and $y_2(0) = -1$.

13.

$$y_1' - y_1 = -2y_2$$
$$y_2' - y_2 = 2y_1$$

with initial conditions $y_1(0) = 2$ and $y_2(0) = -2$.

14.

$$y_1' - 2y_1 = -y_2$$
$$y_2'' - y_2' + y_2 = y_1$$

with initial conditions $y_1(0) = 0$, $y_2(0) = -1$, and $y_2'(0) = 2$.

15.

$$y_1' + 2y_1 = 5y_2$$
$$y_2'' - 2y_2' + 5y_2 = 2y_1$$

with initial conditions $y_1(0) = 1$, $y_2(0) = 0$, and $y_2'(0) = 3$.

16.

$$y_1'' + 2y_1 = -3y_2$$
$$y_2'' + 2y_2' - 9y_2 = 6y_1$$

with initial conditions $y_1(0) = 10$, $y_1'(0) = 0$, $y_2(0) = 10$, and $y_2'(0) = 0$.

4.5 System Modeling

Mathematical modeling involves understanding how a system works mathematically. By a system, we mean something that takes inputs and produces outputs such as might be found in the biological, chemical, engineering, and physical sciences. The core of modeling thus involves expressing how the outputs of a system can be mathematically described as a function of the inputs. The following *system diagram* represents the inputs coming in from the left of the system and outputs going out on the right.

$$f(t) \longrightarrow \boxed{\text{System}} \longrightarrow y(t)$$

For the most part, inputs and outputs will be quantities that are time dependent; they will be represented as functions of t. There are occasions, however, where other parameters such as position or frequency are used in place of time. An input-output pair, $(f(t), y(t))$, implies a relationship which we denote by

$$y(t) = \Phi(f)(t).$$

Notice that Φ is an operation on the input function f and produces the output function y. In a certain sense, understanding Φ is equivalent to understanding the workings of the system. Frequently, we identify the system under study with Φ itself. Our goal in modeling then is to give an explicit mathematical description of Φ.

In many settings, a mathematical model can be described implicitly by a constant coefficient linear differential equation and its solution gives an explicit description. For example, the mixing problems in Sect. 1.5, spring systems in Sect. 3.6, RCL circuits in Sect. 3.7, and the coupled spring systems in Sect. 4.4 are each modeled by constant coefficient linear differential equations and have common features which we explore in this section. To get a better idea of what we have in mind, let us reconsider the mixing problem as an example.

Example 1. Suppose a tank holds 10 liters of a brine solution with initial concentration $a/10$ grams of salt per liter. (Thus, there are a grams of salt in the tank, initially.) Pure water flows in an intake tube at a rate of b liters per minute and the well-mixed solution flows out of the tank at the same rate. Attached to the intake is a hopper containing salt. The amount of salt entering the intake is controlled by a valve and thus varies as a function of time. Let $f(t)$ be the rate (in grams of salt per minute) at which salt enters the system. Let $y(t)$ represent the amount of salt in the tank at time t. Find a mathematical model that describes y.

▶ **Solution.** As in Chap. 1, our focus is on the way y changes. Observe that

$$y' = \text{input rate} - \text{output rate}.$$

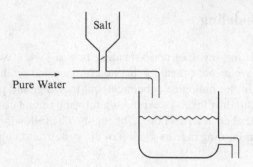

By input rate, we mean the rate at which salt enters the system and this is just $f(t)$. The output rate is the rate at which salt leaves the system and is given by the product of the concentration of salt in the tank, $y(t)/10$, and the flow rate b. We are thus led to the following initial value problem:

$$y' + \frac{b}{10}y = f(t), \quad y(0) = a. \tag{1}$$

In this system, $f(t)$ is the input function and $y(t)$ is the output function. Incorporating this mathematical model in a system diagram gives the following implicit description:

$$f(t) \longrightarrow \boxed{\begin{array}{c} \text{Solve } y' + \frac{b}{10}\ y = f(t) \\ y(0) = a \end{array}} \longrightarrow y(t).$$

Using Algorithm 3 in Sect. 1.5 to solve (1) gives

$$y(t) = a\mathrm{e}^{\frac{-bt}{10}} + \mathrm{e}^{-\frac{bt}{10}} \int_0^t f(x)\mathrm{e}^{\frac{bx}{10}}\,\mathrm{d}x$$

$$= a\mathrm{e}^{\frac{-bt}{10}} + \int_0^t f(x)\mathrm{e}^{-\frac{b(t-x)}{10}}\,\mathrm{d}x$$

$$= ah(t) + f * h(t),$$

where $h(t) = \mathrm{e}^{\frac{-bt}{10}}$ and $f * h$ denotes the convolution of f with h. We therefore arrive at an explicit mathematical model Φ for the mixing system:

$$\Phi(f)(t) = ah(t) + f * h(t). \qquad \blacktriangleleft$$

As we shall see, this simple example illustrates many of the main features shared by all systems modeled by a constant coefficient differential equation. You notice that the output consists of two pieces: $ah(t)$ and $f * h(t)$. If $a = 0$, then the initial state of the system is zero, that is, there is no salt in the tank at time $t = 0$. In this case, the output is

$$y(t) = \Phi(f)(t) = f * h(t). \tag{2}$$

This output is called the ***zero-state response***; it represents the response of the system by purely external forces of the system and not on any nonzero initial condition or state. The zero-state response is the particular solution to $q(D)y = f$, $y(0) = 0$, where $q(s) = s + b/10$.

On the other hand, if $f = 0$, the output is $ah(t)$. This output is called the ***zero-input response***; it represents the response based on purely internal conditions of the system and not on any external inputs. The zero-input response is the homogeneous solution to $q(D)y = 0$ with initial condition $y(0) = a$. The ***total response*** is the sum of the zero-state response and the zero-input response.

The function h is called the ***unit impulse response function***. It is the zero-input response with $a = 1$ and completely characterizes the system. Alternatively, we can understand h in terms of the mixing system by opening the hopper for a very brief moment just before $t = 0$ and letting 1 gram of salt enter the intake. At $t = 0$, the amount of salt in the tank is 1 gram and no salt enters the system thereafter. The expression "unit impulse" reflects the fact that a unit of salt (1 gram) enters the system and does so instantaneously, that is, as an impulse. Such impulsive inputs are discussed more thoroughly in Chap. 6. Over time, the amount of salt in the tank will diminish according to the unit impulse response $h = e^{\frac{-bt}{10}}$ as illustrated in the graph below:

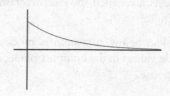

Once h is known, the zero-state response for an input f is completely determined by (2). One of the main points of this section will be to show that in all systems modeled by a constant coefficient differential equation, the zero-state response is given by this same formula, for some h. We will also show how to find h.

With a view to a more general setting, let $q(D)$ be an nth order differential operator with leading coefficient one and let $\mathbf{a} = (a_0, a_1, \ldots, a_{n-1})$ be a vector of n scalars. Suppose Φ is a system. We say Φ is ***modeled by $q(D)$ with initial state $\mathbf{a}$*** if for each input function f, the output function $y = \Phi(f)$ satisfies $q(D)y = f$, and $y(0) = a_0$, $y'(0) = a_1$, $\ldots$, $y^{(n-1)}(0) = a_{n-1}$. Sometimes we say that the ***initial state*** of Φ is $\mathbf{a}$. By the existence and uniqueness theorem, Theorem 5 of Sect. 4.1, y is unique. In terms of a system diagram we have

$$f(t) \longrightarrow \boxed{\begin{array}{c} \text{Solve } q(D)y = f \\ \text{with initial state } \mathbf{a} \end{array}} \longrightarrow y(t).$$

The definitions of zero-state response, zero-input response, and total response given above naturally extend to this more general setting.

The Zero-Input Response

First, we consider the response of the system with no external inputs, that is, $f(t) = 0$. This is the zero-input response and is a result of internal initial conditions of the system only. For example, the initial conditions for a spring system are the initial position and velocity, and for RCL circuits, they are the initial charge and current. In the mixing system we described earlier, the only initial condition is the amount of salt a in the tank at time $t = 0$.

The zero-input response is the solution to

$$q(D)y = 0, \quad y(0) = a_0, \; y'(0) = a_1, \ldots, y^{(n-1)}(0) = a_{n-1}.$$

By Theorem 2 of Sect. 4.2, y is a linear combination of functions in the standard basis $\mathcal{B}_q$. The roots $\lambda_1, \ldots, \lambda_k \in \mathbb{C}$ of q are called the ***characteristic values*** and the functions in the standard basis are called the ***characteristic modes*** of the system. Consider the following example.

Example 2. For each problem below, a system is modeled by $q(D)$ with initial state $\mathbf{a}$. Plot the characteristic values in the complex plane, determine the zero-input response, and graph the response.

a. $q(D) = D + 3, \mathbf{a} = 2$
b. $q(D) = (D - 3)(D + 3), \mathbf{a} = (0, 1)$
c. $q(D) = (D + 1)^2 + 4, \mathbf{a} = (1, 3)$
d. $q(D) = D^2 + 4, \mathbf{a} = (1, 2)$
e. $q(D) = (D^2 + 4)^2, \mathbf{a} = (0, 0, 0, 1)$.

▶ **Solution.** The following table summarizes the calculations that we ask the reader to verify.

Characteristic Values	Zero-input Response	Graph

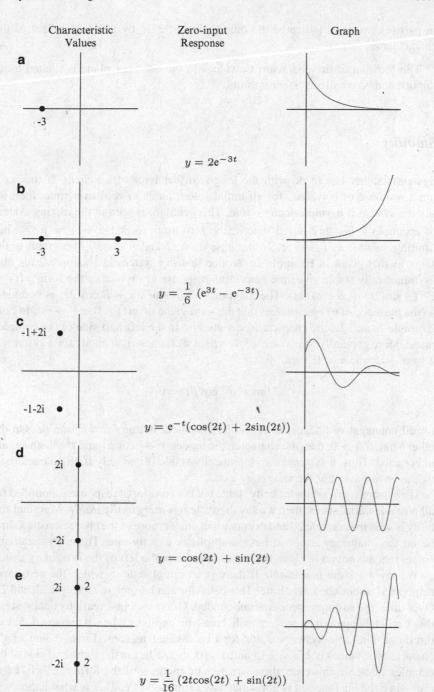

a

$$y = 2e^{-3t}$$

b

$$y = \frac{1}{6}\,(e^{3t} - e^{-3t})$$

c

$$y = e^{-t}(\cos(2t) + 2\sin(2t))$$

d

$$y = \cos(2t) + \sin(2t)$$

e

$$y = \frac{1}{16}\,(2t\cos(2t) + \sin(2t))$$

In part (e), we have indicated the multiplicity of $\pm 2i$ by a 2 to the right of the characteristic value.   ◄

The location of the characteristic values in the complex plane is related to an important notion called system stability.

Stability

System stability has to do with the long-term behavior of a system. If the zero-input response of a system, for all initial states, tends to zero over time, then we say the system is *asymptotically stable*. This behavior is seen in the mixing system of Example 1. For any initial state a, the zero-input response $y(t) = ae^{\frac{-bt}{10}}$ has limiting value 0 as $t \to \infty$. In the case $a = 2$ and $b = 30$, the graph is the same as that given in Example 2a. Notice that the system in Example 2c is also asymptotically stable since the zero-input response always takes the form $y(t) = e^{-t}(A\sin(2t) + B\cos(2t))$. The function $t \to A\sin(2t) + B\cos(2t)$ is bounded, so the presence of e^{-t} guarantees that the limit value of $y(t)$ is 0 as $t \to \infty$. In both Example 2a and 2c, the characteristic values lie in the left-half side of the complex plane. More generally, suppose $t^k e^{\alpha t} \cos \beta t$ is a characteristic mode for a system. If $\lambda = \alpha + i\beta$ and $\alpha < 0$, then

$$\lim_{t \to \infty} t^k e^{\alpha t} \cos \beta t = 0,$$

for all nonnegative integers k. A similar statement is true for $t^k e^{\alpha t} \sin \beta t$. On the other hand, if $\alpha > 0$, then the characteristic modes, $t^k e^{\alpha t} \cos \beta t$ and $t^k e^{\alpha t} \sin \beta t$, are unbounded. Thus, a system is asymptotically stable if and only if all characteristic values lie to the left of the imaginary axis.

If a system is not asymptotically stable but the zero-input response is bounded for all possible initial states, then we say the system is *marginally stable*. Marginal stability is seen in Example 2d and occurs when one or more of the characteristic values lie on the imaginary axis and have multiplicity exactly one. Those characteristic values that are not on the imaginary axis must be to the left of the imaginary axis.

We say a system is *unstable* if there is an initial state in which the zero-input response is unbounded over time. This behavior can be seen in Example 2b and 2e. Over time, the response becomes unbounded. Of course, in a real physical system, this cannot happen. The system will break or explode when it passes a certain threshold. Unstable systems occur for two distinct reasons. First, if one of the characteristic values is $\lambda = \alpha + i\beta$ and $\alpha > 0$, then λ lies in the right-half side of the complex plane. In this case, the characteristic mode is of the form $t^k e^{\alpha t} \cos \beta t$ or $t^k e^{\alpha t} \sin \beta t$. This function is unbounded as a function of t. This is what happens in Example 2b. Second, if one of the characteristic values $\lambda = i\beta$ lies on the imaginary axis and, in addition, the multiplicity is greater than one, then the characteristic modes are of the form $t^k \cos(\beta t)$ or $t^k \sin(\beta t)$, $k \geq 1$. These modes oscillate unboundedly as a function of $t > 0$, as in Example 2e. Remember, it only takes one unbounded characteristic mode for the whole system to be unstable.

Example 3. Determine the stability of each system modeled by $q(D)$ below:

1. $q(D) = (D + 1)^2(D + 3)$
2. $q(D) = (D^2 + 9)(D + 4)$
3. $q(D) = (D + 4)^2(D - 5)$
4. $q(D) = (D^2 + 1)(D^2 + 9)^2$

▶ **Solution.** 1. The characteristic values are $\lambda = -1$ with multiplicity 2 and $\lambda = -3$. The system is asymptotically stable.
2. The characteristic values are $\lambda = \pm 3i$ and $\lambda = -4$. The system is marginally stable.
3. The characteristic values are $\lambda = -4$ with multiplicity 2 and $\lambda = 5$. The system is unstable.
4. The characteristic values are $\lambda = \pm i$ and $\lambda = \pm 3i$ with multiplicity 2. The system is unstable.   ◄

The Unit Impulse Response Function

Suppose Φ is a system modeled by $q(D)$, an nth order constant coefficient differential operator. The **unit impulse response function** $h(t)$ is the zero-input response to Φ when the initial state of the system is $\mathbf{a} = (0, \ldots, 0, 1)$. More specifically, $h(t)$ is the solution to

$$q(D)y = 0 \qquad y(0) = 0, \ldots, y^{(n-2)}(0) = 0, y^{(n-1)}(0) = 1.$$

If $n = 1$, then $y(0) = 1$ as in the mixing system discussed in the beginning of this section. In this simple case, $h(t)$ is a multiple of a single characteristic mode. For higher order systems, however, the unit impulse response function is a homogeneous solution to $q(D)y = 0$ and, hence, a linear combination of the characteristic modes of the system.

Example 4. Find the unit impulse response function for a system Φ modeled by

$$q(D) = (D + 1)(D^2 + 1).$$

▶ **Solution.** It is an easy matter to apply the Laplace transform method to $q(D)y = 0$ with initial condition $y(0) = 0$, $y'(0) = 0$, and $y''(0) = 1$. We get $q(s)Y(s) - 1 = 0$. A short calculation gives

$$Y(s) = \frac{1}{q(s)} = \frac{1}{(s+1)(s^2+1)} = \frac{1}{4}\left(\frac{2}{s+1} + \frac{-2s+2}{s^2+1}\right).$$

The inverse Laplace transform is $y(t) = \frac{1}{2}(e^{-t} + \sin t - \cos t)$. The unit impulse response function is thus

$$h(t) = \frac{1}{2}(e^{-t} + \sin t - \cos t).$$ ◄

Observe in this example that $h(t) = \mathcal{L}^{-1}\left\{\frac{1}{q(s)}\right\}(t)$. It is not hard to see that this formula extends to the general case. We record this in the following theorem. The proof is left to the reader.

Theorem 5. *Suppose a system Φ is modeled by a constant coefficient differential operator $q(D)$. The unit impulse response function, $h(t)$, of the system Φ is given by*

$$h(t) = \mathcal{L}^{-1}\left\{\frac{1}{q(s)}\right\}(t).$$

The Zero-State Response

Let us now turn our attention to a system Φ in the zero-state and consider the zero-state response. This occurs precisely when $\mathbf{a} = \mathbf{0}$, that is, all initial conditions of the system are zero. We should think of the system as initially being at rest. We continue to assume that Φ is modeled by an nth order constant coefficient differential operator $q(D)$. Thus, for each input $f(t)$, the output $y(t)$ satisfies $q(D)y = f$ with y and its higher derivatives up to order $n-1$ all zero at $t = 0$. An important feature of Φ in this case is its linearity.

Proposition 6. *Suppose Φ is a system modeled by $q(D)$ in the zero-state. Then Φ is linear. Specifically, if f, f_1, and f_2 are input functions and c is a scalar, then*

1. $\Phi(f_1 + f_2) = \Phi(f_1) + \Phi(f_2)$
2. $\Phi(cf) = c\Phi(f)$.

Proof. If f is any input function, then $\Phi(f) = y$ if and only if $q(D)y = f$ and $y(0) = y'(0) = \cdots = y^{(n-1)}(0) = 0$. If y_1 and y_2 are the zero-state response functions to f_1 and f_2, respectively, then the linearity of $q(D)$ implies

$$q(D)(y_1 + y_2) = q(D)y_1 + q(D)y_2 = f_1 + f_2.$$

Furthermore, since the initial state of both y_1 and y_2 are zero, so is the initial state of $y_1 + y_2$. This implies $\Phi(f_1 + f_2) = y_1 + y_2$. In a similar way,

$$q(D)(cy) = cq(D)y = cf,$$

by the linearity of $q(D)$. The initial state of cy is clearly zero. So $\Phi(cf) = c\Phi(f)$.

□

The following remarkable theorem gives an explicit formula for Φ in terms of convolution with the unit-impulse response function.

Theorem 7. *Suppose Φ is a system modeled by $q(D)$ in the zero-state. If f is a continuous input function on an interval which includes zero, then the zero-state response is given by the convolution of f with the unit impulse response function h. That is,*

$$\Phi(f)(t) = f * h(t) = \int_0^t f(x)h(t - x)\,dx.$$

If, in addition, we were to assume that f has a Laplace transform, then the proof would be straightforward. Indeed, if $y(t)$ is the zero-state response, then the Laplace transform method would give $q(s)\mathcal{L}\{y\} = \mathcal{L}\{f\} := F(s)$ and therefore

$$\mathcal{L}\{y\}(s) = \frac{1}{q(s)}F(s).$$

The convolution theorem then gives $y(t) = \Phi(f)(t) = h * f$.

For the more general case, let us introduce the following helpful lemma.

Lemma 8. *Suppose f is continuous on a interval I containing 0. Suppose h is differentiable on I. Then*

$$(f * h)'(t) = f(t)h(0) + f * h'.$$

Proof. Let $y(t) = f * h(t)$. Then

$$\frac{\Delta y}{\Delta t} = \frac{y(t + \Delta t) - y(t)}{\Delta t}$$

$$= \frac{1}{\Delta t}\left(\int_0^{t+\Delta t} f(x)h(t + \Delta t - x)\,dx - \int_0^t f(x)h(t - x)\,dx\right)$$

$$= \int_0^t f(x)\frac{h(t + \Delta t - x) - h(t - x)}{\Delta t}\,dx$$

$$+ \frac{1}{\Delta t}\int_t^{t+\Delta t} f(x)h(t + \Delta t - x)\,dx.$$

We now let Δt go to 0. The first summand has limit $\int_0^t f(x)h'(t - x)\,dx = f * h'(t)$. By the fundamental theorem of calculus, the limit of the second summand is obtained by evaluating the integrand at $x = t$, thus getting $f(t)h(0)$. The lemma now follows by adding these two terms.   □

Proof (of Theorem 7). Let $h(t)$ be the unit impulse response function. Then $h(0) = h'(0) = \cdots = h^{(n-2)}(0) = 0$ and $h^{(n-1)}(0) = 1$. Set $y(t) = f * h(t)$. Repeated applications of Lemma 8 to y gives

$$y' = f * h' + h(0)f = f * h'$$
$$y'' = f * h'' + h'(0)f = f * h''$$

$$\vdots$$

$$y^{(n-1)} = f * h^{(n-1)} + h^{(n-2)}(0)f = f * h^{(n-1)}$$
$$y^{(n)} = f * h^{(n)} + h^{(n-1)}(0)f = f * h^{(n)} + f.$$

From this it follows that

$$q(\boldsymbol{D})y = f * q(\boldsymbol{D})h + f = f,$$

since $q(\boldsymbol{D})h = 0$. It is easy to check that y is in the zero-state. Therefore,

$$\Phi(f) = y = f * h.$$ □

At this point, let us make a few remarks on what this theorem tells us. The most remarkable thing is the fact that Φ is precisely determined by the unit impulse response function h. From a mathematical point of view, knowing h means you know how the system Φ works in the zero-state. Once h is determined, all output functions, that is, system responses, are given by the convolution product, $f * h$, for an input function f. Admittedly, convolution is an unusual product. It is not at all like the usual product of functions where the value (or state) at time t is determined by knowing just the value of each factor at time t. Theorem 7 tells us that the state of a system response at time t depends on knowing the values of the input function f for all x between 0 and t. The system "remembers" the whole of the input f up to time t and "meshes" those inputs with the internal workings of the system, as represented by the impulse response function h, to give $f * h(t)$.

Since the zero-state response, $f * h$, is a solution to $q(\boldsymbol{D})y = f$, with zero initial state, it is a particular solution. In practice, computing convolutions can be time consuming and tedious. In the following examples, we will limit the inputs to functions in $\mathcal{E}$ and use Table C.7.

Example 9. A system Φ is modeled by $q(\boldsymbol{D})$. Find the zero-state response for the given input function:

(a) $q(D) = D + 2$ and $f(t) = 1$
(b) $q(D) = D^2 + 4D + 3$ and $f(t) = e^{-t}$
(c) $q(D) = D^2 + 4$ and $f(t) = \cos(2t)$

▶ **Solution.** (a) The characteristic polynomial is $q(s) = s + 2$, and therefore, the characteristic mode is e^{-2t}. It follows that $h(t) = Ae^{-2t}$ and with initial condition $h(0) = 1$, we get $h(t) = e^{-2t}$. The system response, $y(t)$, for the input 1 is

$$y(t) = e^{-2t} * 1(t) = \frac{1}{2}(1 - e^{-2t}).$$

(b) The characteristic polynomial is $q(s) = s^2 + 4s + 3 = (s + 1)(s + 3)$. The characteristic modes are e^{-3t} and e^{-t}. Thus, $h(t)$ has the form $h(t) = Ae^{-t} + Be^{-3t}$. The initial conditions $h(0) = 0$ and $h'(0) = 1$ imply

$$h(t) = \frac{1}{2}(e^{-t} - e^{-3t}).$$

The system response to the input function $f(t) = e^{-t}$ is

$$y(t) = \frac{1}{2}(e^{-t} - e^{-3t}) * e^{-t} = \frac{1}{2}te^{-t} - \frac{1}{4}e^{-t} + \frac{1}{4}e^{-3t}.$$

(c) It is easy to verify that the impulse response function is $h(t) = \frac{1}{2}\sin(2t)$. The system response to the input function $f(t) = \cos(2t)$ is

$$y(t) = \frac{1}{2}\sin(2t) * \cos(2t) = \frac{1}{4}t\sin(2t). \qquad ◀$$

Bounded-In Bounded-Out

In Example 9a, we introduce a bounded input, $f(t) = 1$, and the response $y(t) = \frac{1}{2}(1 - e^{-2t})$ is also bounded, by $\frac{1}{2}$, in fact. On the other hand, in Example 9c, we introduce a bounded input, $f(t) = \cos 2t$, yet the response $y(t) = \frac{1}{4}t\sin 2t$ oscillates unboundedly. We say that a system Φ is **BIBO-stable** if for every bounded input $f(t)$ the response function $y(t)$ is likewise bounded. (*BIBO* stands for "bounded input bounded output.") Note the following theorem. An outline of the proof is given in the exercises.

Theorem 10. *Suppose Φ is an asymptotically stable system. Then Φ is BIBO-stable.*

Unstable systems are of little practical value to an engineer designing a "safe" system. In an unstable system, a set of unintended initial states can lead to an unbounded response that destroys the system entirely. Even marginally stable systems can have bounded input functions that produce unbounded output functions.

This is seen in Example 9c, where the system response to the bounded input $f(t) = \cos(2t)$ is the unbounded function $y(t) = \frac{1}{4}t \sin(2t)$. Asymptotically stable systems are thus the "safest" systems since they are *BIBO*-stable. They produce at worst a bounded response to a bounded input. However, this is not to say that the response cannot be destructive. We will say more about this in the following topic.

Resonance

We now come to a very interesting and important phenomenon called *resonance*. Loosely speaking, resonance is the phenomenon that occurs when a system reacts very energetically to a relatively mild input. Resonance can sometimes be catastrophic for the system. For example, a wine glass has a characteristic frequency at which it will vibrate. You can hear this frequency by rubbing your moistened finger around the rim of the glass to cause it to vibrate. An opera singer who sings a note at this same frequency with sufficient intensity can cause the wine glass to vibrate so much that it shatters. Resonance can also be used to our advantage as is familiar to a musician tuning a musical instrument to a standard frequency given, for example, by a tuning fork. Resonance occurs when the instrument is "in tune."

The characteristic values of a system Φ are sometimes referred to as the *characteristic frequencies* As we saw earlier, the internal workings of a system are governed by these frequencies. A system that is energized tends to operate at these frequencies. Thus, when an input function matches an internal frequency, the system response will generally be quite energetic, even explosive.

A dramatic example of this occurs when a system is marginally stable. Consider the following example.

Example 11. A zero-state system Φ is modeled by the differential equation

$$(D^2 + 1)y = f.$$

Determine the impulse response function h and the system response to the following inputs:

1. $f(t) = \sin(\pi t)$
2. $f(t) = \sin(1.25t)$
3. $f(t) = \sin(t)$.

Discuss the resonance that occurs.

▶ **Solution.** The characteristic values are $\pm i$ with multiplicity one. Thus, the system is marginally stable. The unit impulse response $h(t)$ is the solution to $(D^2 + 1)y = 0$ with initial conditions $y(0) = 0$ and $y'(0) = 1$. The Laplace transform gives $H(s) = \frac{1}{s^2+1}$, and hence, $h(t) = \sin(t)$.

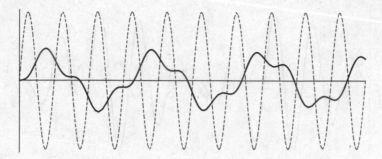

Fig. 4.1 Input and response functions with dissimilar frequencies

The three inputs all have amplitude 1 but different frequencies. We will use the convolution formula

$$\sin(at) * \sin(bt) = \begin{cases} \dfrac{a\sin(bt) - b\sin(at)}{a^2 - b^2} & \text{if } b \neq a \\[2ex] \dfrac{\sin(at) - at\cos(at)}{2a} & \text{if } a = b \end{cases}$$

from Table 2.11.

1. Consider the input function $f(t) = \sin(\pi t)$; its frequency is π. The system response is

$$y(t) = \sin(\pi t) * \sin(t) = \frac{\pi \sin(t) - \sin(\pi t)}{\pi^2 - 1}.$$

The graph of the input function together with the response is given in Fig. 4.1. The graph of the input function is dashed and has amplitude 1 while the response function has an amplitude less than 1. No resonance is occurring here, and this is reflected in the fact that the inputs characteristic frequency, π, is far from the systems characteristic frequency 1. We also note that the response function is not periodic. This is reflected in the fact that the quotient of the frequencies $\frac{\pi}{1} = \pi$ is not rational. We will say more about periodic functions in Chap. 6.

2. Next we take the input function to be $f(t) = \sin(1.25t)$. Its frequency is 1.25 and is significantly closer to the characteristic frequency. The system response is

$$y(t) = \sin(1.25t) * \sin(t) = \frac{1.25\sin(t) - \sin(1.1t)}{0.5625}.$$

The graph of the input function together with the response is given in Fig. 4.2. In this graph, we needed to scale back significantly to see the response function. Notice how the amplitude of the response is significantly higher than that of the input. Also notice how the response comes in pulses. This phenomenon is known as ***beats*** and is familiar to musicians who try to tune an instrument. When the

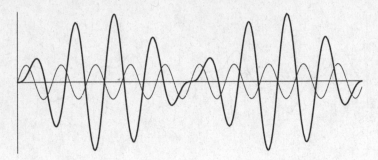

Fig. 4.2 Input and response functions with similar yet unequal frequencies. Beats occur

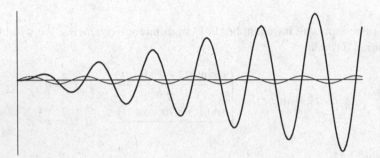

Fig. 4.3 Input and response functions with equal frequencies. Resonance occurs

frequency of vibration of the string is close but not exactly equal to that of the tuning fork, one hears a pulsating beat. The instrument is out of tune.

3. We now consider the input function $f(t) = \sin t$. Here the input frequency matches exactly the characteristic frequency. The system response is

$$y(t) = \sin(t) * \sin(t) = \frac{\sin(t) - t\cos(t)}{2}.$$

The presence of t in $t\cos(t)$ implies that the response will oscillate without bound as seen in Fig. 4.3.

Again in this graph, we needed to scale back to see the enormity of the response function. This is resonance in action. In physical systems, resonance can be so energetic that the system may fall apart. Because of the significant damage that can occur, systems designers must be well aware of the internal characteristic values or frequencies of their system and the likely kinds of inputs it may need to deal with.   ◄

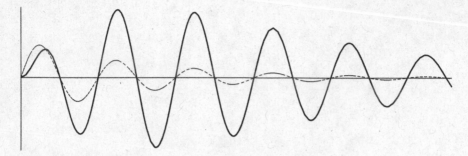

Fig. 4.4 Asymptotically stable system with resonance

As a final example, we consider resonance in an asymptotically stable system.

Example 12. A zero-state system Φ is modeled by the differential equation

$$((D + 0.1)^2 + 1)y = f.$$

Determine the impulse response function h and the system response to the input function $f(t) = e^{-0.1t}\cos(t)$. Discuss the resonance that occurs.

▶ **Solution.** The characteristic values are $-0.1 \pm i$ and lie in the left-hand side of the complex plane. Thus, the system is asymptotically stable. The characteristic modes are $\{e^{-0.1t}\sin t, e^{-0.1t}\cos t\}$. A straightforward calculation gives the unit impulse response function

$$h(t)) = e^{-0.1t}\sin(t).$$

The input function $f(t) = e^{-0.1t}\cos(t)$ is a characteristic mode, and the response function is

$$y(t) = h * f(t) = \frac{1}{2}te^{-0.1t}\sin(t).$$

Figure 4.4 shows the graph. Notice the initial energetic response. This is a manifestation of resonance even though the response dies out in time. If the response passes a certain threshold, the system may break. On the other hand, resonance can be used in a positive way as in tuning a radio to a particular frequency. Again, system designers must be well aware of the resonance effect. ◀

Exercises

1–12. For each problem, a system is modeled by $q(D)$ with initial state **a**. Determine the zero-input response. Determine whether the system is asymptotically stable, marginally stable, or unstable.

1. $q(D) = D + 5, \mathbf{a} = 10$
2. $q(D) = D - 2, \mathbf{a} = 2$
3. $q(D) = D^2 - 4D + 3, \mathbf{a} = (2, 4)$
4. $q(D) = D^2 + 5D + 4, \mathbf{a} = (0, 3)$
5. $q(D) = D^2 + 4D + 5, \mathbf{a} = (0, 1)$
6. $q(D) = D^2 + 9, \mathbf{a} = (1, 1)$
7. $q(D) = D^2 + 6D + 9, \mathbf{a} = (1, 1)$
8. $q(D) = D^2 + D - 2, \mathbf{a} = (1, -2)$
9. $q(D) = D^2 - 2D + 2, \mathbf{a} = (1, 2)$
10. $q(D) = D^3 + D^2, \mathbf{a} = (1, -1, 1)$
11. $q(D) = (D + 1)(D^2 + 1), \mathbf{a} = (1, -1, 1)$
12. $q(D) = D^4 - 1, \mathbf{a} = (0, 1, 0, -1)$

13–17. For each problem, a system is modeled by $q(D)$. Determine the unit impulse response function.

13. $q(D) = D + 1$
14. $q(D) = D^2 + 4$
15. $q(D) = D^2 - 4$
16. $q(D) = D^2 + 2D + 5$
17. $q(D) = D^3 + D$

18–20. In this set of problems, we establish that an asymptotically stable system modeled by $q(D)$, for some constant coefficient differential operator, is *BIBO*-stable.

18. Let k be a nonnegative integer and $\alpha \in \mathbb{R}$. Show that

$$\int_0^t x^k e^{\alpha x}\, dx = C + p(t)e^{\alpha t},$$

where C is a constant and $p(t)$ is a polynomial of degree k. Show that $C + p(t)e^{\alpha t}$ is a bounded function of $[0, \infty)$ if $\alpha < 0$.
19. Suppose $\lambda = \alpha + i\beta$ is a complex number and $\alpha < 0$. Let k be a nonnegative integer and suppose f is a bounded function on $[0, \infty)$. Show that $t^k e^{\alpha t} \cos \beta t * f$ and $t^k e^{\alpha t} \sin \beta t * f$ are bounded functions.
20. Suppose a system modeled by a constant coefficient differential operator is asymptotically stable. Show it is *BIBO*-stable.

Chapter 5
Second Order Linear Differential Equations

In this chapter, we consider the broader class of second order linear differential equations that includes the constant coefficient case. In particular, we will consider differential equations of the following form:

$$a_2(t)y'' + a_1(t)y' + a_0(t)y = f(t). \tag{1}$$

Notice that the coefficients $a_0(t)$, $a_1(t)$, and $a_2(t)$ are functions of the independent variable t and not necessarily constants. This difference has many important consequences, the main one being that there is no general solution method as in the constant coefficient case. Nevertheless, it is still linear and, as we shall see, this implies that the solution set has a structure similar to the constant coefficient case.

In order to find solution methods, one must put some rather strong restrictions on the *coefficient functions* $a_0(t)$, $a_1(t)$ and $a_2(t)$. For example, in the following list, the coefficient functions are polynomial of a specific form. The equations in this list are classical and have important uses in the physical and engineering sciences.

$t^2 y'' + t y' + (t^2 - \nu^2)y = 0$	Bessel's equation of index ν
$t y'' + (1 - t)y' + \lambda y = 0$	Laguerre's equation of index λ
$(1 - t^2)y'' - 2t y' + \alpha(\alpha + 1)y = 0$	Legendre's equation of index α
$(1 - t^2)y'' - t y' + \alpha^2 y = 0$	Chebyshev's equation of index α
$y'' - 2t y' + 2\lambda y = 0$	Hermite's equation of index λ.

Unlike the constant coefficient case, the solutions to these equations are not, in general, expressible in terms of algebraic combinations of polynomials, trigonometric, or exponential functions, nor their inverses. Nevertheless, the general theory implies that solutions exist and traditionally have been loosely categorized as *special functions*. In addition to satisfying the differential equation for the given index, there are other interesting and important functional relations as the index varies. We will explore some of these relations.

W.A. Adkins and M.G. Davidson, *Ordinary Differential Equations*,
Undergraduate Texts in Mathematics, DOI 10.1007/978-1-4614-3618-8_5,
© Springer Science+Business Media New York 2012

5.1   The Existence and Uniqueness Theorem

In this chapter, we will assume that the coefficient functions $a_0(t)$, $a_1(t)$, and $a_2(t)$ and the *forcing function* $f(t)$ are continuous functions on some common interval I. We also assume that $a_2(t) \neq 0$ for all $t \in I$. By dividing by $a_2(t)$, when convenient, we may assume that the leading coefficient function is 1. In this case, we say that the differential equation is in *standard form*. We will adopt much of the notation that we used in Sects. 3.1 and 4.1. In particular, let D denote the derivative operator and let

$$L = a_2(t)D^2 + a_1(t)D + a_0(t). \tag{1}$$

Then (1) in the introductory paragraph can be written as $Ly = f$. If $f = 0$, then the equation $Ly = f = 0$ is called *homogeneous*. Otherwise, it is *nonhomogeneous*. We can think of L as an operation on functions. If $y \in C^2(I)$, in other words if y is a function on an interval I having a second order continuous derivative, then Ly produces a continuous function.

Example 1. Suppose $L = tD^2 + 2D + t$. Find

$$L\left(\frac{\cos t}{t}\right), \quad L\left(\frac{\sin t}{t}\right), \quad L(\sin t), \quad \text{and} \quad L(e^{-t}).$$

▶ **Solution.** The following table gives the first and second derivatives for each function:

$$y = \frac{\cos t}{t} \qquad y' = \frac{-t\sin t - \cos t}{t^2} \qquad y'' = \frac{-t^2\cos t + 2t\sin t + 2\cos t}{t^3}$$

$$y = \frac{\sin t}{t} \qquad y' = \frac{t\cos t - \sin t}{t^2} \qquad y'' = \frac{-t^2\sin t - 2t\cos t + 2\sin t}{t^3}$$

$$y = \sin t \qquad y' = \cos t \qquad y'' = -\sin t$$

$$y = e^{-t} \qquad y' = -e^{-t} \qquad y'' = e^{-t}.$$

It now follows that

- $L\left(\dfrac{\cos t}{t}\right) = t\dfrac{-t^2\cos t + 2t\sin t + 2\cos t}{t^3} + 2\dfrac{-t\sin t - \cos t}{t^2} + t\dfrac{\cos t}{t}$

   $= \dfrac{-t^3\cos t + 2t^2\sin t + 2t\cos t - 2t^2\sin t - 2t\cos t + t^3\cos t}{t^3}$

   $= 0$

- $L\left(\dfrac{\sin t}{t}\right) = t\dfrac{-t^2\sin t - 2t\cos t + 2\sin t}{t^3} + 2\dfrac{t\cos t - \sin t}{t^2} + t\dfrac{\sin t}{t}$

   $= \dfrac{-t^3\sin t - 2t^2\cos t + 2t\sin t + 2t^2\cos t - 2t\sin t + t^3\sin t}{t^3}$

   $= 0$

- $L(\sin t) = t(-\sin t) + 2(\cos t) + t \sin t = 2 \cos t$
- $L(e^{-t}) = te^{-t} + 2(-e^{-t}) + te^{-t} = (2t - 2)e^{-t}$ ◀

The most important general property that we can say about L is that it is linear.

Proposition 2. *The operator*

$$L = a_2(t)D^2 + a_1(t)D + a_0(t)$$

given by (1) *is linear. Specifically,*

1. If y_1 and y_2 have second order continuous derivatives, then

$$L(y_1 + y_2) = L(y_1) + L(y_2).$$

2. If y has a second order continuous derivative and c is a scalar, then

$$L(cy) = cL(y).$$

Proof. The proof of this proposition is essentially the same as the proof of Proposition 2 of Sect. 3.3. We only need to remark that multiplication by a function $a_k(t)$ preserves addition and scalar multiplication in the same way as multiplication by a constant.   □

We call L a **second order linear differential operator**. Proposition 4 of Sect. 3.3 and Theorem 6 of Sect. 3.3 are two important consequences of linearity for the constant coefficient case. The statement and proof are essentially the same. We consolidate these results and Algorithm 7 of Sect. 3.3 into the following theorem:

Theorem 3. *Suppose L is a second order linear differential operator and y_1 and y_2 are solutions to $Ly = 0$. Then $c_1 y_1 + c_2 y_2$ is a solution to $Ly = 0$, for all scalars c_1 and c_2. Suppose f is a continuous function. If y_p is a fixed particular solution to $Ly = f$ and y_h is any solution to the associated homogeneous differential equation $Ly = 0$, then*

$$y = y_p + y_h$$

is a solution to $Ly = f$. Furthermore, any solution to $Ly = f$ has this same form. Thus, to solve $Ly = f$, we proceed as follows:

1. Find all the solutions to the associated homogeneous differential equation $Ly = 0$.
2. Find one particular solution y_p.
3. Add the particular solution to the homogeneous solutions.

As an application of Theorem 3, consider the following example.

Example 4. Let $L = tD^2 + 2D + t$. Use Theorem 3 and the results of Example 1 to write the most general solution to

$$Ly = 2 \cos t.$$

▶ **Solution.** In Example 1, we showed that $y_p = \sin t$ was a particular solution and $y_1 = \cos t / t$ and $y_2 = \sin t / t$ are homogeneous solutions. By Theorem 3, we have that $y_h = c_1 \left(\frac{\cos t}{t} \right) + c_2 \left(\frac{\sin t}{t} \right)$ is also a homogeneous solution and

$$y = y_p + y_h = \sin t + c_1 \left(\frac{\cos t}{t} \right) + c_2 \left(\frac{\sin t}{t} \right)$$

are solutions to $Ly = 2 \cos t$. We will soon see it is the general solution. ◀

Again, suppose L is a second order linear differential operator and f is a function defined on an interval I. Let $t_0 \in I$. To the equation

$$Ly = f$$

we can associate *initial conditions* of the form

$$y(t_0) = y_0, \quad \text{and} \quad y'(t_0) = y_1.$$

We refer to the initial conditions and the differential equation $Ly = f$ as an *initial value problem*.

Example 5. Let $L = tD^2 + 2D + t$. Solve the initial value problem

$$Ly = 2 \cos t, \quad y(\pi) = 1, \; y'(\pi) = -1.$$

▶ **Solution.** By Example 4, all functions of the form

$$y = \sin t + c_1 \frac{\cos t}{t} + c_2 \frac{\sin t}{t}$$

are solutions. Thus, we only need to find constants c_1 and c_2 that satisfy the initial conditions. Since

$$y' = \cos t + c_1 \left(\frac{-t \sin t - \cos t}{t^2} \right) + c_2 \left(\frac{t \cos t - \sin t}{t^2} \right),$$

we have

$$0 + c_1 \left(\frac{-1}{\pi} \right) + c_2(0) = y(\pi) = 1$$

$$-1 + c_1 \left(\frac{1}{\pi^2} \right) + c_2 \left(\frac{-\pi}{\pi^2} \right) = y'(\pi) = -1,$$

which imply $c_1 = -\pi$ and $c_2 = -1$. The solution to the initial value problem is

$$y = \sin t - \pi \frac{\cos t}{t} - \frac{\sin t}{t}.$$ ◀

In the case where the coefficient functions of L are constant, we have an existence and uniqueness theorem. (See Theorem 10 of Sect. 3.1.) In the present case, we still have the existence and uniqueness theorem; however, its proof is beyond the scope of this book.[1]

Theorem 6 (Uniqueness and Existence).   *Suppose $a_0(t)$, $a_1(t)$, $a_2(t)$, and f are continuous functions on an open interval I and $a_2(t) \neq 0$ for all $t \in I$. Suppose $t_0 \in I$ and y_0 and y_1 are fixed real numbers. Let $L = a_2(t)D^2 + a_1(t)D + a_0(t)$. Then there is one and only one solution to the initial value problem*

$$Ly = f, \quad y(t_0) = y_0, \; y'(t_0) = y_1.$$

Theorem 6 does not tell us how to find any solution. We must develop procedures for this. Let us explain in more detail what this theorem does say. Under the conditions stated, the existence and uniqueness theorem says that there always is a solution to the given initial value problem. The solution is at least twice differentiable on I and there is no other solution. In Example 5, we found $y = \sin t - \pi \frac{\cos t}{t} - \frac{\cos t}{t}$ is a solution to $ty'' + 2y' + ty = 2\cos t$ with initial conditions $y(\pi) = 1$ and $y'(\pi) = -1$. Notice, in this case, that y is, in fact, infinitely differentiable on any interval not containing 0. The uniqueness part of Theorem 6 implies that there are no other solutions. In other words, there are no potentially hidden solutions, so that if we can find enough solutions to take care of all possible initial values, then Theorem 6 provides the theoretical underpinnings to know that we have found *all* possible solutions and need look no further. Compare this theorem with the discussion in Sect. 1.7 where we saw examples (in the nonlinear case) of initial value problems which had infinitely many distinct solutions.

---

[1]For a proof, see Theorems 1 and 3 on pages 104–105 of the text *An Introduction to Ordinary Differential Equations* by Earl Coddington, published by Prentice Hall, (1961).

Exercises

1–11. For each of the following differential equations, determine if it is linear (yes/no). For each of those which are linear, further determine if the equation is homogeneous (homogeneous/nonhomogeneous) and constant coefficient (yes/no). Do *not* solve the equations.

1. $y'' + y'y = 0$
2. $y'' + y' + y = 0$
3. $y'' + y' + y = t^2$
4. $y'' + ty' + (1 + t^2)y^2 = 0$
5. $3t^2 y'' + 2ty' + y = e^{2t}$
6. $y'' + \sqrt{y'} + y = t$
7. $y'' + \sqrt{t}\,y' + y = \sqrt{t}$
8. $y'' - 2y = ty$
9. $y'' + 2y + t \sin y = 0$
10. $y'' + 2y' + (\sin t)y = 0$
11. $t^2 y'' + ty' + (t^2 - 5)y = 0$

12–13. For the given differential operator L, compute $L(y)$ for each given y.

12. $L = tD^2 + 1$

1. $y(t) = 1$
2. $y(t) = t$
3. $y(t) = e^{-t}$
4. $y(t) = \cos 2t$

13. $L = t^2 D^2 + tD - 1$

1. $y(t) = \frac{1}{t}$
2. $y(t) = 1$
3. $y(t) = t$
4. $y(t) = t^r$

14. The differential equation $t^2 y'' + ty' - y = t^{\frac{1}{2}}$, $\quad t > 0$ has a solution of the form $y_p(t) = Ct^{\frac{1}{2}}$. Find C.
15. The differential equation $ty'' + (t-1)y' - y = t^2 e^{-t}$ has a solution of the form $y_p(t) = Ct^2 e^{-t}$. Find C.
16. Let $L(y) = (1 + t^2)y'' - 4ty' + 6y$

1. Check that $y(t) = t$ is a solution to the differential equation $L(y) = 2t$.
2. Check that $y_1(t) = 1 - 3t^2$ and $y_2(t) = t - \frac{t^3}{3}$ are two solutions to the differential equation $L(y) = 0$.

3. Using the results of Parts (1) and (2), find a solution to each of the following initial value problems:

   a. $(1+t^2)y'' - 4ty + 6y = 2t$, $y(0) = 1$, $y'(0) = 0$.

   b. $(1+t^2)y'' - 4ty + 6y = 2t$, $y(0) = 0$, $y'(0) = 1$.

   c. $(1+t^2)y'' - 4ty + 6y = 2t$, $y(0) = -1$, $y'(0) = 4$.

   d. $(1+t^2)y'' - 4ty + 6y = 2t$, $y(0) = a$, $y'(0) = b$, where $a, b \in \mathbb{R}$.

17. Let $L(y) = (t-1)y'' - ty' + y$:

  1. Check that $y(t) = e^{-t}$ is a solution to the differential equation $L(y) = 2te^{-t}$.
  2. Check that $y_1(t) = e^t$ and $y_2(t) = t$ are two solutions to the differential equation $L(y) = 0$.
  3. Using the results of Parts (1) and (2), find a solution to each of the following initial value problems:

   a. $(t-1)y'' - ty' + y = 2te^{-t}$ $y(0) = 0$, $y'(0) = 0$.

   b. $(t-1)y'' - ty' + y = 2te^{-t}$ $y(0) = 1$, $y'(0) = 0$.

   c. $(t-1)y'' - ty' + y = 2te^{-t}$ $y(0) = 0$, $y'(0) = 1$.

   d. $(t-1)y'' - ty' + y = 2te^{-t}$, $y(0) = a$, $y'(0) = b$, where $a, b \in \mathbb{R}$.

18. Let $L(y) = t^2 y'' - 4ty' + 6y$:

  1. Check that $y(t) = \frac{1}{6}t^5$ is a solution to the differential equation $L(y) = t^5$.
  2. Check that $y_1(t) = t^2$ and $y_2(t) = t^3$ are two solutions to the differential equation $L(y) = 0$.
  3. Using the results of Parts (1) and (2), find a solution to each of the following initial value problems:

   a. $t^2 y'' - 4ty' + 6y = t^5$, $y(1) = 1$, $y'(1) = 0$.

   b. $t^2 y'' - 4ty' + 6y = t^5$, $y(1) = 0$, $y'(1) = 1$.

   c. $t^2 y'' - 4ty' + 6y = t^5$, $y(1) = -1$, $y'(1) = 3$.

   d. $t^2 y'' - 4ty' + 6y = t^5$, $y(1) = a$, $y'(1) = b$, where $a, b \in \mathbb{R}$.

19–24. For each of the following differential equations, find the largest interval on which a unique solution of the initial value problem

$$a_2(t)y'' + a_1(t)y' + a_0(t)y = f(t)$$

is guaranteed by Theorem 6, if initial conditions, $y(t_0) = y_1$, $y'(t_0) = y_1$, are given at t_0.

19. $t^2 y'' + 3ty' - y = t^4$ $(t_0 = -1)$

20. $y'' - 2y' - 2y = \dfrac{1+t^2}{1-t^2}$ $(t_0 = 2)$

21. $(\sin t)y'' + y = \cos t \qquad (t_0 = \frac{\pi}{2})$
22. $(1 + t^2)y'' - ty' + t^2 y = \cos t \qquad (t_0 = 0)$
23. $y'' + \sqrt{t}\, y' - \sqrt{t - 3}\, y = 0 \qquad (t_0 = 10)$
24. $t(t^2 - 4)y'' + y = e^t \qquad (t_0 = 1)$

25. The functions $y_1(t) = t^2$ and $y_2(t) = t^3$ are two distinct solutions of the initial value problem

$$t^2 y'' - 4ty' + 6y = 0, \quad y(0) = 0, \ y'(0) = 0.$$

Why does this not violate the uniqueness part of Theorem 6?

26. Let $y(t)$ be a solution of the differential equation

$$y'' + a_1(t)y' + a_0(t)y = 0.$$

We assume that $a_1(t)$ and $a_0(t)$ are continuous functions on an interval I, so that Theorem 6 implies that a solution y is defined on I. Show that if the graph of $y(t)$ is tangent to the t-axis at some point t_0 of I, then $y(t) = 0$ for all $t \in I$. *Hint:* If the graph of $y(t)$ is tangent to the t-axis at $(t_0, 0)$, what does this say about $y(t_0)$ and $y'(t_0)$?

27. More generally, let $y_1(t)$ and $y_2(t)$ be two solutions of the differential equation

$$y'' + a_1(t)y' + a_0(t)y = f(t),$$

where, as usual, we assume that $a_1(t)$, $a_0(t)$, and $f(t)$ are continuous functions on an interval I, so that Theorem 6 implies that y_1 and y_2 are defined on I. Show that if the graphs of $y_1(t)$ and $y_2(t)$ are tangent at some point t_0 of I, then $y_1(t) = y_2(t)$ for all $t \in I$.

5.2 The Homogeneous Case

In this section, we are concerned with a concise description of the solution set of the homogeneous linear differential equation

$$L(y) = a_2(t)y'' + a_1(t)y' + a_0(t)y = 0 \tag{1}$$

The main result, Theorem 2 given below, shows that we will in principle be able to find two linearly independent functions y_1 and y_2 such that all solutions to (1) are of the form $c_1 y_1 + c_2 y_2$, for some constants c_1 and c_2. This is just like the second order constant coefficient case. In fact, if $q(s)$ is a characteristic polynomial of degree 2, then $\mathcal{B}_q = \{y_1, y_2\}$ is a set of two linearly independent functions that span the solution set of the corresponding homogeneous constant coefficient differential equation. In Sect. 3.2, we introduced the concept of linear independence for a set of n functions. Let us recall this important concept in the case $n = 2$.

Linear Independence

Two functions y_1 and y_2 defined on some interval I are said to be **linearly independent** if the equation

$$c_1 y_1 + c_2 y_2 = 0 \tag{2}$$

implies that c_1 and c_2 are both 0. Otherwise, we call y_1 and y_2 **linearly dependent**.

Example 1. Show that the functions

$$y_1(t) = \frac{\cos t}{t} \quad \text{and} \quad y_2(t) = \frac{\sin t}{t},$$

defined on the interval $(0, \infty)$ are linearly independent.

▶ **Solution.** The equation $c_1 \frac{\cos t}{t} + c_2 \frac{\sin t}{t} = 0$ on $(0, \infty)$ implies $c_1 \cos t + c_2 \sin t = 0$. Evaluating at $t = \frac{\pi}{2}$ and $t = \pi$ gives

$$c_2 = 0$$
$$-c_1 = 0.$$

It follows that y_1 and y_2 are independent.   ◀

In Sect. 3.2, we provided several other examples that we encourage the student to reconsider by way of review.

The Main Theorem for the Homogeneous Case

Theorem 2. *Let $L = a_2(t)D^2 + a_1(t)D + a_0(t)$, where $a_0(t)$, $a_1(t)$, and $a_2(t)$ are continuous functions on an interval I. Assume $a_2(t) \neq 0$ for all $t \in I$.*

1. *There are two linearly independent solutions to $Ly = 0$.*
2. *If y_1 and y_2 are any two linearly independent solutions to $Ly = 0$, then any homogeneous solution y can be written $y = c_1 y_1 + c_2 y_2$, for some $c_1, c_2 \in \mathbb{R}$.*

Proof. Let $t_0 \in I$. By Theorem 6 of Sect. 5.1, there are functions, ψ_1 and ψ_2, that are solutions to the initial value problems $L(y) = 0$, with initial conditions $y(t_0) = 1$, $y'(t_0) = 0$, and $y(t_0) = 0$, $y'(t_0) = 1$, respectively. Suppose $c_1 \psi_1 + c_2 \psi_2 = 0$. Then

$$c_1 \psi_1(t_0) + c_2 \psi_2(t_0) = 0.$$

Since $\psi_1(t_0) = 1$ and $\psi_2(t_0) = 0$, it follows that $c_1 = 0$. Similarly, we have

$$c_1 \psi_1'(t_0) + c_2 \psi_2'(t_0) = 0.$$

Since $\psi_1'(t_0) = 0$ and $\psi_2'(t_0) = 1$, it follows that $c_2 = 0$. Therefore, ψ_1 and ψ_2 are linearly independent. This proves (1).

Suppose y is a homogeneous solution. Let $r = y(t_0)$ and $s = y'(t_0)$. By Theorem 3 of Sect. 5.1, the function $r\psi_1 + s\psi_2$ is a solution to $Ly = 0$. Furthermore,

$$r\psi_1(t_0) + s\psi_2(t_0) = r$$

$$\text{and} \quad r\psi_1'(t_0) + s\psi_2'(t_0) = s.$$

This means the $r\psi_1 + s\psi_2$ and y satisfy the same initial conditions. By the uniqueness part of Theorem 6 of Sect. 5.1, they are equal. Thus, $y = r\psi_1 + s\psi_2$, that is, every homogeneous solution is a linear combination of ψ_1 and ψ_2.

Now suppose y_1 and y_2 are any two linearly independent homogeneous solutions and suppose y is any other solution. From the argument above, we can write

$$y_1 = a\psi_1 + b\psi_2,$$

$$y_2 = c\psi_1 + d\psi_2,$$

which in matrix form can be written

$$\begin{bmatrix} y_1 \\ y_2 \end{bmatrix} = \begin{bmatrix} a & b \\ c & d \end{bmatrix} \begin{bmatrix} \psi_1 \\ \psi_2 \end{bmatrix}.$$

We multiply both sides of this matrix equation by the adjoint $\begin{bmatrix} d & -b \\ -c & a \end{bmatrix}$ to obtain

$$\begin{bmatrix} d & -b \\ -c & a \end{bmatrix} \begin{bmatrix} y_1 \\ y_2 \end{bmatrix} = \begin{bmatrix} ad - bc & 0 \\ 0 & ad - bc \end{bmatrix} \begin{bmatrix} \psi_1 \\ \psi_2 \end{bmatrix} = (ad - bc) \begin{bmatrix} \psi_1 \\ \psi_2 \end{bmatrix}.$$

Suppose $ad - bc = 0$. Then

$$dy_1 - by_2 = 0$$

$$\text{and} \quad -cy_1 + ay_2 = 0.$$

But since y_1 and y_2 are independent, this implies that $a, b, c,$ and d are zero, which in turn implies that y_1 and y_2 are both zero. But this cannot be. We conclude that $ad - bc \neq 0$. We can now write ψ_1 and ψ_2 each as a linear combination of y_1 and y_2. Specifically,

$$\begin{bmatrix} \psi_1 \\ \psi_2 \end{bmatrix} = \frac{1}{ad - bc} \begin{bmatrix} d & -b \\ -c & a \end{bmatrix} \begin{bmatrix} y_1 \\ y_2 \end{bmatrix}.$$

Since y is a linear combination of ψ_1 and ψ_2, it follows that y is a linear combination of y_1 and y_2. $\square$

Remark 3. The matrix $\begin{bmatrix} a & b \\ c & d \end{bmatrix}$ that appears in the proof above appears in other contexts as well. If y_1 and y_2 are functions, recall from Sect. 3.2 that we defined the **Wronskian matrix** by

$$W(y_1, y_2)(t) = \begin{bmatrix} y_1(t) & y_2(t) \\ y_1'(t) & y_2'(t) \end{bmatrix}$$

and the **Wronskian** by

$$w(y_1, y_2)(t) = \det W(y_1, y_2)(t).$$

If y_1 and y_2 are as in the proof above, then the relations

$$y_1 = a\psi_1 + b\psi_2,$$

$$y_2 = c\psi_1 + d\psi_2,$$

in the proof, when evaluated at t_0 imply that

$$\begin{bmatrix} a & c \\ b & d \end{bmatrix} = \begin{bmatrix} y_1(t_0) & y_2(t_0) \\ y_1'(t_0) & y_2'(t_0) \end{bmatrix} = W(y_1, y_2)(t_0).$$

Since it was shown that $ad - bc \neq 0$, we have $w(y_1, y_2) \neq 0$. On the other hand, given any two differentiable functions, y_1 and y_2 (not necessarily homogeneous solutions of a linear differential equation), whose Wronskian is a nonzero function, then it is easy to see that y_1 and y_2 are independent. For suppose t_0 is chosen so that $w(y_1, y_2)(t_0) \neq 0$ and $c_1 y_1 + c_2 y_2 = 0$. Then $c_1 y_1' + c_2 y_2' = 0$, and we have

$$\begin{bmatrix} 0 \\ 0 \end{bmatrix} = \begin{bmatrix} c_1 y_1(t_0) + c_2 y_2(t_0) \\ c_1 y_1'(t_0) + c_2 y_2'(t_0) \end{bmatrix} = W(y_1, y_2) \begin{bmatrix} c_1 \\ c_2 \end{bmatrix}.$$

Simple matrix algebra[2] gives $c_1 = 0$ and $c_2 = 0$. Hence, y_1 and y_2 are linearly independent.

We have thus shown the following proposition.

Proposition 4. *Suppose L satisfies the conditions of Theorem 2. Suppose y_1 and y_2 are solutions to $L y = 0$. Then y_1 and y_2 are linearly independent if and only if*

$$w(y_1, y_2) \neq 0.$$

The following theorem extends Abel's theorem given in Theorem 8 of Sect. 3.3. You are asked to prove this in the exercises.

Theorem 5 (Abel's Formula). *Suppose f_1 and f_2 are solutions to the second order linear differential equation*

$$y'' + a_1(t)y' + a_0(t)y = 0,$$

where a_0 and a_1 are continuous functions on an interval I. Let $t_0 \in I$. Then

$$w(f_1, f_2)(t) = K e^{-\int_{t_0}^{t} a_1(x)\, dx}, \tag{3}$$

for some constant K. Furthermore, if initial conditions $y(t_0)$ and $y'(t_0)$ are given, then

$$K = w(f_1, f_2)(t_0).$$

Remark 6. Let us now summarize what Theorems 3, 6 of Sect. 5.1, and 2 tell us. In order to solve $L(y) = f$ (satisfying the continuity hypotheses), we first need to find a particular solution y_p, which exists by the existence and uniqueness theorem. Next, Theorem 2 says that if y_1 and y_2 are any two linearly independent solutions of the associated homogeneous equation $L(y) = 0$, then all of the solutions of the associated homogeneous equation are of the form $c_1 y_1 + c_2 y_2$. Theorem 3 of Sect. 5.1 now tells us that the ***general solution*** to $L(y) = f$ is of the form

$$\{y_p + c_1 y_1 + c_2 y_2 : c_1, c_2 \in \mathbb{R}\}.$$

Furthermore, any set of initial conditions uniquely determines the constants c_1 and c_2.

---

[2]cf. Chapter 8 for a discussion of matrices.

A set $\{y_1, y_2\}$ of linearly independent solutions to the homogeneous equation $L(y) = 0$ is called a ***fundamental set*** for the second order linear differential operator L. A fundamental set is a basis of the linear space of homogeneous solutions (cf. Sect. 3.2 for the definition of a basis). Furthermore, the standard basis, $\mathcal{B}_q$, in the context of constant coefficient differential equations, is a fundamental set.

In the following sections, we will develop methods, under suitable assumptions, for finding a fundamental set for L and a particular solution to the differential equation $L(y) = f$. For now, let us illustrate the main theorems with a couple of examples.

Example 7. Consider the differential equation

$$t^2 y'' + t y' + y = 0.$$

Suppose $y_1(t) = \cos(\ln t)$ and $y_2(t) = \sin(\ln t)$ are solutions. Determine the solution set.

▶ **Solution.** We begin by computing the Wronskian of $\{\cos \ln t, \ \sin \ln t\}$:

$$w(\cos \ln t, \ \sin \ln t) = \det \begin{pmatrix} \cos \ln t & \sin \ln t \\ -\dfrac{\sin \ln t}{t} & \dfrac{\cos \ln t}{t} \end{pmatrix}$$

$$= \frac{\cos^2 \ln t + \sin^2 \ln t}{t}$$

$$= \frac{1}{t}.$$

Proposition 4 implies that $\{\cos(\ln t), \sin(\ln t)\}$ is linearly independent on $(0, \infty)$ and thus a fundamental set for $L(y) = 0$. Theorem 2 now implies that

$$\{c_1 \cos(\ln t) + c_2 \sin(\ln t) : c_1, c_2 \in \mathbb{R}\}$$

is the solution set.   ◀

Example 8. Consider the differential equation

$$t y'' + 2 y' + t y = 2 \cos t.$$

Determine the solution set.

▶ **Solution.** By Example 1 of Sect. 5.1, a particular solution is $y_p = \sin t$, and $y_1 = \frac{\cos t}{t}$ and $y_2 = \frac{\sin t}{t}$ are homogeneous solutions. By Example 1, the homogeneous solutions y_1 and y_2 are linearly independent. By Theorem 2, all homogeneous solutions are of the form

$$y_h = c_1 y_1 + c_2 y_2 = c_1 \frac{\cos t}{t} + c_2 \frac{\sin t}{t}.$$

It follows from linearity, Theorem 3 of Sect. 5.1, that the solution set is

$$\left\{ y = y_p + y_h = \sin t + c_1 \frac{\cos t}{t} + c_2 \frac{\sin t}{t}, c_1, c_2 \in \mathbb{R} \right\}.$$

◄

Exercises

1–6. Determine if each of the following pairs of functions are linearly independent or linearly dependent.

1. $y_1(t) = 2t$, $y_2(t) = 5t$
2. $y_1(t) = 2^t$, $y_2(t) = 5^t$
3. $y_1(t) = \ln t$, $y_2(t) = t \ln t$ on the interval $(0, \infty)$
4. $y_1(t) = e^{2t+1}$, $y_2(t) = e^{2t-3}$
5. $y_1(t) = \ln(2t)$, $y_2(t) = \ln(5t)$ on the interval $(0, \infty)$
6. $y_1(t) = \ln t^2$, $y_2(t) = \ln t^5$

7–9. For each exercise below, verify that the functions f_1 and f_2 satisfy the given differential equation. Verify Abel's formula as given in Theorem 5 of Sect. 5.2 for the given initial point t_0. Determine the solution set.

7. $(t - 1)y'' - ty' + y = 0$, $f_1(t) = e^t - t$, $f_2(t) = t$, $t_0 = 0$
8. $(1 + t^2)y'' - 2ty' + 2y = 0$, $f_1(t) = 1 - t^2$, $f_2(t) = t$, $t_0 = 1$
9. $t^2 y'' + ty' + 4y = 0$, $f_1(t) = \cos(2 \ln t)$, $f_2(t) = \sin(2 \ln t)$, $t_0 = 1$
10. Prove Abel's formula as stated in Theorem 5. Hint, carefully look at the proof of Abel's formula given in the second order constant coefficient case (cf. Theorem 8 of Sect. 3.3).

11. 1. Verify that $y_1(t) = t^3$ and $y_2(t) = |t^3|$ are linearly independent on $(-\infty, \infty)$.
    2. Show that the Wronskian, $w(y_1, y_2)(t) = 0$, for all $t \in \mathbb{R}$.
    3. Explain why Parts (a) and (b) do not contradict Proposition 4.
    4. Verify that $y_1(t)$ and $y_2(t)$ are solutions to the linear differential equation
       $t^2 y'' - 2ty' = 0$, $\quad y(0) = 0$, $\quad y'(0) = 0$.
    5. Explain why Parts (a), (b), and (d) do not contradict the existence and uniqueness theorem, Theorem 6 of Sect. 5.1.

5.3 The Cauchy–Euler Equations

When the coefficient functions of a second order linear differential equation are nonconstant, the corresponding equation can become very difficult to solve. In order to expect to find solutions, one must put certain restrictions on the coefficient functions. A class of nonconstant coefficient linear differential equations, known as Cauchy–Euler equations, have solutions that are easy to obtain.

A **Cauchy–Euler equation** is a second order linear differential equation of the following form:

$$at^2 y'' + bt y' + cy = 0, \tag{1}$$

where a, b, and c are real constants and $a \neq 0$. When put in standard form, we obtain

$$y'' + \frac{b}{at} y' + \frac{c}{at^2} y = 0.$$

The functions $\frac{b}{at}$ and $\frac{c}{at^2}$ are continuous everywhere except at 0. Thus, the existence and uniqueness theorem guarantees that solutions exist in either of the intervals $(-\infty, 0)$ or $(0, \infty)$. To work in a specific interval, we will assume $t > 0$. We will refer to $L = at^2 D^2 + bt D + c$ as a **Cauchy–Euler operator**.

The Laplace transform method does not work in any simple fashion here. However, the simple change in variable $t = e^x$ will transform equation (1) into a constant coefficient linear differential equation. To see this, let $Y(x) = y(e^x)$. Then the chain rule gives

$$Y'(x) = e^x y'(e^x)$$

$$\text{and} \quad Y''(x) = e^x y'(e^x) + (e^x)^2 y''(e^x)$$

$$= Y'(x) + (e^x)^2 y''(e^x).$$

Thus,

$$a(e^x)^2 y''(e^x) = aY''(x) - aY'(x)$$

$$be^x y'(e^x) = bY'(x)$$

$$cy(e^x) = cY(x).$$

Addition of these terms gives

$$a(e^x)^2 y''(e^x) + be^x y'(e^x) + cy(e^x) = aY''(x) - aY'(x) + bY'(x) + cY(x)$$

$$= aY''(x) + (b - a)Y'(x) + cY(x).$$

With t replaced by e^x in (1), we now obtain

$$aY''(x) + (b - a)Y'(x) + cY(x) = 0. \tag{2}$$

The polynomial

$$Q(s) = as^2 + (b - a)s + c$$

is the characteristic polynomial of (2) and known as the **indicial polynomial** of (1). Equation (2) is a second order constant coefficient differential equation and by now routine to solve. Its solutions depend on the way $Q(s)$ factors. We consider the three possibilities.

Q Has Distinct Real Roots

Suppose r_1 and r_2 are distinct roots to the indicial polynomial $Q(s)$. Then $e^{r_1 x}$ and $e^{r_2 x}$ are solutions to (2). Solutions to (1) are obtained by the substitution $x = \ln t$: we have $e^{r_1 x} = e^{r_1 \ln t} = t^{r_1}$ and similarly $e^{r_2 x} = t^{r_2}$. Since t^{r_1} is not a multiple of t^{r_2}, they are independent, and hence,

$$\{t^{r_1},\, t^{r_2}\}$$

is a fundamental set for $L(y) = 0$.

Example 1. Find a fundamental set and general solution for the equation $t^2 y'' - 2y = 0$.

▶ **Solution.** The indicial polynomial is $Q(s) = s^2 - s - 2 = (s - 2)(s + 1)$ and it has 2 and -1 as roots, and thus, $\{t^2,\, t^{-1}\}$ is a fundamental set for this Cauchy–Euler equation. The general solution is $y = c_1 t^2 + c_2 t^{-1}$.   ◀

Q Has a Double Root

Suppose r is a double root of Q. Then e^{rx} and xe^{rx} are independent solutions to (2). The substitution $x = \ln t$ then gives t^r and $t^r \ln t$ as independent solutions to (1). Hence,

$$\{t^r, t^r \ln t\}$$

is a fundamental set for $L(y) = 0$.

Example 2. Find a fundamental set and the general solution for the equation $4t^2 y'' + 8ty' + y = 0$.

▶ **Solution.** The indicial polynomial is $Q(s) = 4s^2 + 4s + 1 = (2s + 1)^2$ and has $-\frac{1}{2}$ as a root with multiplicity 2. Thus, $\{t^{-\frac{1}{2}},\, t^{-\frac{1}{2}} \ln t\}$ is a fundamental set. The general solution is $y = c_1 t^{-\frac{1}{2}} + c_2 t^{-\frac{1}{2}} \ln t$.   ◀

Q Has Conjugate Complex Roots

Suppose Q has complex roots $\alpha \pm i\beta$, where $\beta \neq 0$. Then $e^{\alpha x} \cos \beta x$ and $e^{\alpha x} \sin \beta x$ are independent solutions to (2). The substitution $x = \ln t$ then gives

$$\{t^{\alpha} \cos(\beta \ln t), t^{\alpha} \sin(\beta \ln t)\}$$

as a fundamental set for $L y = 0$.

Example 3. Find a fundamental set and the general solution for the equation $t^2 y'' + ty' + y = 0$.

▶ **Solution.** The indicial polynomial is $Q(s) = s^2 + 1$ which has $\pm i$ as complex roots. Theorem 4 implies that $\{\cos \ln t, \sin \ln t\}$ is a fundamental set. The general solution is $y = c_1 \cos \ln t + c_2 \sin \ln t$. ◀

We now summarize the above results into one theorem.

Theorem 4. *Let* $L = at^2 D^2 + bt D + c$, *where* $a, b, c \in \mathbb{R}$ *and* $a \neq 0$. *Let* $Q(s) = as^2 + (b - a)s + c$ *be the indicial polynomial.*

1. If r_1 *and* r_2 *are distinct real roots of* $Q(s)$, *then*

$$\{t^{r_1}, t^{r_2}\}$$

is a fundamental set for $L(y) = 0$.
2. If r *is a double root of* $Q(s)$, *then*

$$\{t^r, t^r \ln t\}$$

is a fundamental set for $L(y) = 0$.
3. If $\alpha \pm i\beta$ *are complex conjugate roots of* $Q(s)$, $\beta \neq 0$, *then*

$$\{t^{\alpha} \sin(\beta \ln t), t^{\alpha} \cos(\beta \ln t)\}$$

is a fundamental set for $L(y) = 0$.

Exercises

1–11. Find the general solution of each of the following homogeneous Cauchy–Euler equations on the interval $(0, \infty)$.

1. $t^2 y'' + 2ty' - 2y = 0$
2. $2t^2 y'' - 5ty' + 3y = 0$
3. $9t^2 y'' + 3ty' + y = 0$
4. $t^2 y'' + ty' - 2y = 0$
5. $4t^2 y'' + y = 0$
6. $t^2 y'' - 3ty' - 21y = 0$
7. $t^2 y'' + 7ty' + 9y = 0$
8. $t^2 y'' + y = 0$
9. $t^2 y'' + ty' - 4y = 0$
10. $t^2 y'' + ty' + 4y = 0$
11. $t^2 y'' - 3ty' + 13y = 0$

12–15. Solve each of the following initial value problems.

12. $t^2 y'' + 2ty' - 2y = 0$,
    $y(1) = 0, \ y'(1) = 1$
13. $4t^2 y'' + y = 0$,
    $y(1) = 2, \ y'(1) = 0$
14. $t^2 y'' + ty' + 4y = 0$,
    $y(1) = -3, \ y'(1) = 4$
15. $t^2 y'' - 4ty' + 6y = 0$,
    $y(0) = 1, \ y'(0) = -1$

5.4   Laplace Transform Methods

In this section, we will develop some further properties of the Laplace transform and use them to solve some linear differential equations with nonconstant coefficient functions. However, we will see that the use of the Laplace transform is limited. Several new Laplace transform rules and formulas are developed in this section. For quick reference Tables 5.1 and 5.2 in Sect. 5.7 summarize these results.

Let us begin by recalling an important definition we saw in Sect. 2.2. A continuous function f on $[0, \infty)$ is said to be of *exponential type with order a* if there is a constant K such that

$$|f(t)| \leq Ke^{at}$$

for all $t \in [0, \infty)$. If the order is not important to the discussion, we will just say f is of *exponential type*. A function of exponential type has limited growth; it cannot grow faster than a multiple of an exponential function. The above inequality means

$$-Ke^{at} \leq f(t) \leq Ke^{at},$$

for all $t \in [0, \infty)$ as illustrated in Fig. 5.1, where the boldfaced curve, $f(t)$, lies between the upper and lower exponential functions. If f is of exponential type, then Proposition 3 of Sect. 2.2 tells us that $F(s) = \mathcal{L}\{f\}(s)$ exists and

$$\lim_{s \to \infty} F(s) = 0.$$

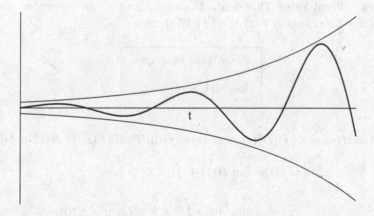

Fig. 5.1 The exponential function Ke^{at} bounds $f(t)$

Asymptotic Values

An interesting property of the Laplace transform is that certain limiting values of $f(t)$ can be deduced from its Laplace transform and vice versa.

Theorem 1 (Initial Value Theorem). *Suppose* f *and its derivative* f' *are of exponential order. Let* $F(s) = \mathcal{L}\{f(t)\}(s)$. *Then*

$$\boxed{\begin{array}{c} \textit{Initial Value Principle} \\[2mm] f(0) = \lim_{s \to \infty} sF(s). \end{array}}$$

Proof. Let $H(s) = \mathcal{L}\{f'(t)\}(s)$. By Proposition 3 of Sect. 2.2, we have

$$0 = \lim_{s \to \infty} H(s) = \lim_{s \to \infty} (sF(s) - f(0)) = \lim_{s \to \infty} (sF(s) - f(0)).$$

This implies the result. □

Example 2. Verify the initial value theorem for $f(t) = \cos at$.

▶ **Solution.** On the one hand, $\cos at|_{t=0} = 1$. On the other hand,

$$s\mathcal{L}\{\cos at\}(s) = \frac{s^2}{s^2 + a^2}$$

which has limit 1 as $s \to \infty$. ◀

Theorem 3 (Final Value Theorem). *Suppose* f *and* f' *are of exponential type and* $\lim_{t \to \infty} f(t)$ *exists. If* $F(s) = \mathcal{L}\{f(t)\}(s)$, *then*

$$\boxed{\begin{array}{c} \textit{Final Value Principle} \\[2mm] \lim_{t \to \infty} f(t) = \lim_{s \to 0} sF(s). \end{array}}$$

Proof. Let $H(s) = \mathcal{L}\{f'(t)\}(s) = sF(s) - f(0)$. Then $sF(s) = H(s) + f(0)$ and

$$\lim_{s \to 0} sF(s) = \lim_{s \to 0} H(s) + f(0)$$

$$= \lim_{s \to 0} \lim_{M \to \infty} \int_0^M e^{-st} f'(t)\, dt + f(0)$$

$$= \lim_{M \to \infty} \int_0^M f'(t)\, dt + f(0)$$

$$= \lim_{M \to \infty} f(M) - f(0) + f(0)$$

$$= \lim_{M \to \infty} f(M).$$

The interchange of the limit operations in line 2 above can be justified for functions of exponential type.   □

Integration in Transform Space

The transform derivative principle, Theorem 20 of Sect. 2.2, tells us that multiplication of an input function by $-t$ induces differentiation of the Laplace transform. One might expect then that division by $-t$ will induce integration in the transform space. This idea is valid but we must be careful about assumptions. First, if $f(t)$ has a Laplace transform, it is not necessarily the case that $f(t)/t$ will likewise. For example, the constant function $f(t) = 1$ has Laplace transform $\frac{1}{s}$ but $\frac{f(t)}{t} = \frac{1}{t}$ does not have a Laplace transform. Second, integration produces an arbitrary constant of integration. What is this constant? The precise statement is as follows:

Theorem 4 (Integration in Transform Space). *Suppose f is of exponential type with order a and $\frac{f(t)}{t}$ has a continuous extension to 0, that is, $\lim_{t \to 0+} \frac{f(t)}{t}$ exists. Then $\frac{f(t)}{t}$ is of exponential type with order a and*

Transform Integral Principle

$$\mathcal{L}\left\{\frac{f(t)}{t}\right\}(s) = \int_s^\infty F(\sigma)\, d\sigma,$$
---

where $s > a$.

Proof. Let $L = \lim_{t \to 0+} \frac{f(t)}{t}$ and define

$$h(t) = \begin{cases} \frac{f(t)}{t} & \text{if } t > 0 \\ L & \text{if } t = 0 \end{cases}.$$

Since f is continuous, so is h. Since f is of exponential type with order a, there is a K so that $|f(t)| \le K e^{at}$. Since $\frac{1}{t} \le 1$ on $[1, \infty)$,

$$|h(t)| = \left|\frac{f(t)}{t}\right| \le |f(t)| \le K e^{at},$$

for all $t \ge 1$. Since h is continuous on $[0, 1]$, it is bounded by B, say. Thus, $|h(t)| \le B \le B e^{at}$ for all $t \in [0, 1]$. If M is the larger of K and B, then

$$|h(t)| \le M \mathrm{e}^{at},$$

for all $t \in [0, \infty)$, and hence h is of exponential type. Let $H(s) = \mathcal{L}\{h(t)\}(s)$ and $F(s) = \mathcal{L}\{f(t)\}(s)$. Then, since $-th(t) = -f(t)$, we have, by Theorem 20 of Sect. 2.2, $H'(s) = -F(s)$. Thus H is an antiderivative of $-F$, and we have

$$H(s) = -\int_a^s F(\sigma)\,\mathrm{d}\sigma + C.$$

Proposition 3 of Sect. 2.2 implies $0 = \lim_{s\to\infty} H(s) = -\int_a^\infty F(\sigma)\,\mathrm{d}\sigma + C$, and hence, $C = \int_a^\infty F(\sigma)\,\mathrm{d}\sigma$. Therefore,

$$\begin{aligned}
\mathcal{L}\left\{\frac{f(t)}{t}\right\}(s) &= H(s) \\
&= \int_s^a F(\sigma)\,\mathrm{d}\sigma + \int_a^\infty F(\sigma)\,\mathrm{d}\sigma \\
&= \int_s^\infty F(\sigma)\,\mathrm{d}\sigma.
\end{aligned}$$

$\square$

The Laplace transform of several new functions can now be deduced from this theorem. Consider an example.

Example 5. Find $\mathcal{L}\left\{\frac{\sin t}{t}\right\}$.

▶ **Solution.** Since $\lim_{t\to 0}\frac{\sin t}{t} = 1$, Theorem 4 applies to give

$$\begin{aligned}
\mathcal{L}\left\{\frac{\sin t}{t}\right\}(s) &= \int_s^\infty \frac{1}{\sigma^2 + 1}\,\mathrm{d}\sigma \\
&= \tan^{-1}\sigma\big|_s^\infty \\
&= \frac{\pi}{2} - \tan^{-1}(s) \\
&= \tan^{-1}\frac{1}{s}.
\end{aligned}$$

The last line can be seen by considering the right triangle

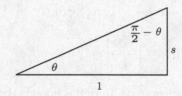

where $\theta = \tan^{-1} s$.   ◀

Solving Linear Differential Equations

We now consider by example how one can use the Laplace transform method to solve some differential equations.

Example 6.   Find a solution of exponential type that solves

$$ty'' - (1+t)y' + y = 0.$$

▶ **Solution.**   Note that the existence and uniqueness theorem implies that solutions exist on intervals that do not contain 0. We presume that such a solution has a continuous extension to $t = 0$ and is of exponential type. Let y be such a solution. Let $y(0) = y_0$, $y'(0) = y_1$, and $Y(s) = \mathcal{L}\{y(t)\}(s)$. Application of transform derivative principle, Theorem 20 of Sect. 2.2, to each component of the differential equation gives

$$\mathcal{L}\{ty''\} = -(s^2 Y(s) - sy(0) - y'(0))'$$

$$= -(2sY(s) + s^2 Y'(s) - y(0))$$

$$\mathcal{L}\{-(1+t)y'\} = \mathcal{L}\{-y'\} + \mathcal{L}\{-ty'\}$$

$$= -sY(s) + y_0 + (sY(s) - y_0)'$$

$$= sY'(s) - (s-1)Y(s) + y_0$$

$$\mathcal{L}\{y\} = Y(s).$$

The sum of the left-hand terms is given to be 0. Thus, adding the right-hand terms and simplifying give

$$(s - s^2)Y'(s) + (-3s + 2)Y(s) + 2y_0 = 0,$$

which can be rewritten in the following way:

$$Y'(s) + \frac{3s - 2}{s(s-1)}Y(s) = \frac{2y_0}{s(s-1)}.$$

This equation is a first order linear differential equation in $Y(s)$. Since $\frac{3s-2}{s(s-1)} = \frac{2}{s} + \frac{1}{s-1}$, it is easy to see that an integrating factor is $I = s^2(s-1)$, and hence,

$$(IY(s))' = 2y_0 s.$$

Integrating and solving for Y give

$$Y(s) = \frac{y_0}{s - 1} + \frac{c}{s^2(s-1)}.$$

The inverse Laplace transform is $y(t) = y_0 e^t + c(e^t - t - 1)$. For simplicity, we can write this solution in the form

$$y(t) = c_1 e^t + c_2 (t + 1),$$

where $c_1 = y_0 + c$ and $c_2 = -c$. It is easy to verify that e^t and $t + 1$ are linearly independent and solutions to the given differential equation.   ◄

Example 7. Find a solution of exponential type that solves

$$ty'' + 2y' + ty = 0.$$

► **Solution.** Again, assume y is a solution of exponential type. Let $Y(s) = \mathcal{L}\{y(t)\}(s)$. As in the preceding example, we apply the Laplace transform and simplify. The result is a simple linear differential equation:

$$Y'(s) = \frac{-y_0}{s^2 + 1},$$

where y_0 is the initial condition $y(0) = y_0$. Integration gives $Y(s) = y_0(-\tan^{-1} s + C)$. By Proposition 3 of Sect. 2.2, we have $0 = \lim_{s \to \infty} Y(s) = y_0 \left(-\frac{\pi}{2} + C\right)$ which implies $C = \frac{\pi}{2}$ and

$$Y(s) = y_0 \left(\frac{\pi}{2} - \tan^{-1} s\right) = y_0 \tan^{-1} \frac{1}{s}.$$

By Example 5, we get

$$y(t) = y_0 \frac{\sin t}{t}.$$ ◄

Theorem 2 of Sect. 5.2 implies that there are two linearly independent solutions. The Laplace transform method has found only one, namely, $\frac{\sin t}{t}$. In Sect. 5.5, we will introduce a technique that will find another independent solution. When applied to this example, we will find $y(t) = \frac{\cos t}{t}$ is another solution. (cf. Example 2 of Sect. 5.5.) It is easy to check that the Laplace transform of $\frac{\cos t}{t}$ does not exist, and thus, the Laplace transform method cannot find it as a solution. Furthermore, the constant of integration, C, in this example cannot be arbitrary because of Proposition 3 of Sect. 2.2. It frequently happens in examples that C must be carefully chosen.

We observe that the presence of the linear factor t in Examples 6 and 7 produces a differential equation of order 1 which can be solved by techniques learned in Chapter 1. Correspondingly, the presence of higher order terms, t^n, produces differential equations of order n. For example, the Laplace transform applied to the differential equation $t^2 y'' + 6y = 0$ gives, after a short calculation, $s^2 Y''(s) + 4s Y'(s) + 8Y(s) = 0$. The resulting differential equation in $Y(s)$ is still second order and no simpler than the original. In fact, both are Cauchy–Euler. Thus, when the coefficient functions are polynomial of order greater than one, the Laplace transform method will generally be of little use. For this reason, we will usually limit our examples to second order linear differential equations with coefficient functions that are linear terms, that is, of the form $at + b$. Even with this restriction, we still will need to solve a first order differential equation in $Y(s)$ and determine its inverse Laplace transform; not always easy problems.

Laguerre Polynomials

The **Laguerre polynomial,** $\boldsymbol{\ell_n(t)}$, **of order n** is the polynomial solution to Laguerre's differential equation

$$ty'' + (1 - t)y' + ny = 0,$$

where $y(0) = 1$ and n is a nonnegative integer.

Proposition 8. *The nth Laguerre polynomial is given by*

$$\ell_n(t) = \sum_{k=0}^{n} (-1)^k \binom{n}{k} \frac{t^k}{k!}$$

and

$$\mathcal{L}\{\ell_n(t)\}(s) = \frac{(s-1)^n}{s^{n+1}}.$$

Proof. Taking the Laplace transform of Laguerre's differential equation gives

$$(s^2 - s)Y'(s) + (s - (1 + n))Y(s) = 0$$

and hence, $Y(s) = C \frac{(s-1)^n}{s^{n+1}}$. By the initial value theorem,

$$1 = y(0) = \lim_{s \to \infty} C \frac{s(s-1)^n}{s^{n+1}} = C.$$

Now using the binomial theorem, we get $(s-1)^n = \sum_{k=0}^{n} \binom{n}{k}(-1)^k s^{n-k}$ and hence $Y(s) = \sum_{k=0}^{n}(-1)^k \binom{n}{k}\frac{1}{s^{k+1}}$. It now follows by inversion that $y(t) = \ell_n(t) = \sum_{k=0}^{n}(-1)^k \binom{n}{k}\frac{t^k}{k!}$. $\qquad \square$

It is easy to see that the first five Laguerre polynomials are:

$$\ell_0(t) = 1$$
$$\ell_1(t) = 1 - t$$

$$\ell_2(t) = 1 - 2t + \frac{t^2}{2}$$

$$\ell_3(t) = 1 - 3t + \frac{3t^2}{2} - \frac{t^3}{6}$$

$$\ell_4(t) = 1 - 4t + 3t^2 - \frac{2t^3}{3} + \frac{t^4}{24}.$$

Below are their graphs on the interval $[0, 6]$.

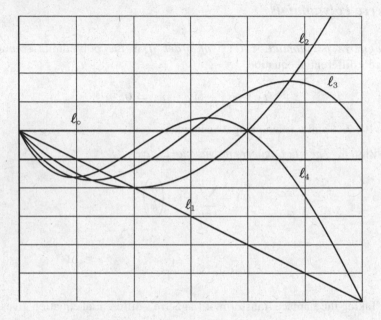

Define the following differential operators:

$$E_\circ = 2t\,\boldsymbol{D}^2 + (2 - 2t)\boldsymbol{D} - 1,$$

$$E_+ = t\,\boldsymbol{D}^2 + (1 - 2t)\boldsymbol{D} + (t - 1),$$

$$E_- = t\boldsymbol{D}^2 + \boldsymbol{D}.$$

Theorem 9. *We have the following differential relationships among the Laguerre polynomials:*

1. $E_\circ \ell_n = -(2n + 1)\ell_n$
2. $E_+ \ell_n = -(n + 1)\ell_{n+1}$
3. $E_- \ell_n = -n\ell_{n-1}$.

Proof.

1. Let $A_n = t\boldsymbol{D}^2 + (1 - t)\boldsymbol{D} + n$ be Laguerre's differential equation. Then from the defining equation of the Laguerre polynomial ℓ_n, we have $A_n\ell_n = 0$.

Multiply this equation by 2 and add $-(1 + 2n)\ell_n$ to both sides. This gives
$E_\circ \ell_n = -(2n + 1)\ell_n$

2. A simple observation gives $E_+ = A_n - t\boldsymbol{D} - (1 - t - n)$. Since $A_n \ell_n = 0$, it
   is enough to verify that $t\ell'_n + (1 - t - n)\ell_n = (n + 1)\ell n + 1$. This we do in
   transform space. Let $L_n = \mathcal{L}\{\ell_n\}$.

$$
\begin{aligned}
\mathcal{L}\left\{t\ell'_n + (1 - t - n)\ell_n\right\}(s) &= -(sL_n(s) - \ell_n(0))' + (1 - n)L_n(s) + L'_n(s) \\
&= -(L_n + sL'_n) + (1 - n)L_n + L'_n \\
&= -(s - 1)L'_n - nL_n \\
&= \frac{-(s - 1)^n}{s^{n+2}}(ns - (n + 1)(s - 1) - ns) \\
&= (n + 1)\mathcal{L}\{\ell_{n+1}\}(s).
\end{aligned}
$$

3. This equation is proved in a similar manner as above. We leave the details to the
   exercises.   ◀

For a differential operator A, let $A^2 y = A(Ay)$, $A^3 y = A(A(Ay))$, etc. It is easy
to verify by induction and the use of Theorem 9 that

$$
\frac{(-1)^n}{n!}E_+^n \ell_\circ = \ell_n.
$$

The operator E_+ is called a ***creation operator*** since successive applications to $\ell_\circ$
creates all the other Laguerre polynomials. In a similar way, it is easy to verify that

$$
E_-^m \ell_n = 0,
$$

for all $m > n$. The operator E_- is called an ***annihilation operator***.

Exercises

1–4. For each of the following functions, show that Theorem 4 applies and use it to find its Laplace transform.

1. $\dfrac{e^{bt} - e^{at}}{t}$

2. $2\dfrac{\cos bt - \cos at}{t}$

3. $2\dfrac{\cos bt - \cos at}{t^2}$

4. $\dfrac{\sin at}{t}$

5–10. Use the Laplace transform to find solutions to each of the following differential equations. In some cases, you may find two independent solutions, and in other cases, you may only find one solution. It may be useful to have the following table for quick reference:

$$
\begin{array}{rcl}
ty & \longleftrightarrow & -Y'(s) \\[4pt]
ty' & \longleftrightarrow & -sY'(s) - Y(s) \\[4pt]
ty'' & \longleftrightarrow & -s^2Y'(s) - 2sY(s) + y_0 \\[4pt]
ty''' & \longleftrightarrow & -s^3Y'(s) - 3s^2Y(s) + 2sy_0 + y_1 \\[12pt]
y & \longleftrightarrow & Y(s) \\[4pt]
y' & \longleftrightarrow & sY(s) - y_0 \\[4pt]
y'' & \longleftrightarrow & s^2Y(s) - sy_0 - y_1
\end{array}
$$

5. $ty'' + (t - 1)y' - y = 0$
6. $ty'' + (1 + t)y' + y = 0$
7. $ty'' + (2 + 4t)y' + (4 + 4t)y = 0$
8. $ty'' - 2y' + ty = 0$
9. $ty'' - 4y' + ty = 0$, assume $y(0) = 0$
10. $ty'' + (2 + 2t)y' + (2 + t)y = 0$

11–16. Use the Laplace transform to find solutions to each of the following differential equations. Use the results of Exercises 1 to 4.

11. $-ty'' + (t - 2)y' + y = 0$
12. $-ty'' - 2y' + ty = 0$
13. $ty'' + (2 - 5t)y' + (6t - 5)y = 0$
14. $ty'' + 2y' + 9ty = 0$

15. $ty''' + 3y'' + ty' + y = 0$
16. $ty'' + (2 + t)y' + y = 0$

17–25. *Laguerre Polynomials:* Each of these problems develops further properties of the Laguerre polynomials.

17. The Laguerre polynomial of order n can be defined in another way: $\ell_n(t) = \frac{1}{n!}e^t \frac{d^n}{dt^n}(e^{-t}t^n)$. Show that this definition is consistent with the definition in the text.

18. Verify (3) in Theorem 9:
$$E_-\ell_n = -n\ell_{n-1}.$$

19. The *Lie bracket* $[A, B]$ of two differential operators A and B is defined by
$$[A, B] = AB - BA.$$

   Show the following:

   • $[E_\circ, E_+] = -2E_+.$
   • $[E_\circ, E_-] = 2E_-.$
   • $[E_+, E_-] = E_\circ.$

20. Show that the Laplace transform of $\ell_n(at)$, $a \in \mathbb{R}$, is $\frac{(s-a)^n}{s^{n+1}}$.

21. Verify that
$$\sum_{k=0}^{n} \binom{n}{k}a^k \ell_k(t)(1 - a)^{n-k} = \ell_n(at).$$

22. Show that
$$\int_0^t \ell_n(x)\,dx = \ell_n(t) - \ell_{n+1}(t).$$

23. Verify the following recursion formula:
$$\ell_{n+1}(t) = \frac{1}{n + 1}\left((2n + 1 - t)\ell_n(t) - n\ell_{n-1}(t)\right).$$

24. Show that
$$\int_0^t \ell_n(x)\ell_m(t - x)\,dx = \ell_{m+n}(t) - \ell_{m+n+1}(t).$$

25. Show that
$$\int_t^\infty e^{-x}\ell_n(x)\,dx = e^{-t}\left(\ell_n(t) - \ell_{n-1}(t)\right).$$

5.5   Reduction of Order

It is a remarkable feature of linear differential equations that one nonzero homogeneous solution can be used to obtain a second independent solution. Suppose $L = a_2(t)D^2 + a_1(t)D + a_0(t)$ and suppose $y_1(t)$ is a known nonzero solution. It turns out that a second independent solution will take the form

$$y_2(t) = u(t)y_1(t), \tag{1}$$

where $u(t)$ is to be determined. By substituting y_2 into $Ly = 0$, we find that $u(t)$ must satisfy a second order differential equation, which, by a simple substitution, can be *reduced* to a first order separable differential equation. After $u(t)$ is found, (1) gives $y_2(t)$, a second independent solution.

The procedure is straightforward. We drop the functional dependence on t to make the notation simpler. The product rule gives

$$y_2' = u'y_1 + uy_1'$$
$$\text{and} \quad y_2'' = u''y_1 + 2u'y_1' + uy_1''.$$

Substituting these equations into Ly_2 gives

$$\begin{aligned}
Ly_2 &= a_2y_2'' + a_1y_2' + a_0y_2 \\
&= a_2(u''y_1 + 2u'y_1' + uy_1'') + a_1(u'y_1 + uy_1') + a_0uy_1 \\
&= u''a_2y_1 + u'(2a_2y_1' + a_1y_1) + u(a_2y_1'' + a_1y_1' + a_0y_1) \\
&= u''a_2y_1 + u'(2a_2y_1' + a_1y_1).
\end{aligned}$$

In the third line above, the coefficient of u is zero because y_1 is assumed to be a solution to $Ly = 0$. The equation $Ly_2 = 0$ implies

$$u''a_2y_1 + u'(2a_2y_1' + a_1y_1) = 0, \tag{2}$$

another second order differential equation in u. One obvious solution to (2) is $u(t)$ a constant, implying y_2 is a multiple of y_1. To find another independent solution, we use the substitution $v = u'$ to get

$$v'a_2y_1 + v(2a_2y_1' + a_1y_1) = 0,$$

a first order separable differential equation in v. This substitution gives this procedure its name: The product **reduction of order**. It is now straightforward to solve for v. In fact, separating variables gives

$$\frac{v'}{v} = \frac{-2y_1'}{y_1} - \frac{a_1}{a_2}.$$

From this, we get

$$v = \frac{1}{y_1^2}e^{-\int a_1/a_2}.$$

Since $v = u'$, we integrate v to get

$$u = \int \frac{1}{y_1^2}e^{-\int a_1/a_2}, \tag{3}$$

which is independent of the constant solution. Substituting (3) into (1) then gives a new solution independent of y_1. Admittedly, (3) is difficult to remember and not very enlightening. In the exercises, we recommend following the procedure we have outlined above. This is what we shall do in the examples to follow.

Example 1. The function $y_1(t) = e^t$ is a solution to

$$(t-1)y'' - ty' + y = 0.$$

Use reduction of order to find another independent solution and write down the general solution.

▶ **Solution.** Let $y_2(t) = u(t)e^t$. Then

$$y_2'(t) = u'(t)e^t + u(t)e^t$$
$$y_2''(t) = u''(t)e^t + 2u'(t)e^t + u(t)e^t.$$

Substitution into the differential equation $(t-1)y'' - ty' + y = 0$ gives

$$(t-1)(u''(t)e^t + 2u'(t)e^t + u(t)e^t) - t(u'(t)e^t + u(t)e^t) + u(t)e^t = 0$$

which simplifies to

$$(t-1)u'' + (t-2)u' = 0.$$

Let $v = u'$. Then we get $(t-1)v' + (t-2)v = 0$. Separating variables gives

$$\frac{v'}{v} = \frac{-(t-2)}{t-1} = -1 + \frac{1}{t-1}$$

with solution $v = e^{-t}(t-1)$. Integration by parts gives $u(t) = \int v(t)\,dt = -te^{-t}$. Substitution gives

$$y_2(t) = u(t)e^t = -te^{-t}e^t = -t.$$

It is easy to verify that this is indeed a solution. Since our equation is homogeneous, we know $-y_2(t) = t$ is also a solution. Clearly t and e^t are independent. By Theorem 2 of Sect. 5.2, the general solution is

$$y(t) = c_1 t + c_2 e^t. \qquad \blacktriangleleft$$

Example 2. In Example 7 of Sect. 5.4, we showed that $y_1 = \frac{\sin t}{t}$ is a solution to

$$ty'' + 2y' + ty = 0.$$

Use reduction of order to find a second independent solution and write down the general solution.

▶ **Solution.** Let $y_2(t) = u(t)\frac{\sin t}{t}$. Then

$$y_2'(t) = u'(t)\frac{\sin t}{t} + u(t)\frac{t\cos t - \sin t}{t^2}$$

$$y_2''(t) = u''(t)\frac{\sin t}{t} + 2u'(t)\frac{t\cos t - \sin t}{t^2} + u(t)\frac{-t^2\sin t - 2t\cos t + 2\sin t}{t^3}.$$

We next substitute y_2 into $ty'' + 2y' + ty = 0$ and simplify to get

$$u''(t)\sin t + 2u'(t)\cos t = 0.$$

Let $v = u'$. Then we get $v'(t)\sin t + 2v(t)\cos t = 0$. Separating variables gives

$$\frac{v'}{v} = \frac{-2\cos t}{\sin t}$$

with solution

$$v(t) = \csc^2(t).$$

Integration gives $u(t) = \int v(t)\,dt = -\cot(t)$, and hence,

$$y_2(t) = -(\cot t)\frac{\sin t}{t} = \frac{-\cos t}{t}.$$

By Theorem 2 of Sect. 5.2, the general solution can be written as

$$c_1\frac{\sin t}{t} + c_2\frac{\cos t}{t}.$$

Compare this result with Examples 4 of Sect. 5.1 and 7 of Sect. 5.4.   ◀

We remark that the constant of integration in the computation of u was chosen to be 0 in both examples. There is no loss in this for if a nonzero constant, c say, is chosen, then $y_2 = uy_1 + cy_1$. But cy_1 is already known to be a homogeneous solution. We gain nothing by adding a multiple of it in y_2.

Exercises

1–16. For each differential equation and the given solution, use reduction of order
to find a second independent solution and write down the general solution.

1. $t^2 y'' - 3ty' + 4y = 0$,
   $y_1(t) = t^2$

2. $t^2 y'' + 2ty' - 2y = 0$,
   $y_1(t) = t$

3. $4t^2 y'' + y = 0$,
   $y_1(t) = \sqrt{t}$

4. $t^2 y'' + 2ty' = 0$,
   $y_1(t) = \frac{1}{t}$

5. $t^2 y'' - t(t + 2)y' + (t + 2)y = 0$,
   $y_1(t) = t$

6. $t^2 y'' - 4ty' + (t^2 + 6)y = 0$,
   $y_1(t) = t^2 \cos t$

7. $ty'' - y' + 4t^3 y = 0$,
   $y_1(t) = \sin t^2$

8. $ty'' - 2(t + 1)y' + 4y = 0$,
   $y_1(t) = e^{2t}$

9. $y'' - 2(\sec^2 t)\, y = 0$,
   $y_1(t) = \tan t$

10. $ty'' + (t - 1)y' - y = 0$,
    $y_1(t) = e^{-t}$

11. $y'' - (\tan t)y' - (\sec^2 t)y = 0$, ·
    $y_1(t) = \tan t$

12. $(1 + t^2)y'' - 2ty' + 2y = 0$,
    $y_1(t) = t$

13. $(\cos 2t + 1)y'' - 4y = 0$, $t \in (-\pi/2, \pi/2)$,
    $y_1(t) = \frac{\sin 2t}{1 + \cos 2t}$

14. $t^2 y'' - 2ty' + (t^2 + 2)y = 0$,
    $y_1(t) = t \cos t$

15. $(1 - t^2)y'' + 2y = 0$, $-1 < t < 1$,
    $y_1(t) = 1 - t^2$

16. $(1 - t^2)y'' - 2ty' + 2y = 0$, $-1 < t < 1$,
    $y_1(t) = t$

5.6 Variation of Parameters

Let L be a second order linear differential operator. In this section, we address the issue of finding a particular solution to a nonhomogeneous linear differential equation $L(y) = f$, where f is continuous on some interval I. It is a pleasant feature of linear differential equations that the homogeneous solutions can be used decisively to find a particular solution. The procedure we use is called *variation of parameters* and, as you shall see, is akin to the method of reduction of order.

Suppose, in particular, that $L = D^2 + a_1(t)D + a_0(t)$, that is, we will assume that the leading coefficient function is 1, and it is important to remember that this assumption is essential for the method we develop below. Suppose $\{y_1, y_2\}$ is a fundamental set for $L(y) = 0$. We know then that all solutions of the homogeneous equation $L(y) = 0$ are of the form $c_1 y_1 + c_2 y_2$. To find a particular solution y_p to $L(y) = f$, the method of variation of parameters makes two assumptions. First, the parameters c_1 and c_2 are allowed to vary (hence the name). We thus replace the constants c_1 and c_2 by functions $u_1(t)$ and $u_2(t)$, and assume that the particular solution y_p, takes the form

$$y_p(t) = u_1(t)y_1(t) + u_2(t)y_2(t). \tag{1}$$

The second assumption is

$$u_1'(t)y_1(t) + u_2'(t)y_2(t) = 0. \tag{2}$$

What is remarkable is that these two assumptions consistently lead to explicit formulas for $u_1(t)$ and $u_2(t)$ and hence a formula for y_p. .

To simplify notation in the calculations that follow, we will drop the "t" in expressions like $u_1(t)$, etc. Before substituting y_p into $L(y) = f$, we first calculate y_p' and y_p'':

$$y_p' = u_1'y_1 + u_1 y_1' + u_2'y_2 + u_2 y_2'$$
$$= u_1 y_1' + u_2 y_2',$$

where we used (2) to simplify. Now for the second derivative

$$y_p'' = u_1'y_1' + u_1 y_1'' + u_2'y_2' + u_2 y_2''.$$

We now substitute y_p into $L(y)$:

$$L(y_p) = y_p'' + a_1 y_p' + a_0 y_p$$
$$= u_1'y_1' + u_1 y_1'' + u_2'y_2' + u_2 y_2'' + a_1(u_1 y_1' + u_2 y_2') + a_0(u_1 y_1 + u_2 y_2)$$
$$= u_1'y_1' + u_2'y_2' + u_1(y_1'' + a_1 y_1' + a_0 y_1) + u_2(y_2'' + a_1 y_2' + a_0 y_2)$$
$$= u_1'y_1' + u_2'y_2'.$$

In the second to the last equation, the coefficients of u_1 and u_2 are zero because y_1 and y_2 are assumed to be homogeneous solutions. The second assumption, (2), and the equation $L(y_p) = f$ now lead to the following system:

$$u_1' y_1 + u_2' y_2 = 0$$
$$u_1' y_1' + u_2' y_2' = f$$

which can be rewritten in matrix form as

$$\begin{bmatrix} y_1 & y_2 \\ y_1' & y_2' \end{bmatrix} \begin{bmatrix} u_1' \\ u_2' \end{bmatrix} = \begin{bmatrix} 0 \\ f \end{bmatrix}. \tag{3}$$

The leftmost matrix in (3) is none other than the Wronskian matrix, $W(y_1, y_2)$, which has a nonzero determinant because $\{y_1, y_2\}$ is a fundamental set (cf. Proposition 4 of Sect. 5.2). By Cramer's rule, we can solve for u_1' and u_2'. We obtain

$$u_1' = \frac{-y_2 f}{w(y_1, y_2)},$$

$$u_2' = \frac{y_1 f}{w(y_1, y_2)}.$$

We now obtain an explicit formula for a particular solution:

$$y_p(t) = u_1 y_1 + u_2 y_2$$
$$= \left(\int \frac{-y_2 f}{w(y_1, y_2)} \right) y_1 + \left(\int \frac{y_1 f}{w(y_1, y_2)} \right) y_2.$$

The following theorem consolidates these results with Theorem 6 of Sect. 5.1.

Theorem 1. *Let $L = D^2 + a_1(t) D + a_0(t)$, where $a_1(t)$ and $a_0(t)$ are continuous on an interval I. Suppose $\{y_1, y_2\}$ is a fundamental set of solutions for $L(y) = 0$. If f is continuous on I, then a particular solution, y_p, to $L(y) = f$ is given by the formula*

$$y_p = \left(\int \frac{-y_2 f}{w(y_1, y_2)} \right) y_1 + \left(\int \frac{y_1 f}{w(y_1, y_2)} \right) y_2. \tag{4}$$

Furthermore, the solution set to $L(y) = f$ becomes

$$\{y_p + c_1 y_1 + c_2 y_2 : c_1, c_2 \in \mathbb{R}\}.$$

Remark 2. Equation (4), which gives an explicit formula for a particular solution, is too complicated to memorize, and we do not recommend students to do this. Rather, the point of variation of parameters is the method that leads to (4), and our recommended starting point is (3). You will see such matrix equations as we proceed in the text.

We will illustrate the method of variation of parameters with two examples.

Example 3. Find the general solution to the following equation:

$$t^2 y'' - 2y = t^2 \ln t.$$

▶ **Solution.** In standard form, this becomes

$$y'' - \frac{2}{t^2} y = \ln t.$$

The associated homogeneous equation is $y'' - (2/t^2)y = 0$ or, equivalently, $t^2 y'' - 2y = 0$ and is a Cauchy–Euler equation. The indicial polynomial is $Q(s) = s^2 - s - 2 = (s-2)(s+1)$, which has 2 and -1 as roots. Thus, $\{t^{-1}, t^2\}$ is a fundamental set to the homogeneous equation $y'' - (2/t^2)y = 0$, by Theorem 4 of Sect. 5.3. Let $y_p = t^{-1} u_1(t) + t^2 u_2(t)$. Our starting point for determining u_1 and u_2 is the matrix equation

$$\begin{bmatrix} t^{-1} & t^2 \\ -t^{-2} & 2t \end{bmatrix} \begin{bmatrix} u_1' \\ u_2' \end{bmatrix} = \begin{bmatrix} 0 \\ \ln t \end{bmatrix}$$

which is equivalent to the system

$$\begin{aligned} t^{-1} u_1' + t^2 u_2' &= 0 \\ -t^{-2} u_1' + 2t u_2' &= \ln t. \end{aligned}$$

Multiplying the bottom equation by t and then adding the equations together give $3t^2 u_2' = t \ln t$, and hence,

$$u_2' = \frac{1}{3t} \ln t.$$

Substituting u_2' into the first equation and solving for u_1' give

$$u_1' = -\frac{t^2}{3} \ln t.$$

Integration by parts leads to

$$u_1 = -\frac{1}{3} \left(\frac{t^3}{3} \ln t - \frac{t^3}{9} \right)$$

and a simple substitution gives

$$u_2 = \frac{1}{6} (\ln t)^2.$$

We substitute u_1 and u_2 into (1) to get

$$y_p(t) = -\frac{1}{3}\left(\frac{t^3}{3}\ln t - \frac{t^3}{9}\right)t^{-1} + \frac{1}{6}(\ln t)^2 t^2 = \frac{t^2}{54}(9(\ln t)^2 - 6\ln t + 2).$$

It follows that the solution set is

$$\left\{\frac{t^2}{54}(9(\ln t)^2 - 6\ln t + 2) + c_1 t^{-1} + c_2 t^2 : c_1, c_2 \in \mathbb{R}\right\}. \qquad \blacktriangleleft$$

Example 4. Find the general solution to

$$ty'' + 2y' + ty = 1.$$

Use the results of Example 2 of Sect. 5.5.

▶ **Solution.** Example 2 of Sect. 5.5 showed that

$$y_1(t) = \frac{\sin t}{t} \quad \text{and} \quad y_2(t) = \frac{\cos t}{t}$$

are homogeneous solutions to $ty'' + 2y' + ty = 0$. Let $y_p = \frac{\sin t}{t}u_1(t) + \frac{\cos t}{t}u_2(t)$. Then

$$\begin{bmatrix} \dfrac{\sin t}{t} & \dfrac{\cos t}{t} \\[2mm] \dfrac{t\cos t - \sin t}{t^2} & \dfrac{-t\sin t - \cos t}{t^2} \end{bmatrix} \begin{bmatrix} u_1'(t) \\[2mm] u_2'(t) \end{bmatrix} = \begin{bmatrix} 0 \\[2mm] \dfrac{1}{t} \end{bmatrix}.$$

(We get $1/t$ in the last matrix because the differential equation in standard form is $y'' + (2/t)y' + y = 1/t$.) From the matrix equation, we get the following system:

$$\frac{\sin t}{t}u_1'(t) + \frac{\cos t}{t}u_2'(t) = 0,$$

$$\frac{t\cos t - \sin t}{t^2}u_1'(t) + \frac{-t\sin t - \cos t}{t^2}u_2'(t) = \frac{1}{t}.$$

The first equation gives

$$u_1'(t) = -(\cot t)u_2'(t).$$

If we multiply the first equation by t, the second equation by t^2, and then add, we get

$$(\cos t)u_1'(t) - (\sin t)u_2'(t) = 1.$$

Substituting in $u_1'(t)$ and solving for $u_2'(t)$ give $u_2'(t) = -\sin t$, and thus, $u_1'(t) = \cos t$. Integration gives

$$u_1(t) = \sin t,$$

$$u_2(t) = \cos t.$$

We now substitute these functions into y_p to get

$$y_p(t) = \frac{\sin t}{t} \sin t + \frac{\cos t}{t} \cos t$$

$$= \frac{\sin^2 t + \cos^2 t}{t}$$

$$= \frac{1}{t}.$$

The general solution is

$$y(t) = \frac{1}{t} + c_1 \frac{\sin t}{t} + c_2 \frac{\cos t}{t}. \qquad \blacktriangleleft$$

Exercises

1–5. Use variation of parameters to find a particular solution and then write down the general solution. Next solve each using the method of undetermined coefficients or the incomplete partial fraction method.

1. $y'' + y = \sin t$
2. $y'' - 4y = e^{2t}$
3. $y'' - 2y' + 5y = e^t$
4. $y'' + 3y' = e^{-3t}$
5. $y'' - 3y' + 2y = e^{3t}$

6–16. Use variation of parameters to find a particular solution and then write down the general solution. In some exercises, a fundamental set $\{y_1, y_2\}$ is given.

6. $y'' + y = \tan t$

7. $y'' - 2y' + y = \dfrac{e^t}{t}$

8. $y'' + y = \sec t$

9. $t^2 y'' - 2ty' + 2y = .t^4$

10. $ty'' - y' = 3t^2 - 1$
    $y_1(t) = 1$ and $y_2(t) = t^2$

11. $t^2 y'' - ty' + y = t$

12. $y'' - 4y' + 4y = \frac{e^{2t}}{t^2+1}$

13. $y'' - (\tan t)y' - (\sec^2 t)y = t$
    $y_1(t) = \tan t$ and $y_2(t) = \sec t$

14. $ty'' + (t - 1)y' - y = t^2 e^{-t}$
    $y_1(t) = t - 1$ and $y_2(t) = e^{-t}$

15. $ty'' - y' + 4t^3 y = 4t^5$
    $y_1 = \cos t^2$ and $y_2(t) = \sin t^2$

16. $y'' - y = \frac{1}{1+e^{-t}}$

17. Show that the constants of integration in the formula for y_p in Theorem 1 can be chosen so that a particular solution can be written in the form:

$$y_p(t) = \int_a^t \frac{\left| \begin{bmatrix} y_1(x) & y_2(x) \\ y_1(t) & y_2(t) \end{bmatrix} \right|}{\left| \begin{bmatrix} y_1(x) & y_2(x) \\ y_1'(x) & y_2'(x) \end{bmatrix} \right|} f(x)\mathrm{d}x,$$

where a and t are in the interval I, and the absolute value signs indicate the determinant.

18–21. For each problem below, use the result of Problem 17, with $a = 0$, to obtain a particular solution to the given differential equation in the form given. Solve the differential equation using the Laplace transform method and compare.

18. $y'' + a^2 y = f(t)$
19. $y'' - a^2 y = f(t)$
20. $y'' - 2ay' + a^2 y = f(t)$
21. $y'' - (a + b)y' + aby = f(t), \ a \neq b$

5.7   Summary of Laplace Transforms

Laplace transforms and rules presented in Chap. 5 are summarized in Tables 5.1 and 5.2.

Table 5.1 Laplace transform rules

Assumptions	Result	Page
Initial value theorem		
1. f, f' of exponential order	$f(0) = \lim\limits_{s \to \infty} sF(s)$	356
Final value theorem		
2. f, f' of exponential order and $\lim_{t \to \infty} f(t)$ exists	$\lim\limits_{t \to \infty} f(t) = \lim\limits_{s \to 0} sF(s)$	356
Transform integral formula		
3. f of exponential order and $\frac{f(t)}{t}$ has a continuous extension to 0	$\mathcal{L}\left\{\frac{f(t)}{t}\right\}(s) = \int_s^\infty F(\sigma)\,d\sigma$	357

Table 5.2 Laplace transforms

$f(t)$	$F(s)$	Page
1. $\dfrac{\sin t}{t}$	$\tan^{-1}\dfrac{1}{s}$	358
2. $\dfrac{\sin at}{t}$	$\tan^{-1}\left(\dfrac{a}{s}\right)$	365
3. $\dfrac{e^{bt} - e^{at}}{t}$	$\ln\left(\dfrac{s-a}{s-b}\right)$	365
4. $2\dfrac{\cos bt - \cos at}{t}$	$\ln\left(\dfrac{s^2+a^2}{s^2+b^2}\right)$	365
5. $2\dfrac{\cos bt - \cos at}{t^2}$	$s\ln\left(\dfrac{s^2+b^2}{s^2+a^2}\right) - 2b\tan^{-1}\left(\dfrac{b}{s}\right)$ $+2a\tan^{-1}\left(\dfrac{a}{s}\right)$	365
Laguerre polynomials		
6. $\ell_n(t) = \sum_{k=0}^n (-1)^k \binom{n}{k}\frac{t^k}{k!}$	$\dfrac{(s-1)^n}{s^{n+1}}$	361
7. $\ell_n(at)$	$\dfrac{(s-a)^n}{s^{n+1}}$	366

Chapter 6
Discontinuous Functions and the Laplace Transform

Our focus in this chapter is a study of first and second order linear constant coefficient differential equations

$$y' + ay = f(t),$$
$$y'' + ay' + by = f(t),$$

where the input or forcing function $f(t)$ is more general than we have studied so far. These types of forcing functions arise in applications only slightly more complicated than those we have already considered. For example, imagine a mixing problem (see Example 11 of Sect. 1.4 and the discussion that followed it for a review of mixing problems) where there are two sources of incoming salt solutions with different concentrations as illustrated in the following diagram.

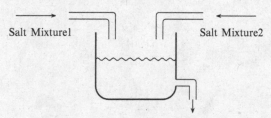

Salt Mixture1   Salt Mixture2

Initially, the first source may be flowing for several minutes. Then the second source is turned on at the same time the first source is turned off. Such a situation will result in a differential equation $y' + ay = f(t)$ where the input function has a graph similar to the one illustrated in Fig. 6.1. The most immediate observation is that the input function is discontinuous. Nevertheless, the Laplace transform methods we will develop will easily handle this situation, leading to a formula for the amount of the salt in the tank as a function of time.

As a second example, imagine that a sudden force is applied to a spring-mass dashpot system (see Sect. 3.6 for a discussion of these systems). For example, hit the mass attached to the spring with a hammer, which is a very good idealization of

W.A. Adkins and M.G. Davidson, *Ordinary Differential Equations*,
Undergraduate Texts in Mathematics, DOI 10.1007/978-1-4614-3618-8_6,
© Springer Science+Business Media New York 2012

Fig. 6.1 The graph of discontinuous input function $f(t)$ where salt of concentration level 1 enters until time $t = 2$, at which time the concentration switches to a different level

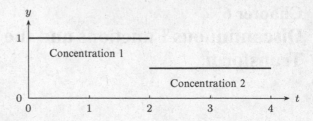

what happens to the shock absorber on a car when the car hits a bump in the road. Modeling this system will lead to a differential equation of the form

$$y'' + ay' + by = f(t),$$

where the forcing function is what we will refer to as an *instantaneous impulse function*. Such a function has a "very large" (or even infinite) value at a single instant $t = t_0$ and is 0 for other times. Such a function is not a true function, but its effect on systems can be analyzed effectively via the Laplace transform methods developed later in this chapter.

This chapter will develop the necessary background on the types of discontinuous functions and impulse functions which arise in basic applications of differential equations. We will start by describing the basic concepts of calculus for these more general classes of functions.

6.1 Calculus of Discontinuous Functions

Piecewise Continuous Functions

A function $f(t)$ has a ***jump discontinuity*** at a point $t = a$ if the left-hand limit $f(a^-) = \lim_{t \to a^-} f(t)$ and the right-hand limit $f(a^+) = \lim_{t \to a^+} f(t)$ both exist (as real numbers, not $\pm\infty$) and

$$f(a^+) \neq f(a^-).$$

The difference $f(a^+) - f(a^-)$ is frequently referred to as the ***jump*** in $f(t)$ at $t = a$. Functions with jump discontinuities are typically described by using different formulas on different subintervals of the domain. For example, the function $f(t)$ defined on the interval $[0, 3]$ by

$$f(t) = \begin{cases} t^3 & \text{if } 0 \le t < 1, \\ 1 - t & \text{if } 1 \le t < 2, \\ 1 & \text{if } 2 \le t \le 3 \end{cases}$$

has a jump discontinuity at $t = 1$ since $f(1^-) = 1 \neq f(1^+) = 0$ and at $t = 2$ since $f(2^-) = -1 \neq f(2^+) = 1$. The jump at $t = 1$ is -1 and the jump at $t = 2$ is 2. The graph of $f(t)$ is given in Fig. 6.2. On the other hand, the function

$$g(t) = \begin{cases} 1/(1-t) & \text{if } 0 \le t < 1, \\ t & \text{if } 1 \le t \le 2. \end{cases}$$

defined on the interval $[0, 2]$ has a discontinuity at $t = 1$, but it is not a jump discontinuity since $\lim_{t \to 1^-} g(t) = \infty$ does not exist.

We will say that a function $f(t)$ is ***piecewise continuous on a closed interval*** $[a, b]$ if $f(t)$ is continuous except for possibly finitely many jump discontinuities.

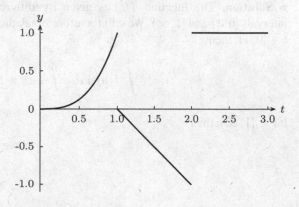

Fig. 6.2 A piecewise continuous function

For convenience, it will not be required that $f(t)$ be defined at the jump discontinuities. Suppose $a_1, \ldots, a_n$ are the locations of the jump discontinuities in the interval $[a, b]$ and assume $a_i < a_{i+1}$, for each i. On the interval (a_i, a_{i+1}), we can extend $f(t)$ to a continuous function on the closed interval $[a_i, a_{i+1}]$ by defining $f(a_i) = \lim_{t \to a_i^+} f(t)$ and $f(a_{i+1}) = \lim_{t \to a_{i+1}^-} f(t)$. Since a continuous function on a closed interval is bounded and there are only finitely many jump discontinuities, we have the following property of piecewise continuous functions.

Proposition 1. *If $f(t)$ is a piecewise continuous function on $[a, b]$, then $f(t)$ is bounded.*

How do we compute the derivative and integral of a piecewise continuous function?

Integration of Piecewise Continuous Functions

If $f(t)$ is a piecewise continuous function on the interval $[a, b]$ and the jump discontinuities are located at $a_1 < \ldots < a_k$, we may let $a_0 = a$ and $a_{k+1} = b$, and, as we observed above, $f(t)$ extends to a continuous function on the each closed interval $[a_i, a_{i+1}]$. Thus, we can define the definite integral of $f(t)$ on $[a, b]$ by the formula

$$\int_a^b f(t)\,dt = \int_{a_0}^{a_1} f(t)\,dt + \int_{a_1}^{a_2} f(t)\,dt + \cdots + \int_{a_k}^{a_{k+1}} f(t)\,dt.$$

Example 2. Find $\int_0^t f(u)\,du$ for all $t \in [0, \infty)$ where $f(t)$ is the piecewise continuous function defined by

$$f(t) = \begin{cases} 1 & \text{if } 0 \le t < 1, \\ 0 & \text{if } 1 \le t < \infty. \end{cases}$$

▶ **Solution.** The function $f(t)$ is given by different formulas on each of the intervals $[0, 1)$ and $[1, \infty)$. We will therefore break the calculation into two cases. If $t \in [0, 1)$, then

$$\int_0^t f(u)\,du = \int_0^t 1\,du = t.$$

It $t \in [1, \infty)$, then

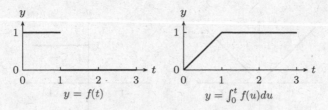

Fig. 6.3   The graph of the piecewise continuous function $f(t)$ and its integral $\int_0^t f(u)\,du$

$$\int_0^t f(u)\,du = \int_0^1 f(u)\,du + \int_1^t f(u)\,du$$

$$= 1 + \int_1^t 0\,du = 1.$$

Piecing these functions together gives

$$\int_0^t f(u)\,du = \begin{cases} t & \text{if } 0 \le t < 1 \\ 1 & \text{if } 1 \le t < \infty. \end{cases}$$

The graph of this function of t is shown in Fig. 6.3.   ◀

·Notice that the function $\int_0^t f(u)\,du$ is a continuous function of t, even though the integrand $f(t)$ is discontinuous. This is always true as long as the function $f(t)$ has only jump discontinuities, which is formalized in the following result.

Proposition 3. *If $f(t)$ is a piecewise continuous function on an interval $[a,b]$ and $c, t \in [a,b]$, then the integral $\int_c^t f(u)\,du$ exists and is a continuous function in the variable t.*

Proof. The integral exists as discussed above. Let $F(t) = \int_c^t f(u)\,du$. Since $f(t)$ is piecewise continuous on $[a, b]$, it is bounded by Proposition 1. We may then suppose $|f(t)| \le B$, for some $B > 0$. Let $\epsilon > 0$. Then

$$|F(t + \epsilon) - F(t)| \le \int_t^{t+\epsilon} |f(u)|\,du \le \int_t^{t+\epsilon} B\,du = B\epsilon.$$

Therefore, $\lim_{\epsilon \to 0} F(t + \epsilon) = F(t)$, and hence, $F(t^+) = F(t)$. In a similar way, $F(t^-) = F(t)$. This establishes the continuity of F.   □

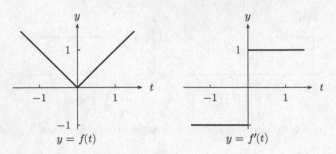

Fig. 6.4 The graph of the piecewise continuous function $f(t)$ and its derivative $f'(t)$

Differentiation of Piecewise Continuous Functions

In the applications, we will consider functions that are differentiable except at finitely many points in any interval $[a, b]$ of finite length. In this case we will use the symbol $f'(t)$ to denote the derivative of $f(t)$ even though it may not be defined at some points. For example, the absolute value function

$$f(t) = |t| = \begin{cases} -t & \text{if } -\infty < t < 0 \\ t & \text{if } 0 \le t < \infty. \end{cases}$$

This function is continuous on $(-\infty, \infty)$ and differentiable at all points except $t = 0$. Then

$$f'(t) = \begin{cases} -1 & \text{if } -\infty < t < 0 \\ 1 & \text{if } 0 < t < \infty. \end{cases}$$

Notice that $f'(t)$ is not defined at $t = 0$, but the derivative of this discontinuous function has produced a function with a jump discontinuity where the derivative does not exist. See Fig. 6.4. Compare this with Fig. 6.3, where we have seen that integrating a function with jump discontinuities produces a continuous function.

Differential Equations and Piecewise Continuous Functions

We now look at some examples of solutions to constant coefficient linear differential equations with piecewise continuous forcing functions. We start with the first order equation $y' + ay = f(t)$ where a is a constant and $f(t)$ is a piecewise continuous function. An equation of this type has a unique solution for each initial condition provided the input function is continuous, which is the situation for $f(t)$ on each subinterval on which it is continuous. To be able to extend the initial condition in a unique manner across each jump discontinuity, we shall define a function $y(t)$ to

Fig. 6.5 The graph of the solution to Example 4

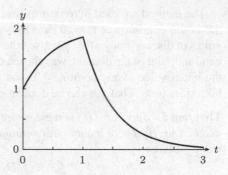

be a *solution* to $y' + ay = f(t)$ if $y(t)$ is *continuous* and satisfies the differential equation except at the jump discontinuities of the of the input function $f(t)$.

Example 4. Find a solution to

$$y' + 2y = f(t) = \begin{cases} 4 & \text{if } 0 \le t < 1 \\ 0 & \text{if } 1 \le t < \infty, \end{cases} \qquad y(0) = 1. \tag{1}$$

▶ **Solution.** The procedure will be to solve the differential equation separately on each of the subintervals where $f(t)$ is continuous and then piece the solution together to make a continuous solution. Start with the interval $[0\ 1)$ which includes the initial time $t = 0$. On this first subinterval, $f(t) = 4$, so the differential equation to be solved is $y' + 2y = 4$. The solution uses the integrating factor technique developed in Sect. 1.4. Multiplying both sides of $y' + 2y = 4$ by the integrating factor, e^{2t}, leads to $(e^{2t} y)' = 4e^{2t}$. Integrating and solving for $y(t)$ gives $y(t) = 2 + ce^{-2t}$, and the initial condition $y(0) = 1$ implies that $c = -1$ so that

$$y = 2 - e^{-2t}, \quad 0 \le t < 1.$$

On the interval $[1, \infty)$, the differential equation to solve is $y' + 2y = 0$, which has the general solution $y(t) = ke^{-2t}$. To produce a continuous function, we need to choose k so that this solution will match up with the solution found for the interval $[0, 1)$. To do this, let $y(1) = y(1^-) = 2 - e^{-2}$. This value must match the value $y(1) = ke^{-2}$ computed from the formula on $[1, \infty)$. Thus, $2 - e^{-2} = ke^{-2}$ and solving for k gives $k = 2e^2 - 1$. Therefore, the solution on the interval $[1, \infty)$ is $y(t) = (2e^2 - 1)e^{-2t}$. Putting the two pieces together gives the solution

$$y(t) = \begin{cases} 2 - e^{-2t} & \text{if } 0 \le t < 1 \\ (2e^2 - 1)e^{-2t} & \text{if } 1 \le t < \infty. \end{cases}$$

The graph of this solution is shown in Fig. 6.5, where the discontinuity of the derivative of $y(t)$ at $t = 1$ is evident by the kink at that point. ◀

The method we used here insures that the solution we obtain is continuous and the initial condition at $t = 0$ determines the subsequent initial conditions at the points of discontinuity of f. We also note that the initial condition at $t = 0$, the left-hand endpoint of the domain, was chosen only for convenience; we could have taken the initial value at any point $t_0 \geq 0$ and pieced together a continuous function on both sides of t_0. That this can be done in general is stated in the following theorem.

Theorem 5. *Suppose $f(t)$ is a piecewise continuous function on an interval $[\alpha, \beta]$ and $t_0 \in [\alpha, \beta]$. There is a unique continuous function $y(t)$ which satisfies the initial value problem*

$$y' + ay = f(t), \qquad y(t_0) = y_0.$$

Proof. Follow the method illustrated in the example above to construct a continuous solution. To prove uniqueness, suppose $y_1(t)$ and $y_2(t)$ are two continuous solutions. If $y(t) = y_1(t) - y_2(t)$, then $y(t_0) = 0$ and $y(t)$ is a continuous solution to $y' + ay = 0$. On the interval containing t_0 on which $f(t)$ is continuous, $y(t) = 0$ by the existence and uniqueness theorem. The initial value at the endpoint of adjacent intervals is thus 0. Continuing in this way, we see that $y(t)$ is identically 0 on $[\alpha, \beta]$ and hence $y_1(t) = y_2(t)$.   □

Now consider a second order constant coefficient differential equation with a piecewise continuous forcing function. Our method is similar to the one above, however, we demand more out of our solution. Since the solution of a second order equation with continuous input function is determined by the initial values of both $y(t)$ and $y'(t)$, it will be necessary to extend *both* of these values across the jump discontinuity in order to obtain a unique solution with a discontinuous input function. Thus, if $f(t)$ is a piecewise continuous function, then we will say that a function $y(t)$ is a **solution** to

$$y'' + ay' + by = f(t),$$

if $y(t)$ is *continuous*, has a *continuous derivative*, and satisfies the differential equation except at the discontinuities of the forcing function $f(t)$.

Example 6. Find a solution y to

$$y'' + y = f(t) = \begin{cases} 1 & \text{if } 0 \leq t < \pi \\ 0 & \text{if } \pi \leq t < \infty, \end{cases} \qquad y(0) = 1, \ y'(0) = 0.$$

▶ **Solution.** The general solution to the differential equation $y'' + y = 1$ on the interval $[0, \pi)$ is $y(t) = 1 + a\cos t + b\sin t$, and the initial conditions $y(0) = 1$, $y'(0) = 0$ imply $a = 0$, $b = 0$, so the solution on $[0, \pi)$ is $y(t) = 1$. Taking limits as $t \to \pi^-$ gives $y(\pi) = 1$, $y'(\pi) = 0$. On the interval $[\pi, \infty)$, the differential equation $y'' + y = f(t)$ becomes $y'' + y = 0$ with the initial conditions $y(\pi) = 1$, $y'(\pi) = 0$. The general solution on this interval is thus $y(t) = a\cos t + b\sin t$, and taking into account the values at $t = \pi$ gives $a = -1$, $b = 0$. Piecing these two solutions together gives

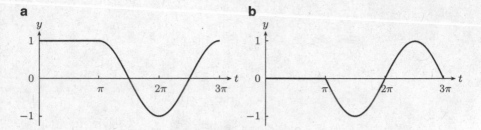

Fig. 6.6 The graph of the solution $y(t)$ to Example 4 is shown in (**a**), while the derivative $y'(t)$ is graphed in (**b**). Note that the derivative $y'(t)$ is continuous, but it is not differentiable at $t = \pi$

$$y(t) = \begin{cases} 1 & \text{if } 0 \le t < \pi \\ -\cos t & \text{if } \pi \le t < \infty. \end{cases}$$

Its derivative is

$$y'(t) = \begin{cases} 0 & \text{if } 0 \le t < \pi \\ \sin t & \text{if } \pi \le t < \infty. \end{cases}$$

Figure 6.6 gives (a) the graph of the solution and (b) the graph of its derivative. The solution is differentiable on the interval $[0, \infty]$, and the derivative is continuous on $[0, \infty)$. However, the kink in the derivative at $t = \pi$ indicates that the second derivative is not continuous. ◄

In direct analogy to the first order case we considered above, we are led to the following theorem. The proof is omitted.

Theorem 7. *Suppose f is a piecewise continuous function on an interval $[\alpha, \beta]$ and $t_0 \in [\alpha, \beta]$. There is a unique continuous function $y(t)$ with continuous derivative which satisfies*

$$y'' + ay' + by = f(t), \qquad y(t_0) = y_0, \ y'(t_0) = y_1.$$

Piecing together solutions in the way that we described above is at best tedious. Later in this chapter, the Laplace transform method for solving differential equations will be extended to provide a simpler alternate method for solving differential equations like the ones above. It is one of the hallmarks of the Laplace transform.

Exercises

1–8. Match the following functions that are given piecewise with their graphs and determine where jump discontinuities occur.

1. $f(t) = \begin{cases} 1 & \text{if } 0 \le t < 4 \\ -1 & \text{if } 4 \le t < 5 \\ 0 & \text{if } 5 \le t \le 6 \end{cases}$

2. $f(t) = \begin{cases} t & \text{if } 0 \le t < 1 \\ 2-t & \text{if } 1 \le t < 2 \\ 1 & \text{if } 2 \le t \le 6 \end{cases}$

3. $f(t) = \begin{cases} t/3 & \text{if } 0 \le t < 3 \\ 2-t/3 & \text{if } 3 \le t \le 6 \end{cases}$

4. $f(t) = t - n$ for $n \le t \le n+1$ and $0 \le n \le 5$

5. $f(t) = \begin{cases} 1 & \text{if } 2n \le t < 2n+1 \\ 0 & \text{if } 2n+1 \le t < 2n+2 \end{cases}$ for $0 \le n \le 2$

6. $f(t) = \begin{cases} t^2 & \text{if } 0 \le t < 2 \\ 4 & \text{if } 2 \le t < 3 \\ 7-t & \text{if } 3 \le t \le 6 \end{cases}$

7. $f(t) = \begin{cases} 1-t & \text{if } 0 \le t < 2 \\ 3-t & \text{if } 2 \le t < 4 \\ 5-t & \text{if } 4 \le t \le 6 \end{cases}$

8. $f(t) = \begin{cases} 1 & \text{if } 0 \le t < 2 \\ 3-t & \text{if } 2 \le t < 3 \\ 2(t-3) & \text{if } 3 \le t < 4 \\ 2 & \text{if } 4 \le t < \infty \end{cases}$

Graphs for problems 1 through 8

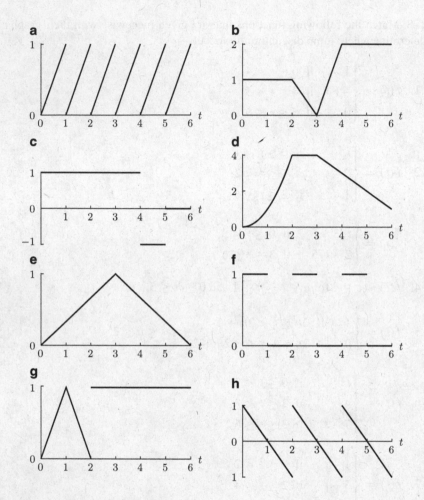

9–12. Compute the indicated integral.

9. $\int_0^5 f(t)\,dt$, where $f(t) = \begin{cases} t^2 - 4 & \text{if } 0 \le t < 2 \\ 0 & \text{if } 2 \le t < 3 \\ -t + 3 & \text{if } 3 \le t < 5 \end{cases}$

10. $\int_0^2 f(u)\,du$, where $f(u) = \begin{cases} 2 - u & \text{if } 0 \le u < 1 \\ u^3 & \text{if } 1 \le u < 2 \end{cases}$

11. $\int_0^{2\pi} |\sin x| \, dx$

12. $\int_0^3 f(w) \, dw$ where $f(w) = \begin{cases} w & \text{if } 0 \le w < 1 \\ \frac{1}{w} & \text{if } 1 \le w < 2 \\ \frac{1}{2} & \text{if } 2 \le w < \infty \end{cases}$

13–16. Compute the indicated integral (See problems 1–8 for the appropriate formula to match with each graph.)

13. $\int_2^5 f(t) \, dt$, where the graph of f is

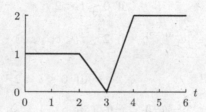

14. $\int_0^6 f(t) \, dt$, where the graph of f is

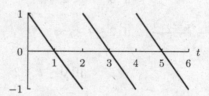

15. $\int_0^6 f(u) \, du$, where the graph of f is

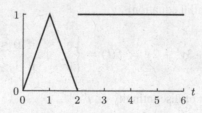

16. $\int_0^6 f(t) \, dt$, where the graph of f is

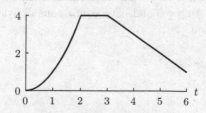

17–20. Of the following four piecewise-defined functions, determine which ones

(a) Satisfy the differential equation

$$y' + 4y = f(t) = \begin{cases} 4 & \text{if } 0 \le t < 2 \\ 8t & \text{if } 2 \le t < \infty, \end{cases}$$

except at the point of discontinuity of f
(b) Are continuous
(c) Are continuous solutions to the differential equation with initial condition $y(0) = 2$. Do not solve the differential equation

17. $y(t) = \begin{cases} 1 & \text{if } 0 \le t < 2 \\ 2t - \frac{1}{2} - \frac{5}{2}e^{-4(t-2)} & \text{if } 2 \le t < \infty \end{cases}$

18. $y(t) = \begin{cases} 1 + e^{-4t} & \text{if } 0 \le t < 2 \\ 2t - \frac{1}{2} - \frac{5}{2}e^{-4(t-2)} + e^{-4t} & \text{if } 2 \le t < \infty \end{cases}$

19. $y(t) = \begin{cases} 1 + e^{-4t} & \text{if } 0 \le t < 2 \\ 2t - \frac{1}{2} - \frac{5e^{-4(t-2)}}{2} & \text{if } 2 \le t < \infty \end{cases}$

20. $y(t) = \begin{cases} 2e^{-4t} & \text{if } 0 \le t < 2 \\ 2t - \frac{1}{2} - \frac{5}{2}e^{-4(t-2)} + e^{-4t} & \text{if } 2 \le t < \infty \end{cases}$

21–24. Of the following four piecewise-defined functions, determine which ones

(a) satisfy the differential equation

$$y'' - 3y' + 2y = f(t) = \begin{cases} e^t & \text{if } 0 \le t < 1 \\ e^{2t} & \text{if } 1 \le t < \infty, \end{cases}$$

except at the point of discontinuity of f
(b) Are continuous
(c) Have continuous derivatives
(d) Are continuous solutions to the differential equation with initial condition $y(0) = 0$ and $y'(0) = 0$ and have continuous derivatives. Do not solve the differential equation.

21. $y(t) = \begin{cases} -te^t - e^t + e^{2t} & \text{if } 0 \le t < 1 \\ te^{2t} - 2e^t & \text{if } 1 \le t < \infty \end{cases}$

22. $y(t) = \begin{cases} -te^t - e^t + e^{2t} & \text{if } 0 \le t < 1 \\ te^{2t} - 3e^t - \frac{1}{2}e^{2t} & \text{if } 1 \le t < \infty \end{cases}$

23. $y(t) = \begin{cases} -te^t - e^t + e^{2t} & \text{if } 0 \le t < 1 \\ te^{2t} + e^{t+1} - e^t - e^{2t} - e^{2t-1} & \text{if } 1 \le t < \infty \end{cases}$

24. $y(t) = \begin{cases} -te^t + e^t - e^{2t} & \text{if } 0 \le t < 1 \\ te^{2t} + e^{t+1} + e^t - e^{2t-1} - 3e^{2t} & \text{if } 1 \le t < \infty \end{cases}$

25–30. Solve the following differential equations.

25. $y' - y = \begin{cases} 1 & \text{if } 0 \le t < 2, \\ -1 & \text{if } 2 \le t < 4, \\ 0 & \text{if } 4 \le t < \infty \end{cases} \qquad y(0) = 0$

26. $y' + 3y = \begin{cases} t & \text{if } 0 \le t < 1 \\ 1 & \text{if } 1 \le t < \infty, \end{cases} \qquad y(0) = 0$

27. $y' - y = \begin{cases} 0 & \text{if } 0 \le t < 1 \\ t - 1 & \text{if } 1 \le t < 2 \\ 3 - t & \text{if } 2 \le t < 3 \\ 0 & \text{if } 3 \le t < \infty, \end{cases} \qquad y(0) = 0$

28. $y' + y = \begin{cases} \sin t & \text{if } 0 \le t < \pi \\ 0 & \text{if } \pi \le t < \infty \end{cases} \qquad y(\pi) = -1$

29. $y'' - y = \begin{cases} t & \text{if } 0 \le t < 1 \\ 0 & \text{if } 1 \le t < \infty, \end{cases} \qquad y(0) = 0, \ y'(0) = 1$

30. $y'' - 4y' + 4y = \begin{cases} 0 & \text{if } 0 \le t < 2 \\ 4 & \text{if } 2 \le t < \infty \end{cases} \qquad y(0) = 1, \ y'(0) = 0$

31. Suppose f is a piecewise continuous function on an interval $[\alpha, \beta]$. Let $a \in [\alpha, \beta]$ and define $y(t) = y_0 + \int_a^t f(u) \, du$. Show that y is a continuous solution to

$$y' = f(t) \qquad y(a) = y_0$$

32. Suppose f is a piecewise continuous function on an interval $[\alpha, \beta]$. Let $a \in [\alpha, \beta]$ and define $y(t) = y_0 + e^{-at} \int_a^t e^{au} f(u)\, du$. Show that y is a continuous solution to

$$y' + ay = f(t) \quad y(a) = y_0$$

33. Let $f(t) = \begin{cases} \sin(1/t) & \text{if } t \neq 0 \\ 0 & \text{if } t = 0 \end{cases}$

    (a) Show that f is bounded.
    (b) Show that f is not continuous at $t = 0$.
    (c) Show that f is not piecewise continuous.

6.2 The Heaviside Class $\mathcal{H}$

In Chap. 2, the Laplace transform was introduced and extensively studied for the class of *continuous* functions of exponential type. We now want to broaden the range of applicability of the Laplace transform method for solving differential equations by allowing for some discontinuous forcing functions. Thus, we say a function $f(t)$ is *piecewise continuous on* $[0, \infty)$ if $f(t)$ is piecewise continuous on each closed subinterval $[0, b]$ for all $b > 0$. In addition, we will maintain the growth condition so that convergence of the integrals defining the Laplace transform is guaranteed. Thus, we will define the *Heaviside class* to be the set $\mathcal{H}$ of all *piecewise continuous functions on* $[0, \infty)$ *of exponential type*. Specifically, $f \in \mathcal{H}$ if:

1. f is piecewise continuous on $[0, \infty)$.
2. There are constants K and a such that $|f(t)| \leq Ke^{at}$ for all $t \geq 0$.

One can show $\mathcal{H}$ is a linear space, that is, closed under addition and scalar multiplication (see Exercises 43–44). It is to this class $\mathcal{H}$ of functions that we extend the Laplace transform. The first observation is that the argument of Proposition 3 of Sect. 2.2 guaranteeing the existence of the Laplace transform extends immediately to functions in $\mathcal{H}$.

Proposition 1. *For* $f \in \mathcal{H}$ *of exponential type of order* a, *the Laplace transform* $F(s) = \int_0^\infty e^{-st} f(t)\, dt$ *exists for* $s > a$, *and* $\lim_{s \to \infty} F(s) = 0$.

Proof. The finite integral $\int_0^N e^{-st} f(t)\, dt$ exists because f is piecewise continuous on $[0, N]$. Since f is also of exponential type, there are constants $K \geq 0$ and a such that $|f(t)| \leq Ke^{at}$ for all $t \geq 0$. Thus, for all $s > a$,

$$|F(s)| \leq \int_0^\infty |e^{-st} f(t)|\, dt \leq \int_0^\infty |e^{-st} Ke^{at}|\, dt = K \int_0^\infty e^{-(s-a)t}\, dt = \frac{K}{s-a}.$$

This shows that the integral converges absolutely, and hence, the Laplace transform exists for $s > a$ and $F(s) \leq K/(s-a)$. It follows that

$$\lim_{s \to \infty} F(s) = 0. \qquad \square$$

Many of the properties of the Laplace transform that were discussed in Chap. 2, and collected in Table 2.3, for continuous functions carry over to the Heaviside class without change in statement or proof. Some of these properties are summarized below.

Linearity $\mathcal{L}\{af + bg\} = a\mathcal{L}\{f\} + b\mathcal{L}\{g\}$.

The first translation principle $\mathcal{L}\{e^{-at} f(t)\}(s) = \mathcal{L}\{f(t)\}(s - a)$.

Differentiation in transform space $\mathcal{L}(-tf(t)) = \dfrac{d}{ds}F(s)$,

$$\mathcal{L}\{(-t)^n f(t)\} = F^{(n)}(s).$$

The dilation principle $\mathcal{L}\{f(bt)\}(s) = \dfrac{1}{b}\mathcal{L}\{f(t)\}(s/b)$.

The important input derivative formula is not on the above list because of a subtlety to be considered in the next section. For now, we will look at some direct computations of Laplace transforms of functions in $\mathcal{H}$ and a useful tool (the second translation theorem) for avoiding most direct calculations.

As might be expected, computations using the definition to compute Laplace transforms of even simple piecewise continuous functions can be tedious.

Example 2. Use the definition to compute the Laplace transform of

$$f(t) = \begin{cases} t^2 & \text{if } 0 \le t < 1, \\ 2 & \text{if } 1 \le t < \infty. \end{cases}$$

▶ **Solution.** Clearly, f is piecewise continuous and bounded; hence, it is in the Heaviside class. We can thus proceed with the definition confident, by Proposition 1, that the improper integral will converge. We have

$$\mathcal{L}\{f(t)\}(s) = \int_0^\infty e^{-st} f(t)\,dt$$

$$= \int_0^1 e^{-st} t^2\,dt + \int_1^\infty e^{-st} 2\,dt$$

For the first integral, we need integration by parts twice:

$$\int_0^1 e^{-st} t^2\,dt = \left.\frac{t^2 e^{-st}}{-s}\right|_0^1 + \frac{2}{s}\int_0^1 e^{-st} t\,dt$$

$$= \frac{e^{-s}}{-s} + \frac{2}{s}\left(\left.\frac{t e^{-st}}{-s}\right|_0^1 + \frac{1}{s}\int_0^1 e^{-st}\,dt\right)$$

Fig. 6.7 The graph of the Heaviside function $h(t)$ (also called the unit step function)

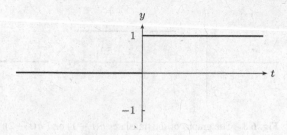

$$= -\frac{e^{-s}}{s} + \frac{2}{s}\left(-\frac{e^{-s}}{s} - \frac{1}{s^2}e^{-st}\Big|_0^1\right)$$

$$= -\frac{e^{-s}}{s} - \frac{2e^{-s}}{s^2} + \frac{2}{s^3} - \frac{2e^{-s}}{s^3}.$$

The second integral is much simpler, and we get

$$\int_1^{\infty} e^{-st}\, 2\, \mathrm{d}t = \frac{2e^{-s}}{s}.$$

Now putting the two pieces together and simplifying gives

$$\mathcal{L}\{f(t)\}(s) = \frac{2}{s^3} + e^{-s}\left(-\frac{2}{s^3} - \frac{2}{s^2} + \frac{1}{s}\right). \qquad \blacktriangleleft$$

As we saw for the Laplace transform of continuous functions, calculations directly from the definition are rarely needed since the Heaviside function that we introduce next will lead to a Laplace transform principle that will allow for the use of our previously derived formulas and make calculations like the one above unnecessary. The **_unit step function_** or **_Heaviside function_** is defined on the real line by

$$h(t) = \begin{cases} 0 & \text{if } t < 0, \\ 1 & \text{if } 0 \le t < \infty. \end{cases}$$

The graph of this function is given in Fig. 6.7.

Clearly, $h(t)$ is piecewise continuous, and it is bounded so it is of exponential type, and hence, $h(t) \in \mathcal{H}$. In addition to $h(t)$ itself, it will be necessary to consider the translations $h(t - c)$ of $h(t)$ for all $c \ge 0$. From the definition of $h(t)$, we see

$$h(t - c) = \begin{cases} 0 & \text{if } -\infty \le t < c, \\ 1 & \text{if } c \le t < \infty. \end{cases}$$

Note that the graph of $h(t - c)$ is just the graph of $h(t)$ translated c units to the right. The graphs of two examples are given in Fig. 6.8.

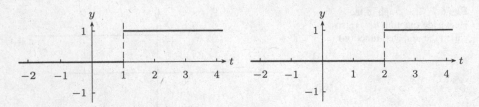

Fig. 6.8 The graphs of the translates $h(t-1)$ and $h(t-2)$

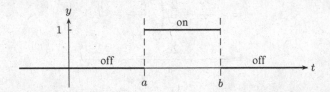

Fig. 6.9 The graph of the on–off switch $\chi_{[a,b)}(t)$

More complicated functions can be built from the Heaviside function. The most important building block is the ***characteristic function*** $\chi_{[a,b)}(t)$ on the interval $[a, b]$ defined by

$$\chi_{[a,b)}(t) = \begin{cases} 1 & \text{if } t \in [a, b), \\ 0 & \text{if } t \notin [a, b). \end{cases}$$

The characteristic function $\chi_{[a,b)}(t)$ serves as the model for an on–off switch at $t = a$ (on) and $t = b$ (off). That is, $\chi_{[a,b)}(t)$ is 1 (the *on* state) for t in the interval $[a, b)$ and 0 (the *off* state) for t not in $[a, b)$. Because of this, we shall also refer to $\chi_{[a,b)}(t)$ as an ***on–off switch***. A graph of a typical on–off switch is given in Fig. 6.9.

Here are some useful relationships between the on-off switches $\chi_{[a,b)}(t)$ and the Heaviside function $h(t)$. All are obtained by direct comparison of the value of the function on the left with that on the right.

1. $\chi_{[0,\infty)}(t) = h(t)$, and if these functions are restricted to the interval $[0, \infty)$ (rather than defined on all of $\mathbb{R}$), then

$$\chi_{[0,\infty)}(t) = h(t) = 1. \tag{1}$$

2. If $0 \leq a < \infty$, then

$$\chi_{[a,\infty)}(t) = h(t - a). \tag{2}$$

3. If $0 \leq a < b < \infty$, then

$$\chi_{[a,b)}(t) = h(t - a) - h(t - b). \tag{3}$$

Using on–off switches, we can easily describe functions defined piecewise. The strategy is to write $f(t)$ as a sum of terms of the form $f_i(t)\chi_{[a_i, a_{i+1})}(t)$ if $f_i(t)$ is the formula used to describe $f(t)$ on the subinterval $[a_i, a_{i+1})$. Then using the relationships listed above, it is possible to write $f(t)$ in terms of translates of the Heaviside function $h(t)$. Here are some examples.

Example 3.   Write the piecewise-defined function

$$f(t) = \begin{cases} t^2 & \text{if } 0 \le t < 1, \\ 2 & \text{if } 1 \le t < \infty. \end{cases}$$

in terms of on–off switches and in terms of translates of the Heaviside function.

▶ **Solution.**   In this piecewise function, t^2 is in the on state only in the interval $[0, 1)$ and 2 is in the on state only in the interval $[1, \infty)$. Thus,

$$f(t) = t^2 \chi_{[0,1)}(t) + 2\chi_{[1,\infty)}(t).$$

Now rewriting the on–off switches in terms of the Heaviside functions using (1)–(3), we obtain

$$\begin{aligned} f(t) &= t^2(h(t) - h(t-1)) + 2h(t-1) \\ &= t^2 h(t) + (2 - t^2)h(t-1) \\ &= t^2 + (2 - t^2)h(t-1). \end{aligned}$$ ◀

Example 4.   Write the piecewise-defined function

$$f(t) = \begin{cases} \cos t & \text{if } 0 \le t < \pi, \\ 1 & \text{if } \pi \le t < 2\pi, \\ 0 & \text{if } 2\pi \le t < \infty. \end{cases}$$

in terms of on–off switches and in terms of translates of the Heaviside function.

▶ **Solution.**   The function is defined by different formulas on each of the intervals $[0, \pi)$, $[\pi, 2\pi)$, and $[2\pi, \infty)$. Thus,

$$\begin{aligned} f(t) &= \cos t \chi_{[0,\pi)}(t) + 1 \chi_{[\pi, 2\pi)}(t) + 0 \chi_{[2\pi,\infty)}(t) \\ &= \cos t(h(t) - h(t-\pi)) + (h(t-\pi) - h(t-2\pi)) \\ &= (\cos t)h(t) + (1 - \cos t)h(t-\pi) - h(t-2\pi) \\ &= \cos t + (1 - \cos t)h(t-\pi) - h(t-2\pi). \end{aligned}$$ ◀

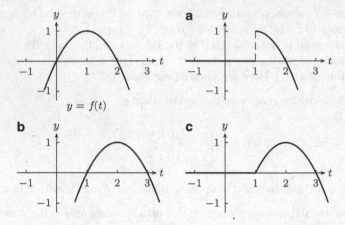

Fig. 6.10 The graphs of (**a**) $h(t-1)f(t)$, (**b**) $f(t-1)$, and (**c**) $h(t-1)f(t-1)$ for $f(t) = 2t - t^2$

In the above descriptions, there are functions of the form $g(t)h(t - c)$ and $g(t)\chi_{[a,b)}(t)$. How do the graphs of these functions correspond to the graph of $g(t)$? Here is an example.

Example 5. Compare the graph of each of the following functions to the graph of $f(t) = 2t - t^2$:

$$\text{(a) } h(t - 1)f(t) \quad \text{(b) } f(t - 1) \quad \text{(c) } h(t - 1)f(t - 1)$$

▶ **Solution.** The graphs are given in Fig. 6.10.
The graph of $f(t)$ is given first. Now, (a) $h(t - 1)f(t)$ simply cuts off the graph of $f(t)$ at $t = 1$ and replaces it with the line $y = 0$ for $t < 1$, (b) $f(t - 1)$ just shifts the graph of $f(t)$ 1 unit to the right, and (c) $h(t - 1)f(t - 1)$ shifts the graph of $f(t)$ 1 unit to the right and then cuts off the resulting graph at $t = 1$ and replaces it with the line $y = 0$ for $t < 1$.   ◀

Functions of the form $h(t - c)f(t - c)$, namely, translation of f by c and then truncation of the resulting graph for $t < c$, as illustrated for $t = 1$ in the previous example, are precisely the special type of functions in the Heaviside class $\mathcal{H}$ for which it is possible to compute the Laplace transform in an efficient manner. Since any piecewise-defined function will be reducible to functions of this form, it will provide an effective method of computation. Let us start by computing the Laplace transform of a translated Heaviside function $h(t - c)$.

Formula 6. If $c \geq 0$ is any nonnegative real number, verify the Laplace transform formula:

Translates of the Heaviside function

$$\mathcal{L}\{h(t - c)\}(s) = \frac{e^{-sc}}{s}, \qquad s > 0.$$

▼ *Verification.* For the Heaviside function $h(t - c)$, we have

$$\mathcal{L}\{h(t - c)\}(s) = \int_0^\infty e^{-st} h(t - c)\, dt = \int_c^\infty e^{-st}\, dt$$

$$= \lim_{r \to \infty} \left. \frac{e^{-ts}}{-s} \right|_c^r$$

$$= \lim_{r \to \infty} \frac{e^{-rs} - e^{-sc}}{-s} = \frac{e^{-sc}}{s} \qquad \text{for } s > 0. \qquad \blacktriangle$$

Combining Formula 6 with linearity gives the following formula.

Formula 7. If $0 \le a < b < \infty$, then

The on–off switch

$$\mathcal{L}\{\chi_{[a,b)}(t)\}(s) = \frac{e^{-as}}{s} - \frac{e^{-bs}}{s}, \qquad s > 0.$$

Formula 6 is a special case of what is known as the second translation principle.

Theorem 8. *Suppose $f(t) \in \mathcal{H}$ is a function with Laplace transform $F(s)$. Then*

Second translation principle

$$\mathcal{L}\{f(t - c)h(t - c)\}(s) = e^{-sc} F(s).$$

In terms of the inverse Laplace transform, this is equivalent to

---

Inverse second translation principle

$$\mathcal{L}^{-1}\{e^{-sc} F(s)\} = f(t - c)h(t - c).$$

Proof. The calculation is straightforward and involves a simple change of variables:

$$\mathcal{L}\{f(t - c)h(t - c)\}(s) = \int_0^\infty e^{-st} f(t - c)h(t - c)\, dt$$

$$= \int_c^\infty e^{-st} f(t - c)\, dt$$

$$= \int_0^\infty e^{-s(t+c)} f(t)\, dt \qquad (t \mapsto t + c)$$

$$= e^{-sc} \int_0^\infty e^{-st} f(t)\, dt$$

$$= e^{-sc} F(s) \qquad\qquad\qquad \square$$

As with the notation used with the first translation theorem, it is frequently convenient to express the inverse second translation theorem in the following format:

$$\mathcal{L}^{-1}\{e^{-sc} F(s)\} = h(t-c)\mathcal{L}^{-1}\{F(s)\}\big|_{t \to t-c}.$$

That is, take the inverse Laplace transform of $F(s)$, replace t by $t-c$, and then multiply by the translated Heaviside function.

In practice, it is more common to encounter expressions written in the form $g(t)h(t-c)$, rather than the nicely arranged format $f(t-c)h(t-c)$. But if $f(t)$ is replaced by $g(t+c)$ in Theorem 8, then we obtain the (apparently) more general version of the second translation principle.

Corollary 9.
$$\mathcal{L}\{g(t)h(t-c)\} = e^{-sc}\mathcal{L}\{g(t+c)\}.$$

A simple example of this occurs when $g = 1$. Then $\mathcal{L}\{h(t-c)\} = e^{-sc}\mathcal{L}\{1\} = e^{-sc}/s$, which agrees with Formula 6 found above. When $c = 0$, then $\mathcal{L}\{h(t-0)\} = 1/s$ which is the same as the Laplace transform of the constant function 1. This is consistent since $h(t-0) = h(t) = 1$ for $t \geq 0$.

Now we give some examples of using these formulas.

Example 10. Find the Laplace transform of

$$f(t) = \begin{cases} t^2 & \text{if } 0 \leq t < 1, \\ 2 & \text{if } 1 \leq t < \infty \end{cases}$$

using the second translation principle. This Laplace transform was previously computed directly from the definition in Example 2.

▶ **Solution.** In Example 3, we found $f(t) = t^2 + (2 - t^2)h(t-1)$. By Corollary 9, we get

$$\mathcal{L}\{f\} = \frac{2}{s^3} + e^{-s}\mathcal{L}\{2 - (t+1)^2\}$$

$$= \frac{2}{s^3} + e^{-s}\mathcal{L}\{-t^2 - 2t + 1\}$$

$$= \frac{2}{s^3} + e^{-s}\left(-\frac{2}{s^3} - \frac{2}{s^2} + \frac{1}{s}\right) \qquad\qquad ◀$$

Example 11. Find the Laplace transform of

$$f(t) = \begin{cases} \cos t & \text{if } 0 \leq t < \pi, \\ 1 & \text{if } \pi \leq t < 2\pi, \\ 0 & \text{if } 2\pi \leq t < \infty. \end{cases}$$

▶ **Solution.** In Example 4, we found

$$f(t) = \cos t + (1 - \cos t)h(t - \pi) - h(t - 2\pi).$$

By Corollary 9, we get

$$F(s) = \frac{s}{s^2 + 1} + e^{-s\pi}\mathcal{L}\{1 - \cos(t + \pi)\} - \frac{e^{-2s\pi}}{s}$$

$$= \frac{s}{s^2 + 1} + e^{-s\pi}\left(\frac{1}{s} + \frac{s}{s^2 + 1}\right) - \frac{e^{-2s\pi}}{s}.$$

In the second line, we have used the fact that $\cos(t + \pi) = -\cos t$.   ◀

Uniqueness of the Inverse Laplace Transform

Theorem 8 gives a formula for the inverse Laplace transform of a function $e^{-sc}F(s)$. Such an explicit formula is suggestive of some type of uniqueness for the Laplace transform, that is, $\mathcal{L}\{f(t)\} = \mathcal{L}\{g(t)\} \implies f(t) = g(t)$, which is Theorem 1 of Sect. 2.5 for continuous functions. By expanding the domain of the Laplace transform $\mathcal{L}$ to include the possibly discontinuous functions in the Heaviside class $\mathcal{H}$, the issue of uniqueness is made somewhat more subtle because changing the value of a function at a single point will not change the integral of the function, and hence, the Laplace transform will not change. Therefore, instead of talking about *equality* of functions, we will instead consider the concept of *essential equality* of functions. Two functions $f_1(t)$ and $f_2(t)$ are said to be ***essentially equal*** on $[0, \infty)$ if for each subinterval $[0, N)$ they are equal as functions except at possibly finitely many points. For example, the functions

$$f_1(t) = \begin{cases} 1 & \text{if } 0 \leq t < 1, \\ 2 & \text{if } 1 \leq t < \infty, \end{cases} \qquad f_2(t) = \begin{cases} 1 & \text{if } 0 \leq t < 1, \\ 3 & \text{if } t = 1, \\ 2 & \text{if } 1 < t < \infty, \end{cases} \qquad \text{and}$$

$$f_3(t) = \begin{cases} 1 & \text{if } 0 \leq t \leq 1, \\ 2 & \text{if } 1 < t < \infty, \end{cases}$$

are essentially equal for they are equal everywhere except at $t=1$, where $f_1(1)=2$, $f_2(1)=3$, and $f_3(1)=1$. Two functions that are essentially equal have the same Laplace transform. This is because the Laplace transform is an integral operator and integration cannot distinguish functions that are essentially equal. The Laplace transforms of $f_1(t)$, $f_2(t)$, and $f_3(t)$ in our example above are all $(1+e^{-s})/s$. Here is our problem: Given a transform, like $(1+e^{-s})/s$, how do we decide what "the" inverse Laplace transform is. It turns out that if $F(s)$ is the Laplace transform of two functions $f_1(t)$, $f_2(t) \in \mathcal{H}$, then $f_1(t)$ and $f_2(t)$ are essentially equal. Since for most practical situations it does not matter which one is chosen, we will make a choice by always choosing the function that is right continuous at each point. A function $f(t)$ in the Heaviside class is said to be **right continuous at a point** a if we have

$$f(a) = f(a^+) = \lim_{t \to a^+} f(t),$$

and it is **right continuous on** $[0, \infty)$ if it is right continuous at each point in $[0, \infty)$. In the example above, $f_1(t)$ is right continuous while $f_2(t)$ and $f_3(t)$ are not. The function $f_3(t)$ is, however, left continuous, using the obvious definition of left continuity. If we decide to use right continuous functions in the Heaviside class, then the correspondence with its Laplace transform is one-to-one. We summarize this discussion as a theorem:

Theorem 12. *If $F(s)$ is the Laplace transform of a function in $\mathcal{H}$, then there is a unique right continuous function $f(t) \in \mathcal{H}$ such that $\mathcal{L}\{f(t)\} = F(s)$. Any two functions in $\mathcal{H}$ with the same Laplace transform are essentially equal.*

All the translates $h(t-c)$ of the Heaviside function $h(t)$ are right continuous, so any piecewise function written as a sum of products of a continuous function and a translated Heaviside function are right continuous. In fact, *the convention of using right continuous functions just means that the inverse transforms of functions in $\mathcal{H}$ will be written as sums of $f(t-c)h(t-c)$, as given in the second translation principle Theorem 8.*

Example 13. Find the inverse Laplace transform of

$$F(s) = \frac{e^{-s}}{s^2} + \frac{e^{-3s}}{s-4}$$

and write it as a right continuous piecewise-defined function.

▶ **Solution.** The inverse Laplace transforms of $1/s^2$ and $1/(s-4)$ are, respectively, t and e^{4t}. By Theorem 8, the inverse Laplace transform of $F(s)$ is

$$
\begin{aligned}
\mathcal{L}^{-1}\{F(s)\} &= h(t-1)\,\mathcal{L}^{-1}\left\{\frac{1}{s^2}\right\}\bigg|_{t \to t-1} + h(t-3)\,\mathcal{L}^{-1}\left\{\frac{1}{s-4}\right\}\bigg|_{t \to t-3} \\
&= h(t-1)\,(t)|_{t \to t-1} + h(t-3)\,(e^{4t})|_{t \to t-3} \\
&= (t-1)h(t-1) + e^{4(t-3)}h(t-3).
\end{aligned}
$$

On the interval $[0, 1)$, both $t - 1$ and $e^{4(t-3)}$ are off. On the interval $[1, 3)$ only $t - 1$ is on. On the interval $[3, \infty)$, both $t - 1$ and $e^{4(t-3)}$ are on. Thus,

$$\mathcal{L}^{-1}\{F(s)\} = \begin{cases} 0 & \text{if } 0 \le t < 1 \\ t - 1 & \text{if } 1 \le t < 3 \\ t - 1 + e^{4(t-3)} & \text{if } 3 \le t < \infty \end{cases} \quad .$$

◄

Exercises

1–8. Graph each of the following functions defined by means of the unit step function $h(t - c)$ and/or the on–off switches $\chi_{[a, b)}$.

1. $f(t) = 3h(t - 2) - h(t - 5)$

2. $f(t) = 2h(t - 2) - 3h(t - 3) + 4h(t - 4)$

3. $f(t) = (t - 1)h(t - 1)$

4. $f(t) = (t - 2)^2 h(t - 2)$

5. $f(t) = t^2 h(t - 2)$

6. $f(t) = h(t - \pi) \sin t$

7. $f(t) = h(t - \pi) \cos 2(t - \pi)$

8. $f(t) = t^2 \chi_{[0, 1)}(t) + (2 - t)\chi_{[1, 3)}(t) + 3\chi_{[3, \infty)}(t)$

9–27. For each of the following functions $f(t)$, (a) express $f(t)$ in terms of on–off switches $\chi_{[a, b)}(t)$, (b) express $f(t)$ in terms of translates $h(t - c)$ of the Heaviside function $h(t)$, and (c) compute the Laplace transform $F(s) = \mathcal{L}\{f(t)\}$.

9. $f(t) = \begin{cases} 0 & \text{if } 0 \le t < 2 \\ t - 2 & \text{if } 2 \le t < \infty \end{cases}$

10. $f(t) = \begin{cases} 0 & \text{if } 0 \le t < 2 \\ t & \text{if } 2 \le t < \infty \end{cases}$

11. $f(t) = \begin{cases} 0 & \text{if } 0 \le t < 2 \\ t + 2 & \text{if } 2 \le t < \infty \end{cases}$

12. $f(t) = \begin{cases} 0 & \text{if } 0 \le t < 4 \\ (t - 4)^2 & \text{if } 4 \le t < \infty \end{cases}$

13. $f(t) = \begin{cases} 0 & \text{if } 0 \le t < 4 \\ t^2 & \text{if } 4 \le t < \infty \end{cases}$

14. $f(t) = \begin{cases} 0 & \text{if } 0 \le t < 4 \\ t^2 - 4 & \text{if } 4 \le t < \infty \end{cases}$

15. $f(t) = \begin{cases} 0 & \text{if } 0 \le t < 2 \\ (t - 4)^2 & \text{if } 2 \le t < \infty \end{cases}$

16. $f(t) = \begin{cases} 0 & \text{if } 0 \leq t < 4 \\ e^{t-4} & \text{if } 4 \leq t < \infty \end{cases}$

17. $f(t) = \begin{cases} 0 & \text{if } 0 \leq t < 4 \\ e^{t} & \text{if } 4 \leq t < \infty \end{cases}$

18. $f(t) = \begin{cases} 0 & \text{if } 0 \leq t < 6 \\ e^{t-4} & \text{if } 6 \leq t < \infty \end{cases}$

19. $f(t) = \begin{cases} 0 & \text{if } 0 \leq t < 4 \\ te^{t} & \text{if } 4 \leq t < \infty \end{cases}$

20. $f(t) = \begin{cases} 1 & \text{if } 0 \leq t < 4 \\ -1 & \text{if } 4 \leq t < 5 \\ 0 & \text{if } 5 \leq t < \infty \end{cases}$

21. $f(t) = \begin{cases} t & \text{if } 0 \leq t < 1 \\ 2-t & \text{if } 1 \leq t < \infty \end{cases}$

22. $f(t) = \begin{cases} t & \text{if } 0 \leq t < 1 \\ 2-t & \text{if } 1 \leq t < 2 \\ 1 & \text{if } 2 \leq t < \infty \end{cases}$

23. $f(t) = \begin{cases} t^2 & \text{if } 0 \leq t < 2 \\ 4 & \text{if } 2 \leq t < 3 \\ 7-t & \text{if } 3 \leq t < \infty \end{cases}$

24. $f(t) = \begin{cases} 1 & \text{if } 0 \leq t < 2 \\ 3-t & \text{if } 2 \leq t < 3 \\ 2(t-3) & \text{if } 3 \leq t < 4 \\ 2 & \text{if } 4 \leq t < \infty \end{cases}$

25. $f(t) = \begin{cases} t & \text{if } 0 \leq t < 1 \\ t-1 & \text{if } 1 \leq t < 2 \\ t-2 & \text{if } 2 \leq t < 3 \\ \vdots \end{cases}$

26. $f(t) = \begin{cases} 1 & \text{if } 2n \leq t < 2n + 1 \\ 0 & \text{if } 2n + 1 \leq t < 2n + 2 \end{cases}$

27. $f(t) = \begin{cases} 1 - t & \text{if } 0 \leq t < 2 \\ 3 - t & \text{if } 2 \leq t < 4 \\ 5 - t & \text{if } 4 \leq t < 6 \\ \vdots \end{cases}$

28–42. Compute the inverse Laplace transform of each of the following functions.

28. $\dfrac{e^{-3s}}{s - 1}$

29. $\dfrac{e^{-3s}}{s^2}$

30. $\dfrac{e^{-3s}}{(s - 1)^3}$

31. $\dfrac{e^{-\pi s}}{s^2 + 1}$

32. $\dfrac{se^{-3\pi s}}{s^2 + 1}$

33. $\dfrac{e^{-\pi s}}{s^2 + 2s + 5}$

34. $\dfrac{e^{-s}}{s^2} + \dfrac{e^{-2s}}{(s - 1)^3}$

35. $\dfrac{e^{-2s}}{s^2 + 4}$

36. $\dfrac{e^{-2s}}{s^2 - 4}$

37. $\dfrac{se^{-4s}}{s^2 + 3s + 2}$

38. $\dfrac{e^{-2s} + e^{-3s}}{s^2 - 3s + 2}$

39. $\dfrac{1 - e^{-5s}}{s^2}$

40. $\dfrac{1 + e^{-3s}}{s^4}$

41. $e^{-\pi s} \dfrac{2s + 1}{s^2 + 6s + 13}$

42. $(1 - e^{-\pi s}) \dfrac{2s + 1}{s^2 + 6s + 13}$

43–44. These exercises show that the Heaviside class $\mathcal{H}$ is a linear space.

43. Suppose f_1 and f_2 are piecewise continuous on $[0, \infty)$ and $c \in \mathbb{R}$. Show $f_1 + cf_2$ is piecewise continuous on $[0, \infty)$.

44. Suppose $f_1, f_2 \in \mathcal{H}$ and $c \in \mathbb{R}$. Show that $f_1 + cf_2 \in \mathcal{H}$ and hence, $\mathcal{H}$ is a linear space.

6.3 Laplace Transform Method for $f(t) \in \mathcal{H}$

The differential equations that we will solve by means of the Laplace transform are first and second order constant coefficient linear differential equations with a forcing function $f(t) \in \mathcal{H}$:

$$y' + ay = f(t),$$
$$y'' + ay' + by = f(t).$$

To employ this method, it will be necessary to compute the Laplace transform of derivatives of functions in $\mathcal{H}$. That is, we need to know the extent to which the input derivative formula

$$\mathcal{L}\{f'(t)\} = s\mathcal{L}\{f(t)\} - f(0)$$

is valid when $f(t)$ is in the Heaviside class $\mathcal{H}$. Recall that for $f(t) \in \mathcal{H}$, the symbol $f'(t)$ is used to denote the derivative of $f(t)$ if $f(t)$ is differentiable except possibly at a finite number of points on each interval of the form $[0, N]$. Thus, $f'(t)$ will be (possibly) undefined for finitely many points in $[0, N]$.

Example 1. Verify that the input derivative formula is not valid for the on–off switch function

$$f(t) = \chi_{[0, 1)}(t) = \begin{cases} 1 & \text{if } 0 \le t < 1, \\ 0 & \text{if } t \ge 1. \end{cases}$$

▼ *Verification.* From the definition, it is clear that $f'(t) = 0$ for all $t \neq 1$; it is not differentiable, or even continuous, at $t = 1$. Thus, $\mathcal{L}\{f'(t)\} = 0$. However, by Formula 7 of Sect. 6.2, $\mathcal{L}\{f(t)\} = F(s) = (1 - e^{-s})/s$ so that

$$sF(s) - f(0) = 1 - e^{-s} - 1 = e^{-s} \neq 0 = \mathcal{L}\{f'(t)\}. \qquad ▲$$

Example 2. Verify that the input derivative formula is valid for the following function

$$f(t) = \begin{cases} 1 & \text{if } 0 \le t < 1, \\ t & \text{if } t \ge 1. \end{cases}$$

See Fig. 6.11

▼ *Verification.* Write $f(t)$ in the standard form using translates of the Heaviside function $h(t)$ to get

$$f(t) = \chi_{[0, 1)}(t) + th(t - 1) = 1 + (t - 1)h(t - 1).$$

Thus,

$$sF(s) - f(0) = s\left(\frac{1}{s} + \frac{e^{-s}}{s^2}\right) - 1 = \frac{e^{-s}}{s} = \mathcal{L}\{h(t - 1)\} = \mathcal{L}\{f'(t)\},$$

since $f'(t) = h(t - 1)$. Therefore, the input derivative principle is satisfied for this function $f(t)$.   ▲

Fig. 6.11 A continuous $f(t)$ with discontinuous derivative

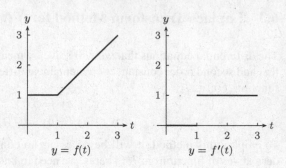

$y = f(t)$ $y = f'(t)$

Thus, we have an example of a piecewise continuous function (Example 1) for which the input derivative principle fails, and another example (Example 2) for which the input derivative principle holds. What is the significant difference between these two simple examples? For one thing, the second example has a *continuous* $f(t)$, while that of the first example is *discontinuous*. It turns out that this is the feature that needs to be included. That is, the derivative need not be continuous, but the function must.

Theorem 3. *Suppose $f(t)$ is a continuous function on $[0, \infty)$ such that both $f(t)$ and $f'(t)$ are in $\mathcal{H}$. Then*

> ### The input derivative principle
> ### The first derivative
>
> $$\mathcal{L}\{f'(t)\}(s) = s\mathcal{L}\{f(t)\}(s) - f(0).$$

Proof. We begin by computing $\int_0^N e^{-st} f'(t)\, dt$. This computation requires a careful analysis of the points where $f'(t)$ is discontinuous. There are only finitely many such discontinuities on the interval $[0, N)$, which will be labeled $a_1, \ldots, a_k$, and we may assume $a_i < a_{i+1}$. If we let $a_0 = 0$ and $a_{k+1} = N$, then we obtain

$$\int_0^N e^{-st} f'(t)\, dt = \sum_{i=0}^{k} \int_{a_i}^{a_{i+1}} e^{-st} f'(t)\, dt,$$

and integration by parts gives

$$\int_0^N e^{-st} f'(t)\, dt = \sum_{i=0}^{k} \left(f(t)e^{-st} \Big|_{a_i}^{a_{i+1}} + s \int_{a_i}^{a_{i+1}} e^{-st} f(t)\, dt \right)$$

$$= \sum_{i=0}^{k} (f(a_{i+1}^-)e^{-sa_{i+1}} - f(a_i^+)e^{-sa_i}) + s \int_0^N e^{-st} f(t)\, dt$$

$$= \sum_{i=0}^{k} (f(a_{i+1})e^{-sa_{i+1}} - f(a_i)e^{-sa_i}) + s \int_{0}^{N} e^{-st} f(t) \, dt$$

$$= f(N)e^{-Ns} - f(0) + s \int_{0}^{N} e^{-st} f(t) \, dt.$$

Since $f(t)$ is continuous, we have $f(a_i^+) = f(a_i)$ and $f(a_{i+1}^-) = f(a_{i+1})$ which allows us to go from the second to the third line. The telescoping nature of the sum in line 3 allows it to collapse to $f(a_{k+1})e^{-sa_{k+1}} - f(a_0)e^{-sa_0} = f(N)e^{-Ns} - f(0)$ which produces the final line. Now take the limit as N goes to infinity and the result follows. □

If $f(t) \in \mathcal{H}$, then the definite integral $g(t) = \int_0^t f(u) \, du$ is continuous, in the Heaviside class, and moreover, $g(0) = 0$. Thus, applying the input derivative principle to the function $g(t) \in \mathcal{H}$ gives the input integral principle:

Corollary 4. *Suppose $f(t)$ is a function defined on $[0, \infty)$ such $f(t) \in \mathcal{H}$, and $F(s) = \mathcal{L}\{f(t)\}(s)$, then*

> **The input integral principle**
>
> $$\mathcal{L}\left\{ \int_{0}^{t} f(u) \, du \right\}(s) = \frac{F(s)}{s}.$$

The second order input derivative principle is an immediate corollary of the first, as long as we are careful to identify the appropriate hypotheses that $f(t)$ must satisfy.

Corollary 5. *If $f(t)$ and $f'(t)$ are continuous and $f(t)$, $f'(t)$, and $f''(t)$ are in $\mathcal{H}$, then*

> **Input derivative principle**
> **The second derivative**
>
> $$\mathcal{L}\{f''(t)\} = s^2 \mathcal{L}\{f(t)\} - sf(0) - f'(0).$$

We are now in a position to illustrate the Laplace transform method for solving first and second order constant coefficient differential equations

$$y' + ay = f(t),$$
$$y'' + ay' + by = f(t)$$

with a forcing function $f(t)$ that is possibly discontinuous. In order to apply the Laplace transform method, we will need to take the Laplace transform of a potential solution $y(t)$. To apply the input derivative formula to $y(t)$, it is necessary to know that $y(t)$ is continuous and $y'(t) \in \mathcal{H}$ for the first order equation and both $y(t)$ and $y'(t)$ are continuous while $y''(t) \in \mathcal{H}$ for the second order equation. These facts were proved in Theorems 5 and 7 of Sect. 6.1, and hence, these theorems provide the theoretical underpinnings for applying the Laplace transform method formally. Here are some examples.

Example 6. Solve the following first order differential equation:

$$y' + 2y = f(t), \qquad y(0) = 1,$$

where

$$f(t) = \begin{cases} 0 & \text{if } 0 \le t < 1, \\ t & \text{if } 1 \le t < \infty. \end{cases}$$

▶ **Solution.** We first rewrite $f(t)$ in terms of Heaviside functions:

$$f(t) = t\, \chi_{[1,\infty)}(t) = t\, h(t-1).$$

By Corollary 9 of Sect. 6.2 of the second translation principle, its Laplace transform is

$$F(s) = \mathcal{L}\{th(t-1)\} = e^{-s}\mathcal{L}\{t+1\} = e^{-s}\left(\frac{1}{s^2} + \frac{1}{s}\right) = e^{-s}\left(\frac{s+1}{s^2}\right).$$

Let $Y(s) = \mathcal{L}\{y(t)\}$ where $y(t)$ is the solution to the differential equation. Since the analysis done above shows that $y(t)$ satisfies the hypotheses of the input derivative principle, we can apply the Laplace transform to the differential equation and conclude

$$sY(s) - y(0) + 2Y(s) = e^{-s}\left(\frac{s+1}{s^2}\right).$$

Solving for $Y(s)$ gives

$$Y(s) = \frac{1}{s+2} + e^{-s}\frac{s+1}{s^2(s+2)}.$$

A partial fraction decomposition gives

$$\frac{s+1}{s^2(s+2)} = \frac{1}{4}\frac{1}{s} + \frac{1}{2}\frac{1}{s^2} - \frac{1}{4}\frac{1}{s+2},$$

Fig. 6.12 The graph of the solution to Example 6

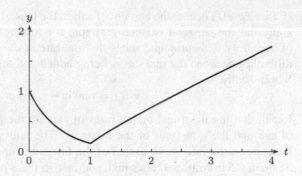

and the second translation principle (Theorem 8 of Sect. 6.2) gives

$$y(t) = \mathcal{L}^{-1}\left\{\frac{1}{s+2}\right\} + \frac{1}{4}\mathcal{L}^{-1}\left\{e^{-s}\frac{1}{s}\right\} + \frac{1}{2}\mathcal{L}^{-1}\left\{e^{-s}\frac{1}{s^2}\right\} - \frac{1}{4}\mathcal{L}^{-1}\left\{e^{-s}\frac{1}{s+2}\right\}$$

$$= e^{-2t} + \frac{1}{4}h(t-1) + \frac{1}{2}(t-1)h(t-1) - \frac{1}{4}e^{-2(t-1)}h(t-1).$$

$$= \begin{cases} e^{-2t} & \text{if } 0 \leq t < 1 \\ e^{-2t} + \frac{1}{4}(2t-1) - \frac{1}{4}e^{-2(t-1)} & \text{if } 1 \leq t < \infty. \end{cases}$$

The graph of $y(t)$ is shown in Fig. 6.12, where the discontinuity of the forcing function $f(t)$ at time $t = 1$ is reflected in the abrupt change in the direction of the tangent line of the graph of $y(t)$ at $t = 1$. ◀

We now consider a mixing problem of the type mentioned in the introduction to this chapter.

Example 7. Suppose a tank holds 10 gal of pure water. There are two input sources of brine solution: the first source has a concentration of 2 lbs of salt per gallon while the second source has a concentration of 3 lbs of salt per gallon. The first source flows into the tank at a rate of 1 gal/min for 5 min after which it is turned off and simultaneously the second source is turned on at a rate of 1 gal/min. The well-mixed solution flows out of the tank at a rate of 1 gal/min. Find the amount of salt in the tank at any time t.

▶ **Solution.** A pictorial representation of the problem is the following diagram.

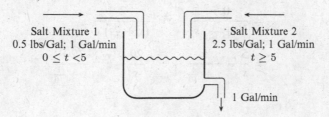

Letting $y(t)$ denote the amount of salt in the tank at time t, measured in pounds, apply the fundamental balance principle for mixing problems (see Example 11 of Sect. 1.4). This principle states that the rate of change of $y(t)$ comes from the difference between the rate salt is being added and the rate salt is being removed. Symbolically,

$$y'(t) = \text{rate in} - \text{rate out}.$$

Recall that the input and output rates of salt are the product of the concentration of salt and the flow rates of the mixtures. The rate at which salt is being added depends on the interval of time. For the first five minutes, source one adds salt at a rate of 0.5 lbs/min, and after that, source two takes over and adds salt at a rate of 2.5 lbs/min. Since the flow rate in is 1 gal/min, the rate at which salt is being added is given by the function

$$f(t) = \begin{cases} 0.5 & \text{if } 0 \leq t < 5, \\ 2.5 & \text{if } 5 \leq t < \infty. \end{cases}$$

The concentration of salt at time t is $y(t)/10$ lbs/gal, and since the flow rate out is 1 gal/min, it follows that the rate at which salt is being removed is $y(t)/10$ lbs/min. Since initially there is pure water, it follows that $y(0) = 0$, and therefore, we have that $y(t)$ satisfies the following initial value problem:

$$y' = f(t) - \frac{y(t)}{10}, \qquad y(0) = 0.$$

Rewriting $f(t)$ in terms of translates of the Heaviside function gives

$$f(t) = 0.5\chi_{[0,5)}(t) + 2.5\chi_{[5,\infty)}(t)$$
$$= 0.5(h(t) - h(t-5)) + 2.5h(t-5)$$
$$= 0.5 + 2h(t-5).$$

Applying the Laplace transform to the differential equation and solving for $Y(s) = \mathcal{L}\{y(t)\}(s)$ gives

$$Y(s) = \left(\frac{1}{s + \frac{1}{10}} \right) \left(\frac{0.5 + 2e^{-5s}}{s} \right)$$

$$= \frac{0.5}{\left(s + \frac{1}{10}\right)s} + e^{-5s}\frac{2}{\left(s + \frac{1}{10}\right)s}$$

$$= \frac{5}{s} - \frac{5}{s + \frac{1}{10}} + e^{-5s}\frac{20}{s} - e^{-5s}\frac{20}{s + \frac{1}{10}}.$$

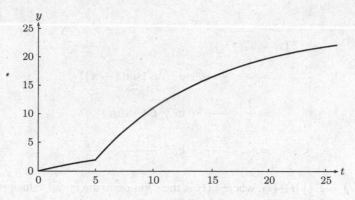

Fig. 6.13 The solution to a mixing problem with discontinuous input function

Taking the inverse Laplace transform of $Y(s)$ gives

$$y(t) = 5 - 5e^{-\frac{t}{10}} + 20h(t-5) - 20e^{-\frac{t-5}{10}}h(t-5)$$

$$= \begin{cases} 5 - 5e^{-\frac{t}{10}} & \text{if } 0 \le t < 5, \\ 25 - 5e^{-\frac{t}{10}} - 20e^{-\frac{t-5}{10}} & \text{if } 5 \le t < \infty. \end{cases}$$

The graph of y is given in Fig. 6.13. As expected, we observe that the solution is continuous, but the kink at $t = 5$ indicates that there is a discontinuity of the derivative at this point. This occurred when the flow of the second source, which had a higher concentration of salt, was turned on.   ◄

Here is an example of a second order equation.

Example 8. Solve the following second order initial value problem

$$y'' + 4y = f(t), \qquad y(0) = 0, \ y'(0) = 1$$

where

$$f(t) = \begin{cases} 1 & \text{if } 0 \le t < \pi, \\ \sin t & \text{if } t \ge \pi. \end{cases}$$

▶ **Solution.** First note that

$$f(t) = \chi_{[0,\pi)}(t) + (\sin t)\chi_{[\pi,\infty)}(t) \quad .$$
$$= 1 - h(t-\pi) + (\sin t)h(t-\pi),$$

so that

$$F(s) = \mathcal{L}\{f(t)\}$$

$$= \frac{1}{s} - \frac{e^{-\pi s}}{s} + e^{-\pi s}\mathcal{L}\{\sin(t + \pi)\}$$

$$= \frac{1}{s} - \frac{e^{-\pi s}}{s} + e^{-\pi s}\mathcal{L}\{-\sin t\}$$

$$= \frac{1}{s} - \frac{e^{-\pi s}}{s} + e^{-\pi s}\frac{1}{s^2 + 1}.$$

Letting $Y(s) = \mathcal{L}\{y(t)\}(s)$, where $y(t)$ is the solution to the initial value problem, and taking the Laplace transform of the differential equation gives

$$s^2 Y(s) - 1 + 4Y(s) = \frac{1}{s} - \frac{e^{-\pi s}}{s} + e^{-\pi s}\frac{1}{s^2 + 1}.$$

Now solve for $Y(s)$ to get

$$Y(s) = \frac{1}{s^2 + 4} + \frac{1}{s(s^2 + 4)}(1 - e^{-\pi s}) + e^{-\pi s}\frac{1}{(s^2 + 1)(s^2 + 4)}.$$

The solution is completed by taking the inverse Laplace transform, using the second translation theorem:

$$y(t) = \mathcal{L}^{-1}\{Y(s)\}$$

$$= \frac{1}{2}\sin 2t + \mathcal{L}^{-1}\left\{\frac{1}{s(s^2 + 4)}\right\} - h(t - \pi)\mathcal{L}^{-1}\left\{\frac{1}{s(s^2 + 4)}\right\}(t - \pi)$$

$$+ h(t - \pi)\mathcal{L}^{-1}\left\{\frac{1}{(s^2 + 1)(s^2 + 4)}\right\}(t - \pi).$$

Computing the partial fractions in the usual manner

$$\frac{1}{s(s^2 + 4)} = \frac{\frac{1}{4}}{s} - \frac{\frac{1}{4}s}{s^2 + 4} \quad \text{and}$$

$$\frac{1}{(s^2 + 1)(s^2 + 4)} = \frac{\frac{1}{3}}{s^2 + 1} - \frac{\frac{1}{3}}{s^2 + 4}$$

and substituting these into the inverse Laplace transforms gives

$$y(t) = \frac{1}{2}\sin 2t + \frac{1}{4} - \frac{1}{4}\cos 2t - \frac{1}{4}h(t - \pi)(1 - \cos 2(t - \pi))$$

$$+ \frac{1}{3}h(t - \pi)\left(\sin(t - \pi) - \frac{1}{2}\sin 2(t - \pi)\right).$$

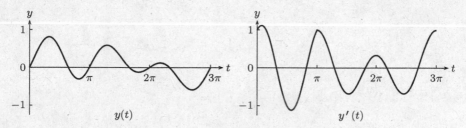

Fig. 6.14 The graph of the solution $y(t)$ to Example 8 is shown along with the graph of the derivative $y'(t)$. Note that the derivative $y'(t)$ is continuous, but it is not differentiable at $t = \pi$, signifying that the second derivative $y''(t)$ is not continuous at $t = \pi$

Evaluating this piecewise gives

$$y(t) = \begin{cases} \frac{1}{2}\sin 2t + \frac{1}{4} - \frac{1}{4}\cos 2t & \text{if } 0 \leq t < \pi, \\ \\ -\frac{1}{3}\sin t + \frac{1}{3}\sin 2t & \text{if } t \geq \pi. \end{cases}$$

The graph of the solution $y(t)$ and its derivative $y'(t)$ are given in Fig. 6.14. ◄

Exercises

1–12. Solve each of the following initial value problems.

1. $y' + 2y = \begin{cases} 0 & \text{if } 0 \le t < 1 \\ -3 & \text{if } t \ge 1 \end{cases}$, $y(0) = 0$

2. $y' + 5y = \begin{cases} -5 & \text{if } 0 \le t < 1 \\ 5 & \text{if } t \ge 1 \end{cases}$, $y(0) = 1$

3. $y' - 3y = \begin{cases} 0 & \text{if } 0 \le t < 2 \\ 2 & \text{if } 2 \le t < 3 , \\ 0 & \text{if } t \ge 3 \end{cases}$ $y(0) = 0$

4. $y' + 2y = \begin{cases} t & \text{if } 0 \le t < 1 \\ 0 & \text{if } 1 \le t < \infty \end{cases}$, $y(0) = 0$

5. $y' - 4y = \begin{cases} 12e^t & \text{if } 0 \le t < 1 \\ 12e & \text{if } 1 \le t < \infty \end{cases}$, $y(0) = 2$

6. $y' + 3y = \begin{cases} 10 \sin t & \text{if } 0 \le t < \pi \\ 0 & \text{if } \pi \le t < \infty \end{cases}$, $y(0) = -1$

7. $y'' + 9y = h(t - 3), \quad y(0) = 0, \, y'(0) = 0$

8. $y'' - 5y' + 4y = \begin{cases} 1 & \text{if } 0 \le t < 5 \\ 0 & \text{if } t \ge 5 \end{cases}$, $y(0) = 0, \, y'(0) = 1$

9. $y'' + 5y' + 6y = \begin{cases} 0 & \text{if } 0 \le t < 1 \\ 6 & \text{if } 1 \le t < 3 , \\ 0 & \text{if } t \ge 3 \end{cases}$ $y(0) = 0, \, y'(0) = 0$

10. $y'' + 9y = h(t - 2\pi) \sin t, \quad y(0) = 1, y'(0) = 0$

11. $y'' + 2y' + y = h(t - 3), \quad y(0) = 0, y'(0) = 1$

12. $y'' + 2y' + y = \begin{cases} e^{-t} & \text{if } 0 \le t < 4 \\ 0 & \text{if } 4 \le t < \infty \end{cases}$, $y(0) = 0, y'(0) = 0$

13–15. *Mixing Problems*

13. Suppose a tank holds 4 gal of pure water. There are two input sources of brine solution: the first source has a concentration of 1 lb of salt per gallon while the second source has a concentration of 5 lbs of salt per gallon. The first source flows into the tank at a rate of 2 gal/min for 3 min after which it is turned off, and simultaneously, the second source is turned on at a rate of 2 gal/min. The well-mixed solution flows out of the tank at a rate of 2 gal/min. Find the amount of salt in the tank at any time t.

14. Suppose a tank holds a brine solution consisting of 1 kg salt dissolved in 4 L of water. There are two input sources of brine solution: the first source has a concentration of 2 kg of salt per liter while the second source has a concentration of 3 kg of salt per liter. The first source flows into the tank at a rate of 4 L/min for 5 min after which it is turned off, and simultaneously, the second source is turned on at a rate of 4 L/min. The well-mixed solution flows out of the tank at a rate of 4 L/min. Find the amount of salt in the tank at any time t.

15. Suppose a tank holds a brine solution consisting of 2 kg salt dissolved in 10 L of water. There are two input sources of brine solution: the first source has a concentration of 1 kg of salt per liter while the second source is pure water. The first source flows into the tank at a rate of 3 L/min for 2 min. Thereafter, it is turned off and simultaneously the second source is turned on at a rate of 3 L/min for 2 min. Thereafter, it is turned off and simultaneously the first source is turned back on at a rate of 3 L/min and remains on. The well-mixed solution flows out of the tank at a rate of 3 L/min. Find the amount of salt in the tank at any time t.

16. Suppose $a \neq 0$. Show that the solution to

$$y' + ay = A\chi_{[\alpha,\beta)}, \ y(0) = y_0$$

is

$$y(t) = y_0 e^{-at} + \frac{A}{a} \begin{cases} 0 & \text{if } 0 \leq t < \alpha, \\ 1 - e^{-a(t-\alpha)} & \text{if } \alpha \leq t < \beta, \\ e^{-a(t-\beta)} - e^{-a(t-\alpha)} & \text{if } \beta \leq t < \infty. \end{cases}$$

6.4   The Dirac Delta Function

In applications, we may encounter an input into a system we wish to study that is very large in magnitude, but applied over a short period of time. Consider, for example, the following mixing problem:

Example 1. A tank holds 10 gal of a brine solution in which each gallon contains 2 lbs of dissolved salt. An input source begins pouring fresh water into the tank at a rate of 1 gal/min and the thoroughly mixed solution flows out of the tank at the same rate. After 5 min, 3 lbs of salt are poured into the tank where it instantly mixes into the solution. Find the amount of salt in the tank at any time t.

This example introduces a sudden action, namely, the sudden input of 3 lbs of salt at time $t = 5$ min. If we imagine that it actually takes 1 second to do this, then the average rate of input of salt would be 3 lbs/s = 180 lbs/min. Thus, we see a high magnitude in the rate of input of salt over a short interval. Moreover, the rate multiplied by the duration of input gives the total input.

More generally, if $r(t)$ represents the rate of input over a time interval $[a, b]$, then $\int_a^b r(t)\,dt$ would represent the total input. A unit input means that this integral is 1. Let $t = c \geq 0$ be fixed and let ϵ be a small positive number. Imagine a constant input rate over the interval $[c, c + \epsilon)$ and 0 elsewhere. The function $d_{c,\epsilon} = \frac{1}{\epsilon}\chi_{[c,c+\epsilon)}$ represents such an input rate with constant input $1/\epsilon$ over the interval $[c, c + \epsilon)$ (cf. Sect. 6.2 where the on–off switch $\chi_{[a,b)}$ is discussed). The constant $1/\epsilon$ is chosen so that the total input is

$$\int_0^\infty d_{c,\epsilon}\,dt = \frac{1}{\epsilon}\int_c^{c+\epsilon} 1\,dt = \frac{1}{\epsilon}\epsilon = 1.$$

For example, if $\epsilon = \frac{1}{60}$ min, then $3d_{5,\epsilon}$ would represent the input of 3 lbs of salt over a 1-s interval beginning at $t = 5$.

Figure 6.15 shows the graphs of $d_{c,\epsilon}$ for $\epsilon = 2, 0.5$, and 0.1. The area of the region under each line segment is 1. The main idea will be to take smaller and smaller values of ϵ, that is, we want to imagine the total input being concentrated at the point c. We would like to define the **Dirac delta function** by $\delta_c(t) = \lim_{\epsilon \to 0^+} d_{c,\epsilon}(t)$. However, the pointwise limit would give

$$\delta_c(t) = \begin{cases} \infty & \text{if } t = c, \\ 0 & \text{elsewhere.} \end{cases}$$

In addition, we would like to have the property that

$$\int_0^\infty \delta_c(t)\,dt = \lim_{\epsilon \to 0}\int_0^\infty d_{c,\epsilon}\,dt = 1.$$

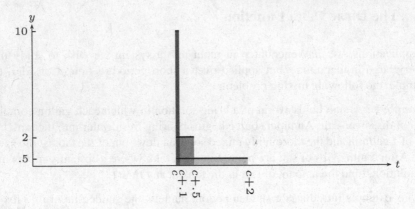

Fig. 6.15 Approximation to a delta function

Of course, there is really no such function with this property. (Mathematically, we can make precise sense out of this idea by extending the Heaviside class to a class that includes **distributions** or **generalized functions**. We will not pursue distributions here as it will take us far beyond the introductory nature of this text.) Nevertheless, this is the idea we want to develop. We will consider first order constant coefficient differential equations of the form

$$y' + ay = f(t),$$

where f involves the Dirac delta function δ_c. It turns out that the main problem lies in the fact that the solution is *not* continuous, so Theorem 3 of Sect. 6.3 does not apply. Nevertheless, we will justify that we can apply the usual Laplace transform method in a formal way to produce the desired solutions. The beauty of doing this is found in the ease in which we can work with the "Laplace transform" of δ_c.

We define the Laplace transform of δ_c by the formula:

$$\mathcal{L}\{\delta_c\} = \lim_{\epsilon \to 0} \mathcal{L}\{d_{c,\epsilon}\}.$$

Theorem 2. *The Laplace transform of δ_c is*

$$\mathcal{L}\{\delta_c\} = e^{-cs}.$$

Proof. We begin with $d_{c,\epsilon}$:

$$\mathcal{L}\{d_{c,\epsilon}\} = \frac{1}{\epsilon}\mathcal{L}\{h(t-c) - h(t-c-\epsilon)\}$$

$$= \frac{1}{\epsilon}\left(\frac{e^{-cs} - e^{-(c+\epsilon)s}}{s}\right)$$

$$= \frac{e^{-cs}}{s}\left(\frac{1 - e^{-\epsilon s}}{\epsilon}\right).$$

We now take limits as ϵ goes to 0 and use L'Hospital's rule to obtain:

$$\mathcal{L}\{\delta_c\} = \lim_{\epsilon \to 0} \mathcal{L}\{d_{c,\epsilon}\} = \frac{e^{-cs}}{s} \left(\lim_{\epsilon \to 0} \frac{1 - e^{-\epsilon s}}{\epsilon} \right) = \frac{e^{-cs}}{s} \cdot s = e^{-cs}. \qquad \square$$

We remark that when $c = 0$ we have $\mathcal{L}\{\delta_0\} = 1$. By Proposition 1 of Sect. 6.2, there is no Heaviside function with this property. Thus, to reiterate, even though $\mathcal{L}\{\delta_c\}$ is a function, δ_c is *not*. We will frequently write $\delta_c(t) = \delta(t - c)$ and $\delta = \delta_0$.

The Dirac delta function allows us to model the mixing problem from Example 1 as a first order linear differential equation.

▶ **Solution.** *Setting up the differential equation in Example 1*: Let $y(t)$ be the amount of salt in the tank at time t. Then $y(0) = 20$ and y' is the difference of the input rate and the output rate. The only input of salt occurs at $t = 5$. If the salt were input over a small interval, $[5, 5 + \epsilon)$ say, then $\frac{3}{\epsilon} \chi_{[5,5+\epsilon)}$ would represent the input of 3 lbs of salt over a period of ϵ minutes. If we let ϵ go to zero, then $3\delta_5$ would represent the input rate. The output rate is $y(t)/10$. We are thus led to the differential equation:

$$y' + \frac{y}{10} = 3\delta_5, \qquad y(0) = 20.$$

The solution to this differential equation will continue below and fall out of the slightly more general discussion we now give.   ◀

Differential Equations of the Form $y' + ay = k\delta_c$

We will present progressively three procedures for solving

$$y' + ay = k\delta_c, \qquad y(0) = y_0. \tag{1}$$

The last one, the *formal Laplace transform method*, is the simplest and is, in part, justified by the methods that precede it. The formal method will thereafter be used to solve (1) and will be extended to second order equations.

Limiting Procedure

In our first approach, we solve the equation

$$y' + ay = \frac{k}{\epsilon} \chi_{[c,c+\epsilon)}, \qquad y(0) = y_0$$

and call the solution y_ϵ. Since $\delta_c = \lim_{\epsilon \to 0} \frac{1}{\epsilon} \chi_{[c,c+\epsilon)}$, we let $y(t) = \lim_{\epsilon \to 0} y_\epsilon$. Then $y(t)$ will be the solution to $y' + ay = k\delta_c$, $y(0) = y_0$. We will assume $a \neq 0$ and leave the case $a = 0$ to the reader. Recall from Exercise 16 of Sect. 6.3 the solution to

$$y' + ay = A\chi[\alpha, \beta], \quad y(0) = y_0,$$

is

$$y(t) = y_0 e^{-at} + \frac{A}{a} \begin{cases} 0 & \text{if } 0 \leq t < \alpha, \\ 1 - e^{-a(t-\alpha)} & \text{if } \alpha \leq t < \beta, \\ e^{-a(t-\beta)} - e^{-a(t-\alpha)} & \text{if } \beta \leq t < \infty. \end{cases}$$

We let $A = \frac{k}{\epsilon}$, $\alpha = c$, and $\beta = c + \epsilon$ to get

$$y_\epsilon(t) = y_0 e^{-at} + \frac{k}{a\epsilon} \begin{cases} 0 & \text{if } 0 \leq t < c, \\ 1 - e^{-a(t-c)} & \text{if } c \leq t < c + \epsilon, \\ e^{-a(t-c-\epsilon)} - e^{-a(t-c)} & \text{if } c + \epsilon \leq t < \infty. \end{cases}$$

The computation of $\lim_{\epsilon \to 0} y_\epsilon$ is done on each interval separately. If $0 \leq t \leq c$, then $y_\epsilon = y_0 e^{-at}$ is independent of ϵ, and hence,

$$\lim_{\epsilon \to 0} y_\epsilon(t) = y_0 e^{-at} \qquad 0 \leq t \leq c.$$

If $c < t < \infty$, then for ϵ small enough, $c + \epsilon < t$ and thus

$$y_\epsilon(t) = y_0 e^{-at} + \frac{k}{a\epsilon}(e^{-a(t-c-\epsilon)} - e^{-a(t-c)}) = y_0 e^{-at} + \frac{k}{a} e^{-a(t-c)} \frac{e^{a\epsilon} - 1}{\epsilon}.$$

Since $\lim_{t \to \epsilon} \frac{e^{a\epsilon} - 1}{\epsilon} = a$, we get

$$\lim_{\epsilon \to 0} y_\epsilon(t) = y_0 e^{-at} + k e^{-a(t-c)}, \qquad c < t < \infty.$$

We thus obtain

$$y(t) = \begin{cases} y_0 e^{-at} & \text{if } 0 \leq t \leq c, \\ y_0 e^{-at} + k e^{-a(t-c)} & \text{if } c < t < \infty. \end{cases}$$

Observe that there is a jump discontinuity in $y(t)$ at $t = c$ with jump k.

Extension of Input Derivative Principle

In this method, we want to focus on the differential equation, $y' + ay = 0$, on the entire interval $[0, \infty)$ with the a priori knowledge that there is a jump discontinuity in $y(t)$ at $t = c$ with jump k. Recall from Theorem 3 of Sect. 6.3 that when y is continuous and both y and y' are in $\mathcal{H}$, we have the formula

$$\mathcal{L}\{y'\}(s) = sY(s) - y(0).$$

We cannot apply this theorem as stated for y is not continuous. But if y has a single jump discontinuity at $t = c$, we can prove a slight generalization.

Theorem 3. *Suppose y and y' are in $\mathcal{H}$ and y is continuous except for one jump discontinuity at $t = c$ with jump k. Then*

$$\mathcal{L}\{y'\}(s) = sY(s) - y(0) - ke^{-cs}.$$

Proof. Let $N > c$. Then integration by parts gives

$$\int_0^N e^{-st} y'(t)\,dt = \int_0^c e^{-st} y'(t)\,dt + \int_c^N e^{-st} y'(t)\,dt$$

$$= e^{-st} y(t)\big|_0^c + s \int_0^c e^{-st} y(t)\,dt$$

$$+ e^{-st} y(t)\big|_c^N + s \int_c^N e^{-st} y(t)\,dt$$

$$= s \int_0^N e^{-st} y(t)\,dt + e^{-sN} y(N) - y(0)$$

$$- e^{-sc}(y(c^+) - y(c^-)).$$

We take the limit as N goes to infinity and obtain

$$\mathcal{L}\{y'\} = s\mathcal{L}\{y\} - y(0) - ke^{-cs},$$

where $k = y(c^+) - y(c^-)$ is the jump of $y(t)$ at $t = c$. $\qquad\square$

We apply this theorem to the initial value problem

$$y' + ay = 0, \qquad y(0) = y_0$$

with the knowledge that the solution y has a jump discontinuity at $t=c$ with jump k. Apply the Laplace transform to the differential equation to obtain

$$sY(s) - y(0) - ke^{-cs} + aY(s) = 0.$$

Solving for Y gives

$$Y(s) = \frac{y_0}{s+a} + k\frac{e^{-cs}}{s+a}.$$

Applying the inverse Laplace transform gives the solution

$$y(t) = y_0 e^{-at} + k e^{-a(t-c)} h(t-c)$$

$$= \begin{cases} y_0 e^{-at} & \text{if } 0 \le t < c, \\ y_0 e^{-at} + k e^{-a(t-c)} & \text{if } c \le t < \infty. \end{cases}$$

The Formal Laplace Transform Method

We now return to the differential equation

$$y' + ay = k\delta_c, \qquad y(0) = y_0$$

and apply the Laplace transform method directly. From Theorem 2, the Laplace transform of $k\delta_c$ is ke^{-cs}. This is precisely the term found in Theorem 3 where the assumption of a single jump discontinuity is assumed. Thus, the presence of $k\delta_c$ automatically encodes the jump discontinuity in the solution. Therefore, we can (formally) proceed without any advance knowledge of jump discontinuities. The Laplace transform of

$$y' + ay = k\delta_c, \qquad y(0) = y_0$$

gives

$$sY(s) - y(0) + aY(s) = ke^{-cs}.$$

Solving for $Y(s)$, we get

$$Y(s) = \frac{y_0}{s+a} + k\frac{e^{-cs}}{s+a}.$$

Apply the inverse Laplace transform as above to get

$$y(t) = \begin{cases} y_0 e^{-at} & \text{if } 0 \le t < c, \\ y_0 e^{-at} + k e^{-a(t-c)} & \text{if } c \le t < \infty. \end{cases}$$

Observe that the same result is obtained in each procedure and justifies the formal Laplace transform method, which is thus the preferred method to use. We now use this method to solve the mixing problem given in Example 1.

▶ **Solution.** We apply the Laplace transform method to

$$y' + \frac{y}{10} = 3\delta_5, \qquad y(0) = 20 .$$

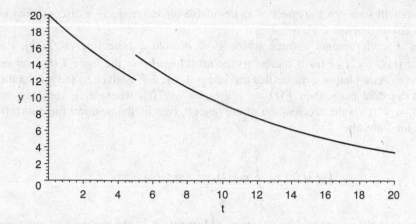

Fig. 6.16 Graph of the solution to the mixing problem

to get

$$sY(s) - 20 + \frac{1}{10}Y(s) = 3e^{-5s}.$$

Solving for $Y(s)$ gives

$$Y(s) = \frac{20}{s + \frac{1}{10}} + \frac{3e^{-5s}}{s + \frac{1}{10}}$$

and Laplace inversion gives

$$y(t) = 20e^{-t/10} + 3e^{-(t-5)/10}h(t-5)$$

$$= \begin{cases} 20e^{-t/10} & \text{if } 0 \le t \le 5, \\ 20e^{-t/10} + 3e^{-(t-5)/10} & \text{if } 5 < t < \infty. \end{cases}$$

The graph is given in Fig. 6.16 where the jump of 3 is clearly seen at $t = 5$. ◄

Impulse Functions

An impulsive force is a force with high magnitude introduced over a short period of time. For example, a bat hitting a ball or a spike in electricity on an electric circuit both involve impulsive forces and are best represented by the Dirac delta function. We consider the effect of the introduction of impulsive forces into spring systems and how they lead to second order differential equations of the form

$$my'' + \mu y' + ky = K\delta_c(t).$$

As we will soon see, the effect of an impulsive force introduces a discontinuity not in y but its derivative y'.

If $F(t)$ represents a force which is 0 outside a time interval $[a, b]$, then $\int_0^\infty F(t)\,dt = \int_a^b F(t)\,dt$ represents the **total impulse** of the force $F(t)$ over that interval. A unit impulse means that this integral is 1. If F is given by the acceleration of a constant mass, then $F(t) = ma(t) = my''(t)$, where m is the mass and $a(t) = y''(t)$ is the acceleration of the object given by the position function $y(t)$. The total impulse

$$\int\limits_a^b F(t)\,dt = \int\limits_a^b ma(t)\,dt = my'(b) - my'(a)$$

represents the change of momentum. (Momentum is the product of mass and velocity). Now imagine a constant force is introduced over a very short period of time with unit impulse. We then model the force by $d_{c,\epsilon} = \frac{1}{\epsilon}\chi_{[c,c+\epsilon)}$. Letting ϵ go to 0 then leads to the Dirac delta function δ_c to represent the instantaneous change of momentum. Since momentum is proportional to velocity, we see that such impacts lead to discontinuities in the derivative y'.

Example 4 (See Sect. 3.6 for a discussion of spring-mass-dashpot systems). A spring is stretched 49 cm when a 1 kg mass is attached. The body is pulled to 10 cm below its spring-body equilibrium and released. We assume the system is frictionless. After 3 s, the mass is suddenly struck by a hammer in a downward direction with total impulse of 1 kg·m/s. Find the motion of the mass.

▶ **Solution.** *Setting up the differential equation*: We will work in units of kg, m, and s. Thus, the spring constant k is given by $1(9.8) = k\frac{49}{100}$, so that $k = 20$. An external force to the system occurs as an impulse at $t = 3$ which may be represented by the Dirac delta function δ_3. The initial conditions are given by $y(0) = 0.10$ and $y'(0) = 0$, and since the system is frictionless, the initial value problem is

$$y'' + 20y = \delta_3, \qquad y(0) = 0.10, \quad y'(0) = 0.$$

We will return to the solution of this problem after we discuss the more general second order case. ◀

Equations of the Form $y'' + ay' + by = K\delta_c$

Our goal is to solve

$$y'' + ay' + by = K\delta_c, \qquad y(0) = y_0, \quad y'(0) = y_1 \tag{2}$$

using the formal Laplace transform method discussed above for first order differential equations. As we discussed, the effect of $K\delta_c$ is to introduce a single jump discontinuity in y' at $t = c$ with jump K. Therefore, the solution to (2) is equivalent to solving

$$y'' + ay' + by = 0$$

with the advanced knowledge that y' has a jump discontinuity at $t = c$. If we apply Theorem 3 to y', we obtain

$$\mathcal{L}\{y''\} = s\mathcal{L}\{y'\} - y'(0) - Ke^{-sc}$$
$$= s^2 Y(s) - sy(0) - y'(0) - Ke^{-sc}.$$

Therefore, the Laplace transform of $y'' + ay' + by = 0$ leads to

$$(s^2 + as + b)Y(s) - sy(0) - y'(0) - Ke^{-sc} = 0.$$

On the other hand, if we (formally) proceed with the Laplace transform of (2) without foreknowledge of discontinuities in y', we obtain the equivalent equation

$$(s^2 + as + b)Y(s) - sy(0) - y'(0) = Ke^{-sc}.$$

Again, the Dirac function δ_c encodes the jump discontinuity automatically. If we proceed as usual, we obtain

$$Y(s) = \frac{sy(0) + y'(0)}{s^2 + as + b} + \frac{Ke^{-sc}}{s^2 + as + b}.$$

The inversion will depend on the way the characteristic polynomial factors.

We now return to Example 4. The equation we wish to solve is

$$y'' + 20y = \delta_3, \qquad y(0) = 0.10, \quad y'(0) = 0.$$

▶ **Solution.** We apply the formal Laplace transform to obtain

$$Y(s) = \frac{0.1s}{s^2 + 20} + \frac{e^{-3s}}{s^2 + 20}.$$

The inversion gives

$$y(t) = \frac{1}{10}\cos\left(\sqrt{20}\,t\right) + \frac{1}{\sqrt{20}}\sin\left(\sqrt{20}\,(t-3)\right)h(t-3)$$

$$= \frac{1}{10}\cos\left(\sqrt{20}\,t\right) + \begin{cases} 0 & \text{if } 0 \le t < 3, \\ \frac{1}{\sqrt{20}}\sin\left(\sqrt{20}\,(t-3)\right) & \text{if } 3 \le t < \infty. \end{cases}$$

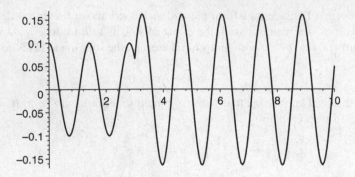

Fig. 6.17 Harmonic motion with impulse function

Figure 6.17 gives the graph of the solution. You will note that y is continuous, but the little kink at $t = 3$ indicates the discontinuity of y'. This is precisely when the impulse to the system was delivered.   ◄

Exercises

1–10. Solve each of the following initial value problems.

1. $y' + 2y = \delta_1(t), \quad y(0) = 0$
2. $y' - 3y = 3 + \delta_2(t), \quad y(0) = -1$
3. $y' - 4y = \delta_4(t), \quad y(0) = 2$
4. $y' + y = \delta_1(t) - \delta_3(t), \quad y(0) = 0$
5. $y'' + 4y = \delta_\pi(t), \quad y(0) = 0, \, y'(0) = 1$
6. $y'' - y = \delta_1(t) - \delta_2(t), \quad y(0) = 0, \, y'(0) = 0$
7. $y'' + 4y' + 3y = 2\delta_2(t), \quad y(0) = 1, \, y'(0) = -1$
8. $y'' + 4y = \delta_\pi(t) - \delta_{2\pi}(t), \quad y(0) = 1, \, y'(0) = 0$
9. $y'' + 4y' + 4y = 3\delta_1(t), \quad y(0) = -1, \, y'(0) = 3$
10. $y'' + 4y' + 5y = 3\delta_\pi(t), \quad y(0) = 0, \, y'(0) = 1$

11–13. *Mixing Problems.*

11. Suppose a tank is filled with 12 gal of pure water. A brine solution with concentration 2 lbs salt per gallon flows into the tank at a rate of 3 gal/min and the well-stirred solution flows out of the tank at the same rate. In addition, at $t = 3$ min, 4 lbs of salt is instantly poured into the tank where it immediately dissolves. Find the amount of salt, $y(t)$, in the tank at any time t.

12. A tank holds 10 L of a brine solution in which each liter contains 1 kg of dissolved salt. A brine solution with concentration 0.5 kg/L is poured into the tank at a rate of 2 L/min, and the thoroughly mixed solution flows out of the tank at the same rate. After 2 min, 1 L of salt is poured into the tank where it instantly mixes into the solution. Find the amount of salt, $y(t)$, in the tank at any time t.

13. A tank holds 1 gal of pure water. Pure water flows into the tank at a rate of 1 gal/min, and the well-stirred mixture flows out of the tank at the same rate. At $t = 0, 2, 4, 6$ min, 1 lb of salt is instantly added to the tank where it immediately dissolves. Find the amount, $y(t)$, of salt in the tank at time t. How much salt is in the tank just after the last addition at $t = 6$ min?

14–17. *Spring Problems*

14. A spring is stretched 6 in. when a 1-lb object is attached. The 1-lb object is pulled to 12 in. below its spring-body equilibrium and released. We assume the system is frictionless. After 5π seconds, the mass is suddenly struck by a hammer in a downward direction with total impulse of 1 slug ·ft/s. Find the motion of the object. Determine the amplitudes before and after the hammer impact.

15. A spring is stretched 1 m by a force of 8 N. A body of mass 2 kg is attached to the spring with accompanying dashpot. Suppose the damping force of the dashpot is 8 N when the velocity of the body is 1 m/s. At $t = 0$, the mass is pulled down from its equilibrium position a distance of 10 cm and given an

initial downward velocity of 5 cm/s. After 4 s, the mass is suddenly struck by a hammer in a downward direction with total impulse of 2 kg·m/s. Determine the resulting motion.

16. A 1-Newton force will stretch a spring 1 m. A body of mass 1 kg is attached to the spring and allowed to come to rest. There is no dashpot. At time $t = 0, 2\pi, 4\pi, 6\pi$, a hammer impacts the mass in the downward direction with a magnitude of 1 kg m/s. Find the equation of motion and provide a graph on the interval $[0, 10\pi)$.

17. A 1-Newton force will stretch a spring 1 m. A body of mass 1 kg is attached to the spring and allowed to come to rest. There is no dashpot. At time $t = 0, \pi, 2\pi, 3\pi, 4\pi, 5\pi$, a hammer impacts the mass in the downward direction with a magnitude of 1 kg m/s. Find the equation of motion and provide a graph on the interval $[0, 6\pi)$. Explain.

18–20. In these problems, we justify the Laplace transform method for solving $y' + ay = k\delta_c$, $y(0) = y_0$ in a way different from the limiting procedure and the extension of the input derivative principle introduced in the text.

18. On the interval $[0, c)$, solve $y' + ay = 0$ with initial value $y(0) = y_0$.
19. On the interval $[c, \infty)$, solve $y' + ay = 0$ with initial value $y(c) + k$, where $y(c)$ is the value of $y(t)$ at $t = c$ obtained from Exercise 18 for the interval $[0, c]$.
20. Piece together the solutions obtained from Exercise 18 for the interval $[0, c)$ and from Exercise 19 for the interval $[c, \infty)$ and verify that it is the same obtained from the formal Laplace transform method.

6.5 Convolution

In this section, we extend to the Heaviside class the definition of convolution that
we introduced in Sect. 2.8. The importance of convolution is that it is precisely the
operation in input space that corresponds via the Laplace transform to the ordinary
product in transform space. This is the essence of the convolution principle stated
in Theorem 1 of Sect. 2.8 and which will be proved here for the Heaviside class.
We will then consider further extensions to the Dirac delta functions δ_c and explore
some very pleasant properties.

Given two functions f and g in $\mathcal{H}$, the function

$$u \mapsto f(u)g(t-u)$$

is continuous except for perhaps finitely many points on each interval of the form
$[0, t]$. Therefore, the integral

$$\int_0^t f(u)g(t-u)\,du$$

exists for each $t > 0$. The **convolution** of f and g is given by

$$f * g(t) = \int_0^t f(u)g(t-u)\,du.$$

We will not make the argument but it can be shown that $f * g$ is in fact continuous.
Since there are numbers K, L, a, and b such that

$$|f(t)| \le Ke^{at} \quad \text{and} \quad |g(t)| \le Le^{bt},$$

it follows that

$$|f * g(t)| \le \int_0^t |f(u)|\,|g(t-u)|\,du$$

$$\le KL \int_0^t e^{au}e^{b(t-u)}\,du$$

$$= KLe^{bt} \int_0^t e^{(a-b)u}\,du$$

$$= KL \begin{cases} te^{bt} & \text{if } a = b \\ \frac{e^{at}-e^{bt}}{a-b} & \text{if } a \ne b \end{cases}.$$

This shows that $f * g$ is of exponential type since both te^{bt} and $\frac{e^{at}-e^{bt}}{a-b}$ are exponential polynomials. It follows now that $f * g \in \mathcal{H}$.

Several important properties we listed in Sect. 2.5 extend to $\mathcal{H}$. For convenience, we restate them here: Suppose f, g, and h are in $\mathcal{H}$. Then

1. $(f + h) * g = f * g + h * g$
2. $(cf) * g = c(f * g)$, for c a scalar
3. $f * g = g * f$
4. $(f * g) * h = f * (g * h)$
5. $f * 0 = 0$.

The Sliding Window

When one of the functions, g say, is an on–off switch, then convolution takes a particularly simple form. Suppose $g = \chi_{[a,b)}$. Then $g(t) = 1$ if and only if $a \leq t < b$. Hence, $g(t - u) = 1$ if and only if $a \leq t - u < b$ which is equivalent to $t - b < u \leq t - a$. Thus, $g(t - u) = \chi_{(t-b,t-a]}(u)$ is an on–off switch on the interval $(t - b, t - a]$ which slides to the right as t increases. It follows that

$$f * g(t) = \int_0^t f(u)\chi_{(t-b,t-a]}(u)\,du.$$

One can think of $g(t - u)$ as a horizontally *sliding window* by which a portion of f is turned on. That portion is then measured by integration. To illustrate this, consider the following example.

Example 1. Let $f(t) = (t - 3)h(t - 3)$ and $g(t) = \chi_{[1,2)}(t)$. Find the convolution $f * g$.

▶ **Solution.** In this case, $g(t - u) = \chi_{[1,2)}(t - u) = \chi_{(t-2,t-1]}(u)$. If $t < 4$, then $t - 2 < t - 1 < 3$ so $g(t - u)$ turns on that part of f which is 0. Thus,

$$f * g(t) = 0, \quad \text{if } 0 \leq t < 4,$$

as illustrated in Fig. 6.18.

If $4 \leq t < 5$, then $t - 2 < 3 \leq t - 1$. Thus, $g(t - u)$ turns of that part of f which is 0 on the interval $(t - 2, 3)$ and which is $u - 3$ on the interval $[3, t - 1)$. Thus,

$$f * g(t) = \int_{t-2}^{t-1} f(u)\,du = \int_3^{t-1} u - 3\,du$$

$$= \frac{(u - 3)^2}{2}\Big|_3^{t-1} = \frac{(t - 4)^2}{2}, \quad 4 \leq t < 5.$$

This is illustrated in Fig. 6.19.

Now when $t \geq 5$, then $3 \leq t - 2$. Thus, $g(t - u)$ turns on that portion of f which is $u - 3$ on the interval $(t - 2, t - 1]$ and

$$f * g(t) = \int_{t-2}^{t-1} (u - 3)\, du$$

$$= \left. \frac{(u - 3)^2}{2} \right|_{t-2}^{t-1} = \frac{2t - 9}{2}, \quad 5 \leq t < \infty.$$

This is illustrated in Fig. 6.20. Finally, we piece together the convolution to get

$$f * g = \begin{cases} 0 & \text{if } 0 \leq t < 4 \\ \frac{(t-2)^2}{2} & \text{if } 4 \leq t < 5 \\ \frac{2t-9}{2} & \text{if } 5 \leq t < \infty \end{cases} \tag{1}$$

The graph of $f * g$ is given in Fig. 6.21.   ◄

Theorem 2 (The Convolution Theorem).   *Suppose f and g are in $\mathcal{H}$ and F and G are their Laplace transforms, respectively. Then*

$$\mathcal{L}\{f * g\}(s) = F(s)G(s)$$

or, equivalently,

$$\mathcal{L}^{-1}\{F(s)G(s)\}(t) = (f * g)(t).$$

Proof. For any $f \in \mathcal{H}$, we will define $f(t) = 0$ for $t < 0$. By Theorem 8 of Sect. 6.2,

$$e^{-st}G(s) = \mathcal{L}\{g(u - t)h(u - t)\}.$$

Therefore,

$$F(s)G(s) = \int_0^\infty e^{-st} f(t)\, dt \; G(s)$$

$$= \int_0^\infty e^{-st} G(s) f(t)\, dt$$

$$= \int_0^\infty \mathcal{L}\{g(u - t)h(u - t)\}(s) f(t)\, dt$$

$$= \int_0^\infty \int_0^\infty e^{-su} g(u - t)h(u - t) f(t)\, du\, dt. \tag{2}$$

Fig. 6.18 When $t - 1 < 3$, the on–off switch $\chi_{(t-2,t-1]}$ turns on that portion of f that is zero. Hence, $f * g(t) = 0$ for all $0 \le t < 4$. Notice how the on–off switch slides to the right as t increases in the following figures

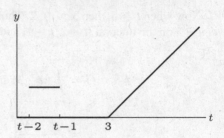

Fig. 6.19 When $4 < t < 5$, then $t - 2 < 3 < t - 1$ and the on–off switch $\chi_{(t-2,t-1]}$ turns on that portion of f that is zero to the left of 3 and the line $u - 3$ to the right of 3. Hence, $f * g(t) = \int_3^{t-1} (u - 3)\, du = \frac{(t-4)^2}{2}$ for $4 \le t < 5$

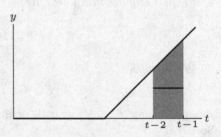

Fig. 6.20 When $5 \le t$, then $3 < t - 2 < t - 1$ and the on–off switch $\chi_{(t-2,t-1]}$ turns on that portion of f that is the line $u - 3$. Hence, $f * g(t) = \int_{t-2}^{t-1} (u - 3)\, du = \frac{2t-9}{2}$ for $5 \le t < \infty$

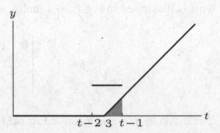

Fig. 6.21 The graph of the convolution $f * g(t) =$
$$\begin{cases} 0 & \text{if } 0 \le t < 4 \\ \frac{(t-2)^2}{2} & \text{if } 4 \le t < 5 \\ \frac{2t-9}{2} & \text{if } 5 \le t < \infty \end{cases}$$

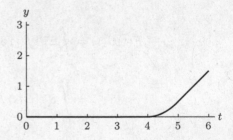

A theorem in calculus[1] tells us that we can switch the order of integration in (2) when f and g are in $\mathcal{H}$. Thus, we obtain

$$F(s)G(s) = \int_0^\infty \int_0^\infty e^{-su} g(u-t)h(u-t)f(t)\,dt\,du$$

$$= \int_0^\infty \int_0^u e^{-su} g(u-t)f(t)\,dt\,du$$

$$= \int_0^\infty e^{-su} (f * g)(u)\,du$$

$$= \mathcal{L}\{f * g\}(s). \qquad \qquad \square$$

There are a variety of uses for the convolution theorem. For one, it is sometimes a convenient way to compute the convolution of two functions f and g using the formula $(f * g)(t) = \mathcal{L}^{-1}\{F(s)G(s)\}$. In the following example, we rework Example 1 in this way.

Example 3. Compute the convolution $f * g$ where

$$f(t) = (t-3)h(t-3) \quad \text{and} \quad g(t) = \chi_{[1,2)}.$$

▶ **Solution.** The Laplace transforms of f and g are, respectively,

$$F(s) = \frac{e^{-3s}}{s^2} \quad \text{and} \quad G(s) = \frac{e^{-s}}{s} - \frac{e^{-2s}}{s}.$$

The product simplifies to

$$F(s)G(s) = \frac{e^{-4s}}{s^3} - \frac{e^{-5s}}{s^3}.$$

Its inverse Laplace transform is

$$(f * g)(t) = \mathcal{L}^{-1}\{F(s)G(s)\}(t)$$

$$= \frac{(t-4)^2}{2} h(t-4) - \frac{(t-5)^2}{2} h(t-5)$$

---

[1]cf. *Vector Calculus, Linear Algebra, and Differential Forms*, J.H. Hubbard and B.B Hubbard, page 444.

$$= \begin{cases} 0 & \text{if } 0 \le t < 4 \\ \frac{(t-2)^2}{2} & \text{if } 4 \le t < 5 \\ \frac{2t-9}{2} & \text{if } 5 \le t < \infty \end{cases} \quad\blacktriangleleft$$

Convolution and the Dirac Delta Function

We would like to extend the definition of convolution to include the Dirac delta functions δ_c, $c \ge 0$. Recall that we formally defined the Dirac delta function by

$$\delta_c(t) = \lim_{\epsilon \to 0} d_{c,\epsilon}(t),$$

where $d_{c,\epsilon} = \frac{1}{\epsilon}\chi_{[c,c+\epsilon)}$. In like manner, for $f \in \mathcal{H}$, we define

$$f * \delta_c(t) = \lim_{\epsilon \to 0} f * d_{c,\epsilon}(t).$$

Theorem 4. *For $f \in \mathcal{H}$,*

$$f * \delta_c(t) = f(t-c)h(t-c),$$

where the equality is understood to mean essentially equal.

Proof. Let $f \in \mathcal{H}$. Then

$$f * d_{c,\epsilon}(t) = \int_0^t f(u)d_{c,\epsilon}(t-u)\,dt$$

$$= \frac{1}{\epsilon} \int_0^t f(u)\chi_{[c,c+\epsilon)}(t-u)\,du$$

$$= \frac{1}{\epsilon} \int_0^t f(u)\chi_{[t-c-\epsilon,t-c)}(u)\,du$$

Now suppose $t < c$. Then $\chi_{[t-c-\epsilon,t-c)}(u) = 0$, for all $u \in [0,t)$. Thus, $f * d_{c,\epsilon} = 0$. On the other hand, if $t > c$, then for ϵ small enough, we have

$$f * d_{c,\epsilon}(t) = \frac{1}{\epsilon} \int_{t-c-\epsilon}^{t-c} f(u)\,du.$$

Let t be such that $t - c$ is a point of continuity of f. Then by the fundamental theorem of calculus

$$\lim_{\epsilon \to 0} \frac{1}{\epsilon} \int\limits_{t-c-\epsilon}^{t-c} f(u)\, du = f(t-c).$$

Since f has only finitely many jump discontinuities on any finite interval, it follows that $f * \delta_c$ is essentially equal to $f(t-c)h(t-c)$. $\square$

The special case $c = 0$ produces the following pleasant corollary.

Corollary 5. *For $f \in \mathcal{H}$, we have*

$$f * \delta_0 = f.$$

This corollary tells us that this extension to the Dirac delta function gives an identity under the convolution product. We thus have a correspondence between the multiplicative identities in input space and transform space under the Laplace transform since $\mathcal{L}\{\delta_0\} = 1$.

Remark 6. Notice that when $f(t) = 1$, in Theorem 4, we get

$$1 * \delta_c = h(t-c).$$

This can be reexpressed as

$$\int\limits_0^t \delta_c(u)\, du = h(t-c).$$

Thus, the integral of the Dirac delta function is the unit step function shifted by c. Put another way, the derivative of the unit step function, $h(t-c)$, is the Dirac delta function, δ_c.

The Impulse Response Function

Suppose $q(s)$ is a polynomial of degree n. The solution $\zeta(t)$ to the initial value problem

$$q(\mathbf{D})y = \delta_0 \quad y(0) = 0,\ y'(0) = 0,\ \ldots,\ y^{(n-1)}(0) = 0, \qquad (3)$$

is called the **unit impulse response function**. In the case where $\deg q(s) = 2$, the unit impulse response function may be viewed as the response to a mass-spring dashpot system initially at rest but hit with a hammer of impulse 1 at $t = 0$ as represented by the Dirac delta function δ_0. Applying the Laplace transform to (3), we get $Y(s) = \frac{1}{q(s)}$, and thus the unit impulse response function is given by the Laplace inversion formula

$$\zeta(t) = \mathcal{L}^{-1}\left\{\frac{1}{q(s)}\right\}.$$

Let $f \in \mathcal{H}$ (or f could be a Dirac delta function) and let us consider the general differential equation

$$q(D)y = f, \quad y(0) = y_0, \ y'(0) = y_1, \ y^{(n-1)}(0) = y_{n-1}. \tag{4}$$

Let $F(s) = \mathcal{L}\{f(t)\}$. Applying the Laplace transform to both sides gives

$$q(s)Y(s) - p(s) = F(s),$$

where $p(s)$ is a polynomial that depends on the initial conditions and has degree at most $n - 1$. Solving for $Y(s)$ gives

$$Y(s) = \frac{p(s)}{q(s)} + \frac{F(s)}{q(s)}. \tag{5}$$

Let

$$Y_h(s) = \frac{p(s)}{q(s)} \quad \text{and} \quad Y_p(s) = \frac{F(s)}{q(s)}.$$

If $y_h(t) = \mathcal{L}^{-1}\{Y_h(s)\}$, then y_h is the homogeneous solution to (4). Specifically, y_h is the solution to (4) when $f = 0$ but with the same initial conditions. We sometimes refer to y_h as the **zero-input solution**.

On the other hand, let $y_p(t) = \mathcal{L}^{-1}\{Y_p(s)\}$. Then y_p is the solution to (4) when all the initial conditions are zero, sometimes referred to as the **zero-state**. We refer to y_p as the **zero-state solution**. Since $Y(s) = \frac{1}{q(s)}F(s)$, we get by the convolution theorem

$$y_p(t) = \zeta * f(t). \tag{6}$$

This tells us that the solution to a system in the zero-state is completely determined by convolution of the input function with the unit impulse response function.

We summarize this discussion in the following theorem:

Theorem 7. *Let $f \in \mathcal{H}$ (or a linear combination of Dirac delta functions). The solution to (4) can be expressed as*

$$y_h + y_p,$$

where y_h is the zero-input solution, that is, the homogeneous solution to (4) (with the same initial conditions), and $y_p = \zeta * f$ is the zero-state solution given by convolution of the unit impulse response function ζ and the input function f.

Remark 8. In Exercises 15–21, we outline a proof that the homogeneous solution y_h can be expressed as a linear combination of

$$\{\zeta, \zeta', \ldots, \zeta^{(n-1)}\}$$

and provide formulas for the coefficients in terms of the initial conditions. The reader is encouraged to explore these exercises. It follows then that both y_h and y_p are directly determined by the unit impulse response function, ζ.

As an example of the techniques we have developed in this section, consider the differential equation that solved the mixing problem given in Example 1 of Sect. 6.4.

Example 9. Solve the following differential equation:

$$y' + \frac{1}{10}y = 3\delta_5 \quad y(0) = 20.$$

▶ **Solution.** The characteristic polynomial is $q(s) = s + 1/10$, and therefore, the unit impulse response function is $\zeta(t) = \mathcal{L}^{-1}\left\{\frac{1}{s+1/10}\right\} = e^{-1/10t}$. The zero-state solution is $y_p = \zeta * 3\delta_5 = 3e^{-(t-5)/10}h(t - 5)$ by Theorem 4. It is straightforward to see that the homogeneous solution is $y_h = 20e^{-1/10t}$. We thus get

$$y(t) = 20e^{-t/10} + 3e^{-(t-5)/10}h(t - 5)$$

$$= \begin{cases} 20e^{-t/10} & \text{if } 0 \leq t \leq 5, \\ 20e^{-t/10} + 3e^{-(t-5)/10} & \text{if } 5 < t < \infty. \end{cases}$$ ◀

Example 10. Solve the following differential equation:

$$y'' + 4y = \chi_{[0,1)} \quad y(0) = 0 \text{ and } y'(0) = 0.$$

▶ **Solution.** The homogeneous solution to

$$y'' + 4y = 0 \quad y(0) = 0 \text{ and } y'(0) = 0$$

is the trivial solution $y_h = 0$. The unit impulse response function is

$$\zeta(t) = \mathcal{L}^{-1}\left\{\frac{1}{s^2 + 4}\right\} = \frac{1}{2}\sin 2t.$$

By Theorem 7, the solution is

$$y(t) = \zeta * \chi_{[0,1)}$$

$$= \int_0^t \frac{1}{2}\sin(2u)\chi_{[0,1)}(t - u)\,du$$

$$= \frac{1}{2} \int_0^t \sin(2u)\chi_{(t-1,t]}(u)\,du$$

$$= \frac{1}{2} \begin{cases} \int_0^t \sin 2u\,du & \text{if } 0 \le t < 1 \\ \int_{t-1}^t \sin 2u\,du & \text{if } 1 \le t < \infty \end{cases}$$

$$= \frac{1}{4} \begin{cases} 1 - \cos 2t & \text{if } 0 \le t < 1 \\ \cos 2(t-1) - \cos 2t & \text{if } 1 \le t < \infty \end{cases}. \qquad \blacktriangleleft$$

In the following, we revisit Example 4 of Sect. 6.4.

Example 11. Solve the differential equation

$$y'' + 20y = \delta_3, \quad y(0) = 0.1, \ y'(0) = 0,$$

that models the spring problem given in Exercise 4 of Sect. 6.4.

▶ **Solution.** The characteristic polynomial is $q(s) = s^2 + 20$. So the homogeneous solution is $y_h = c_1 \cos \sqrt{20}t + c_2 \sin \sqrt{20}t$. The initial conditions easily imply that $c_1 = \frac{1}{10}$ and $c_2 = 0$. So $y_h = \frac{1}{10} \cos \sqrt{20}t$. The unit impulse response function is

$$\zeta(t) = \mathcal{L}^{-1}\left\{\frac{1}{s^2 + 20}\right\} = \frac{1}{\sqrt{20}} \sin \sqrt{20}t.$$

It follows from Theorem 4 and 7 that

$$y_p(t) = \zeta * \delta_3(t)$$

$$= \frac{1}{\sqrt{20}} (\sin \sqrt{20}(t-3))h(t-3).$$

It follows now that the solution to the spring problem in Example 4 of Sect. 6.4 is

$$y(t) = \frac{1}{10} \cos\left(\sqrt{20}\,t\right) + \frac{1}{\sqrt{20}} \sin\left(\sqrt{20}\,(t-3)\right) h(t-3)$$

$$= \frac{1}{10} \cos\left(\sqrt{20}\,t\right) + \begin{cases} 0 & \text{if } 0 \le t < 3, \\ \frac{1}{\sqrt{20}} \sin\left(\sqrt{20}\,(t-3)\right) & \text{if } 3 \le t < \infty. \end{cases} \qquad \blacktriangleleft$$

Exercises

1–8. Find the convolution of the following pairs of functions.

1. $f(t) = e^t$ and $g(t) = \chi_{[0,1)}(t)$
2. $f(t) = \sin t$ and $g(t) = h(t - \pi)$
3. $f(t) = t h(t - 1)$ and $g(t) = \chi_{[3,4)}(t)$
4. $f(t) = t$ and $g(t) = (t - 1)h(t - 1)$
5. $f(t) = \chi_{[0,2)}$ and $g(t) = \chi_{[0,2)}$
6. $f(t) = \cos t$ and $g(t) = \delta_{\pi/2}$
7. $f(t) = \sin t$ and $g(t) = \delta_0 + \delta_\pi$
8. $f(t) = t e^{2t}$ and $g(t) = \delta_1 - \delta_2$

9–14. Find the unit impulse response function ζ and use Theorem 7 to solve the following differential equations.

9. $y' - 3y = h(t - 2), \quad y(0) = 2$
10. $y' + 4y = \delta_3, \quad y(0) = 1$
11. $y' + 8y = \chi_{[3,5)}, \quad y(0) = -2$
12. $y'' - y = \delta_1(t) - \delta_2(t), \quad y(0) = 0, \ y'(0) = 0$
13. $y'' + 9y = \chi_{[0,2\pi)}, \quad y(0) = 1, y'(0) = 0.$
14. $y'' - 6y' + 9y = \delta_3, \quad y(0) = -1, \ y'(0) = -3$

15–21. Suppose $q(s) = a_n s^n + a_{n-1} s^{n-1} + \cdots + a_1 s + a_0, a_n \neq 0$. In these exercises, we explore properties of the unit impulse response function ζ for $q(D)y = \delta_0$.

15. Show that ζ is the solution to the homogeneous differential equation

$$q(D)y = 0$$

with initial conditions

$$y(0) = 0$$
$$y'(0) = 0$$
$$\vdots$$
$$y^{(n-2)}(0) = 0$$
$$y^{(n-1)}(0) = 1/a_n.$$

16. Show that $\mathcal{L}\{\zeta^{(k)}\} = \frac{s^k}{q(s)}, 0 \leq k < n$, and hence $\zeta^{(k)} \in \mathcal{E}_q$.

17. Show that $\{\zeta, \zeta', \ldots, \zeta^{(n-1)}\}$ is a linearly independent subset of $\mathcal{E}_q$.

18. Show that $\{\zeta, \zeta', \dots, \zeta^{(n-1)}\}$ spans $\mathcal{E}_q$.

19. Show that $\{\zeta, \zeta', \dots, \zeta^{(n-1)}\}$ is a basis of $\mathcal{E}_q$.

20. Show that the Wronskian of $\zeta, \zeta', \dots, \zeta^{(n-1)}$ is given by the following formula:

$$w(\zeta, \zeta', \dots, \zeta^{(n-1)}) = \frac{(-1)^{\lfloor \frac{n}{2} \rfloor}}{a_n^n} e^{\frac{-a_{n-1}}{a_n}t},$$

where $\lfloor x \rfloor$ is the largest integer less than or equal to x.

21. Let y be the solution to $q(D)y = 0$ with initial conditions

$$y(0) = y_0$$
$$y'(0) = y_1$$
$$\vdots$$
$$y^{(n-1)}(0) = y_{n-1}.$$

Since $y \in \mathcal{E}_q$, we may write

$$y = c_0 \zeta + c_1 \zeta' + \cdots + c_{n-1} \zeta^{(n-1)}.$$

Show that

$$c_l = \sum_{k=0}^{n-l-1} a_{k+l+1} y_k.$$

22–26. Use the result of Exercise 21 to find the homogeneous solution to each of the following differential equations. It may be helpful to organize the computation of the coefficients in the following way: Let $0 \leq l \leq n - 1$ and write

$$
\begin{array}{llllllll}
a_0 & a_1 & \cdots & a_l & a_{l+1} & a_{l+2} & \cdots & a_n \\
 & & & y_0 & y_1 & \cdots & y_{n-l-1} & \cdots & y_n \\
c_0 & c_1 & \cdots & c_l & \cdots & & &
\end{array}
$$

In the first row, put the coefficients of $q(s)$ starting with the constant coefficient on the left. In the second row, put the initial conditions with y_0 under a_{l+1}, y_1 under a_{l+2}, etc. Multiply terms that overlap in the first two rows and add. Put the result in c_l. Now shift the second row of initial conditions to the right one place and repeat to get c_{l+1}. Repeating this will give all the coefficients $c_0, \dots, c_{n-1}$ needed in $y = \sum_{l=0}^{n-1} c_l \zeta^{(l)}$.

22. $y'' + 9y = 0, \quad y(0) = 1, \; y'(0) = 2$
23. $y'' - 2y' + y = 0, \quad y(0) = 2, \; y'(0) = -3$
24. $y'' + 4y' + 3y = 0, \quad y(0) = -1, \; y'(0) = 1$
25. $y''' + y' = 0, \quad y(0) = 1, \; y'(0) = 0, \; y''(0) = 4$
26. $q(D)y = 0$ where $q(s) = (s-1)^4$ and $y(0) = 0, \; y'(0) = 1, \; y''(0) = 0$, and $y'''(0) = -1$

6.6   Periodic Functions

In modeling mechanical and other systems, it frequently happens that the forcing function repeats over time. Periodic functions best model such repetition.

A function f defined on $[0, \infty)$ is said to be **periodic** if there is a positive number p such that $f(t + p) = f(t)$ for all t in the domain of f. We say p is **a period** of f. If $p > 0$ is a period of f and there is no smaller period, then we say p is the **fundamental period** of f although we will usually just say **the period**. The interval $[0, p)$ is called the **fundamental interval**. If there is no such smallest positive p for a periodic function, then the period is defined to be 0. The constant function $f(t) = 1$ is an example of a periodic function with period 0. The sine function is periodic with period 2π: $\sin(t + 2\pi) = \sin(t)$. Knowing the sine on the interval $[0, 2\pi)$ implies knowledge of the function everywhere. Similarly, if we know f is periodic with period $p > 0$ and we know the function on the fundamental interval, then we know the function everywhere. Figure 6.22 illustrates this point.

The Sawtooth Function

A particularly useful periodic function is the **sawtooth** function. With it, we may express other periodic functions simply by composition. Let $p > 0$. The saw tooth function is given by

$$\langle t \rangle_{p} = \begin{cases} t & \text{if } 0 \leq t < p \\ t - p & \text{if } p \leq t < 2p \\ t - 2p & \text{if } 2p \leq t < 3p \\ \vdots \end{cases}.$$

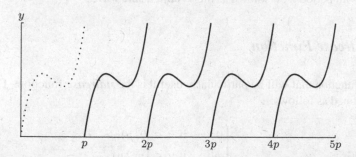

Fig. 6.22 An example of a periodic function with period p. Notice how the interval $[0, p)$ determines the function everywhere

Fig. 6.23 The sawtooth function $\langle t \rangle_p$ with period p

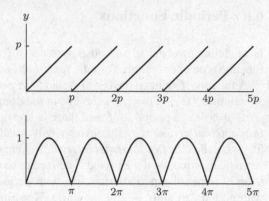

Fig. 6.24 The rectified sine wave: $\sin(\langle t \rangle_\pi)$

It is periodic with period p. Its graph is given in Fig. 6.23. The sawtooth function $\langle t \rangle_p$ is obtained by extending the function $y = t$ on the interval $[0, p)$ periodically to $[0, \infty)$. More generally, given a function f defined on the interval $[0, p)$, then the composition of f and $\langle t \rangle_p$ is the periodic extension of f to $[0, \infty)$. It is given by the formula

$$
f(\langle t \rangle_p) = \begin{cases} f(t) & \text{if } 0 \le t < p \\ f(t - p) & \text{if } p \le t < 2p \\ f(t - 2p) & \text{if } 2p \le t < 3p \\ \vdots & \vdots \end{cases}
$$

In applications, it is useful to rewrite this piecewise function as

$$
\sum_{n=0}^{\infty} f(t - np)\chi_{[np,(n+1)p)}(t).
$$

For example, Fig. 6.24 is the graph of $y = \sin(\langle t \rangle_\pi)$. This function, which is periodic with period π, is known as the *rectified sine wave*.

The Staircase Function

Another function that will be particularly useful is the *staircase* function. For $p > 0$, it is defined as follows:

$$
[t]_p = \begin{cases} 0 & \text{if } t \in [0, p) \\ p & \text{if } t \in [p, 2p) \\ 2p & \text{if } t \in [2p, 3p) \\ \vdots & \end{cases}
$$

Its graph is given in Fig. 6.25.

Fig. 6.25 The staircase
function: $[t]_p$

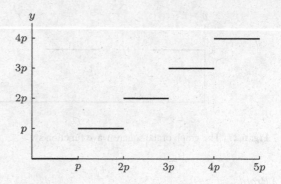

Fig. 6.26 The graph of
$1 - e^{-t}$ and $1 - e^{-[t]_s}$

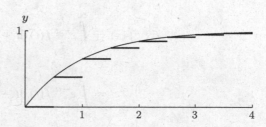

The staircase function is *not* periodic. It is useful in expressing piecewise
functions that are like steps on intervals of length p. For example, if f is a function
on $[0, \infty)$, then $f([t]_p)$ is a function whose value on $[np, (n + 1)p)$ is the constant
$f(np)$. Thus,

$$f([t]_p) = \sum_{n=0}^{\infty} f(np) \chi_{[np,(n+1)p)}(t).$$

Figure 6.26 illustrates this idea with the function $f(t) = 1 - e^{-t}$ and $p = 0.5$.
Observe that the staircase function and the sawtooth function are related by

$$< t >_p = t - [t]_p.$$

The Laplace Transform of Periodic Functions

Not surprisingly, the formula for the Laplace transform of a periodic function is
determined by the fundamental interval.

Theorem 1. *Let f be a periodic function in $\mathcal{H}$ and $p > 0$ a period of f. Then*

$$\mathcal{L}\{f\}(s) = \frac{1}{1 - e^{-sp}} \int_0^p e^{-st} f(t)\, dt.$$

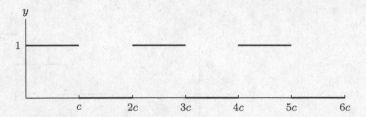

Fig. 6.27 The graph of the square-wave function sw_c

Proof.

$$\mathcal{L}\{f\}(s) = \int\limits_{0}^{\infty} e^{-st} f(t)\, dt$$

$$= \int\limits_{0}^{p} e^{-st} f(t)\, dt + \int\limits_{p}^{\infty} e^{-st} f(t)\, dt.$$

However, the change of variables $t \to t+p$ in the second integral and the periodicity of f gives

$$\int\limits_{p}^{\infty} e^{-st} f(t)\, dt = \int\limits_{0}^{\infty} e^{-s(t+p)} f(t+p)\, dt$$

$$= e^{-sp} \int\limits_{0}^{\infty} e^{-st} f(t)\, dt$$

$$= e^{-sp} \mathcal{L}\{f\}(s).$$

Therefore,

$$\mathcal{L}\{f\}(s) = \int\limits_{0}^{p} e^{-st} f(t)\, dt + e^{-sp} \mathcal{L}\{f\}(s).$$

Solving for $\mathcal{L}\{f\}$ gives the desired result.   □

Example 2.   Find the Laplace transform of the **square-wave** function sw_c given by

$$\text{sw}_c(t) = \begin{cases} 1 & \text{if } t \in [2nc, (2n+1)c) \\ 0 & \text{if } t \in [(2n+1)c, (2n+2)c) \end{cases} \qquad \text{for each integer } n.$$

▶ **Solution.**   The square-wave function sw_c is periodic with period $2c$. Its graph is given in Fig. 6.27 and, by Theorem 1, its Laplace transform is

$$\mathcal{L}\{sw_c\}(s) = \frac{1}{1 - e^{-2cs}} \int_0^{2c} e^{-st} \, sw_c(t) \, dt$$

$$= \frac{1}{1 - e^{-2cs}} \int_0^{c} e^{-st} \, dt$$

$$= \frac{1}{1 - (e^{-cs})^2} \frac{1 - e^{-cs}}{s}$$

$$= \frac{1}{1 + e^{-cs}} \frac{1}{s}. \qquad \blacktriangleleft$$

Example 3. Find the Laplace transform of the sawtooth function $<t>_p$.

▶ **Solution.** Since the sawtooth function is periodic with period p and since $<t>_p = t$ for $0 \le t < p$, Theorem 1 gives

$$\mathcal{L}\{\langle t \rangle_p\}(s) = \frac{1}{1 - e^{-sp}} \int_0^{p} e^{-st} t \, dt.$$

Integration by parts gives

$$\int_0^{p} e^{-st} t \, dt = \frac{t e^{-st}}{-s}\Big|_0^p - \frac{1}{-s} \int_0^{p} e^{-st} \, dt$$

$$= -\frac{p e^{-sp}}{s} - \frac{1}{s^2} e^{-st}\Big|_0^p$$

$$= -\frac{p e^{-sp}}{s} - \frac{e^{-sp} - 1}{s^2}.$$

With a little algebra, we obtain

$$\mathcal{L}\{<t>_p\}(s) = \frac{1}{s^2}\left(1 - \frac{s p e^{-sp}}{1 - e^{-sp}}\right). \qquad \blacktriangleleft$$

As mentioned above, it frequently happens that we build periodic functions by restricting a given function f to the interval $[0, p)$ and then extending it to be periodic with period p: $f(\langle t \rangle_p)$. Suppose now that $f \in \mathcal{H}$. We can then express the Laplace transform of $f(\langle t \rangle_p)$ in terms of the Laplace transform of f. The following corollary expresses this relationship and simplifies unnecessary calculations like the integration by parts that we did in the previous example.

Corollary 4. *Let $p > 0$. Suppose $f \in \mathcal{H}$. Then*

$$\mathcal{L}\{f(\langle t \rangle_p)\}(s) = \frac{1}{1 - e^{-sp}} \mathcal{L}\{f(t) - f(t)h(t - p)\}.$$

Proof. The function $f(t) - f(t)h(t - p) = f(t)(1 - h(t - p))$ is the same as f on the interval $[0, p)$ and 0 on the interval $[p, \infty)$. Therefore,

$$\int_0^p e^{-st} f(t)\, dt = \int_0^\infty e^{-st} (f(t) - f(t)h(t - p))\, dt = \mathcal{L}\{f(t) - f(t)h(t - p)\}.$$

The result now follows from Theorem 1. □

Let us return to the sawtooth function in Example 3 and see how Corollary 4 simplifies the calculation of its Laplace transform.

$$\mathcal{L}\{\langle t \rangle_p\}(s) = \frac{1}{1 - e^{-sp}} \mathcal{L}\{t - th(t - p)\}$$

$$= \frac{1}{1 - e^{-sp}} \left(\frac{1}{s^2} - e^{-sp} \mathcal{L}\{t + p\} \right)$$

$$= \frac{1}{1 - e^{-sp}} \left(\frac{1}{s^2} - e^{-sp} \frac{1 + sp}{s^2} \right)$$

$$= \frac{1}{s^2} \left(1 - \frac{spe^{-sp}}{1 - e^{-sp}} \right).$$

The last line requires a few algebraic steps.

Example 5. Find the Laplace transform of the rectified sine wave

$$\sin(\langle t \rangle_\pi).$$

See Fig. 6.24.

▶ **Solution.** Corollary 4 gives

$$\mathcal{L}\{\sin(\langle t \rangle_\pi)\} = \frac{1}{1 - e^{-\pi s}} \mathcal{L}\{\sin t - \sin t\, h(t - \pi)\}$$

$$= \frac{1}{1 - e^{-\pi s}} \left(\frac{1}{s^2 + 1} - e^{-\pi s} \mathcal{L}\{\sin(t + \pi)\} \right)$$

$$= \frac{1}{1 - e^{-\pi s}} \left(\frac{1 + e^{-\pi s}}{s^2 + 1} \right),$$

where we use the fact that $\sin(t + \pi) = -\sin(t)$. ◀

Periodic Extensions of the Dirac Delta Function

We will also consider in the applications inputs that are periodic extensions of the Dirac delta function. For example,

$$\delta_c(\langle t \rangle_p) = \delta_c + \delta_{c+p} + \delta_{c+2p} + \delta_{c+3p} + \cdots,$$

where $0 \leq c < p$ is the periodic extension of the Dirac delta function, δ_c. An important case is when $c = 0$. Then $\delta_0(\langle t \rangle_p)$, is the periodic extension of δ_0 with period p and represents a unit impulse at each multiple of $t = p$. Another important example is

$$(\delta_0 - \delta_p)(\langle t \rangle_{2p}) = \delta_0 - \delta_p + \delta_{2p} - \delta_{3p} + \cdots,$$

the periodic extension of $\delta_0 - \delta_p$ with period $2p$ which represents a unit impulse at each even multiple of p and a negative unit impulse at odd multiples of p.

Proposition 6. *The Laplace transforms of $\delta_0(\langle t \rangle_p)$ and $(\delta_0 - \delta_p)(\langle t \rangle_{2p})$ are given by the following formulas:*

$$\mathcal{L}\left\{\delta_0(\langle t \rangle_p)\right\} = \frac{1}{1 - e^{-ps}},$$

$$\mathcal{L}\left\{(\delta_0 - \delta_p)(\langle t \rangle_{2p})\right\} = \frac{1}{1 + e^{-ps}}.$$

Proof. Let r be a fixed real or complex number. Recall that the geometric series

$$\sum_{n=0}^{\infty} r^n = 1 + r + r^2 + r^3 + \cdots$$

converges to $\frac{1}{1-r}$ when $|r| < 1$. We can now compute the Laplace transforms.

$$\mathcal{L}\left\{\delta_0(\langle t \rangle_p)\right\} = 1 + e^{-ps} + e^{-2ps} + e^{-3ps} + \cdots$$

$$= \sum_{n=0}^{\infty} (e^{-ps})^n$$

$$= \frac{1}{1 - e^{-ps}}.$$

Similarly,

$$\mathcal{L}\left\{(\delta_0 - \delta_p)(\langle t \rangle_{2p})\right\} = 1 - e^{-ps} + e^{-2ps} - e^{-3ps} + \cdots$$

$$= \sum_{n=0}^{\infty} (-e^{-ps})^n$$

$$= \frac{1}{1 + e^{-ps}}. \qquad \Box$$

The Inverse Laplace Transform

The inverse Laplace transform of functions of the form

$$\frac{1}{1 - e^{-sp}} F(s)$$

is not always a straightforward matter to find unless, of course, $F(s)$ is of the form $\mathcal{L}\{f(t) - f(t)h(t - p)\}$ so that Corollary 4 can be used. Usually, though, this is not the case. Since

$$\frac{1}{1 - e^{-sp}} = \sum_{n=0}^{\infty} e^{-snp},$$

we can write

$$\frac{1}{1 - e^{-sp}} F(s) = \sum_{n=0}^{\infty} e^{-snp} F(s).$$

If $f = \mathcal{L}^{-1}\{F\}$, then a termwise computation gives

$$\mathcal{L}^{-1}\left\{\frac{1}{1 - e^{-sp}} F(s)\right\} = \sum_{n=0}^{\infty} \mathcal{L}^{-1}\{e^{-snp} F(s)\} = \sum_{n=0}^{\infty} f(t - np)h(t - np).$$

For t in an interval of the form $[Np, (N + 1)p)$, the function $h(t - np)$ is 1 for $n = 0, \ldots, N$ and 0 otherwise. We thus obtain

$$\mathcal{L}^{-1}\left\{\frac{1}{1 - e^{-sp}} F(s)\right\} = \sum_{N=0}^{\infty} \left(\sum_{n=0}^{N} f(t - np)\right) \chi_{[Np,(N+1)p)}.$$

A similar argument gives

$$\mathcal{L}^{-1}\left\{\frac{1}{1 + e^{-sp}} F(s)\right\} = \sum_{N=0}^{\infty} \left(\sum_{n=0}^{N} (-1)^n f(t - np)\right) \chi_{[Np,(N+1)p)}.$$

For reference, we record these results in the following theorem:

Theorem 7. *Let $p > 0$ and suppose $\mathcal{L}\{f(t)\} = F(s)$. Then*

1. $\mathcal{L}^{-1}\left\{\frac{1}{1-e^{-sp}}F(s)\right\} = \sum_{N=0}^{\infty}\left(\sum_{n=0}^{N}f(t-np)\right)\chi_{[Np,(N+1)p)}.$

2. $\mathcal{L}^{-1}\left\{\frac{1}{1+e^{-sp}}F(s)\right\} = \sum_{N=0}^{\infty}\left(\sum_{n=0}^{N}(-1)^{n}f(t-np)\right)\chi_{[Np,(N+1)p)}.$

Example 8.   Find the inverse Laplace transform of

$$\frac{1}{(1-e^{-2s})s}.$$

▶ **Solution.** If $f(t) = 1$, then $F(s) = \frac{1}{s}$ is its Laplace transform. We thus have

$$\mathcal{L}^{-1}\left\{\frac{1}{(1-e^{-2s})s}\right\} = \sum_{N=0}^{\infty}\left(\sum_{n=0}^{N}f(t-2n)\right)\chi_{[2N,2(N+1))}$$

$$= \sum_{N=0}^{\infty}(N+1)\chi_{[2N,2(N+1))}$$

$$= 1 + \frac{1}{2}\sum_{N=0}^{\infty}2N\chi_{[2N,2(N+1))}$$

$$= 1 + \frac{1}{2}[t]_{2}. \qquad \blacktriangleleft$$

Remark 9. Generally, it will not be possible to express the final answer in a nice closed form as in Example 8; one may have to settle for an infinite sum as given in Theorem 7.

Exercises

1–5. Reexpress each sum in terms of the sawtooth function $\langle t \rangle_p$ and/or the staircase function $[t]_p$. (Since $t - \langle t \rangle_p = [t]_p$, there are more than one equivalent answer.)

1. $\sum_{n=0}^{\infty} (t - n)^2 \chi_{[n,(n+1))}(t)$

2. $\sum_{n=0}^{\infty} (t - n)^2 \chi_{[2n,2(n+1))}(t)$

3. $\sum_{n=0}^{\infty} n^2 \chi_{[3n,3(n+1))}(t)$

4. $\sum_{n=0}^{\infty} e^{8n} \chi_{[4n,4(n+1))}(t)$

5. $\sum_{n=0}^{\infty} (t + n) \chi_{[2n,2(n+1))}(t)$

6–10. Find the Laplace transform of each periodic function.

6. $f(\langle t \rangle_2)$ where $f(t) = t^2$

7. $f(\langle t \rangle_3)$ where $f(t) = e^t$

8. $f(\langle t \rangle_2)$ where $f(t) = \begin{cases} t & \text{if } 0 \le t < 1 \\ 2 - t & \text{if } 1 \le t < 2 \end{cases}$

9. $f(\langle t \rangle_{2p})$ where $f(t) = \begin{cases} 1 & \text{if } 0 \le t < p \\ -1 & \text{if } p \le t < 2p \end{cases}$

10. $f(\langle t \rangle_{\pi})$ where $f(t) = \cos t$

11–14. Find the Laplace transform of $f([t]_p)$, where $[t]_p$ is the staircase function.

11. $f([t]_p)$ where $f(t) = t$. That is, find the Laplace transform of the staircase function $[t]_p$.

12. $f([t]_1)$ where $f(t) = e^t$

13. $f([t]_2)$ where $f(t) = e^{-t}$

14. $f([t]_3)$ where $f(t) = t^2$. Hint: Use the fact that $\sum_{n=0}^{\infty} n^2 x^n = \frac{x(1+x)}{(1-x)^3}$ for $|x| < 1$.

15. Suppose $f \in \mathcal{H}$. Show that

$$\mathcal{L}\{f([t]_p)\} = \frac{1 - e^{-ps}}{s} \sum_{n=0}^{\infty} f(np) e^{-nps}.$$

16–18. Find the inverse Laplace transform of each function.

16. $\dfrac{e^{-2s}}{s(1-e^{-2s})}$

17. $\dfrac{1-e^{-4(s-2)}}{(1-e^{-4s})(s-2)}$

18. $\dfrac{1}{(1-e^{-4s})(s-2)}$

19. Let $F(s) = \frac{1}{(s+a)(1+e^{-ps})}$. Show that

$$\mathcal{L}^{-1}\{F(s)\} = e^{-at}\begin{cases} \frac{1+e^{a(N+1)p}}{1+e^{ap}} & \text{if } t \in [Np,(N+1)p),\ (N \text{ even}) \\[2ex] \frac{1-e^{a(N+1)p}}{1+e^{ap}} & \text{if } t \in [Np,(N+1)p),\ (N \text{ odd}) \end{cases}$$

$$= e^{-at}\left(\frac{1+(-1)^{\frac{[t]p}{p}}e^{a([t]_p+p)}}{1+e^{ap}} \right).$$

(Use the fact that $1 + x + x^2 + \cdots + x^N = \frac{1-x^{N+1}}{1-x}$.)

20. In the text, we stated that the constant function 1 is periodic with period 0. Here is another example: Let $\mathbb{Q}$ denote the set of rational numbers. Let

$$\chi_{\mathbb{Q}}(t) = \begin{cases} 1 & \text{if } t \in \mathbb{Q} \\ 0 & \text{if } t \notin \mathbb{Q} \end{cases}.$$

Show that χ is periodic with period q for each positive $q \in \mathbb{Q}$. Conclude that $\chi_{\mathbb{Q}}$ has fundamental period 0.

6.7 First Order Equations with Periodic Input

We now turn our attention to two examples of mixing problems with periodic input functions. Each example will be modeled by a first order differential equation of the form

$$y' + ay = f(t),$$

where $f(t)$ is a periodic function.

Example 1. Suppose a tank contains 10 gal of pure water. Two input sources alternately flow into the tank for 1-min intervals. The first input source begins flowing at $t = 0$. It consists of a brine solution with concentration 1 lb salt per gallon and flows (when on) at a rate of 5 gal/min. The second input source is pure water and flows (when on) at a rate of 5 gal/min. The tank has a drain with a constant outflow of 5 gal/min. Let $y(t)$ denote the total amount of salt at time t. Find $y(t)$ and for large values of t determine how $y(t)$ fluctuates.

▶ **Solution.** The input rate of salt is given piecewise by the formula

$$5\,\text{sw}_1(t) = \begin{cases} 5 & \text{if } 2n \le t < 2n + 1) \\ 0 & \text{if } 2n + 1 \le t < 2n + 2 \end{cases}.$$

The output rate is given by

$$\frac{y(t)}{10} \cdot 5.$$

This leads to the first order differential equation

$$y' + \frac{1}{2}y = 5\,\text{sw}_1(t) \qquad y(0) = 0.$$

A calculation using Example 2 of Sect. 6.6 shows that the Laplace transform is

$$Y(s) = 5\frac{1}{1 + e^{-s}} \frac{1}{s\left(s + \frac{1}{2}\right)},$$

and a partial fraction decomposition of $\frac{1}{s(s+1/2)}$ gives

$$Y(s) = 10\frac{1}{1 + e^{-s}} \frac{1}{s} - 10\frac{1}{1 + e^{-s}} \frac{1}{s + \frac{1}{2}}.$$

Now apply the inverse Laplace transform. To simplify the calculations, let

$$Y_1(s) = 10\frac{1}{1 + e^{-s}} \frac{1}{s},$$

$$Y_2(s) = 10\frac{1}{1 + e^{-s}} \frac{1}{s + \frac{1}{2}}.$$

Then $Y(s) = Y_1(s) - Y_2(s)$. By Example 2 of Sect. 6.6, we have

$$\mathcal{L}^{-1}\{Y_1(s)\} = 10\,sw_1(t).$$

By Theorem 7 of Sect. 6.6, the inverse Laplace transform of the second expression is

$$\mathcal{L}^{-1}\{Y_2(s)\} = 10 \sum_{N=0}^{\infty} \sum_{n=0}^{N} (-1)^n e^{-\frac{1}{2}(t-n)} \chi_{[N,N+1)}$$

$$= 10 e^{-\frac{1}{2}t} \sum_{N=0}^{\infty} \sum_{n=0}^{N} \left(-e^{\frac{1}{2}}\right)^n \chi_{[N,N+1)}$$

$$= 10 e^{-\frac{1}{2}t} \sum_{N=0}^{\infty} \frac{1 - \left(-e^{\frac{1}{2}}\right)^{N+1}}{1 + e^{\frac{1}{2}}} \chi_{[N,N+1)}$$

$$= \frac{10 e^{-\frac{1}{2}t}}{1 + e^{\frac{1}{2}}} \begin{cases} 1 + e^{\frac{N+1}{2}} & \text{if } t \in [N, N+1) \text{ (N even)} \\ 1 - e^{\frac{N+1}{2}} & \text{if } t \in [N, N+1) \text{ (N odd)} \end{cases}.$$

Finally, we put these two expression together to get our solution

$$y(t) = 10\,sw_1(t) - \frac{10 e^{-\frac{1}{2}t}}{1 + e^{\frac{1}{2}}} \begin{cases} 1 + e^{\frac{N+1}{2}} & \text{if } t \in [N, N+1) \text{ (N even)} \\ 1 - e^{\frac{N+1}{2}} & \text{if } t \in [N, N+1) \text{ (N odd)} \end{cases}$$

$$= \begin{cases} 10 - 10\dfrac{e^{-\frac{1}{2}t} + e^{\frac{-t+N+1}{2}}}{1 + e^{\frac{1}{2}}} & \text{if } t \in [N, N+1) \text{ (N even)} \\[3mm] -10\dfrac{e^{-\frac{1}{2}t} - e^{\frac{-t+N+1}{2}}}{1 + e^{\frac{1}{2}}} & \text{if } t \in [N, N+1) \text{ (N odd)} \end{cases}$$

The graph of $y(t)$, obtained with the help of a computer, is presented in Fig. 6.28. The solution is sandwiched in between a lower and upper curve. The lower curve, $l(t)$, is obtained by setting $t = m$ to be an even integer in the formula for the solution and then continuing it to all reals. We obtain

$$l(m) = 10 - 10\frac{e^{-\frac{1}{2}m} + e^{\frac{-m+m+1}{2}}}{1 + e^{\frac{1}{2}}} = 10 - 10\frac{e^{-\frac{1}{2}m} + e^{\frac{1}{2}}}{1 + e^{\frac{1}{2}}}$$

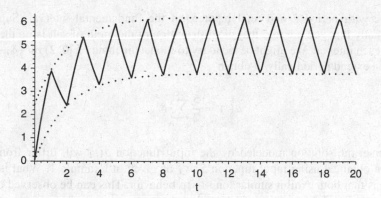

Fig. 6.28   A mixing problem with square-wave input function

and thus

$$l(t) = 10 - 10\frac{e^{-\frac{1}{2}t} + e^{\frac{1}{2}}}{1 + e^{\frac{1}{2}}}.$$

In a similar way, the upper curve, $u(t)$, is obtained by setting $t = m^-$ to be slight smaller than an odd integer and continuing to all reals. We obtain

$$u(t) = -10\frac{e^{-\frac{1}{2}t} - e^{\frac{1}{2}}}{1 + e^{\frac{1}{2}}}.$$

An easy calculation gives

$$\lim_{t\to\infty} l(t) = 10 - \frac{10e^{\frac{1}{2}}}{1 + e^{\frac{1}{2}}} \simeq 3.78 \quad \text{and} \quad \lim_{t\to\infty} u(t) = \frac{10e^{\frac{1}{2}}}{1 + e^{\frac{1}{2}}} \simeq 6.22.$$

This means that the salt fluctuation in the tank varies between 3.78 and 6.22 lbs for large values of t.   ◀

In practice, it is not always possible to know the input function, $f(t)$, precisely. Suppose though that it is known that f is periodic with period p. Then the total input on all intervals of the form $[np, (n+1)p)$ is $\int_{np}^{(n+1)p} f(t)\, dt = h$, a constant. On the interval $[0, p)$, we could model the input with a Dirac delta function concentrated at a point, c say, and then extend it periodically. We would then obtain a sum of Dirac delta functions of the form

$$a(t) = h(\delta_c + \delta_{c+p} + \delta_{c+2p} + \cdots)$$

that may adequately represent the input for the system we are trying to model. Additional information may justify distributing the total input over two or more points in the interval and extend periodically. Whatever choices are made, the solution will need to be analyzed in the light of empirical data known about the system. Consider the example above. Suppose that it is known that the input is periodic with period 2 and total input 5 on the fundamental interval. Suppose additionally that you are told that the distribution of the input of salt is on the first half of each interval. We might be led to try to model the input on $[0, 2)$ by $\frac{5}{2}\delta_0 + \frac{5}{2}\delta_1$ and then extend periodically to obtain

$$a(t) = \frac{5}{2} \sum_{n=0}^{\infty} \delta_n.$$

Of course, the solution modeled by the input function $a(t)$ will differ from the solution obtained using input function $5\,\mathrm{sw}_1$ as given in Example 1. What is true though is that both exhibit similar long-term behavior. This can be observed in the following example.

Example 2. Suppose a tank contains 10 gal of pure water. Pure water flows into the tank at a rate of 5 gal/min. The tank has a drain with a constant outflow of 5 gal/min. Suppose $\frac{5}{2}$ lbs of salt is put in the tank each minute whereupon it instantly and uniformly dissolves. Assume the level of fluid in the tank is always 10 gal. Let $y(t)$ denote the total amount of salt at time t. Find $y(t)$ and for large values of t determine how $y(t)$ fluctuates.

▶ **Solution.** As discussed above, the input function is $\frac{5}{2} \sum_{n=1}^{\infty} \delta_n$, and therefore, the differential equation that models this system is

$$y' + \frac{1}{2}y = \frac{5}{2} \sum_{n=1}^{\infty} \delta_n, \quad y(0) = 0.$$

The Laplace transform leads to

$$Y(s) = \frac{5}{2} \sum_{n=0}^{\infty} e^{-sn} \frac{1}{s + \frac{1}{2}},$$

and inverting the Laplace transform and using Theorem 7 of Sect. 6.6 give

$$y(t) = \frac{5}{2} e^{-\frac{1}{2}t} \sum_{N=0}^{\infty} \left(\sum_{n=0}^{N} \left(e^{-\frac{1}{2}(t-n)} \right) \right) \chi_{[N,N+1)}$$

$$= \frac{5}{2} e^{-\frac{1}{2}t} \sum_{N=0}^{\infty} \left(\sum_{n=0}^{N} \left(e^{\frac{1}{2}} \right)^n \right) \chi_{[N,N+1)}$$

$$= \frac{5}{2} e^{-\frac{1}{2}t} \sum_{N=0}^{\infty} \frac{1 - e^{\frac{N+1}{2}}}{1 - e^{\frac{1}{2}}} \chi_{[N,N+1)}$$

$$= \frac{5 \left(e^{-\frac{1}{2}t} - e^{-\frac{1}{2}(t-[t]_1-1)} \right)}{2 \left(1 - e^{\frac{1}{2}} \right)}.$$

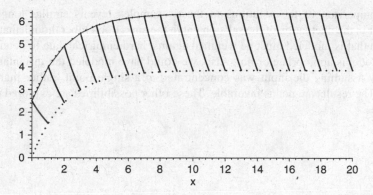

Fig. 6.29 A mixing problem with a periodic Dirac delta function: The solution to the differential equation $y' + \frac{1}{2}y = \frac{5}{2}\sum_{n=1}^{\infty}\delta_n$, $\quad y(0) = 0$

The graph of this equation is given in Fig. 6.29. The solution is sandwiched in between a lower and upper curve. The upper curve, $u(t)$, is obtained by setting $t = m$ to be an integer in the formula for the solution and then continuing it to all reals. We obtain

$$u(m) = \frac{5}{2\left(1 - e^{-\frac{1}{2}}\right)}\left(e^{-\frac{m}{2}} - e^{\frac{-m+m+1}{2}}\right) = \frac{5}{2\left(1 - e^{-\frac{1}{2}}\right)}\left(e^{-\frac{m}{2}} - e^{\frac{1}{2}}\right)$$

and thus

$$u(t) = \frac{5}{2\left(1 - e^{-\frac{1}{2}}\right)}\left(e^{-\frac{t}{2}} - e^{\frac{1}{2}}\right).$$

In a similar way, the upper curve, $l(t)$, is obtained by setting $t = (m+1)^-$ (slightly less than the integer $m + 1$) and continuing to all reals. We obtain

$$l(t) = \frac{5}{2\left(1 - e^{-\frac{1}{2}}\right)}\left(e^{-\frac{t}{2}} - 1\right).$$

An easy calculation gives

$$\lim_{t\to\infty} u(t) = \frac{-5e^{\frac{1}{2}}}{2\left(1 - e^{\frac{1}{2}}\right)} \simeq 6.35 \text{ and } \lim_{t\to\infty} l(t) = \frac{-5}{2\left(1 - e^{\frac{1}{2}}\right)} \simeq 3.85.$$

This means that the salt fluctuation in the tank varies between 3.85 and 6.35 lbs for large values of t. ◄

A comparison of the solutions in these examples reveals similar long-term behavior in the fluctuation of the salt content in the tank. Remember though that each problem that is modeled must be weighed against hard empirical data to determine if the model is appropriate or not. Also, we could have modeled the instantaneous input by assuming the input was concentrated at a single point, rather than two points. The results are not as favorable. These other possibilities are explored in the exercises.

Exercises

1–3. Solve the following mixing problems.

1. Suppose a tank contains 10 gal of pure water. Two input sources alternately flow into the tank for 2-min intervals. The first input source begins flowing at $t = 0$. It is a brine solution with concentration 2 lbs salt per gallon and flows (when on) at a rate of 4 gal/min. The second input source is a brine solution with concentration 1 lb salt per gallon and flows (when on) at a rate of 4 gal/min. The tank has a drain with a constant outflow of 4 gal/min. Let $y(t)$ denote the total amount of salt at time t. Find $y(t)$ and for large values of t determine how $y(t)$ fluctuates.

2. Suppose a tank contains 10 gal of brine in which 20 lbs of salt are dissolved. Two input sources alternately flow into the tank for 1-min intervals. The first input source begins flowing at $t = 0$. It is a brine solution with concentration 1 lb salt per gallon and flows (when on) at a rate of 2 gal/min. The second input source is a pure water and flows (when on) at a rate of 2 gal/min. The tank has a drain with a constant outflow of 2 gal/min. Let $y(t)$ denote the total amount of salt at time t. Find $y(t)$ and for large values of t determine how $y(t)$ fluctuates.

3. Suppose a tank contains 10 gal of pure water. Pure water flows into the tank at a rate of 5 gal/min. The tank has a drain with a constant outflow of 5 gal/min. Suppose 5 lbs of salt is put in the tank every other minute beginning at $t = 0$ whereupon it instantly and uniformly dissolves. Assume the level of fluid in the tank is always 10 gal. Let $y(t)$ denote the total amount of salt at time t. Find $y(t)$ and for large values of t determine how $y(t)$ fluctuates.

4–5. Solve the following harvesting problems.

4. In a certain area of the Louisiana swamp, a population of 2,500 alligators is observed. Given adequate amounts of food and space, their population will follow the Malthusian growth model. After 12 months, scientists observe that there are 3,000 alligators. Alarmed by their rapid growth, the Louisiana Wildlife and Fisheries institutes the following hunting policy for a specialized group of alligator hunters: Hunting is allowed in only an odd-numbered month, and the total number of alligators taken is limited to 80. Assuming the limit is attained in each month allowed and uniformly over the month, determine a model that gives the population of alligators. Solve that model. How many alligators are there at the beginning of the fifth year? (Assume a population of 3,000 alligators at the beginning of the initial year.)

5. Assume the premise of Exercise 4, but instead of the stated hunting policy in odd-numbered months, assume that the Louisiana Wildlife and Fisheries contracts an elite force of Cajun alligator hunters to take out 40 alligators at the beginning of each month. (You may assume this is done instantly on the first day of each month.) Determine a model that gives the population of alligators. Solve that model. How many alligators are there at the beginning of the fifth year?

6.8   Undamped Motion with Periodic Input

In Sect. 3.6, we discussed various kinds of motion of a spring-body-dashpot system modeled by the differential equation

$$my'' + \mu y' + ky = f(t).$$

Undamped motion led to the differential equation

$$my'' + ky = f(t). \tag{1}$$

In particular, we explored the case where $f(t) = F_0 \cos \omega t$ and were led to the solution

$$y(t) = \begin{cases} \frac{F_0}{a(\beta^2 - \omega^2)} (\cos \omega t - \cos \beta t) & \text{if } \beta \neq \omega, \\[2mm] \frac{F_0}{2a\omega} t \sin \omega t & \text{if } \beta = \omega, \end{cases} \tag{2}$$

where $\beta = \sqrt{\frac{k}{m}}$. The case where β is close to but not equal to ω gave rise to the notion of beats, while the case $\beta = \omega$ gave us resonance. Since $\cos \omega t$ is periodic, the system that led to (1) is an example of *undamped motion with periodic input*. In this section, we will explore this phenomenon with two further examples: a square-wave periodic function, sw_c, and a periodic impulse function, $\delta_0(\langle t \rangle_c) = \sum_{n=0}^{\infty} \delta_{nc}$. Both examples are algebraically tedious, so you will be asked to fill in some of the algebraic details in the exercises. To simplify the notation, we will rewrite (1) as

$$y'' + \beta^2 y = g(t)$$

and assume $y(0) = y'(0) = 0$.

Undamped Motion with Square-Wave Forcing Function

Example 1. A constant force of r units for c units of time is applied to a mass-spring system with no damping force that is initially at rest. The force is then released for c units of time. This on–off force is extended periodically to give a periodic forcing function with period $2c$. Describe the motion of the mass.

▶ **Solution.** The differential equation which describes this system is

$$y'' + \beta^2 y = r\, sw_c(t), \quad y(0) = 0,\ y'(0) = 0, \tag{3}$$

where sw_c is the square-wave function with period $2c$ and β^2 is the spring constant. By Example 2 of Sect. 6.6, the Laplace transform leads to the equation

$$Y(s) = r\frac{1}{1+e^{-sc}}\frac{1}{s(s^2+\beta^2)} = \frac{r}{\beta^2}\frac{1}{1+e^{-sc}}\left(\frac{1}{s} - \frac{s}{s^2+\beta^2}\right).$$

$$= \frac{r}{\beta^2}\frac{1}{1+e^{-sc}}\frac{1}{s} - \frac{r}{\beta^2}\frac{1}{1+e^{-sc}}\frac{s}{s^2+\beta^2}. \tag{4}$$

Let

$$F_1(s) = \frac{r}{\beta^2}\frac{1}{1+e^{-sc}}\frac{1}{s} \quad \text{and} \quad F_2(s) = \frac{r}{\beta^2}\frac{1}{1+e^{-sc}}\frac{s}{s^2+\beta^2}.$$

Then $Y(s) = F_1(s) - F_2(s)$. Again, by Example 2 of Sect. 6.6, we have

$$f_1(t) = \frac{r}{\beta^2}\,sw_c(t). \tag{5}$$

By Theorem 7 of Sect. 6.6, we have

$$f_2(t) = \frac{r}{\beta^2}\sum_{N=0}^{\infty}\left(\sum_{n=0}^{N}(-1)^n\cos(\beta t - n\beta c)\right)\chi_{[Nc,(N+1)c)}. \tag{6}$$

We consider two cases.

βc Is not an Odd Multiple of π

Lemma 2. *Suppose v is not an odd multiple of π and let $\alpha = \frac{\sin(v)}{1+\cos(v)}$. Then*

1. $\displaystyle\sum_{n=0}^{N}(-1)^n\cos(u+nv) = \frac{1}{2}(\cos u + \alpha \sin u)$

$$+\frac{(-1)^N}{2}(\cos(u+Nv) - \alpha\sin(u+Nv)).$$

2. $\displaystyle\sum_{n=0}^{N}(-1)^n\sin(u+nv) = \frac{1}{2}(\sin u - \alpha\cos(u))$

$$+\frac{(-1)^N}{2}(\sin(u+Nv) + \alpha\cos(u+Nv)).$$

Proof. The proof of the lemma is left as an exercise.   □

Let $u = \beta t$ and $v = -\beta c$. Then $\alpha = \frac{-\sin(\beta c)}{1+\cos(\beta c)}$. In this case, we can apply part (1) of the lemma to (6) to get

$$f_2(t) = \frac{r}{2\beta^2} \sum_{N=0}^{\infty} (\cos \beta t + \alpha \sin \beta t) \, \chi_{[Nc, N+1)c}$$

$$+ \frac{r}{2\beta^2} \sum_{N=0}^{\infty} (-1)^N (\cos \beta(t - Nc) - \alpha \sin \beta(t - Nc)) \, \chi_{[Nc, N+1)c}$$

$$= \frac{r}{2\beta^2} (\cos \beta t + \alpha \sin \beta t)$$

$$+ \frac{r}{2\beta^2} (-1)^{[t/c]_1} (\cos \beta \langle t \rangle_c - \alpha \sin \beta \langle t \rangle_c).$$

Let

$$y_1(t) = \frac{r}{\beta^2} \, \mathrm{sw}_c(t) - \frac{r}{2\beta^2} (-1)^{[t/c]_1} (\cos \beta \langle t \rangle_c - \alpha \sin \beta \langle t \rangle_c)$$

$$= \frac{r}{2\beta^2} \left(2 \, \mathrm{sw}_c(t) - (-1)^{[t/c]_1} (\cos \beta \langle t \rangle_c - \alpha \sin \beta \langle t \rangle_c) \right)$$

and

$$y_2(t) = -\frac{r}{2\beta^2} (\cos \beta t + \alpha \sin \beta t).$$

Then

$$y(t) = f_1(t) - f_2(t) = y_1(t) + y_2(t)$$

$$= \frac{r}{2\beta^2} \left(2 \, \mathrm{sw}_c(t) - (-1)^{[t/c]_1} (\cos \beta \langle t \rangle_c - \alpha \sin \beta \langle t \rangle_c) \right)$$

$$- \frac{r}{2\beta^2} (\cos \beta t + \alpha \sin \beta t). \tag{7}$$

A quick check shows that y_1 is periodic with period $2c$ and y_2 is periodic with period $\frac{2\pi}{\beta}$. Clearly, y_2 is continuous, and since the solution $y(t)$ is continuous by Theorem 7 of Sect. 6.1, so is y_1. The following lemma will help us determine when y is a periodic solution.

Lemma 3. *Suppose g_1 and g_2 are continuous periodic functions with periods $p_1 > 0$ and $p_2 > 0$, respectively. Then $g_1 + g_2$ is periodic if and only if $\frac{p_1}{p_2}$ is a rational number.*

Proof. If $\frac{p_1}{p_2} = \frac{m}{n}$ is rational, then $np_1 = mp_2$ is a common period of g_1 and g_2 and hence is a period of $g_1 + g_2$. It follows that $g_1 + g_2$ is periodic. The opposite implication, namely, that the periodicity of $g_1 + g_2$ implies $\frac{p_1}{p_2}$ is rational, is a nontrivial fact. We do not include a proof. □

Using this lemma, we can determine precisely when the solution $y = y_1 + y_2$ is periodic. Namely, y is periodic precisely when $\frac{2c}{2\pi/\beta} = \frac{c\beta}{\pi}$ is rational. Consider

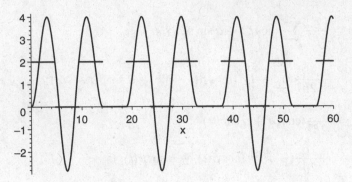

Fig. 6.30 The graph of (8) with $c = \frac{3\pi}{2}$: a periodic solution

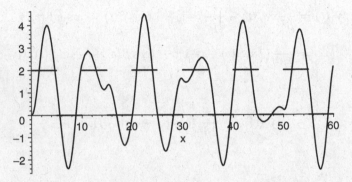

Fig. 6.31 The graph of (9): a nonperiodic solution

the following illustrative example. Set $r = 2$, $c = \frac{3\pi}{2}$, and $\beta = 1$. Then $\frac{c\beta}{\pi} = \frac{3}{2}$ is rational. Further, α, defined in Lemma 2, is 1 and

$$y(t) = 2\,\mathrm{sw}_c(t) - (-1)^{[t/c]_1}(\cos\langle t\rangle_c - \sin\langle t\rangle_c) - (\cos t + \sin t). \qquad (8)$$

This function is graphed simultaneously with the forcing function in Fig. 6.30. The solution is periodic with period $4c = 6\pi$. Notice that there is an interval where the motion of the mass is stopped. This occurs in the interval $[3c, 4c)$. The constant force applied on the interval $[2c, 3c)$ gently stops the motion of the mass by the time $t = 3c$. Since the force is 0 on $[3c, 4c)$, there is no movement. At $t = 4c$, the force is reapplied and the process thereafter repeats itself. This phenomenon occurs in all cases where the solution y is periodic.

Now consider the following example that illustrates a nonperiodic solution. Set $r = 2$, $c = 5$, and $\beta = 1$ in (7). Then

$$y(t) = 2\,\mathrm{sw}_5(t) - (-1)^{[t/5]_1}(\cos\langle t\rangle_5 - \alpha\sin\langle t\rangle_5) - \cos t - \alpha\sin t, \qquad (9)$$

where $\alpha = \frac{-\sin 5}{1 + \cos 5}$. Further, $\frac{c\beta}{\pi} = \frac{5}{\pi}$ is irrational so $y(t)$ is not periodic. This is clearly seen in the rather erratic motion given by the graph of $y(t)$ in Fig. 6.31.

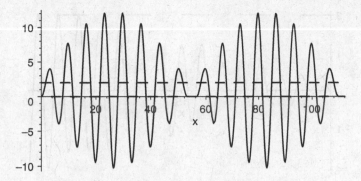

Fig. 6.32 The graph of (8) with $c = \frac{9\pi}{8}$: the beats are evident here

In Sect. 3.6, we observed that when the characteristic frequency of the spring is close to but not equal to the frequency of the forcing function, $\cos(\omega t)$, then one observes vibrations that exhibit a beat. This phenomenon likewise occurs for the square-wave forcing function. Let $r = 2$, $c = \frac{9\pi}{8}$, and $\beta = 1$. Recall that *frequency* is merely the reciprocal of the period so when these frequencies are close, so are their periods. The period of the spring is $\frac{2\pi}{\beta} = 2\pi$ while the period of the forcing function is $2c = \frac{9\pi}{4}$: their periods are close and likewise their frequencies. Figure 6.32 gives a graph of y in this case. Again it is evident that the motion of the mass stops on the last subinterval before the end of its period. More interesting is the fact that y oscillates with an amplitude that varies with time and produces "beats".

βc Is an Odd Multiple of π

We now return to equation (6) in the case βc is an odd multiple of π. Things reduce substantially because $\cos(\beta t - N\beta c) = (-1)^N \cos(\beta t)$ and we get

$$f_2(t) = \frac{r}{\beta^2} \sum_{N=0}^{\infty} \sum_{n=0}^{N} \cos(\beta t) \chi_{[Nc,(N+1)c)}$$

$$= \frac{r}{\beta^2} \sum_{N=0}^{\infty} (N+1) \chi_{[Nc,(N+1)c)} \cos(\beta t)$$

$$= \frac{r}{\beta^2} ([t/c]_1 + 1) \cos(\beta t).$$

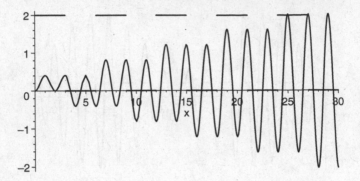

Fig. 6.33 The graph of (10) with $r = 2$, $\beta = \pi$, and $c = 3$: resonance is evident here

The solution now is

$$
\begin{aligned}
y(t) &= f_1(t) - f_2(t) \\
&= \frac{r}{\beta^2} \left(\text{sw}_c(t) - [t/c]_1 \cos(\beta t) - \cos(\beta t) \right) \\
&= \frac{r}{\beta^2} \begin{cases} 1 - (n+1)\cos(\beta t) & \text{if } t \in [cn, c(n+1)), \ n \text{ even} \\ -(n+1)\cos(\beta t) & \text{if } t \in [cn, c(n+1)), \ n \text{ odd} \end{cases}.
\end{aligned} \tag{10}
$$

The presence of the factor $n + 1$ implies that $y(t)$ is unbounded. Figure 6.33 gives the graph of this in the case where $r = 2$, $\beta = \pi$, and $c = 3$. Resonance becomes clearly evident. Of course, this is an idealized situation; the system would eventually fail. ◄

Undamped Motion with Periodic Impulses

Example 4. A mass-spring system with no damping force is acted upon at rest by an impulse force of r units at all multiples of c units of time starting at $t = 0$. (Imagine a hammer exerting blows to the mass at regular intervals.) Describe the motion of the mass.

► **Solution.** The differential equation that describes this system is given by

$$
y'' + \beta^2 y = r \sum_{n=0}^{\infty} \delta_{nc} \quad y(0) = 0, \ y'(0) = 0,
$$

where, again, β^2 is the spring constant. The Laplace transform gives

$$Y(s) = \frac{r}{\beta} \sum_{n=0}^{\infty} e^{-ncs} \frac{\beta}{s^2 + \beta^2}.$$

By Theorem 7 of Sect. 6.6,

$$y(t) = \frac{r}{\beta} \sum_{N=0}^{\infty} \sum_{n=0}^{N} \sin(\beta t - n\beta c) \chi_{[Nc,(N+1)c)}. \tag{11}$$

Again we will consider two cases.   ◀

βc Is not a Multiple of 2π

Lemma 5. *Suppose v is not a multiple of 2π. Let $\gamma = \frac{\sin v}{1-\cos v}$. Then*

1. $\sum_{n=0}^{N} \sin(u + nv) = \frac{1}{2}(\sin u + \gamma \cos u + \sin(u + Nv) - \gamma \cos(u + Nv)).$

2. $\sum_{n=0}^{N} \cos(u + nv) = \frac{1}{2}(\cos u - \gamma \sin u + \cos(u + Nv) + \gamma \sin(u + Nv)).$

Let $u = \beta t$ and $v = -\beta c$. By the first part of Lemma 5, we get

$$y(t) = \frac{r}{2\beta} \sum_{N=0}^{\infty} (\sin \beta t + \gamma \cos \beta t)\, \chi_{[Nc,(N+1)c)}$$

$$+ \frac{r}{2\beta} \sum_{N=0}^{\infty} (\sin \beta(t - Nc) - \gamma \cos \beta(t - Nc))\, \chi_{[Nc,(N+1)c)}$$

$$= \frac{r}{2\beta}(\sin \beta t + \gamma \cos \beta t + \sin \beta \langle t \rangle_c - \gamma \cos \beta \langle t \rangle_c), \tag{12}$$

where $\gamma = \frac{-\sin \beta c}{1 - \cos \beta c}$. Lemma 3 implies that the solution will be periodic when $\frac{c}{2\pi/\beta} = \frac{\beta c}{2\pi}$ is rational. Consider the following example. Let $r = 2$, $\beta = 1$, and $c = \frac{3\pi}{2}$. Equation (12) becomes

$$y(t) = \sin t + \cos t + \sin \langle t \rangle_c - \cos \langle t \rangle_c \tag{13}$$

and its graph is given in Fig. 6.34. The period is $6\pi = 4c$. Observe that on the interval $[3c, 4c)$, the motion of the mass is completely stopped. At $t = 3c$, the hammer strikes and imparts a velocity that stops the mass dead in its track. At $t = 4c$, the process begins to repeat itself. As in the previous example, this phenomenon occurs in all cases where the solution y is periodic, that is, when $\frac{c}{2\pi/(\beta)} = \frac{\beta c}{2\pi}$ is rational.

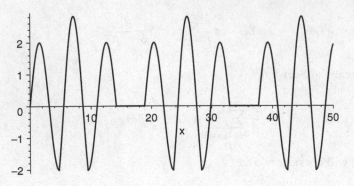

Fig. 6.34 The graph of (13): $c = \frac{3\pi}{2}$

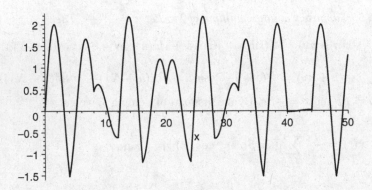

Fig. 6.35 The graph of (14): a nonperiodic solution

Now consider the following example that illustrates a nonperiodic solution. Set $r = 2$, $c = 4$, and $\beta = 1$ in (12). Then

$$y(t) = \sin t - \gamma \cos t + \sin\langle t\rangle_4 - \gamma \cos\langle t\rangle_4, \tag{14}$$

where $\gamma = \frac{-\sin 4}{1 - \cos 4}$. Further, $\frac{c\beta}{2\pi} = \frac{2}{\pi}$ is irrational so $y(t)$ is not periodic. The graph of $y(t)$ in is given Fig. 6.35. Observe that the impulses given every 4 units suddenly changes the direction of the motion in a most erratic way.

When the period of the forcing function is close to that of the period of the spring, the beats in the solution can again be seen. For example, if $c = \frac{9}{8}(2\pi) = \frac{9\pi}{4}$, $\beta = 1$, and $r = 2$, then (12) becomes

$$y(t) = \sin t - \left(\sqrt{2} + 1\right)\cos t + \sin\langle t\rangle_c + \left(\sqrt{2} + 1\right)\cos\langle t\rangle_c \tag{15}$$

and Fig. 6.36 shows its graph.

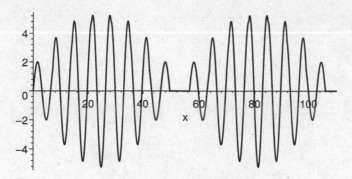

Fig. 6.36   The graph of (15) with $c = \frac{9\pi}{4}$. A solution that demonstrates beats

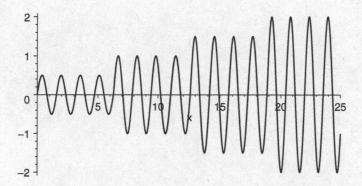

Fig. 6.37   The graph of (16) with $r = 2$, $c = 2\pi$, and $\beta = 4$. Resonance is evident

βc Is a Multiple of 2π

In this case, (11) simplifies to

$$y(t) = \frac{r}{\beta}\left(\sin\beta t + [t/c]_1 \sin\beta t\right) \tag{16}$$

$$= \frac{r}{\beta}(n+1)\sin\beta t \quad t \in [cn, c(n+1)). \tag{17}$$

The presence of the factor $n+1$ implies that $y(t)$ is unbounded; resonance is present. Figure 6.37 gives a graph of the solution when $c = 2\pi$, $\beta = 4$, and $r = 2$.

Exercises

1–6. For the parameters β, r, and c given in each problem below, determine the solution $y(t)$ to the differential equation

$$y'' + \beta^2 y = r\, \mathrm{sw}_c(t), \quad y(0) = 0, \ y'(0) = 0,$$

which models undamped motion with square-wave forcing function. Is the solution periodic or nonperiodic? Does it exhibit resonance?

1. $r = 4, c = 1, \beta = \sqrt{2}$
2. $r = 2, c = 2\pi, \beta = 1/\pi$
3. $r = 2, c = 2, \beta = \pi$
4. $r = 4, c = 1, \beta = \pi/2$
5. $r = 2, c = 1, \beta = \pi$
6. $r = 3, c = 5\pi, \beta = 1$

7–12. For the parameters β, r and c given in each problem below determine the solution $y(t)$ to the differential equation

$$y'' + \beta^2 y = r \sum_{n=0}^{\infty} \delta_{nc}, \quad y(0) = 0, \ y'(0) = 0,$$

which models undamped motion with a periodic impulse function. Is the solution periodic or non periodic? Does it exhibit resonance?

7. $r = 2, c = \pi, \beta = 1$
8. $r = 1, c = 2, \beta = \pi/4$
9. $r = 2, c = 1, \beta = 1$
10. $r = 2, c = \sqrt{2}, \beta = \pi/2$
11. $r = 2, c = 2\pi, \beta = 1$
12. $r = 2, c = 4, \beta = \pi$

13–16. Euler's formula

$$e^{i\theta} = \cos\theta + i\sin\theta$$

is very useful in establishing Lemmas 2 and 5. These exercises guide you through the verifications.

13. Suppose θ is not a multiple of 2π. Let $\gamma = \gamma(\theta) = \frac{\sin\theta}{1-\cos\theta}$. Use Euler's formula to show that

$$\sum_{n=0}^{N} \cos n\theta = \frac{1}{2}\left(1 + \cos N\theta + \gamma \sin N\theta\right),$$

$$\sum_{n=0}^{N} \sin n\theta = \frac{1}{2}\left(\sin N\theta + \gamma(1 - \cos N\theta)\right).$$

(Hint: Expand $\sum_{n=0}^{N} \left(e^{i\theta}\right)^n$ in two different ways and equate real and imaginary parts, use the formula $1 + x + x^2 + \cdots x^N = \frac{x^{N+1}-1}{x-1}$, and use the trigonometric sum and difference formulas.)

14. Suppose θ is not an odd multiple of π. Let $\alpha = \alpha(\theta) = \frac{\sin\theta}{1+\cos\theta}$. Show that

$$\sum_{n=0}^{N} (-1)^n \cos n\theta = \frac{1}{2}\left(1 + (-1)^N(\cos N\theta - \alpha \sin N\theta)\right),$$

$$\sum_{n=0}^{N} (-1)^n \sin n\theta = \frac{1}{2}\left(-\alpha + (-1)^N(\sin N\theta + \alpha \cos N\theta)\right).$$

15. Prove Lemma 5. Namely, suppose v is not a multiple of 2π. Let $\gamma = \frac{\sin v}{1-\cos v}$. Then

   1. $\sum_{n=0}^{N} \cos(u + nv) = \frac{1}{2}(\cos u - \gamma \sin u + \cos(u + Nv) + \gamma \sin(u + Nv))$.

   2. $\sum_{n=0}^{N} \sin(u + nv) = \frac{1}{2}(\sin u + \gamma \cos u + \sin(u + Nv) - \gamma \cos(u + Nv))$.

16. Prove Lemma 2. Namely, suppose v is not an odd multiple of π and let $\alpha = \frac{\sin(v)}{1+\cos(v)}$. Then

   1. $$\sum_{n=0}^{N} (-1)^n \cos(u + nv) = \frac{1}{2}(\cos u + \alpha \sin u)$$
   $$+ \frac{(-1)^N}{2}(\cos(u + Nv) - \alpha \sin(u + Nv)).$$

   2. $$\sum_{n=0}^{N} (-1)^n \sin(u + nv) = \frac{1}{2}(\sin u - \alpha \cos(u))$$
   $$+ \frac{(-1)^N}{2}(\sin(u + Nv) + \alpha \cos(u + Nv)).$$

6.9   Summary of Laplace Transforms

Laplace transforms and convolutions presented in Chap. 6 are summarized in Tables 6.1–6.3.

Table 6.1 Laplace transform rules

$f(t)$	$F(s)$	Page
Second translation principle		
1. $f(t-c)h(t-c)$	$e^{-sc}F(s)$	405
Corollary to the second translation principle		
2. $g(t)h(t-c)$	$e^{-sc}\mathcal{L}\{g(t+c)\})$	405
Periodic functions		
3. $f(t)$, periodic with period p	$\dfrac{1}{1-e^{-sp}}\int_0^p e^{-st}f(t)\,dt$	455
4. $f(\langle t\rangle_p)$	$\dfrac{1}{1-e^{-sp}}\mathcal{L}\{f(t)-f(t)h(t-p)\}$	458
Staircase functions		
5. $f([t]_p)$	$\dfrac{1-e^{-ps}}{s}\sum_{n=0}^{\infty}f(np)e^{-nps}$	463
Transforms involving $\dfrac{1}{1\pm e^{-sp}}$		
6. $\displaystyle\sum_{N=0}^{\infty}\sum_{n=0}^{N}f(t-np)\chi_{[Np,(N+1)p]}$	$\dfrac{1}{1-e^{-sp}}F(s)$	461
7. $\displaystyle\sum_{N=0}^{\infty}\sum_{n=0}^{N}(-1)^n f(t-np)\chi_{[Np,(N+1)p]}$	$\dfrac{1}{1+e^{-sp}}F(s)$	461

Table 6.2 Laplace transforms

$f(t)$	$F(s)$	Page
The Heaviside function		
1. $h(t-c)$	$\dfrac{e^{-sc}}{s}$	404
The on–off switch		
2. $\chi_{[a,b)}$	$\dfrac{e^{-as}}{s} - \dfrac{e^{-bs}}{s}$	405
The Dirac delta function		
3. δ_c	e^{-cs}	428
The square-wave function		
4. sw_c	$\dfrac{1}{1+e^{-cs}}\dfrac{1}{s}$	456
The sawtooth function		
5. $\langle t \rangle_p$	$\frac{1}{s^2}\left(1 - \frac{spe^{-sp}}{1-e^{-sp}}\right)$	457
Periodic Dirac delta functions		
6. $\delta_0(\langle t \rangle_p)$	$\frac{1}{1-e^{-ps}}$	459
Alternating periodic Dirac delta functions		
7. $(\delta_0 - \delta_p)(\langle t \rangle_{2p})$	$\frac{1}{1+e^{-ps}}$	459

Table 6.3 Convolutions

$f(t)$	$g(t)$	$(f * g)(t)$	Page
1. $f(t)$	$g(t)$	$f * g(t) = \int_0^t f(u)g(t-u)\,du$	439
2. f	$\delta_c(t)$	$f(t-c)h(t-c)$	444
3. f	$\delta_0(t)$	$f(t)$	445

Chapter 7
Power Series Methods

Thus far in our study of linear differential equations, we have imposed severe restrictions on the coefficient functions in order to find solution methods. Two special classes of note are the constant coefficient and Cauchy–Euler differential equations. The Laplace transform method was also useful in solving some differential equations where the coefficients were linear. Outside of special cases such as these, linear second order differential equations with variable coefficients can be very difficult to solve.

In this chapter, we introduce the use of power series in solving differential equations. Here is the main idea. Suppose a second order differential equation

$$a_2(t)y'' + a_1(t)y' + a_0(t)y = f(t)$$

is given. Under the right conditions on the coefficient functions, a solution can be expressed in terms of a power series which takes the form

$$y(t) = \sum_{n=0}^{\infty} c_n(t - t_0)^n,$$

for some fixed t_0. Substituting the power series into the differential equation gives relationships among the coefficients $\{c_n\}_{n=0}^{\infty}$, which when solved gives a power series solution. This technique is called the ***power series method***. While we may not enjoy a closed form solution, as in the special cases thus far considered, power series methods imposes the least restrictions on the coefficient functions.

W.A. Adkins and M.G. Davidson, *Ordinary Differential Equations*,
Undergraduate Texts in Mathematics, DOI 10.1007/978-1-4614-3618-8_7,
© Springer Science+Business Media New York 2012

7.1 A Review of Power Series

We begin with a review of the main properties of power series that are usually learned in a first year calculus course.

Definitions and Convergence

A *power series centered at t_0 in the variable t* is a series of the form

$$\sum_{n=0}^{\infty} c_n(t - t_0)^n = c_0 + c_1(t - t_0) + c_2(t - t_0)^2 + \cdots. \tag{1}$$

The *center* of the power series is t_0, and the *coefficients* are the constants $\{c_n\}_{n=0}^{\infty}$. Frequently, we will simply refer to (1) as a *power series*. Let I be the set of real numbers where the series converges. Obviously, t_0 is in I, so I is nonempty. It turns out that I is an interval and is called the *interval of convergence*. It contains an open interval of the form $(t_0 - R, t_0 + R)$ and possibly one or both of the endpoints. The number R is called the *radius of convergence* and can frequently be determined by the ratio test.

The Ratio Test for Power Series Let $\sum_{n=0}^{\infty} c_n(t - t_0)^n$ be a given power series and suppose $L = \lim_{n \to \infty} \left| \frac{c_{n+1}}{c_n} \right|$. *Define R in the following way:*

$$\begin{aligned} R &= 0 & &\text{if } L = \infty, \\ R &= \infty & &\text{if } L = 0, \\ R &= \tfrac{1}{L} & &\text{if } 0 < L < \infty. \end{aligned}$$

Then

1 The power series converges only at $t = t_0$ if $R = 0$.

2 The power series converges absolutely for all $t \in \mathbb{R}$ if $R = \infty$.

3 The power series converges absolutely when $|t - t_0| < R$ and diverges when $|t - t_0| > R$ if $0 < R < \infty$.

If $R = 0$, then I is the degenerate interval $[t_0, t_0]$, and if $R = \infty$, then $I = (-\infty, \infty)$. If $0 < R < \infty$, then I is the interval $(t_0 - R, t_0 + R)$ and possibly the endpoints, $t_0 - R$ and $t_0 + R$, which one must check separately using other tests of convergence.

Recall that *absolute convergence* means that $\sum_{n=0}^{\infty} |c_n(t - t_0)^n|$ converges and implies the original series converges. One of the important advantages absolute convergence gives us is that we can add up the terms in a series in any order we

please and still get the same result. For example, we can add all the even terms and
then the odd terms separately. Thus,

$$\sum_{n=0}^{\infty} c_n(t - t_0)^n = \sum_{n \text{ odd}} c_n(t - t_0)^n + \sum_{n \text{ even}} c_n(t - t_0)^n$$

$$= \sum_{n=0}^{\infty} c_{2n+1}(t - t_0)^{2n+1} + \sum_{n=0}^{\infty} c_{2n}(t - t_0)^{2n}.$$

Example 1. Find the interval of convergence of the power series

$$\sum_{n=1}^{\infty} \frac{(t - 4)^n}{n2^n}.$$

▶ **Solution.** The ratio test gives

$$\left| \frac{c_{n+1}}{c_n} \right| = \frac{n2^n}{(n + 1)2^{n+1}} = \frac{n}{2(n + 1)} \to \frac{1}{2}$$

as $n \to \infty$. The radius of convergence is 2. The interval of convergence has 4 as the
center, and thus, the endpoints are 2 and 6. When $t = 2$, the power series reduces to
$\sum_{n=1}^{\infty} \frac{(-1)^n}{n}$, which is the alternating harmonic series and known to converge. When
$t = 6$, the power series reduces to $\sum_{n=0}^{\infty} \frac{1}{n}$, which is the harmonic series and known
to diverge. The interval of convergence is thus $I = [2, 6)$.   ◀

Example 2. Find the interval of convergence of the power series

$$J_0(t) = \sum_{n=0}^{\infty} \frac{(-1)^n t^{2n}}{2^{2n}(n!)^2}.$$

▶ **Solution.** Let $u = t^2$. We apply the ratio test to $\sum_{n=0}^{\infty} \frac{(-1)^n u^n}{2^{2n}(n!)^2}$ to get

$$\left| \frac{c_{n+1}}{c_n} \right| = \frac{2^{2n}(n!)^2}{2^{2(n+1)}((n + 1)!)^2} = \frac{1}{4(n + 1)^2} \to 0$$

as $n \to \infty$. It follows that $R = \infty$ and the series converges for all u. Hence,
$\sum_{n=0}^{\infty} \frac{(-1)^n t^{2n}}{2^{2n}(n!)^2}$ converges for all t and $I = (-\infty, \infty)$.   ◀

In each example, the power series defines a function on its interval of convergence. In Example 2, the function $J_0(t)$ is known as the **Bessel function of
order** 0 and plays an important role in many physical problems. More generally,
let $f(t) = \sum_{n=0}^{\infty} c_n(t - t_0)^n$ for all $t \in I$. Then f is a function on the interval
of convergence I, and (1) is its **power series representation**. A simple example

of a power series representation is a polynomial defined on $\mathbb{R}$. In this case, the coefficients are all zero except for finitely many. Other well-known examples from calculus are:

$$e^t = \sum_{n=0}^{\infty} \frac{t^n}{n!} = 1 + t + \frac{t^2}{2} + \frac{t^3}{3!} + \cdots, \tag{2}$$

$$\cos t = \sum_{n=0}^{\infty} \frac{(-1)^n t^{2n}}{(2n)!} = 1 - \frac{t^2}{2} + \frac{t^4}{4!} - \cdots, \tag{3}$$

$$\sin t = \sum_{n=0}^{\infty} \frac{(-1)^n t^{2n+1}}{(2n+1)!} = t - \frac{t^3}{3!} + \frac{t^5}{5!} - \cdots, \tag{4}$$

$$\frac{1}{1-t} = \sum_{n=0}^{\infty} t^n = 1 + t + t^2 + \cdots, \tag{5}$$

$$\ln t = \sum_{n=1}^{\infty} \frac{(-1)^{(n+1)}(t-1)^n}{n!} = (t-1) - \frac{(t-1)^2}{2} + \frac{(t-1)^3}{3} - \cdots. \tag{6}$$

Equations (2), (3), and (4) are centered at 0 and have interval of convergence $(-\infty, \infty)$. Equation (5), known as the **geometric series**, is centered at 0 and has interval of convergence $(-1, 1)$. Equation (6) is centered at 1 and has interval of convergence $(0, 2]$.

Index Shifting

In calculus, the variable x in a definite integral $\int_a^b f(x)\,dx$ is called a dummy variable because the value of the integral is independent of x. Sometimes it is convenient to change the variable. For example, if we replace x by $x - 1$ in the integral $\int_1^2 \frac{1}{x+1}\,dx$, we obtain

$$\int_1^2 \frac{1}{x+1}\,dx = \int_{x-1=1}^{x-1=2} \frac{1}{x-1+1}\,d(x-1) = \int_2^3 \frac{1}{x}\,dx.$$

In like manner, the index n in a power series is referred to as a dummy variable because the sum is independent of n. It is also sometimes convenient to make a change of variable, which, for series, is called an **index shift**. For example, in the series $\sum_{n=0}^{\infty}(n+1)t^{n+1}$, we replace n by $n - 1$ to obtain

$$\sum_{n=0}^{\infty}(n+1)t^{n+1} = \sum_{n-1=0}^{n-1=\infty}(n-1+1)t^{n-1+1} = \sum_{n=1}^{\infty} nt^n.$$

The lower limit $n = 0$ is replaced by $n - 1 = 0$ or $n = 1$. The upper limit $n = \infty$ is replace by $n - 1 = \infty$ or $n = \infty$. The terms $(n + 1)t^{n+1}$ in the series go to $(n - 1 + 1)t^{n-1+1} = nt^n$.

When a power series is given is such a way that the index n in the sum is the power of $(t - t_0)$, we say the power series is written in **standard form**. Thus, $\sum_{n=1}^{\infty} nt^n$ is in standard form while $\sum_{n=0}^{\infty}(n + 1)t^{n+1}$ is not.

Example 3. Make an index shift so that the series $\sum_{n=2}^{\infty} \frac{t^{n-2}}{n^2}$ is expressed as a series in standard form.

▶ **Solution.** We replace n by $n + 2$ and get

$$\sum_{n=2}^{\infty} \frac{t^{n-2}}{n^2} = \sum_{n+2=2}^{n+2=\infty} \frac{t^{n+2-2}}{(n+2)^2} = \sum_{n=0}^{\infty} \frac{t^n}{(n+2)^2}. \qquad ◀$$

Differentiation and Integration of Power Series

If a function can be represented by a power series, then we can compute its derivative and integral by differentiating and integrating each term in the power series as noted in the following theorem.

Theorem 4. *Suppose*

$$f(t) = \sum_{n=0}^{\infty} c_n(t - t_0)^n$$

is defined by a power series with radius of convergence $R > 0$. Then f is differentiable and integrable on $(t_0 - R, t_0 + R)$ and

$$f'(t) = \sum_{n=1}^{\infty} nc_n(t - t_0)^{n-1} \qquad (7)$$

and

$$\int f(t)\,dt = \sum_{n=0}^{\infty} c_n \frac{(t - t_0)^{n+1}}{n + 1} + C. \qquad (8)$$

Furthermore, the radius of convergence for the power series representations of f' and $\int f$ are both R.

We note that the presence of the factor n in $f'(t)$ allows us to write

$$f'(t) = \sum_{n=0}^{\infty} nc_n(t - t_0)^{n-1}$$

since the term at $n = 0$ is zero. This observation is occasionally used. Consider the following examples.

Example 5. Find a power series representation for $\frac{1}{(1-t)^2}$ in standard form.

▶ **Solution.** If $f(t) = \frac{1}{1-t}$, then $f'(t) = \frac{1}{(1-t)^2}$. If follows from Theorem 4 that

$$\frac{1}{(1-t)^2} = \frac{d}{dt} \sum_{n=0}^{\infty} t^n = \sum_{n=1}^{\infty} n t^{n-1} = \sum_{n=0}^{\infty} (n+1) t^n. \qquad \blacktriangleleft$$

Example 6. Find the power series representation for $\ln(1-t)$ in standard form.

▶ **Solution.** For $t \in (-1,1)$, $\ln(1-t) = -\int \frac{1}{1-t} + C$. Thus,

$$\ln(1-t) = C - \int \sum_{n=0}^{\infty} t^n \, dt = C - \sum_{n=0}^{\infty} \frac{t^{n+1}}{n+1} = C - \sum_{n=1}^{\infty} \frac{t^n}{n}.$$

Evaluating both side at $t = 0$ gives $C = 0$. It follows that

$$\ln(1-t) = -\sum_{n=1}^{\infty} \frac{t^n}{n}. \qquad \blacktriangleleft$$

The Algebra of Power Series

Suppose $f(t) = \sum_{n=0}^{\infty} a_n (t - t_0)^n$ and $g(t) = \sum_{n=0}^{\infty} b_n (t - t_0)^n$ are power series representation of f and g and converge on the interval $(t_0 - R, t_0 + R)$ for some $R > 0$. Then

$$f(t) = g(t) \text{ if and only if } a_n = b_n,$$

for all $n = 1, 2, 3, \ldots$. Let $c \in \mathbb{R}$. Then the power series representation of $f \pm g$, cf, fg, and f/g are given by

$$f(t) \pm g(t) = \sum_{n=0}^{\infty} (a_n \pm b_n)(t - t_0)^n, \qquad (9)$$

$$cf(t) = \sum_{n=0}^{\infty} c a_n (t - t_0)^n, \qquad (10)$$

$$f(t)g(t) = \sum_{n=0}^{\infty} c_n (t - t_0)^n \quad \text{where } c_n = a_0 b_n + a_1 b_{n-1} + \cdots + a_n b_0, \quad (11)$$

and $\quad \dfrac{f(t)}{g(t)} = \displaystyle\sum_{n=0}^{\infty} d_n (t - t_0)^n, \quad g(a) \neq 0, \qquad (12)$

where each d_n is determined by the equation $f(t) = g(t) \sum_{n=0}^{\infty} d_n(t - t_0)^n$. In (9), (10), and (11), the series converges on the interval $(t_0 - R, t_0 + R)$. For division of power series, (12), the radius of convergence is positive but may not be as large as R.

Example 7. Compute the power series representations of

$$\cosh t = \frac{e^t + e^{-t}}{2} \quad \text{and} \quad \sinh t = \frac{e^t - e^{-t}}{2}.$$

▶ **Solution.** We write out the terms in each series, e^t and e^{-t}, and get

$$e^t = 1 + t + \frac{t^2}{2!} + \frac{t^3}{3!} + \frac{t^4}{4!} + \cdots,$$

$$e^{-t} = 1 - t + \frac{t^2}{2!} - \frac{t^3}{3!} + \frac{t^4}{4!} - \cdots,$$

$$e^t + e^{-t} = 2 + 2\frac{t^2}{2!} + 2\frac{t^4}{4!} + \cdots,$$

$$e^t - e^{-t} = 2t + 2\frac{t^3}{3!} + 2\frac{t^5}{5!} + \cdots.$$

It follows that

$$\cosh t = \sum_{n=0}^{\infty} \frac{t^{2n}}{(2n)!} \quad \text{and} \quad \sinh t = \sum_{n=0}^{\infty} \frac{t^{2n+1}}{(2n+1)!}. \qquad \blacktriangleleft$$

Example 8. Let $y(t) = \sum_{n=0}^{\infty} c_n t^n$. Compute

$$(1 + t^2)y'' + 4ty' + 2y$$

as a power series.

▶ **Solution.** We differentiate y twice to get

$$y'(t) = \sum_{n=1}^{\infty} c_n n t^{n-1} \quad \text{and} \quad y''(t) = \sum_{n=2}^{\infty} c_n n(n - 1)t^{n-2}.$$

In the following calculations, we shift indices as necessary to obtain series in standard form:

$$t^2 y'' = \sum_{n=2}^{\infty} c_n n(n - 1)t^n = \sum_{n=0}^{\infty} c_n n(n - 1)t^n,$$

$$y'' = \sum_{n=2}^{\infty} c_n n(n - 1)t^{n-2} = \sum_{n=0}^{\infty} c_{n+2}(n + 2)(n + 1)t^n,$$

$$4ty' = \sum_{n=1}^{\infty} 4c_n n t^n = \sum_{n=0}^{\infty} 4c_n n t^n,$$

$$2y = \sum_{n=0}^{\infty} 2c_n t^n.$$

Notice that the presence of the factors n and $n-1$ in the first series allows us to write it with a starting point $n=0$ instead of $n=2$, similarly for the third series. Adding these results and simplifying gives

$$(1+t^2)y'' + 4ty' + 2y = \sum_{n=0}^{\infty} ((c_{n+2} + c_n)(n+2)(n+1))\, t^n. \quad \blacktriangleleft$$

A function f is said to be an **odd** function if $f(-t) = -f(t)$ and **even** if $f(-t) = f(t)$. If f is odd and has a power series representation with center 0, then all coefficients of even powers of t are zero. Similarly, if f is even, then all the coefficients of odd powers are zero. Thus, f has the following form:

$$f(t) = a_0 + a_2 t^2 + a_4 t^4 + \ldots = \sum_{n=0}^{\infty} a_{2n} t^{2n} \qquad f\text{ -even,}$$

$$f(t) = a_1 t + a_3 t^3 + a_5 t^5 + \ldots = \sum_{n=0}^{\infty} a_{2n+1} t^{2n+1} \quad f\text{ -odd.}$$

For example, the power series representations of $\cos t$ and $\cosh t$ reflect that they are even, while those of $\sin t$ and $\sinh t$ reflect that they are odd functions.

Example 9. Compute the first four nonzero terms in the power series representation of

$$\tanh t = \frac{\sinh t}{\cosh t}.$$

▶ **Solution.** Division of power series is generally complicated. To make things a little simpler, we observe that $\tanh t$ is an odd function. Thus, its power series expansion is of the form $\tanh t = \sum_{n=1}^{\infty} d_{2n+1} t^{2n+1}$ and satisfies $\sinh t = \cosh t \sum_{n=0}^{\infty} d_{2n+1} t^{2n+1}$. By Example 7, this means

$$\left(t + \frac{t^3}{3!} + \frac{t^5}{5!} + \cdots\right) = \left(1 + \frac{t^2}{2!} + \frac{t^4}{4!} + \cdots\right)\left(d_1 t + d_3 t^3 + d_5 t^5 + \cdots\right)$$

$$= \left(d_1 t + \left(d_3 + \frac{d_1}{2!}\right) t^3 + \left(d_5 + \frac{d_3}{2!} + \frac{d_1}{4!}\right) t^5 + \cdots\right).$$

We now equate coefficients to get the following sequence of equations:

$$d_1 = 1$$

$$d_3 + \frac{d_1}{2!} = \frac{1}{3!}$$

$$d_5 + \frac{d_3}{2!} + \frac{d_1}{4!} = \frac{1}{5!}$$

$$d_7 + \frac{d_5}{2!} + \frac{d_3}{4!} + \frac{d_1}{6!} = \frac{1}{7!}$$

$$\vdots$$

Recursively solving these equations gives $d_1 = 1$, $d_3 = \frac{-1}{3}$, $d_5 = \frac{2}{15}$, and $d_7 = \frac{-17}{315}$. The first four nonzero terms in the power series expansion for $\tanh t$ is thus

$$\tanh t = t - \frac{1}{3}t^3 + \frac{2}{15}t^5 - \frac{17}{315}t^7 + \cdots .$$ ◀

Identifying Power Series

Given a power series, with positive radius of convergence, it is sometimes possible to identify it with a known function. When we can do this, we will say that it is written in **closed form**. Usually, such identifications come by using a combination of differentiation, integration, or the algebraic properties of power series discussed above. Consider the following examples.

Example 10. Write the power series

$$\sum_{n=0}^{\infty} \frac{t^{2n+1}}{n!}$$

in closed form.

▶ **Solution.** Observe that we can factor out t and associate the term t^2 to get

$$\sum_{n=0}^{\infty} \frac{t^{2n+1}}{n!} = t \sum_{n=0}^{\infty} \frac{(t^2)^n}{n!} = te^{t^2},$$

from (2).   ◀

Example 11. Write the power series

$$\sum_{n=1}^{\infty} n(-1)^n t^{2n}$$

in closed form.

▶ **Solution.** Let $z(t) = \sum_{n=1}^{\infty} n(-1)^n t^{2n}$. Then dividing both sides by t gives

$$\frac{z(t)}{t} = \sum_{n=1}^{\infty} n(-1)^n t^{2n-1}.$$

Integration will now simplify the sum:

$$\int \frac{z(t)}{t} \, dt = \sum_{n=1}^{\infty} \int n(-1)^n t^{2n-1} \, dt$$

$$= \sum_{n=1}^{\infty} (-1)^n \frac{t^{2n}}{2}$$

$$= \frac{1}{2} \sum_{n=1}^{\infty} (-t^2)^n$$

$$= \frac{1}{2} \left(\frac{1}{1+t^2} - 1 \right) + c,$$

where the last line is obtained from the geometric series by adding and subtracting the $n = 0$ term and c is a constant of integration. We now differentiate this equation to get

$$\frac{z(t)}{t} = \frac{1}{2} \frac{d}{dt} \left(\frac{1}{1+t^2} - 1 \right)$$

$$= \frac{-t}{(1+t^2)^2}.$$

It follows now that

$$z(t) = \frac{-t^2}{(1+t^2)^2}.$$

It is straightforward to check that the radius of convergence is 1 so we get the equality

$$\frac{-t^2}{(1+t^2)^2} = \sum_{n=1}^{\infty} n(-1)^n t^{2n},$$

on the interval $(-1, 1)$.   ◀

Taylor Series

Suppose $f(t) = \sum_{n=0}^{\infty} c_n (t - t_0)^n$, with positive radius of converge. Theorem 4 implies that the derivatives, $f^{(n)}$, exist for all $n = 0, 1, \ldots$. Furthermore, it is easy to check that $f^{(n)}(t_0) = n!c_n$ and thus $c_n = \frac{f^{(n)}(t_0)}{n!}$. Therefore, if f is represented by a power series, then it must be that

$$f(t) = \sum_{n=0}^{\infty} \frac{f^{(n)}(t_0)}{n!}(t - t_0)^n. \tag{13}$$

This series is called the *Taylor series of f centered at t_0*.

Now let us suppose that f is a function on some domain D and we wish to find a power series representation centered at $t_0 \in D$. By what we have just argued, f will have to be given by its Taylor Series, which, of course, means that all higher order derivatives of f at t_0 must exist. However, it can happen that the Taylor series may not converge to f on any interval containing t_0 (see Exercises 28–29 where such an example is considered). When (13) is valid on an open interval containing t_0, we call f *analytic at t_0*. The properties of power series listed above shows that the sum, difference, scalar multiple, and product of analytic functions is again analytic. The quotient of analytic functions is likewise analytic at points where the denominator is not zero. Derivatives and integrals of analytic functions are again analytic.

Example 12. Verify that the Taylor series of $\sin t$ centered at 0 is that given in (4).

▶ **Solution.** The first four derivatives of $\sin t$ and their values at 0 are as follows:

order n	$\sin^{(n)}(t)$	$\sin^{(n)}(0)$
$n = 0$	$\sin t$	0
$n = 1$	$\cos t$	1
$n = 2$	$-\sin t$	0
$n = 3$	$-\cos t$	-1
$n = 4$	$\sin t$	0

The sequence $0, 1, 0, -1$ thereafter repeats. Hence, the Taylor series of $\sin t$ is

$$0 + 1t + 0\frac{t^2}{2!} - \frac{t^3}{3!} + 0\frac{t^4}{4!} + 1\frac{t^5}{5!} + \cdots = \sum_{n=0}^{\infty} \frac{t^{2n+1}}{(2n + 1)!},$$

as in (4).   ◀

Given a function f, it can sometimes be difficult to compute the Taylor series by computing $f^{(n)}(t_0)$ for all $n = 0, 1, \ldots$. For example, compute the first few derivatives of $\tanh t$, considered in Example 9, to see how complicated the derivatives become. Additionally, determining whether the Taylor series converges to f requires some additional information, for example, the Taylor remainder theorem. We will not include this in our review. Rather we will stick to examples where we derive new power series representations from existing ones as we did in Examples 5–9.

Rational Functions

A *rational function* is the quotient of two polynomials and is analytic at all points where the denominator is nonzero. Rational functions will arise in many examples. It will be convenient to know what the radius of convergence about a point t_0. The following theorem allows us to determine this without going through the work of determining the power series. The proof is beyond the scope of this text.

Theorem 13. *Suppose $\frac{p(t)}{q(t)}$ is a quotient of two polynomials p and q. Suppose $q(t_0) \neq 0$. Then the power series expansion for $\frac{p}{q}$ about t_0 has radius of convergence equal to the closest distance from t_0 to the roots (including complex roots) of q.*

Example 14.   Find the radius of convergence for each rational function about the given point.

1. $\frac{t}{4-t}$ about $t_0 = 1$

2. $\frac{1-t}{9-t^2}$ about $t_0 = 2$

3. $\frac{t^3}{t^2+1}$ about $t_0 = 2$

▶ **Solution.**

1. The only root of $4 - t$ is 4. Its distance to $t_0 = 1$ is 3. The radius of convergence is 3.

2. The roots of $9-t^2$ are 3 and -3. Their distances to $t_0 = 2$ is 1 and 5, respectively. The radius of convergence is 1.

3. The roots of $t^2 + 1$ are i and $-$i. Their distances to $t_0 = 2$ are $|2 - i| = \sqrt{5}$ and $|2 + i| = \sqrt{5}$. The radius of convergence is $\sqrt{5}$.   ◀

Exercises

1–9. Compute the radius of convergence for the given power series.

1. $\sum_{n=0}^{\infty} n^2(t-2)^n$

2. $\sum_{n=1}^{\infty} \dfrac{t^n}{n}$

3. $\sum_{n=0}^{\infty} \dfrac{(t-1)^n}{2^n n!}$

4. $\sum_{n=0}^{\infty} \dfrac{3^n(t-3)^n}{n+1}$

5. $\sum_{n=0}^{\infty} n! t^n$

6. $\sum_{n=0}^{\infty} \dfrac{(-1)^n t^{2n}}{(2n+1)!}$

7. $\sum_{n=0}^{\infty} \dfrac{(-1)^n t^{2n+1}}{(2n)!}$

8. $\sum_{n=0}^{\infty} \dfrac{n^n t^n}{n!}$

9. $\sum_{n=0}^{\infty} \dfrac{n! t^n}{1 \cdot 3 \cdot 5 \cdots (2n+1)}$

10–16. Find the Taylor series for each function with center $t_0 = 0$.

10. $\dfrac{1}{1+t^2}$

11. $\dfrac{1}{t-a}$

12. e^{at}

13. $\dfrac{\sin t}{t}$

14. $\dfrac{e^t - 1}{t}$

15. $\tan^{-1} t$

16. $\ln(1 + t^2)$

17–20. Find the first four nonzero terms in the Taylor series with center 0 for each function.

17. $\tan t$

18. $\sec t$

19. $e^t \sin t$

20. $e^t \cos t$

21–25. Find a closed form expression for each power series.

21. $\sum_{n=0}^{\infty} (-1)^n \frac{n+1}{n!} t^n$

22. $\sum_{n=0}^{\infty} \frac{3^n - 2}{n!} t^n$

23. $\sum_{n=0}^{\infty} (n+1) t^n$

24. $\sum_{n=0}^{\infty} \frac{t^{2n+1}}{2n-1}$

25. $\sum_{n=0}^{\infty} \frac{t^{2n+1}}{(2n+1)(2n-1)}$

26. Redo Exercises 19 and 20 in the following way. Recall Euler's formula $e^{it} = \cos t + i \sin t$ and write $e^t \cos t$ and $e^t \sin t$ as the real and imaginary parts of $e^t e^{it} = e^{(1+i)t}$ expanded as a power series.

27. Use the power series (with center 0) for the exponential function and expand both sides of the equation $e^{at} e^{bt} = e^{(a+b)t}$. What well-known formula arises when the coefficients of $\frac{t^n}{n!}$ are equated?

28–29. A test similar to the ratio test is the root test.

The Root Test for Power Series. Let $\sum_{n=0}^{\infty} c_n (t - t_0)^n$ *be a given power series and suppose* $L = \lim_{n \to \infty} \sqrt[n]{|c_n|}$. *Define R in the following way:*

$$
\begin{aligned}
R = \tfrac{1}{L} \quad &\text{if} \quad 0 < L < \infty, \\
R = 0 \quad &\text{if} \quad L = \infty, \\
R = \infty \quad &\text{if} \quad L = 0.
\end{aligned}
$$

Then

i. The power series converges only at $t = t_0$ if $R = 0$.

ii. The power series converges for all $t \in \mathbb{R}$ if $R = \infty$.

iii. The power series converges if $|t - t_0| < R$ and diverges if $|t - t_0| > R$.

28. Use the root test to determine the radius of convergence of $\sum_{n=0}^{\infty} \frac{t^n}{n^n}$.
29. Let $c_n = 1$ if n is odd and $c_n = 2$ if n is even. Consider the power series $\sum_{n=0}^{\infty} c_n t^n$. Show that the ratio test does not apply. Use the root test to determine the radius of convergence.

30–34. In this sequence of exercises, we consider a function that is infinitely differentiable but not analytic. Let

$$f(t) = \begin{cases} 0 & \text{if } t \leq 0 \\ e^{\frac{-1}{t}} & \text{if } t > 0 \end{cases}.$$

30. Compute $f'(t)$ and $f''(t)$ and observe that $f^{(n)}(t) = e^{\frac{-1}{t}} p_n(\frac{1}{t})$ where p_n is a polynomial, $n = 1, 2$. Find p_1 and p_2.
31. Use mathematical induction to show that $f^{(n)}(t) = e^{\frac{-1}{t}} p_n(\frac{1}{t})$ where p_n is a polynomial.
32. Show that $\lim_{t \to 0+} f^{(n)}(t) = 0$. To do this, let $u = \frac{1}{t}$ in $f^{(n)}(t) = e^{\frac{-1}{t}} p_n(\frac{1}{t})$ and let $u \to \infty$. Apply L'Hospital's rule.
33. Show that $f^n(0) = 0$ for all $n = 0, 1, \ldots$.
34. Conclude that f is not analytic at $t = 0$ though all derivatives at $t = 0$ exist.

7.2   Power Series Solutions About an Ordinary Point

A point t_0 is called an *ordinary point* of $Ly = 0$ if we can write the differential equation in the form

$$y'' + a_1(t)y' + a_0(t)y = 0, \tag{1}$$

where $a_0(t)$ and $a_1(t)$ are analytic at t_0. If t_0 is not an ordinary point, we call it a *singular point*.

Example 1. Determine the ordinary and singular points for each of the following differential equations:

1. $y'' + \frac{1}{t^2-9}y' + \frac{1}{t+1}y = 0.$
2. $(1 - t^2)y'' - 2ty' + n(n + 1)y = 0$, where n is an integer.
3. $ty'' + (\sin t)y' + (e^t - 1)y = 0.$

▶ **Solution.**
1. Here $a_1(t) = \frac{1}{t^2-9}$ is analytic except at $t = \pm3$. The function $a_0 = \frac{1}{t+1}$ is analytic except at $t = -1$. Thus, the singular points are -3, 3, and -1. All other points are ordinary.
2. This is Legendre's equation. In standard form, we find $a_1(t) = \frac{-2t}{1-t^2}$ and $a_0(t) = \frac{n(n+1)}{1-t^2}$. They are analytic except at 1 and -1. These are the singular points and all other points are ordinary.
3. In standard form, $a_1(t) = \frac{\sin t}{t}$ and $a_0(t) = \frac{e^t - 1}{t}$. Both of these are analytic everywhere. (See Exercises 13 and 14 of Sect. 7.1.) It follows that all points are ordinary.   ◀

In this section, we restrict our attention to ordinary points. Their importance is underscored by the following theorem. It tells us that there is always a power series solution about ordinary points.

Theorem 2. *Suppose $a_0(t)$ and $a_1(t)$ are analytic at t_0, both of which converge for $|t - t_0| < R$. Then there is a unique solution $y(t)$, analytic at t_0, to the initial value problem*

$$y'' + a_1(t)y' + a_0(t)y = 0, \quad y(t_0) = \alpha, \ y'(t_0) = \beta. \tag{2}$$

If

$$y(t) = \sum_{n=0}^{\infty} c_n(t - t_0)^n$$

then $c_0 = \alpha$, $c_1 = \beta$, and all other c_k, $k = 2, 3, \ldots,$ are determined by c_0 and c_1. Furthermore, the power series for y converges for $|t - t_0| < R$.

Of course the uniqueness and existence theorem, Theorem 6 of Sect. 5.1, implies there is a unique solution. What is new here is that the solution is analytic at t_0. Since the solution is necessarily unique, it is not at all surprising that the coefficients are determined by the initial conditions. The only hard part about the proof, which we omit, is showing that the solution converges for $|t - t_0| < R$. Let y_1 be the solution with initial conditions $y(t_0) = 1$ and $y'(t_0) = 0$ and y_2 the solution with initial condition $y(t_0) = 0$ and $y'(t_0) = 1$. Then it is easy to see that y_1 and y_2 are independent solutions, and hence, all solutions are of the form $c_1 y_1 + c_2 y_2$. (See Theorem 2 of Sect. 5.2) The **power series method** refers to the use of this theorem by substituting $y(t) = \sum_{n=0}^{\infty} c_n(t - t_0)^n$ into (2) and determining the coefficients.

We illustrate the use of Theorem 2 with a few examples. Let us begin with a familiar constant coefficient differential equation.

Example 3. Use the power series method to solve

$$y'' + y = 0.$$

▶ **Solution.** Of course, this is a constant coefficient differential equation. Since $q(s) = s^2 + 1$ and $\mathcal{B}_q = \{\cos t, \sin t\}$, we get solution $y(t) = c_1 \sin t + c_2 \cos t$. Let us see how the power series method gives the same answer. Since the coefficients are constant, they are analytic everywhere with infinite radius of convergence. Theorem 2 implies that the power series solutions converge everywhere. Let $y(t) = \sum_{n=0}^{\infty} c_n t^n$ be a power series about $t_0 = 0$. Then

$$y'(t) = \sum_{n=1}^{\infty} c_n n t^{n-1}$$

$$\text{and} \quad y''(t) = \sum_{n=2}^{\infty} c_n n(n-1) t^{n-2}.$$

An index shift, $n \to n + 2$, gives $y''(t) = \sum_{n=0}^{\infty} c_{n+2}(n + 2)(n + 1)t^n$. Therefore, the equation $y'' + y = 0$ gives

$$\sum_{n=0}^{\infty} (c_n + c_{n+2}(n + 2)(n + 1))t^n = 0,$$

which implies $c_n + c_{n+2}(n + 2)(n + 1) = 0$, or, equivalently,

$$c_{n+2} = \frac{-c_n}{(n + 2)(n + 1)} \quad \text{for all } n = 0, 1, \ldots. \tag{3}$$

Equation (3) is an example of a **recurrence relation**: terms of the sequence are determined by earlier terms. Since the difference in indices between c_n and c_{n+2} is 2, it follows that even terms are determined by previous even terms and odd terms are determined by previous odd terms. Let us consider these two cases separately.

	The Even Case		*The Odd Case*
$n = 0$	$c_2 = \frac{-c_0}{2\cdot 1}$	$n = 1$	$c_3 = \frac{-c_1}{3\cdot 2}$
$n = 2$	$c_4 = \frac{-c_2}{4\cdot 3} = \frac{c_0}{4\cdot 3\cdot 2\cdot 1} = \frac{c_0}{4!}$	$n = 3$	$c_5 = \frac{-c_3}{5\cdot 4} = \frac{c_1}{5\cdot 4\cdot 3\cdot 2} = \frac{c_1}{5!}$
$n = 4$	$c_6 = \frac{-c_4}{6\cdot 5} = \frac{-c_0}{6!}$	$n = 5$	$c_7 = \frac{-c_5}{7\cdot 6} = \frac{-c_1}{7!}$
$n = 6$	$c_8 = \frac{-c_6}{8\cdot 7} = \frac{c_0}{8!}$	$n = 7$	$c_9 = \frac{-c_7}{9\cdot 8} = \frac{c_1}{9!}$
$\vdots$	$\vdots$	$\vdots$	$\vdots$

More generally, we can see that $\qquad\qquad$ Similarly, we see that

$$c_{2n} = (-1)^n \frac{c_0}{(2n)!}. \qquad\qquad c_{2n+1} = (-1)^n \frac{c_1}{(2n+1)!}.$$

Now, as we mentioned in Sect. 7.1, we can change the order of absolutely convergent sequences without affecting the sum. Thus, let us rewrite $y(t) = \sum_{n=0}^{\infty} c_n t^n$ in terms of odd and even indices to get

$$y(t) = \sum_{n=0}^{\infty} c_{2n} t^{2n} + \sum_{n=0}^{\infty} c_{2n+1} t^{2n+1}$$

$$= c_0 \sum_{n=0}^{\infty} \frac{(-1)^n}{(2n)!} t^{2n} + c_1 \sum_{n=0}^{\infty} \frac{(-1)^n}{(2n+1)!} t^{2n+1}$$

$$= c_0 \cos t + c_1 \sin t,$$

where the first power series in the second line is that of $\cos t$ and the second power series is that of $\sin t$ (See (3) and (4) of Sect. 7.1). ◀

Example 4. Use the power series method with center $t_0 = 0$ to solve

$$(1 + t^2)y'' + 4ty' + 2y = 0.$$

What is a lower bound on the radius of convergence?

▶ **Solution.** We write the given equation in standard form to get

$$y'' + \frac{4t}{1 + t^2} y' + \frac{2}{1 + t^2} y = 0.$$

Since the coefficient functions $a_1(t) = \frac{4t}{1+t^2}$ and $a_2(t) = \frac{2}{1+t^2}$ are rational functions with nonzero denominators, they are analytic at all points. By Theorem 13 of Sect. 7.1, it is not hard to see that they have power series expansions about $t_0 = 0$ with radius of convergence 1. By Theorem 2, the radius of convergence for a solution, $y(t) = \sum_{n=0}^{\infty} c_n t^n$ is at least 1. To determine the coefficients, it is easier to substitute $y(t)$ directly into $(1 + t^2)y'' + 4ty' + 2y = 0$ instead of its equivalent standard form. The details were worked out in Example 8 of Sect. 7.1. We thus obtain

$$(1+t^2)y'' + 4ty' + 2y = \sum_{n=0}^{\infty} ((c_{n+2} + c_n)(n+2)(n+1)) t^n = 0.$$

From this equation, we get $c_{n+2} + c_n = 0$ for all $n = 0, 1, \cdots$. Again we consider even and odd cases.

	The Even Case		*The Odd Case*
$n = 0$	$c_2 = -c_0$	$n = 1$	$c_3 = -c_1$
$n = 2$	$c_4 \doteq -c_2 = c_0$	$n = 3$	$c_5 = -c_3 = c_1$
$n = 4$	$c_6 = -c_0$	$n = 5$	$c_7 = -c_1$

More generally, we can see that

$$c_{2n} = (-1)^n c_0.$$

Similarly, we see that

$$c_{2n+1} = (-1)^n c_1.$$

It follows now that

$$y(t) = c_0 \sum_{n=0}^{\infty} (-1)^n t^{2n} + c_1 \sum_{n=0}^{\infty} (-1)^n t^{2n+1}.$$

As we observed earlier, each of these series has radius of convergence at least 1. In fact, the radius of convergence of each is 1. ◄

A couple of observations are in order for this example. First, we can relate the power series solutions to the geometric series, (5) of Sect. 7.1, and write them in closed form. Thus,

$$\sum_{n=0}^{\infty} (-1)^n t^{2n} = \sum_{n=0}^{\infty} (-t^2)^n = \frac{1}{1+t^2},$$

$$\sum_{n=0}^{\infty} (-1)^n t^{2n+1} = t \sum_{n=0}^{\infty} (-t^2)^n = \frac{t}{1+t^2}.$$

It follows now that the general solution is $y(t) = c_0 \frac{1}{1+t^2} + c_1 \frac{t}{1+t^2}$. Second, since $a_1(t)$ and $a_0(t)$ are continuous on $\mathbb{R}$, the uniqueness and existence theorem, Theorem 6 of Sect. 5.1, guarantees the existence of solutions defined on all of $\mathbb{R}$. It is easy to check that these closed forms, $\frac{1}{1+t^2}$ and $\frac{t}{1+t^2}$, are defined on all of $\mathbb{R}$ and satisfy the given differential equation.

This example illustrates that there is some give and take between the uniqueness and existence theorem, Theorem 6 of Sect. 5.1, and Theorem 2 above. On the one hand, Theorem 6 of Sect. 5.1 may guarantee a solution, but it may be difficult or impossible to find without the power series method. The power series method, Theorem 2, on the other hand, may only find a series solution on the interval of convergence, which may be quite smaller than that guaranteed by Theorem 6 of Sect. 5.1. Further analysis of the power series may reveal a closed form solution valid on a larger interval as in the example above. However, it is not always possible to do this. Indeed, some recurrence relations can be difficult to solve and we must be satisfied with writing out only a finite number of terms in the power series solution.

Example 5. Discuss the radius of convergence of the power series solution about $t_0 = 0$ to

$$(1 - t)y'' + y = 0.$$

Write out the first five terms given the initial conditions

$$y(0) = 1 \quad \text{and} \quad y'(0) = 0.$$

▶ **Solution.** In standard form, the differential equation is

$$y'' + \frac{1}{1-t}y = 0.$$

Thus, $a_1(t) = 0$, $a_0(t) = \frac{1}{1-t}$, and $t_0 = 0$ is an ordinary point. Since $a_0(t)$ is represented by the geometric series, which has radius of convergence 1, it follows that any solution will have radius of convergence at least 1. Let $y(t) = \sum_{n=0}^{\infty} c_n t^n$. Then y'' and $-ty''$ are given by

$$y''(t) = \sum_{n=0}^{\infty} c_{n+2}(n + 2)(n + 1)t^n,$$

$$-ty''(t) = \sum_{n=2}^{\infty} -c_n n(n - 1)t^{n-1},$$

$$= \sum_{n=0}^{\infty} -c_{n+1}(n + 1)nt^n.$$

It follows that

$$(1 - t)y'' + y = \sum_{n=0}^{\infty} (c_{n+2}(n + 2)(n + 1) - c_{n+1}(n + 1)n + c_n)t^n,$$

which leads to the recurrence relations

$$c_{n+2}(n + 2)(n + 1) - c_{n+1}(n + 1)n + c_n = 0,$$

for all $n = 0, 1, 2, \ldots$. This recurrence relation is not easy to solve generally. We can, however, compute any finite number of terms. First, we solve for c_{n+2}:

$$c_{n+2} = c_{n+1} \frac{n}{n+2} - c_n \frac{1}{(n+2)(n+1)}. \tag{4}$$

The initial conditions $y(0) = 1$ and $y'(0) = 0$ imply that $c_0 = 1$ and $c_1 = 0$. Recursively applying (4), we get

$$
\begin{array}{lll}
n = 0 & c_2 = c_1 \cdot 0 - c_0 \frac{1}{2} = -\frac{1}{2}, \\
n = 1 & c_3 = c_2 \frac{1}{3} - c_1 \frac{1}{6} = -\frac{1}{6}, \\
n = 2 & c_4 = c_3 \frac{1}{2} - c_2 \frac{1}{12} = -\frac{1}{24}, \\
n = 3 & c_5 = c_4 \frac{3}{5} - c_3 \frac{1}{20} = -\frac{1}{60}.
\end{array}
$$

It now follows that the first five terms of $y(t)$ is

$$y(t) = 1 - \frac{1}{2}t^2 - \frac{1}{6}t^3 - \frac{1}{24}t^4 - \frac{1}{60}t^5. \qquad \blacktriangleleft$$

In general, it may not be possible to find a closed form description of c_n. Nevertheless, we can use the recurrence relation to find as many terms as we desire. Although this may be tedious, it may suffice to give an approximate solution to a given differential equation.

We note that the examples we gave are power series solutions about $t_0 = 0$. We can always reduce to this case by a substitution. To illustrate, consider the differential equation

$$t y'' - (t-1)y' - ty = 0. \tag{5}$$

It has $t_0 = 1$ as an ordinary point. Suppose we wish to derive a power series solution about $t_0 = 1$. Let $y(t)$ be a solution and let $Y(x) = y(x + 1)$. Then $Y'(x) = y'(x + 1)$ and $Y''(x) = y''(x + 1)$. In the variable x, (5) becomes $(x+1)Y''(x) - xY'(x) - (x+1)Y(x) = 0$ and $x_0 = 0$ is an ordinary point. We solve $Y(x) = \sum_{n=0}^{\infty} c_n x^n$ as before. Now let $x = t - 1$. That is, $y(t) = Y(t-1) = \sum_{n=0}^{\infty} c_n (t-1)^n$ is the series solution to (5) about $t_0 = 1$.

Chebyshev Polynomials

We conclude this section with the following two related problems: For a nonnegative integer n, expand $\cos nx$ and $\sin nx$ in terms of just $\cos x$ and $\sin x$. It is an easy exercise (see Exercises 12 and 13) to show that we can write $\cos nx$ as a polynomial in $\cos x$ and we can write $\sin nx$ as a product of $\sin x$ and a polynomial in $\cos x$. More specifically, we will find polynomials T_n and U_n such that

$$\cos nx = T_n(\cos x),$$
$$\sin(n + 1)x = \sin x\, U_n(\cos x). \tag{6}$$

(The shift by 1 in the formula defining U_n is intentional.) The polynomials T_n and U_n are called the **Chebyshev polynomials** of the first and second kind, respectively. They each have degree n. For example, if $n = 2$, we have $\cos 2x = \cos^2 x - \sin^2 x = 2\cos^2 x - 1$. Thus, $T_2(t) = 2t^2 - 1$; if $t = \cos x$, we have

$$\cos 2x = T_2(\cos x).$$

Similarly, $\sin 2x = 2 \sin x \cos x$. Thus, $U_1(t) = 2t$ and

$$\sin 2x = \sin x U_1(\cos x).$$

More generally, we can use the trigonometric summation formulas

$$\sin(x + y) = \sin x \cos y + \cos x \sin y,$$
$$\cos(x + y) = \cos x \cos y - \sin x \sin y,$$

and the basic identity $\sin^2 x + \cos^2 x = 1$ to expand

$$\cos nx = \cos((n - 1)x + x) = \cos((n - 1)x) \sin x - \sin((n - 1)x) \sin x.$$

Now expand $\cos((n - 1)x)$ and $\sin((n - 1)x)$ and continue inductively to the point where all occurrences of $\cos kx$ and $\sin kx$, $k > 1$, are removed. Whenever $\sin^2 x$ occurs, replace it by $1 - \cos^2 x$. In the table below, we have done just that for some small values of n. We include in the table the resulting Chebyshev polynomials of the first kind, T_n.

n	$\cos nx$	$T_n(t)$
0	$\cos 0x = 1$	$T_0(t) = 1$
1	$\cos 1x = \cos x$	$T_1(t) = t$
2	$\cos 2x = 2\cos^2 x - 1$	$T_2(t) = 2t^2 - 1$
3	$\cos 3x = 4\cos^3 x - 3\cos x$	$T_3(t) = 4t^3 - 3t$
4	$\cos 4x = 8\cos^4 x - 8\cos^2 x + 1$	$T_4(t) = 8t^4 - 8t^2 + 1$

In a similar way, we expand $\sin(n+1)x$. The following table gives the Chebyshev polynomials of the second kind, U_n, for some small values of n.

n	$\sin(n+1)x$	$U_n(t)$
0	$\sin 1x = \sin x$	$U_0(t) = 1$
1	$\sin 2x = \sin x(2\cos x)$	$U_1(t) = 2t$
2	$\sin 3x = \sin x(4\cos^2 x - 1)$	$U_2(t) = 4t^2 - 1$
3	$\sin 4x = \sin x(8\cos^3 x - 4\cos x)$	$U_3(t) = 8t^3 - 4t$
4	$\sin 5x = \sin x(16\cos^4 x - 12\cos^2 x + 1)$	$U_4(t) = 16t^4 - 12t^2 + 1$

The method we used for computing the tables is not very efficient. We will use the interplay between the defining equations, (6), to derive second order differential equations that will determine T_n and U_n. This theme of using the interplay between two related families of functions will come up again in Sect. 7.4.

Let us begin by differentiating the equations that define T_n and U_n in (6). For the first equation, we get

LHS: $\dfrac{d}{dx}\cos nx = -n\sin nx = -n\sin x\, U_{n-1}(\cos x),$

RHS: $\dfrac{d}{dx}T_n(\cos x) = T_n'(\cos x)(-\sin x).$

Equating these results, simplifying, and substituting $t = \cos x$ gives

$$T_n'(t) = nU_{n-1}(t). \tag{7}$$

For the second equation, we get

LHS: $\dfrac{d}{dx}\sin(n+1)x = (n+1)\cos(n+1)x = (n+1)T_{n+1}(\cos x),$

RHS: $\dfrac{d}{dx}\sin x\, U_n(\cos x) = \cos x\, U_n(\cos x) + \sin x\, U_n'(\cos x)(-\sin x)$

$$= \cos x\, U_n(\cos x) - (1 - \cos^2 x)U_n'(\cos x).$$

It now follows that $(n+1)T_{n+1}(t) = tU_n(t) - (1-t^2)U_n'(t)$. Replacing n by $n-1$ gives

$$nT_n(t) = tU_{n-1}(t) - (1-t^2)U_{n-1}'(t). \tag{8}$$

We now substitute (7) and its derivative $T_n''(t) = nU_{n-1}'(t)$ into (8). After simplifying, we get that T_n satisfies

$$(1 - t^2)T_n''(t) - tT_n' + n^2T_n(t) = 0. \qquad (9)$$

By substituting (7) into the derivative of (8) and simplifying we get that U_n satisfies

$$(1 - t^2)U_n''(t) - 3tU_n' + n(n + 2)U_n(t) = 0. \qquad (10)$$

The differential equations

$$(1 - t^2)y''(t) - ty' + \alpha^2 y(t) = 0 \qquad (11)$$

$$(1 - t^2)y''(t) - 3ty' + \alpha(\alpha + 2)y(t) = 0 \qquad (12)$$

are known as **Chebyshev's differential equations**. Each have $t_0 = \pm 1$ as singular points and $t_0 = 0$ is an ordinary point. The Chebyshev polynomial T_n is a polynomial solution to (11) and U_n is a polynomial solution to (12), when $\alpha = n$.

Theorem 6. *We have the following explicit formulas for T_n and U_n;*

$$T_{2n}(t) = n(-1)^n \sum_{k=0}^{n}(-1)^k \frac{(n + k - 1)!}{(n - k)!} \frac{(2t)^{2k}}{(2k)!},$$

$$T_{2n+1}(t) = \frac{2n + 1}{2}(-1)^n \sum_{k=0}^{n}(-1)^k \frac{(n + k)!}{(n - k)!} \frac{(2t)^{2k+1}}{(2k + 1)!},$$

$$U_{2n}(t) = (-1)^n \sum_{k=0}^{n}(-1)^k \frac{(n + k)!}{(n - k)!} \frac{(2t)^{2k}}{(2k)!},$$

$$U_{2n+1}(t) = (-1)^n \sum_{k=0}^{n}(-1)^k \frac{(n + k + 1)!}{(n - k)!} \frac{(2t)^{2k+1}}{(2k + 1)!}.$$

Proof. Let us first consider Chebyshev's first differential equation, for general α. Let $y(t) = \sum_{k=0}^{\infty} c_k t^k$. Substituting $y(t)$ into (11), we get the following relation for the coefficients:

$$c_{k+2} = \frac{-(\alpha^2 - k^2)c_k}{(k + 2)(k + 1)}.$$

Let us consider the even and odd cases.

The Even Case

$$c_2 = -\frac{\alpha^2 - 0^2}{2 \cdot 1} c_0$$

$$c_4 = -\frac{\alpha^2 - 2^2}{4 \cdot 3} c_2 = \frac{(\alpha^2 - 2^2)(\alpha^2 - 0^2)}{4!} c_0$$

$$c_6 = -\frac{\alpha^2 - 4^2}{6 \cdot 5} c_4 = -\frac{(\alpha^2 - 4^2)(\alpha^2 - 2^2)(\alpha^2 - 0^2)}{6!} c_0$$

$$\vdots$$

More generally, we can see that

$$c_{2k} = (-1)^k \frac{(\alpha^2 - (2k-2)^2) \cdots (\alpha^2 - 0^2)}{(2k)!} c_0.$$

By factoring each expression $\alpha^2 - (2j)^2$ that appears in the numerator into $2^2 \left(\frac{\alpha}{2} + j\right)\left(\frac{\alpha}{2} - j\right)$ and rearranging factors, we can write

$$c_{2k} = (-1)^k \frac{\alpha}{2} \frac{\left(\frac{\alpha}{2} + k - 1\right) \cdots \left(\frac{\alpha}{2} - 1\right)\left(\frac{\alpha}{2}\right)\left(\frac{\alpha}{2} + 1\right) \cdots \left(\frac{\alpha}{2} - k + 1\right)}{(2k)!} 2^{2k} c_0.$$

The Odd Case

$$c_3 = -\frac{(\alpha^2 - 1^2)}{3 \cdot 2} c_1$$

$$c_5 = -\frac{(\alpha^2 - 3^2)}{5 \cdot 4} c_3 = \frac{(\alpha^2 - 3^2)(\alpha^2 - 1^2)}{5!} c_1$$

$$c_7 = -\frac{(\alpha^2 - 5^2)}{7 \cdot 6} c_5 = -\frac{(\alpha^2 - 5^2)(\alpha^2 - 3^2)(\alpha^2 - 1^2)}{7!} c_1$$

$$\vdots$$

Similarly, we see that

$$c_{2k+1} = (-1)^k \frac{(\alpha^2 - (2k-1)^2) \cdots (\alpha^2 - 1^2)}{(2k+1)!} c_1.$$

By factoring each expression $\alpha^2 - (2j-1)^2$ that appears in the numerator into $2^2 \left(\frac{\alpha-1}{2} + j\right)\left(\frac{\alpha-1}{2} - (j-1)\right)$ and rearranging factors, we get

$$c_{2k+1} = (-1)^k \frac{\left(\frac{\alpha-1}{2} + k\right) \cdots \left(\frac{\alpha-1}{2} - (k-1)\right)}{(2k+1)!} 2^{2k} c_1.$$

Let

$$y_0 = \frac{\alpha}{2} \sum_{k=0}^{\infty} (-1)^k \frac{\left(\frac{\alpha}{2} + k - 1\right) \cdots \left(\frac{\alpha}{2} - k + 1\right)}{(2k)!} (2t)^{2k}$$

and

$$y_1 = \frac{1}{2} \sum_{k=0}^{\infty} (-1)^k \frac{\left(\frac{\alpha-1}{2} + k\right) \cdots \left(\frac{\alpha-1}{2} - (k-1)\right)}{(2k+1)!} (2t)^{2k+1}.$$

Then the general solution to Chebyshev's first differential equation is

$$y = c_0 y_0 + c_1 y_1.$$

It is clear that neither y_0 nor y_1 is a polynomial if α is not an integer.

The Case $\alpha = 2n$

In this case, y_0 is a polynomial while y_1 is not. In fact, for $k > n$, the numerator in the sum for y_0 is zero, and hence,

$$y_0(t) = n \sum_{k=0}^{n} (-1)^k \frac{(n+k-1) \cdots (n-k+1)}{(2k)!} (2t)^{2k}$$

$$= n \sum_{k=0}^{n} (-1)^k \frac{(n+k-1)!}{(n-k)!} \frac{(2t)^{2k}}{(2k)!}.$$

It follows $T_{2n}(t) = c_0 y_0(t)$, where $c_0 = T_{2n}(0)$. To determine $T_{2n}(0)$, we evaluate the defining equation $T_{2n}(\cos x) = \cos 2nx$ at $x = \frac{\pi}{2}$ to get $T_{2n}(0) = \cos n\pi = (-1)^n$. The formula for T_{2n} now follows.

The Case $\alpha = 2n + 1$

In this case, y_1 is a polynomial while y_0 is not. Further,

$$y_1(t) = \frac{1}{2} \sum_{k=0}^{n} (-1)^k \frac{(n+k) \cdots (n-k+1)}{(2k+1)!} (2t)^{2k+1}$$

$$= \frac{1}{2} \sum_{k=0}^{n} (-1)^k \frac{(n+k)!}{(n-k)!} \frac{(2t)^{2k+1}}{(2k+1)!}.$$

It now follows that $T_{2n+1}(t) = c_1 y_1(t)$. There is no constant coefficient term in y_1. However, $y_1'(0) = 1$ is the coefficient of t in y_1. Differentiating the defining equation $T_{2n+1}(\cos x) = \cos((2n+1)x)$ at $x = \frac{\pi}{2}$ gives $T_{2n+1}'(0) = (2n+1)(-1)^n$. Let $c_1 = (2n+1)(-1)^n$. The formula for T_{2n+1} now follows. The formulas for U_n follow from (7) which can be written $U_n = \frac{1}{n+1}T_{n+1}'$. $\qquad\square$

Exercises

1–4. Use the power series method about $t_0 = 0$ to solve the given differential equation. Identify the power series with known functions. Since each is constant coefficient, use the characteristic polynomial to solve and compare.

1. $y'' - y = 0$
2. $y'' - 2y' + y = 0$
3. $y'' + k^2 y = 0$, where $k \in \mathbb{R}$
4. $y'' - 3y' + 2y = 0$

5–10. Use the power series method about $t_0 = 0$ to solve each of the following differential equations. Write the solution in the form $y(t) = c_0 y_0(t) + c_1 y_1(t)$, where $y_0(0) = 1$, $y_0'(0) = 0$ and $y_1(0) = 0$, $y_1'(t) = 1$. Find y_0 and y_1 in closed form.

5. $(1 - t^2)y'' + 2y = 0 \quad -1 < t < 1$
6. $(1 - t^2)y'' - 2ty' + 2y = 0 \quad -1 < t < 1$
7. $(t - 1)y'' - ty' + y = 0$
8. $(1 + t^2)y'' - 2ty' + 2y = 0$
9. $(1 + t^2)y'' - 4ty' + 6y = 0$
10. $(1 - t^2)y'' - 6ty' - 4y = 0$

11–19. *Chebyshev Polynomials:*

11. Use Euler's formula to derive the following identity known as de Moivre's formula:
$$(\cos x + i \sin x)^n = \cos nx + i \sin nx,$$
for any integer n.

12. Assume n is a nonnegative integer. Use the binomial theorem on de Moivre's formula to show that $\cos nx$ is a polynomial in $\cos x$ and that
$$T_n(t) = \sum_{k=0}^{\lfloor \frac{n}{2} \rfloor} \binom{n}{2k} t^{n-2k} (1 - t^2)^k.$$

13. Assume n is a nonnegative integer. Use the binomial theorem on de Moivre's formula to show that $\sin(n + 1)x$ is a product of $\sin x$ and a polynomial in $\cos x$ and that
$$U_n(t) = \sum_{k=0}^{\lfloor \frac{n}{2} \rfloor} \binom{n + 1}{2k + 1} t^{n-2k} (1 - t^2)^k.$$

14. Show that
$$(1 - t^2)U_n(t) = t T_{n+1}(t) - T_{n+2}(t).$$

15. Show that

$$U_{n+1}(t) = t U_n(t) + T_{n+1}(t).$$

16. Show that

$$T_{n+1}(t) = 2t T_n(t) - T_{n-1}(t).$$

17. Show that

$$U_{n+1}(t) = 2t U_n(t) - U_{n-1}(t).$$

18. Show that

$$T_n(t) = \frac{1}{2}\left(U_n(t) - U_{n-2}(t)\right).$$

19. Show that

$$U_n(t) = \frac{1}{2}\left(T_n(t) - T_{n+2}(t)\right).$$

7.3 Regular Singular Points and the Frobenius Method

Recall that the singular points of a differential equation

$$y'' + a_1(t)y' + a_0(t)y = 0 \tag{1}$$

are those points for which either $a_1(t)$ or $a_0(t)$ is not analytic. Generally, they are few in number but tend to be the most important and interesting. In this section, we will describe a modified power series method, called the **Frobenius Method**, that can be applied to differential equations with certain kinds of singular points.

We say that the point t_0 is a **regular singular point** of (1) if

1. t_0 is a singular point.
2. $A_1(t) = (t - t_0)a_1(t)$ and $A_0(t) = (t - t_0)^2 a_0(t)$ are analytic at t_0.

Note that by multiplying $a_1(t)$ by $t - t_0$ and $a_0(t)$ by $(t - t_0)^2$, we "restore" the analyticity at t_0. In this sense, a regular singularity at t_0 is not too bad. A singular point that is not regular is called **irregular**.

Example 1. Show $t_0 = 0$ is a regular singular point for the differential equation

$$t^2 y'' + t \sin t \, y' - 2(t + 1)y = 0. \tag{2}$$

▶ **Solution.** Let us rewrite (2) by dividing by t^2. We get

$$y'' + \frac{\sin t}{t} y' - \frac{2(t + 1)}{t^2}.$$

While $a_1(t) = \frac{\sin t}{t}$ is analytic at 0 (see Exercise 13, of Sect. 7.1) the coefficient function $a_0(t) = \frac{-2(t+1)}{t^2}$ is not. However, both $t a_1(t) = t \frac{\sin t}{t} = \sin t$ and $t^2 a_0(t) = t^2 \frac{-2(t+1)}{t^2} = -2(1 + t)$ are analytic at $t_0 = 0$. It follows that $t_0 = 0$ is a regular singular point. ◀

In the case of a regular singular point, we will rewrite (1): multiply both sides by $(t - t_0)^2$ and note that

$$(t - t_0)^2 a_1(t) = (t - t_0)A_1(t) \quad \text{and} \quad (t - t_0)^2 a_0(t) = A_0(t).$$

We then get
$$(t - t_0)^2 y'' + (t - t_0)A_1(t)y' + A_0(t)y = 0. \tag{3}$$

We will refer to this equation as the **standard form** of the differential equation when t_0 is a regular singular point. By making a change of variable, if necessary, we can assume that $t_0 = 0$. We will restrict our attention to this case. Equation (3) then becomes

$$t^2 y'' + t A_1(t)y' + A_0(t)y = 0. \tag{4}$$

When $A_1(t)$ and $A_0(t)$ are constants then (4) is a Cauchy–Euler equation. We would expect that any reasonable adjustment to the power series method should able to handle this simplest case. Before we describe the adjustments let us explore, by an example, what goes wrong when we apply the power series method to a Cauchy–Euler equation. Consider the differential equation

$$2t^2 y'' + 5t y' - 2y = 0. \tag{5}$$

Let $y(t) = \sum_{n=0}^{\infty} c_n t^n$. Then

$$2t^2 y'' = \sum_{n=0}^{\infty} 2n(n-1)c_n t^n,$$

$$5t y' = \sum_{n=0}^{\infty} 5n c_n t^n,$$

$$-2y = \sum_{n=0}^{\infty} -2c_n t^n.$$

Thus,

$$2t^2 y'' + 5t y' - 2y = \sum_{n=0}^{\infty} (2n(n-1) + 5n - 2)c_n t^n = \sum_{n=0}^{\infty} (2n-1)(n+2)c_n t^n.$$

Equation (5) now implies $(2n-1)(n+2)c_n = 0$, and hence $c_n = 0$, for all $n = 0, 1, \ldots$. The power series method has failed; it has only given us the trivial solution. With a little forethought, we could have seen the problem. The indicial polynomial for (5) is $2s^2 + 5s - 2 = (2s-1)(s+2)$. The roots are $\frac{1}{2}$ and -2. Thus, a fundamental set is $\left\{ t^{\frac{1}{2}}, t^{-2} \right\}$ and neither of these functions is analytic at $t_0 = 0$. Our assumption that there was a power series solution centered at 0 was wrong!

Any modification of the power series method must take into account that solutions to differential equations about regular singular points can have fractional or negative powers of t, as in the example above. It is thus natural to consider solutions of the form

$$y(t) = t^r \sum_{n=0}^{\infty} c_n t^n, \tag{6}$$

where r is a constant to be determined. This is the starting point for the Frobenius method. We may assume that c_0 is nonzero for if $c_0 = 0$, we could factor out a power of t and incorporate it into r. Under this assumption, we call (6) a **Frobenius series**. Of course, if r is a nonnegative integer, then a Frobenius series is a power series.

Recall that the fundamental sets for Cauchy–Euler equations take the form

$$\{t^{r_1}, t^{r_2}\}, \quad \{t^r, t^r \ln t\}, \quad \text{and} \quad \{t^\alpha \cos \beta \ln t, t^\alpha \sin \beta \ln t\}.$$

(cf. Sect. 5.3). The power of t depends on the roots of the indicial polynomial. For differential equations with regular singular points, something similar occurs. Suppose $A_1(t)$ and $A_0(t)$ have power series expansions about $t_0 = 0$ given by

$$A_1(t) = a_0 + a_1 t + a_2 t^2 + \cdots \quad \text{and} \quad A_0(t) = b_0 + b_1 t + b_2 t^2 + \cdots.$$

The polynomial
$$q(s) = s(s-1) + a_0 s + b_0 \tag{7}$$

is called the **indicial polynomial** associated to (4) and extends the definition given in the Cauchy–Euler case.[1] Its roots are called the **exponents of singularity** and, as in the Cauchy–Euler equations, indicate the power to use in the Frobenius series. A Frobenius series that solves (4) is called a **Frobenius series solution**.

Theorem 2 (The Frobenius Method).   *Suppose $t_0 = 0$ is a regular singular point of the differential equation*

$$t^2 y'' + t A_1(t) y' + A_0(t) y = 0.$$

Suppose r_1 and r_2 are the exponents of singularity.

The Real Case: *Assume r_1 and r_2 are real and $r_1 \geq r_2$. There are three cases to consider:*

1. *If $r_1 - r_2$ is not an integer, then there are two Frobenius solutions of the form*

$$y_1(t) = t^{r_1} \sum_{n=0}^{\infty} c_n t^n \quad \text{and} \quad y_2(t) = t^{r_2} \sum_{n=0}^{\infty} C_n t^n.$$

2. *If $r_1 - r_2$ is a positive integer, then there is one Frobenius series solution of the form*

$$y_1(t) = t^{r_1} \sum_{n=0}^{\infty} c_n t^n$$

and a second independent solution of the form

$$y_2(t) = \epsilon y_1(t) \ln t + t^{r_2} \sum_{n=0}^{\infty} C_n t^n.$$

---

[1]If the coefficient of $t^2 y''$ is a number c other than 1, we take the indicial polynomial to be $q(s) = cs(s-1) + a_0 s + b_0$.

It can be arranged so that ϵ is either 0 or 1. When $\epsilon = 0$, there are two Frobenius series solution. When $\epsilon = 1$, then a second independent solution is the sum of a Frobenius series and a logarithmic term. We refer to these as the nonlogarithmic and logarithmic cases, respectively.

3. *If $r_1 - r_2 = 0$, let $r = r_1 = r_2$. Then there is one Frobenius series solution of the form*

$$y_1(t) = t^r \sum_{n=0}^{\infty} c_n t^n$$

and a second independent solution of the form

$$y_2(t) = y_1(t) \ln t + t^r \sum_{n=0}^{\infty} C_n t^n.$$

The second solution y_2 is also referred to as a logarithmic case.

The Complex Case: *If the roots of the indicial polynomial are distinct complex numbers, r and $\bar{r}$ say, then there is a complex-valued Frobenius series solution of the form*

$$y(t) = t^r \sum_{n=0}^{\infty} c_n t^n,$$

where the coefficients c_n may be complex. The real and imaginary parts of $y(t)$, $y_1(t)$, and $y_2(t)$, respectively, are linearly independent solutions.

Each series given for all five different cases has a positive radius of convergence.

Remark 3. You will notice that in each case, there is at least one Frobenius solution. When the roots are real, there is a Frobenius solution for the larger of the two roots. If y_1 is a Frobenius solution and there is not a second Frobenius solution, then a second independent solution is the sum of a logarithmic expression $y_1(t) \ln t$ and a Frobenius series. This fact is obtained by applying reduction of order. We will not provide the proof as it is long and not very enlightening. However, we will consider an example of each case mentioned in the theorem. Read these examples carefully. They will reveal some of the subtleties involved in the general proof and, of course, are a guide through the exercises.

Example 4 (Real Roots Not Differing by an Integer). Use Theorem 2 to solve the following differential equation:

$$2t^2 y'' + 3t(1 + t) y' - y = 0. \tag{8}$$

▶ **Solution.** We can identify $A_1(t) = 3 + 3t$ and $A_0(t) = -1$. It is easy to see that $t_0 = 0$ is a regular singular point and the indicial equation

$$q(s) = 2s(s - 1) + 3s - 1 = (2s^2 + s - 1) = (2s - 1)(s + 1).$$

The exponents of singularity are thus $\frac{1}{2}$ and -1, and since their difference is not an integer, Theorem 2 tells us there are two Frobenius solutions: one for each exponent of singularity. Before we specialize to each case, we will first derive the general recurrence relation from which the indicial equation falls out. Let

$$y(t) = t^r \sum_{n=0}^{\infty} c_n t^n = \sum_{n=0}^{\infty} c_n t^{n+r}.$$

Then

$$y'(t) = \sum_{n=0}^{\infty} (n+r)c_n t^{n+r-1} \quad \text{and} \quad y''(t) = \sum_{n=0}^{\infty} (n+r)(n+r-1)c_n t^{n+r-2}.$$

It follows that

$$2t^2 y''(t) = t^r \sum_{n=0}^{\infty} 2(n+r)(n+r-1)c_n t^n,$$

$$3ty'(t) = t^r \sum_{n=0}^{\infty} 3(n+r)c_n t^n,$$

$$3t^2 y'(t) = t^r \sum_{n=0}^{\infty} 3(n+r)c_n t^{n+1} = t^r \sum_{n=1}^{\infty} 5(n-1+r)c_{n-1} t^n,$$

$$-y(t) = t^r \sum_{n=0}^{\infty} -c_n t^n.$$

The sum of these four expressions is $2t^2 y'' + 3t(1+t)y' - 1y = 0$. Notice that each term has t^r as a factor. It follows that the sum of each corresponding power series is 0. They are each written in standard form so the sum of the coefficients with the same powers must likewise be 0. For $n = 0$, only the first, second, and fourth series contribute constant coefficients (t^0), while for $n \geq 1$, all four series contribute coefficients for t^n. We thus get

$$
\begin{aligned}
n = 0 \quad & (2r(r-1) + 3r - 1)c_0 = 0, \\
n \geq 1 \quad & (2(n+r)(n+r-1) + 3(n+r) - 1)c_n + 3(n-1+r)c_{n-1} = 0.
\end{aligned}
$$

Now observe that for $n = 0$, the coefficient of c_0 is the indicial polynomial $q(r) = 2r(r-1) + 3r - 1 = (2r-1)(r+1)$, and for $n \geq 1$, the coefficient of c_n is $q(n+r)$. This will happen routinely. We can therefore rewrite these equations in the form

$$
\begin{aligned}
n = 0 \quad & q(r)c_0 = 0 \\
n \geq 1 \quad & q(n+r)c_n + 3(n-1+r)c_{n-1} = 0.
\end{aligned}
\tag{9}
$$

Since a Frobenius series has a nonzero constant term, it follows that $q(r) = 0$ implies $r = \frac{1}{2}$ and $r = -1$, the exponents of singularity derived in the beginning. Let us now specialize to these cases individually. We start with the larger of the two.

The Case $r = \frac{1}{2}$. Let $r = \frac{1}{2}$ in the recurrence relation given in (9). Observe that $q(n + \frac{1}{2}) = 2n(n + \frac{3}{2}) = n(2n+3)$ and is nonzero for all positive n since the only roots are $\frac{1}{2}$ and -1. We can therefore solve for c_n in the recurrence relation and get

$$c_n = \frac{-3(n - \frac{1}{2})}{n(2n+3)}c_{n-1} = \left(\frac{-3}{2}\right)\frac{(2n-1)}{n(2n+3)}c_{n-1}. \qquad (10)$$

Recursively applying (10), we get

$$n = 1 \qquad c_1 = \left(\frac{-3}{2}\right)\frac{1}{5}c_0 = \left(\frac{-3}{2}\right)\frac{3}{1\cdot(5\cdot3)}c_0,$$

$$n = 2 \qquad c_2 = \left(\frac{-3}{2}\right)\frac{3}{2\cdot7}c_1 = \left(\frac{-3}{2}\right)^2\frac{3}{2\cdot(7\cdot5)}c_0,$$

$$n = 3 \qquad c_3 = \left(\frac{-3}{2}\right)\frac{5}{3\cdot9}c_2 = \left(\frac{-3}{2}\right)^3\frac{5\cdot3}{(3\cdot2)(9\cdot7\cdot5)}c_0 = \left(\frac{-3}{2}\right)^3\frac{3}{(3!)(9\cdot7)}c_0,$$

$$n = 4 \qquad c_4 = \left(\frac{-3}{2}\right)\frac{7}{4\cdot11}c_2 = \left(\frac{-3}{2}\right)^4\frac{3}{(4!)(11\cdot9)}c_0.$$

Generally, we have $c_n = \left(\frac{-3}{2}\right)^n\frac{3}{n!(2n+3)(2n+1)}c_0$. We let $c_0 = 1$ and substitute c_n into the Frobenius series with $r = \frac{1}{2}$ to get

$$y_1(t) = t^{\frac{1}{2}}\sum_{n=0}^{\infty}\left(\frac{-3}{2}\right)^n\frac{3}{n!(2n+3)(2n+1)}t^n.$$

The Case $r = -1$. In this case, $q(n+r) = q(n-1) = (2n-3)(n)$ is again nonzero for all positive integers n. The recurrence relation in (9) simplifies to

$$c_n = -3\frac{n-2}{(2n-3)(n)}c_{n-1}. \qquad (11)$$

Recursively applying (11), we get

$$n = 1 \qquad c_1 = -3\frac{-1}{-1\cdot1}c_0 = -3c_0$$

$$n = 2 \qquad c_2 = 0c_1 = 0$$

$$n = 3 \qquad c_3 = 0$$

$$\vdots \qquad \vdots$$

Again, we let $c_0 = 1$ and substitute c_n into the Frobenius series with $r = -1$ to get

$$y_2(t) = t^{-1}(1 - 3t) = \frac{1 - 3t}{t}.$$

Since y_1 and y_2 are linearly independent, the solutions to (8) is the set of all linear combinations.   ◄

Before proceeding to our next example let us make a couple of observations that will apply in general. It is not an accident that the coefficient of c_0 is the indicial polynomial q. This will happen in general, and since we assumed from the outset that $c_0 \neq 0$, it follows that if a Frobenius series solution exists, then $q(r) = 0$; that is, r must be an exponent of singularity. It is also not an accident that the coefficient of c_n is $q(n + r)$ in the recurrence relation. This will happen in general as well. If we can guarantee that $q(n + r)$ is not zero for all positive integers n, then we obtain a consistent recurrence relation, that is, we can solve for c_n to obtain a Frobenius series solution. This will always happen for r_1, the larger of the two roots. For the smaller root r_2, we need to be more careful. In fact, in the previous example, we were careful to point out that $q(n + r_2) \neq 0$ for $n > 0$. However, if the roots differ by an integer, then the consistency of the recurrence relation comes into question in the case of the smaller root. The next two examples consider this situation.

Example 5 (Real Roots Differing by an Integer: The Nonlogarithmic Case). Use Theorem 2 to solve the following differential equation:

$$ty'' + 2y' + ty = 0.$$

► **Solution.** We first multiply both sides by t to put in standard form. We get

$$t^2 y'' + 2ty' + t^2 y = 0 \tag{12}$$

and it is easy to verify that $t_0 = 0$ is a regular singular point. The indicial polynomial is $q(s) = s(s - 1) + 2s = s^2 + s = s(s + 1)$. It follows that 0 and -1 are the exponents of singularity. They differ by an integer so there may or may not be a second Frobenius solution. Let

$$y(t) = t^r \sum_{n=0}^{\infty} c_n t^n.$$

Then

$$t^2 y''(t) = t^r \sum_{n=0}^{\infty} (n + r)(n + r - 1)c_n t^n,$$

$$2ty'(t) = t^r \sum_{n=0}^{\infty} 2(n + r)c_n t^n,$$

$$t^2 y(t) = t^r \sum_{n=2}^{\infty} c_{n-2} t^n.$$

The sum of the left side of each of these equations is $t^2 y'' + 2ty' + t^2 y = 0$, and, therefore, the sum of the series is zero. We separate the $n = 0$, $n = 1$, and $n \geq 2$ cases and simplify to get

$$n = 0 \quad (r(r+1))c_0 = 0,$$
$$n = 1 \quad ((r+1)(r+2))c_1 = 0,$$
$$n \geq 2 \quad ((n+r)(n+r+1))c_n + c_{n-2} = 0. \tag{13}$$

The $n = 0$ case tells us that $r = 0$ or $r = -1$, the exponents of singularity.

The Case $r = 0$. If $r = 0$ is the larger of the two roots, then the $n = 1$ case in (13) implies $c_1 = 0$. Also $q(n+r) = q(n) = n(n+1)$ is nonzero for all positive integers n. The recurrence relation simplifies to

$$c_n = \frac{-c_{n-2}}{(n+1)n}.$$

Since the difference in indices in the recurrence relation is 2 and $c_1 = 0$, it follows that all the odd terms, c_{2n+1}, are zero. For the even terms, we get

$$n = 2 \quad c_2 = \frac{-c_0}{3 \cdot 2}$$
$$n = 4 \quad c_4 = \frac{-c_2}{5 \cdot 4} = \frac{c_0}{5!}$$
$$n = 6 \quad c_6 = \frac{-c_4}{7 \cdot 6} = \frac{-c_0}{7!},$$

and generally, $c_{2n} = \frac{(-1)^n}{(2n+1)!} c_0$. If we choose $c_0 = 1$, then

$$y_1(t) = \sum_{n=0}^{\infty} \frac{(-1)^n t^{2n}}{(2n+1)!}$$

is a Frobenius solution (with exponent of singularity 0).

The Case $r = -1$. In this case, we see something different in the recurrence relation. For in the $n = 1$ case, (13) gives the equation

$$0 \cdot c_1 = 0.$$

This equation is satisfied for all c_1. There is no restriction on c_1 so we will choose $c_1 = 0$, as this is most convenient. The recurrence relation becomes

$$c_n = \frac{-c_{n-2}}{(n-1)n}, \quad n \geq 2.$$

A calculation similar to what we did above gives all the odd terms c_{2n+1} zero and

$$c_{2n} = \frac{(-1)^n}{(2n)!} c_0.$$

If we set $c_0 = 1$, we find the Frobenius solution with exponent of singularity -1

$$y_2(t) = \sum_{n=0}^{\infty} \frac{(-1)^n t^{2n-1}}{(2n)!}.$$

It is easy to verify that $y_1(t) = \frac{\sin t}{t}$ and $y_2(t) = \frac{\cos t}{t}$. Since y_1 and y_2 are linearly independent, the solutions to (12) is the set of all linear combinations of y_1 and y_2. ◀

The main difference that we saw in the previous example from that of the first example was in the case of the smaller root $r = -1$. We had $q(n-1) = 0$ when $n = 1$, and this leads to the equation $c_1 \cdot 0 = 0$. We were fortunate in that any c_1 is a solution and choosing $c_1 = 0$ leads to a second Frobenius solution. The recurrence relation remained consistent. In the next example, we will not be so fortunate. (If c_1 were chosen to be a fixed nonzero number, then the odd terms would add up to a multiple of y_1; nothing is gained.)

Example 6 (Real Roots Differing by an Integer: The Logarithmic Case). Use Theorem 2 to solve the following differential equation:

$$t^2 y'' - t^2 y' - (3t+2)y = 0. \tag{14}$$

▶ **Solution.** It is easy to verify that $t_0 = 0$ is a regular singular point. The indicial polynomial is $q(s) = s(s-1) - 2 = s^2 - s - 2 = (s-2)(s+1)$. It follows that 2 and -1 are the exponents of singularity. They differ by an integer so there may or may not be a second Frobenius solution. Let

$$y(t) = t^r \sum_{n=0}^{\infty} c_n t^n.$$

Then

$$t^2 y''(t) = t^r \sum_{n=0}^{\infty} (n+r)(n+r-1)c_n t^n,$$

$$-t^2 y'(t) = t^r \sum_{n=1}^{\infty} -(n-1+r)c_{n-1}t^n,$$

$$-3ty(t) = t^r \sum_{n=1}^{\infty} -3c_{n-1}t^n,$$

$$-2y(t) = t^r \sum_{n=0}^{\infty} -2c_n t^n.$$

As in the previous examples, the sum of the series is zero. We separate the $n = 0$
and $n \geq 1$ cases and simplify to get

$$n = 0 \quad (r-2)(r+1)c_0 = 0,$$
$$n \geq 1 \quad ((n+r-2)(n+r+1))c_n - (n+r+2)c_{n-1} = 0. \tag{15}$$

The $n = 0$ case tells us that $r = 2$ or $r = -1$, the exponents of singularity.

The Case $r = 2$. Since $r = 2$ is the larger of the two roots, the $n \geq 1$ case in
(15) implies $c_n = \frac{n+4}{n(n+3)}c_{n-1}$. We then get

$$n = 1 \quad c_1 = \frac{5}{1 \cdot 4}c_0$$

$$n = 2 \quad c_2 = \frac{6}{2 \cdot 5}c_1 = \frac{6}{2! \cdot 4}c_0$$

$$n = 3 \quad c_3 = \frac{7}{3 \cdot 6}c_2 = \frac{7}{3! \cdot 4}c_0,$$

and generally, $c_n = \frac{n+4}{n! \cdot 4}c_0$. If we choose $c_0 = 4$, then

$$y_1(t) = t^2 \sum_{n=0}^{\infty} \frac{n+4}{n!}t^n = \sum_{n=0}^{\infty} \frac{n+4}{n!}t^{n+2} \tag{16}$$

is a Frobenius series solution; the exponent of singularity is 2. (It is easy to see that
$y_1(t) = (t^3 + 4t^2)e^t$ but we will not use this fact.)

The Case $r = -1$. The recurrence relation in (15) simplifies to

$$n(n-3)c_n = -(n+1)c_{n-1}.$$

In this case, there is a problem when $n = 3$. Observe

$$n = 1 \quad -2c_1 = 2c_0 \quad \text{hence} \quad c_1 = -c_0,$$
$$n = 2 \quad -2c_2 = 3c_1 \quad \text{hence} \quad c_2 = \tfrac{3}{2}c_0,$$
$$n = 3 \quad 0 \cdot c_3 = 4c_2 = 6c_0, \quad \Rightarrow\Leftarrow$$

In the $n = 3$ case, there is no solution since $c_0 \neq 0$. The recurrence relation is inconsistent and there is no second Frobenius series solution. However, Theorem 2 tells us there is a second independent solution of the form

$$y(t) = y_1(t) \ln t + t^{-1} \sum_{n=0}^{\infty} c_n t^n. \tag{17}$$

Although the calculations that follow are straightforward, they are more involved than in the previous examples. The idea is simple though: substitute (17) into (14) and solve for the coefficients c_n, $n = 0, 1, 2, \ldots$. If $y(t)$ is as in (17), then a calculation gives

$$t^2 y'' = t^2 y_1'' \ln t + 2t y_1' - y_1 + t^{-1} \sum_{n=0}^{\infty} (n-1)(n-2) c_n t^n,$$

$$-t^2 y' = -t^2 y_1' \ln t - t y_1 + t^{-1} \sum_{n=1}^{\infty} -(n-2) c_{n-1} t^n,$$

$$-3t y = -3t y_1 \ln t + t^{-1} \sum_{n=1}^{\infty} -3 c_{n-1} t^n,$$

$$-2y = -2 y_1 \ln t + t^{-1} \sum_{n=0}^{\infty} -2 c_n t^n.$$

The sum of the terms on the left is zero since we are assuming a y is a solution. The sum of the terms with $\ln t$ as a factor is also zero since y_1 is the solution, (16), we found in the case $r = 2$. Observe also that the $n = 0$ term only occurs in the first and fourth series. In the first series, the constant term is $(-1)(-2) c_0 = 2 c_0$, and in the fourth series, the constant term is $(-2) c_0 = -2 c_0$. Since the $n = 0$ terms cancel, we can thus start all the series at $n = 1$. Adding these terms together and simplifying gives

$$0 = 2t y_1' - (t+1) y_1$$
$$+ t^{-1} \sum_{n=1}^{\infty} (n(n-3)) c_n - (n+1) c_{n-1}) t^n. \tag{18}$$

Now let us calculate the power series for $2t y_1' - (t+1) y_1$ and factor t^{-1} out of the sum. A short calculation and some index shifting gives

$$2t y_1' = \sum_{n=0}^{\infty} \frac{2(n+4)(n+2)}{n!} t^{n+2} = t^{-1} \sum_{n=3}^{\infty} \frac{2(n+1)(n-1)}{(n-3)!} t^n,$$

$$-y_1 = \sum_{n=0}^{\infty} -\frac{n+4}{n!} t^{n+2} = t^{-1} \sum_{n=3}^{\infty} -\frac{n+1}{(n-3)!} t^n,$$

$$-ty_1 = \sum_{n=0}^{\infty} -\frac{n+4}{n!} t^{n+3} = t^{-1} \sum_{n=3}^{\infty} -\frac{n(n-3)}{(n-3)!} t^n.$$

Adding these three series and simplifying gives

$$2ty_1' - (t+1)y_1 = t^{-1} \sum_{n=3}^{\infty} \frac{(n+3)(n-1)}{(n-3)!} t^n.$$

We now substitute this calculation into (18), cancel out the common factor t^{-1}, and get

$$\sum_{n=3}^{\infty} \frac{(n+3)(n-1)}{(n-3)!} t^n + \sum_{n=1}^{\infty} (n(n-3)c_n - (n+1)c_{n-1}) t^n = 0.$$

We separate out the $n = 1$ and $n = 2$ cases to get:

$$n = 1 \qquad -2c_1 - 2c_0 = 0 \text{ hence } c_1 = -c_0,$$
$$n = 2 \qquad -2c_2 - 3c_1 = 0 \text{ hence } c_2 = \tfrac{3}{2}c_0.$$

For $n \geq 3$, we get

$$\frac{(n+3)(n-1)}{(n-3)!} + n(n-3)c_n - (n+1)c_{n-1} = 0$$

which we can rewrite as

$$n(n-3)c_n = (n+1)c_{n-1} - \frac{(n+3)(n-1)}{(n-3)!} \qquad n \geq 3. \qquad (19)$$

You should notice that the coefficient of c_n is zero when $n = 3$. As observed earlier, this led to an inconsistency of the recurrence relation for a Frobenius series. However, the additional term $\frac{(n+3)(n-1)}{(n-3)!}$ that comes from the logarithmic term results in consistency but only for a specific value of c_2. To see this, let $n = 3$ in (19) to get $0 = 0c_3 = 4c_2 - 12$ and hence $c_2 = 3$. Since $c_2 = \tfrac{3}{2}c_0$, we have that $c_0 = 2$ and $c_1 = -2$. We now have $0c_3 = 4c_2 - 12 = 0$, and we can choose c_3 to be any number. It is convenient to let $c_3 = 0$. For $n \geq 4$, we can write (19) as

$$c_n = \frac{n+1}{n(n-3)} c_{n-1} - \frac{(n+3)(n-1)}{n(n-3)(n-3)!} \qquad n \geq 4. \qquad (20)$$

Such recurrence relations are generally very difficult to solve in a closed form. However, we can always solve any finite number of terms. In fact, the following terms are easy to check:

$n = 0$	$c_0 = 2,$	$n = 4$	$c_4 = \frac{-21}{4}$
$n = 1$	$c_1 = -2,$	$n = 5$	$c_5 = \frac{-19}{4}$
$n = 2$	$c_2 = 3,$	$n = 6$	$c_6 = \frac{-163}{72}$
$n = 3$	$c_3 = 0,$	$n = 7$	$c_7 = \frac{-53}{72}.$

We substitute these values into (17) to obtain (an approximation to) a second linearly independent solution

$$y_2(t) = y_1(t) \ln t + t^{-1} \left(2 - 2t + 3t^2 - \frac{21}{4}t^4 - \frac{19}{4}t^5 - \frac{163}{72}t^6 - \frac{53}{72}t^7 \cdots \right). \quad \blacktriangleleft$$

A couple of remarks are in order. By far the logarithmic cases are the most tedious. In the case just considered, the difference in the roots is $2 - (-1) = 3$ and it was precisely at $n = 3$ in the recurrence relation where c_0 is determined in order to achieve consistency. After $n = 3$, the coefficients are nonzero so the recurrence relation is consistent. In general, it is at the difference in the roots where this junction occurs. If we choose c_3 to be a nonzero fixed constant, the terms that would arise with c_3 as a coefficient would be a multiple of y_1, and thus, nothing is gained. Choosing $c_3 = 0$ does not exclude any critical part of the solution.

Example 7 (Real roots: Coincident). Use Theorem 2 to solve the following differential equation:

$$t^2 y'' - t(t + 3)y' + 4y = 0. \tag{21}$$

▶ **Solution.** It is easy to verify that $t_0 = 0$ is a regular singular point. The indicial polynomial is $q(s) = s(s - 1) - 3s + 4 = s^2 - 4s + 4 = (s - 2)^2$. It follows that 2 is a root with multiplicity two. Hence, $r = 2$ is the only exponent of singularity. There will be only one Frobenius series solution. A second solution will involve a logarithmic term. Let

$$y(t) = t^r \sum_{n=0}^{\infty} c_n t^n.$$

Then

$$t^2 y''(t) = t^r \sum_{n=0}^{\infty} (n+r)(n+r-1)c_n t^n,$$

$$-t^2 y'(t) = t^r \sum_{n=1}^{\infty} (n+r-1)c_{n-1} t^n,$$

$$-3t y'(t) = t^r \sum_{n=0}^{\infty} -3c_n(n+r) t^n,$$

$$4y(t) = t^r \sum_{n=0}^{\infty} 4c_n t^n.$$

As in previous examples, the sum of the series is zero. We separate the $n = 0$ and $n \geq 1$ cases and simplify to get

$$\begin{aligned} n &= 0 \qquad (r-2)^2 c_0 = 0, \\ n &\geq 1 \qquad (n+r-2)^2 c_n = (n+r-1)c_{n-1}. \end{aligned} \tag{22}$$

The Case $r = 2$. Equation (22) implies $r = 2$ and

$$c_n = \frac{n+1}{n^2} c_{n-1} \quad n \geq 1.$$

We then get

$$\begin{aligned} n &= 1 \qquad c_1 = \tfrac{2}{1^2} c_0, \\ n &= 2 \qquad c_2 = \tfrac{3}{2^2} c_1 = \tfrac{3 \cdot 2}{2^2 \cdot 1^2} c_0, \\ n &= 3 \qquad c_3 = \tfrac{4}{3^2} c_2 = \tfrac{4 \cdot 3 \cdot 2}{3^2 \cdot 2^2 \cdot 1^2} c_0, \end{aligned}$$

and generally,

$$c_n = \frac{(n+1)!}{(n!)^2} c_0 = \frac{n+1}{n!} c_0.$$

If we choose $c_0 = 1$, then

$$y_1(t) = t^2 \sum_{n=0}^{\infty} \frac{n+1}{n!} t^n \tag{23}$$

is a Frobenius series solution (with exponent of singularity 2). This is the only Frobenius series solution.

A second independent solution takes the form

$$y(t) = y_1(t) \ln t + t^2 \sum_{n=0}^{\infty} c_n t^n. \tag{24}$$

The ideas and calculations are very similar to the previous example. A straightforward calculation gives

$$t^2 y'' = t^2 y_1'' \ln t + 2t y_1' - y_1 + t^2 \sum_{n=0}^{\infty} (n+2)(n+1) c_n t^n,$$

$$-t^2 y' = -t^2 y_1' \ln t - t y_1 + t^2 \sum_{n=1}^{\infty} -(n+1) c_{n-1} t^n,$$

$$-3t y' = -3t y_1' \ln t - 3 y_1 + t^2 \sum_{n=0}^{\infty} -3(n+2) c_n t^n,$$

$$4y = 4 y_1 \ln t + t^2 \sum_{n=0}^{\infty} 4 c_n t^n.$$

The sum of the terms on the left is zero since we are assuming a y is a solution. The sum of the terms with $\ln t$ as a factor is also zero since y_1 is a solution. Observe also that the $n = 0$ terms occur in the first, third, and fourth series. In the first series the coefficient is $2c_0$, in the third series the coefficient is $-6c_0$, and in the fourth series the coefficient is $4c_0$. We can thus start all the series at $n = 1$ since the $n = 0$ terms cancel. Adding these terms together and simplifying gives

$$0 = 2t y_1' - (t+4) y_1 + t^2 \sum_{n=1}^{\infty} \left(n^2 c_n - (n+1) c_{n-1} \right) t^n. \tag{25}$$

Now let us calculate the power series for $2t y_1' - (t+4) y_1$ and factor t^2 out of the sum. A short calculation and some index shifting gives

$$2t y_1' - (t+4) y_1 = t^2 \sum_{n=1}^{\infty} \frac{n+2}{(n-1)!} t^n. \tag{26}$$

We now substitute this calculation into (25), cancel out the common factor t^2, and equate coefficients to get

$$\frac{n+2}{(n-1)!} + n^2 c_n - (n+1) c_{n-1} = 0.$$

Since $n \geq 1$, we can solve for c_n to get the following recurrence relation:

$$c_n = \frac{n+1}{n^2} c_{n-1} - \frac{n+2}{n(n!)}, \qquad n \geq 1.$$

As in the previous example, such recurrence relations are difficult to solve in a closed form. There is no restriction on c_0 so we may assume it is zero. The first few terms thereafter are as follows:

$n = 0$	$c_0 = 0$,	$n = 4$	$c_4 = \dfrac{-173}{288}$,
$n = 1$	$c_1 = -3$,	$n = 5$	$c_5 = \dfrac{-187}{1200}$,
$n = 2$	$c_2 = \dfrac{-13}{4}$,	$n = 6$	$c_6 = \dfrac{-463}{14400}$,
$n = 3$	$c_3 = \dfrac{-31}{18}$,	$n = 7$	$c_7 = \dfrac{-971}{176400}$.

We substitute these values into (24) to obtain (an approximation to) a second linearly independent solution

$$y_2(t) = y_1(t) \ln t - \left(3t^3 + \frac{13}{4}t^4 + \frac{31}{18}t^5 + \frac{173}{288}t^6 + \frac{187}{1200}t^7 + \frac{463}{14400}t^8 + \cdots \right).$$

◀

Since the roots in this example are coincident, their difference is 0. The juncture mentioned in the example that proceeded this one thus occurs at $n = 0$ and so we can make the choice $c_0 = 0$. If c_0 is chosen to be nonzero, then y_2 will include an extra term $c_0 y_1$. Thus nothing is gained.

Example 8 (Complex Roots). Use Theorem 2 to solve the following differential equation:

$$t^2(t + 1)y'' + ty' + (t + 1)^3 y = 0. \tag{27}$$

▶ **Solution.** It is easy to see that $t_0 = 0$ is a regular singular point. The indicial polynomial is $q(s) = s(s - 1) + s + 1 = s^2 + 1$ and has roots $r = \pm i$. Thus, there is a complex-valued Frobenius solution, and its real and imaginary parts will be two linear independent solutions to (27). A straightforward substitution gives

$$t^3 y'' = t^r \sum_{n=1}^{\infty} (n + r - 1)(n + r - 2)c_{n-1}t^n,$$

$$t^2 y'' = t^r \sum_{n=0}^{\infty} (n + r)(n + r - 1)c_n t^n,$$

$$ty' = t^r \sum_{n=0}^{\infty} (n + r)c_n t^n,$$

$$y = t^r \sum_{n=0}^{\infty} c_n t^n,$$

$$3ty = t^r \sum_{n=1}^{\infty} 3c_{n-1}t^n,$$

$$3t^2 y = t^r \sum_{n=2}^{\infty} 3c_{n-2}t^n,$$

$$t^3 y = t^r \sum_{n=3}^{\infty} c_{n-3}t^n.$$

As usual, the sum of these series is zero. Since the index of the sums have different starting points, we separate the cases $n = 0$, $n = 1$, $n = 2$, and $n \geq 3$ to get, after some simplification, the following:

$$n = 0 \qquad (r^2 + 1)c_0 = 0,$$

$$n = 1 \qquad ((r+1)^2 + 1)c_1 + (r(r-1)+3)c_0 = 0,$$

$$n = 2 \qquad ((r+2)^2 + 1)c_2 + ((r+1)r + 3)c_1 + 3c_0 = 0,$$

$$n \geq 3 \qquad ((r+n)^2 + 1)c_n + ((r+n-1)(r+n-2)+3)c_{n-1}$$
$$+ \, 3c_{n-2} + c_{n-3} = 0.$$

The $n = 0$ case implies that $r = \pm i$. We will let $r = i$ (the $r = -i$ case will give equivalent results). As usual, c_0 is arbitrary but nonzero. For simplicity, let us fix $c_0 = 1$. Substituting these values into the cases, $n = 1$ and $n = 2$ above, gives $c_1 = i$ and $c_2 = \frac{-1}{2}$. The general recursion relation is

$$((i+n)^2 + 1)c_n + ((i+n-1)(i+n-2)+3)c_{n-1} + 3c_{n-2} + c_{n-3} = 0 \quad (28)$$

from which we see that c_n is determined as long as we know the previous three terms. Since c_0, c_1, and c_2 are known, it follows that we can determine all c_n's. Although (28) is somewhat tedious to work with, straightforward calculations give the following values:

$n = 0$	$c_0 = 1,$		$n = 3$	$c_3 = \frac{-i}{6} = \frac{i^3}{3!},$
$n = 1$	$c_1 = i,$		$n = 4$	$c_4 = \frac{1}{24} = \frac{i^4}{4!},$
$n = 2$	$c_2 = \frac{-1}{2} = \frac{i^2}{2!},$		$n = 5$	$c_5 = \frac{i}{120} = \frac{i^5}{5!}.$

We will leave it as an exercise to verify by mathematical induction that $c_n = \frac{i^n}{n!}$. It follows now that

$$y(t) = t^i \sum_{n=0}^{\infty} \frac{i^n}{n!} t^n = t^i e^{it}.$$

We assume $t > 0$. Therefore, we can write $t^i = e^{i \ln t}$ and

$$y(t) = e^{i(t + \ln t)}.$$

By Euler's formula, the real and imaginary parts are

$$y_1(t) = \cos(t + \ln t) \quad \text{and} \quad y_2(t) = \sin(t + \ln t).$$

It is easy to see that these functions are linearly independent solutions. We remark that the $r = -i$ case gives the solution $y(t) = e^{-i(t + \ln t)}$. Its real and imaginary parts are, up to sign, the same as y_1 and y_2 given above. ◄

Exercises

1–5. For each problem, determine the singular points. Classify them as regular or irregular.

1. $y'' + \frac{t}{1-t^2} y' + \frac{1}{1+t} y = 0$

2. $y'' + \frac{1-t}{t} y' + \frac{1-\cos t}{t^3} y = 0$

3. $y'' + 3t(1-t)y' + \frac{1-e^t}{t} y = 0$

4. $y'' + \frac{1}{t} y' + \frac{1-t}{t^3} y = 0$

5. $ty'' + (1-t)y' + 4ty = 0$

6–10. Each differential equation has a regular singular point at $t = 0$. Determine the indicial polynomial and the exponents of singularity. How many Frobenius solutions are guaranteed by Theorem 2? How many could there be?

6. $2ty'' + y' + ty = 0$

7. $t^2 y'' + 2ty' + t^2 y = 0$

8. $t^2 y'' + te^t y' + 4(1 - 4t)y = 0$

9. $ty'' + (1-t)y' + \lambda y = 0$

10. $t^2 y'' + 3t(1 + 3t)y' + (1 - t^2)y = 0$

11–14. Verify the following claims that were made in the text.

11. In Example 5, verify the claims that $y_1(t) = \frac{\sin t}{t}$ and $y_2(t) = \frac{\cos t}{t}$.

12. In Example 8, we claimed that the solution to the recursion relation

$$((i + n)^2 + 1)c_n + ((i + n - 1)(i + n - 2) + 3)c_{n-1} + 3c_{n-2} + c_{n-3} = 0$$

was $c_n = \frac{i^n}{n!}$. Use mathematical induction to verify this claim.

13. In Remark 3, we stated that the logarithmic case could be obtained by a reduction of order argument. Consider the Cauchy–Euler equation

$$t^2 y'' + 5ty' + 4y = 0.$$

One solution is $y_1(t) = t^{-2}$. Use reduction of order to show that a second independent solution is $y_2(t) = t^{-2} \ln t$, in harmony with the statement for the appropriate case of the theorem.

14. Verify the claim made in Example 6 that $y_1(t) = (t^3 + 4t^2)e^t$

15–26. Use the Frobenius method to solve each of the following differential equations. For those problems marked with a (*), one of the independent solutions can easily be written in closed form. For those problems marked with a (**), both independent solutions can easily be written in closed form. In each case below we let $y = t^r \sum_{n=0}^{\infty} c_n t^n$ where we assume $c_0 \neq 0$ and r is the exponent of singularity.

15. $ty'' - 2y' + ty = 0$ (**) (real roots, differ by integer, two Frobenius solutions)

16. $2t^2 y'' - ty' + (1+t)y = 0$ (**) (real roots, do not differ by an integer, two Frobenius solutions)

17. $t^2 y'' - t(1+t)y' + y = 0$, (*) (real roots, coincident, logarithmic case)

18. $2t^2 y'' - ty' + (1-t)y = 0$ (**) (real roots, do not differ by an integer, two Frobenius solutions)

19. $t^2 y'' + t^2 y' - 2y = 0$ (**) (real roots, differ by integer, two Frobenius solutions)

20. $t^2 y'' + 2ty' - a^2 t^2 y = 0$ (**) (real roots, differ by integer, two Frobenius Solutions)

21. $ty'' + (t-1)y' - 2y = 0$, (*) (real roots, differ by integer, logarithmic case)

22. $ty'' - 4y = 0$ (real roots, differ by an integer, logarithmic case)

23. $t^2(-t+1)y'' + (t+t^2)y' + (-2t+1)y = 0$ (**) (complex)

24. $t^2 y'' + t(1+t)y' - y = 0$, (**) (real roots, differ by an integer, two Frobenius solutions)

25. $t^2 y'' + t(1-2t)y' + (t^2 - t + 1)y = 0$ (**) (complex)

26. $t^2(1+t)y'' - t(1+2t)y' + (1+2t)y = 0$ (**) (real roots, equal, logarithmic case)

7.4   Application of the Frobenius Method: Laplace Inversion Involving Irreducible Quadratics

In this section, we return to the question of determining formulas for the Laplace inversion of

$$\frac{b}{(s^2 + b^2)^{k+1}} \quad \text{and} \quad \frac{s}{(s^2 + b^2)^{k+1}}, \tag{1}$$

for k a nonnegative integer. Recall that in Sect. 2.5, we developed reduction of order formulas so that each inversion could be recursively computed. In this section, we will derive a closed formula for the inversion by solving a distinguished second order differential equation, with a regular singular point at $t = 0$, associated to each simple rational function given above. Table 7.1 of Sect. 7.5 summarizes the Laplace transform formulas we obtain in this section.

To begin with, we use the dilatation principle to reduce the simple quadratic rational functions given in (1) to the case $b = 1$. Recall the dilation principle, Theorem 23 of Sect. 2.2. For an input function $f(t)$ and b a positive number, we have

$$\mathcal{L}\{f(bt)\}(s) = \frac{1}{b}\mathcal{L}\{f(t)\}(s/b).$$

The corresponding inversion formula gives

$$\mathcal{L}^{-1}\{F(s/b)\} = bf(bt), \tag{2}$$

where as usual $\mathcal{L}\{f\} = F$.

Proposition 1. *Suppose $b > 0$. Then*

$$\mathcal{L}^{-1}\left\{\frac{b}{(s^2 + b^2)^{k+1}}\right\}(t) = \frac{1}{b^{2k}}\mathcal{L}^{-1}\left\{\frac{1}{(s^2 + 1)^{k+1}}\right\}(bt),$$

$$\mathcal{L}^{-1}\left\{\frac{s}{(s^2 + b^2)^{k+1}}\right\}(t) = \frac{1}{b^{2k}}\mathcal{L}^{-1}\left\{\frac{s}{(s^2 + 1)^{k+1}}\right\}(bt).$$

Proof. We simply apply (2) to get

$$\mathcal{L}^{-1}\left\{\frac{b}{(s^2 + b^2)^{k+1}}\right\}(t) = \frac{b}{b^{2(k+1)}}\mathcal{L}^{-1}\left\{\frac{1}{((s/b)^2 + 1)^{k+1}}\right\}(t)$$

$$= \frac{1}{b^{2k}}\mathcal{L}^{-1}\left\{\frac{1}{(s^2 + 1)^{k+1}}\right\}(bt).$$

A similar calculation holds for the second simple rational function.   □

It follows from Proposition 1 that we only need to consider the cases

$$\frac{s}{(s^2 + 1)^{k+1}} \quad \text{and} \quad \frac{1}{(s^2 + 1)^{k+1}}. \tag{3}$$

We now define $A_k(t)$ and $B_k(t)$ as follows:

$$A_k(t) = 2^k k! \mathcal{L}^{-1}\left\{\frac{1}{(s^2+1)^{k+1}}\right\}$$

$$B_k(t) = 2^k k! \mathcal{L}^{-1}\left\{\frac{s}{(s^2+1)^{k+1}}\right\}. \tag{4}$$

The factor $2^k k!$ will make the resulting formulas a little simpler.

Lemma 2. *For $k \geq 1$, we have*

$$A_k(t) = -t^2 A_{k-2}(t) + (2k-1)A_{k-1}(t), \quad k \geq 2, \tag{5}$$

$$B_k(t) = t A_{k-1}(t), \tag{6}$$

$$A_k(0) = 0, \tag{7}$$

$$B_k(0) = 0. \tag{8}$$

Proof. Let $b = 1$ in Proposition 8 of Sect. 2.5, use the definition of A_k and B_k, and simplify to get

$$A_k(t) = -t B_{k-1}(t) + (2k-1)A_{k-1}(t),$$

$$B_k(t) = t A_{k-1}(t).$$

Equation (6) is the second of the two above. Now replace k by $k-1$ in (6) and substitute into the first equation above to get (5). By the initial value theorem, Theorem 1 of Sect. 5.4, we have

$$A_k(0) = 2^k k! \lim_{s\to\infty} \frac{s}{(s^2+1)^{k+1}} = 0,$$

$$B_k(0) = 2^k k! \lim_{s\to\infty} \frac{s^2}{(s^2+1)^{k+1}} = 0. \qquad \square$$

Proposition 3. *Suppose $k \geq 1$. Then*

$$A_k'(t) = B_k(t), \tag{9}$$

$$B_k'(t) = 2k A_{k-1}(t) - A_k(t). \tag{10}$$

Proof. By the input derivative principle and the Lemma above, we have

$$\frac{1}{2^k k!}\mathcal{L}\{A_k'(t)\} = \frac{1}{2^k k!}(s\mathcal{L}\{A_k(t)\} - A_k(0))$$

$$= \frac{s}{(s^2+1)^{k+1}}$$

$$= \frac{1}{2^k k!}\mathcal{L}\{B_k(t)\}.$$

Equation (9) now follows. In a similar way, we have

$$\frac{1}{2^k k!} \mathcal{L}\{B'_k(t)\} = \frac{1}{2^k k!}(s\mathcal{L}\{B_k(t)\} - B_k(0))$$

$$= \frac{s^2}{(s^2+1)^{k+1}}$$

$$= \frac{1}{(s^2+1)^k} - \frac{1}{(s^2+1)^{k+1}}$$

$$= \frac{2k}{2^k k!}\mathcal{L}\{A_{k-1}(t)\} - \frac{1}{2^k k!}\mathcal{L}\{A_k(t)\}.$$

Equation (10) now follows.   □

Proposition 4. *With notation as above, we have*

$$tA''_k - 2kA'_k + tA_k = 0, \tag{11}$$

$$t^2 B''_k - 2ktB'_k + (t^2 + 2k)B_k = 0. \tag{12}$$

Proof. We first differentiate equation (9) and then substitute in (10) to get

$$A''_k = B'_k = 2kA_{k-1} - A_k.$$

Now multiply this equation by t and simplify using (6) and (9) to get

$$tA''_k = 2ktA_{k-1} - tA_k = 2kB_k - tA_k = 2kA'_k - tA_k.$$

Equation (11) now follows. To derive equation (12), first differentiate equation (11) and then multiply by t to get

$$t^2 A'''_k + (1 - 2k)tA''_k + t^2 A'_k + tA_k = 0.$$

From (11), we get $tA_k = -tA''_k + 2kA'_k$ which we substitute into the equation above to get

$$t^2 A'''_k - 2ktA''_k + (t^2 + 2k)A'_k = 0.$$

Equation (12) follows now from (9).   □

Proposition 4 tells us that A_k and B_k both satisfy differential equations with a regular singular point at $t_0 = 0$. We know from Corollary 11 of Sect. 2.5 that A_k and B_k are sums of products of polynomials with $\sin t$ and $\cos t$ (see Table 2.5 for the cases $k = 0, 1, 2, 3$). Specifically, we can write

$$A_k(t) = p_1(t)\cos t + p_2(t)\sin t, \tag{13}$$

$$B_k(t) = q_1(t)\cos t + q_2(t)\sin t, \tag{14}$$

where p_1, p_2, q_1, and q_2 are polynomials of degree at most k. Because of the presence of the sin and cos functions, the Frobenius method will produce rather complicated power series and it will be very difficult to recognize these polynomial factors. Let us introduce a simplifying feature that gets to the heart of the polynomial coefficients. We will again assume some familiarity of complex number arithmetic. By Euler's formula, $e^{it} = \cos t + i \sin t$ and $e^{-it} = \cos t - i \sin t$. Adding these formulas together and dividing by 2 gives a formula for $\cos t$. Similarly, subtracting these formula and dividing by $2i$ gives a formula for $\sin t$. Specifically, we get

$$\cos t = \frac{e^{it} + e^{-it}}{2},$$

$$\sin t = \frac{e^{it} - e^{-it}}{2i}.$$

Substituting these formulas into (13) and simplifying gives

$$\begin{aligned}
A_k(t) &= p_1(t) \cos t + p_2(t) \sin t \\
&= p_1(t) \frac{e^{it} + e^{-it}}{2} + p_2(t) \frac{e^{it} - e^{-it}}{2i} \\
&= \frac{p_1(t) - i p_2(t)}{2} e^{it} + \frac{p_1(t) + i p_2(t)}{2} e^{-it} \\
&= \frac{a_k(t)}{2} e^{it} + \frac{\overline{a_k(t)}}{2} e^{it} \\
&= \operatorname{Re}(a_k(t) e^{it}),
\end{aligned}$$

where $a_k(t) = p_1(t) - i p_2(t)$ is a complex-valued polynomial, which we determine below. Observe that since p_1 and p_2 have degrees at most k, it follows that $a_k(t)$ is a polynomial of degree at most k. In a similar way, we can write

$$B_k(t) = \operatorname{Re}(b_k(t) e^{it}),$$

for some complex-valued polynomial $b_k(t)$ whose degree is at most k. We summarize the discussion above for easy reference.

Proposition 5. *There are complex-valued polynomials $a_k(t)$ and $b_k(t)$ so that*

$$A_k(t) = \operatorname{Re}(a_k(t) e^{it}),$$

$$B_k(t) = \operatorname{Re}(b_k(t) e^{it}).$$

We now proceed to show that $a_k(t)$ and $b_k(t)$ satisfy second order differential equations with a regular singular point at $t_0 = 0$. The Frobenius method will give only one polynomial solution in each case, which we identify with $a_k(t)$ and $b_k(t)$. From there, it is an easy matter to use Proposition 5 to find $A_k(t)$ and $B_k(t)$.

First a little Lemma.

Lemma 6. *Suppose $p(t)$ is a complex-valued polynomial and $\mathrm{Re}(p(t)e^{it}) = 0$ for all $t \in \mathbb{R}$. Then $p(t) = 0$ for all t.*

Proof. Write $p(t) = \alpha(t) + i\beta(t)$, where $\alpha(t)$ and $\beta(t)$ are real-valued polynomials. Then the assumption that $\mathrm{Re}(p(t)e^{it}) = 0$ becomes

$$\alpha(t)\cos t - \beta(t)\sin t = 0. \tag{15}$$

Let $t = 2\pi n$ in (15). Then we get $\alpha(2\pi n) = 0$ for each integer n. This means $\alpha(t)$ has infinitely many roots, and this can only happen when a polynomial is zero. Similarly, if $t = \frac{\pi}{2} + 2\pi n$ is substituted into 15, then we get $\beta\left(\frac{\pi}{2} + 2\pi n\right) = 0$, for all n. We similarly get $\beta(t) = 0$. It now follows that $p(t) = 0$. $\qquad\square$

Proposition 7. *The polynomials $a_k(t)$ and $b_k(t)$ satisfy*

$$ta_k'' + 2(it - k)a_k' - 2ki a_k = 0,$$
$$t^2 b_k'' + 2t(it - k)b_k' - 2k(it - 1)b_k = 0.$$

Proof. Let us start with A_k. Since differentiation respects the real and imaginary parts of complex-valued functions, we have

$$A_k(t) = \mathrm{Re}(a_k(t)e^{it}),$$
$$A_k'(t) = \mathrm{Re}((a_k(t)e^{it})') = \mathrm{Re}((a_k'(t) + ia_k(t))e^{it}),$$
$$A_k''(t) = \mathrm{Re}((a_k''(t) + 2ia_k'(t) - a_k(t))e^{it}).$$

It follows now from Proposition 4 that

$$
\begin{aligned}
0 &= tA_k''(t) - 2kA_k'(t) + tA_k(t) \\
&= t\,\mathrm{Re}(a_k(t)e^{it})'' - 2k\,\mathrm{Re}(a_k(t)e^{it})' + t\,\mathrm{Re}(a_k(t)e^{it}) \\
&= \mathrm{Re}((t(a_k''(t) + 2ia_k'(t) - a_k(t)) - 2k(a_k'(t) + ia_k(t)) + ta_k(t))e^{it}) \\
&= \mathrm{Re}\left((ta_k''(t) + 2(it - k)a_k'(t) - 2ki a_k(t))e^{it}\right).
\end{aligned}
$$

Now, Lemma 6 implies

$$ta_k''(t) + 2(it - k)a_k'(t) - 2ki a_k(t) = 0.$$

The differential equation in b_k is done similarly and left as an exercise (see Exercise 3). $\qquad\square$

Each differential equation involving a_k and b_k in Proposition 7 is second order and has $t_0 = 0$ as a regular singular point. Do not be troubled by the presence of complex coefficients; the Frobenius method applies over the complex numbers as well.

The following lemma will be useful in determining the coefficient needed in the Frobenius solutions given below for $a_k(t)$ and $b_k(t)$.

Lemma 8. *The constant coefficient of $a_k(t)$ is given by*

$$a_k(0) = -i \frac{(2k)!}{k!2^k}.$$

The coefficient of t in b_k is given by

$$b_k'(0) = -i \frac{(2(k-1))!}{(k-1)!2^{(k-1)}}.$$

Proof. Replace k by $k+2$ in (5) to get

$$A_{k+2} = (2k+3)A_{k+1} - t^2 A_k,$$

for all t and $k \geq 1$. Lemma 6 gives that a_k satisfies the same recursion relation:

$$a_{k+2}(t) = (2k+3)a_{k+1}(t) - t^2 a_k(t).$$

If $t = 0$, we get $a_{k+2}(0) = (2k+3)a_{k+1}(0)$. Replace k by $k-1$ to get

$$a_{k+1}(0) = (2k+1)a_k(0).$$

By Table 2.5, we have

$$A_1(t) = 2^1(1!)\mathcal{L}^{-1}\left\{\frac{1}{(s^2+1)^2}\right\}$$

$$= \sin t - t \cos t$$

$$= \text{Re}((-t-i)e^{it}).$$

Thus, $a_1(t) = -t - i$ and $a_1(0) = -i$. The above recursion relation gives the following first four terms:

$a_1(0) = -i,$	$a_3(0) = 5a_2(0) = -5 \cdot 3 \cdot i,$
$a_2(0) = 3a_1(0) = -3i,$	$a_4(0) = 7a_3(0) = -7 \cdot 5 \cdot 3 \cdot i.$

Inductively, we get

$$a_k(0) = -(2k - 1) \cdot (2k - 3) \cdots 5 \cdot 3 \cdot i$$

$$= -i \frac{(2k)!}{k!2^k}.$$

For any polynomial $p(t)$, the coefficient of t is given by $p'(0)$. Thus, $b_k'(0)$ is the coefficient of t in $b_k(t)$. On the other hand, (6) implies $b_{k+1}(t) = ta_k(t)$, and hence, the coefficient of t in $b_{k+1}(t)$ is the same as the constant coefficient, $a_k(0)$ of $a_k(t)$. Replacing k by $k - 1$ and using the formula for $a_k(0)$ derived above, we get

$$b_k'(0) = a_{k-1}(0) = -i \frac{(2(k - 1))!}{(k - 1)!2^{(k-1)}}. \qquad \square$$

Proposition 9. *With the notation as above, we have*

$$a_k(t) = \frac{-i}{2^k} \sum_{n=0}^{k} \frac{(2k - n)!}{n!(k - n)!}(-2it)^n$$

$$and \qquad b_k(t) = \frac{-it}{2^{k-1}} \sum_{n=0}^{k-1} \frac{(2(k - 1) - n)!}{n!(k - 1 - n)!}(-2it)^n.$$

Proof. By Proposition 7, $a_k(t)$ is the polynomial solution to the differential equation

$$ty'' + 2(it - k)y' - 2kiy = 0 \tag{16}$$

with constant coefficient $a_k(0)$ as given in Lemma 8. Multiplying equation (16) by t gives $t^2y'' + 2t(it - k)y' - 2kity = 0$ which is easily seen to have a regular singular point at $t = 0$. The indicial polynomial is given by $q(s) = s(s-1)-2ks = s(s - (2k + 1))$. It follows that the exponents of singularity are 0 and $2k + 1$. We will show below that the $r = 0$ case gives a polynomial solution. We leave it to the exercises (see Exercise 1) to verify that the $r = 2k + 1$ case gives a nonpolynomial Frobenius solution. We thus let

$$y(t) = \sum_{n=0}^{\infty} c_n t^n.$$

Then

$$ty''(t) = \sum_{n=1}^{\infty} (n)(n + 1)c_{n+1} t^n,$$

$$2ity'(t) = \sum_{n=1}^{\infty} 2inc_n t^n,$$

$$-2ky'(t) = \sum_{n=0}^{\infty} -2k(n+1)c_{n+1}t^n,$$

$$-2iky(t) = \sum_{n=0}^{\infty} -2ikc_n t^n.$$

By assumption, the sum of the series is zero. We separate the $n = 0$ and $n \geq 1$ cases and simplify to get

$$n = 0 \qquad -2kc_1 - 2ikc_0 = 0,$$

$$n \geq 1 \qquad (n - 2k)(n + 1)c_{n+1} + 2i(n - k)c_n = 0. \tag{17}$$

The $n = 0$ case tells us that $c_1 = -ic_0$. For $1 \leq n \leq 2k - 1$, we have

$$c_{n+1} = \frac{-2i(k - n)}{(2k - n)(n + 1)}c_n.$$

Observe that $c_{k+1} = 0$, and hence, $c_n = 0$ for all $k + 1 \leq n \leq 2k - 1$. For $n = 2k$ we get from the recursion relation $0c_{2k+1} = 2ikc_k = 0$. This implies that c_{2k+1} can be arbitrary. We will choose $c_{2k+1} = 0$. Then $c_n = 0$ for all $n \geq k + 1$, and hence, the solution y is a polynomial. We make the usual comment that if c_{k+1} is chosen to be nonzero, then those terms with c_{k+1} as a factor will make up the Frobenius solution for exponent of singularity $r = 2k + 1$. Let us now determine the coefficients c_n for $0 \leq n \leq k$. From the recursion relation, (17), we get

$$n = 0 \qquad c_1 = -ic_0,$$

$$n = 1 \qquad c_2 = \frac{-2i(k - 1)}{(2k - 1)2}c_1 = \frac{2(-i)^2(k - 1)}{(2k - 1)2}c_0 = \frac{(-2i)^2k(k - 1)}{(2k)(2k - 1)2}c_0,$$

$$n = 2 \qquad c_3 = \frac{-2i(k - 2)}{(2k - 2)3}c_2 = \frac{(-2i)^3k(k - 1)(k - 2)}{2k(2k - 1)(2k - 2)3!}c_0,$$

$$n = 3 \qquad c_4 = \frac{-2i(k - 3)}{(2k - 3)4}c_3 = \frac{(-2i)^4k(k - 1)(k - 2)(k - 3)}{2k(2k - 1)(2k - 2)(2k - 3)4!}c_0.$$

and generally,

$$c_n = \frac{(-2i)^n k(k - 1) \cdots (k - n + 1)}{2k(2k - 1) \cdots (2k - n + 1)n!}c_0 \qquad n = 1, \ldots, k.$$

We can write this more compactly in terms of binomial coefficients as

$$c_n = \frac{(-2i)^n \binom{k}{n}}{\binom{2k}{n}n!}c_0.$$

It follows now that

$$y(t) = c_0 \sum_{n=0}^{k} \frac{(-2i)^n \binom{k}{n}}{\binom{2k}{n} n!} t^n.$$

The constant coefficient is c_0. Thus, we choose $c_0 = a_k(0)$ as given in Lemma 8. Then $y(t) = a_k(t)$ is the polynomial solution we seek. We get

$$a_k(t) = a_k(0) \sum_{n=0}^{k} \frac{(-2i)^n \binom{k}{n}}{\binom{2k}{n} n!} t^n$$

$$= \sum_{n=0}^{k} -i \frac{(2k)!}{k! 2^k} \frac{(-2i)^n \binom{k}{n}}{\binom{2k}{n} n!} t^n$$

$$= \frac{-i}{2^k} \sum_{n=0}^{k} \frac{(2k-n)!}{(k-n)! n!} (-2it)^n.$$

It is easy to check that (6) has as an analogue the equation $b_{k+1}(t) = t a_k(t)$. Replacing k by $k-1$ in the formula for $a_k(t)$ and multiplying by t thus establishes the formula for $b_k(t)$.   □

For $x \in \mathbb{R}$, we let $\lfloor x \rfloor$ denote the greatest integer function of x. It is defined to be the greatest integer less than or equal to x. We are now in a position to give a closed formula for the inverse Laplace transforms of the simple rational functions given in (4).

Theorem 10. *For the simple rational functions, we have*

$$\mathcal{L}^{-1}\left\{\frac{1}{(s^2+1)^{k+1}}\right\}(t) = \frac{\sin t}{2^{2k}} \sum_{m=0}^{\lfloor \frac{k}{2} \rfloor} (-1)^m \binom{2k-2m}{k} \frac{(2t)^{2m}}{(2m)!}$$

$$- \frac{\cos t}{2^{2k}} \sum_{m=0}^{\lfloor \frac{k-1}{2} \rfloor} (-1)^m \binom{2k-2m-1}{k} \frac{(2t)^{2m+1}}{(2m+1)!},$$

$$\mathcal{L}^{-1}\left\{\frac{s}{(s^2+1)^{k+1}}\right\}(t) = \frac{2t \sin t}{k \cdot 2^{2k}} \sum_{m=0}^{\lfloor \frac{k-1}{2} \rfloor} (-1)^m \binom{2k-2m-2}{k-1} \frac{(2t)^{2m}}{(2m)!}$$

$$- \frac{2t \cos t}{k \cdot 2^{2k}} \sum_{m=0}^{\lfloor \frac{k-2}{2} \rfloor} (-1)^m \binom{2k-2m-3}{k-1} \frac{(2t)^{2m+1}}{(2m+1)!}.$$

The first formula is valid for $k \geq 0$, and the second formula is valid for $k \geq 1$. Sums where the upper limit is less than 0 (which occur in the cases $k = 0$ and 1) should be understood to be 0.

Proof. From Proposition 9, we can write

$$a_k(t) = \frac{1}{2^k} \sum_{n=0}^{k} \frac{(2k-n)!}{n!(k-n)!}(-1)^{n+1}(i)^{n+1}(2t)^n.$$

It is easy to see that the real part of a_k consists of those terms where n is odd. The imaginary part consists of those terms where n is even. The odd integers from $n = 0, \ldots, k$ can be written $n = 2m + 1$ where $m = 0, \ldots, \lfloor \frac{k-1}{2} \rfloor$. Similarly, the even integers can be written $n = 2m$, where $m = 0, \ldots, \lfloor \frac{k}{2} \rfloor$. We thus have

$$\mathrm{Re}(a_k(t)) = \frac{1}{2^k} \sum_{m=0}^{\lfloor \frac{k-1}{2} \rfloor} \frac{(2k-2m-1)!}{(2m+1)!(k-2m-1)!}(-1)^{2m+2}(i)^{2m+2}(2t)^{2m+1}$$

$$= \frac{-1}{2^k} \sum_{m=0}^{\lfloor \frac{k-1}{2} \rfloor} \frac{(2k-2m-1)!}{(2m+1)!(k-2m-1)!}(-1)^m(2t)^{2m+1}$$

and

$$\mathrm{Im}(a_k(t)) = \frac{1}{i}\frac{1}{2^k} \sum_{m=0}^{\lfloor \frac{k}{2} \rfloor} \frac{(2k-2m)!}{(2m)!(k-2m)!}(-1)^{2m+1}(i)^{2m+1}(2t)^{2m}$$

$$= \frac{-1}{2^k} \sum_{m=0}^{\lfloor \frac{k}{2} \rfloor} \frac{(2k-2m)!}{(2m)!(k-2m)!}(-1)^m(2t)^{2m}.$$

Now $\mathrm{Re}(a_k(t)e^{it}) = \mathrm{Re}(a_k(t))\cos t - \mathrm{Im}(a_k(t))\sin t$. It follows from Propositions 5 that

$$\mathcal{L}^{-1}\left\{\frac{1}{(s^2+1)^{k+1}}\right\} = \frac{1}{2^k k!}\mathrm{Re}(a_k(t)e^{it})$$

$$= \frac{1}{2^k k!}\mathrm{Re}(a_k(t))\cos t - \frac{1}{2^k k!}\mathrm{Im}(a_k(t))\sin t$$

$$= \frac{-\cos t}{2^{2k}} \sum_{m=0}^{\lfloor \frac{k-1}{2} \rfloor} \frac{(2k-2m-1)!}{k!(2m+1)!(k-2m-1)!}(-1)^m(2t)^{2m+1}$$

$$+ \frac{\sin t}{2^{2k}} \sum_{m=0}^{\lfloor \frac{k}{2} \rfloor} \frac{(2k-2m)!}{k!(2m)!(k-2m)!}(-1)^m(2t)^{2m}$$

$$= \frac{-\cos t}{2^{2k}} \sum_{m=0}^{\lfloor \frac{k-1}{2} \rfloor} \binom{2k-2m-1}{k} (-1)^m \frac{(2t)^{2m+1}}{(2m+1)!}$$

$$+ \frac{\sin t}{2^{2k}} \sum_{m=0}^{\lfloor \frac{k}{2} \rfloor} \binom{2k-2m}{k} (-1)^m \frac{(2t)^{2m}}{(2m)!}.$$

A similar calculation gives the formula for $\mathcal{L}^{-1}\left\{ \frac{s}{(s^2+1)^{k+1}} \right\}$. $\qquad$ □

We conclude with the following corollary which is immediate from Proposition 1.

Corollary 11. *Let* $b > 0$, *then*

$$\mathcal{L}^{-1}\left\{ \frac{b}{(s^2+b^2)^{k+1}} \right\}(t) = \frac{\sin bt}{(2b)^{2k}} \sum_{m=0}^{\lfloor \frac{k}{2} \rfloor} (-1)^m \binom{2k-2m}{k} \frac{(2bt)^{2m}}{(2m)!}$$

$$- \frac{\cos bt}{(2b)^{2k}} \sum_{m=0}^{\lfloor \frac{k-1}{2} \rfloor} (-1)^m \binom{2k-2m-1}{k} \frac{(2bt)^{2m+1}}{(2m+1)!},$$

$$\mathcal{L}^{-1}\left\{ \frac{s}{(s^2+b^2)^{k+1}} \right\}(t) = \frac{2bt \sin bt}{k \cdot (2b)^{2k}} \sum_{m=0}^{\lfloor \frac{k-1}{2} \rfloor} (-1)^m \binom{2k-2m-2}{k-1} \frac{(2bt)^{2m}}{(2m)!}$$

$$- \frac{2bt \cos bt}{k \cdot (2b)^{2k}} \sum_{m=0}^{\lfloor \frac{k-2}{2} \rfloor} (-1)^m \binom{2k-2m-3}{k-1} \frac{(2bt)^{2m+1}}{(2m+1)!}.$$

Exercises

1–3. Verify the following unproven statements made in this section.

1. Verify the statement made that the Frobenius solution to

$$ty'' + 2(it - k)y' - 2kiy = 0$$

with exponent of singularity $r = 2k + 1$ is not a polynomial.

2. Verify the second inversion formula

$$\mathcal{L}^{-1}\left\{\frac{s}{(s^2 + b^2)^{k+1}}\right\}(t) = \frac{1}{b^{2k}}\mathcal{L}^{-1}\left\{\frac{s}{(s^2 + 1)^{k+1}}\right\}(bt)$$

given in Proposition 1.

3. Verify the second differential equation formula

$$t^2 b_k'' + 2t(it - k)b_k' - 2k(it - 1)b_k = 0$$

given in Proposition 7.

4–15. *This series of exercises leads to closed formulas for the inverse Laplace transform of*

$$\frac{1}{(s^2 - 1)^{k+1}} \quad and \quad \frac{s}{(s^2 - 1)^{k+1}}.$$

Define $C_k(t)$ and $D_k(t)$ by the formulas

$$\frac{1}{2^k k!}\mathcal{L}\{C_k(t)\}(s) = \frac{1}{(s^2 - 1)^{k+1}}$$

and

$$\frac{1}{2^k k!}\mathcal{L}\{D_k(t)\}(s) = \frac{s}{(s^2 - 1)^{k+1}}.$$

4. Show that C_k and D_k are related by

$$D_{k+1}(t) = tC_k(t).$$

5. Show that C_k and D_k satisfy the recursion formula

$$C_k(t) = tD_{k-1}(t) - (2k - 1)C_{k-1}(t).$$

6. Show that C_k satisfies the recursion formula

$$C_{k+2}(t) = t^2 C_k(t) - (2k + 3)C_{k+1}(t).$$

7. Show that D_k satisfies the recursion formula

$$D_{k+2}(t) = t^2 D_k(t) - (2k + 1)D_{k+1}(t).$$

8. For $k \geq 1$, show that

$$C_k(0) = 0,$$
$$D_k(0) = 0.$$

9. Show that for $k \geq 1$,

1. $C_k'(t) = D_k(t)$.
2. $D_k'(t) = 2kC_{k-1}(t) + C_k(t)$.

10. Show the following:

1. $tC_k''(t) - 2kC_k'(t) - tC_k(t) = 0$.
2. $t^2 D_k''(t) - 2ktD_k'(t) + (2k - t^2)D_k(t) = 0$.

11. Show that there are polynomials $c_k(t)$ and $d_k(t)$, each of degree at most k, such that

1. $C_k(t) = c_k(t)e^t - c_k(-t)e^{-t}$.
2. $D_k(t) = d_k(t)e^t + d_k(-t)e^{-t}$.

12. Show the following:

1. $tc_k''(t) + (2t - 2k)c_k'(t) - 2kc_k(t) = 0$.
2. $td_k''(t) + 2t(t - k)d_k'(t) - 2k(t - 1)d_k(t) = 0$.

13. Show the following:

1. $c_k(0) = \dfrac{(-1)^k(2k)!}{2^{k+1}k!}$.

2. $d_k'(0) = \dfrac{(-1)^{k-1}(2(k-1))!}{2^k(k-1)!}$.

14. Show the following:

1. $c_k(t) = \dfrac{(-1)^k}{2^{k+1}} \displaystyle\sum_{n=0}^{k} \dfrac{(2k-n)!}{n!(k-n)!}(-2t)^n$.

2. $d_k(t) = \dfrac{(-1)^k}{2^{k+1}} \displaystyle\sum_{n=1}^{k} \dfrac{(2k-n-1)!}{(n-1)!(k-n)!}(-2t)^n$.

15. Show the following:

1. $\mathcal{L}^{-1}\left\{\dfrac{1}{(s^2-1)^{k+1}}\right\}(t) = \dfrac{(-1)^k}{2^{2k+1}k!} \displaystyle\sum_{n=0}^{k} \dfrac{(2k-n)!}{n!(k-n)!}\left((-2t)^n e^t - (2t)^n e^{-t}\right).$

2. $\mathcal{L}^{-1}\left\{\dfrac{s}{(s^2-1)^{k+1}}\right\}(t) = \dfrac{(-1)^k}{2^{2k+1}k!} \displaystyle\sum_{n=1}^{k} \dfrac{(2k-n-1)!}{(n-1)!(k-n)!}\left((-2t)^n e^t + (2t)^n e^{-t}\right).$

7.5 Summary of Laplace Transforms

Table 7.1 Laplace transforms

$f(t)$	$F(s)$	Page
Laplace transforms involving the quadratic $s^2 + b^2$		
1. $\dfrac{\sin bt}{(2b)^{2k}} \displaystyle\sum_{m=0}^{\lfloor \frac{k}{2} \rfloor} (-1)^m \binom{2k-2m}{k} \dfrac{(2bt)^{2m}}{(2m)!}$ $\quad -\dfrac{\cos bt}{(2b)^{2k}} \displaystyle\sum_{m=0}^{\lfloor \frac{k-1}{2} \rfloor} (-1)^m \binom{2k-2m-1}{k} \dfrac{(2bt)^{2m+1}}{(2m+1)!}$	$\dfrac{b}{(s^2+b^2)^{k+1}}$	549
2. $\dfrac{2bt \sin bt}{k \cdot (2b)^{2k}} \displaystyle\sum_{m=0}^{\lfloor \frac{k-1}{2} \rfloor} (-1)^m \binom{2k-2m-2}{k-1} \dfrac{(2bt)^{2m}}{(2m)!}$ $\quad -\dfrac{2bt \cos bt}{k \cdot (2b)^{2k}} \displaystyle\sum_{m=0}^{\lfloor \frac{k-2}{2} \rfloor} (-1)^m \binom{2k-2m-3}{k-1} \dfrac{(2bt)^{2m+1}}{(2m+1)!}$	$\dfrac{s}{(s^2+b^2)^{k+1}}$	549
Laplace transforms involving the quadratic $s^2 - b^2$		
3. $\dfrac{(-1)^k}{2^{2k+1}k!} \displaystyle\sum_{n=0}^{k} \dfrac{(2k-n)!}{n!(k-n)!} \left((-2t)^n e^t - (2t)^n e^{-t}\right)$	$\dfrac{1}{(s^2-1)^{k+1}}$	553
4. $\dfrac{(-1)^k}{2^{2k+1}k!} \displaystyle\sum_{n=1}^{k} \dfrac{(2k-n-1)!}{(n-1)!(k-n)!} \left((-2t)^n e^t + (2t)^n e^{-t}\right)$	$\dfrac{s}{(s^2-1)^{k+1}}$	553

Chapter 8
Matrices

Most students by now have been exposed to the language of matrices. They arise naturally in many subject areas but mainly in the context of solving a simultaneous system of linear equations. In this chapter, we will give a review of matrices, systems of linear equations, inverses, determinants, and eigenvectors and eigenvalues. The next chapter will apply what is learned here to linear systems of differential equations.

W.A. Adkins and M.G. Davidson, *Ordinary Differential Equations*,
Undergraduate Texts in Mathematics, DOI 10.1007/978-1-4614-3618-8_8,
© Springer Science+Business Media New York 2012

8.1 Matrix Operations

A *matrix* is a rectangular array of entities and is generally written in the following way:

$$X = \begin{bmatrix} x_{11} & \cdots & x_{1n} \\ \vdots & \ddots & \vdots \\ x_{m1} & \cdots & x_{mn} \end{bmatrix}.$$

We let $\mathcal{R}$ denote the set of entities that will be in use at any particular time. Each x_{ij} is in $\mathcal{R}$, and in this text, $\mathcal{R}$ can be one of the following sets:

$\mathbb{R}$ or $\mathbb{C}$	The scalars
$\mathbb{R}[t]$ or $\mathbb{C}[t]$	Polynomials with real or complex entries
$\mathbb{R}(s)$ or $\mathbb{C}(s)$	The real or complex rational functions
$C^k(I, \mathbb{R})$ or $C^k(I, \mathbb{C})$	Real- or complex-valued functions
	with k continuous derivatives

Notice that addition and multiplication are defined on $\mathcal{R}$. Below we will extend these operations to matrices.

The following are examples of matrices.

Example 1.

$$A = \begin{bmatrix} 1 & 0 & 3 \\ 2 & -1 & 4 \end{bmatrix}, \quad B = \begin{bmatrix} 1 & -1 & 9 \end{bmatrix}, \quad C = \begin{bmatrix} i & 2-i \\ 1 & 0 \end{bmatrix},$$

$$D = \begin{bmatrix} \sin t \\ \cos t \\ \tan t \end{bmatrix}, \quad E = \begin{bmatrix} \dfrac{s}{s^2-1} & \dfrac{1}{s^2-1} \\ \dfrac{-1}{s^2-1} & \dfrac{s+2}{s^2-1} \end{bmatrix}.$$

It is a common practice to use capital letters, like A, B, C, D, and E, to denote matrices. The *size* of a matrix is determined by the number of rows m and the number of columns n and written $m \times n$. In Example 1, A is a 2×3 matrix, B is a 1×3 matrix, C and E are 2×2 matrices, and D is a 3×1 matrix. A matrix is *square* if the number of rows is the same as the number of columns. Thus, C and E are square matrices. An entry in a matrix is determined by its position. If X is a matrix, the (i, j) *entry* is the entry that appears in the ith row and jth column. We denote it in two ways: $\text{ent}_{ij}(X)$ or more simply X_{ij}. Thus, in Example 1, $A_{13} = 3$, $B_{12} = -1$, and $C_{22} = 0$. We say that two matrices X and Y are *equal* if the corresponding entries are equal, that is, $X_{ij} = Y_{ij}$, for all indices i and j. Necessarily, X and Y must be the same size. The *main diagonal* of a square $n \times n$ matrix X is the vector formed from the entries X_{ii}, for $i = 1, \ldots, n$. The main diagonal of C is $(i, 0)$ and the main diagonal of E is $(\frac{s}{s^2-1}, \frac{s+2}{s^2-1})$. A matrix is said to be a ***real matrix*** if each entry is real and a ***complex matrix*** if each entry is complex. Since every real number is also complex, every real matrix is also a complex matrix. Thus, A and B are real (and complex) matrices while C is a complex matrix.

Even though a matrix is a structured array of entities in $\mathcal{R}$, it should be viewed as a single object just as a word is a single object though made up of many letters. We let $M_{m,n}(\mathcal{R})$ denote the set of all $m \times n$ matrices with entries in $\mathcal{R}$. If the focus is on matrices of a certain size and not on the entries, we will sometimes write $M_{m,n}$. The following definitions highlight various kinds of matrices that commonly arise:

1. A ***diagonal*** matrix D is a square matrix in which all entries off the main diagonal are 0. We can say this in another way:

$$D_{ij} = 0 \text{ if } i \neq j.$$

Examples of diagonal matrices are:

$$\begin{bmatrix} 1 & 0 \\ 0 & 4 \end{bmatrix} \qquad \begin{bmatrix} e^t & 0 & 0 \\ 0 & e^{4t} & 0 \\ 0 & 0 & 1 \end{bmatrix} \qquad \begin{bmatrix} \frac{1}{s} & 0 & 0 & 0 \\ 0 & \frac{2}{s-1} & 0 & 0 \\ 0 & 0 & 0 & 0 \\ 0 & 0 & 0 & \frac{-1}{s-2} \end{bmatrix}.$$

It is convenient to write $\operatorname{diag}(d_1, \ldots, d_n)$ to represent the diagonal $n \times n$ matrix with $(d_1, \ldots, d_n)$ on the diagonal. Thus, the diagonal matrices listed above are $\operatorname{diag}(1, 4)$, $\operatorname{diag}(e^t, e^{4t}, 1)$, and $\operatorname{diag}(\frac{1}{s}, \frac{2}{s-1}, 0, -\frac{1}{s-2})$, respectively.

2. The ***zero*** matrix $\mathbf{0}$ is the matrix with each entry 0. The size is usually determined by the context. If we need to be specific, we will write $\mathbf{0}_{m,n}$ to mean the $m \times n$ zero matrix. Note that the square zero matrix, $\mathbf{0}_{n,n}$ is diagonal and is $\operatorname{diag}(0, \ldots, 0)$.

3. The ***identity matrix***, I, is the square matrix with ones on the main diagonal and zeros elsewhere. The size is usually determined by the context, but if we want to be specific, we write I_n to denote the $n \times n$ identity matrix. The 2×2 and the 3×3 identity matrices are

$$I_2 = \begin{bmatrix} 1 & 0 \\ 0 & 1 \end{bmatrix}, \qquad I_3 = \begin{bmatrix} 1 & 0 & 0 \\ 0 & 1 & 0 \\ 0 & 0 & 1 \end{bmatrix}.$$

4. We say a square matrix is ***upper triangular*** if each entry below the main diagonal is zero. We say a square matrix is ***lower triangular*** if each entry above the main diagonal is zero. The matrices

$$\begin{bmatrix} 1 & 2 \\ 0 & 3 \end{bmatrix} \quad \text{and} \quad \begin{bmatrix} 1 & 3 & 5 \\ 0 & 0 & 3 \\ 0 & 0 & -4 \end{bmatrix}$$

are upper triangular, and

$$\begin{bmatrix} 4 & 0 \\ 1 & 1 \end{bmatrix} \quad \text{and} \quad \begin{bmatrix} 0 & 0 & 0 \\ 2 & 0 & 0 \\ 1 & 1 & -7 \end{bmatrix}$$

are lower triangular.

5. Suppose A is an $m \times n$ matrix. The **transpose** of A, denoted A^t, is the $n \times m$ matrix obtained by turning the rows of A into columns. In terms of the entries we have, more explicitly,

$$(A^t)_{ij} = A_{ji}.$$

This expression reverses the indices of A. Simple examples are

$$\begin{bmatrix} 2 & 3 \\ 9 & 0 \\ 1 & 4 \end{bmatrix}^t = \begin{bmatrix} 2 & 9 & 1 \\ 3 & 0 & 4 \end{bmatrix} \quad \begin{bmatrix} e^t \\ e^{-t} \end{bmatrix}^t = \begin{bmatrix} e^t & e^{-t} \end{bmatrix} \quad \begin{bmatrix} \frac{1}{s} & \frac{2}{s^3} \\ \frac{2}{s^2} & \frac{3}{s} \end{bmatrix}^t = \begin{bmatrix} \frac{1}{s} & \frac{2}{s^2} \\ \frac{2}{s^3} & \frac{3}{s} \end{bmatrix}.$$

Matrix Algebra

There are three matrix operations that make up the algebraic structure of matrices: addition, scalar multiplication, and matrix multiplication.

Addition

Suppose A and B are two matrices of the same size. We define **matrix addition**, $A + B$, entrywise by the following formula:

$$(A + B)_{ij} = A_{ij} + B_{ij}.$$

Thus, if

$$A = \begin{bmatrix} 1 & -2 & 0 \\ 4 & 5 & -3 \end{bmatrix} \quad \text{and} \quad B = \begin{bmatrix} 4 & -1 & 0 \\ -3 & 8 & 1 \end{bmatrix},$$

then

$$A + B = \begin{bmatrix} 1+4 & -2-1 & 0+0 \\ 4-3 & 5+8 & -3+1 \end{bmatrix} = \begin{bmatrix} 5 & -3 & 0 \\ 1 & 13 & -2 \end{bmatrix}.$$

Corresponding entries are added. Addition preserves the size of matrices. We can symbolize this in the following way:

$$+ : M_{m,n}(\mathcal{R}) \times M_{m,n}(\mathcal{R}) \to M_{m,n}(\mathcal{R}).$$

Addition satisfies the following properties:

Proposition 2. *Suppose A, B, and C are m × n matrices. Then*

$$A + B = B + A, \qquad\qquad\qquad \text{(commutative)}$$

$$(A + B) + C = A + (B + C), \qquad\qquad \text{(associative)}$$

$$A + \mathbf{0} = A, \qquad\qquad\qquad \text{(additive identity)}$$

$$A + (-A) = \mathbf{0}. \qquad\qquad\qquad \text{(additive inverse)}$$

Scalar Multiplication

Suppose A is an matrix and $c \in \mathcal{R}$. We define *scalar multiplication*, $c \cdot A$, (but usually we will just write cA), entrywise by the following formula

$$(cA)_{ij} = cA_{ij}.$$

For example,

$$-2 \begin{bmatrix} 1 & 9 \\ -3 & 0 \\ 2 & 5 \end{bmatrix} = \begin{bmatrix} -2 & -18 \\ 6 & 0 \\ -4 & -10 \end{bmatrix}.$$

Scalar multiplication preserves the size of matrices. Thus,

$$\cdot : \mathcal{R} \times M_{m,n}(\mathcal{R}) \to M_{m,n}(\mathcal{R}).$$

Scalar multiplication satisfies the following properties:

Proposition 3. *Suppose A and B are matrices of the same size. Suppose $c_1, c_2 \in \mathcal{R}$. Then*

$$c_1(A + B) = c_1 A + c_1 B, \qquad\qquad \text{(distributive)}$$

$$(c_1 + c_2)A = c_1 A + c_2 A, \qquad\qquad \text{(distributive)}$$

$$c_1(c_2 A) = (c_1 c_2)A, \qquad \text{(associative)}$$

$$1A = A, \qquad \text{(1 is a multiplicative identity)}$$

$$0A = \mathbf{0}.$$

Matrix Multiplication

Matrix multiplication is more complicated than addition and scalar multiplication. We will define it in two stages: first on row and column matrices and then on general matrices.

A *row matrix* or *row vector* is a matrix which has only one row. Thus, row vectors are in $M_{1,n}$. Similarly, a *column matrix* or *column vector* is a matrix which has only one column. Thus, column vectors are in $M_{m,1}$. We frequently will denote column and row vectors by lower case boldface letters like $\mathbf{v}$ or $\mathbf{x}$ instead of capital letters. It is unnecessary to use double subscripts to indicate the entries of a row or column matrix: if $\mathbf{v}$ is a row vector, then we write $\mathbf{v}_i$ for the ith entry instead of $\mathbf{v}_{1i}$. Similarly for column vectors. Suppose $\mathbf{v} \in M_{1,n}$ and $\mathbf{w} \in M_{n,1}$. We define the product $\mathbf{v} \cdot \mathbf{w}$ (or preferably $\mathbf{vw}$) to be the scalar given by

$$\mathbf{vw} = v_1 w_1 + \cdots + v_n w_n.$$

Even though this formula looks like the scalar product or dot product that you likely have seen before, keep in mind that $\mathbf{v}$ is a row vector while $\mathbf{w}$ is a column vector. For example, if

$$\mathbf{v} = \begin{bmatrix} 1 & 3 & -2 & 0 \end{bmatrix} \text{ and } \mathbf{w} = \begin{bmatrix} 1 \\ 3 \\ 0 \\ 9 \end{bmatrix},$$

then

$$\mathbf{vw} = 1 \cdot 1 + 3 \cdot 3 + (-2) \cdot 0 + 0 \cdot 9 = 10.$$

Now suppose that A is any matrix. It is often convenient to distinguish the rows of A in the following way. Let $\text{Row}_i(A)$ denotes the ith row of A. Then

$$A = \begin{bmatrix} \text{Row}_1(A) \\ \text{Row}_2(A) \\ \vdots \\ \text{Row}_m(A) \end{bmatrix}.$$

In a similar way, if B is another matrix, we can distinguish the columns of B. Let $\text{Col}_j(B)$ denote the jth column of B, then

$$B = \left[\text{Col}_1(B)\ \text{Col}_2(B)\ \cdots\ \text{Col}_p(B)\right].$$

Now let $A \in M_{mn}$ and $B \in M_{np}$. We define the **matrix product** of A and B to be the $m \times p$ matrix given entrywise by $\text{ent}_{i\,j}(AB) = \text{Row}_i(A)\,\text{Col}_j(B)$. In other words, the (i, j)-entry of the product of A and B is the ith row of A times the jth column of B. We thus have

$$AB = \begin{bmatrix} \text{Row}_1(A)\,\text{Col}_1(B) & \text{Row}_1(A)\,\text{Col}_2(B) & \cdots & \text{Row}_1(A)\,\text{Col}_p(B) \\ \text{Row}_2(A)\,\text{Col}_1(B) & \text{Row}_2(A)\,\text{Col}_2(B) & \cdots & \text{Row}_2(A)\,\text{Col}_p(B) \\ \vdots & \vdots & \ddots & \vdots \\ \text{Row}_m(A)\,\text{Col}_1(B) & \text{Row}_m(A)\,\text{Col}_2(B) & \cdots & \text{Row}_m(A)\,\text{Col}_p(B) \end{bmatrix}.$$

Notice that each entry of AB is given as a product of a row vector and a column vector. Thus, it is necessary that the number of columns of A (the first matrix) match the number of rows of B (the second matrix). This common number is n. The resulting product is an $m \times p$ matrix. We thus write

$$\cdot : M_{m,n}(\mathcal{R}) \times M_{n,p}(\mathcal{R}) \to M_{m,p}(\mathcal{R}).$$

In terms of the entries of A and B, we have

$$\text{ent}_{i\,j}(AB) = \text{Row}_i(A)\,\text{Col}_j(B) = \sum_{k=1}^{n} \text{ent}_{i\,k}(A)\text{ent}_{k\,j}(B) = \sum_{k=1}^{n} A_{i,k}B_{k,j}.$$

Example 4.

1. If $A = \begin{bmatrix} 2 & 1 \\ -1 & 3 \\ 4 & -2 \end{bmatrix}$ and $B = \begin{bmatrix} 2 & 1 \\ 2 & -2 \end{bmatrix}$, then AB is defined because the number of columns of A is the number of rows of B. Further, AB is a 3×2 matrix and

$$AB = \begin{bmatrix} \begin{bmatrix} 2 & 1 \end{bmatrix}\begin{bmatrix} 2 \\ 2 \end{bmatrix} & \begin{bmatrix} 2 & 1 \end{bmatrix}\begin{bmatrix} 1 \\ -2 \end{bmatrix} \\ \begin{bmatrix} -1 & 3 \end{bmatrix}\begin{bmatrix} 2 \\ 2 \end{bmatrix} & \begin{bmatrix} -1 & 3 \end{bmatrix}\begin{bmatrix} 1 \\ -2 \end{bmatrix} \\ \begin{bmatrix} 4 & -2 \end{bmatrix}\begin{bmatrix} 2 \\ 2 \end{bmatrix} & \begin{bmatrix} 4 & -2 \end{bmatrix}\begin{bmatrix} 1 \\ -2 \end{bmatrix} \end{bmatrix} = \begin{bmatrix} 6 & 0 \\ 4 & -7 \\ 4 & 8 \end{bmatrix}.$$

2. If $A = \begin{bmatrix} e^t & 2e^t \\ e^{2t} & 3e^{2t} \end{bmatrix}$ and $B = \begin{bmatrix} -2 \\ 1 \end{bmatrix}$, then

$$AB = \begin{bmatrix} -2e^t + 2e^t \\ -2e^{2t} + 3e^{2t} \end{bmatrix} = \begin{bmatrix} 0 \\ e^{2t} \end{bmatrix}.$$

Notice in the definition (and the example) that in a given column of AB, the corresponding column of B appears as the second factor. Thus,

$$\mathrm{Col}_j(AB) = A\,\mathrm{Col}_j(B). \tag{1}$$

Similarly, in each row of AB, the corresponding row of A appears and we get

$$\mathrm{Row}_i(A)B = \mathrm{Row}_i(AB). \tag{2}$$

Notice too that even though the product AB is defined, it is not necessarily true that BA is defined. This is the case in part 1 of the above example due to the fact that the number of columns of B (2) does not match the number of rows of A (3). Even when AB and BA are defined, it is not necessarily true that they are equal. Consider the following example:

Example 5. Suppose

$$A = \begin{bmatrix} 1 & 2 \\ 0 & 3 \end{bmatrix} \text{ and } B = \begin{bmatrix} 2 & 1 \\ 4 & -1 \end{bmatrix}.$$

Then

$$AB = \begin{bmatrix} 1 & 2 \\ 0 & 3 \end{bmatrix} \begin{bmatrix} 2 & 1 \\ 4 & -1 \end{bmatrix} = \begin{bmatrix} 10 & -1 \\ 12 & -3 \end{bmatrix}$$

while

$$BA = \begin{bmatrix} 2 & 1 \\ 4 & -1 \end{bmatrix} \begin{bmatrix} 1 & 2 \\ 0 & 3 \end{bmatrix} = \begin{bmatrix} 2 & 7 \\ 4 & 5 \end{bmatrix}.$$

These products are not the same. This example shows that matrix multiplication is *not* commutative. On the other hand, there can be two special matrices A and B for which $AB = BA$. In this case, we say A and B **commute**. For example, if

$$A = \begin{bmatrix} 1 & -1 \\ 1 & 1 \end{bmatrix} \text{ and } B = \begin{bmatrix} 1 & 1 \\ -1 & 1 \end{bmatrix},$$

then

$$AB = \begin{bmatrix} 2 & 0 \\ 0 & 2 \end{bmatrix} = BA.$$

Thus A and B commute. However, such occurrences are special. The other properties that we are used to in an algebra are valid. We summarize them in the following proposition.

Proposition 6. *Suppose A, B, and C are matrices whose sizes are such that each line below is defined. Suppose $c_1, c_2 \in \mathcal{R}$. Then*

$A(BC) = (AB)C,$ (associatvie)

$A(cB) = (cA)B = c(AB),$ (commutes with scalar multiplication)

$(A + B)C = AC + BC,$ (distributive)

$A(B + C) = AB + AC,$ (distributive)

$IA = AI = A.$ (I is a multiplicative identity)

We highlight two useful formulas that follow from these algebraic properties. If A is an $m \times n$ matrix, then

$$A\mathbf{x} = x_1 \operatorname{Col}_1(A) + \cdots x_n \operatorname{Col}_n(A), \quad \text{where } \mathbf{x} = \begin{bmatrix} x_1 \\ \vdots \\ x_n \end{bmatrix} \tag{3}$$

and

$$\mathbf{y}A = y_1 \operatorname{Row}_1(A) + \cdots y_m \operatorname{Row}_m(A), \quad \text{where } \mathbf{y} = [y_1 \cdots y_m]. \tag{4}$$

Henceforth, we will use these algebraic properties without explicit reference. The following result expresses the relationship between multiplication and transposition of matrices

Theorem 7. *Let A and B be matrices such that AB is defined. Then $B^t A^t$ is defined and*

$$B^t A^t = (AB)^t.$$

Proof. The number of columns of B^t is the same as the number of rows of B while the number of rows of A^t is the number of columns of A. These numbers agree since AB is defined so $B^t A^t$ is defined. If n denotes these common numbers, then

$$(B^t A^t)_{ij} = \sum_{k=1}^{n} (B^t)_{ik} (A^t)_{kj} = \sum_{k=1}^{n} A_{jk} B_{ki} = (AB)_{ji} = ((AB)^t)_{ij}.$$

All entries of $A^t B^t$ and $(AB)^t$ agree so they are equal.   □

Exercises

1–3. Let $A = \begin{bmatrix} 2 & -1 & 3 \\ 1 & 0 & 4 \end{bmatrix}$, $B = \begin{bmatrix} 1 & -1 \\ 2 & 3 \\ -1 & 2 \end{bmatrix}$, and $C = \begin{bmatrix} 0 & 2 \\ -3 & 4 \\ 1 & 1 \end{bmatrix}$. Compute the following matrices.

1. $B + C$, $B - C$, $2B - 3C$
2. AB, AC, BA, CA
3. $A(B + C)$, $AB + AC$, $(B + C)A$

4. Let $A = \begin{bmatrix} 2 & 1 \\ 3 & 4 \\ -1 & 0 \end{bmatrix}$ and $B = \begin{bmatrix} 1 & 2 \\ -1 & 1 \\ 1 & 0 \end{bmatrix}$. Find C so that $3A + C = 4B$.

5–9. Let $A = \begin{bmatrix} 3 & -1 \\ 0 & -2 \\ 1 & 2 \end{bmatrix}$, $B = \begin{bmatrix} 2 & 1 & 1 & -3 \\ 0 & -1 & 4 & -1 \end{bmatrix}$, and $C = \begin{bmatrix} 2 & 1 & 2 \\ 1 & 3 & 1 \\ 0 & 1 & 8 \\ 1 & 1 & 7 \end{bmatrix}$. Find the following products.

5. AB
6. BC
7. CA
8. $B^t A^t$
9. ABC

10. Let $A = \begin{bmatrix} 1 & 4 & 3 & 1 \end{bmatrix}$ and $B = \begin{bmatrix} 1 \\ 0 \\ -1 \\ -2 \end{bmatrix}$. Find AB and BA.

11–13. Let $A = \begin{bmatrix} 1 & 2 & 5 \\ 2 & 4 & 10 \\ -1 & -2 & -5 \end{bmatrix}$, $B = \begin{bmatrix} 1 & 0 \\ 4 & -1 \end{bmatrix}$, $C = \begin{bmatrix} 3 & -2 \\ 3 & -2 \end{bmatrix}$. Verify the following facts:

11. $A^2 = 0$
12. $B^2 = I_2$
13. $C^2 = C$

14–15. Compute $AB - BA$ in each of the following cases.

14. $A = \begin{bmatrix} 0 & 1 \\ 1 & 1 \end{bmatrix}$, $B = \begin{bmatrix} 1 & 0 \\ 1 & 1 \end{bmatrix}$

15. $A = \begin{bmatrix} 2 & 1 & 0 \\ 1 & 1 & 1 \\ -1 & 2 & 1 \end{bmatrix}$, $B = \begin{bmatrix} 3 & 1 & -2 \\ 3 & -2 & 4 \\ -3 & 5 & -1 \end{bmatrix}$

16. Let $A = \begin{bmatrix} 1 & a \\ 0 & 1 \end{bmatrix}$ and $B = \begin{bmatrix} 1 & 0 \\ b & 1 \end{bmatrix}$. Show that there are no numbers a and b so that $AB - BA = I$, where I is the 2×2 identity matrix.

17. Suppose that A and B are 2×2 matrices.

1. Show by example that it need not be true that $(A + B)^2 = A^2 + 2AB + B^2$.
2. Find conditions on A and B to insure that the equation in Part (a) is valid.

18. If $A = \begin{bmatrix} 0 & 1 \\ 1 & 1 \end{bmatrix}$, compute A^2 and A^3.

19. If $B = \begin{bmatrix} 1 & 1 \\ 0 & 1 \end{bmatrix}$, compute B^n for all n.

20. If $A = \begin{bmatrix} a & 0 \\ 0 & b \end{bmatrix}$, compute A^2, A^3, and more generally, A^n for all n.

21. Let $A = \begin{bmatrix} v_1 \\ v_2 \end{bmatrix}$ be a matrix with two rows v_1 and v_2. (The number of columns of A is not relevant for this problem.) Describe the effect of multiplying A on the left by the following matrices:

(a) $\begin{bmatrix} 0 & 1 \\ 1 & 0 \end{bmatrix}$ (b) $\begin{bmatrix} 1 & c \\ 0 & 1 \end{bmatrix}$ (c) $\begin{bmatrix} 1 & 0 \\ c & 1 \end{bmatrix}$ (d) $\begin{bmatrix} a & 0 \\ 0 & 1 \end{bmatrix}$ (e) $\begin{bmatrix} 1 & 0 \\ 0 & a \end{bmatrix}$

22. Let $E(\theta) = \begin{bmatrix} \cos \theta & \sin \theta \\ -\sin \theta & \cos \theta \end{bmatrix}$. Show that $E(\theta_1 + \theta_2) = E(\theta_1)E(\theta_2)$.

23. Let $F(\theta) = \begin{bmatrix} \cosh \theta & \sinh \theta \\ \sinh \theta & \cosh \theta \end{bmatrix}$. Show that $F(\theta_1 + \theta_2) = F(\theta_1)F(\theta_2)$.

24. Let $D = \text{diag}(d_1, \ldots, d_n)$ and $E = \text{diag}(e_1, \ldots, e_n)$. Show that

$$DE = \text{diag}(d_1 e_1, \ldots, d_n e_n).$$

8.2   Systems of Linear Equations

Most students have learned various techniques for finding the solution of a system of linear equations. They usually include various forms of elimination and substitutions. In this section, we will learn the Gauss-Jordan elimination method. It is essentially a highly organized method involving elimination and substitution that always leads to the solution set. This general method has become the standard for solving systems. At first reading, it may seem to be a bit complicated because of its description for general systems. However, with a little practice on a few examples, it is quite easy to master. We will as usual begin with our definitions and proceed with examples to illustrate the needed concepts. To make matters a bit cleaner, we will stick to the case where $\mathcal{R} = \mathbb{R}$. Everything we do here will work for $\mathcal{R} = \mathbb{C}$, $\mathbb{R}(s)$, or $\mathbb{C}(s)$ as well. (A technical difficulty for general $\mathcal{R}$ is the lack of inverses.)

If $x_1, \ldots, x_n$ are variables, then the equation

$$a_1 x_1 + \cdots + a_n x_n = b$$

is called a *linear equation* in the unknowns $x_1, \ldots, x_n$. A *system of linear equations* is a set of m linear equations in the unknowns $x_1, \ldots, x_n$ and is written in the form

$$
\begin{aligned}
a_{11}x_1 + a_{12}x_2 + \cdots + a_{1n}x_n &= b_1 \\
a_{21}x_1 + a_{22}x_2 + \cdots + a_{2n}x_n &= b_2 \\
\vdots \qquad \vdots \qquad\qquad \vdots \qquad\quad &\ \ \vdots \\
a_{m1}x_1 + a_{m2}x_2 + \cdots + a_{mn}x_n &= b_m.
\end{aligned}
\tag{1}
$$

The entries a_{ij} are in $\mathbb{R}$ and are called *coefficients*. Likewise, each b_j is in $\mathbb{R}$. A key observation is that (1) can be rewritten in matrix form as:

$$A\mathbf{x} = \mathbf{b}, \tag{2}$$

where

$$
A = \begin{bmatrix} a_{11} & a_{12} & \cdots & a_{1n} \\ a_{21} & a_{22} & \cdots & a_{2n} \\ \vdots & \vdots & & \vdots \\ a_{m1} & a_{m2} & \cdots & a_{mn} \end{bmatrix}, \quad
\mathbf{x} = \begin{bmatrix} x_1 \\ x_2 \\ \vdots \\ x_n \end{bmatrix}, \quad \text{and} \quad
\mathbf{b} = \begin{bmatrix} b_1 \\ b_2 \\ \vdots \\ b_m \end{bmatrix}.
$$

We call A the *coefficient matrix*, $\mathbf{x}$ the *variable matrix*, and $\mathbf{b}$ the *output matrix*. Any column vector $\mathbf{x}$ with entries in $\mathbb{R}$ that satisfies equation (1) (or, equivalently, equation (2)) is called a *solution*. If a system has a solution, we say it is *consistent*; otherwise, it is *inconsistent*. The *solution set* is the set of all solutions. The system is said to be **homogeneous** if $\mathbf{b} = \mathbf{0}$, otherwise it is called **nonhomogeneous**. The homogeneous case is especially important. We call the solution set to

$$A\mathbf{x} = \mathbf{0}$$

the **null space** of A and denote it by $\mathrm{NS}(A)$. Another important matrix associated with (2) is the **augmented matrix**:

$$[A \mid b] = \begin{bmatrix} a_{11} & a_{12} & \cdots & a_{1n} & b_1 \\ a_{21} & a_{22} & \cdots & a_{2n} & b_2 \\ \vdots & \vdots & & \vdots & \vdots \\ a_{m1} & a_{m2} & \cdots & a_{mn} & b_m \end{bmatrix},$$

where the vertical line only serves to separate A from $\mathbf{b}$.

Example 1. Write the coefficient, variable, output, and augmented matrices for the following system:

$$\begin{aligned} -2x_1 + 3x_2 - x_3 &= 4 \\ x_1 - 2x_2 + 4x_3 &= 5. \end{aligned}$$

Determine whether the following vectors are solutions:

$$\mathbf{v_1} = \begin{bmatrix} -3 \\ 0 \\ 2 \end{bmatrix}, \quad \mathbf{v_2} = \begin{bmatrix} 7 \\ 7 \\ 3 \end{bmatrix}, \quad \mathbf{v_3} = \begin{bmatrix} 10 \\ 7 \\ 1 \end{bmatrix}, \quad \mathbf{v_4} = \begin{bmatrix} 2 \\ 1 \end{bmatrix}.$$

▶ **Solution.** The coefficient matrix is $A = \begin{bmatrix} -2 & 3 & -1 \\ 1 & -2 & 4 \end{bmatrix}$, the variable matrix is $\mathbf{x} = \begin{bmatrix} x_1 \\ x_2 \\ x_3 \end{bmatrix}$, the output matrix is $\mathbf{b} = \begin{bmatrix} 4 \\ 5 \end{bmatrix}$, and the augmented matrix is $\begin{bmatrix} -2 & 3 & -1 & 4 \\ 1 & -2 & 4 & 5 \end{bmatrix}$. The system is nonhomogeneous. Notice that

$$A\mathbf{v_1} = \begin{bmatrix} 4 \\ 5 \end{bmatrix} \quad \text{and} \quad A\mathbf{v_2} = \begin{bmatrix} 4 \\ 5 \end{bmatrix} \text{ while } A\mathbf{v_3} = \begin{bmatrix} 0 \\ 0 \end{bmatrix}.$$

Therefore, $\mathbf{v_1}$ and $\mathbf{v_2}$ are solutions, $\mathbf{v_3}$ is not a solution but since $A\mathbf{v_3} = \mathbf{0}$, we have $\mathbf{v_3}$ is in the null space of A. Finally, $\mathbf{v_4}$ is not the right size and thus cannot be a solution.   ◀

Remark 2. When only 2 or 3 variables are involved in an example, we will frequently use the variables x, y, and z instead of the subscripted variables x_1, x_2, and x_3.

Linearity

It is convenient to think of $\mathbb{R}^n$ as the set of column vectors $M_{n,1}(\mathbb{R})$. If A is an $m \times n$ real matrix, then for each column vector $\mathbf{x} \in \mathbb{R}^n$, the product, $A\mathbf{x}$, is a column vector in $\mathbb{R}^m$. Thus, the matrix A induces a map which we also denote just by $A : \mathbb{R}^n \to \mathbb{R}^m$ given by matrix multiplication. It satisfies the following important property.

Proposition 3. *The map $A : \mathbb{R}^n \to \mathbb{R}^m$ is linear. In other words,*

1. $A(\mathbf{x} + \mathbf{y}) = A(\mathbf{x}) + A(\mathbf{y})$
2. $A(c\mathbf{x}) = cA(\mathbf{x})$

for all $\mathbf{x}, \mathbf{y} \in \mathbb{R}^n$ and $c \in \mathbb{R}$.

Proof. This follows directly from Propositions 3 and 6 of Sect. 8.1.   □

Linearity is an extremely important property for it allows us to describe the structure of the solution set to $A\mathbf{x} = \mathbf{b}$ in a particularly nice way.

Proposition 4. *With A, $\mathbf{x}$, and $\mathbf{b}$ as above, we have two possibilities. Either the solution set to $A\mathbf{x} = \mathbf{b}$ is the empty set or we can write all solutions in the following form:*

$$\mathbf{x}_p + \mathbf{x}_h,$$

where $\mathbf{x}_p$ is a fixed particular solution and $\mathbf{x}_h$ is any vector in $\mathrm{NS}(A)$.

Remark 5. We will write the solution set to $A\mathbf{x} = \mathbf{b}$, when it is nonempty, as

$$x_p + \mathrm{NS}(A).$$

Proof. Suppose $\mathbf{x}_p$ is a fixed particular solution and $\mathbf{x}_h \in \mathrm{NS}(A)$. Then $A(\mathbf{x}_p + \mathbf{x}_h) = A\mathbf{x}_p + A\mathbf{x}_h = \mathbf{b} + \mathbf{0} = \mathbf{b}$. This implies that $\mathbf{x}_p + \mathbf{x}_h$ is a solution. On the other hand, suppose $\mathbf{x}$ is a solution to $A\mathbf{x} = \mathbf{b}$. Let $\mathbf{x}_h = \mathbf{x} - \mathbf{x}_p$. Then $A\mathbf{x}_h = A(\mathbf{x} - \mathbf{x}_p) = A\mathbf{x} - A\mathbf{x}_p = \mathbf{b} - \mathbf{b} = 0$. This means that $\mathbf{x}_h$ is in the null space of A and we get $\mathbf{x} = \mathbf{x}_p + \mathbf{x}_h$.   □

Remark 6. The solution set being empty is a legitimate possibility. For example, the simple equation $0x = 1$ has empty solution set. The system of equations $A\mathbf{x}=0$ is called the *associated homogeneous system*. It should be mentioned that the particular solution $\mathbf{x}_p$ is not necessarily unique. In Chap. 5, we saw a similar theorem for a second order differential equation $Ly = f$. That theorem provided a strategy for solving such differential equations: First we solved the homogeneous equation $Ly = 0$ and second found a particular solution (using variation of parameters or undetermined coefficients). For a linear system of equations, the matter is much simpler; the Gauss-Jordan method will give the whole solution set at one time. We will see that it has the above form.

Homogeneous Systems

The homogeneous case, $A\mathbf{x} = \mathbf{0}$, is of particular interest. Observe that $\mathbf{x} = \mathbf{0}$ is always a solution so NS(A) is never the empty set. But much more is true.

Proposition 7. *The solution set,* NS(A), *to a homogeneous system* $A\mathbf{x} = \mathbf{0}$ *is a linear space. In other words, if* $\mathbf{x}$ *and* $\mathbf{y}$ *are solutions to the homogeneous system and c is a scalar, then* $\mathbf{x} + \mathbf{y}$ *and* $c\mathbf{x}$ *are also solutions.*

Proof. Suppose $\mathbf{x}$ and $\mathbf{y}$ are in NS(A). Then $A(\mathbf{x}+\mathbf{y}) = A\mathbf{x}+A\mathbf{y} = \mathbf{0}+\mathbf{0} = \mathbf{0}$. This shows that $\mathbf{x} + \mathbf{y} \in$ NS(A). Now suppose $c \in \mathbb{R}$. Then $A(c\mathbf{x}) = cA\mathbf{x} = c\mathbf{0} = \mathbf{0}$. Hence $c\mathbf{x} \in$ NS(A). Thus NS(A) is a linear space.   □

Corollary 8. *The solution set to a general system of linear equations,* $A\mathbf{x} = \mathbf{b}$, *is either*

1. *Empty*
2. *Unique*
3. *Infinite*

Proof. The associated homogeneous system $A\mathbf{x} = \mathbf{0}$ has solution set, NS(A), that is either equal to the trivial set $\{\mathbf{0}\}$ or an infinite set. To see this suppose that $\mathbf{x}$ is a nonzero solution to $A\mathbf{x} = \mathbf{0}$, then by Proposition 7, all multiples, $c\mathbf{x}$, are in NS(A) as well. Therefore, by Proposition 4, if there is a solution to $A\mathbf{x} = \mathbf{b}$, it is unique or there are infinitely many.   □

The Elementary Equation and Row Operations

We say that two systems of equations are *equivalent* if their solution sets are the same. This definition implies that the variable matrix is the same for each system.

Example 9. Consider the following systems of equations:

$$2x + 3y = 5 \qquad\qquad x = 1$$
$$\qquad\qquad\qquad\text{and}$$
$$x - y = 0 \qquad\qquad y = 1.$$

The solution set to the second system is transparent. For the first system, there are some simple operations that easily lead to the solution: First, switch the two equations around. Next, multiply the equation $x - y = 1$ by -2 and add the result to the second equation. We then obtain

$$x - y = 0$$
$$5y = 5$$

Next, multiply the second equation by $\frac{1}{5}$ to get $y = 1$. Then add this equation to the first. We get $x = 1$ and $y = 1$. Thus, they both have the same solution set, namely, the single vector $\begin{bmatrix} 1 \\ 1 \end{bmatrix}$. They are thus equivalent. When used in the right way, these kinds of operations can transform a complicated system into a simpler one. We formalize these operations in the following definition:

Suppose $A\mathbf{x} = \mathbf{b}$ is a given system of linear equations. The following three operations are called *elementary equation operations*:

1. Switch the order in which two equations are listed.
2. Multiply an equation by a nonzero scalar.
3. Add a multiple of one equation to another.

Notice that each operation produces a new system of linear equations but leaves the size of the system unchanged. Furthermore, we have the following proposition.

Proposition 10. *An elementary equation operation applied to a system of linear equations is an equivalent system of equations.*

Proof. Suppose S is a system of equations and S' is a system obtained from S by switching two equations. A vector $\mathbf{x}$ is a solution to S if and only if it satisfies each equation in the system. If we switch the order of two of the equations, then $\mathbf{x}$ still satisfies each equation, and hence is a solution to S'. Notice that applying the same elementary equation operation to S' produces S. Hence, a solution to S' is a solution to S. It follows that S and S' are equivalent. The proof for the second and third elementary equation operations is similar. $\qquad\square$

The main idea in solving a system of linear equations is to perform a finite sequence of elementary equation operations to transform a system into simpler system where the solution set is transparent. Proposition 10 implies that the solution set of the simpler system is the same as original system. Let us consider our example above.

Example 11. Use elementary equation operations to transform

$$2x + 3y = 5,$$
$$x - y = 0$$

into

$$x = 1,$$
$$y = 1.$$

► **Solution.**

$$2x + 3y = 5$$
$$x - y = 0$$

Switch the order of the two equations.

$$x - y = 0$$
$$2x + 3y = 5$$

Add -2 times the first equation to the second equation.

$$x - y = 0$$
$$5y = 5$$

Multiply the second equation by $\frac{1}{5}$.

$$x - y = 0$$
$$y = 1$$

Add the second equation to the first.

$$x \quad = 1$$
$$y = 1$$

◄

Each operation produces a new system equivalent to the first by Proposition 10. The end result is a system where the solution is transparent. Since $y = 1$ is apparent in the fourth system, we could have stopped and used the method of **back substitution**, that is, substitute $y = 1$ into the first equation and solve for x. However, it is in accord with the Gauss–Jordan elimination method to continue as we did to eliminate the variable y in the first equation of the fourth system.

You will notice that the variables x and y play no prominent role in any of the calculations. They merely serve as placeholders for the coefficients, some of which change with each operation. We thus simplify the notation by performing the elementary operations on just the augmented matrix. The elementary equation operations become the *elementary row operations* which act on the augmented matrix of the system.

The elementary row operations on a matrix are:

1. Switch two rows
2. Multiply a row by a nonzero constant
3. Add a multiple of one row to another

The following notations for these operations will be useful:

1. p_{ij} - switch rows i and j
2. $m_i(a)$ - multiply row i by $a \neq 0$
3. $t_{ij}(a)$ - add to row j the value of a times row i

The effect of p_{ij} on a matrix A is denoted by $p_{ij}(A)$. Similarly for the other elementary row operations.

The corresponding operations when applied to the augmented matrix for the system in Example 11 becomes:

$$
\begin{bmatrix} 2 & 3 & | & 5 \\ 1 & -1 & | & 0 \end{bmatrix}
\xrightarrow{p_{12}}
\begin{bmatrix} 1 & -1 & | & 0 \\ 2 & 3 & | & 5 \end{bmatrix}
\xrightarrow{t_{12}(-2)}
\begin{bmatrix} 1 & -1 & | & 0 \\ 0 & 5 & | & 5 \end{bmatrix}
$$

$$
\xrightarrow{m_2(1/5)}
\begin{bmatrix} 1 & -1 & | & 0 \\ 0 & 1 & | & 1 \end{bmatrix}
\xrightarrow{t_{21}(1)}
\begin{bmatrix} 1 & 0 & | & 1 \\ 0 & 1 & | & 1 \end{bmatrix}.
$$

Above each arrow is the notation for the elementary row operation performed to produce the next augmented matrix. The sequence of elementary row operations chosen follows a certain strategy: Starting from left to right and top down, one tries to isolate a 1 in a given column and produce 0's above and below it. This corresponds to isolating and eliminating variables.

Let us consider three illustrative examples. The sequence of elementary row operation we perform is in accord with the Gauss–Jordan method which we will discuss in detail later on in this section. For now, verify each step. The end result will be an equivalent system for which the solution set will be transparent.

Example 12. Consider the following system of linear equations:

$$
\begin{aligned}
2x + 3y + 4z &= 9 \\
x + 2y - z &= 2
\end{aligned}.
$$

Find the solution set and write it in the form $\mathbf{x}_p + \mathrm{NS}(A)$.

▶ **Solution.** We first will write the augmented matrix and perform a sequence of elementary row operations:

$$
\begin{bmatrix} 2 & 3 & 4 & | & 9 \\ 1 & 2 & -1 & | & 2 \end{bmatrix}
\xrightarrow{p_{12}}
\begin{bmatrix} 1 & 2 & -1 & | & 2 \\ 2 & 3 & 4 & | & 9 \end{bmatrix}
\xrightarrow{t_{12}(-2)}
\begin{bmatrix} 1 & 2 & -1 & | & 2 \\ 0 & -1 & 6 & | & 5 \end{bmatrix}
$$

$$
\xrightarrow{m_2(-1)}
\begin{bmatrix} 1 & 2 & -1 & | & 2 \\ 0 & 1 & -6 & | & -5 \end{bmatrix}
\xrightarrow{t_{21}(-2)}
\begin{bmatrix} 1 & 0 & 11 & | & 12 \\ 0 & 1 & -6 & | & -5 \end{bmatrix}.
$$

The last augmented matrix corresponds to the system

$$
\begin{aligned}
x + \quad\;\; 11z &= 12 \\
y - 6z &= -5.
\end{aligned}
$$

In the first equation, we can solve for x in terms of z, and in the second equation, we can solve for y in terms of z. We refer to z as a *free variable* and let $z = \alpha$ be a parameter in $\mathbb{R}$. Then we obtain

$$x = 12 - 11\alpha$$
$$y = -5 + 6\alpha$$
$$z = \quad \alpha.$$

In vector form, we write

$$\mathbf{x} = \begin{bmatrix} x \\ y \\ z \end{bmatrix} = \begin{bmatrix} 12 - 11\alpha \\ -5 + 6\alpha \\ \alpha \end{bmatrix} = \begin{bmatrix} 12 \\ -5 \\ 0 \end{bmatrix} + \alpha \begin{bmatrix} -11 \\ 6 \\ 1 \end{bmatrix}.$$

The vector, $\mathbf{x}_p = \begin{bmatrix} 12 \\ -5 \\ 0 \end{bmatrix}$ is a particular solution (corresponding to $\alpha = 0$) while all

multiples of the vector $\begin{bmatrix} -11 \\ 6 \\ 1 \end{bmatrix}$ gives the null space of A. We have thus written the

solution in the form $\mathbf{x}_p + \text{NS}(A)$. In this case, there are infinitely many solutions. ◀

Example 13. Find the solution set for the system

$$3x + 2y + z = 4$$
$$2x + 2y + z = 3$$
$$x + y + z = 0.$$

▶ **Solution.** Again we start with the augmented matrix and apply elementary row operations. Occasionally, we will apply more than one operation at a time. When this is so, we stack the operations above the arrow with the topmost operation performed first followed in order by the ones below it:

$$\begin{bmatrix} 3 & 2 & 1 & 4 \\ 2 & 2 & 1 & 3 \\ 1 & 1 & 1 & 0 \end{bmatrix} \xrightarrow{p_{13}} \begin{bmatrix} 1 & 1 & 1 & 0 \\ 2 & 2 & 1 & 3 \\ 3 & 2 & 1 & 4 \end{bmatrix} \xrightarrow[\;t_{13}(-3)\;]{t_{12}(-2)} \begin{bmatrix} 1 & 1 & 1 & 0 \\ 0 & 0 & -1 & 3 \\ 0 & -1 & -2 & 4 \end{bmatrix}$$

$$\xrightarrow{p_{23}} \begin{bmatrix} 1 & 1 & 1 & 0 \\ 0 & -1 & -2 & 4 \\ 0 & 0 & -1 & 3 \end{bmatrix} \xrightarrow[\;m_3(-1)\;]{m_2(-1)} \begin{bmatrix} 1 & 1 & 1 & 0 \\ 0 & 1 & 2 & -4 \\ 0 & 0 & 1 & -3 \end{bmatrix}$$

$$\xrightarrow[\;t_{31}(-1)\;]{t_{32}(-2)} \begin{bmatrix} 1 & 1 & 0 & 3 \\ 0 & 1 & 0 & 2 \\ 0 & 0 & 1 & -3 \end{bmatrix} \xrightarrow{t_{21}(-1)} \begin{bmatrix} 1 & 0 & 0 & 1 \\ 0 & 1 & 0 & 2 \\ 0 & 0 & 1 & -3 \end{bmatrix}.$$

The last augmented matrix corresponds to the system

$$
\begin{aligned}
x &= 1 \\
y &= 2 \\
z &= -3.
\end{aligned}
$$

The solution set is transparent: $\mathbf{x} = \begin{bmatrix} 1 \\ 2 \\ -3 \end{bmatrix}$. In this example, we note that $NS(A) =$ $\{\mathbf{0}\}$, and the system has a unique solution.   ◄

Example 14. Solve the following system of linear equations:

$$
\begin{aligned}
x + 2y + 4z &= -2 \\
x + y + 3z &= 1 \\
2x + y + 5z &= 2.
\end{aligned}
$$

► **Solution.** Again we begin with the augmented matrix and perform elementary row operations.

$$
\begin{bmatrix} 1 & 2 & 4 & -2 \\ 1 & 1 & 3 & 1 \\ 2 & 1 & 5 & 2 \end{bmatrix}
\xrightarrow[\;t_{13}(-2)\;]{t_{12}(-1)}
\begin{bmatrix} 1 & 2 & 4 & -2 \\ 0 & -1 & -1 & 3 \\ 0 & -3 & -3 & 6 \end{bmatrix}
$$

$$
\xrightarrow{m_2(-1)}
\begin{bmatrix} 1 & 2 & 4 & -2 \\ 0 & 1 & 1 & -3 \\ 0 & -3 & -3 & 6 \end{bmatrix}
\xrightarrow{t_{23}(3)}
\begin{bmatrix} 1 & 2 & 4 & -2 \\ 0 & 1 & 1 & -3 \\ 0 & 0 & 0 & -3 \end{bmatrix}
$$

$$
\xrightarrow{m_3(-1/3)}
\begin{bmatrix} 1 & 2 & 4 & -2 \\ 0 & 1 & 1 & -3 \\ 0 & 0 & 0 & 1 \end{bmatrix}
\xrightarrow[\;t_{21}(-2)\;]{\substack{t_{31}(2) \\ t_{32}(3)}}
\begin{bmatrix} 1 & 0 & 2 & 0 \\ 0 & 1 & 1 & 0 \\ 0 & 0 & 0 & 1 \end{bmatrix}.
$$

The system that corresponds to the last augmented matrix is

$$
\begin{aligned}
x + 2z &= 0 \\
y + z &= 0 \\
0 &= 1.
\end{aligned}
$$

The last equation, which is shorthand for $0x + 0y + 0z = 1$, clearly has no solution. Thus, the system has no solution.   ◄

Reduced Matrices

These last three examples typify what happens in general and illustrate the three possible outcomes discussed in Corollary 8: infinitely many solutions, a unique solution, or no solution at all. The most involved case is when the solution set has infinitely many solutions. In Example 12, a single parameter α was needed to parameterize the set of solutions. However, in general, there may be many parameters needed. We will always want to use the least number of parameters possible, without dependencies among them. In each of the three preceding examples, it was transparent what the solution was by considering the system determined by the last listed augmented matrix. The last matrix was in a certain sense reduced as simple as possible.

We say that a matrix is in   *row echelon form (REF)*   if the following three conditions are satisfied:

1. The nonzero rows lie above the zero rows.
2. The first nonzero entry in a nonzero row is 1. (We call such a 1 a *leading one*.)
3. For any two adjacent nonzero rows, the leading one of the upper row is to the left of the leading one of the lower row. (We say the leading ones are in echelon form.)

We say a matrix is in *row reduced echelon form (RREF)* if it also satisfies

4. The entries above each leading one are zero.

Example 15. Determine which of the following matrices are row echelon form, row reduced echelon form, or neither. For the matrices in row echelon form, determine the columns (**C**) of the leading ones. If a matrix is not in row reduced echelon form, explain which conditions are violated.

$$
(1) \begin{bmatrix} 1 & 0 & -3 & 11 & 2 \\ 0 & 0 & 1 & 0 & 3 \\ 0 & 0 & 0 & 1 & 4 \end{bmatrix} \quad (2) \begin{bmatrix} 0 & 1 & 0 & 1 & 4 \\ 0 & 0 & 1 & 0 & 2 \\ 0 & 0 & 0 & 0 & 0 \end{bmatrix} \quad (3) \begin{bmatrix} 0 & 1 & 0 \\ 0 & 0 & 0 \\ 0 & 0 & 1 \end{bmatrix}
$$

$$
(4) \begin{bmatrix} 1 & 0 & 0 & 4 & 3 & 0 \\ 0 & 2 & 1 & 2 & 0 & 2 \\ 0 & 0 & 0 & 0 & 0 & 0 \end{bmatrix} \quad (5) \begin{bmatrix} 1 & 1 & 2 & 4 & -7 \\ 0 & 0 & 0 & 0 & 1 \end{bmatrix} \quad (6) \begin{bmatrix} 0 & 1 & 0 & 2 \\ 1 & 0 & 0 & -2 \\ 0 & 0 & 1 & 0 \end{bmatrix}
$$

▶ **Solution.**

1. (REF): Leading ones are in the first, third, and fourth columns. It is not reduced because there is a nonzero entry above the leading one in the third column.
2. (RREF): The leading ones are in the second and third columns.
3. Neither: The zero row is not at the bottom.
4. Neither: The first nonzero entry in the second row is not 1.
5. (REF): Leading ones are in the first and fifth columns. It is not reduced because there is a nonzero entry above the leading one in the fifth column.
6. Neither: The leading ones are not in echelon form.   ◀

Suppose a matrix A transforms by elementary row operations into a matrix A' which is in row reduced echelon form. We will sometimes say that A *row reduces* to A'. The *rank* of A, denoted Rank$A$, is the number of nonzero rows in A'. The definition of row reduced echelon form is valid for arbitrary matrices and not just matrices that come from a system of linear equations, that is, the augmented matrix. Suppose though that we consider a system $A\mathbf{x} = \mathbf{b}$, where A is an $m \times n$ matrix. Suppose the augmented matrix $[A|\mathbf{b}]$ is row reduced to a matrix $[A'|\mathbf{b}']$. Let $r = $ Rank$A$ and $r_\mathbf{b} = $ Rank$[A|\mathbf{b}]$. The elementary row operations that row reduce $[A|\mathbf{b}]$ to $[A'|\mathbf{b}']$ are the same elementary row operations the row reduce A to A'. Further, A' is row reduced echelon form. Hence, r is the number of nonzero rows in A' and $r \le r_\mathbf{b}$. Consider the following possibilities:

1. If $r < r_\mathbf{b}$, then there is a row of the form $[0 \cdots 0|1]$ in $[A'|\mathbf{b}']$ in which case $r_\mathbf{b} = r + 1$. Such a row translates into the equation

$$0x_1 + \cdots + 0x_n = 1,$$

which means the system is inconsistent. This is what occurs in Example 14. Recall that the augmented matrix row reduces as follows:

$$\begin{bmatrix} 1 & 2 & 4 & -2 \\ 1 & 1 & 3 & 1 \\ 2 & 1 & 5 & 2 \end{bmatrix} \rightarrow \cdots \rightarrow \begin{bmatrix} 1 & 0 & 2 & 0 \\ 0 & 1 & 1 & 0 \\ 0 & 0 & 0 & 1 \end{bmatrix},$$

where $\rightarrow \cdots \rightarrow$ denotes the sequence of elementary row operations used. Notice that Rank$(A) = 2$ while the presence of $[000|1]$ in the last row of $[A'|\mathbf{b}']$ gives Rank$[A|\mathbf{b}] = 3$. There are no solutions.

2. Suppose $r = r_\mathbf{b}$. The variables that correspond to the columns where the leading ones occur are called the *leading variables* or *dependent variables*. Since each nonzero row has a leading one, there are r leading variables. All of the other variables are called *free variables*, and we are able to solve each leading variable in terms of the free variables and so there are solutions. The system $A\mathbf{x} = \mathbf{b}$ is consistent. Consider two subcases:

a. Suppose $r < n$. Since there are a total of n (the number of columns of A) variables, there are $n - r$ free variables and hence infinitely many solutions. This is what occurs in Example 12. Recall

$$\begin{bmatrix} 2 & 3 & 4 & | & 9 \\ 1 & 2 & -1 & | & 2 \end{bmatrix} \rightarrow \cdots \rightarrow \begin{bmatrix} 1 & 0 & 11 & | & 12 \\ 0 & 1 & -6 & | & -5 \end{bmatrix}.$$

Here, we have $r = \text{Rank}A = \text{Rank}[A|\mathbf{b}]=2$ and $n = 3$. There is exactly $n - r = 3 - 2 = 1$ free variable.

b. Now suppose $r = n$. Then every variable is a leading variable. There are no free variables so there is a unique solution. This is what occurs in Example 13. Recall

$$\begin{bmatrix} 3 & 2 & 1 & | & 4 \\ 2 & 2 & 1 & | & 3 \\ 1 & 1 & 1 & | & 0 \end{bmatrix} \rightarrow \cdots \rightarrow \begin{bmatrix} 1 & 0 & 0 & | & 1 \\ 0 & 1 & 0 & | & 2 \\ 0 & 0 & 1 & | & -3 \end{bmatrix}.$$

Here we have $r = \text{Rank}A = \text{Rank}[A|\mathbf{b}] = 3$ and $n = 3$. There are no free variables. The solution is unique.

We summarize our discussion in the following proposition.

Proposition 16. *Let A be an $m \times n$ matrix and $\mathbf{b}$ an $n \times 1$ column vector. Let $r = \text{Rank}A$ and $r_\mathbf{b} = \text{Rank}[A|\mathbf{b}]$:*

1. *If $r < r_\mathbf{b}$, then $A\mathbf{x} = \mathbf{b}$ is inconsistent.*
2. *If $r = r_\mathbf{b}$, then $A\mathbf{x} = \mathbf{b}$ is consistent. Further,*

   a. *if $r < n$, there are $n - r > 0$ free variables and hence infinitely many solutions.*
   b. *if $r = n$, there is exactly one solution.*

Example 17. Suppose the following matrices are obtained by row reducing the augmented matrix of a system of linear equations. Identify the leading and free variables and write down the solution set. Assume the variables are $x_1, x_2, \ldots$.

$$(1) \begin{bmatrix} 1 & 1 & 4 & 0 & | & 2 \\ 0 & 0 & 0 & 1 & | & 3 \\ 0 & 0 & 0 & 0 & | & 0 \end{bmatrix} \quad (2) \begin{bmatrix} 1 & 1 & 0 & | & 1 \\ 0 & 0 & 0 & | & 0 \end{bmatrix}$$

$$(3) \begin{bmatrix} 1 & 0 & 0 & | & 3 \\ 0 & 1 & 0 & | & 4 \\ 0 & 0 & 1 & | & 5 \end{bmatrix} \quad (4) \begin{bmatrix} 1 & 0 & | & 1 \\ 0 & 1 & | & 2 \\ 0 & 0 & | & 1 \\ 0 & 0 & | & 0 \\ 0 & 0 & | & 0 \end{bmatrix}$$

▶ **Solution.** 1. The zero row provides no information and can be ignored. The variables are x_1, x_2, x_3, and x_4. The leading ones occur in the first and fourth column. Therefore, x_1 and x_4 are the leading variables. The free variables are x_2 and x_3. Let $\alpha = x_2$ and $\beta = x_3$. The first row implies the equation $x_1 + x_2 + 4x_3 = 2$. We solve for x_1 and obtain $x_1 = 2 - x_2 - 4x_3 = 2 - \alpha - 4\beta$. The second row implies the equation $x_4 = 3$. Thus,

$$\mathbf{x} = \begin{bmatrix} x_1 \\ x_2 \\ x_3 \\ x_4 \end{bmatrix} = \begin{bmatrix} 2 - \alpha - 4\beta \\ \alpha \\ \beta \\ 3 \end{bmatrix} = \begin{bmatrix} 2 \\ 0 \\ 0 \\ 3 \end{bmatrix} + \alpha \begin{bmatrix} -1 \\ 1 \\ 0 \\ 0 \end{bmatrix} + \beta \begin{bmatrix} -4 \\ 0 \\ 1 \\ 0 \end{bmatrix},$$

where α and β are arbitrary parameters in $\mathbb{R}$.

2. x_1 is the leading variable. $\alpha = x_2$ and $\beta = x_3$ are free variables. The first row implies $x_1 = 1 - \alpha$. The solution is

$$\mathbf{x} = \begin{bmatrix} 1 - \alpha \\ \alpha \\ \beta \end{bmatrix} = \begin{bmatrix} 1 \\ 0 \\ 0 \end{bmatrix} + \alpha \begin{bmatrix} -1 \\ 1 \\ 0 \end{bmatrix} + \beta \begin{bmatrix} 0 \\ 0 \\ 1 \end{bmatrix},$$

where α and β are in $\mathcal{R}$.

3. The leading variables are x_1, x_2, and x_3. There are no free variables. The solution set is

$$\mathbf{x} = \begin{bmatrix} 3 \\ 4 \\ 5 \end{bmatrix}.$$

4. The row $\begin{bmatrix} 0 & 0 & 1 \end{bmatrix}$ implies the solution set is empty.   ◀

The Gauss–Jordan Elimination Method

Now that you have seen several examples, we present the Gauss-Jordan elimination method for any matrix. It is an algorithm to transform any matrix to row reduced echelon form using a finite number of elementary row operations. When applied to an augmented matrix of a system of linear equations, the solution set can be readily discerned. It has other uses as well so our description will be for an arbitrary matrix.

Algorithm 18. The Gauss–Jordan Elimination Method Let A be a matrix. There is a finite sequence of elementary row operations that transform A to a matrix in row reduced echelon form. There are two stages of the process: (1) The first stage is called *Gaussian elimination* and transforms a given matrix to row echelon form, and (2) the second stage is called *Gauss–Jordan elimination* and transforms a matrix in row echelon form to row reduced echelon form.

From A to REF: Gaussian Elimination

1. Let $A_1 = A$. If $A_1 = \mathbf{0}$, then A is in row echelon form.
2. If $A_1 \neq \mathbf{0}$, then in the first nonzero column from the left (say the jth column), locate a nonzero entry in one of the rows (say the ith row with entry a):

    a. Multiply that row by the reciprocal of that nonzero entry: $m_i(1/a)$.
    b. Permute that row with the top row: p_{1i}. There is now a 1 in the $(1, j)$ entry.
    c. If b is a nonzero entry in the (i, j) position for $i \neq 1$, add $-b$ times the first row to the ith row: $t_{1j}(-b)$. Do this for each row below the first.

    The transformed matrix will have the following form:

$$
\begin{bmatrix}
0 & \cdots & 0 & 1 & * & \cdots & * \\
0 & \cdots & 0 & 0 & & & \\
\vdots & \ddots & \vdots & & & A_2 & \\
0 & \cdots & 0 & 0 & & &
\end{bmatrix}.
$$

   The *'s in the first row are unknown entries, and A_2 is a matrix with fewer rows and columns than A_1.

3. If $A_2 = \mathbf{0}$, we are done. The above matrix is in row echelon form.
4. If $A_2 \neq \mathbf{0}$, apply step (2) to A_2. Since there are zeros to the left of A_2 and the only elementary row operations we apply affect the rows of A_2 (and not all of A), there will continue to be zeros to the left of A_2. The result will be a matrix of the form

$$
\begin{bmatrix}
0 & \cdots & 0 & 1 & * & \cdots & * & * & * & \cdots & * \\
& & 0 & 0 & \cdots & 0 & 1 & * & \cdots & * \\
\vdots & \ddots & \vdots & 0 & 0 & \cdots & 0 & 0 & & \\
& & & \vdots & \vdots & & \vdots & \vdots & A_3 & \\
0 & \cdots & 0 & 0 & 0 & \cdots & 0 & 0 & &
\end{bmatrix}.
$$

5. If $A_3 = 0$, we are done. Otherwise, continue repeating step (2) until a matrix $A_k = \mathbf{0}$ is obtained.

From REF to RREF: Gauss–Jordan Elimination

1. The leading ones now become apparent in the previous process. We begin with the rightmost leading one. Suppose it is in the kth row and lth column. If there

is a nonzero entry (b say) above that leading one, we add $-b$ times the kth row to it: $t_{kj}(-b)$. We do this for each nonzero entry in the lth column. The result is zeros above the rightmost leading one. (The entries to the left of a leading one are zeros. This process preserves that property.)

2. Now repeat the process described above to each leading one moving right to left. The result will be a matrix in row reduced echelon form.

Example 19. Use the Gauss–Jordan method to row reduce the following matrix to row reduced echelon form:

$$\begin{bmatrix} 2 & 3 & 8 & 0 & 4 \\ 3 & 4 & 11 & 1 & 8 \\ 1 & 2 & 5 & 1 & 6 \\ -1 & 0 & -1 & 0 & 1 \end{bmatrix}.$$

▶ **Solution.**   The Gauss–Jordan algorithm produces

$$\begin{bmatrix} 2 & 3 & 8 & 0 & 4 \\ 3 & 4 & 11 & 1 & 8 \\ 1 & 2 & 5 & 1 & 6 \\ -1 & 0 & -1 & 0 & 1 \end{bmatrix} \xrightarrow{p_{13}} \begin{bmatrix} 1 & 2 & 5 & 1 & 6 \\ 3 & 4 & 11 & 1 & 8 \\ 2 & 3 & 8 & 0 & 4 \\ -1 & 0 & -1 & 0 & 1 \end{bmatrix} \begin{array}{l} t_{12}(-3) \\ t_{13}(-2) \\ \xrightarrow{t_{14}(1)} \end{array}$$

$$\begin{bmatrix} 1 & 2 & 5 & 1 & 6 \\ 0 & -2 & -4 & -2 & -10 \\ 0 & -1 & -2 & -2 & -8 \\ 0 & 2 & 4 & 1 & 7 \end{bmatrix} \xrightarrow{m_2(-1/2)} \begin{bmatrix} 1 & 2 & 5 & 1 & 6 \\ 0 & 1 & 2 & 1 & 5 \\ 0 & -1 & -2 & -2 & -8 \\ 0 & 2 & 4 & 1 & 7 \end{bmatrix} \begin{array}{l} t_{23}(1) \\ t_{24}(-2) \\ \xrightarrow{} \end{array}$$

$$\begin{bmatrix} 1 & 2 & 5 & 1 & 6 \\ 0 & 1 & 2 & 1 & 5 \\ 0 & 0 & 0 & -1 & -3 \\ 0 & 0 & 0 & -1 & -3 \end{bmatrix} \begin{array}{l} m_3(-1) \\ t_{34}(1) \\ \xrightarrow{} \end{array} \begin{bmatrix} 1 & 2 & 5 & 1 & 6 \\ 0 & 1 & 2 & 1 & 5 \\ 0 & 0 & 0 & 1 & 3 \\ 0 & 0 & 0 & 0 & 0 \end{bmatrix} \begin{array}{l} t_{32}(-1) \\ t_{31}(-1) \\ \xrightarrow{} \end{array}$$

$$\begin{bmatrix} 1 & 2 & 5 & 0 & 3 \\ 0 & 1 & 2 & 0 & 2 \\ 0 & 0 & 0 & 1 & 3 \\ 0 & 0 & 0 & 0 & 0 \end{bmatrix} \xrightarrow{t_{21}(-2)} \begin{bmatrix} 1 & 0 & 1 & 0 & -1 \\ 0 & 1 & 2 & 0 & 2 \\ 0 & 0 & 0 & 1 & 3 \\ 0 & 0 & 0 & 0 & 0 \end{bmatrix}.$$

In the first step, we observe that the first column is nonzero so it is possible to produce a 1 in the upper left-hand corner. This is most easily accomplished by $p_{1,3}$. The next set of operations produces 0's below this leading one. We repeat this procedure on the submatrix to the right of the zeros. We produce a one in the 2, 2 position by $m_2(-\frac{1}{2})$, and the next set of operations produce zeros below this second leading one. Now notice that the third column below the second leading one is zero. There are no elementary row operations that can produce a leading one in the (3, 3) position that involve just the third and fourth row. We move over to the fourth column and observe that the entries below the second leading one are not both zero. The elementary row operation $m_3(-1)$ produces a leading one in the (3, 4) position and the subsequent operation produces a zero below it. At this point, A has been transformed to row echelon form. Now starting at the rightmost leading one, the 1 in the (3, 4) position, we use operations of the form $t_{3i}(a)$ to produce zeros above that leading one. This is applied to each column that contains a leading one. The result is in row reduced echelon form.   ◄

The student is encouraged to go carefully through Examples 12–14. In each of those examples, the Gauss–Jordan Elimination method was used to transform the augmented matrix to the matrix in row reduced echelon form.

A Basis of the Null Space

When the Gauss-Jordan elimination method is used to compute the null space of A, the solution space takes the form

$$\{\alpha_1 \mathbf{v}_1 + \cdots + \alpha_k \mathbf{v}_k : \alpha_1, \ldots, \alpha_k \in \mathbb{R}\}. \tag{3}$$

We simplify the notation and write $\text{Span}\{\mathbf{v}_1, \ldots, \mathbf{v}_k\}$ for the set of all linear combinations of $\mathbf{v}_1, \ldots, \mathbf{v}_k$ given by (3). The vectors $\mathbf{v}_1, \ldots, \mathbf{v}_k$ turn out to also be linearly independent. The notion of linear independence for linear spaces of functions was introduced earlier in the text and that same notion extends to vectors in $\mathbb{R}^n$.

Before we give the definition of linear independence, consider the following example.

Example 20. Find the null space of

$$A = \begin{bmatrix} 2 & 3 & 1 & 4 \\ 1 & 1 & -1 & 2 \\ 3 & 5 & 3 & 6 \\ 4 & 5 & -1 & 8 \end{bmatrix}.$$

▶ **Solution.** We augment A with the zero vector and row reduce:

$$
\begin{bmatrix} 2 & 3 & 1 & 4 & -2 & | & 0 \\ 1 & 1 & -1 & 2 & 3 & | & 0 \\ 3 & 5 & 3 & 6 & -7 & | & 0 \\ 4 & 5 & -1 & 8 & 4 & | & 0 \end{bmatrix}
\overset{p_{12}}{\to}
\begin{bmatrix} 1 & 1 & -1 & 2 & 3 & | & 0 \\ 2 & 3 & 1 & 4 & -2 & | & 0 \\ 3 & 5 & 3 & 6 & -7 & | & 0 \\ 4 & 5 & -1 & 8 & 4 & | & 0 \end{bmatrix}
$$

$$
\overset{\substack{t_{12}(-2) \\ t_{13}(-3) \\ t_{14}(-4)}}{\longrightarrow}
\begin{bmatrix} 1 & 1 & -1 & 2 & 3 & | & 0 \\ 0 & 1 & 3 & 0 & -8 & | & 0 \\ 0 & 2 & 6 & 0 & -16 & | & 0 \\ 0 & 1 & 3 & 0 & -8 & | & 0 \end{bmatrix}
\overset{\substack{t_{23}(-2) \\ t_{24}(-1)}}{\longrightarrow}
\begin{bmatrix} 1 & 1 & -1 & 2 & 3 & | & 0 \\ 0 & 1 & 3 & 0 & -8 & | & 0 \\ 0 & 0 & 0 & 0 & 0 & | & 0 \\ 0 & 0 & 0 & 0 & 0 & | & 0 \end{bmatrix}
$$

$$
\overset{t_{21}(-1)}{\longrightarrow}
\begin{bmatrix} 1 & 0 & -4 & 2 & 11 & | & 0 \\ 0 & 1 & 3 & 0 & -8 & | & 0 \\ 0 & 0 & 0 & 0 & 0 & | & 0 \\ 0 & 0 & 0 & 0 & 0 & | & 0 \end{bmatrix}.
$$

If the variables are $x_1, \ldots, x_5$, then x_1 and x_2 are the leading variables and x_3, x_4, and x_5 are the free variables. Let $\alpha = x_3$, $\beta = x_4$, and $\gamma = x_5$. Then

$$
\begin{bmatrix} x_1 \\ x_2 \\ x_3 \\ x_4 \\ x_5 \end{bmatrix} = \begin{bmatrix} 4\alpha - 2\beta - 11\gamma \\ -3\alpha + 8\gamma \\ \alpha \\ \beta \\ \gamma \end{bmatrix} = \alpha \begin{bmatrix} 4 \\ -3 \\ 1 \\ 0 \\ 0 \end{bmatrix} + \beta \begin{bmatrix} -2 \\ 0 \\ 0 \\ 1 \\ 0 \end{bmatrix} + \gamma \begin{bmatrix} -11 \\ 8 \\ 0 \\ 0 \\ 1 \end{bmatrix}.
$$

It follows that

$$
\mathrm{NS}(A) = \mathrm{Span} \left\{ \begin{bmatrix} 4 \\ -3 \\ 1 \\ 0 \\ 0 \end{bmatrix}, \begin{bmatrix} -2 \\ 0 \\ 0 \\ 1 \\ 0 \end{bmatrix}, \begin{bmatrix} -11 \\ 8 \\ 0 \\ 0 \\ 1 \end{bmatrix} \right\}.
$$
◀

We say that vectors $\mathbf{v}_1, \ldots, \mathbf{v}_k$ in $\mathbb{R}^n$ are *linearly independent* if the equation

$$
a_1 \mathbf{v}_1 + \cdots + a_k \mathbf{v}_k = \mathbf{0}
$$

implies that the coefficients $a_1, \ldots, a_k$ are all zero. Otherwise, they are said to be *linearly dependent*.

Example 21. Show that the vectors

$$\mathbf{v}_1 = \begin{bmatrix} 4 \\ -3 \\ 1 \\ 0 \\ 0 \end{bmatrix}, \ \mathbf{v}_2 = \begin{bmatrix} -2 \\ 0 \\ 0 \\ 1 \\ 0 \end{bmatrix}, \ \text{and} \ \mathbf{v}_3 = \begin{bmatrix} -11 \\ 8 \\ 0 \\ 0 \\ 1 \end{bmatrix}$$

that span the null space in Example 20 are linearly independent.

▶ **Solution.** The equation $a_1\mathbf{v}_1 + a_2\mathbf{v}_2 + a_3\mathbf{v}_3 = \mathbf{0}$ means

$$\begin{bmatrix} 4a_1 - 2a_3 - 11a_3 \\ -3a_1 + 8a_3 \\ a_1 \\ a_2 \\ a_3 \end{bmatrix} = \begin{bmatrix} 0 \\ 0 \\ 0 \\ 0 \\ 0 \end{bmatrix}.$$

From the last three rows, it is immediate that $a_1 = a_2 = a_3 = 0$. This implies the linear independence. ◀

Suppose V is a linear subspace of $\mathbb{R}^n$. By this we mean that V is in $\mathbb{R}^n$ and is closed under addition and scalar multiplication. That is, if

1. $\mathbf{v}_1, \mathbf{v}_2$ are in V, then so is $\mathbf{v}_1 + \mathbf{v}_2$
2. $c \in \mathbb{R}$ and $\mathbf{v} \in V$, then $c\mathbf{v} \in V$.

We say that the set of vectors $\mathcal{B} = \{\mathbf{v}_1, \ldots, \mathbf{v}_k\}$ forms a **basis** of V if $\mathcal{B}$ is linearly independent and spans V. Thus, in Example 20, the set $\{\mathbf{v}_1, \mathbf{v}_2, \mathbf{v}_3\}$ is a basis of $\mathrm{NS}(A)$. We observe that the number of vectors is three: one for each free variable.

Theorem 22. *Suppose A is an $m \times n$ matrix of rank r then there are $n - r$ vectors that form a basis of the null space of A.*

Proof. Suppose $[A|\mathbf{0}]$ is row reduced to $[R|\mathbf{0}]$, in row reduced echelon form. By Theorem 16, there are $n - r$ free variables. Let $k = n - r$. Suppose $f_1, \ldots, f_k$ are the columns of R that do not contain leading ones. Then $x_{f_1}, \ldots, x_{f_k}$ are the free variables. Let $\alpha_j = x_{f_j}, \ j = 1, \ldots, k$ be parameters. Solving for $\mathbf{x}$ in $R\mathbf{x} = \mathbf{0}$ in terms of the free variables, we get vectors $v_1, \ldots, v_k$ such that

$$\mathbf{x} = \alpha_1\mathbf{v}_1 + \cdots + \alpha_k\mathbf{v}_k.$$

It follows that $\mathbf{v}_1, \ldots, \mathbf{v}_k$ span $\mathrm{NS}(A)$. Since R is in row reduced echelon form, the f_j^{th} entry of $\mathbf{v}_j$ is one while the f_j^{th} entry of $\mathbf{v}_i$ is zero, when $i \neq j$. If $a_1\mathbf{v}_1 + \cdots + a_k\mathbf{v}_k = \mathbf{0}$, then the f_j^{th} entry of the left-hand side is a_j, and hence, $a_j = 0$, $j = 1, \ldots k$. It follows that $v_1, \ldots, v_k$ are linearly independent and hence form a basis of $\mathrm{NS}(A)$. □

Exercises

1–2. For each system of linear equations, identify the coefficient matrix A, the variable matrix $\mathbf{x}$, the output matrix $\mathbf{b}$, and the augmented matrix $[A|\mathbf{b}]$.

1.
$$\begin{aligned} x + 4y + 3z &= 2 \\ x + y - z &= 4 \\ 2x + z &= 1 \\ y - z &= 6 \end{aligned}$$

2.
$$\begin{aligned} 2x_1 - 3x_2 + 4x_3 + x_4 &= 0 \\ 3x_1 + 8x_2 - 3x_3 - 6x_4 &= 1 \end{aligned}$$

3. Suppose $A = \begin{bmatrix} 1 & 0 & -1 & 4 & 3 \\ 5 & 3 & -3 & -1 & -3 \\ 3 & -2 & 8 & 4 & -3 \\ -8 & 2 & 0 & 2 & 1 \end{bmatrix}$, $\mathbf{x} = \begin{bmatrix} x_1 \\ x_2 \\ x_3 \\ x_4 \\ x_5 \end{bmatrix}$, and $\mathbf{b} = \begin{bmatrix} 2 \\ 1 \\ 3 \\ -4 \end{bmatrix}$. Write out the system of linear equations that corresponds to $A\mathbf{x} = \mathbf{b}$.

4–9. In the following, matrices identify those that are in row reduced echelon form. If a matrix is not in row reduced echelon form, find a single elementary row operation that will transform it to row reduced echelon form and write the new matrix.

4. $A = \begin{bmatrix} 1 & 0 & 1 \\ 0 & 0 & 0 \\ 0 & 1 & -4 \end{bmatrix}$

5. $A = \begin{bmatrix} 1 & 0 & 4 \\ 0 & 1 & 2 \end{bmatrix}$

6. $A = \begin{bmatrix} 1 & 2 & 1 & 0 & 1 \\ 0 & 1 & 3 & 1 & 1 \end{bmatrix}$

7. $A = \begin{bmatrix} 0 & 1 & 0 & 3 \\ 0 & 0 & 2 & 6 \\ 0 & 0 & 0 & 0 \end{bmatrix}$

8. $A = \begin{bmatrix} 0 & 1 & 1 & 0 & 3 \\ 0 & 0 & 0 & 1 & 2 \\ 0 & 0 & 0 & 0 & 0 \end{bmatrix}$

9. $A = \begin{bmatrix} 1 & 0 & 1 & 0 & 3 \\ 0 & 1 & 3 & 4 & 1 \\ 3 & 0 & 3 & 0 & 9 \end{bmatrix}$

10–18. Use the Gauss-Jordan elimination method to row reduce each matrix.

10. $\begin{bmatrix} 1 & 2 & 3 & 1 \\ -1 & 0 & 3 & -5 \\ 0 & 1 & 1 & 0 \end{bmatrix}$

11. $\begin{bmatrix} 2 & 1 & 3 & 1 & 0 \\ 1 & -1 & 1 & 2 & 0 \\ 0 & 2 & 1 & 1 & 2 \end{bmatrix}$

12. $\begin{bmatrix} 0 & -2 & 3 & 2 & 1 \\ 0 & 2 & -1 & 4 & 0 \\ 0 & 6 & -7 & 0 & -2 \\ 0 & 4 & -6 & -4 & -2 \end{bmatrix}$

13. $\begin{bmatrix} 1 & 2 & 1 & 1 & 5 \\ 2 & 4 & 0 & 0 & 6 \\ 1 & 2 & 0 & 1 & 3 \\ 0 & 0 & 1 & 1 & 2 \end{bmatrix}$

14. $\begin{bmatrix} -1 & 0 & 1 & 1 & 0 & 0 \\ -3 & 1 & 3 & 0 & 1 & 0 \\ 7 & -1 & -4 & 0 & 0 & 1 \end{bmatrix}$

15. $\begin{bmatrix} 1 & 2 & 4 \\ 2 & 4 & 8 \\ -1 & 2 & 0 \\ 1 & 6 & 8 \\ 0 & 4 & 4 \end{bmatrix}$

16.
$$\begin{bmatrix} 5 & 1 & 8 & 1 \\ 1 & 1 & 4 & 0 \\ 2 & 0 & 2 & 1 \\ 4 & 1 & 7 & 1 \end{bmatrix}$$

17.
$$\begin{bmatrix} 2 & 8 & 0 & 0 & 6 \\ 1 & 4 & 1 & 1 & 7 \\ -1 & -4 & 0 & 1 & 0 \end{bmatrix}$$

18.
$$\begin{bmatrix} 1 & -1 & 1 & -1 & 1 \\ 1 & 1 & -1 & -1 & 1 \\ -1 & -1 & 1 & 1 & -1 \\ 1 & 1 & -1 & 1 & -1 \end{bmatrix}$$

19–25. Solve the following systems of linear equations:

19.
$$\begin{aligned} x + 3y &= 2 \\ 5x \quad\quad + 3z &= -5 \\ 3x - y + 2z &= -4 \end{aligned}$$

20.
$$\begin{aligned} 3x_1 + 2x_2 + 9x_3 + 8x_4 &= 10 \\ x_1 + \quad\quad x_3 + 2x_4 &= 4 \\ -2x_1 + x_2 + x_3 - 3x_4 &= -9 \\ x_1 + x_2 + 4x_3 + 3x_4 &= 3 \end{aligned}$$

21.
$$\begin{aligned} -x + 4y &= -3x \\ x - y &= -3y \end{aligned}$$

22.
$$\begin{aligned} -2x_1 - 8x_2 - x_3 - x_4 &= -9 \\ -x_1 - 4x_2 \quad\quad - x_4 &= -8 \\ x_1 + 4x_2 + x_3 + x_4 &= 6 \end{aligned}$$

23.
$$\begin{aligned} 2x + 3y + 8z &= 5 \\ 2x + y + 10z &= 3 \\ 2x \quad\quad + 8z &= 4 \end{aligned}$$

24.
$$x_1 + x_2 + x_3 + 5x_4 = 3$$
$$x_2 + x_3 + 4x_4 = 1$$
$$x_1 + x_3 + 2x_4 = 2$$
$$2x_1 + 2x_2 + 3x_3 + 11x_4 = 8$$
$$2x_1 + x_2 + 2x_3 + 7x_4 = 7$$

25.
$$x_1 + x_2 = 3 + x_1$$
$$x_2 + 2x_3 = 4 + x_2 + x_3$$
$$x_1 + 3x_2 + 4x_3 = 11 + x_1 + 2x_2 + 2x_3$$

26–32. For each of the following matrices A, find a basis of the null space.

26. $A = \begin{bmatrix} 1 & 2 \\ 2 & 4 \end{bmatrix}$

27. $A = \begin{bmatrix} 1 & 3 \\ 3 & 5 \end{bmatrix}$

28. $A = \begin{bmatrix} 2 & 6 \\ 1 & 3 \\ -1 & -3 \end{bmatrix}$

29. $A = \begin{bmatrix} 1 & 3 & 1 \\ 1 & 4 & 1 \end{bmatrix}$

30. $A = \begin{bmatrix} 1 & 1 & 3 & -2 \\ 1 & 4 & 1 & 1 \\ 4 & 7 & 10 & -5 \end{bmatrix}$

31. $A = \begin{bmatrix} -1 & 2 & 1 & 1 \\ 6 & -2 & 3 & 1 \\ 2 & -1 & 0 & 4 \\ 5 & 1 & 7 & -9 \end{bmatrix}$

32. $A = \begin{bmatrix} 2 & 3 & 1 & 9 & 3 \\ 0 & 1 & -3 & 4 & 0 \\ 2 & 1 & 7 & 1 & 3 \\ 4 & 4 & 8 & 10 & 6 \end{bmatrix}$

33. Suppose the homogeneous system $A\mathbf{x} = \mathbf{0}$ has the following two solutions: $\begin{bmatrix} 1 \\ 1 \\ 2 \end{bmatrix}$ and $\begin{bmatrix} 1 \\ -1 \\ 0 \end{bmatrix}$. Is $\begin{bmatrix} 5 \\ -1 \\ 4 \end{bmatrix}$ a solution? Why or why not?

34. For what value of k will the following system have a solution:

$$\begin{aligned} x_1 + x_2 - x_3 &= 2 \\ 2x_1 + 3x_2 + x_3 &= 4 \\ x_1 - 2x_2 - 10x_3 &= k \end{aligned}$$

35–36. Let $A = \begin{bmatrix} 1 & 3 & 4 \\ -2 & 1 & 7 \\ 1 & 1 & 0 \end{bmatrix}$, $\mathbf{b}_1 = \begin{bmatrix} 1 \\ 0 \\ 0 \end{bmatrix}$, $\mathbf{b}_2 = \begin{bmatrix} 1 \\ 1 \\ 0 \end{bmatrix}$ and $\mathbf{b}_3 = \begin{bmatrix} 1 \\ 1 \\ 1 \end{bmatrix}$.

35. Solve $A\mathbf{x} = \mathbf{b}_i$, for each $i = 1, 2, 3$.
36. Solve the above systems simultaneously by row reducing

$$[A|b_1|b_2|b_3] = \begin{bmatrix} 1 & 3 & 4 & 1 & 1 & 1 \\ -2 & 1 & 7 & 0 & 1 & 1 \\ 1 & 1 & 0 & 0 & 0 & 1 \end{bmatrix}$$

8.3   Invertible Matrices

Let A be a square matrix. A matrix B is said to be an ***inverse*** of A if $BA = AB = I$. In this case, we say A is ***invertible*** or ***nonsingular***. If A is not invertible, we say A is ***singular***.

Example 1. Suppose

$$A = \begin{bmatrix} 3 & 1 \\ -4 & -1 \end{bmatrix}.$$

Show that A is invertible and an inverse is

$$B = \begin{bmatrix} -1 & -1 \\ 4 & 3 \end{bmatrix}.$$

▶ **Solution.** Observe that

$$AB = \begin{bmatrix} 3 & 1 \\ -4 & -1 \end{bmatrix}\begin{bmatrix} -1 & -1 \\ 4 & 3 \end{bmatrix} = \begin{bmatrix} 1 & 0 \\ 0 & 1 \end{bmatrix}$$

and

$$BA = \begin{bmatrix} -1 & -1 \\ 4 & 3 \end{bmatrix}\begin{bmatrix} 3 & 1 \\ -4 & -1 \end{bmatrix} = \begin{bmatrix} 1 & 0 \\ 0 & 1 \end{bmatrix}. \qquad ◀$$

The following proposition says that when A has an inverse, there can only be one.

Proposition 2. *Let A be an invertible matrix. Then the inverse is unique.*

Proof. Suppose B and C are inverses of A. Then

$$B = BI = B(AC) = (BA)C = IC = C. \qquad \square$$

Because of uniqueness, we can properly say *the inverse* of A when A is invertible. In Example 1, the matrix $B = \begin{pmatrix} -1 & -1 \\ 4 & 3 \end{pmatrix}$ is the inverse of A; there are no others. It is standard convention to denote the inverse of A by A^{-1}.

We say that B is a ***left inverse*** of A if $BA = I$ and a ***right inverse*** if $AB = I$. For square matrices, we have the following proposition which we will not prove. This proposition tells us that it is enough to check that either $AB = I$ or $BA = I$. It is not necessary to check both products.

Proposition 3. *Suppose A is a square matrix and B is a left or right inverse of A. Then A is invertible and $A^{-1} = B$.*

For many matrices, it is possible to determine their inverse by inspection. For example, the identity matrix I_n is invertible and its inverse is I_n: $I_n I_n = I_n$. A diagonal matrix $\text{diag}(a_1, \ldots, a_n)$ is invertible if each $a_i \neq 0$, $i = 1, \ldots, n$. The inverse then is simply $\text{diag}\left(\frac{1}{a_1}, \ldots, \frac{1}{a_n}\right)$. However, if one of the a_i is zero, then the matrix in not invertible. Even more is true. If A has a zero row, say the ith row, then A is not invertible. To see this, we get from (2) of Sect. 8.1 that $\text{Row}_i(AB) = \text{Row}_i(A)B = 0$. Hence, there is no matrix B for which $AB = I$. Similarly, a matrix with a zero column cannot be invertible.

Proposition 4. *Let A and B be invertible matrices. Then*

1. A^{-1} is invertible and $(A^{-1})^{-1} = A$.
2. AB is invertible and $(AB)^{-1} = B^{-1}A^{-1}$.

Proof. Suppose A and B are invertible. The symmetry of the equation $A^{-1}A = AA^{-1} = I$ says that A^{-1} is invertible and $(A^{-1})^{-1} = A$. We also have

$$(B^{-1}A^{-1})(AB) = B^{-1}(A^{-1}A)B = B^{-1}IB = B^{-1}B = I.$$

This shows $(AB)^{-1} = B^{-1}A^{-1}$. $\qquad\qquad\qquad\qquad\qquad\qquad\qquad\qquad$ □

The following corollary easily follows by induction:

Corollary 5. *If $A = A_1 \cdots A_k$ is the product of invertible matrices, then A is invertible and $A^{-1} = A_k^{-1} \cdots A_1^{-1}$.*

The Elementary Matrices

When an elementary row operation is applied to the identity matrix I, the resulting matrix is called an ***elementary matrix***.

Example 6. Show that each of the following matrices are elementary matrices:

$$E_1 = \begin{bmatrix} 1 & 0 \\ 0 & 3 \end{bmatrix}, \quad E_2 = \begin{bmatrix} 1 & 0 & 0 \\ 0 & 1 & 0 \\ 0 & 2 & 1 \end{bmatrix}, \quad E_3 = \begin{bmatrix} 0 & 1 & 0 & 0 \\ 1 & 0 & 0 & 0 \\ 0 & 0 & 1 & 0 \\ 0 & 0 & 0 & 1 \end{bmatrix}.$$

▶ **Solution.** We have

$$m_2(3)I = \begin{bmatrix} 1 & 0 \\ 0 & 3 \end{bmatrix} = E_1,$$

$$t_{23}(2)I = \begin{bmatrix} 1 & 0 & 0 \\ 0 & 1 & 0 \\ 0 & 2 & 1 \end{bmatrix} = E_2,$$

$$p_{12}I = \begin{bmatrix} 0 & 1 & 0 & 0 \\ 1 & 0 & 0 & 0 \\ 0 & 0 & 1 & 0 \\ 0 & 0 & 0 & 1 \end{bmatrix} = E_3,$$

where the size of the identity matrix matches the size of the given matrix.   ◀

The following example shows a useful relationship between an elementary row operation and left multiplication by the corresponding elementary matrix.

Example 7. Use the elementary matrices in Example 6 to show that multiplying a matrix A on the left by E_i produces the same effect as applying the corresponding elementary row operation to A.

▶ **Solution.** 1. Let $A = \begin{bmatrix} A_1 \\ A_2 \end{bmatrix}$. Then $E_1 A = \begin{bmatrix} A_1 \\ 3A_2 \end{bmatrix} = m_2(3)A$.

2. Let $A = \begin{bmatrix} A_1 \\ A_2 \\ A_3 \end{bmatrix}$. Then $E_2 A = \begin{bmatrix} A_1 \\ A_2 \\ 2A_2 + A_3 \end{bmatrix} = t_{23}(2)A$.

3. Let $A = \begin{bmatrix} A_1 \\ A_2 \\ A_3 \\ A_4 \end{bmatrix}$. Then $E_3 A = \begin{bmatrix} A_2 \\ A_1 \\ A_3 \\ A_4 \end{bmatrix} = p_{12}A$.   ◀

This important relationship is summarized in general as follows:

Proposition 8. *Let e be an elementary row operation. That is, let e be one of p_{ij}, $t_{ij}(a)$, or $m_i(a)$. Let $E = eI$ be the elementary matrix obtained by applying e to the identity matrix. Then*

$$e(A) = EA.$$

In other words, when A is multiplied on the left by the elementary matrix E, it is the same as applying the elementary row operation e to A.

If $E = eI$ is an elementary matrix, then we can apply a second elementary row operation to E to get I back. For example, consider the elementary matrices in Example 6. It is easy to see that

$$m_2(1/3)E_1 = I, \quad t_{23}(-2)E_2 = I, \quad \text{and} \quad p_{12}E_3 = I.$$

In general, each elementary row operation is reversible by the another elementary row operation. Switching two rows is reversed by switching those two rows again, thus $p_{ij}^{-1} = p_{ij}$. Multiplying a row by a nonzero constant is reversed by multiplying that same row by the reciprocal of that nonzero constant, thus $(m_i(a))^{-1} = m_i(1/a)$. Finally, adding a multiple of a row to another is reversed by adding the negative multiple of first row to the second, thus $(t_{ij}(a))^{-1} = t_{ij}(-a)$. If e is an elementary row operation, we let e^{-1} denote the inverse row operation. Thus,

$$ee^{-1}A = A,$$

for any matrix A. These statements imply

Proposition 9. *An elementary matrix is invertible and the inverse is an elementary matrix.*

Proof. Let e be an elementary row operation and e^{-1} its inverse elementary row operation. Let $E = e(I)$ and $B = e^{-1}I$. By Proposition 8, we have

$$EB = E(e^{-1}I) = ee^{-1}I = I.$$

It follows the E is invertible and $E^{-1} = B = e^{-1}(I)$ is an elementary matrix. $\quad\square$

We now have a useful result that will be used later.

Theorem 10. *Let A by an $n \times n$ matrix. Then the following are equivalent:*

1. *A is invertible.*
2. *The null space of A, NS(A), consists only of the zero vector.*
3. *A row reduces (by Gauss–Jordan) to the identity matrix.*

Proof. Suppose A is invertible and $\mathbf{c} \in \text{NS}(A)$. Then $A\mathbf{c} = \mathbf{0}$. Multiply both sides by A^{-1} to get

$$\mathbf{c} = A^{-1}A\mathbf{c} = A^{-1}\mathbf{0} = \mathbf{0}.$$

Thus, the null space of A consists only of the zero vector.

Suppose now that the null space of A is trivial. Then the system $A\mathbf{x} = \mathbf{0}$ is equivalent to

$$x_1 \qquad\quad = 0$$
$$x_2 \qquad = 0$$
$$\ddots$$
$$x_n = 0$$

Thus, the augmented matrix $[A \mid \mathbf{0}]$ reduces to $[I \mid \mathbf{0}]$. The same elementary row operations row reduces A to the identity.

Now suppose A row reduces to the identity. Then there is a sequence of elementary matrices, $E_1, \ldots, E_k$, (corresponding to the elementary row operations) such that $E_1 \cdots E_k A = I$. Let $B = E_1 \cdots E_k$. Then $BA = I$ and this implies A is invertible.   □

Corollary 11. *Suppose A is an $n \times n$ matrix. If the null space of A is not the zero vector, then A row reduces to a matrix that has a zero row.*

Proof. Suppose A transforms by elementary row operations to R which is in row reduced echelon form. The system $A\mathbf{x} = \mathbf{0}$ has infinitely many solutions, and this only occurs if there are one or more free variables in the system $R\mathbf{x} = \mathbf{0}$. The number of leading variables which is the same as the number of nonzero rows is thus less than n. Hence, there are some zero rows.   □

Inversion Computations

Let $\mathbf{e}_i$ be the column vector with 1 in the ith position and 0's elsewhere. By Equation 8.1.(1), the equation $AB = I$ implies that $A\,\mathrm{Col}_i(B) = \mathrm{Col}_i(I) = \mathbf{e}_i$. This means that the solution to $A\mathbf{x} = \mathbf{e}_i$ is the ith column of the inverse of A, when A is invertible. We can thus compute the inverse of A one column at a time using the Gauss–Jordan elimination method on the augmented matrix $[A|\mathbf{e}_i]$. Better yet, though, is to perform the Gauss–Jordan elimination method on the matrix $[A|I]$, that is, the matrix A augmented with I. If A is invertible, it will reduce to a matrix of the form $[I|B]$ and B will be A^{-1}. If A is not invertible, it will not be possible to produce the identity in the first slot.

We illustrate this in the following two examples.

Example 12. Determine whether the matrix

$$A = \begin{bmatrix} 2 & 0 & 3 \\ 0 & 1 & 1 \\ 3 & -1 & 4 \end{bmatrix}$$

is invertible. If it is, compute the inverse.

▶ **Solution.** We will augment A with I and follow the procedure outlined above:

$$\begin{bmatrix} 2 & 0 & 3 & 1 & 0 & 0 \\ 0 & 1 & 1 & 0 & 1 & 0 \\ 3 & -1 & 4 & 0 & 0 & 1 \end{bmatrix} \begin{matrix} t_{13}(-1) \\ \xrightarrow{p_{13}} \end{matrix} \begin{bmatrix} 1 & -1 & 1 & -1 & 0 & 1 \\ 0 & 1 & 1 & 0 & 1 & 0 \\ 2 & 0 & 3 & 1 & 0 & 0 \end{bmatrix} \begin{matrix} t_{13}(-2) \\ \xrightarrow{t_{23}(-2)} \end{matrix}$$

$$\begin{bmatrix} 1 & -1 & 1 & -1 & 0 & 1 \\ 0 & 1 & 1 & 0 & 1 & 0 \\ 0 & 0 & -1 & 3 & -2 & -2 \end{bmatrix} \begin{matrix} m_3(-1) \\ t_{32}(-1) \\ \xrightarrow{t_{31}(-1)} \end{matrix} \begin{bmatrix} 1 & -1 & 0 & 2 & -2 & -1 \\ 0 & 1 & 0 & 3 & -1 & -2 \\ 0 & 0 & 1 & -3 & 2 & 2 \end{bmatrix} \begin{matrix} t_{21}(1) \\ \xrightarrow{} \end{matrix}$$

$$\begin{bmatrix} 1 & 0 & 0 & 5 & -3 & -3 \\ 0 & 1 & 0 & 3 & -1 & -2 \\ 0 & 0 & 1 & -3 & 2 & 2 \end{bmatrix}.$$

It follows that A is invertible and $A^{-1} = \begin{bmatrix} 5 & -3 & -3 \\ 3 & -1 & -2 \\ -3 & 2 & 2 \end{bmatrix}.$ ◀

Example 13. Let $A = \begin{bmatrix} 1 & -4 & 0 \\ 2 & 1 & 3 \\ 0 & -7 & 3 \end{bmatrix}$. Determine whether A is invertible. If it is, find its inverse.

▶ **Solution.** Again, we augment A with I and row reduce:

$$\begin{bmatrix} 1 & -4 & 0 & 1 & 0 & 0 \\ 2 & 1 & 3 & 0 & 1 & 0 \\ 0 & 9 & 3 & 0 & 0 & 1 \end{bmatrix} \begin{matrix} t_{12}(-2) \\ \xrightarrow{t_{23}(-1)} \end{matrix} \begin{bmatrix} 1 & -4 & 0 & 1 & 0 & 0 \\ 0 & 9 & 3 & -2 & 1 & 0 \\ 0 & 0 & 0 & 2 & -1 & 1 \end{bmatrix}.$$

We can stop at this point. Notice that the row operations produced a **0** row in the reduction of A. This implies A cannot be invertible. ◀

Solving a System of Equations

Suppose A is a square matrix with a known inverse. Then the equation $A\mathbf{x} = \mathbf{b}$ implies $\mathbf{x} = A^{-1}A\mathbf{x} = A^{-1}\mathbf{b}$ and thus gives the solution.

Example 14. Solve the following system:

$$
\begin{aligned}
2x + + 3z &= 1 \\
y + z &= 2 \\
3x - y + 4z &= 3.
\end{aligned}
$$

▶ **Solution.** The coefficient matrix is

$$
A = \begin{bmatrix} 2 & 0 & 3 \\ 0 & 1 & 1 \\ 3 & -1 & 4 \end{bmatrix}
$$

whose inverse we computed in the example above:

$$
A^{-1} = \begin{bmatrix} 5 & -3 & -3 \\ 3 & -1 & -2 \\ -3 & 2 & 2 \end{bmatrix}.
$$

The solution to the system is thus

$$
\mathbf{x} = A^{-1}\mathbf{b} = \begin{bmatrix} 5 & -3 & -3 \\ 3 & -1 & -2 \\ -3 & 2 & 2 \end{bmatrix} \begin{bmatrix} 1 \\ 2 \\ 3 \end{bmatrix} = \begin{bmatrix} -10 \\ -5 \\ 7 \end{bmatrix}.
$$

◀

Exercises

1–12. Determine whether the following matrices are invertible. If so, find the inverse:

1. $\begin{bmatrix} 1 & 1 \\ 3 & 4 \end{bmatrix}$

2. $\begin{bmatrix} 3 & 2 \\ 4 & 3 \end{bmatrix}$

3. $\begin{bmatrix} 1 & -2 \\ 2 & -4 \end{bmatrix}$

4. $\begin{bmatrix} 1 & -2 \\ 3 & -4 \end{bmatrix}$

5. $\begin{bmatrix} 1 & 2 & 4 \\ 0 & 1 & -3 \\ 2 & 5 & 5 \end{bmatrix}$

6. $\begin{bmatrix} 1 & 1 & 1 \\ 0 & 1 & 2 \\ 0 & 0 & 1 \end{bmatrix}$

7. $\begin{bmatrix} 1 & 2 & 3 \\ 4 & 5 & 1 \\ -1 & -1 & 1 \end{bmatrix}$

8. $\begin{bmatrix} 1 & 0 & -2 \\ 2 & -2 & 0 \\ 1 & 2 & -1 \end{bmatrix}$

9. $\begin{bmatrix} 1 & 3 & 0 & 1 \\ 2 & 2 & -2 & 0 \\ 1 & -1 & 0 & 4 \\ 1 & 2 & 3 & 9 \end{bmatrix}$

10. $\begin{bmatrix} -1 & 1 & 1 & -1 \\ 1 & -1 & 1 & -1 \\ 1 & 1 & -1 & -1 \\ -1 & -1 & -1 & 1 \end{bmatrix}$

11. $\begin{bmatrix} 0 & 1 & 0 & 0 \\ 1 & 0 & 1 & 0 \\ 0 & 1 & 1 & 1 \\ 1 & 1 & 1 & 1 \end{bmatrix}$

12. $\begin{bmatrix} -3 & 2 & -8 & 2 \\ 0 & 2 & -3 & 5 \\ 1 & 2 & 3 & 5 \\ 1 & -1 & 1 & -1 \end{bmatrix}$

13–18. Solve each system $A\mathbf{x} = \mathbf{b}$, where A and $\mathbf{b}$ are given below, by first computing A^{-1} and applying it to $A\mathbf{x} = \mathbf{b}$ to get $\mathbf{x} = A^{-1}\mathbf{b}$.

13. $A = \begin{bmatrix} 1 & 1 \\ 3 & 4 \end{bmatrix}$ $\mathbf{b} = \begin{bmatrix} 2 \\ 3 \end{bmatrix}$

14. $A = \begin{bmatrix} 1 & 1 & 1 \\ 0 & 1 & 2 \\ 0 & 0 & 1 \end{bmatrix}$ $\mathbf{b} = \begin{bmatrix} 1 \\ 0 \\ -3 \end{bmatrix}$

15. $A = \begin{bmatrix} 1 & 0 & -2 \\ 2 & -2 & 0 \\ 1 & 2 & -1 \end{bmatrix}$ $\mathbf{b} = \begin{bmatrix} -2 \\ 1 \\ 2 \end{bmatrix}$

16. $A = \begin{bmatrix} 1 & -1 & 1 \\ -2 & 5 & -2 \\ 0 & 2 & -1 \end{bmatrix}$ $\mathbf{b} = \begin{bmatrix} 1 \\ 1 \\ 1 \end{bmatrix}$

17. $A = \begin{bmatrix} 1 & 3 & 0 & 1 \\ 2 & 2 & -2 & 0 \\ 1 & -1 & 0 & 4 \\ 1 & 2 & 3 & 9 \end{bmatrix}$ $\mathbf{b} = \begin{bmatrix} 1 \\ 0 \\ -1 \\ 2 \end{bmatrix}$

18. $A = \begin{bmatrix} 0 & 1 & 0 & 0 \\ 1 & 0 & 1 & 0 \\ 0 & 1 & 1 & 1 \\ 1 & 1 & 1 & 1 \end{bmatrix}$ $\mathbf{b} = \begin{bmatrix} 1 \\ -1 \\ -2 \\ 1 \end{bmatrix}$

19. Suppose A is an invertible matrix. Show that A^t is invertible and give a formula for the inverse.

20. Let $E(\theta) = \begin{bmatrix} \cos\theta & \sin\theta \\ -\sin\theta & \cos\theta \end{bmatrix}$. Show $E(\theta)$ is invertible and find its inverse.

21. Let $F(\theta) = \begin{bmatrix} \sinh\theta & \cosh\theta \\ \cosh\theta & \sinh\theta \end{bmatrix}$. Show $F(\theta)$ is invertible and find its inverse.

22. Suppose A is invertible and $AB = AC$. Show that $B = C$. Give an example of a nonzero matrix A (not invertible) with $AB = AC$, for some B and C, but $B \neq C$.

8.4 Determinants

In this section, we will discuss the definition of the determinant and some of its properties. For our purposes, the determinant is a very useful number that we can associate to a square matrix. The determinant has an wide range of applications. It can be used to determine whether a matrix is invertible. Cramer's rule gives the unique solution to a system of linear equations as the quotient of determinants. In multidimensional calculus, the Jacobian is given by a determinant and expresses how area or volume changes under a transformation. Most students by now are familiar with the definition of the determinant for a 2×2 matrix: Let $A = \begin{bmatrix} a & b \\ c & d \end{bmatrix}$.

The *determinant* of A is given by

$$\det(A) = ad - bc.$$

It is the product of the diagonal entries minus the product of the off diagonal entries. For example, $\det \begin{bmatrix} 1 & 3 \\ 5 & -2 \end{bmatrix} = 1 \cdot (-2) - 5 \cdot 3 = -17$.

The definition of the determinant for an $n \times n$ matrix is decidedly more complicated. We will present an inductive definition. Let A be an $n \times n$ matrix and let $A(i, j)$ be the matrix obtained from A by deleting the ith row and jth column. Since $A(i, j)$ is an $(n - 1) \times (n - 1)$ matrix, we can inductively define the (i, j) *minor*, $\text{Minor}_{i\,j}(A)$, to be the determinant of $A(i, j)$:

$$\text{Minor}_{i\,j}(A) = \det(A(i, j)).$$

The following theorem, whose proof we omit, is the basis for the definition of the determinant.

Theorem 1 (Laplace Expansion Formulas). *Suppose A is an $n \times n$ matrix. Then the following numbers are all equal and we call this number the **determinant** of A:*

$$\det A = \sum_{j=1}^{n} (-1)^{i+j} a_{i,j} \text{Minor}_{i\,j}(A) \quad \textit{for each } i$$

and

$$\det A = \sum_{i=1}^{n} (-1)^{i+j} a_{i,j} \text{Minor}_{i\,j}(A) \quad \textit{for each } j.$$

Any of these formulas can thus be taken as the definition of the determinant. In the first formula, the index i is fixed and the sum is taken over all j. The entries $a_{i,j}$ thus fill out the ith row. We therefore call this formula the *Laplace expansion of the determinant along the ith row* or simply a *row expansion*. Since the index i can range from 1 to n, there are n row expansions. In a similar way, the second formula is called the **Laplace expansion of the determinant along the jth column** or simply a **column expansion** and there are n column expansions. The presence of the factor $(-1)^{i+j}$ alternates the signs along the row or column according as $i+j$ is even or odd. The **sign matrix**

$$\begin{bmatrix} + & - & + & \cdots \\ - & + & - & \cdots \\ + & - & + & \cdots \\ \vdots & \vdots & \vdots & \ddots \end{bmatrix}$$

is a useful tool to organize the signs in an expansion.

It is common to use the absolute value sign $|A|$ to denote the determinant of A. This should not cause confusion unless A is a 1×1 matrix, in which case we will not use this notation.

Example 2. Find the determinant of the matrix

$$A = \begin{bmatrix} 1 & 2 & -2 \\ 3 & -2 & 4 \\ 1 & 0 & 5 \end{bmatrix}.$$

▶ **Solution.** For purposes of illustration, we compute the determinant in two ways. First, we expand along the first row:

$$\det A = 1 \cdot \begin{vmatrix} -2 & 4 \\ 0 & 5 \end{vmatrix} - 2 \begin{vmatrix} 3 & 4 \\ 1 & 5 \end{vmatrix} + (-2) \begin{vmatrix} 3 & -2 \\ 1 & 0 \end{vmatrix} = 1 \cdot (-10) - 2 \cdot (11) - 2(2) = -36.$$

Second, we expand along the second column:

$$\det A = (-)2 \begin{vmatrix} 3 & 4 \\ 1 & 5 \end{vmatrix} + (-2) \begin{vmatrix} 1 & -2 \\ 1 & 5 \end{vmatrix} (-)0 \begin{vmatrix} 1 & -2 \\ 3 & 4 \end{vmatrix} = (-2) \cdot 11 - 2 \cdot (7) = -36.$$

Of course, we get the same answer; that is what the theorem guarantees. Observe though that the second column has a zero entry which means that we really only needed to compute two minors. In practice, we usually try to use an expansion along a row or column that has a lot of zeros. Also note that we use the sign matrix to adjust the signs on the appropriate terms.   ◀

Properties of the Determinant

The determinant has many important properties. The three listed below show how the elementary row operations affect the determinant. They are used extensively to simplify many calculations.

Proposition 3. *Let A be an $n \times n$ matrix. Then*

1. $\det p_{i,j}(A) = -\det A$.
2. $\det m_i(a)(A) = a \det A$.
3. $\det t_{i,j}(A) = \det A$.

Proof. We illustrate the proof for the 2×2 case. Let $A = \begin{bmatrix} r & s \\ t & u \end{bmatrix}$. We then have

1. $|p_{1,2}(A)| = \begin{vmatrix} t & u \\ r & s \end{vmatrix} = ts - ru = -|A|$.

2. $|m_1(a)(A)| = \begin{vmatrix} ar & as \\ t & u \end{vmatrix} = aru - ast = a|A|$.

3. $|t_{1,2}(a)(A)| = \begin{vmatrix} r & s \\ t+ar & u+as \end{vmatrix} = r(u+as) - s(t+ar) = ru - st = |A|$. $\quad\square$

Another way to express 2. is

$2'$. If $A' = m_i(a)A$ then $\det A = \frac{1}{a} \det A'$.

Corollary 4. *Let E be an elementary matrix. Consider the three cases: If*

1. $E = p_{ij}I$, *then* $\det E = -1$.
2. $E = m_i(a)I$, *then* $\det E = a$.
3. $E = t_{ij}I$, *then* $\det E = 1$.

Furthermore,
$$\det EA = \det E \det A.$$

Proof. Let $A = I$ in Proposition 3 to get the stated formulas for $\det E$. Now let A be an arbitrary square matrix. The statement $\det EA = \det E \det A$ is now just a restatement of Proposition 3 and the fact that $eA = EA$ for an elementary row operation e and its associated elementary matrix E. $\quad\square$

Further important properties include:

Proposition 5. *1.* $\det A = \det A^t$.
2. If A has a zero row (or column), then $\det A = 0$.
3. If A has two equal rows (or columns), then $\det A = 0$.
4. If A is upper or lower triangular, then the determinant of A is the product of the diagonal entries.

Proof. 1. The transpose changes row expansions to column expansions and column expansions to row expansions. By Theorem 1, they are all the same.

2. All coefficients of the minors in an expansion along a zero row (column) are zero so the determinant is zero.

3. If the ith and jth rows are equal, then $\det A = \det(p_{i,j} A) = -\det A$ and this implies $\det A = 0$. If A has two equal columns, then A^t has two equal rows. Thus, $\det A = \det A^t = 0$.

4. Suppose A is upper triangular. Expansion along the first column gives $a_{11} Minor_{11}(A)$. But $Minor_{11}(A)$ is an upper triangular matrix of size one less than A. By induction, $\det A$ is the product of the diagonal entries. Since the transpose changes lower triangular matrices to upper triangular matrices, we get that the determinant of a lower triangular matrix is likewise the product of the diagonal entries.   □

Example 6. Use elementary row operations to find $\det A$ if

$$(1)\ A = \begin{bmatrix} 2 & 4 & 2 \\ -1 & 3 & 5 \\ 0 & 1 & 1 \end{bmatrix} \quad \text{and} \quad (2)\ A = \begin{bmatrix} 1 & 0 & 5 & 1 \\ -1 & 2 & 1 & 3 \\ 2 & 2 & 16 & 6 \\ 3 & 1 & 0 & 1 \end{bmatrix}.$$

▶ **Solution.** Again we will write the elementary row operation that we have used above the equal sign.

$$(1) \quad \begin{vmatrix} 2 & 4 & 2 \\ -1 & 3 & 5 \\ 0 & 1 & 1 \end{vmatrix} \overset{m_1(\frac{1}{2})}{=} 2\begin{vmatrix} 1 & 2 & 1 \\ -1 & 3 & 5 \\ 0 & 1 & 1 \end{vmatrix} \overset{t_{12}(1)}{=} 2\begin{vmatrix} 1 & 2 & 1 \\ 0 & 5 & 6 \\ 0 & 1 & 1 \end{vmatrix}$$

$$\overset{p_{23}}{=} -2\begin{vmatrix} 1 & 2 & 1 \\ 0 & 1 & 1 \\ 0 & 5 & 6 \end{vmatrix} \overset{t_{23}(-5)}{=} -2\begin{vmatrix} 1 & 2 & 1 \\ 0 & 1 & 1 \\ 0 & 0 & 1 \end{vmatrix} = -2$$

For the first equality, we have used (3′) above, and in the last equality, we have used the fact that the last matrix is upper triangular and its determinant is the product of the diagonal entries:

$$(2) \quad \begin{vmatrix} 1 & 0 & 5 & 1 \\ -1 & 2 & 1 & 3 \\ 2 & 2 & 16 & 6 \\ 3 & 1 & 0 & 1 \end{vmatrix} \begin{matrix} t_{12}(1) \\ t_{13}(-2) \\ t_{14}(-3) \\ = \end{matrix} \begin{vmatrix} 1 & 0 & 5 & 1 \\ 0 & 2 & 6 & 4 \\ 0 & 2 & 6 & 4 \\ 0 & 1 & -15 & -2 \end{vmatrix} = 0,$$

with the last equality because two rows are equal. ◄

In the following example, we use elementary row operations to zero out entries in a column and then use a Laplace expansion formula.

Example 7. Find the determinant of

$$A = \begin{bmatrix} 1 & 4 & 2 & -1 \\ 2 & 2 & 3 & 0 \\ -1 & 1 & 2 & 4 \\ 0 & 1 & 3 & 2 \end{bmatrix}.$$

► **Solution.**

$$\det(A) = \begin{vmatrix} 1 & 4 & 2 & -1 \\ 2 & 2 & 3 & 0 \\ -1 & 1 & 2 & 4 \\ 0 & 1 & 3 & 2 \end{vmatrix} \begin{matrix} t_{1,2}(-2) \\ t_{1,3}(1) \\ = \end{matrix} \begin{vmatrix} 1 & 4 & 2 & -1 \\ 0 & -6 & -1 & 2 \\ 0 & 5 & 4 & 3 \\ 0 & 1 & 3 & 2 \end{vmatrix}$$

$$= \begin{vmatrix} -6 & -1 & 2 \\ 5 & 4 & 3 \\ 1 & 3 & 2 \end{vmatrix} \begin{matrix} t_{3,1}(6) \\ t_{3,2}(-5) \\ = \end{matrix} \begin{vmatrix} 0 & 17 & 14 \\ 0 & -11 & -7 \\ 1 & 3 & 2 \end{vmatrix}$$

$$= \begin{vmatrix} 17 & 14 \\ -11 & -7 \end{vmatrix} = -119 + 154 = 35, \quad ◄$$

The following two theorems state very important properties about the determinant.

Theorem 8. *A square matrix A is invertible if and only if* $\det A \neq 0$.

Proof. Apply Gauss–Jordan to A to get R, a matrix in row reduce echelon form. There is a sequence of elementary row operations $e_1, \ldots, e_k$ such that

$$e_1 \cdots e_k A = R.$$

Let $E_i = e_i I$, $i = 1, \ldots, k$ be the corresponding elementary matrices. Then

$$E_1 \cdots E_k A = R.$$

By repeatedly using Corollary 4, we get

$$\det E_1 \det E_2 \cdots \det E_k \det A = \det R.$$

Now suppose A is invertible. By Theorem 10 of Sect. 8.3, $R = I$. Since each factor $\det E_i \neq 0$ by Corollary 4, it follows that $\det A \neq 0$. On the other hand, if A is not invertible, then R has a zero row by Theorem 10 of Sect. 8.3 and Corollary 11 of Sect. 8.3. By Proposition 5, we have $\det R = 0$. Since each factor E_i has a nonzero determinant, it follows that $\det A = 0$.   □

Theorem 9. *If A and B are square matrices of the same size, then*

$$\det(AB) = \det A \det B.$$

Proof. We consider two cases.

1. Suppose A is invertible. Then there is a sequence of elementary matrices such that $A = E_1 \cdots E_k$. Now repeatedly use Corollary 4 to get

$$\det AB = \det E_1 \cdots E_k B = \det E_1 \cdots \det E_k \det B = \det A \det B.$$

2. Now suppose A is not invertible. Then AB is not invertible for otherwise there would be a C such that $(AB)C = I$. But by associativity of the product, we have $A(BC) = I$, and this implies A is invertible (with inverse BC). Now by Theorem 8, we have $\det AB = 0$ and $\det A = 0$ and the result follows.   □

The Cofactor and Adjoint Matrices

Again, let A be a square matrix. We define the ***cofactor*** matrix, $\mathrm{Cof}(A)$, of A to be the matrix whose (i, j)-entry is $(-1)^{i+j} \mathrm{Minor}_{i,j}$. We define the ***adjoint*** matrix, $\mathrm{Adj}(A)$, of A by the formula $\mathrm{Adj}(A) = (\mathrm{Cof}(A))^t$. The important role of the adjoint matrix is seen in the following theorem and its corollary.

Theorem 10. *For A a square matrix, we have*

$$A \, \mathrm{Adj}(A) = \mathrm{Adj}(A) \, A = \det(A)I.$$

Proof. The (i, j) entry of $A \, \mathrm{Adj}(A)$ is

$$\sum_{k=0}^{n} A_{i\,k}(\mathrm{Adj}(A))_{k\,j} = \sum_{k=0}^{n} (-1)^{k+j} A_{i\,k} \mathrm{Minor}_{j\,k}(A).$$

When $i = j$, this is a Laplace expansion formula and is hence det A by Theorem 1. When $i \neq j$, this is the expansion of a determinant for a matrix with two equal rows and hence is zero by Proposition 5. □

The following corollary immediately follows.

Corollary 11 (The Adjoint Inversion Formula). *If* det $A \neq 0$, *then*

$$A^{-1} = \frac{1}{\det A} \text{Adj}(A).$$

The inverse of a 2×2 matrix is a simple matter: Let $A = \begin{bmatrix} a & b \\ c & d \end{bmatrix}$. Then $\text{Adj}(A) = \begin{bmatrix} d & -b \\ -c & a \end{bmatrix}$ and if $\det(A) = ad - bd \neq 0$, then

$$A^{-1} = \frac{1}{ad - bc} \begin{bmatrix} d & -b \\ -c & a \end{bmatrix}. \tag{1}$$

For an example, suppose $A = \begin{bmatrix} 1 & -3 \\ -2 & 1 \end{bmatrix}$. Then $\det(A) = 1 - (6) = -5 \neq 0$ so A is invertible and $A^{-1} = \frac{-1}{5} \begin{bmatrix} 1 & 3 \\ 2 & 1 \end{bmatrix} = \begin{bmatrix} \frac{-1}{5} & \frac{-3}{5} \\ \frac{-2}{5} & \frac{-1}{5} \end{bmatrix}$.

The general formula for the inverse of a 3×3 is substantially more complicated and difficult to remember. Consider though an example.

Example 12. Let

$$A = \begin{bmatrix} 1 & 2 & 0 \\ 1 & 4 & 1 \\ -1 & 0 & 3 \end{bmatrix}.$$

Find its inverse if it is invertible.

▶ **Solution.** We expand along the first row to compute the determinant and get

$$\det(A) = 1 \det \begin{bmatrix} 4 & 1 \\ 0 & 3 \end{bmatrix} - 2 \det \begin{bmatrix} 1 & 1 \\ -1 & 3 \end{bmatrix} = 1(12) - 2(4) = 4. \text{ Thus, } A \text{ is invertible.}$$

The cofactor of A is $\text{Cof}(A) = \begin{bmatrix} 12 & -4 & 4 \\ -6 & 3 & -2 \\ 2 & -1 & 2 \end{bmatrix}$ and $\text{Adj}(A) = \text{Cof}(A)^t =$

$\begin{bmatrix} 12 & -6 & 2 \\ -4 & 3 & -1 \\ 4 & -2 & 2 \end{bmatrix}$. The inverse of A is thus

$$A^{-1} = \frac{1}{4} \begin{bmatrix} 12 & -6 & 2 \\ -4 & 3 & -1 \\ 4 & -2 & 2 \end{bmatrix} = \begin{bmatrix} 3 & \frac{-3}{2} & \frac{1}{2} \\ -1 & \frac{3}{4} & \frac{-1}{4} \\ 1 & \frac{-1}{2} & \frac{1}{2} \end{bmatrix}. \qquad \blacktriangleleft$$

In our next example, we will consider a matrix with entries in $\mathcal{R} = \mathbb{R}[s]$. Such matrices will arise naturally in Chap. 9.

Example 13. Let

$$A = \begin{bmatrix} 1 & 2 & 1 \\ 0 & 1 & 3 \\ 1 & 1 & 2 \end{bmatrix}.$$

Find the inverse of the matrix

$$sI - A = \begin{bmatrix} s-1 & -2 & -1 \\ 0 & s-1 & -3 \\ -1 & -1 & s-2 \end{bmatrix}.$$

▶ **Solution.** A straightforward computation gives

$$\det(sI - A) = (s-4)(s^2 + 1).$$

The matrix of minors for $sI - A$ is

$$\begin{bmatrix} (s-1)(s-2)-3 & -3 & s-1 \\ -2(s-2)-1 & (s-1)(s-2)-1 & -(s-1)-2 \\ 6+(s-1) & -3(s-1) & (s-1)^2 \end{bmatrix}.$$

After simplifying, we obtain the cofactor matrix

$$\begin{bmatrix} s^2 - 3s - 1 & 3 & s-1 \\ 2s - 3 & s^2 - 3s + 1 & s+1 \\ s+5 & 3s - 3 & (s-1)^2 \end{bmatrix}.$$

The adjoint matrix is

$$\begin{bmatrix} s^2 - 3s - 1 & 2s - 3 & s+5 \\ 3 & s^2 - 3s + 1 & 3s - 3 \\ s - 1 & s+1 & (s-1)^2 \end{bmatrix}.$$

Finally, we obtain the inverse:

$$(sI - A)^{-1} = \begin{bmatrix} \frac{s^2-3s-1}{(s-4)(s^2+1)} & \frac{2s-3}{(s-4)(s^2+1)} & \frac{s+5}{(s-4)(s^2+1)} \\ \frac{3}{(s-4)(s^2+1)} & \frac{s^2-3s+1}{(s-4)(s^2+1)} & \frac{3s-3}{(s-4)(s^2+1)} \\ \frac{s-1}{(s-4)(s^2+1)} & \frac{s+1}{(s-4)(s^2+1)} & \frac{(s-1)^2}{(s-4)(s^2+1)} \end{bmatrix}. \qquad \blacktriangleleft$$

Cramer's Rule

We finally consider a well-known theoretical tool used to solve a system $A\mathbf{x} = \mathbf{b}$ when A is invertible. Let $A(i, \mathbf{b})$ denote the matrix obtained by replacing the ith column of A with the column vector $\mathbf{b}$. We then have the following theorem:

Theorem 14. *Suppose* $\det A \neq 0$. *Then the solution to* $A\mathbf{x} = \mathbf{b}$ *is given coordinate-wise by the formula*

$$\mathbf{x}_i = \frac{\det A(i, \mathbf{b})}{\det A}.$$

Proof. Since A is invertible, we have

$$\mathbf{x}_i = (A^{-1}\mathbf{b})_i = \sum_{k=1}^{n} (A^{-1})_{ik} \mathbf{b}_k$$

$$= \frac{1}{\det A} \sum_{k=1}^{n} (-1)^{i+k} \text{Minor}_{ki}(A) \mathbf{b}_k$$

$$= \frac{1}{\det(A)} \sum_{k=1}^{n} (-1)^{i+k} \mathbf{b}_k \text{Minor}_{ki}(A) = \frac{\det A(i, \mathbf{b})}{\det A}. \qquad \square$$

The following example should convince you that Cramer's rule is mainly a theoretical tool and not a practical one for solving a system of linear equations. The Gauss–Jordan elimination method is usually far more efficient than computing $n + 1$ determinants for a system $A\mathbf{x} = \mathbf{b}$, where A is $n \times n$.

Example 15. Solve the following system of linear equations using Cramer's rule:

$$
\begin{aligned}
x + y + z &= 0 \\
2x + 3y - z &= 11 \\
x + z &= -2
\end{aligned}
$$

▶ **Solution.** We have

$$
\det A = \begin{vmatrix} 1 & 1 & 1 \\ 2 & 3 & -1 \\ 1 & 0 & 1 \end{vmatrix} = -3,
$$

$$
\det A(1, \mathbf{b}) = \begin{vmatrix} 0 & 1 & 1 \\ 11 & 3 & -1 \\ -2 & 0 & 1 \end{vmatrix} = -3,
$$

$$
\det A(2, \mathbf{b}) = \begin{vmatrix} 1 & 0 & 1 \\ 2 & 11 & -1 \\ 1 & -2 & 1 \end{vmatrix} = -6,
$$

$$
\text{and } \det A(3, \mathbf{b}) = \begin{vmatrix} 1 & 1 & 0 \\ 2 & 3 & 11 \\ 1 & 0 & -2 \end{vmatrix} = 9,
$$

where $\mathbf{b} = \begin{bmatrix} 0 \\ 11 \\ -2 \end{bmatrix}$. Since $\det A \neq 0$, Cramer's rule gives

$$
x_1 = \frac{\det A(1, \mathbf{b})}{\det A} = \frac{-3}{-3} = 1,
$$

$$
x_2 = \frac{\det A(2, \mathbf{b})}{\det A} = \frac{-6}{-3} = 2,
$$

and

$$
x_3 = \frac{\det A(3, \mathbf{b})}{\det A} = \frac{9}{-3} = -3.
$$
◀

Exercises

1–9. Find the determinant of each matrix given below in three ways: a row expansion, a column expansion, and using row operations to reduce to a triangular matrix.

1. $\begin{bmatrix} 1 & 4 \\ 2 & 9 \end{bmatrix}$

2. $\begin{bmatrix} 1 & 1 \\ 4 & 4 \end{bmatrix}$

3. $\begin{bmatrix} 3 & 4 \\ 2 & 6 \end{bmatrix}$

4. $\begin{bmatrix} 1 & 1 & -1 \\ 1 & 4 & 0 \\ 2 & 3 & 1 \end{bmatrix}$

5. $\begin{bmatrix} 4 & 0 & 3 \\ 8 & 1 & 7 \\ 3 & 4 & 1 \end{bmatrix}$

6. $\begin{bmatrix} 3 & 98 & 100 \\ 0 & 2 & 99 \\ 0 & 0 & 1 \end{bmatrix}$

7. $\begin{bmatrix} 0 & 1 & -2 & 4 \\ 2 & 3 & 9 & 2 \\ 1 & 4 & 8 & 3 \\ -2 & 3 & -2 & 4 \end{bmatrix}$

8. $\begin{bmatrix} -4 & 9 & -4 & 1 \\ 2 & 3 & 0 & -4 \\ -2 & 3 & 5 & -6 \\ -3 & 2 & 0 & 1 \end{bmatrix}$

9. $\begin{bmatrix} 2 & 4 & 2 & 3 \\ 1 & 2 & 1 & 4 \\ 4 & 8 & 4 & 6 \\ 1 & 9 & 11 & 13 \end{bmatrix}$

10–15. Find the inverse of $(sI - A)$ and determine for which values of s det$(sI - A) = 0$.

10. $\begin{bmatrix} 1 & 2 \\ 1 & 2 \end{bmatrix}$

11. $\begin{bmatrix} 3 & 1 \\ 1 & 3 \end{bmatrix}$

12. $\begin{bmatrix} 1 & 1 \\ -1 & 1 \end{bmatrix}$

13. $\begin{bmatrix} 1 & 0 & 1 \\ 0 & 1 & 0 \\ 0 & 3 & 1 \end{bmatrix}$

14. $\begin{bmatrix} 1 & -3 & 3 \\ -3 & 1 & 3 \\ 3 & -3 & 1 \end{bmatrix}$

15. $\begin{bmatrix} 0 & 4 & 0 \\ -1 & 0 & 0 \\ 1 & 4 & -1 \end{bmatrix}$

16–24. Use the adjoint formula for the inverse for the matrices given below.

16. $\begin{bmatrix} 1 & 4 \\ 2 & 9 \end{bmatrix}$

17. $\begin{bmatrix} 1 & 1 \\ 4 & 4 \end{bmatrix}$

18. $\begin{bmatrix} 3 & 4 \\ 2 & 6 \end{bmatrix}$

19. $\begin{bmatrix} 1 & 1 & -1 \\ 1 & 4 & 0 \\ 2 & 3 & 1 \end{bmatrix}$

20. $\begin{bmatrix} 4 & 0 & 3 \\ 8 & 1 & 7 \\ 3 & 4 & 1 \end{bmatrix}$

21. $\begin{bmatrix} 3 & 98 & 100 \\ 0 & 2 & 99 \\ 0 & 0 & 1 \end{bmatrix}$

22. $\begin{bmatrix} 0 & 1 & -2 & 4 \\ 2 & 3 & 9 & 2 \\ 1 & 4 & 8 & 3 \\ -2 & 3 & -2 & 4 \end{bmatrix}$

23. $\begin{bmatrix} -4 & 9 & -4 & 1 \\ 2 & 3 & 0 & -4 \\ -2 & 3 & 5 & -6 \\ -3 & 2 & 0 & 1 \end{bmatrix}$

24. $\begin{bmatrix} 2 & 4 & 2 & 3 \\ 1 & 2 & 1 & 4 \\ 4 & 8 & 4 & 6 \\ 1 & 9 & 11 & 13 \end{bmatrix}$

25–28. Use Cramer's rule to solve the system $A\mathbf{x} = \mathbf{b}$ for the given matrices A and $\mathbf{b}$.

25. $A = \begin{bmatrix} 1 & 1 \\ 3 & 4 \end{bmatrix} \quad \mathbf{b} = \begin{bmatrix} 2 \\ 3 \end{bmatrix}$

26. $A = \begin{bmatrix} 1 & 1 & 1 \\ 0 & 1 & 2 \\ 0 & 0 & 1 \end{bmatrix} \quad \mathbf{b} = \begin{bmatrix} 1 \\ 0 \\ -3 \end{bmatrix}$

27. $A = \begin{bmatrix} 1 & 0 & -2 \\ 2 & -2 & 0 \\ 1 & 2 & -1 \end{bmatrix}$ $\mathbf{b} = \begin{bmatrix} -2 \\ 1 \\ 2 \end{bmatrix}$

28. $A = \begin{bmatrix} 1 & 3 & 0 & 1 \\ 2 & 2 & -2 & 0 \\ 1 & -1 & 0 & 4 \\ 1 & 2 & 3 & 9 \end{bmatrix}$ $\mathbf{b} = \begin{bmatrix} 1 \\ 0 \\ -1 \\ 2 \end{bmatrix}$

8.5   Eigenvectors and Eigenvalues

Suppose A is a square $n \times n$ matrix. Again, it is convenient to think of $\mathbb{R}^n$ as the set of column vectors $M_{n,1}(\mathbb{R})$. If $\mathbf{v} \in \mathbb{R}^n$, then A transforms $\mathbf{v}$ to a new vector $A\mathbf{v}$ as seen below.

As illustrated, A rotates and compresses (or stretches) a given vector $\mathbf{v}$. However, if a vector $\mathbf{v}$ points in the right direction, then A acts in a much simpler manner.

   We say that s is an *eigenvalue* of A if there is a *nonzero* vector $\mathbf{v}$ in $\mathbb{R}^n$ such that

$$A\mathbf{v} = s\mathbf{v}. \qquad (1)$$

The vector $\mathbf{v}$ is called an *eigenvector*[1] associated to s. The pair $(s, \mathbf{v})$ is called an *eigenpair*. One should think of eigenvectors as the directions for which A acts by stretching or compressing vectors by the length determined by the eigenvalue s: if $s > 1$, the eigenvector is stretched; if $0 < s < 1$, then the eigenvector is compressed; if $s < 0$, the direction of the eigenvector is reversed; and if $s = 0$, the eigenvector is in the null space of A. See the illustration below.

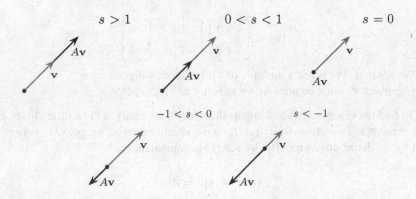

This notion is very important and has broad applications in mathematics, computer science, physics, engineering, and economics. For example, the Google page rank algorithm is based on this concept.

---

[1]Eigenvectors and eigenvalues are also called characteristic vectors and characteristic values, respectively.

In this section, we discuss how to find the eigenpairs for a given matrix A. We begin with a simple example.

Example 1. Suppose $A = \begin{bmatrix} -3 & 1 \\ -4 & 2 \end{bmatrix}$. Let

$$\mathbf{v}_1 = \begin{bmatrix} 2 \\ 8 \end{bmatrix}, \ \mathbf{v}_2 = \begin{bmatrix} 1 \\ 1 \end{bmatrix}, \ \mathbf{v}_3 = \begin{bmatrix} 1 \\ -1 \end{bmatrix}, \ \text{and} \ \mathbf{v}_4 = \begin{bmatrix} 0 \\ 0 \end{bmatrix}.$$

Show that $\mathbf{v}_1$ and $\mathbf{v}_2$ are eigenvectors for A. What are the associated eigenvalues? Show that $\mathbf{v}_3$ and $\mathbf{v}_4$ are not eigenvectors.

▶ **Solution.** We simply observe that

1.
$$A\mathbf{v}_1 = \begin{bmatrix} -3 & 1 \\ -4 & 2 \end{bmatrix} \begin{bmatrix} 2 \\ 8 \end{bmatrix} = \begin{bmatrix} 2 \\ 8 \end{bmatrix} = 1 \begin{bmatrix} 2 \\ 8 \end{bmatrix}.$$

Thus, $\mathbf{v}_1$ an eigenvector with eigenvalue 1.

2.
$$A\mathbf{v}_2 = \begin{bmatrix} -3 & 1 \\ -4 & 2 \end{bmatrix} \begin{bmatrix} 1 \\ 1 \end{bmatrix} = \begin{bmatrix} -2 \\ -2 \end{bmatrix} = -2 \begin{bmatrix} 1 \\ 1 \end{bmatrix}.$$

Thus, $\mathbf{v}_2$ an eigenvector with eigenvalue -2.

3.
$$A\mathbf{v}_3 = \begin{bmatrix} -3 & 1 \\ -4 & 2 \end{bmatrix} \begin{bmatrix} 1 \\ -1 \end{bmatrix} = \begin{bmatrix} -4 \\ -6 \end{bmatrix}.$$

We see that $A\mathbf{v}_3$ is not a multiple of $\mathbf{v}_3$. It is not an eigenvector.
4. Eigenvectors must be nonzero so $\mathbf{v}_4$ is not an eigenvector.   ◀

To find the eigenvectors and eigenvalues for A, we analyze (1) a little closer. Let us rewrite it as $sv - Av = 0$. By inserting the identity matrix, we get $sIv - Av = 0$, and by the distributive property, we see (1) is equivalent to

$$(sI - A)v = 0. \tag{2}$$

Said another way, a nonzero vector $\mathbf{v}$ is an eigenvector for A if and only if it is in the null space of $sI - A$ for an eigenvalue s. Let E_s be the null space of $sI - A$; it is called the **eigenspace** for A with eigenvalue s. So once an eigenvalue is known, the corresponding eigenspaces are easily computed. How does one determine the eigenvalues? By Theorems 10 of Sect. 8.3 and 8 of Sect. 8.4, s is an eigenvalue if and only if

$$\det(sI - A) = 0. \tag{3}$$

As a function of s, we let $c_A(s) = \det(sI - A)$; it is called the **characteristic polynomial of A**, for it is a polynomial of degree n (assuming A is an $n \times n$ matrix). The matrix $sI - A$ is called the **characteristic matrix**, and (3) is called the **characteristic equation**. By solving the characteristic equation, we determine the eigenvalues.

Example 2. Determine the characteristic polynomial, $c_A(s)$, for

$$A = \begin{bmatrix} -3 & 1 \\ -4 & 2 \end{bmatrix},$$

as given in Example 1. Find the eigenvalues and corresponding eigenspaces.

▶ **Solution.** The characteristic matrix is

$$sI - A = \begin{bmatrix} s+3 & -1 \\ 4 & s-2 \end{bmatrix},$$

and the characteristic polynomial is given as follows:

$$c_A(s) = \det(sI - A) = \det \begin{bmatrix} s+3 & -1 \\ 4 & s-2 \end{bmatrix}$$

$$= (s+3)(s-2) + 4 = s^2 + s - 2 = (s+2)(s-1).$$

The characteristic equation is $(s+2)(s-1) = 0$, and hence, the eigenvalues are -2 and 1. The eigenspaces, E_s, are computed as follows:

E_{-2}: We let $s = -2$ in the characteristic matrix and row reduce the corresponding augmented matrix $[-2I - A | \mathbf{0}]$ to get the null space of $-2I - A$. We get

$$\begin{bmatrix} 1 & -1 & | & 0 \\ 4 & -4 & | & 0 \end{bmatrix} \xrightarrow{t_{12}(-4)} \begin{bmatrix} 1 & -1 & | & 0 \\ 0 & 0 & | & 0 \end{bmatrix}.$$

If x and y are the variables, then y is a free variable. Let $y = \alpha$. Then $x = y = \alpha$. From this, we see that

$$E_{-2} = \text{Span}\left\{ \begin{bmatrix} 1 \\ 1 \end{bmatrix} \right\}.$$

E_1: We let $s = 1$ in the characteristic matrix and row reduce the corresponding augmented matrix $[1I - A | \mathbf{0}]$ to get the null space of $1I - A$. We get

$$\begin{bmatrix} 4 & -1 & | & 0 \\ 4 & -1 & | & 0 \end{bmatrix} \xrightarrow{m_1(1/4)} \begin{bmatrix} 1 & -1/4 & | & 0 \\ 4 & -1 & | & 0 \end{bmatrix} \xrightarrow{t_{12}(-4)} \begin{bmatrix} 1 & -1/4 & | & 0 \\ 0 & 0 & | & 0 \end{bmatrix}.$$

Again $y = \alpha$ is the free variable and $x = \frac{1}{4}\alpha$. It follows that the null space of $1I - A$ is all multiples of the vector $\begin{bmatrix} 1/4 \\ 1 \end{bmatrix}$. Since we are considering all multiples of a vector, we can clear the fraction (by multiplying by 4) and write

$$E_1 = \text{Span} \left\{ \begin{bmatrix} 1 \\ 4 \end{bmatrix} \right\}.$$

We will routinely do this.   ◀

Remark 3. In Example 1, we found that the vector $\mathbf{v}_1 = \begin{bmatrix} 2 \\ 8 \end{bmatrix}$ was an eigenvector with eigenvalue 1. We observe that $\mathbf{v}_1 = 2 \begin{bmatrix} 1 \\ 4 \end{bmatrix} \in E_1$. In like manner, $\mathbf{v}_2 = \begin{bmatrix} 1 \\ 1 \end{bmatrix} \in E_{-2}$.

Example 4. Determine the characteristic polynomial for

$$A = \begin{bmatrix} 3 & -1 & 1 \\ 3 & -1 & 3 \\ 1 & -1 & 3 \end{bmatrix}.$$

Find the eigenvalues and corresponding eigenspaces.

▶ **Solution.** The characteristic matrix is

$$sI - A = \begin{bmatrix} -3 & 1 & -1 \\ -3 & s+1 & -3 \\ -1 & 1 & s-3 \end{bmatrix},$$

and the characteristic polynomial is calculated by expanding along the first row as follows:

$$c_A(s) = \det(sI - A) = \det \begin{bmatrix} s-3 & 1 & -1 \\ -3 & s+1 & -3 \\ -1 & 1 & s-3 \end{bmatrix}$$

$$= (s-3)((s+1)(s-3)+3) - (-3(s-3)-3) - (-3+s+1)$$

$$= s^3 - 5s^2 + 8s - 4 = (s - 1)(s - 2)^2.$$

It follows that the eigenvalues are 1 and 2. The eigenspaces, E_s, are computed as follows:

E_1: We let $s = 1$ in the characteristic matrix and row reduce the corresponding augmented matrix $[I - A | \mathbf{0}]$. We forego the details but we get

$$\begin{bmatrix} 1 & 0 & -1 & | & 0 \\ 0 & 1 & -3 & | & 0 \\ 0 & 0 & 0 & | & 0 \end{bmatrix}.$$

If x, y, and z are the variables, then z is the free variable. Let $z = \alpha$. Then $x = \alpha$, $y = 3\alpha$, and $z = \alpha$. From this, we see that

$$E_1 = \text{Span} \left\{ \begin{bmatrix} 1 \\ 3 \\ 1 \end{bmatrix} \right\}.$$

E_2: We let $s = 2$ in the characteristic matrix and row reduce the corresponding augmented matrix $[2I - A | \mathbf{0}]$ to get

$$\begin{bmatrix} 1 & -1 & 1 & | & 0 \\ 0 & 0 & 0 & | & 0 \\ 0 & 0 & 0 & | & 0 \end{bmatrix}.$$

Here we see that $y = \alpha$ and $z = \beta$ are free variable and $x = \alpha - \beta$. It follows that the null space of $2I - A$ is

$$E_2 = \text{Span} \left\{ \begin{bmatrix} 1 \\ 1 \\ 0 \end{bmatrix}, \begin{bmatrix} -1 \\ 0 \\ 1 \end{bmatrix} \right\}. \qquad \blacktriangleleft$$

For a diagonal matrix, the eigenpairs are simple to find. Let

$$A = \text{diag}(a_1, \dots, a_n) = \begin{bmatrix} a_1 & 0 & \cdots & 0 \\ 0 & a_2 & & 0 \\ \vdots & & \ddots & 0 \\ 0 & 0 & \cdots & a_n \end{bmatrix}.$$

Let $\mathbf{e}_i$ be the column vector in $\mathbb{R}^n$ with 1 in the ith position and zeros elsewhere; $\mathbf{e}_i$ is the ith column of the identity matrix I_n. A simple calculation gives

$$A\mathbf{e}_i = a_i\mathbf{e}_i.$$

Thus, for a diagonal matrix, the eigenvalues are the diagonal entries and the eigenvectors are the coordinate axes in $\mathbb{R}^n$. In other words, the coordinate axes point in the directions for which A scales vectors.

If A is a real matrix, then it can happen that the characteristic polynomial have complex roots. In this case, we view A a complex matrix and make all computation over the complex numbers. If $\mathbb{C}^n$ denotes the $n \times 1$ column vectors with entries in $\mathbb{C}$, then we view A as transforming vectors $\mathbf{v} \in \mathbb{C}^n$ to vectors $A\mathbf{v} \in \mathbb{C}^n$.

Example 5. Determine the characteristic polynomial for

$$A = \begin{bmatrix} 2 & -5 \\ 1 & -2 \end{bmatrix}.$$

Find the eigenvalues and corresponding eigenspaces.

▶ **Solution.** The characteristic matrix is

$$sI - A = \begin{bmatrix} s-2 & 5 \\ -1 & s+2 \end{bmatrix},$$

and the characteristic polynomial is given as follows:

$$\begin{aligned} c_A(s) = \det(sI - A) &= \det \begin{bmatrix} s-2 & 5 \\ -1 & s+2 \end{bmatrix} \\ &= (s-2)(s+2) + 5 = s^2 + 1. \end{aligned}$$

The eigenvalues are $\pm i$. The eigenspaces, E_s, are computed as follows:

E_i: We let $s = i$ in the characteristic matrix and row reduce the corresponding augmented matrix $[iI - A | \mathbf{0}]$. We get

$$\begin{bmatrix} i-2 & 5 & | & 0 \\ -1 & i+2 & | & 0 \end{bmatrix} \xrightarrow[m_1(-1)]{p_{12}} \begin{bmatrix} 1 & -i-2 & | & 0 \\ i-2 & 5 & | & 0 \end{bmatrix}$$

$$\xrightarrow{t_{12}(-(i-2))} \begin{bmatrix} 1 & -i-2 & | & 0 \\ 0 & 0 & | & 0 \end{bmatrix}.$$

If x and y are the variables, then y is a free variable. Let $y = \alpha$. Then $x = \alpha(i+2)$. From this, we see that

$$E_i = \text{Span} \left\{ \begin{bmatrix} i+2 \\ 1 \end{bmatrix} \right\}.$$

E_{-i}: We let $s = -i$ in the characteristic matrix and row reduce the corresponding augmented matrix $[-iI - A | \mathbf{0}]$. We get

$$\begin{bmatrix} -i-2 & 5 & | & 0 \\ -1 & -i+2 & | & 0 \end{bmatrix} \xrightarrow[m_1(-1)]{p_{12}} \begin{bmatrix} 1 & i-2 & | & 0 \\ -i-2 & 5 & | & 0 \end{bmatrix}$$

$$\xrightarrow{t_{12}(i+2)} \begin{bmatrix} 1 & i-2 & | & 0 \\ 0 & 0 & | & 0 \end{bmatrix}.$$

Again $y = \alpha$ is the free variable and $x = (2-i)\alpha$. It follows that

$$E_{-i} = \text{Span} \left\{ \begin{bmatrix} 2-i \\ 1 \end{bmatrix} \right\}. \qquad \blacktriangleleft$$

Remark 6. It should be noted in this example that the eigenvalues and eigenvectors are complex conjugate of one another. In other words, if $s = i$ is an eigenvalue, then $\bar{s} = -i$ is another eigenvalue. Further, if $\mathbf{v} = \alpha \begin{bmatrix} i+2 \\ 1 \end{bmatrix}$ is an eigenvector with eigenvalue i, then $\bar{\mathbf{v}} = \alpha \begin{bmatrix} -i+2 \\ 1 \end{bmatrix}$ is an eigenvector with eigenvalue $\bar{i} = -i$ and vice versa. The following theorem shows that this happens in general as long as A is a real matrix. Thus, E_{-i} may be computed by simply taking the complex conjugate of E_i.

Theorem 7. *Suppose A is a real $n \times n$ matrix with a complex eigenvalue s. Then $\bar{s}$ is an eigenvalue and*

$$E_{\bar{s}} = \overline{E_s}.$$

Proof. Suppose s is a complex eigenvalue and $\mathbf{v} \in E_s$ is a corresponding eigenvector. Then $A\mathbf{v} = s\mathbf{v}$. Taking complex conjugates and keeping in mind that $\overline{A} = A$ since A is real, we get $A\bar{\mathbf{v}} = \overline{A\mathbf{v}} = \overline{s\mathbf{v}} = \bar{s}\,\bar{\mathbf{v}}$. It follows that $\bar{\mathbf{v}} \in E_{\bar{s}}$ and $\overline{E_s} \subset E_{\bar{s}}$. This argument is symmetric so $\overline{E_{\bar{s}}} \subset E_s$. These two statements imply $\overline{E_s} = E_{\bar{s}}$. $\qquad \square$

Exercises

1–7. For each of the following matrices A determine the characteristic polynomial, the eigenvalues, and the eigenspaces.

1. $A = \begin{bmatrix} 2 & 0 \\ -1 & 1 \end{bmatrix}$

2. $A = \begin{bmatrix} -6 & -5 \\ 7 & 6 \end{bmatrix}$

3. $A = \begin{bmatrix} 0 & -1 \\ 1 & 2 \end{bmatrix}$

4. $A = \begin{bmatrix} 7 & 4 \\ -16 & -9 \end{bmatrix}$

5. $A = \begin{bmatrix} 9 & 12 \\ -8 & -11 \end{bmatrix}$

6. $A = \begin{bmatrix} 1 & -2 \\ 2 & 1 \end{bmatrix}$

7. $A = \begin{bmatrix} 20 & -15 \\ 30 & -22 \end{bmatrix}$

8–13. For the following problems, A and its characteristic polynomial are given. Find the eigenspaces for each eigenvalue.

8. $A = \begin{bmatrix} 3 & -2 & 2 \\ 9 & -7 & 9 \\ 5 & -4 & 6 \end{bmatrix}, c_A(s) = (s+1)(s-1)(s-2)$

9. $A = \begin{bmatrix} 8 & -5 & 5 \\ 0 & -2 & 0 \\ -10 & 5 & -7 \end{bmatrix}, c_A(s) = (s+2)^2(s-3)$

10. $A = \begin{bmatrix} 1 & 1 & -1 \\ 0 & -1 & 4 \\ 0 & -2 & 5 \end{bmatrix}, c_A(s) = (s-1)^2(s-3)$

11. $A = \begin{bmatrix} 2 & 0 & 0 \\ 2 & -3 & 6 \\ 1 & -3 & 6 \end{bmatrix}, c_A(s) = s(s-2)(s-3)$

12. $A = \begin{bmatrix} -6 & 11 & -16 \\ 4 & -4 & 8 \\ 7 & -10 & 16 \end{bmatrix}, c_A(s) = (s-2)^3$

13. $A = \begin{bmatrix} -8 & 13 & -19 \\ 8 & -8 & 14 \\ 11 & -14 & 22 \end{bmatrix}, c_A(s) = (s-2)(s^2-4s+5)$

Chapter 9
Linear Systems of Differential Equations

9.1 Introduction

In previous chapters, we have discussed ordinary differential equations in a single unknown function, $y(t)$. These are adequate to model real-world systems as they evolve in time, provided that only one state, that is, the number $y(t)$, is needed to describe the system. For instance, we might be interested in the temperature of an object, the concentration of a pollutant in a lake, or the displacement of a weight attached to a spring. In each of these cases, the system we wish to describe is adequately represented by a single function of time. However, a single ordinary differential equation is inadequate for describing the evolution over time of a system with interdependent subsystems, each with its own state. Consider such an example.

Example 1. Two tanks are interconnected as illustrated below.

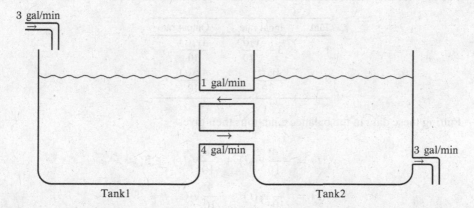

Assume that Tank 1 contains 10 gallons of brine in which 2 pounds of salt are initially dissolved and Tank 2 initially contains 10 gallons of pure water. Moreover, the mixtures are pumped between the two tanks, 4 gal/min from Tank 1 to Tank

W.A. Adkins and M.G. Davidson, *Ordinary Differential Equations*,
Undergraduate Texts in Mathematics, DOI 10.1007/978-1-4614-3618-8_9,
© Springer Science+Business Media New York 2012

2 and 1 gal/min going from Tank 2 back to Tank 1. Assume that a brine mixture containing 1 lb salt/gal enters Tank 1 at a rate of 3 gal/min, and the well-stirred mixture is removed from Tank 2 at the same rate of 3 gal/min. Let $y_1(t)$ be the amount of salt in Tank 1 at time t and let $y_2(t)$ be the amount of salt in Tank 2 at time t. Determine how y_1 and y_2 and their derivatives are related.

▶ **Solution.** The underlying principle is the same as that of the single tank mixing problem. Namely, we apply the balance equation

$$y'(t) = \text{input rate} - \text{output rate}$$

to the amount of salt in *each* tank. If $y_1(t)$ denotes the amount of salt at time t in Tank 1, then the ***concentration*** of salt at time t in Tank 1 is $c_1(t) = (y_1(t)/10)$ lb/gal. Similarly, the concentration of salt in Tank 2 at time t is $c_2(t) = (y_2(t)/10)$ lb/gal. The input and output rates are determined by the product of the concentration and the flow rate of the fluid at time t. The relevant rates of change are summarized in the following table.

From	To	Rate
Outside	Tank 1	$(1 \text{ lb/gal}) \cdot (3 \text{ gal/min}) = 3 \text{ lb/min}$
Tank 1	Tank2	$\dfrac{y_1(t)}{10} \text{lb/gal} \cdot 4 \text{ gal/min} = \dfrac{4y_1(t)}{10} \text{ lb/min}$
Tank 2	Tank 1	$\dfrac{y_2(t)}{10} \text{lb/gal} \cdot 1 \text{ gal/min} = \dfrac{y_2(t)}{10} \text{ lb/min}$
Tank 2	Outside	$\dfrac{y_2(t)}{10} \text{lb/gal} \cdot 3 \text{ gal/min} = \dfrac{3y_2(t)}{10} \text{ lb/min}$

The data for the balance equations can then be read from the following table:

Tank	Input rate	Output rate
1	$3 + \dfrac{y_2(t)}{10}$	$\dfrac{4y_1(t)}{10}$
2	$\dfrac{4y_1(t)}{10}$	$\dfrac{4y_2(t)}{10}$

Putting these data in the balance equations then gives

$$y_1'(t) = -\frac{4}{10}y_1(t) + \frac{1}{10}y_2(t) + 3,$$

$$y_2'(t) = \frac{4}{10}y_1(t) - \frac{4}{10}y_2(t).$$

These equations thus describe the relationship between y_1 and y_2 and their derivatives. We observe also that the statement of the problem includes initial conditions, namely, $y_1(0) = 2$ and $y_2(0) = 0$.   ◀

These two equations together comprise an example of a (constant coefficient) linear system of ordinary differential equations. Notice that two states are involved: the amount of salt in each tank, $y_1(t)$ and $y_2(t)$. Notice also that y_1' depends not only on y_1 but also y_2 and likewise for y_2'. When such occurs, we say that y_1 and y_2 are coupled. In order to find one, we need the other. This chapter is devoted to developing theory and solution methods for such equations. Before we discuss such methods, let us lay down the salient definitions, notation, and basic facts.

9.2   Linear Systems of Differential Equations

A system of equations of the form

$$y_1'(t) = a_{11}(t)y_1(t) + \cdots + a_{1n}(t)y_n(t) + f_1(t)$$

$$y_2'(t) = a_{21}(t)y_1(t) + \cdots + a_{2n}(t)y_n(t) + f_2(t)$$

$$\vdots \tag{1}$$

$$y_n'(t) = a_{n1}(t)y_1(t) + \cdots + a_{nn}(t)y_n(t) + f_n(t),$$

where $a_{ij}(t)$ and $f_i(t)$ are functions defined on some common interval, is called a *first order linear system of ordinary differential equations* or just *linear differential system*, for short. If

$$y(t) = \begin{bmatrix} y_1(t) \\ \vdots \\ y_n(t) \end{bmatrix}, \quad A(t) = \begin{bmatrix} a_{11}(t) & \cdots & a_{1n}(t) \\ \vdots & \ddots & \vdots \\ a_{n1}(t) & \cdots & a_{nn}(t) \end{bmatrix}, \quad \text{and} \quad f(t) = \begin{bmatrix} f_1(t) \\ \vdots \\ f_n(t) \end{bmatrix}$$

then (1) can be written more succinctly in matrix form as

$$y'(t) = A(t)y(t) + f(t). \tag{2}$$

If $A(t) = A$ is a matrix of constants, the linear differential system is said to be *constant coefficient*. A linear differential system is *homogeneous* if $f = 0$; otherwise it is *nonhomogeneous*. The homogeneous linear differential system obtained from (2) (equivalently (1)) by setting $f = 0$ (equivalently setting each $f_i(t) = 0$), namely,

$$y' = A(t)y, \tag{3}$$

is known as the *associated homogeneous equation* for the system (2).

As was the case for a single differential equation, it is conventional to suppress the independent variable t in the unknown functions $y_i(t)$ and their derivatives $y_i'(t)$. Thus, (1) and (2) would normally be written as

$$y_1' = a_{11}(t)y_1 + \cdots + a_{1n}(t)y_n + f_1(t)$$

$$y_2' = a_{21}(t)y_1 + \cdots + a_{2n}(t)y_n + f_2(t)$$

$$\vdots \tag{4}$$

$$y_n' = a_{n1}(t)y_1 + \cdots + a_{nn}(t)y_n + f_n(t),$$

and

$$y' = A(t)y + f(t), \tag{5}$$

respectively.

Example 1. For each of the following systems, determine which is a linear differential system. If it is, write it in matrix form and determine whether it is homogeneous and whether it is constant coefficient:

1.

$$
\begin{aligned}
y_1' &= (1-t)y_1 + e^t y_2 + (\sin t)y_3 + \quad 1 \\
y_2' &= \qquad 3y_1 + \ ty_2 + (\cos t)y_3 + te^{2t} \\
y_3' &= \qquad\qquad\qquad\qquad\qquad\qquad y_3
\end{aligned}
$$

2.

$$
\begin{aligned}
y_1' &= \ ay_1 - by_1y_2 \\
y_2' &= -cy_1 + dy_1y_2
\end{aligned}
$$

3.

$$
\begin{aligned}
y_1' &= -\frac{4}{10}y_1 + \frac{1}{10}y_2 + 3 \\
y_2' &= \ \ \frac{4}{10}y_1 - \frac{4}{10}y_2
\end{aligned}
$$

▶ **Solution.** 1. This is a linear differential system with

$$
A(t) = \begin{bmatrix} 1-t & e^t & \sin t \\ 3 & t & \cos t \\ 0 & 0 & 1 \end{bmatrix} \quad \text{and} \quad f(t) = \begin{bmatrix} 1 \\ te^{2t} \\ 0 \end{bmatrix}.
$$

Since $f \neq 0$, this system is nonhomogeneous, and since $A(t)$ is not a constant function, this system is not a constant coefficient linear differential system.

2. This system is not a *linear* differential system because of the presence of the products $y_1 y_2$.

3. This is a linear differential system with

$$
A(t) = A = \begin{bmatrix} -\frac{4}{10} & \frac{1}{10} \\ \frac{4}{10} & -\frac{4}{10} \end{bmatrix} \quad \text{and} \quad f = \begin{bmatrix} 3 \\ 0 \end{bmatrix}.
$$

It is constant coefficient but nonhomogeneous. This is the linear differential system we introduced in Example 1 of Sect. 9.1.   ◄

If all the entry functions of $A(t)$ and $f(t)$ are defined on a common interval, I, then a vector function y, defined on I, that satisfies (2) (or, equivalently, (1)) is a *solution*. A solution of the associated homogeneous equation $y' = A(t)y$ of (2) is referred to as a *homogeneous solution* to (2). Note that a homogeneous solution to (2) is not a solution of the given equation $y' = A(t)y + f(t)$, but rather a solution of the related equation $y' = A(t)y$.

Example 2. Consider the following first order system of ordinary differential equations:

$$
\begin{aligned}
y_1' &= 3y_1 - y_2 \\
y_2' &= 4y_1 - 2y_2.
\end{aligned} \tag{6}
$$

Let

$$y(t) = \begin{bmatrix} e^{2t} \\ e^{2t} \end{bmatrix} \quad \text{and} \quad z(t) = \begin{bmatrix} e^{-t} \\ 4e^{-t} \end{bmatrix}.$$

Show that $y(t)$, $z(t)$, and $w(t) = c_1 y(t) + c_2 z(t)$, where c_1 and c_2 are scalars, are solutions to (6).

▶ **Solution.** Let $A = \begin{bmatrix} 3 & -1 \\ 4 & -2 \end{bmatrix}$. Then (6) can be written

$$y' = Ay.$$

This system is a constant coefficient homogeneous linear differential system. For $y(t) = \begin{bmatrix} e^{2t} \\ e^{2t} \end{bmatrix}$, we have on the one hand

$$y'(t) = \begin{bmatrix} 2e^{2t} \\ 2e^{2t} \end{bmatrix}$$

and on the other hand,

$$Ay(t) = \begin{bmatrix} 3 & -1 \\ 4 & -2 \end{bmatrix} \begin{bmatrix} e^{2t} \\ e^{2t} \end{bmatrix} = \begin{bmatrix} 3e^{2t} - e^{2t} \\ 4e^{2t} - 2e^{2t} \end{bmatrix} = \begin{bmatrix} 2e^{2t} \\ 2e^{2t} \end{bmatrix}.$$

It follows that $y' = Ay$, and hence, y is a solution.

For $z(t) = \begin{bmatrix} e^{-t} \\ 4e^{-t} \end{bmatrix}$, we have on the one hand

$$z'(t) = \begin{bmatrix} -e^{-t} \\ -4e^{-t} \end{bmatrix}$$

and on the other hand,

$$Az(t) = \begin{bmatrix} 3 & -1 \\ 4 & -2 \end{bmatrix} \begin{bmatrix} e^{-t} \\ 4e^{-t} \end{bmatrix} = \begin{bmatrix} 3e^{-t} - 4e^{-t} \\ 4e^{-t} - 8e^{-t} \end{bmatrix} = \begin{bmatrix} -e^{-t} \\ -4e^{-t} \end{bmatrix}.$$

Again, it follows that z is a solution.

Suppose c_1 and c_2 are *any* constants and $w(t) = c_1 y(t) + c_2 z(t)$. Since differentiation is linear, we have $w'(t) = c_1 y'(t) + c_2 z'(t)$. Since matrix multiplication is linear, we have

$$Aw(t) = c_1 Ay(t) + c_2 Az(t) = c_1 y'(t) + c_2 z'(t) = w'(t).$$

It follows that $w(t)$ is a solution.   ◀

Example 3. Consider the following first order system of ordinary differential equations:

$$y_1' = 3y_1 - y_2 + 2t$$
$$y_2' = 4y_1 - 2y_2 + 2. \tag{7}$$

1. Verify that $y_p(t) = \begin{bmatrix} -2t + 1 \\ -4t + 5 \end{bmatrix}$ is a solution of (7).

2. Verify that $z_p(t) = 2y_p(t) = \begin{bmatrix} -4t + 2 \\ -8t + 10 \end{bmatrix}$ is *not* a solution to (7).

3. Verify that $y_g(t) = w(t) + y_p(t)$ is a solution of (7), where $w(t)$ is the general solution of (6) from the previous example.

▶ **Solution.** We begin by writing (7) in matrix form as:

$$y' = Ay + f,$$

where $A = \begin{bmatrix} 3 & -1 \\ 4 & -2 \end{bmatrix}$ and $f = \begin{bmatrix} 2t \\ 2 \end{bmatrix}$. Note that the associated homogeneous linear differential system, that is, the equation $y' = Ay$ obtained by setting $f = 0$, is the system from the previous example.

1. On the one hand,

$$y_p'(t) = \begin{bmatrix} -2 \\ -4 \end{bmatrix},$$

and on the other hand,

$$A y_p + f = \begin{bmatrix} 3 & -1 \\ 4 & -2 \end{bmatrix} \begin{bmatrix} -2t + 1 \\ -4t + 5 \end{bmatrix} + \begin{bmatrix} 2t \\ 2 \end{bmatrix} = \begin{bmatrix} 3(-2t + 1) - (-4t + 5) + 2t \\ 4(-2t + 1) - 2(-4t + 5) + 2 \end{bmatrix}$$

$$= \begin{bmatrix} -2 \\ -4 \end{bmatrix}.$$

Since $y_p' = A y_p + f$, it follows that y_p is a solution to (7).

2. On the one hand,

$$z_p' = \begin{bmatrix} -4 \\ -8 \end{bmatrix},$$

and on the other hand,

$$A z_p + f = \begin{bmatrix} 3 & -1 \\ 4 & -2 \end{bmatrix} \begin{bmatrix} -4t + 2 \\ -8t + 10 \end{bmatrix} + \begin{bmatrix} 2t \\ 2 \end{bmatrix}$$

$$= \begin{bmatrix} 3(-4t+2) - (-8t+10) + 2t \\ 4(-4t+2) - 2(-8t+10) + 2 \end{bmatrix}$$

$$= \begin{bmatrix} -2t - 4 \\ -10 \end{bmatrix}.$$

Since $z'_p \neq Az_p + f$, z_p is *not* a solution.

3. Since $y' = Ay$ is the homogeneous linear differential system associated to (7), we know from the previous example that w is a solution to the homogeneous equation $y' = Ay$. Since differentiation is linear, we have

$$\begin{aligned} y'_g &= w' + y'_p \\ &= Aw + Ay_p + f \\ &= A(w + y_p) + f \\ &= Ay_g + f. \end{aligned}$$

It follows that y_g is a solution to (7). ◀

These two examples illustrate the power of linearity, a concept that we have repeatedly encountered, and are suggestive of the following general statement.

Theorem 4. *Consider the linear differential system*

$$y' = A(t)y + f(t). \tag{8}$$

1. If y_1 and y_2 are solutions to the associated homogeneous linear differential system

$$y' = A(t)y, \tag{9}$$

and c_1 and c_2 are constants, then $c_1 y_1 + c_2 y_2$ is also a solution to (9).

2. If y_p is a fixed particular solution to (8) and y_h is any homogeneous solution (i.e., any solution to (9)), then

$$y_h + y_p$$

is also a solution to (8) and all solutions to (8) are of this form.

Proof. Let $L = D - A(t)$ be the operator on vector-valued functions given by

$$Ly = (D - A(t))y = y' - A(t)y.$$

Then L is linear since differentiation and matrix multiplication are linear. The rest of the proof follows the same line of argument given in the proof of Theorem 6 of Sect. 3.1. □

Linear differential systems also satisfy the superposition principle, an analogue to Theorem 8 of Sect. 3.4.

Theorem 5. *Suppose* y_{p_1} *is a solution to* $y' = A(t)y + f_1(t)$ *and* y_{p_2} *is a solution to* $y' = A(t)y + f_2(t)$. *Then* $y = y_{p_1}(t) + y_{p_2}(t)$ *is a solution to* $y' = A(t)y + f_1(t) + f_2(t)$.

Proof. Let $L = D - A(t)$ be as above. Then linearity implies

$$L(y_{p_1} + y_{p_2}) = Ly_{p_1} + Ly_{p_2} = f_1(t) + f_2(t).$$ □

Initial Value Problems

For a linear differential system $y' = A(t)y + f(t)$, assume the entries of $A(t)$ and $f(t)$ are defined on a common interval I. Let $t_0 \in I$. When we associate to a linear differential system an initial value $y(t_0) = y_0$, we call the resulting problem an *initial value problem*. The mixing problem we discussed in the introduction to this chapter is an example of an initial value problem:

$$y' = Ay + f, \quad y(0) = y_0,$$

where $A = \begin{bmatrix} -\frac{4}{10} & \frac{1}{10} \\ \frac{4}{10} & -\frac{4}{10} \end{bmatrix}$, $f(t) = \begin{bmatrix} 3 \\ 0 \end{bmatrix}$, and $y(0) = \begin{bmatrix} 2 \\ 0 \end{bmatrix}$.

Linear Differential Equations and Systems

For each linear *ordinary* differential equation $Ly = f$ with initial conditions, we can construct a corresponding linear system with initial condition. The solution of one will imply the solution of the other. The following example will illustrate the procedure.

Example 6. Construct a first order linear differential system with initial value from the second order differential equation

$$y'' + ty' + y = \sin t \quad y(0) = 1, \ y'(0) = 2.$$

▶ **Solution.** If $y(t)$ is a solution to the second order equation, form a vector function $y(t) = \begin{bmatrix} y_1(t) \\ y_2(t) \end{bmatrix}$ by setting $y_1(t) = y(t)$ and $y_2(t) = y'(t)$. Then

$$y_1'(t) = y'(t) = y_2(t),$$

$$y_2'(t) = y''(t) = -y(t) - ty'(t) + \sin t = -y_1(t) - ty_2(t) + \sin t.$$

The second equation is obtained by solving $y'' + ty' + y = \sin t$ for y'' and substituting y_1 for y and y_2 for y'. In vector form, this equation becomes

$$y' = A(t)y + f,$$

where $A(t) = \begin{bmatrix} 0 & 1 \\ -1 & -t \end{bmatrix}$, $f(t) = \begin{bmatrix} 0 \\ \sin t \end{bmatrix}$, and $y(0) = \begin{bmatrix} 1 \\ 2 \end{bmatrix}$.   ◄

The solution to the first order differential system implies a solution to the original second order differential equation and vice versa. Specifically, if $y = \begin{bmatrix} y_1 \\ y_2 \end{bmatrix}$ is a solution of the system, then the first entry of y, namely, y_1, is a solution of the second order equation, and conversely, as illustrated in the above example, if y is a solution of the second order equation, then $y = \begin{bmatrix} y \\ y' \end{bmatrix}$ is a solution of the linear system.

Linear differential equations of order n are transformed into linear systems in a similar way.

Extension of Basic Definitions and Operations to Matrix-Valued Functions

It is convenient to state most of our results on linear systems of ordinary differential equations in the language of matrices. To this end, we extend several definitions familiar for real-valued functions to matrix-(or vector-)valued functions. Let $v(t)$ be an $n \times m$ matrix-valued function with entries $v_{i,j}(t)$, for $i = 1 \ldots n$ and $j = 1 \ldots m$.

1. $v(t)$ is **defined** on an interval I of $\mathbb{R}$ if each $v_{ij}(t)$ is defined on I.
2. $v(t)$ is **continuous** on an interval I of $\mathbb{R}$ if each $v_{ij}(t)$ is continuous on I. For instance, the matrix

$$v(t) = \begin{bmatrix} \dfrac{1}{t+2} & \cos 2t \\ e^{-2t} & \dfrac{1}{(2t-3)^2} \end{bmatrix}$$

is continuous on each of the intervals $I_1 = (-\infty, -2)$, $I_2 = (-2, 3/2)$ and $I_3 = (3/2, \infty)$, but it is not continuous on the interval $I_4 = (0, 2)$.
3. $v(t)$ is **differentiable** on an interval I of $\mathbb{R}$ if each $v_{ij}(t)$ is differentiable on I. Moreover, $v'(t) = [a'_{ij}(t)]$. That is, the matrix $v(t)$ is differentiated by differentiating each entry of the matrix. For instance, for the matrix

$$v(t) = \begin{bmatrix} e^t & \sin t & t^2 + 1 \\ \ln t & \cos t & \sinh t \end{bmatrix},$$

we have

$$v'(t) = \begin{bmatrix} e^t & \cos t & 2t \\ 1/t & -\sin t & \cosh t \end{bmatrix}.$$

4. An *antiderivative* of $v(t)$ is a matrix-valued function $V(t)$ (necessarily of the same size) such that $V'(t) = v(t)$. Since the derivative is calculated entry by entry, so likewise is the antiderivative. Thus, if

$$v(t) = \begin{bmatrix} e^{4t} & \sin t \\ 2t & \ln t \\ \cos 2t & 5 \end{bmatrix},$$

then an antiderivative is

$$V(t) = \int v(t)dt = \begin{bmatrix} \frac{1}{4}e^{4t} + c_{11} & -\cos t + c_{12} \\ t^2 + c_{21} & t\ln t - t + c_{22} \\ \frac{1}{2}\sin 2t + c_{31} & 5t + c_{32} \end{bmatrix}$$

$$= \begin{bmatrix} \frac{1}{4}e^{4t} & -\cos t \\ t^2 & t\ln t - t \\ \frac{1}{2}\sin 2t & 5t \end{bmatrix} + C,$$

where C is the matrix of constants $[c_{ij}]$. Thus, if $v(t)$ is defined on an interval I and $V_1(t)$ and $V_2(t)$ are two antiderivatives of $v(t)$, then they differ by a constant matrix.

5. The integral of $v(t)$ on the interval $[a, b]$ is computed by computing the integral of each entry of the matrix over $[a, b]$, that is, $\int_a^b v(t)\, dt = \left[\int_a^b v_{ij}(t)\, dt\right]$. For the matrix $v(t)$ of item 2 above, this gives

$$\int_0^1 v(t)\, dt = \begin{bmatrix} \int_0^1 \frac{1}{t+2}\, dt & \int_0^1 \cos 2t\, dt \\ \int_0^1 e^{-2t}\, dt & \int_0^1 \frac{1}{(2t-3)^2}\, dt \end{bmatrix} = \begin{bmatrix} \ln\frac{3}{2} & \frac{1}{2}\sin 2 \\ \frac{1}{2}(1-e^{-2}) & \frac{1}{3} \end{bmatrix}.$$

6. If each entry $v_{ij}(t)$ of $v(t)$ is of exponential type (recall the definition on page 111), we can take the *Laplace transform of* $v(t)$, by taking the Laplace transform of each entry. That is, $\mathcal{L}(v(t))(s) = \left[\mathcal{L}(v_{ij}(t))(s)\right]$. For example, if $v(t) = \begin{bmatrix} te^{-2t} & \cos 2t \\ e^{3t}\sin t & (2t-3)^2 \end{bmatrix}$, this gives

$$\mathcal{L}(v(t))(s) = \begin{bmatrix} \mathcal{L}\left(te^{-2t}\right)(s) & \mathcal{L}(\cos 2t)(s) \\ \mathcal{L}(e^{3t}\sin t)(s) & \mathcal{L}\left((2t-3)^2\right)(s) \end{bmatrix} = \begin{bmatrix} \frac{1}{(s+2)^2} & \frac{2}{s^2+4} \\ \frac{1}{(s-3)^2+1} & \frac{9s^2-12s+8}{s^3} \end{bmatrix}.$$

7. We define the *inverse Laplace transform* entry by entry as well. For example, if

$$V(s) = \begin{bmatrix} \frac{1}{s-1} & \frac{1}{(s-1)^2} \\ \frac{s}{s^2+4} & \frac{2}{(s-1)^2+4} \end{bmatrix},$$

then

$$\mathcal{L}^{-1}\{V(s)\}(t) = \begin{bmatrix} e^t & te^t \\ \cos 2t & e^t \sin 2t \end{bmatrix}.$$

8. Finally, we extend convolution to matrix products. Suppose $v(t)$ and $w(t)$ are matrix-valued functions such that the usual matrix product $v(t)w(t)$ is defined. We define the convolution $v(t) * w(t)$ as follows:

$$(v(t) * w(t))_{i,j} = \sum_k v_{i,k} * w_{k,j}(t).$$

Thus, in the matrix product, we replace each product of terms by the convolution product. For example, if

$$v(t) = \begin{bmatrix} e^t & e^{2t} \\ -e^{2t} & 2e^t \end{bmatrix} \quad \text{and} \quad w(t) = \begin{bmatrix} 3e^t \\ -e^t \end{bmatrix},$$

then

$$v * w(t) = \begin{bmatrix} e^t & e^{2t} \\ -e^{2t} & 2e^t \end{bmatrix} * \begin{bmatrix} 3e^t \\ -e^t \end{bmatrix}$$

$$= \begin{bmatrix} 3e^t * e^t - e^{2t} * e^t \\ -3e^{2t} * e^t - 2e^t * e^t \end{bmatrix}$$

$$= \begin{bmatrix} 3te^t - (e^{2t} - e^t) \\ -3(e^{2t} - e^t) - 2te^t \end{bmatrix}$$

$$= te^t \begin{bmatrix} 3 \\ -2 \end{bmatrix} + (e^{2t} - e^t) \begin{bmatrix} -1 \\ -3 \end{bmatrix}.$$

Alternately, if we write

$$v(t) = e^t \begin{bmatrix} 1 & 0 \\ 0 & 2 \end{bmatrix} + e^{2t} \begin{bmatrix} 0 & 1 \\ -1 & 0 \end{bmatrix} \quad \text{and} \quad w(t) = e^t \begin{bmatrix} 3 \\ -1 \end{bmatrix},$$

then the preceding calculations can be performed as follows:

$$v * w(t) = \left(e^t \begin{bmatrix} 1 & 0 \\ 0 & 2 \end{bmatrix} + e^{2t} \begin{bmatrix} 0 & 1 \\ -1 & 0 \end{bmatrix} \right) * e^t \begin{bmatrix} 3 \\ -1 \end{bmatrix}$$

$$= e^t * e^t \begin{bmatrix} 1 & 0 \\ 0 & 2 \end{bmatrix} \begin{bmatrix} 3 \\ -1 \end{bmatrix} + e^{2t} * e^t \begin{bmatrix} 0 & 1 \\ -1 & 0 \end{bmatrix} \begin{bmatrix} 3 \\ -1 \end{bmatrix}$$

$$= t e^t \begin{bmatrix} 3 \\ -2 \end{bmatrix} + (e^{2t} - e^t) \begin{bmatrix} -1 \\ -3 \end{bmatrix}.$$

The following theorem extends basic operations of calculus and Laplace transforms to matrix-valued functions.

Theorem 7. *Assume $v(t)$ and $w(t)$ are matrix-valued functions, $g(t)$ is a real-valued function, and A is a matrix of constants.*

1. *Suppose v is differentiable and the product $Av(t)$ is defined. Then $Av(t)$ is differentiable and*

$$(Av(t))' = Av'(t).$$

2. *Suppose v is differentiable and the product $v(t)A$ is defined. Then $v(t)A$ is differentiable and*

$$(v(t)A)' = v'(t)A.$$

3. *Suppose the product, $v(t)w(t)$, is defined and v and w are both differentiable. Then*

$$(v(t)w(t))' = v'(t)w(t) + v(t)w'(t).$$

4. *Suppose the composition $v(g(t))$ is defined. Then $(v(g(t)))' = v'(g(t))g'(t)$.*
5. *Suppose v is integrable over the interval $[a, b]$ and the product $Av(t)$ is defined. Then $Av(t)$ is integrable over the interval $[a, b]$ and*

$$\int_a^b Av(t)\, dt = A \int_a^b v(t)\, dt.$$

6. *Suppose v is integrable over the interval $[a, b]$ and the product $v(t)A$ is defined. The $v(t)A$ is integrable over the interval $[a, b]$ and*

$$\int_a^b v(t)A\, dt = \left(\int_a^b v(t)\, dt \right) A.$$

7. *Suppose $v(t)$ is defined on $[0, \infty)$, has a Laplace transform, and the product $Av(t)$ is defined. Then $Av(t)$ has a Laplace transform and*

$$\mathcal{L}\{Av(t)\}(s) = A\mathcal{L}\{v(t)\}(s).$$

8. *Suppose $v(t)$ is defined on $[0, \infty)$, has a Laplace transform, and the product $v(t)A$ is defined. Then $v(t)A$ has a Laplace transform and*

$$\mathcal{L}\{v(t)A\}(s) = (\mathcal{L}\{v(t)\}(s))A.$$

9. *Suppose $v(t)$ is defined on $[0, \infty)$ and $v'(t)$ exists and has a Laplace transform. Then*

$$\mathcal{L}\{v'(t)\}(s) = s\mathcal{L}\{v(t)\}(s) - v(0).$$

10. *The convolution theorem extends as well: Suppose $v(t)$ and $w(t)$ have Laplace transforms and $v * w(t)$ is defined. Then*

$$\mathcal{L}\{v * w(t)\}(s) = \mathcal{L}\{v(t)\}(s) \cdot \mathcal{L}\{w(t)\}(s).$$

Remark 8. Where matrix multiplication is involved in these formulas it is important to preserve the order of the multiplication. It is particularly worth emphasizing this dependency on the order of multiplication in the product rule for the derivative of the product of matrix-valued functions (formula (3) above). Also note that formula (4) is just the chain rule in the context of matrix-valued functions.

Exercises

1-6. For each of the following systems of differential equations, determine if it is linear. For each of those which are linear, write it in matrix form; determine if the equation is (1) homogeneous or nonhomogeneous and (2) constant coefficient. Do *not* try to solve the equations.

1. $\begin{aligned} y_1' &= y_2 \\ y_2' &= y_1 y_2 \end{aligned}$

2. $\begin{aligned} y_1' &= y_1 + y_2 + t^2 \\ y_2' &= -y_1 + y_2 + 1 \end{aligned}$

3. $\begin{aligned} y_1' &= (\sin t) y_1 \\ y_2' &= y_1 + (\cos t) y_2 \end{aligned}$

4. $\begin{aligned} y_1' &= t \sin y_1 - y_2 \\ y_2' &= y_1 + t \cos y_2 \end{aligned}$

5. $\begin{aligned} y_1' &= y_1 \\ y_2' &= 2y_1 + y_4 \\ y_3' &= y_4 \\ y_4' &= y_2 + 2y_3 \end{aligned}$

6. $\begin{aligned} y_1' &= \frac{1}{2} y_1 - y_2 + 5 \\ y_2' &= -y_1 + \frac{1}{2} y_2 - 5 \end{aligned}$

7–10. Verify that the given vector function $y(t) = \begin{bmatrix} y_1(t) \\ y_2(t) \end{bmatrix}$ is a solution to the given linear differential system with the given initial value.

7.
$$y' = \begin{bmatrix} 5 & -2 \\ 4 & -1 \end{bmatrix} y; \quad y(0) = \begin{bmatrix} 0 \\ 1 \end{bmatrix}, \quad y(t) = \begin{bmatrix} e^t - e^{3t} \\ 2e^t - e^{3t} \end{bmatrix}.$$

8.
$$y' = \begin{bmatrix} 3 & -1 \\ 4 & -1 \end{bmatrix} y; \quad y(0) = \begin{bmatrix} 1 \\ 0 \end{bmatrix}, \quad y(t) = \begin{bmatrix} e^t + 2te^t \\ 4te^t \end{bmatrix}.$$

9.
$$y' = \begin{bmatrix} 2 & -1 \\ 3 & -2 \end{bmatrix} y + \begin{bmatrix} e^t \\ e^t \end{bmatrix}; \quad y(0) = \begin{bmatrix} 1 \\ 3 \end{bmatrix}, \quad y(t) = \begin{bmatrix} e^{-t} + te^t \\ 3e^{-t} + te^t \end{bmatrix}.$$

10.

$$y = \begin{bmatrix} 0 & 1 \\ -1 & 0 \end{bmatrix} y + \begin{bmatrix} t \\ -t \end{bmatrix}, \quad y(0) = \begin{bmatrix} 1 \\ 1 \end{bmatrix}, \quad y(t) = \begin{bmatrix} 1 - t + 2\sin t \\ -1 - t + 2\cos t \end{bmatrix}.$$

11–15. Rewrite each of the following initial value problems for an ordinary differential equation as an initial value problem for a first order system of ordinary differential equations.

11. $y'' + 5y' + 6y = e^{2t}$, $y(0) = 1, y'(0) = -2$.
12. $y'' + k^2 y = 0$, $y(0) = -1, y'(0) = 0$
13. $y'' - k^2 y = A\cos\omega t$, $y(0) = 0, y'(0) = 0$
14. $ay'' + by' + cy = 0$, $y(0) = \alpha, y'(0) = \beta$
15. $t^2 y'' + 2ty' + y = 0$, $y(1) = -2, y'(1) = 3$

16–21. Compute the derivative of each of the following matrix functions.

16. $A(t) = \begin{bmatrix} \cos 2t & \sin 2t \\ -\sin 2t & \cos 2t \end{bmatrix}$

17. $A(t) = \begin{bmatrix} e^{-3t} & t \\ t^2 & e^{2t} \end{bmatrix}$

18. $A(t) = \begin{bmatrix} e^{-t} & te^{-t} & t^2 e^{-t} \\ 0 & e^{-t} & te^{-t} \\ 0 & 0 & e^{-t} \end{bmatrix}$

19. $y(t) = \begin{bmatrix} t \\ t^2 \\ \ln t \end{bmatrix}$

20. $A(t) = \begin{bmatrix} 1 & 2 \\ 3 & 4 \end{bmatrix}$

21. $v(t) = \begin{bmatrix} e^{-2t} & \ln(t^2 + 1) & \cos 3t \end{bmatrix}$

22–25. For each of the following matrix functions, compute the requested integral.

22. Compute $\int_0^{\frac{\pi}{2}} A(t)\, dt$ if $A(t) = \begin{bmatrix} \cos 2t & \sin 2t \\ -\sin 2t & \cos 2t \end{bmatrix}$.

23. Compute $\int_0^1 A(t)\, dt$ if $A(t) = \dfrac{1}{2} \begin{bmatrix} e^{2t} + e^{-2t} & e^{2t} - e^{-2t} \\ e^{-2t} - e^{2t} & e^{2t} + e^{-2t} \end{bmatrix}$.

24. Compute $\int_1^2 y(t)\,dt$ if $y(t) = \begin{bmatrix} t \\ t^2 \\ \ln t \end{bmatrix}$.

25. Compute $\int_1^5 A(t)\,dt$ if $A(t) = \begin{bmatrix} 1 & 2 \\ 3 & 4 \end{bmatrix}$.

26. On which of the following intervals is the matrix function $A(t) = \begin{bmatrix} t & (t+1)^{-1} \\ (t-1)^{-2} & t+6 \end{bmatrix}$ continuous?

    (a) $I_1 = (-1,1)$ (b) $I_2 = (0,\infty)$ (c) $I_3 = (-1,\infty)$
    (d) $I_4 = (-\infty,-1)$ (e) $I_5 = (2,6)$

27–32. Compute the Laplace transform of each of the following matrix functions.

27. $A(t) = \begin{bmatrix} 1 & t \\ t^2 & e^{2t} \end{bmatrix}$

28. $A(t) = \begin{bmatrix} \cos t & \sin t \\ -\sin t & \cos t \end{bmatrix}$

29. $A(t) = \begin{bmatrix} t^3 & t\sin t & te^{-t} \\ t^2 - t & e^{3t}\cos 2t & 3 \end{bmatrix}$

30. $A(t) = \begin{bmatrix} t \\ t^2 \\ t^3 \end{bmatrix}$

31. $A(t) = e^t \begin{bmatrix} 1 & -1 \\ -1 & 1 \end{bmatrix} + e^{-t} \begin{bmatrix} -1 & 1 \\ 1 & -1 \end{bmatrix}$

32. $A(t) = \begin{bmatrix} 1 & \sin t & 1-\cos t \\ 0 & \cos t & \sin t \\ 0 & -\sin t & \cos t \end{bmatrix}$

33–36. Compute the inverse Laplace transform of each matrix function:

33. $\begin{bmatrix} \dfrac{1}{s} & \dfrac{2}{s^2} & \dfrac{6}{s^3} \end{bmatrix}$

34. $\begin{bmatrix} \dfrac{1}{s} & \dfrac{1}{s^2} \\ \dfrac{s}{s^2-1} & \dfrac{s}{s^2+1} \end{bmatrix}$

35. $\begin{bmatrix} \dfrac{2s}{s^2 - 1} & \dfrac{2}{s^2 - 1} \\ \dfrac{2}{s^2 - 1} & \dfrac{2s}{s^2 - 1} \end{bmatrix}$.

36. $\begin{bmatrix} \dfrac{1}{s - 1} & \dfrac{1}{s^2 - 2s + 1} \\ \dfrac{4}{s^3 + 2s^2 - 3s} & \dfrac{1}{s^2 + 1} \\ \dfrac{3s}{s^2 + 9} & \dfrac{1}{s - 3} \end{bmatrix}$

9.3   The Matrix Exponential and Its Laplace Transform

One of the most basic Laplace transforms that we learned early on was that of the exponential function:

$$\mathcal{L}\{e^{at}\} = \frac{1}{s-a}. \tag{1}$$

This basic formula has proved to be a powerful tool for solving constant coefficient linear differential equations of order n. Our goal here is to extend (1) to the case where the constant a is replaced by an $n \times n$ matrix A. The resulting extension will prove to be an equally powerful tool for solving linear systems of differential equations with constant coefficients.

Let A be an $n \times n$ matrix of scalars. Formally, we define the *matrix exponential*, e^{At}, by the formula

$$e^{At} = I + At + \frac{A^2 t^2}{2!} + \frac{A^3 t^3}{3!} + \cdots \tag{2}$$

When A is a scalar, this definition is the usual power series expansion of the exponential function. Equation (2) is an infinite sum of $n \times n$ matrices: The first term I is the $n \times n$ identity matrix, the second term is the $n \times n$ matrix At, the third term is the $n \times n$ matrix $\frac{A^2 t^2}{2!}$, and so forth. To compute the sum, one must compute each (i, j) entry and add the corresponding terms. Thus, the (i, j) entry of e^{At} is

$$(e^{At})_{i,j} = (I)_{i,j} + t(A)_{i,j} + \frac{t^2}{2!}(A^2)_{i,j} + \cdots, \tag{3}$$

which is a power series centered at the origin. To determine this sum, one must be able to calculate the (i, j) entry of all the powers of A. This is easy enough for $I = A^0$ and $A = A^1$. For the (i, j) entry of A^2, we get the ith row of A times the jth column of A. For the higher powers, A^3, A^4, etc., the computations become more complicated and the resulting power series is difficult to identify, unless A is very simple. However, one can see that each entry is some power series in the variable t and thus defines a function (if it converges). In Appendix A.4, we show that the series in (3) converges absolutely for all $t \in \mathbb{R}$ and for all matrices A so that the matrix exponential is a well-defined matrix function defined on $\mathbb{R}$. However, knowing that the series converges is a far cry from knowing the sum of the series.

In the following examples, A is simple enough to allow the computation of e^{At}.

Example 1.   Let $A = \begin{bmatrix} 2 & 0 & 0 \\ 0 & 3 & 0 \\ 0 & 0 & -1 \end{bmatrix}$. Compute e^{At}.

▶ **Solution.** In this case, the powers of the matrix A are easy to compute. In fact

$$A = \begin{bmatrix} 2 & 0 & 0 \\ 0 & 3 & 0 \\ 0 & 0 & -1 \end{bmatrix}, \quad A^2 = \begin{bmatrix} 4 & 0 & 0 \\ 0 & 9 & 0 \\ 0 & 0 & 1 \end{bmatrix}, \quad \cdots, \quad A^n = \begin{bmatrix} 2^n & 0 & 0 \\ 0 & 3^n & 0 \\ 0 & 0 & (-1)^n \end{bmatrix},$$

so that

$$e^{At} = I + At + \frac{1}{2}A^2 t^2 + \frac{1}{3!}A^3 t^3 + \cdots$$

$$= \begin{bmatrix} 1 & 0 & 0 \\ 0 & 1 & 0 \\ 0 & 0 & 1 \end{bmatrix} + \begin{bmatrix} 2 & 0 & 0 \\ 0 & 3 & 0 \\ 0 & 0 & -1 \end{bmatrix} t + \frac{1}{2!} \begin{bmatrix} 4 & 0 & 0 \\ 0 & 9 & 0 \\ 0 & 0 & 1 \end{bmatrix} t^2 + \frac{1}{3!} \begin{bmatrix} 8 & 0 & 0 \\ 0 & 27 & 0 \\ 0 & 0 & -1 \end{bmatrix} t^3 + \cdots$$

$$= \begin{bmatrix} 1 + 2t + \frac{4t^2}{2} + \cdots & 0 & 0 \\ 0 & 1 + 3t + \frac{9t^2}{2} + \cdots & 0 \\ 0 & 0 & 1 - t + \frac{t^2}{2} + \cdots \end{bmatrix}$$

$$= \begin{bmatrix} e^{2t} & 0 & 0 \\ 0 & e^{3t} & 0 \\ 0 & 0 & e^{-t} \end{bmatrix}. \qquad \blacktriangleleft$$

Example 2. Let $A = \begin{bmatrix} 0 & 1 \\ 0 & 0 \end{bmatrix}$. Compute e^{At}.

▶ **Solution.** In this case, $A^2 = \begin{bmatrix} 0 & 1 \\ 0 & 0 \end{bmatrix} \begin{bmatrix} 0 & 1 \\ 0 & 0 \end{bmatrix} = \begin{bmatrix} 0 & 0 \\ 0 & 0 \end{bmatrix} = \mathbf{0}$ and $A^n = \mathbf{0}$ for all $n \geq 2$. Hence,

$$e^{At} = I + At + \frac{1}{2}A^2 t^2 + \frac{1}{3!}A^3 t^3 + \cdots$$

$$= I + At$$

$$= \begin{bmatrix} 1 & t \\ 0 & 1 \end{bmatrix}.$$

Note that in this case, the individual entries of e^{At} are not exponential functions.   ◀

Example 3. Let $A = \begin{bmatrix} 0 & 1 \\ -1 & 0 \end{bmatrix}$. Compute e^{At}.

▶ **Solution.** The first few powers of A are $A^2 = \begin{bmatrix} -1 & 0 \\ 0 & -1 \end{bmatrix}$, $A^3 = \begin{bmatrix} 0 & -1 \\ 1 & 0 \end{bmatrix}$,

$A^4 = \begin{bmatrix} 1 & 0 \\ 0 & 1 \end{bmatrix} = I_2$, $A^5 = A$, $A^6 = A^2$, etc. That is, the powers repeat with period 4. Thus,

$$
e^{At} = I + At + \frac{1}{2}A^2t^2 + \frac{1}{3!}A^3t^3 + \cdots
$$

$$
= \begin{bmatrix} 1 & 0 \\ 0 & 1 \end{bmatrix} + \begin{bmatrix} 0 & t \\ -t & 0 \end{bmatrix} + \frac{1}{2}\begin{bmatrix} -t^2 & 0 \\ 0 & -t^2 \end{bmatrix} + \frac{1}{3!}\begin{bmatrix} 0 & -t^3 \\ t^3 & 0 \end{bmatrix} + \frac{1}{4!}\begin{bmatrix} t^4 & 0 \\ 0 & t^4 \end{bmatrix} + \cdots
$$

$$
= \begin{bmatrix} 1 - \frac{1}{2}t^2 + \frac{1}{4!}t^4 + \cdots & t - \frac{1}{3!}t^3 + \frac{1}{5!}t^5 + \cdots \\ -t + \frac{1}{3!}t^3 - \frac{1}{5!}t^5 + \cdots & 1 - \frac{1}{2}t^2 + \frac{1}{4!}t^4 + \cdots \end{bmatrix}
$$

$$
= \begin{bmatrix} \cos t & \sin t \\ -\sin t & \cos t \end{bmatrix}.
$$

(cf. (3) and (4) of Sect. 7.1). In this example also, the individual entries of e^{At} are not themselves exponential functions. ◀

Do not let Examples 1–3 fool you. Unless A is very special, it is difficult to directly determine the entries of A^n and use this to compute e^{At}. In the following subsection, we will compute the Laplace transform of the matrix exponential function. The resulting inversion formula provides an effective method for explicitly computing e^{At}.

The Laplace Transform of the Matrix Exponential

Let A be an $n \times n$ matrix of scalars. As discussed above, each entry of e^{At} converges absolutely on $\mathbb{R}$. From a standard theorem in calculus, we have that e^{At} is differentiable and the derivative can be computed termwise. Thus,

$$
\frac{d}{dt}e^{At} = \frac{d}{dt}\left(I + At + \frac{A^2t^2}{2!} + \frac{A^3t^3}{3!} + \cdots\right)
$$

$$
= A + \frac{A^2t}{1!} + \frac{A^3t^2}{2!} + \frac{A^4t^3}{3!} + \cdots
$$

$$
= A\left(I + At + \frac{A^2t^2}{2!} + \frac{A^3t^3}{3!} + \cdots\right)
$$

$$
= Ae^{At}.
$$

By factoring A out on the right-hand side in the second line, we also get

$$\frac{d}{dt}e^{At} = e^{At}A.$$

Appendix A.4 shows that each entry is of exponential type and thus has a Laplace transform. Now apply the derivative formula

$$\frac{d}{dt}e^{At} = Ae^{At},$$

and the first derivative principle for the Laplace transform of matrix-valued functions (Theorem 7 of Sect. 9.2) applied to $v(t) = e^{At}$, to get

$$A\mathcal{L}\left\{e^{At}\right\} = \mathcal{L}\left\{Ae^{At}\right\} = \mathcal{L}\left\{\frac{d}{dt}e^{At}\right\} = s\mathcal{L}\left\{e^{At}\right\} - I,$$

where we have used $v(0) = e^{At}|_{t=0} = I$. Combining terms gives

$$(sI - A)\mathcal{L}\left\{e^{At}\right\} = I$$

and thus,

$$\mathcal{L}\left\{e^{At}\right\} = (sI - A)^{-1}.$$

This is the extension of (1) mentioned above. We summarize this discussion in the following theorem.

Theorem 4. *Let A be an $n \times n$ matrix. Then e^{At} is a well-defined matrix-valued function and*

$$\mathcal{L}\left\{e^{At}\right\} = (sI - A)^{-1}. \tag{4}$$

The Laplace inversion formula is given by

$$e^{At} = \mathcal{L}^{-1}\left\{(sI - A)^{-1}\right\}. \tag{5}$$

The matrix $(sI - A)^{-1}$ is called the **resolvent matrix** of A. It is a function of s, defined for all s for which the inverse exists. Let $c_A(s) = \det(sI - A)$. It is not hard to see that $c_A(s)$ is a polynomial of degree n. We call $c_A(s)$ the **characteristic polynomial of** A. As a polynomial of degree n, it has at most n roots. The roots are called **characteristic values** or **eigenvalues of** A. Thus, if s is larger than the absolute value of all the eigenvalues of A then $sI - A$ is invertible and the resolvent matrix is defined. By the adjoint formula for matrix inversion, Corollary 11 of Sect. 8.4 each entry of $(sI - A)^{-1}$ is the quotient of a cofactor of $sI - A$ and the characteristic polynomial $c_A(s)$, hence, a proper rational function. Thus, e^{At} is a matrix of exponential polynomials. The Laplace inversion formula given in Theorem 4 now provides a method to compute explicitly the matrix exponential without appealing to the power series expansion given by (2). It will frequently involve partial fraction decompositions of each entry of the resolvent matrix. Consider the following examples.

Example 5. Let $A = \begin{bmatrix} 3 & 4 \\ -2 & -3 \end{bmatrix}$. Compute the resolvent matrix $(sI - A)^{-1}$ and use the Laplace inversion formula to compute e^{At}.

▶ **Solution.** The characteristic polynomial is

$$c_A(s) = \det(sI - A)$$

$$= \det \begin{bmatrix} s - 3 & -4 \\ 2 & s + 3 \end{bmatrix}$$

$$= (s - 3)(s + 3) + 8$$

$$= s^2 - 1 = (s - 1)(s + 1).$$

The adjoint formula for the inverse thus gives

$$(sI - A)^{-1} = \frac{1}{(s - 1)(s + 1)} \begin{bmatrix} s + 3 & 4 \\ -2 & s - 3 \end{bmatrix}$$

$$= \begin{bmatrix} \dfrac{s + 3}{(s - 1)(s + 1)} & \dfrac{4}{(s - 1)(s + 1)} \\ \dfrac{-2}{(s - 1)(s + 1)} & \dfrac{s - 3}{(s - 1)(s + 1)} \end{bmatrix}$$

$$= \frac{1}{s - 1} \begin{bmatrix} 2 & 2 \\ -1 & -1 \end{bmatrix} + \frac{1}{s + 1} \begin{bmatrix} -1 & -2 \\ 1 & 2 \end{bmatrix},$$

where the third line is obtained by computing partial fractions of each entry in the second line.

To compute the matrix exponential, we use the Laplace inversion formula from Theorem 4 to get

$$e^{At} = \mathcal{L}^{-1} \left\{ (sI - A)^{-1} \right\}$$

$$= e^t \begin{bmatrix} 2 & 2 \\ -1 & -1 \end{bmatrix} + e^{-t} \begin{bmatrix} -1 & -2 \\ 1 & 2 \end{bmatrix}$$

$$= \begin{bmatrix} 2e^t - e^{-t} & 2e^t - 2e^{-t} \\ -e^t + e^{-t} & -e^t + 2e^{-t} \end{bmatrix}. \qquad ◀$$

As a second example, we reconsider Example 3.

Example 6. Let $A = \begin{bmatrix} 0 & 1 \\ -1 & 0 \end{bmatrix}$. Compute the resolvent matrix $(sI - A)^{-1}$ and use the Laplace inversion formula to compute e^{At}.

▶ **Solution.** The characteristic polynomial is

$$c_A(s) = \det(sI - A)$$

$$= \det \begin{bmatrix} s & -1 \\ 1 & s \end{bmatrix}$$

$$= s^2 + 1.$$

The adjoint formula for the inverse thus gives

$$(sI - A)^{-1} = \frac{1}{s^2 + 1} \begin{bmatrix} s & 1 \\ -1 & s \end{bmatrix}$$

$$= \begin{bmatrix} \dfrac{s}{s^2 + 1} & \dfrac{1}{s^2 + 1} \\ \dfrac{-1}{s^2 + 1} & \dfrac{s}{s^2 + 1} \end{bmatrix}.$$

Using the inversion formula, we get

$$e^{At} = \mathcal{L}^{-1} \left\{ (sI - A)^{-1} \right\}$$

$$= \begin{bmatrix} \cos t & \sin t \\ -\sin t & \cos t \end{bmatrix}, \qquad \blacktriangleleft$$

Theorem 4 thus gives an effective method for computing the matrix exponential. There are many other techniques. In the next section, we discuss a useful alternative that circumvents the need to compute partial fraction decompositions.

Exercises

1–7. Use the power series definition of the matrix exponential to compute e^{At} for the given matrix A.

1. $A = \begin{bmatrix} 1 & 0 \\ 0 & -2 \end{bmatrix}$

2. $A = \begin{bmatrix} 3 & -1 \\ 9 & -3 \end{bmatrix}$

3. $A = \begin{bmatrix} 0 & 1 \\ 1 & 0 \end{bmatrix}$

4. $A = \begin{bmatrix} 1 & -1 \\ 0 & 1 \end{bmatrix}$

5. $A = \begin{bmatrix} 1 & 1 \\ 1 & 1 \end{bmatrix}$

6. $A = \begin{bmatrix} 0 & 2 & 0 \\ 1 & 0 & -1 \\ 0 & 2 & 0 \end{bmatrix}$

7. $A = \begin{bmatrix} 0 & 1 & 0 \\ -1 & 0 & 0 \\ 0 & 0 & 2 \end{bmatrix}$

8–13. For each matrix A given below

(i) Compute the resolvent matrix $(sI - A)^{-1}$.

(ii) Compute the matrix exponential $e^{At} = \mathcal{L}^{-1}\left\{(sI - A)^{-1}\right\}$.

8. $A = \begin{bmatrix} 1 & 0 \\ 0 & 2 \end{bmatrix}$

9. $A = \begin{bmatrix} 1 & -1 \\ -2 & 2 \end{bmatrix}$

10. $A = \begin{bmatrix} 1 & 1 \\ 1 & 1 \end{bmatrix}$

11. $A = \begin{bmatrix} 3 & 5 \\ -1 & -1 \end{bmatrix}$

12. $A = \begin{bmatrix} 4 & -10 \\ 1 & -2 \end{bmatrix}$

13. $\begin{bmatrix} 0 & 1 & 1 \\ 0 & 0 & 1 \\ 0 & 0 & 0 \end{bmatrix}$

14. Suppose $A = \begin{bmatrix} M & 0 \\ 0 & N \end{bmatrix}$ where M is an $r \times r$ matrix and N is an $s \times s$ matrix.

Show that $e^{At} = \begin{bmatrix} e^{Mt} & 0 \\ 0 & e^{Nt} \end{bmatrix}$.

15–16. Use Exercise 14 to compute the matrix exponential for each A.

15. $A = \begin{bmatrix} 0 & 1 & 0 \\ -1 & 0 & 0 \\ 0 & 0 & 2 \end{bmatrix}$

16. $A = \begin{bmatrix} 1 & 1 & 0 & 0 \\ 1 & 1 & 0 & 0 \\ 0 & 0 & 1 & -1 \\ 0 & 0 & -2 & 2 \end{bmatrix}$

9.4 Fulmer's Method for Computing e^{At}

The matrix exponential is fundamental to much of what we do in this chapter. It is therefore useful to have efficient techniques for calculating it. Here we will present a small variation on a technique[1] due to Fulmer[2] for computing the matrix exponential, e^{At}. It is based on the knowledge of what *types* of functions are included in the individual entries of e^{At}. This knowledge is derived from our understanding of the Laplace transform formula

$$e^{At} = \mathcal{L}^{-1}\left\{(sI - A)^{-1}\right\}.$$

Assume that A is an $n \times n$ constant matrix. Let $c_A(s) = \det(sI - A)$ be the characteristic polynomial of A. The characteristic polynomial has degree n, and by the adjoint formula for matrix inversion, $c_A(s)$ is in the denominator of each term of the inverse of $sI - A$. Therefore, each entry in $(sI - A)^{-1}$ belongs to $\mathcal{R}_{c_A}$, and hence, each entry of the matrix exponential, $e^{At} = \mathcal{L}^{-1}\left\{(sI - A)^{-1}\right\}$, is in $\mathcal{E}_{c_A}$. Recall from Sects. 2.6 and 2.7 that if $q(s)$ is a polynomial, then $\mathcal{R}_q$ is the set of proper rational functions that can be written with denominator $q(s)$, and $\mathcal{E}_q$ is the set of all exponential polynomials $f(t)$ with $\mathcal{L}\{f(t)\} \in \mathcal{R}_q$. If $\mathcal{B}_{c_A} = \{\phi_1, \phi_2, \ldots, \phi_n\}$ is the standard basis of $\mathcal{E}_{c_A}$, then it follows that

$$e^{At} = M_1\phi_1 + \cdots + M_n\phi_n,$$

where M_i is an $n \times n$ matrix for each index $i = 1, \ldots, n$. Fulmer's method is a procedure to determine the coefficient matrices $M_1, \ldots, M_n$.

Before considering the general procedure and its justification, we illustrate Fulmer's method with a simple example. If

$$A = \begin{bmatrix} 3 & 5 \\ -1 & -1 \end{bmatrix},$$

then the characteristic polynomial is

$$c_A(s) = \det(sI - A) = s^2 - 2s + 2 = (s - 1)^2 + 1$$

---

[1] There are many other techniques. For example, see the articles "Nineteen Dubious Ways to Compute the Exponential of a Matrix" in *Siam Review*, Vol 20, no. 4, pp 801-836, October 1978 and "Nineteen Dubious Ways to Compute the Exponential of a Matrix, Twenty-Five Years Later" in *Siam Review*, Vol 45, no. 1, pp 3-49, 2003.

[2] Edward P. Fulmer, Computation of the Matrix Exponential, *American Mathematical Monthly*, **82** (1975) 156–159.

and the standard basis is $\mathcal{B}_{c_A} = \{e^t \cos t, \, e^t \sin t\}$. It follows that

$$e^{At} = M_1 e^t \cos t + M_2 e^t \sin t. \tag{1}$$

Differentiate (1) to get

$$A e^{At} = M_1(e^t \cos t - e^t \sin t) + M_2(e^t \sin t + e^t \cos t) \tag{2}$$

and evaluate (1) and (2) at $t = 0$ to get the system

$$\begin{aligned} I &= M_1 \\ A &= M_1 + M_2. \end{aligned} \tag{3}$$

It is immediate that $M_1 = I = \begin{bmatrix} 1 & 0 \\ 0 & 1 \end{bmatrix}$ and $M_2 = A - M_1 = A - I = \begin{bmatrix} 2 & 5 \\ -1 & -2 \end{bmatrix}$.
Substituting these matrices into (1) gives

$$\begin{aligned} e^{At} &= \begin{bmatrix} 1 & 0 \\ 0 & 1 \end{bmatrix} e^t \cos t + \begin{bmatrix} 2 & 5 \\ -1 & -2 \end{bmatrix} e^t \sin t \\[2mm] &= \begin{bmatrix} e^t \sin t + 2e^t \cos t & 5e^t \sin t \\ -e^t \sin t & e^t \sin t - 2e^t \cos t \end{bmatrix}. \end{aligned}$$

Compare the results here to those obtained in Exercise 11 in Sect. 9.3.

The General Case. Let A be an $n \times n$ matrix and $c_A(s)$ its characteristic polynomial. Suppose $\mathcal{B}_{c_A} = \{\phi_1, \ldots, \phi_n\}$. Reasoning as we did above, there are matrices $M_1, \ldots, M_n$ so that

$$e^{At} = M_1 \phi_1(t) + \ldots + M_n \phi_n(t). \tag{4}$$

We need to find these matrices. By taking $n - 1$ derivatives, we obtain a system of linear equations (with matrix coefficients)

$$\begin{aligned} e^{At} &= M_1 \phi_1(t) + \cdots + M_n \phi_n(t) \\ A e^{At} &= M_1 \phi_1'(t) + \cdots + M_n \phi_n'(t) \\ &\;\;\vdots \\ A^{n-1} e^{At} &= M_1 \phi_1^{(n-1)}(t) + \cdots + M_n \phi_n^{(n-1)}(t). \end{aligned}$$

Now we evaluate this system at $t = 0$ to obtain

$$I = M_1 \phi_1(0) + \cdots + M_n \phi_n(0)$$
$$A = M_1 \phi_1'(0) + \cdots + M_n \phi_n'(0)$$

$$\tag{5}$$

$$\vdots$$

$$A^{n-1} = M_1 \phi_1^{(n-1)}(0) + \cdots + M_n \phi_n^{(n-1)}(0).$$

At this point, we want to argue that it is always possible to solve (5) by showing that the coefficient matrix is invertible. However, in the examples and exercises, it is usually most efficient to solve (5) by elimination of variables. Let

$$W = \begin{bmatrix} \phi_1(0) & \cdots & \phi_n(0) \\ \vdots & \ddots & \vdots \\ \phi_1^{(n-1)}(0) & \cdots & \phi_n^{(n-1)}(0) \end{bmatrix}.$$

Then W is the Wronskian of $\phi_1, \ldots, \phi_n$ at $t = 0$. Since $\phi_1, \ldots, \phi_n$ are solutions to the linear homogeneous constant coefficient differential equation $c_A(D)(y) = 0$ (by Theorem 2 of Sect. 4.2) and since they are linearly independent ($\mathcal{B}_{c_A}$ is a *basis* of $\mathcal{E}_{c_A}$), Abel's formula, Theorem 6 of Sect. 4.2, applies to show the determinant of W is nonzero so W is invertible. The above system of equations can now be written:

$$\begin{bmatrix} I \\ A \\ \vdots \\ A^{n-1} \end{bmatrix} = W \begin{bmatrix} M_1 \\ M_2 \\ \vdots \\ M_n \end{bmatrix}.$$

Therefore,

$$\begin{bmatrix} M_1 \\ M_2 \\ \vdots \\ M_n \end{bmatrix} = W^{-1} \begin{bmatrix} I \\ A \\ \vdots \\ A^{n-1} \end{bmatrix}.$$

Having solved for $M_1, \ldots, M_n$, we obtain e^{At}.

Remark 1. Note that this last equation implies that each matrix M_i is a polynomial in the matrix A since W^{-1} is a constant matrix. Specifically, $M_i = p_i(A)$ where

$$p_i(s) = \text{Row}_i(W^{-1}) \begin{bmatrix} 1 \\ s \\ \vdots \\ s^{n-1} \end{bmatrix}.$$

The following algorithm outlines Fulmer's method.

Algorithm 2 (Fulmer's Method).

Computation of e^{At} Where A Is a Given $n \times n$ Constant Matrix.

1. Compute the characteristic polynomial $c_A(s) = \det(sI - A)$.
2. Determine the standard basis $\mathcal{B}_{c_A} = \{\phi_1, \ldots, \phi_n\}$ of $\mathcal{E}_{c_A}$.
3. We then have

$$e^{At} = M_1\phi_1(t) + \cdots + M_n\phi_n(t) \tag{6}$$

   where $M_i \; i = 1, \ldots, n$ are $n \times n$ matrices.
4. Take the derivative of (6) $n - 1$ times and evaluate each resulting equation at $t = 0$ to get a system of matrix equations.
5. Solve the matrix equations for $M_1, \ldots, M_n$.

Example 3. Find the matrix exponential, e^{At}, if $A = \begin{bmatrix} 2 & 1 \\ -4 & 6 \end{bmatrix}$.

▶ **Solution.** The characteristic polynomial is $c_A(s) = (s - 2)(s - 6) + 4 = s^2 - 8s + 16 = (s - 4)^2$. Hence, $\mathcal{B}_{c_A} = \{e^{4t}, te^{4t}\}$ and it follows that

$$e^{At} = M_1 e^{4t} + M_2 t e^{4t}.$$

Differentiating and evaluating at $t = 0$ gives

$$I = M_1$$

$$A = 4M_1 + M_2.$$

It follows that

$$M_1 = I = \begin{bmatrix} 1 & 0 \\ 0 & 1 \end{bmatrix} \text{ and } M_2 = A - 4I = \begin{bmatrix} -2 & 1 \\ -4 & 2 \end{bmatrix}.$$

Thus,

$$e^{At} = M_1 e^{4t} + M_2 t e^{4t}$$

$$= \begin{bmatrix} 1 & 0 \\ 0 & 1 \end{bmatrix} e^{4t} + \begin{bmatrix} -2 & 1 \\ -4 & 2 \end{bmatrix} t e^{4t}$$

$$= \begin{bmatrix} e^{4t} - 2te^{4t} & te^{4t} \\ -4te^{4t} & e^{4t} + 2te^{4t} \end{bmatrix}. \qquad \blacktriangleleft$$

As a final example, consider the following 3×3 matrix.

Example 4. Find the matrix exponential, e^{At}, if $A = \begin{bmatrix} 0 & 0 & 1 \\ 0 & 2 & 0 \\ 1 & 0 & 0 \end{bmatrix}$.

▶ **Solution.** The characteristic polynomial is

$$c_A(s) = \det \begin{bmatrix} s & 0 & -1 \\ 0 & s-2 & 0 \\ -1 & 0 & s \end{bmatrix}$$

$$= (s-2) \det \begin{bmatrix} s & -1 \\ -1 & s \end{bmatrix}$$

$$= (s-2)(s^2-1) = (s-2)(s-1)(s+1).$$

It follows that $\mathcal{B}_{c_A} = \{e^{2t}, e^t, e^{-t}\}$ and

$$e^{At} = M_1 e^{2t} + M_2 e^t + M_3 e^{-t}.$$

Differentiating twice gives

$$Ae^{At} = 2M_1 e^{2t} + M_2 e^t - M_3 e^{-t}$$
$$A^2 e^{At} = 4M_1 e^{2t} + M_2 e^t + M_3 e^{-t}$$

and evaluating at $t = 0$ gives

$$I = M_1 + M_2 + M_3$$
$$A = 2M_1 + M_2 - M_3$$
$$A^2 = 4M_1 + M_2 + M_3.$$

It is an easy exercise to solve for M_1, M_2, and M_3. We get

$$M_1 = \frac{A^2 - I}{3} = \frac{(A-I)(A+I)}{3} = \begin{bmatrix} 0 & 0 & 0 \\ 0 & 1 & 0 \\ 0 & 0 & 0 \end{bmatrix},$$

$$M_2 = -\frac{A^2 - A - 2I}{2} = -\frac{(A-2I)(A+I)}{2} = \begin{bmatrix} \frac{1}{2} & 0 & \frac{1}{2} \\ 0 & 0 & 0 \\ \frac{1}{2} & 0 & \frac{1}{2} \end{bmatrix},$$

$$M_3 = \frac{A^2 - 3A + 2I}{6} = \frac{(A-2I)(A-I)}{6} = \begin{bmatrix} \frac{1}{2} & 0 & -\frac{1}{2} \\ 0 & 0 & 0 \\ -\frac{1}{2} & 0 & \frac{1}{2} \end{bmatrix}.$$

It follows now that

$$e^{At} = e^{2t} M_1 + e^t M_2 + e^{-t} M_3$$

$$= \begin{bmatrix} \cosh t & 0 & \sinh t \\ 0 & e^{2t} & 0 \\ \sinh t & 0 & \cosh t \end{bmatrix}.$$

◀

Exercises

1–19. Use Fulmer's method to compute the matrix exponential e^{At}.

1. $A = \begin{bmatrix} 2 & -1 \\ 1 & 0 \end{bmatrix}$

2. $A = \begin{bmatrix} -1 & 2 \\ -3 & 4 \end{bmatrix}$

3. $A = \begin{bmatrix} 2 & 1 \\ -4 & -2 \end{bmatrix}$

4. $A = \begin{bmatrix} 1 & 1 \\ -2 & 4 \end{bmatrix}$

5. $A = \begin{bmatrix} 4 & -10 \\ 1 & -2 \end{bmatrix}$

6. $A = \begin{bmatrix} 4 & -1 \\ 1 & 2 \end{bmatrix}$

7. $A = \begin{bmatrix} -9 & 11 \\ -7 & 9 \end{bmatrix}$

8. $A = \begin{bmatrix} -5 & -8 \\ 4 & 3 \end{bmatrix}$

9. $A = \begin{bmatrix} 26 & 39 \\ -15 & -22 \end{bmatrix}$

10. $A = \begin{bmatrix} 3 & -3 \\ 2 & -2 \end{bmatrix}$

11. $A = \begin{bmatrix} -3 & 1 \\ -1 & -1 \end{bmatrix}$

12. $A = \begin{bmatrix} 6 & -4 \\ 2 & 0 \end{bmatrix}$

13. $A = \begin{bmatrix} 1 & 0 & -1 \\ 0 & 1 & 0 \\ 2 & 0 & -2 \end{bmatrix}$, where $c_A(s) = s(s-1)(s+1)$

14. $A = \begin{bmatrix} 3 & -1 & -1 \\ -1 & 1 & 1 \\ 2 & -1 & 0 \end{bmatrix}$, where $c_A(s) = (s-1)^2(s-2)$

15. $A = \begin{bmatrix} 1 & -\frac{1}{2} & 0 \\ 1 & 1 & -1 \\ 0 & \frac{1}{2} & 1 \end{bmatrix}$, where $c_A(s) = (s-1)(s^2 - 2s + 2)$

16. $A = \begin{bmatrix} 2 & 1 & 0 \\ -1 & 2 & 1 \\ 0 & 1 & 2 \end{bmatrix}$, where $c_A(s) = (s-2)^3$

17. $A = \begin{bmatrix} -1 & -2 & 1 \\ 4 & 0 & -2 \\ -2 & -2 & 2 \end{bmatrix}$, where $c_A(s) = (s-1)(s^2+4)$

18. $A = \begin{bmatrix} 4 & 1 & -2 & 0 \\ 0 & 4 & 0 & -2 \\ 4 & 0 & -2 & 1 \\ 0 & 4 & 0 & -2 \end{bmatrix}$, where $c_A(s) = s^2(s-2)^2$

19. $A = \begin{bmatrix} -1 & 1 & 1 & 0 \\ -1 & -1 & 0 & 1 \\ -1 & 0 & 1 & 1 \\ 0 & -1 & -1 & 1 \end{bmatrix}$, where $c_A(s) = (s^2+1)^2$

20–22. Suppose A is a 2×2 real matrix with characteristic polynomial $c_A(s) = \det(sI - A) = s^2 + bs + c$. In these exercises, you are asked to derive a general formula for the matrix exponential e^{At}. We distinguish three cases.

20. **Distinct Real Roots:** Suppose $c_A(s) = (s-r_1)(s-r_2)$ with r_1 and r_2 distinct real numbers. Show that

$$ e^{At} = \frac{A - r_2 I}{r_1 - r_2} e^{r_1 t} + \frac{A - r_1 I}{r_2 - r_1} e^{r_2 t}. $$

21. **Repeated Root:** Suppose $c_A(s) = (s-r)^2$. Show that

$$ e^{At} = (I + (A - rI)t)\, e^{rt}. $$

22. **Complex Roots:** Suppose $c_A(s) = (s-\alpha)^2 + \beta^2$ where $\beta \neq 0$. Show that

$$ e^{At} = I e^{\alpha t} \cos \beta t + \frac{(A - \alpha I)}{\beta} e^{\alpha t} \sin \beta t. $$

9.5 Constant Coefficient Linear Systems

In this section, we turn our attention to solving a first order constant coefficient linear differential system

$$y' = Ay + f, \qquad y(t_0) = y_0. \tag{1}$$

The solution method parallels that of Sect. 1.4. Specifically, the matrix exponential e^{-At} serves as an integrating factor to simplify the equivalent system

$$y' - Ay = f. \tag{2}$$

The result is the existence and uniqueness theorem for such systems. As a corollary, we obtain the existence and uniqueness theorems for ordinary constant coefficient linear differential equations, as stated in Sects. 3.1 and 4.1.

We begin with a lemma that lists the necessary properties of the matrix exponential to implement the solution method.

Lemma 1. *Let A be an $n \times n$ matrix. The following statements then hold:*

1. $e^{At}\big|_{t=0} = I$.
2. $\frac{d}{dt} e^{At} = A e^{At} = e^{At} A$ *for all $t \in \mathbb{R}$.*
3. $e^{A(t+a)} = e^{At} e^{Aa} = e^{Aa} e^{At}$ *for all $t, a \in \mathbb{R}$.*
4. e^{At} *is an invertible matrix with inverse $\left(e^{At}\right)^{-1} = e^{-At}$ for all $t \in \mathbb{R}$.*

Proof. Items 1. and 2. were proved in Sect. 9.3. Fix $a \in \mathbb{R}$ and let

$$\Phi(t) = e^{-At} e^{A(t+a)}.$$

Then

$$\begin{aligned}
\Phi'(t) &= -A e^{-At} e^{A(t+a)} + e^{-At} A e^{A(t+a)} \\
&= -A e^{-At} e^{A(t+a)} + A e^{-At} e^{A(t+a)} = 0,
\end{aligned}$$

which follows from the product rule and part 2. It follows that Φ is a constant matrix and since $\Phi(0) = e^{-A0} e^{Aa} = e^{Aa}$ by part 1, we have

$$e^{-At} e^{A(t+a)} = e^{Aa}, \tag{3}$$

for all $t, a \in \mathbb{R}$. Now let $a = 0$ then $e^{-At} e^{At} = I$. From this, it follows that e^{At} is invertible and $\left(e^{At}\right)^{-1} = e^{-At}$. This proves item 4. Further, from (3), we have

$$e^{A(t+a)} = (e^{-At})^{-1} e^{Aa} = e^{At} e^{Aa}.$$

This proves item 3. $\qquad \square$

To solve (1), multiply (2) by e^{-At} to get

$$e^{-At} y'(t) - e^{-At} A y(t) = e^{-At} f(t). \tag{4}$$

By the product rule and Lemma 1, part 2, we have

$$(e^{-At} y)'(t) = -e^{-At} A y(t) + e^{-At} y'(t).$$

We can thus rewrite (4) as

$$(e^{-At} y)'(t) = e^{-At} f(t).$$

Now change the variable from t to u and integrate both sides from t_0 to t to get

$$e^{-At} y(t) - e^{-At_0} y(t_0) = \int_{t_0}^{t} (e^{-Au} y)'(u)\, du = \int_{t_0}^{t} e^{-Au} f(u)\, du.$$

Now add $e^{-At_0} y(t_0)$ to both sides and multiply by the inverse of e^{-At}, which is e^{At} by Lemma 1, part 4, to get

$$y(t) = e^{At} e^{-At_0} y(t_0) + e^{At} \int_{t_0}^{t} e^{-Au} f(u)\, du.$$

Now use Lemma 1, part 3, to simplify. We get

$$y(t) = e^{A(t-t_0)} y_0 + \int_{t_0}^{t} e^{A(t-u)} f(u)\, du. \tag{5}$$

This argument shows that if there is a solution, it must take this form. However, it is a straightforward calculation to verify that (5) is a solution to (1). We thereby obtain

Theorem 2 (Existence and Uniqueness Theorem).   *Let A be an $n \times n$ constant matrix and $f(t)$ an $\mathbb{R}^n$-valued continuous function defined on an interval I. Let $t_0 \in I$ and $y_0 \in \mathbb{R}^n$. Then the unique solution to the initial value problem*

$$y'(t) = A y(t) + f(t), \quad y(t_0) = y_0, \tag{6}$$

is the function $y(t)$ defined for $t \in I$ by

> **Solution to a First Order Differential System**
>
> $$y(t) = e^{A(t-t_0)} y_0 + \int_{t_0}^{t} e^{A(t-u)} f(u)\, du.$$

$$\tag{7}$$

Let us break up this general solution into its two important parts. First, when $f = 0$, (6) reduces to $y'(t) = Ay(t)$, the *associated homogeneous equation*. Its solution is the *homogeneous solution* given simply by

$$y_h = e^{A(t-t_0)} y_0.$$

Let $y_0 = e_i$, the column vector with 1 in the ith position and 0's elsewhere. Define $y_i = e^{A(t-t_0)} e_i$. Then y_i is the ith column of $e^{A(t-t_0)}$. This means that each column of $e^{A(t-t_0)}$ is a solution to the associated homogeneous equation. Furthermore, if

$$y_0 = \begin{bmatrix} a_1 \\ a_2 \\ \vdots \\ a_n \end{bmatrix},$$

then

$$y_h = e^{A(t-t_0)} y_0 = a_1 y_1 + a_2 y_2 + \cdots + a_n y_n$$

is a linear combination of the columns of e^{At} and all homogeneous solutions are of this form.

The other piece of the general solution is the *particular solution*:

$$y_p(t) = \int_{t_0}^{t} e^{A(t-u)} f(u) \, du. \tag{8}$$

We then get the familiar formula

$$y = y_h + y_p.$$

The *general solution* to $y' = Ay + f$ is thus obtained by adding all possible homogeneous solutions to one fixed particular solution.

When $t_0 = 0$, the particular solution y_p of (8) becomes the convolution product of e^{At} and $f(t)$. We record this important special case as a corollary.

Corollary 3. *Let A be an $n \times n$ constant matrix and $f(t)$ an $\mathbb{R}^n$-valued continuous function defined on an interval I containing the origin. Let $y_0 \in \mathbb{R}^n$. The unique solution to*

$$y'(t) = Ay(t) + f(t), \quad y(0) = y_0,$$

is the function defined for $t \in I$ by

$$y(t) = e^{At} y_0 + e^{At} * f(t). \tag{9}$$

When each entry of f is of exponential type, then the convolution theorem, Theorem 1 of Sect. 2.8, can be used to compute $y_p = e^{At} * f(t)$. Let us consider a few examples.

Example 4. Solve the following linear system of differential equations:

$$
\begin{aligned}
y_1' &= -y_1 + 2y_2 + e^t \\
y_2' &= -3y_1 + 4y_2 - 2e^t,
\end{aligned}
\tag{10}
$$

with initial conditions $y_1(0) = 1$ and $y_2(0) = 1$.

▶ **Solution.** We begin by writing the given system in matrix form:

$$
y' = Ay + f, \quad y(0) = y_0
\tag{11}
$$

where

$$
A = \begin{bmatrix} -1 & 2 \\ -3 & 4 \end{bmatrix}, \quad
f(t) = \begin{bmatrix} e^t \\ -2e^t \end{bmatrix} = e^t \begin{bmatrix} 1 \\ -2 \end{bmatrix}, \quad \text{and} \quad
y_0 = \begin{bmatrix} 1 \\ 1 \end{bmatrix}.
$$

The characteristic polynomial is

$$
c_A(s) = \det(sI - A) = \det \begin{bmatrix} s+1 & -2 \\ 3 & s-4 \end{bmatrix} = s^2 - 3s + 2 = (s-1)(s-2).
\tag{12}
$$

Therefore,

$$
\begin{aligned}
(sI - A)^{-1} &= \begin{bmatrix}
\dfrac{s-4}{(s-1)(s-2)} & \dfrac{2}{(s-1)(s-2)} \\[2ex]
\dfrac{-3}{(s-1)(s-2)} & \dfrac{s+1}{(s-1)(s-2)}
\end{bmatrix} \\[2ex]
&= \frac{1}{s-1} \begin{bmatrix} 3 & -2 \\ 3 & -2 \end{bmatrix} + \frac{1}{s-2} \cdot \begin{bmatrix} -2 & 2 \\ -3 & 3 \end{bmatrix}.
\end{aligned}
$$

It now follows that

$$
e^{At} = \mathcal{L}^{-1} \left\{ (sI - A)^{-1} \right\} = e^t \begin{bmatrix} 3 & -2 \\ 3 & -2 \end{bmatrix} + e^{2t} \begin{bmatrix} -2 & 2 \\ -3 & 3 \end{bmatrix}.
$$

By Corollary 3, the homogeneous part of the solution is given by

$$
\begin{aligned}
y_h &= e^{At} y_0 \\[1ex]
&= \left(e^t \begin{bmatrix} 3 & -2 \\ 3 & -2 \end{bmatrix} + e^{2t} \begin{bmatrix} -2 & 2 \\ -3 & 3 \end{bmatrix} \right) \begin{bmatrix} 1 \\ 1 \end{bmatrix} \\[1ex]
&= e^t \begin{bmatrix} 1 \\ 1 \end{bmatrix}.
\end{aligned}
$$

To compute y_p we will use two simple convolution formulas: $e^t * e^t = te^t$ and $e^{2t} * e^t = e^{2t} - e^t$. By Corollary 3, we have

$$y_p(t) = e^{At} * f(t)$$

$$= \left(e^t \begin{bmatrix} 3 & -2 \\ 3 & -2 \end{bmatrix} + e^{2t} \begin{bmatrix} -2 & 2 \\ -3 & 3 \end{bmatrix} \right) * e^t \begin{bmatrix} 1 \\ -2 \end{bmatrix}$$

$$= e^t * e^t \begin{bmatrix} 3 & -2 \\ 3 & -2 \end{bmatrix} \begin{bmatrix} 1 \\ -2 \end{bmatrix} + e^{2t} * e^t \begin{bmatrix} -2 & 2 \\ -3 & 3 \end{bmatrix} \begin{bmatrix} 1 \\ -2 \end{bmatrix}$$

$$= te^t \begin{bmatrix} 7 \\ 7 \end{bmatrix} + (e^{2t} - e^t) \begin{bmatrix} -6 \\ -9 \end{bmatrix}$$

$$= \begin{bmatrix} 7te^t + 6e^t - 6e^{2t} \\ 7te^t + 9e^t - 9e^{2t} \end{bmatrix}.$$

Now, adding the homogeneous and particular solutions together leads to the solution:

$$y(t) = y_h(t) + y_p(t) = e^{At} y(0) + e^{At} * f(t)$$

$$= \begin{bmatrix} e^t \\ e^t \end{bmatrix} + \begin{bmatrix} 7te^t + 6e^t - 6e^{2t} \\ 7te^t + 9e^t - 9e^{2t} \end{bmatrix}$$

$$= \begin{bmatrix} 7te^t + 7e^t - 6e^{2t} \\ 7te^t + 10e^t - 9e^{2t} \end{bmatrix}. \tag{13}$$

◀

Example 5. Find the general solution to the following system of differential equations:

$$\begin{aligned} y_1' &= y_2 + t \\ y_2' &= -y_1 - t \end{aligned}$$

▶ **Solution.** If

$$A = \begin{bmatrix} 0 & 1 \\ -1 & 0 \end{bmatrix}, \quad f(t) = \begin{bmatrix} t \\ -t \end{bmatrix},$$

the given system can be expressed as

$$y' = Ay + f.$$

By Example 6 of Sect. 9.3, we have $e^{At} = \begin{bmatrix} \cos t & \sin t \\ -\sin t & \cos t \end{bmatrix}$. If $y_0 = \begin{bmatrix} a_1 \\ a_2 \end{bmatrix}$ then

$$y_h = e^{At} y_0 = \begin{bmatrix} \cos t & \sin t \\ -\sin t & \cos t \end{bmatrix} \begin{bmatrix} a_1 \\ a_2 \end{bmatrix} = a_1 \begin{bmatrix} \cos t \\ -\sin t \end{bmatrix} + a_2 \begin{bmatrix} \sin t \\ \cos t \end{bmatrix}.$$

Further,

$$y_p = e^{At} * f(t) = \begin{bmatrix} \cos t & \sin t \\ -\sin t & \cos t \end{bmatrix} * \begin{bmatrix} t \\ -t \end{bmatrix} = \begin{bmatrix} (\cos t) * t - (\sin t) * t \\ -(\sin t) * t - (\cos t) * t \end{bmatrix}.$$

Table 2.11 gives formulas for the convolutions $(\cos t) * t$ and $(\sin t) * t$. However, we will use the convolution principle to make these computations. First,

$$\mathcal{L}\{(\cos t) * t\} = \frac{s}{(s^2 + 1)} \frac{1}{s^2} = \frac{1}{(s^2 + 1)s} = \frac{1}{s} - \frac{s}{s^2 + 1},$$

$$\text{and}\quad \mathcal{L}\{(\sin t) * t\} = \frac{1}{(s^2 + 1)s^2} = \frac{1}{s^2} - \frac{1}{s^2 + 1},$$

so that $(\cos t) * t = 1 - \cos t$ and $(\sin t) * t = t - \sin t$. Therefore,

$$y_p = \begin{bmatrix} 1 - \cos t - (t - \sin t) \\ -(t - \sin t) - (1 - \cos t) \end{bmatrix}.$$

By Corollary 3, the general solution is thus

$$y(t) = a_1 \begin{bmatrix} \cos t \\ -\sin t \end{bmatrix} + a_2 \begin{bmatrix} \sin t \\ \cos t \end{bmatrix} + \begin{bmatrix} 1 - t - \cos t + \sin t \\ -1 - t + \sin t + \cos t \end{bmatrix},$$

which we can rewrite more succinctly as

$$y(t) = (a_1 - 1) \begin{bmatrix} \cos t \\ -\sin t \end{bmatrix} + (a_2 + 1) \begin{bmatrix} \sin t \\ \cos t \end{bmatrix} + \begin{bmatrix} 1 - t \\ -1 - t \end{bmatrix}$$

$$= \alpha_1 \begin{bmatrix} \cos t \\ -\sin t \end{bmatrix} + \alpha_2 \begin{bmatrix} \sin t \\ \cos t \end{bmatrix} + \begin{bmatrix} 1 - t \\ -1 - t \end{bmatrix}$$

after relabeling the coefficients.   ◄

Example 6. Solve the following system of equations:

$$y_1' = 2y_1 + y_2 + 1/t \qquad y_1(1) = 2$$
$$y_2' = -4y_1 - 2y_2 + 2/t \qquad y_2(1) = 1$$

on the interval $(0, \infty)$.

► **Solution.** We can write the given system as $y' = Ay + f$, where

$$A = \begin{bmatrix} 2 & 1 \\ -4 & -2 \end{bmatrix}, \quad f(t) = \begin{bmatrix} 1/t \\ 2/t \end{bmatrix}, \quad \text{and} \quad y(1) = \begin{bmatrix} 2 \\ 1 \end{bmatrix}.$$

Observe that f is continuous on $(0, \infty)$. The characteristic polynomial is

$$c_A(s) = \det \begin{bmatrix} s-2 & -1 \\ 4 & s+2 \end{bmatrix} = s^2 \qquad (14)$$

and $\mathcal{B}_{c_A} = \{1, t\}$. Therefore, $e^{At} = M_1 + M_2 t$ and its derivative is $Ae^{At} = M_2$. Evaluating at $t = 0$ gives $M_1 = I$ and $M_2 = A$. Fulmer's method now gives

$$e^{At} = I + At = \begin{bmatrix} 1+2t & t \\ -4t & 1-2t \end{bmatrix}.$$

It follows that

$$y_h(t) = e^{A(t-1)} y(1) = \begin{bmatrix} 1+2(t-1) & t-1 \\ -4(t-1) & 1-2(t-1) \end{bmatrix} \begin{bmatrix} 2 \\ 1 \end{bmatrix} = \begin{bmatrix} 2+5(t-1) \\ 1-10(t-1) \end{bmatrix}.$$

The following calculation gives the particular solution:

$$y_p(t) = \int_1^t e^{A(t-u)} f(u)\, du$$

$$= e^{At} \int_1^t e^{-Au} f(u)\, du$$

$$= e^{At} \int_1^t \begin{bmatrix} 1-2u & -u \\ 4u & 1+2u \end{bmatrix} \begin{bmatrix} 1/u \\ 2/u \end{bmatrix} du$$

$$= e^{At} \int_1^t \begin{bmatrix} \frac{1}{u} - 4 \\ \frac{2}{u} + 8 \end{bmatrix} du$$

$$= \begin{bmatrix} 1+2t & t \\ -4t & 1-2t \end{bmatrix} \begin{bmatrix} \ln t - 4(t-1) \\ 2\ln t + 8(t-1) \end{bmatrix}$$

$$= \begin{bmatrix} (1+4t)\ln t - 4t + 4 \\ (2-8t)\ln t + 8t - 8 \end{bmatrix}.$$

Where in line 2 we used Lemma 1 to write $e^{A(t-u)} = e^{At} e^{-Au}$. We now get

$$y(t) = y_h(t) + y_p(t)$$

$$= \begin{bmatrix} 2+5(t-1) \\ 1-10(t-1) \end{bmatrix} + \begin{bmatrix} (1+4t)\ln t - 4t + 4 \\ (2-8t)\ln t + 8t - 8 \end{bmatrix}$$

$$= \begin{bmatrix} 1+t+(1+4t)\ln t \\ 3-2t+(2-8t)\ln t \end{bmatrix},$$

valid for all $t > 0$. ◀

Example 7. Solve the mixing problem introduced at the beginning of this chapter in Example 1 of Sect. 9.1. Namely,

$$y_1'(t) = \frac{-4}{10}y_1(t) + \frac{1}{10}y_2(t) + 3$$

$$y_2'(t) = \frac{4}{10}y_1(t) - \frac{4}{10}y_2(t),$$

with initial condition $y_1(0) = 2$ and $y_2(0) = 0$. Also, determine the amount of salt in each tank at time $t = 10$.

▶ **Solution.** In matrix form, this system can be written as

$$y'(t) = Ay(t) + f(t),$$

where

$$A = \begin{bmatrix} -4/10 & 1/10 \\ 4/10 & -4/10 \end{bmatrix}, \quad f(t) = \begin{bmatrix} 3 \\ 0 \end{bmatrix}, \quad \text{and} \quad y_0 = y(0) = \begin{bmatrix} 2 \\ 0 \end{bmatrix}.$$

We let the reader verify that

$$e^{At} = e^{-2t/10}\begin{bmatrix} 1/2 & 1/4 \\ 1 & 1/2 \end{bmatrix} + e^{-6t/10}\begin{bmatrix} 1/2 & -1/4 \\ -1 & 1/2 \end{bmatrix}.$$

The homogeneous solution is

$$y_h(t) = e^{At}y_0 = \left(e^{-2t/10}\begin{bmatrix} 1/2 & 1/4 \\ 1 & 1/2 \end{bmatrix} + e^{-6t/10}\begin{bmatrix} 1/2 & -1/4 \\ -1 & 1/2 \end{bmatrix} \right)\begin{bmatrix} 2 \\ 0 \end{bmatrix}$$

$$= e^{-2t/10}\begin{bmatrix} 1 \\ 2 \end{bmatrix} + e^{-6t/10}\begin{bmatrix} 1 \\ -2 \end{bmatrix}.$$

We use the fact that $e^{-2t/10} * 1 = 5 - 5e^{-2t/10}$ and $e^{-6t/10} * 1 = \frac{1}{3}(5 - 5e^{-6t/10})$ to get

$$y_p = e^{At} * \begin{bmatrix} 3 \\ 0 \end{bmatrix} = 3e^{At} * \begin{bmatrix} 1 \\ 0 \end{bmatrix}$$

$$= 3\left(e^{-2t/10} * 1 \right)\begin{bmatrix} 1/2 & 1/4 \\ 1 & 1/2 \end{bmatrix}\begin{bmatrix} 1 \\ 0 \end{bmatrix} + 3\left(e^{-6t/10} * 1 \right)\begin{bmatrix} 1/2 & -1/4 \\ -1 & 1/2 \end{bmatrix}\begin{bmatrix} 1 \\ 0 \end{bmatrix}$$

$$= 3\left(5 - 5e^{-2t/10}\right)\begin{bmatrix} 1/2 \\ 1 \end{bmatrix} + \left(5 - 5e^{-6t/10}\right)\begin{bmatrix} 1/2 \\ -1 \end{bmatrix}$$

$$= \begin{bmatrix} 10 \\ 10 \end{bmatrix} - e^{-2t/10}\begin{bmatrix} 15/2 \\ 15 \end{bmatrix} - e^{-6t/10}\begin{bmatrix} 5/2 \\ -5 \end{bmatrix}.$$

We now obtain the solution:

$$y(t) = y_h(t) + y_p(t)$$

$$= \begin{bmatrix} 10 \\ 10 \end{bmatrix} + e^{-2t/10}\begin{bmatrix} -13/2 \\ -13 \end{bmatrix} + e^{-6t/10}\begin{bmatrix} -3/2 \\ 3 \end{bmatrix}.$$

At time $t = 10$, we have

$$y(10) = \begin{bmatrix} 10 - (13/2)e^{-2} - (3/2)e^{-6} \\ 10 - 13e^{-2} + 3e^{-6} \end{bmatrix} = \begin{bmatrix} 9.117 \\ 8.248 \end{bmatrix}.$$

At $t = 10$ minutes, Tank 1 contains 9.117 pounds of salt and Tank 2 contains 8.248 pounds of salt.   ◄

We now summarize in the following algorithm the procedure for computing the solution to a constant coefficient first order system.

Algorithm 8. Given a constant coefficient first order system,

$$y' = Ay + f, \quad y(t_0) = y_0$$

we proceed as follows to determine the solution set.

Solution Method for a Constant Coefficient First Order System

1. Determine e^{At}: This may be done by the inverse Laplace transform formula $e^{At} = \mathcal{L}^{-1}\left\{(sI - A)^{-1}\right\}$ or by Fulmer's method.

2. Determine the homogeneous part $y_h(t) = e^{A(t-t_0)}y(t_0)$.

3. Determine the particular solution $y_p(t) = \int_{t_0}^{t} e^{A(t-u)}f(u)\,du$. It is sometimes useful to use $e^{A(t-u)} = e^{At}e^{-Au}$.

4. The general solution is $y_g = y_h + y_p$.

Eigenvectors and Eigenvalues

When the initial condition $y_0 = v$ is an eigenvector for A, then the solution to $y' = Ay$, $y(0) = v$, takes on a very simple form.

Lemma 9. *Suppose A is an $n \times n$ matrix and v is an eigenvector with eigenvalue λ. Then the solution to*

$$y' = Ay, \quad y(0) = v$$

is

$$y = e^{\lambda t} v.$$

In other words,

$$e^{At} v = e^{\lambda t} v.$$

Proof. Let $y(t) = e^{\lambda t} v$. Then $y'(t) = \lambda e^{\lambda t} v$ and $Ay(t) = e^{\lambda t} Av = \lambda e^{\lambda t} v$. Therefore, $y'(t) = Ay(t)$. By the uniqueness and existence theorem, we have

$$e^{At} v = e^{\lambda t} v. \qquad \square$$

Example 10. Let $A = \begin{bmatrix} 1 & 2 \\ 2 & 4 \end{bmatrix}$. Find the solution to

$$y'(t) = Ay(t), \quad y(0) = \begin{bmatrix} 1 \\ 2 \end{bmatrix}.$$

▶ **Solution.** We observe that $\begin{bmatrix} 1 \\ 2 \end{bmatrix}$ is an eigenvector of A with eigenvalue 5:

$$\begin{bmatrix} 1 & 2 \\ 2 & 4 \end{bmatrix} \begin{bmatrix} 1 \\ 2 \end{bmatrix} = \begin{bmatrix} 5 \\ 10 \end{bmatrix} = 5 \begin{bmatrix} 1 \\ 2 \end{bmatrix}.$$

Thus,

$$y(t) = e^{5t} \begin{bmatrix} 1 \\ 2 \end{bmatrix},$$

is the unique solution to $y' = Ay$, $y(0) = \begin{bmatrix} 1 \\ 2 \end{bmatrix}$.   ◀

More generally, we have

Theorem 11. *Suppose $v = c_1 v_1 + \cdots + c_k v_k$, where $v_1, \ldots, v_k$ are eigenvectors of A with corresponding eigenvalues $\lambda_1, \ldots, \lambda_k$. Then the solution to*

$$y' = Ay, \quad y(0) = v$$

is

$$y = c_1 e^{\lambda_1 t} v_1 + \cdots + c_2 e^{\lambda_k t} v_k.$$

Proof. By Theorem 2, the solution is $y(t) = e^{At}v$. By linearity and Lemma 9, we get

$$\begin{aligned} y(t) = e^{At}v &= c_1 e^{At} v_1 + \cdots + c_k e^{At} v_k \\ &= c_1 e^{\lambda_1 t} v_1 + \cdots + c_k e^{\lambda_k t} v_k. \end{aligned} \qquad \square$$

Existence and Uniqueness Theorems

We conclude this section with the existence and uniqueness theorems referred to earlier in the text, namely, Theorem 10 of Sect. 3.1 and Theorem 5 of Sect. 4.1.

For convenience of expression, we will say that an $\mathbb{R}^n$-valued function f is an *exponential polynomial* if each component f_i of f is an exponential polynomial. Similarly, we say that an $\mathbb{R}^n$-valued function f is of *exponential type* if each component f_i of f is of exponential type.

Corollary 12. *Suppose f is an exponential polynomial. Then the solution to*

$$y'(t) = Ay(t) + f(t), \quad y(t_0) = y_0$$

is an exponential polynomial defined on $\mathbb{R}$.

Proof. The formula for the solution is given by (7) in Theorem 2. Each entry in e^{At} is an exponential polynomial. Therefore, each entry of $e^{A(t-t_0)} = e^{At}e^{-At_0}$ is a linear combination of entries of e^{At}, hence an exponential polynomial. It follows that $e^{A(t-t_0)}y_0$ is an exponential polynomial. The function $u \to e^{A(t-u)}f(u)$ is a translation and product of exponential polynomials. Thus, by Exercises 34 and 35 of Sect. 2.7 it is in $\mathcal{E}$, and by Exercise 37 of Sect. 2.7 we have $\int_{t_0}^{t} e^{A(t-u)}f(u)\,du$ is in $\mathcal{E}$. Thus, each piece in (7) is an exponential polynomial so the solution $y(t)$ is an exponential polynomial. $\qquad \square$

Corollary 13. *Suppose f is of exponential type. Then the solution to*

$$y'(t) = Ay(t) + f(t), \quad y(t_0) = y_0$$

is of exponential type.

Proof. By Proposition 1 of Sect. 2.2, Exercise 37 of Sect. 2.2, and Lemma 4 of Sect. 2.2, we find that sums, products, and integrals of functions that are of exponential type are again of exponential type. Reasoning as above, we obtain the result. $\qquad \square$

Theorem 14 (The Existence and Uniqueness Theorem for Constant Coefficient Linear Differential Equations). *Suppose $f(t)$ is a continuous real-valued function on an interval I. Let $t_0 \in I$. Then there is a unique real-valued function y*

defined on I satisfying

$$a_n y^{(n)} + a_{n-1} y^{(n-1)} + \cdots a_1 y' + a_0 y = f(t), \tag{15}$$

with initial conditions $y(t_0) = y_0$, $y'(t_0) = y_1$, $\ldots y^{(n-1)} = y_{n-1}$. If $f(t)$ is of exponential type, so is the solution $y(t)$ and its derivatives $y^{(i)}(t)$, for $i = 1, \ldots, n - 1$. Furthermore, if $f(t)$ is in $\mathcal{E}$, then $y(t)$ is also in $\mathcal{E}$.

Proof. We may assume $a_n = 1$ by dividing by a_n, if necessary. Let $y_1 = y$, $y_2 = y'$, $\ldots$, $y_n = y^{(n-1)}$, and let y be the column vector with entries $y_1, \ldots, y_n$. Then $y_1' = y' = y_2, \ldots, y_{n-1}' = y^{(n-1)} = y_n$, and

$$y_n' = y^{(n)} = -a_0 y - a_1 y' - a_2 y'' - \cdots - a_{n-1} y^{(n-1)} + f$$
$$= -a_0 y_1 - a_1 y_2 - a_2 y_3 - \cdots - a_{n-1} y_n + f.$$

It is simple to check that y is a solution to (15) if and only if y is a solution to (1), where

$$A = \begin{bmatrix} 0 & 1 & 0 & \cdots & 0 \\ 0 & 0 & 1 & \cdots & 0 \\ \vdots & & & & \vdots \\ 0 & 0 & \cdots & 0 & 1 \\ -a_0 & -a_1 & \cdots & -a_{n-2} & -a_{n-1} \end{bmatrix}, \quad f = \begin{bmatrix} 0 \\ 0 \\ \vdots \\ 0 \\ f \end{bmatrix}, \quad \text{and} \quad y_0 = \begin{bmatrix} y_0 \\ y_1 \\ \vdots \\ y_{n-2} \\ y_{n-1} \end{bmatrix}.$$

By Theorem 2, there is a unique solution y. The first entry, y, in y is the unique solution to (15). If f is an exponential polynomial, then f is likewise, and Corollary 12 implies that y is an exponential polynomial. Hence, $y_1 = y$, $y_2 = y'$, $\ldots$, $y_n = y^{(n-1)}$ are all exponential polynomials. If f is of exponential type, then so is f, and Corollary 13 implies y is of exponential type. This, in turn, implies y, $y', \ldots, y^{n-1}$ are each of exponential type. $\square$

Exercises

1–9. Solve the homogeneous systems $y' = Ay$, $y(0) = y_0$, for the given A and y_0.

1. $A = \begin{bmatrix} -1 & 0 \\ 0 & 3 \end{bmatrix}$; $y(0) = \begin{bmatrix} 1 \\ -2 \end{bmatrix}$

2. $A = \begin{bmatrix} 0 & 2 \\ -2 & 0 \end{bmatrix}$; $y(0) = \begin{bmatrix} 1 \\ -1 \end{bmatrix}$

3. $A = \begin{bmatrix} 2 & 1 \\ 0 & 2 \end{bmatrix}$; $y(0) = \begin{bmatrix} -1 \\ 2 \end{bmatrix}$

4. $A = \begin{bmatrix} -1 & 2 \\ -2 & -1 \end{bmatrix}$; $y(0) = \begin{bmatrix} 1 \\ 0 \end{bmatrix}$

5. $A = \begin{bmatrix} 2 & -1 \\ 3 & -2 \end{bmatrix}$; $y(0) = \begin{bmatrix} 1 \\ 3 \end{bmatrix}$

6. $A = \begin{bmatrix} 2 & -5 \\ 1 & -2 \end{bmatrix}$; $y(0) = \begin{bmatrix} 1 \\ -1 \end{bmatrix}$

7. $A = \begin{bmatrix} 3 & -4 \\ 1 & -1 \end{bmatrix}$; $y(0) = \begin{bmatrix} 1 \\ 1 \end{bmatrix}$

8. $A = \begin{bmatrix} -1 & 0 & 3 \\ 0 & 2 & 0 \\ 0 & 0 & 1 \end{bmatrix}$; $y(0) = \begin{bmatrix} 1 \\ 1 \\ 2 \end{bmatrix}$

9. $A = \begin{bmatrix} 0 & 4 & 0 \\ -1 & 0 & 0 \\ 1 & 4 & -1 \end{bmatrix}$; $y(0) = \begin{bmatrix} 2 \\ 1 \\ 2 \end{bmatrix}$

10–17. Use Corollary 3 to solve $y' = Ay + f$ for the given matrix A, forcing function f, and initial condition $y(0)$.

10. $A = \begin{bmatrix} 2 & -1 \\ 3 & -2 \end{bmatrix}$, $f(t) = \begin{bmatrix} e^t \\ e^t \end{bmatrix}$, $y(0) = \begin{bmatrix} 1 \\ 3 \end{bmatrix}$

11. $A = \begin{bmatrix} -1 & 2 \\ -2 & -1 \end{bmatrix}$, $f(t) = \begin{bmatrix} 5 \\ 0 \end{bmatrix}$, $y(0) = \begin{bmatrix} 1 \\ 0 \end{bmatrix}$

12. $A = \begin{bmatrix} 2 & -5 \\ 1 & -2 \end{bmatrix}$, $f(t) = \begin{bmatrix} 2\cos t \\ \cos t \end{bmatrix}$, $y(0) = \begin{bmatrix} 1 \\ -1 \end{bmatrix}$

13. $A = \begin{bmatrix} -1 & -4 \\ 1 & -1 \end{bmatrix}$, $f(t) = \begin{bmatrix} 4 \\ 1 \end{bmatrix}$, $y(0) = \begin{bmatrix} 2 \\ -1 \end{bmatrix}$

14. $A = \begin{bmatrix} 2 & 1 \\ 1 & 2 \end{bmatrix}$, $f(t) = \begin{bmatrix} e^t \\ -e^t \end{bmatrix}$, $y(0) = \begin{bmatrix} 1 \\ 1 \end{bmatrix}$

15. $A = \begin{bmatrix} 5 & 2 \\ -8 & -3 \end{bmatrix}$, $f(t) = \begin{bmatrix} t \\ -2t \end{bmatrix}$, $y(0) = \begin{bmatrix} 0 \\ 1 \end{bmatrix}$

16. $A = \begin{bmatrix} -2 & 2 & 1 \\ 0 & -1 & 0 \\ 2 & -2 & -1 \end{bmatrix}, f(t) = \begin{bmatrix} e^{-2t} \\ 0 \\ -e^{-2t} \end{bmatrix}, y(0) = \begin{bmatrix} 2 \\ 1 \\ 1 \end{bmatrix}$

17. $A = \begin{bmatrix} 0 & 1 & 1 \\ 1 & 1 & -1 \\ -2 & 1 & 3 \end{bmatrix}, f(t) = \begin{bmatrix} e^{2t} \\ e^{2t} \\ -e^{2t} \end{bmatrix}, y(0) = \begin{bmatrix} 0 \\ 0 \\ 0 \end{bmatrix}$

18–21. Solve each mixing problem.

18. Two tanks are interconnected as illustrated below.

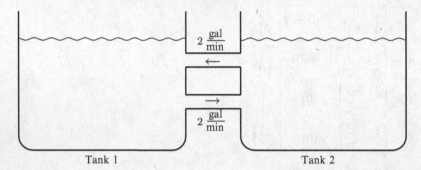

Assume that Tank 1 contains 1 gallon of brine in which 4 pounds of salt are initially dissolved and Tank 2 initially contains 1 gallon of pure water. Moreover, at time $t = 0$, the mixtures are pumped between the two tanks, each at a rate of 2 gal/min. Assume the mixtures are well stirred. Let $y_1(t)$ be the amount of salt in Tank 1 at time t and let $y_2(t)$ be the amount of salt in Tank 2 at time t. Determine y_1, y_2. Find the amount of salt in each tank after 30 seconds.

19. Two tanks are interconnected as illustrated below.

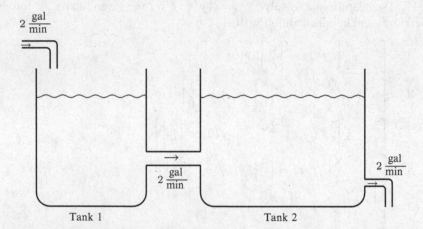

Assume that Tank 1 contains 1 gallon of brine in which 4 pounds of salt is initially dissolved and Tank 2 contains 2 gallons of pure water. Moreover, the

6. A curve C that lies in a region R in the (u, v) phase plane is transformed to a curve $P(C)$ that lies in the region $P(R)$ in the (x, y) phase plane.

You will be guided through a proof of these statements in the exercises. In view of the discussion above, we say that the phase portraits of $z' = Az$ and $z' = Bz$ are *affine equivalent* if A and B are similar.

To illustrate the value of affine equivalence, consider the following example.

Example 1. Discuss the phase portrait for the linear differential system

$$z'(t) = Az(t),$$

where

$$A = \begin{bmatrix} 4 & 1 \\ 2 & 5 \end{bmatrix}.$$

▶ **Solution.** The characteristic polynomial of A is $c_A(s) = s^2 - 9s + 18 = (s - 3)(s - 6)$. The eigenvalues of A are thus 3 and 6. By Fulmer's method, we have $e^{At} = Me^{3t} + Ne^{6t}$ from which we get

$$I = M + N$$

$$A = 3M + 6N.$$

From these equations, it follows that

$$M = \frac{1}{3}(6I - A) = \frac{1}{3}\begin{bmatrix} 2 & -1 \\ -2 & 1 \end{bmatrix}$$

$$N = \frac{1}{3}(A - 3I) = \frac{1}{3}\begin{bmatrix} 1 & 1 \\ 2 & 2 \end{bmatrix}.$$

Hence,

$$e^{At} = \frac{1}{3}\begin{bmatrix} 2 & -1 \\ -2 & 1 \end{bmatrix}e^{3t} + \frac{1}{3}\begin{bmatrix} 1 & 1 \\ 2 & 2 \end{bmatrix}e^{6t}.$$

A short calculation gives the solution to $z'(t) = Az(t)$ with initial value $z(0) = \begin{bmatrix} c_1 \\ c_2 \end{bmatrix}$ as

$$x(t) = \frac{1}{3}(2c_1 - c_2)e^{3t} + \frac{1}{3}(c_1 + c_2)e^{6t}$$

$$y(t) = -\frac{1}{3}(2c_1 - c_2)e^{3t} + \frac{2}{3}(c_1 + c_2)e^{6t} \tag{5}$$

The orbits $\{(x(t), y(t)) : t \in \mathbb{R}\}$ are difficult to describe in general except for a few carefully chosen initial values. Notice that

1. If $c_1 = 0$ and $c_2 = 0$, then $x(t) = y(t) = 0$ so that the origin $(0,0)$ is a trajectory.
2. If $c_2 = -c_1 \neq 0$, then $x(t) = c_1 e^{3t}$ and $y(t) = -c_1 e^{3t}$. This means that the trajectory is of the form

$$(x(t), y(t)) = (c_1 e^{3t}, -c_1 e^{3t}) = c_1 e^{3t}(1, -1).$$

The function e^{3t} is positive and increasing as a function of t. Thus, if c_1 is positive, then the trajectory is a half-line in the fourth quadrant consisting of all positive multiples of the vector $(1, -1)$ and pointing away from the origin. This is the trajectory marked **A** in Fig. 9.1 in the (x, y) phase plane. If c_1 is negative, then the trajectory is a half-line in the second quadrant consisting of all positive multiples of the vector $(-1, 1)$ and pointing away from the origin.
3. If $c_2 = 2c_1 \neq 0$, then $x(t) = c_1 e^{6t}$ and $y(t) = 2c_1 e^{6t}$. This means that the trajectory is of the form

$$(x(t), y(t)) = (c_1 e^{6t}, 2c_1 e^{6t}) = c_1 e^{6t}(1, 2).$$

Again e^{6t} is positive and increasing as a function of t. Thus, if c_1 is positive, the trajectory is a half-line in the first quadrant consisting of all positive multiples of the vector $(1, 2)$ and pointing away from the origin. This is the trajectory marked **D** in Fig. 9.1 in the (x, y) phase plane. If c_1 is negative, then the trajectory is a half-line in the third quadrant consisting of all positive multiples of the vector $(-1, -2)$ and pointing away from the origin.

For initial values other than the ones listed above, it is rather tedious to directly describe the trajectories. Notice though how a change in coordinates simplifies matters significantly. In (5), we can eliminate e^{6t} by subtracting $y(t)$ from twice $x(t)$ and we can eliminate e^{3t} by adding $x(t)$ and $y(t)$. We then get

$$2x(t) - y(t) = (2c_1 - c_2) e^{3t} = k_1 e^{3t},$$
$$x(t) + y(t) = (c_1 + c_2) e^{6t} = k_2 e^{6t}, \tag{6}$$

where $k_1 = 2c_1 - c_2$ and $k_2 = c_1 + c_2$. Now let

$$\begin{aligned} u &= 2x - y \\ v &= x + y \end{aligned} = \begin{bmatrix} 2 & -1 \\ 1 & 1 \end{bmatrix} \begin{bmatrix} x \\ y \end{bmatrix}.$$

The matrix $\begin{bmatrix} 2 & -1 \\ 1 & 1 \end{bmatrix}$ is invertible and we can thus solve for $\begin{bmatrix} x \\ y \end{bmatrix}$ to get

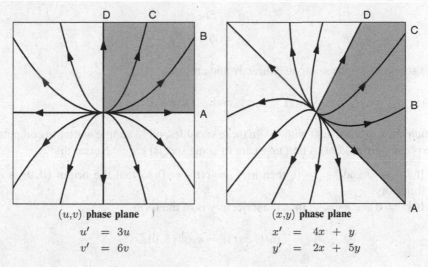

(u,v) **phase plane**
$$u' = 3u$$
$$v' = 6v$$

(x,y) **phase plane**
$$x' = 4x + y$$
$$y' = 2x + 5y$$

Fig. 9.1 Affine equivalent phase portraits

$$\begin{bmatrix} x \\ y \end{bmatrix} = \frac{1}{3} \begin{bmatrix} 1 & 1 \\ -1 & 2 \end{bmatrix} \begin{bmatrix} u \\ v \end{bmatrix}.$$

Let $P = \dfrac{1}{3} \begin{bmatrix} 1 & 1 \\ -1 & 2 \end{bmatrix}$. This is the affine transformation that implements the change in variables that we need. A simple calculation gives

$$B = P^{-1}AP = \frac{1}{3} \begin{bmatrix} 2 & -1 \\ 1 & 1 \end{bmatrix} \begin{bmatrix} 4 & 1 \\ 2 & 5 \end{bmatrix} \begin{bmatrix} 1 & 1 \\ -1 & 2 \end{bmatrix} = \begin{bmatrix} 3 & 0 \\ 0 & 6 \end{bmatrix}, \tag{7}$$

a diagonal matrix consisting of the eigenvalues 3 and 6.

We make an important observation about P here: Since A has two distinct eigenvalues, there are two linearly independent eigenvectors. Notice that the first column of P is an eigenvector with eigenvalue 3 and the second column is an eigenvector with eigenvalue 6. The importance of this will be made clear when we talk about Jordan canonical forms below.

If $w = \begin{bmatrix} u \\ v \end{bmatrix}$, then

$$w' = Bw,$$

and the initial condition is given by

$$w(0) = P^{-1}z(0) = \begin{bmatrix} 2 & -1 \\ 1 & 1 \end{bmatrix} \begin{bmatrix} c_1 \\ c_1 \end{bmatrix} = \begin{bmatrix} 2c_1 - c_2 \\ c_1 + c_2 \end{bmatrix} = \begin{bmatrix} k_1 \\ k_2 \end{bmatrix}.$$

The equations for $w' = Bw$ are simply

$$u' = 3u,$$

$$v' = 6v.$$

The solutions can be computed directly and are

$$u(t) = k_1 e^{3t} \text{ and } v(t) = k_2 e^{6t},$$

which are also consistent with (6). In these variables, it is a simple matter to compute the phase portrait. Let us first take care of some special cases. Notice that

1. If $k_1 = 0$ and $k_2 = 0$, then $u(t) = v(t) = 0$ so that the origin $(0,0)$ is a trajectory.
2. If $k_1 \neq 0$ and $k_2 = 0$, then the trajectory is of the form

$$(u(t), v(t)) = k_1 e^{3t}(1, 0).$$

As before, we observe that e^{3t} is positive and increasing. Thus, if k_1 is positive, the trajectory is the positive u-axis pointing away from the origin. This is the trajectory marked **A** in Fig. 9.1 in the (u, v) phase plane. If k_1 is negative, then the trajectory is the negative u-axis pointing away from the origin.
3. If $k_1 = 0$ and $k_2 \neq 0$ then the trajectory is of the form

$$(u(t), v(t)) = k_2 e^{6t}(0, 1).$$

Again e^{6t} is positive and increasing. Thus if k_2 is positive, then the trajectory is the positive y-axis pointing away from the origin. This is the trajectory marked **D** in Fig. 9.1 in the (u, v) phase plane. If k_2 is negative, then the trajectory is the negative y-axis pointing away from the origin.

Now assume $k_1 \neq 0$ and $k_2 \neq 0$. Since e^{3t} and e^{6t} take on all positive real numbers and are increasing, the trajectories $(u(t), v(t)) = (k_1 e^{3t}, k_2 e^{6t})$ are located in the quadrant determined by the initial value (k_1, k_2) and are pointed in the direction away from the origin. To see what kinds of curves arise, let us determine how $u(t)$ and $v(t)$ are related. For notation's sake, we drop the "t" in $u(t)$ and $v(t)$. Observe that $k_2 u^2 = k_1^2 k_2 e^{6t} = k_1^2 v$, and hence,

$$v = \frac{k_2}{k_1^2} u^2.$$

Hence, a trajectory is that portion of a parabola that lies in the quadrant determined by the initial value (k_1, k_2). In Fig. 9.1, the two trajectories marked **B** and **C** are those trajectories that go through $(4, 3)$ and $(1, 1)$, respectively. The other unmarked trajectories are obtained from initial values, $(\pm 1, \pm 1)$ and $(\pm 4, \pm 3)$.

Now to determine the trajectories in the (x, y) phase plane for $z' = Az$, we utilize the affine transformation P. Since

$$P \begin{bmatrix} u \\ v \end{bmatrix} = \frac{u}{3} \begin{bmatrix} 1 \\ -1 \end{bmatrix} + \frac{v}{3} \begin{bmatrix} 1 \\ 2 \end{bmatrix},$$

it follows that the region determined by the first quadrant in the (u, v) phase plane transforms to the region consisting of all sums of nonnegative multiples of $(1, -1)$ and nonnegative multiples of $(1, 2)$. We have shaded those regions in Fig. 9.1. An affine transformation such as P transforms parabolas to parabolas and preserves tangent lines. The parabola marked **B** in the (u, v) plane lies in the first quadrant and is tangent to the trajectory marked **A** at the origin. Therefore, the transformed trajectory in the (x, y) phase plane must (1) lie in the shaded region, (2) be a parabola, and (3) be tangent to trajectory **A** at the origin. It is similarly marked **B**. Now consider the region between trajectory **B** and **D** in the (u, v) phase plane. Trajectory **C** lies in this region and must therefore transform to a trajectory which (1) lies in the region between trajectory **B** and **D** in the (x, y) phase plane, (2) is a parabola, and (3) is tangent to trajectory **A** at the origin. We have marked it correspondingly **C**. The analysis of the other trajectories in the (u, v) phase plane is similar; they are each transformed to parabolically shaped trajectories in the (x, y) phase plane.   ◄

Jordon Canonical Forms

In the example above, we showed by (7) that A is similar to a diagonal matrix. In general, this cannot always be done. However, in the theorem below, we show that A is similar to one of four forms, called the *Jordan Canonical Forms*.

Theorem 2. *Let A be a real 2×2 matrix. Then there is an invertible matrix P so that $P^{-1}AP$ is one of the following matrices:*

1. $J_1 = \begin{bmatrix} \lambda_1 & 0 \\ 0 & \lambda_2 \end{bmatrix}, \begin{matrix} \lambda_1, \lambda_2 \in \mathbb{R} \\ \lambda_1 \neq \lambda_2 \end{matrix},$ 3. $J_3 = \begin{bmatrix} \lambda & 0 \\ 0 & \lambda \end{bmatrix}, \lambda \in \mathbb{R},$

2. $J_2 = \begin{bmatrix} \lambda & 0 \\ 1 & \lambda \end{bmatrix}, \lambda \in \mathbb{R},$ 4. $J_4 = \begin{bmatrix} \alpha & -\beta \\ \beta & \alpha \end{bmatrix}, \alpha \in \mathbb{R}, \beta > 0.$

Furthermore, the affine transformation P may be determined correspondingly as follows:

1. *If A has two distinct real eigenvalues, λ_1 and λ_2, then the first column of P is an eigenvector for λ_1 and the second column of P is an eigenvector for λ_2.*
2. *If A has only one real eigenvalue λ with eigenspace of dimension 1, then the first column of P may be chosen to be any vector v that is not an eigenvector and the second column of P is $(A - \lambda I)v$.*

3. *If A has only one real eigenvalue with eigenspace of dimension 2 then A is J_3.*
   Hence, P may be chosen to be the identity.
4. *If A has a complex eigenvalue, then one of them is of the form $\alpha - i\beta$ with $\beta > 0$.*
   If w is a corresponding eigenvector, then the first column of P is the real part of
   w and the second column of P, is the imaginary part of w.

Remark 3. Any one of the four matrices, $J_1, \ldots, J_4$ is called a **Jordan matrix**.
Note that the affine transformation P is not unique.

Proof. We consider the eigenvalues of A. There are four possibilities.

1. Suppose A has two distinct real eigenvalues λ_1 and λ_2. Let v_1 be an eigenvector
   with eigenvalue λ_1 and v_2 an eigenvector with eigenvalue λ_2. Let $P = \begin{bmatrix} v_1 & v_2 \end{bmatrix}$,
   the matrix 2×2 matrix with v_1 the first column and v_2 the second column. Then

$$AP = \begin{bmatrix} Av_1 & Av_2 \end{bmatrix} = \begin{bmatrix} \lambda_1 v_1 & \lambda_2 v_2 \end{bmatrix} = \begin{bmatrix} v_1 & v_2 \end{bmatrix} \begin{bmatrix} \lambda_1 & 0 \\ 0 & \lambda_2 \end{bmatrix} = PJ_1.$$

   Now multiply both sides on the left by P^{-1} to get $P^{-1}AP = J_1$.
2. Suppose A has only a single real eigenvalue λ. Then the characteristic polynomial is $c_A(s) = (s - \lambda)^2$. Let E_λ be the eigenspace for λ and suppose further that
   $E_\lambda \neq \mathbb{R}^2$. Let v_1 be a vector outside of E_λ. Then $(A - \lambda I)v_1 \neq 0$. However, by
   the Cayley-Hamilton theorem (see Appendix A.5), $(A - \lambda I)(A - \lambda I)v_1 = (A - \lambda I)^2 v_1 = 0$. It follows that $(A - \lambda I)v_1$ is an eigenvector. Let $v_2 = (A - \lambda I)v_1$.
   Then $Av_1 = ((A - \lambda I) + \lambda I)v_1 = v_2 + \lambda v_1$ and $Av_2 = \lambda v_2$. Let $P = \begin{bmatrix} v_1 & v_2 \end{bmatrix}$
   be the matrix with v_1 in the first column and v_2 in the second column. Then

$$AP = \begin{bmatrix} Av_1 & Av_2 \end{bmatrix} = \begin{bmatrix} v_2 + \lambda v_1 & \lambda v_2 \end{bmatrix} = \begin{bmatrix} v_1 & v_2 \end{bmatrix} \begin{bmatrix} \lambda & 0 \\ 1 & \lambda \end{bmatrix} = PJ_2.$$

   Now multiply both sides on the left by P^{-1} to get $P^{-1}AP = J_2$.
3. Suppose A has only a single real eigenvalue λ and the eigenspace $E_\lambda = \mathbb{R}^2$.
   Then $Av = \lambda v$ for all $v \in \mathbb{R}^2$. This means A must already be $J_3 = \lambda I$.
4. Suppose A does not have a real eigenvalue. Since A is real, the two complex
   eigenvalues are of the form $\alpha + i\beta$ and $\alpha - i\beta$, with $\beta > 0$. Let w be an eigenvector
   in the complex plane $\mathbb{C}^2$ with eigenvalue $\alpha - i\beta$. Let v_1 be the real part of w and
   v_2 the imaginary part of w. Then $w = v_1 + iv_2$ and since

$$A(v_1 + iv_2) = (\alpha - i\beta)(v_1 + iv_2) = (\alpha v_1 + \beta v_2) + i(-\beta v_1 + \alpha v_2)$$

we get

$$Av_1 = \alpha v_1 + \beta v_2,$$
$$Av_2 = -\beta v_1 + \alpha v_2.$$

Let $P = \begin{bmatrix} v_1 & v_2 \end{bmatrix}$. Then

$$AP = \begin{bmatrix} Av_1 & Av_2 \end{bmatrix} = \begin{bmatrix} \alpha v_1 + \beta v_2 & -\beta v_1 + \alpha v_2 \end{bmatrix} = \begin{bmatrix} v_1 & v_2 \end{bmatrix} \begin{bmatrix} \alpha & -\beta \\ \beta & \alpha \end{bmatrix} = PJ_4.$$

Now multiply both sides on the left by P^{-1} to get $P^{-1}AP = J_4$.   □

Example 4. For each of the following matrices, determine an affine transformation P so the $P^{-1}AP$ is a Jordan matrix:

1. $A = \begin{bmatrix} 6 & -2 \\ -3 & 7 \end{bmatrix}$ 2. $A = \begin{bmatrix} -5 & 2 \\ -2 & -1 \end{bmatrix}$ 3. $A = \begin{bmatrix} -5 & -8 \\ 4 & 3 \end{bmatrix}$

▶ **Solution.** 1. The characteristic polynomial is $c_A(s) = (s - 6)(s - 7) - 6 = s^2 - 13s + 36 = (s - 4)(s - 9)$. There are two distinct eigenvalues, 4 and 9. It is an easy calculation to see that $v_1 = \begin{bmatrix} 1 \\ 1 \end{bmatrix}$ is an eigenvector with eigenvalue 4 and $v_2 = \begin{bmatrix} -2 \\ 3 \end{bmatrix}$ is an eigenvector with eigenvalue 9. Let

$$P = \begin{bmatrix} v_1 & v_2 \end{bmatrix} = \begin{bmatrix} 1 & -2 \\ 1 & 3 \end{bmatrix}.$$

Then an easy calculation gives

$$P^{-1}AP = \begin{bmatrix} 4 & 0 \\ 0 & 9 \end{bmatrix}.$$

2. The characteristic polynomial is $c_A(s) = (s + 3)^2$. Thus, $\lambda = -3$ is the only eigenvalue. Since $A - -3I = \begin{bmatrix} -2 & 2 \\ -2 & 2 \end{bmatrix}$, it is easy to see that all eigenvectors are multiples of $\begin{bmatrix} 1 \\ 1 \end{bmatrix}$. Let $v_1 = \begin{bmatrix} 1 \\ -1 \end{bmatrix}$. Then v_1 is not an eigenvector. Let

$$v_2 = (A - \lambda I)v = \begin{bmatrix} -2 & 2 \\ -2 & 2 \end{bmatrix}\begin{bmatrix} 1 \\ -1 \end{bmatrix} = \begin{bmatrix} -4 \\ -4 \end{bmatrix}.$$

Let

$$P = \begin{bmatrix} v_1 & v_2 \end{bmatrix} = \begin{bmatrix} 1 & -4 \\ -1 & -4 \end{bmatrix}.$$

Then an easy calculation gives

$$P^{-1}AP = \begin{bmatrix} -3 & 0 \\ 1 & -3 \end{bmatrix}.$$

3. The characteristic polynomial is $c_A(s) = s^2 + 2s + 17 = (s+1)^2 + 16$ so the eigenvalues are $-1 \pm 4i$. We compute an eigenvector with eigenvalue $-1 - 4i$. To do this, we solve

$$((-1-4i)I - A)\begin{bmatrix} a \\ b \end{bmatrix} = \begin{bmatrix} 0 \\ 0 \end{bmatrix}$$

for a and b. This is equivalent to

$$(4 - 4i)a + 8b = 0,$$
$$-4a + (-4 - 4i)b = 0.$$

If we choose $b = 1$, then $a = \dfrac{-8}{4-4i} = \dfrac{-8(4+4i)}{(4-4i)(4+4i)} = -1 - i$. Therefore,

$$v = \begin{bmatrix} -1-i \\ 1 \end{bmatrix} = \begin{bmatrix} -1 \\ 1 \end{bmatrix} + i \begin{bmatrix} -1 \\ 0 \end{bmatrix}$$

is an eigenvector for A with eigenvalue $-1 - 4i$. Let $v_1 = \begin{bmatrix} -1 \\ 1 \end{bmatrix}$ and $v_2 = \begin{bmatrix} -1 \\ 0 \end{bmatrix}$ be the real and imaginary parts of v. Let

$$P = \begin{bmatrix} v_1 & v_2 \end{bmatrix} = \begin{bmatrix} -1 & -1 \\ 1 & 0 \end{bmatrix}.$$

Then

$$P^{-1}AP = \begin{bmatrix} -1 & -4 \\ 4 & -1 \end{bmatrix}. \qquad \blacktriangleleft$$

Notice in the following example the direct use of affine equivalence.

Example 5. Discuss the phase portrait for the linear differential system

$$z'(t) = Az(t),$$

where

$$A = \begin{bmatrix} -5 & -8 \\ 4 & 3 \end{bmatrix}.$$

▶ **Solution.** The characteristic polynomial is

$$c_A(s) = s^2 + 2s + 17 = (s+1)^2 + 16.$$

It is straightforward to determine that

$$e^{At} = e^{-t} \begin{bmatrix} \cos 4t - \sin 4t & 2\sin 4t \\ \sin 4t & \cos 4t + \sin 4t \end{bmatrix}.$$

For a given initial condition $z(0) = \begin{bmatrix} c_1 \\ c_2 \end{bmatrix}$, we have

$$x(t) = e^{-t}(c_1 \cos 4t + (2c_2 - c_1)\sin 4t)$$
$$y(t) = e^{-t}((c_1 + c_2)\sin 4t + c_2 \cos 4t).$$

The phase plane portrait for this system is very difficult to directly deduce without the help of an affine transformation. In Example 4, part 3, we determined

$$P = \begin{bmatrix} -1 & -1 \\ 1 & 0 \end{bmatrix}$$

and

$$B = P^{-1}AP = \begin{bmatrix} -1 & -4 \\ 4 & -1 \end{bmatrix}.$$

In the new variables $z = Pw$, we get $w' = Bw$, and the initial condition is given by

$$w(0) = P^{-1}z(0) = \begin{bmatrix} 0 & 1 \\ -1 & -1 \end{bmatrix}\begin{bmatrix} c_1 \\ c_2 \end{bmatrix} = \begin{bmatrix} c_2 \\ -c_1 - c_2 \end{bmatrix}.$$

Let $\begin{bmatrix} k_1 \\ k_2 \end{bmatrix} = \begin{bmatrix} c_2 \\ -c_1 - c_2 \end{bmatrix}$. A straightforward calculation gives

$$e^{Bt} = e^{-t} \begin{bmatrix} \cos 4t & -\sin 4t \\ \sin 4t & \cos 4t \end{bmatrix}$$

and

$$\begin{bmatrix} u \\ v \end{bmatrix} = e^{Bt}\begin{bmatrix} k_1 \\ k_2 \end{bmatrix} = e^{-t}\begin{bmatrix} k_1 \cos 4t - k_2 \sin 4t \\ k_1 \sin 4t + k_2 \cos 4t \end{bmatrix}.$$

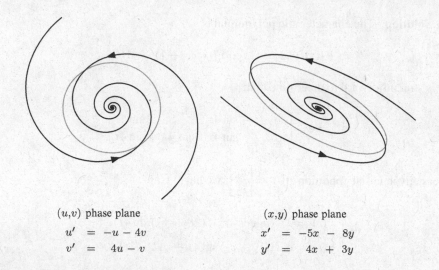

(u,v) phase plane (x,y) phase plane

$$u' = -u - 4v \qquad\qquad x' = -5x - 8y$$
$$v' = 4u - v \qquad\qquad\quad\; y' = 4x + 3y$$

Fig. 9.2 Spiral phase portraits

If ϕ is the angle made by the vector (k_1, k_2) and the positive u-axis, then

$$\cos\phi = \frac{k_1}{\sqrt{k_1^2 + k_2^2}} \quad \text{and} \quad \sin\phi = \frac{k_2}{\sqrt{k_1^2 + k_2^2}}.$$

We can then express u and v as

$$u(t) = \sqrt{k_1^2 + k_2^2}\,e^{-t}(\cos\phi\cos 4t - \sin\phi\sin 4t) = \sqrt{k_1^2 + k_2^2}\,e^{-t}(\cos(4t + \phi)),$$

$$v(t) = \sqrt{k_1^2 + k_2^2}\,e^{-t}(\cos\phi\sin 4t + \sin\phi\cos 4t) = \sqrt{k_1^2 + k_2^2}\,e^{-t}(\sin(4t + \phi)),$$

Now observe that

$$u^2(t) + v^2(t) = (k_1^2 + k_2^2)e^{-2t}.$$

If $u^2(t) + v^2(t)$ were constant, then the trajectories would be circles. However, the presence of the factor e^{-2t} shrinks the distance to the origin as t increases. The result is that the trajectory is a spiral pointing toward the origin. We show two such trajectories in Fig. 9.2 in the (u, v) phase plane.

Notice that the trajectories in the (u, v) phase plane rotate one onto another. By this we mean if we rotate a fixed trajectory, you will get another trajectory. In fact, all trajectories can be obtained by rotating a fixed one. Specifically, if we rotate a trajectory by an angle θ. Then the matrix that implements that rotation is given by

$$R(\theta) = \begin{bmatrix} \cos\theta & -\sin\theta \\ \sin\theta & \cos\theta \end{bmatrix}.$$

It is a nice property about rotation matrices that they commute. In other words, if θ_1 and θ_2 are two angles, then

$$R(\theta_1)R(\theta_2) = R(\theta_2)R(\theta_1).$$

Now observe that $e^{Bt} = e^{-t}R(4t)$. When we apply $R(\theta)$ to w, we get

$$R(\theta)w(t) = e^{-t}R(\theta)R(4t)k = e^{-t}R(4t)(R(\theta)k) = e^{Bt}(R(\theta)k). \qquad (8)$$

Thus, $R(\theta)w(t)$ is the solution to $w' = Bw$ with just a different initial condition, namely, $R(\theta)k$.

Now to see what is going on in (x, y) phase plane, we use the affine map P. In the (u, v) phase plane, we have drawn a gray circle centered at the origin (it is not a trajectory). Suppose its radius is k. Recall that an affine transformation maps circles to ellipses. Specifically, since $w = P^{-1}z$, we have $u = y$ and $v = -x - y$. From this we get

$$k^2 = u^2 + v^2 = y^2 + x^2 + 2xy + y^2 = x^2 + 2xy + 2y^2.$$

This equation defines an ellipse. That ellipse is drawn in gray in the (x, y) phase plane. The trajectories in the (x, y) phase plane are still spirals that point toward the origin but elongate in the direction of the semimajor axis of the ellipse. ◄

Critical Points

A solution for which the associated path

$$\{(x(t), y(t)), t \in \mathbb{R}\}$$

is just a point is called an **equilibrium solution** or **critical point**. This means then that $x(t) = c_1$ and $y(t) = c_2$ for all $t \in \mathbb{R}$ and occurs if and only if

$$Ac = 0, \quad c = \begin{bmatrix} c_1 \\ c_2 \end{bmatrix}. \qquad (9)$$

If A is nonsingular, that is, $\det A \neq 0$, then (9) implies $c = 0$. In the phase portrait, the origin is an orbit and it is the only orbit consisting of a single point. On the other hand, if A is singular, then the solutions to (9) consists of the whole plane in the case $A = 0$ and a line through the origin in the case $A \neq 0$. If $A = 0$, then the phase portrait is trivial; each point in the plane is an orbit. If $A \neq 0$ (but singular), then each point on the line is an orbit and off that line there will be nontrivial orbits. We will assume A is nonsingular for the remainder of this section and develop the case where A is nonzero but singular in the exercises.

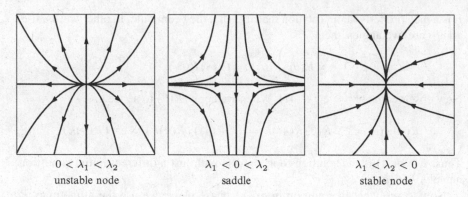

$$0 < \lambda_1 < \lambda_2 \qquad\qquad \lambda_1 < 0 < \lambda_2 \qquad\qquad \lambda_1 < \lambda_2 < 0$$

unstable node   saddle   stable node

Fig. 9.3 Phase portrait for the canonical system J_1

The Canonical Phase Portraits

In view of Theorem 2, all phase portraits are affine equivalent to just a few simple types. These types are referred to as the *canonical phase portraits*. There are four of them (if we do not take into account directions associated with the paths) corresponding to each of the four Jordan canonical forms. Let us consider each case.

J_1 : In J_1, we will order the eigenvalues λ_1 and λ_2 to satisfy $\lambda_1 < \lambda_2$. Since $\det A \neq 0$, neither λ_1 nor λ_2 are zero. The solutions to $z' = J_1 z$ are $x(t) = c_1 e^{\lambda_1 t}$ and $y(t) = c_2 e^{\lambda_2 t}$. Therefore, the trajectories lie on the power curve defined by $|y(t)| = K |x(t)|^{\frac{\lambda_2}{\lambda_1}}$ (a similar calculation was done in Example 1). The shape of the trajectories are determined by $p = \lambda_2/\lambda_1$. Refer to Fig. 9.3 for the phase portrait for each of the following three subcases:

1. Suppose $0 < \lambda_1 < \lambda_2$. Then $x(t)$ and $y(t)$ become infinite as t gets large. The trajectories lie on the curve $|y| = K |x|^p$, $p > 1$. All of the trajectories point away from the origin; the origin is said to be an ***unstable node***.
2. Suppose $\lambda_1 < 0 < \lambda_2$. Then $x(t)$ approaches zero while $y(t)$ becomes infinite as t gets large. The trajectories lie on the curve $|y| = K / |x|^q$, where $q = -p > 0$. The origin is said to be a ***saddle***.
3. Suppose $\lambda_1 < \lambda_2 < 0$. Then $x(t)$ and $y(t)$ approach zero as t gets large. The trajectories lie on the curve $|y| = K |x|^p$, $0 < p < 1$. In this case, all the trajectories point toward the origin; the origin is said to be a ***stable node***.

J_2 : It is straightforward to see that the solutions to $z' = J_2 z$ are $x(t) = c_1 e^{\lambda t}$ and $y(t) = (c_1 t + c_2) e^{\lambda t}$. Observe that if $c_1 = 0$, then $x(t) = 0$ and $y(t) = c_2 e^{\lambda t}$. If $c_2 > 0$, then the positive y-axis is a trajectory, and if $c_2 < 0$, then the negative y-axis is a trajectory. Now assume $c_1 \neq 0$. We can solve for y in terms of x as

Fig. 9.4 Phase portrait for the canonical system J_2

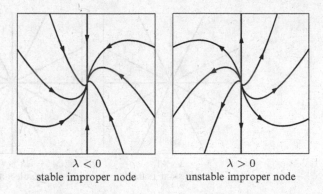

$\lambda < 0$
stable improper node

$\lambda > 0$
unstable improper node

follows. In the equation, $x(t) = c_1 e^{\lambda t}$, we get $t = \frac{1}{\lambda} \ln \left(\frac{x}{c_1} \right)$. Substituting into $y(t)$ gives

$$y = \frac{x}{\lambda} \ln \left(\frac{x}{c_1} \right) + \frac{c_2 x}{c_1}. \tag{10}$$

Note that if $c_1 > 0$, then $x > 0$ and the trajectory lies in the right half plane, and if $c_1 < 0$, then $x < 0$ and the trajectory lies in the left half plane. An easy exercise shows that the graph of Equation (10) has a vertical tangent at the origin; the origin is called an **improper node**. Now refer to Fig. 9.4 for the phase portrait for each of the following two subcases:

1. Suppose $\lambda < 0$. Then $x(t)$ and $y(t)$ approach zero as t gets large. Thus, all trajectories point toward the origin. In this case, the origin is **stable**. If $c_1 < 0$, then the trajectory is concave upward and has a single local minimum. If $c_1 > 0$, then the trajectory is concave downward and has a single local maximum.
2. Suppose $\lambda > 0$. Then $x(t)$ and $y(t)$ approach infinity as t gets large. Thus, all trajectories point away from the origin. In this case, the origin is **unstable**. If $c_1 < 0$, then the trajectory is concave downward and has a single local maximum. If $c_1 > 0$, then the trajectory is concave upward and has a single local minimum.

J_3 : It is straightforward to see that the solutions to $z' = J_3 z$ are $x(t) = c_1 e^{\lambda t}$ and $y(t) = c_2 e^{\lambda t}$. Thus, the orbits are of the form $(x(t), y(t)) = e^{\lambda t}(c_1, c_2)$ and hence are rays from the origin through the initial condition (c_1, c_2). The origin is called a **star node**. Now refer to Fig. 9.5 for the phase portrait for each of the following two subcases:

1. If $\lambda < 0$, then $x(t)$ and $y(t)$ approach zero as x gets large. All trajectories point toward the origin. In this case, the origin is a **stable star node**.
2. If $\lambda > 0$, then $x(t)$ and $y(t)$ approach infinity as x gets large. All trajectories point away from the origin. In this case, the origin is an **unstable star node**.

J_4 : By a calculation similar to the one done in Example 5, it is easy to see that the solutions to $z' = J_3 z$ are $x(t) = e^{\alpha t}(c_1 \cos \beta t - c_2 \sin \beta t) = |c| e^{\alpha t} \cos(\beta t + \phi)$

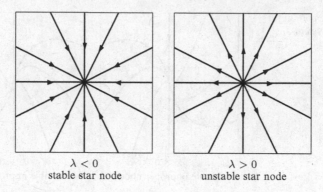

$$\lambda < 0$$
stable star node

$$\lambda > 0$$
unstable star node

Fig. 9.5 Phase portrait for the canonical system J_3

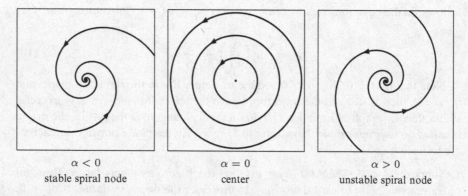

$$\alpha < 0$$
stable spiral node

$$\alpha = 0$$
center

$$\alpha > 0$$
unstable spiral node

Fig. 9.6 Phase portrait for the canonical system J_4

and $y(t) = e^{\alpha t}(c_1 \sin \beta t + c_2 \cos \beta t) = |c| e^{\alpha t} \sin(\beta t + \phi)$, where $|c| = \sqrt{c_1^2 + c_2^2}$ and ϕ is the angle made by the vector (c_1, c_2) and the x-axis. From this it follows that

$$x^2 + y^2 = |c|^2 e^{2\alpha t}$$

and the trajectories are spirals if $\alpha \neq 0$. Now refer to Fig. 9.6 for the phase portrait for each of the following three subcases:

1. If $\alpha < 0$, then $x(t)$ and $y(t)$ approach zero as t gets large. Thus, the trajectories point toward the origin; the origin is called a **stable spiral node**.
2. If $\alpha = 0$, then $x^2 + y^2 = |c|^2$ and the trajectories are circles with center at the origin; the origin is simply called a **center**.
3. If $\alpha > 0$, then $x(t)$ and $y(t)$ approach infinity as t gets large. Thus, the trajectories point away from the origin; the origin is called a **unstable spiral node**.

Classification of Critical Points

Now let A be any nonsingular matrix. The critical point $(0, 0)$ is classified as a ***node, center, or saddle*** according to whether the critical point in its affine equivalent is a node, center, or saddle. In like manner, we extend the adjectives proper, improper, stable, unstable, spiral, and star. Thus, in Example 1, the origin is an unstable node, and in Example 5, the origin is stable spiral node. Below we summarize the classification of the critical points in terms of the eigenvalues of A.

Classification of critical points

Jordan form	Eigenvalues of A	Critical point
J_1 :	$\lambda_1 \neq \lambda_2$	
	$\lambda_1 < \lambda_2 < 0$	Stable node
	$\lambda_1 < 0 < \lambda_2$	Saddle
	$0 < \lambda_1 < \lambda_2$	Unstable node
J_2 :	$\lambda \neq 0$	
	$\lambda < 0$	Stable improper node
	$\lambda > 0$	Unstable improper node
J_3 :	$\lambda \neq 0$	
	$\lambda < 0$	Stable star node
	$\lambda > 0$	Unstable star node
J_4 :	$\alpha \pm i\beta, \ \beta > 0$	
	$\alpha < 0$	Stable spiral node
	$\alpha = 0$	Center
	$\alpha > 0$	Unstable spiral node

Example 6. Classify the critical points for the system $z' = Az$ where

$$1. \ A = \begin{bmatrix} 6 & -2 \\ -3 & 7 \end{bmatrix} \qquad 2. \ A = \begin{bmatrix} -5 & 2 \\ -2 & -1 \end{bmatrix} \qquad 3. \ A = \begin{bmatrix} -5 & -8 \\ 4 & 3 \end{bmatrix}$$

▶ **Solution.** 1. In Example 4 part (1), we found that A is of type J_1 with positive eigenvalues, $\lambda_1 = 4$ and $\lambda_2 = 9$. The origin is an unstable node.

2. In Example 4 part (2), we found that A is of type J_2 with single eigenvalue, $\lambda = -3$. Since it is negative, the origin is an improper stable node.

3. In Example 4 part (3), we found that A is of type J_4 with eigenvalue, $\lambda = -1 \pm 4i$. Since the real part is negative, the origin is a stable star node. ◀

Exercises

1–10. For each of the following matrices A determine an affine transformation P and a Jordan matrix J so that $J = P^{-1}AP$. Then classify the critical point of A.

1. $A = \begin{bmatrix} 2 & 3 \\ -1 & -2 \end{bmatrix}$

2. $A = \begin{bmatrix} -3 & 2 \\ -2 & 1 \end{bmatrix}$

3. $A = \begin{bmatrix} -5 & -2 \\ 5 & 1 \end{bmatrix}$

4. $A = \begin{bmatrix} 2 & -2 \\ 4 & -2 \end{bmatrix}$

5. $A = \begin{bmatrix} 2 & 0 \\ 0 & 2 \end{bmatrix}$

6. $A = \begin{bmatrix} 4 & 1 \\ -1 & 2 \end{bmatrix}$

7. $A = \begin{bmatrix} 5 & 3 \\ -1 & 1 \end{bmatrix}$

8. $A = \begin{bmatrix} 1 & -2 \\ 4 & -5 \end{bmatrix}$

9. $A = \begin{bmatrix} 3 & 1 \\ -8 & -1 \end{bmatrix}$

10. $A = \begin{bmatrix} -3 & 0 \\ 0 & -3 \end{bmatrix}$

11–14. In the following exercises we examine how an affine transformation preserves basic kinds of shapes. Let P be an affine transformation.

11. The general equation of a line is $Du + Ev + F = 0$, with D and E not both zero. Show that the change of variable $z = Pw$ transforms a line L in the (u, v) plane to a line $P(L)$ in the (x, y) plane. If the line goes through the origin show that the transformed line also goes through the origin.

12. The general equation of a conic section is given by

$$Au^2 + Buv + Cv^2 + Du + Ev + F = 0,$$

where A, B, and C are not all zero. Let $\Delta = B^2 - 4AC$ be the *discriminant*. If

1. $\Delta < 0$ the graph is an ellipse.
2. $\Delta = 0$ the graph is a parabola.
3. $\Delta > 0$ the graph is a hyperbola.

Show that the change of variable $z = Pw$ transforms an ellipse to an ellipse, a parabola to a parabola, and a hyperbola to a hyperbola.

13. Suppose C is a power curve, i.e., the graph of a relation $Au + Bv = (Cu + Dv)^p$, where p is a real number and the variables are suitably restricted so the power is well defined. Show that $P(C)$ is again a power curve.

14. Suppose C is a differentiable curve with tangent line L at the point (u_0, v_0) in the (u, v) plane. Show that $P(L)$ is a tangent line to the curve $P(C)$ at the point $P(u_0, v_0)$.

15–19. In this set of exercises we consider the phase portraits when A is non zero but singular. Thus assume $\det A = 0$ and $A \neq 0$.

15. Show that A is similar to one of the following matrices:

$$J_1 = \begin{bmatrix} 0 & 0 \\ 0 & \lambda \end{bmatrix}, \lambda \neq 0 \text{ or } J_2 = \begin{bmatrix} 0 & 0 \\ 1 & 0 \end{bmatrix}.$$

(Hint: Since 0 is an eigenvalue consider two cases: the second eigenvalue is nonzero or it is 0. Then mimic what was done for the cases J_1 and J_2 in Theorem 2.)

16. Construct the Phase Portrait for $J_1 = \begin{bmatrix} 0 & 0 \\ 0 & \lambda \end{bmatrix}$.

17. Construct the Phase Portrait for $J_2 = \begin{bmatrix} 0 & 0 \\ 1 & 0 \end{bmatrix}$.

18. Suppose $\det A = 0$ and A is similar to J_1. Let λ be the nonzero eigenvalue. Show that the phase portrait for A consists of equilibrium points on the zero eigenspace and half lines parallel to the eigenvector for the nonzero eigenvalue λ with one end at an equilibrium point. If $\lambda > 0$ then the half lines point away from the equilibrium point and if $\lambda < 0$ they point toward the equilibrium point.

19. Suppose $\det A = 0$ and A is similar to J_2. Show that the phase portrait for A consists of equilibrium points on the zero eigenspace and lines parallel to the eigenspace.

20–23. In this set of problems we consider some the properties mentioned in the text about the canonical phase portrait J_2. Let (c_1, c_2) be a point in the plane but not on the y axis and let $\lambda \neq 0$. Let $y = \dfrac{x}{\lambda} \ln\left(\dfrac{x}{c_1}\right) + \dfrac{c_2 x}{c_1}$.

20. If $c_1 > 0$ show $\lim\limits_{x \to 0^+} y = 0$ and if $c_1 < 0$ $\lim\limits_{x \to 0^-} y = 0$.

21. Show that y has a vertical tangent at the origin.

22. Show that y has a single critical point.

23. Assume $c_1 > 0$ and hence $x > 0$. Show y is concave upward on $(0, \infty)$ if $\lambda > 0$ and y is concave downward on $(0, \infty)$ if $\lambda < 0$.

9.7 General Linear Systems

In this section, we consider the broader class of linear differential systems where the coefficient matrix $A(t)$ in

$$y'(t) = A(t)y(t) + f(t) \tag{1}$$

is a function of t and not necessarily a constant matrix. Under rather mild conditions on $A(t)$ and the *forcing function* $f(t)$, there is an existence and uniqueness theorem. The linearity of (1) implies that the structure of the solution set is very similar to that of the constant coefficient case. However, it can be quite difficult to find solution methods unless rather strong conditions are imposed on $A(t)$.

An Example: The Mixing Problem

The following example shows that a simple variation of the mixing problem produces a linear differential system with nonconstant coefficient matrix.

Example 1. Two tanks are interconnected as illustrated below.

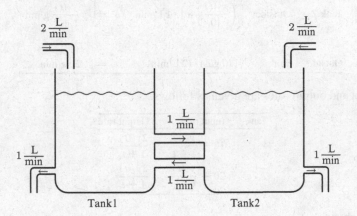

Assume that Tank 1 initially contains 1 liter of brine in which 8 grams of salt are dissolved and Tank 2 initially contains 1 liter of pure water. The mixtures are then pumped between the two tanks, 1 L/min from Tank 1 to Tank 2 and 1 L/min from Tank 2 back to Tank 1. Assume that a brine mixture containing 6 grams salt/L enters Tank 1 at a rate of 2 L/min, and the well-stirred mixture is removed from Tank 1 at the rate of 1 L/min. Assume that a brine mixture containing 10 grams salt/L enters Tank 2 at a rate of 2 L/min and the well-stirred mixture is removed from Tank 2 at the rate of 1 L/min. Let $y_1(t)$ be the amount of salt in Tank 1 at time t and let $y_2(t)$ be the amount of salt in Tank 2 at time t. Determine a linear differential system that describes how y_1, y_2, and their derivatives are related.

▶ **Solution.** In this example, the amount of brine solution is not constant. The net increase of brine in each tank is one liter per minute. Thus, if $v_1(t)$ and $v_2(t)$ represent the amount of brine in Tank 1 and Tank 2, respectively, then $v_1(t) = v_2(t) = 1 + t$. Further, the concentration of salt in each tank is given by $\frac{y_1(t)}{1+t}$ and $\frac{y_2(t)}{1+t}$. As usual, y_1' and y_2' are the differences between the input rate of salt and the output rate of salt, and each rate is the product of the flow rate and the concentration. The relevant rates of change are summarized in the following table.

From	To	Rate		
Outside	Tank 1	(6 g/L)·(2 L/min)	=	12 g/min
Tank 1	Outside	$\left(\frac{y_1(t)}{1+t}\ \text{g/L}\right)\cdot(1\text{L/min})$	=	$\frac{y_1(t)}{1+t}$ g/min
Tank 1	Tank2	$\left(\frac{y_1(t)}{1+t}\ \text{g/L}\right)\cdot(1\ \text{L/min})$	=	$\frac{y_1(t)}{1+t}$ g/min
Tank 2	Tank 1	$\left(\frac{y_2(t)}{1+t}\text{g/L}\right)\cdot(1\ \text{L/min})$	=	$\frac{y_2(t)}{1+t}$ g/min
Tank 2	Outside	$\left(\frac{y_2(t)}{10}\text{g/L}\right)\cdot 1\ \text{L/min}$	=	$\frac{y_2(t)}{1+t}$ g/min
Outside	Tank 2	(10 g/L)·(2 L/min)	=	20 g/min

The input and output rates are given as follows:

Tank	Input rate	Output rate
1	$12 + \dfrac{y_2(t)}{1+t}$	$\dfrac{2y_1(t)}{1+t}$
2	$20 + \dfrac{y_1(t)}{1+t}$	$\dfrac{2y_2(t)}{1+t}$

We thus obtain

$$y_1'(t) = \frac{-2}{1+t}y_1(t) + \frac{1}{1+t}y_2(t) + 12,$$

$$y_2'(t) = \frac{1}{1+t}y_1(t) - \frac{2}{1+t}y_2(t) + 20.$$

The initial conditions are $y_1(0) = 8$ and $y_2(0) = 0$. We may now write the linear differential system in the form

$$y' = A(t)y(t) + f(t), \quad y(0) = y_0,$$

where

$$A(t) = \begin{bmatrix} \dfrac{-2}{1+t} & \dfrac{1}{1+t} \\ \dfrac{1}{1+t} & \dfrac{-2}{1+t} \end{bmatrix}, \quad f(t) = \begin{bmatrix} 12 \\ 20 \end{bmatrix}, \quad \text{and} \quad y(0) = \begin{bmatrix} 8 \\ 0 \end{bmatrix}.$$

We will show how to solve this nonconstant linear differential system later in this section.   ◄

The Existence and Uniqueness Theorem

The following theorem is the fundamental result for linear systems. It guarantees that solutions exist, and if we can find a solution to a initial value problem by any means whatsoever, then we know that we have found the only possible solution.

Theorem 2 (Existence and Uniqueness). [7]*Suppose that the $n \times n$ matrix function $A(t)$ and the $n \times 1$ matrix function $f(t)$ are both continuous on an interval I in $\mathbb{R}$. Let $t_0 \in I$. Then for every choice of the vector y_0, the initial value problem*

$$y' = A(t)y + f(t), \quad y(t_0) = y_0,$$

has a unique solution $y(t)$ which is defined on the same interval I.

Remark 3. How is this theorem related to existence and uniqueness theorems we have stated previously?

- If $A(t) = A$ is a constant matrix, then Theorem 2 is precisely Theorem 2 of Sect. 9.5 where we have actually provided a solution method and a formula.
- If $n = 1$, then this theorem is just Corollary 8 of Sect. 1.5. In this case, we have actually proved the result by exhibiting a formula for the unique solution. However, for general n, there is no formula like (15) of Sect. 1.5, unless $A(t)$ satisfies certain stronger conditions.
- Theorem 6 of Sect. 5.1 is a corollary of Theorem 2. Indeed, if $n = 2$,

$$A(t) = \begin{bmatrix} 0 & 1 \\ -\dfrac{a_0(t)}{a_2(t)} & -\dfrac{a_1(t)}{a_2(t)} \end{bmatrix}, \quad f(t) = \begin{bmatrix} 0 \\ f(t) \end{bmatrix}, \quad y_0 = \begin{bmatrix} y_0 \\ y_1 \end{bmatrix}, \quad \text{and} \quad y = \begin{bmatrix} y \\ y' \end{bmatrix},$$

---

[7]A proof of this result can be found in the text *An Introduction to Ordinary Differential Equations* by Earl Coddington, Prentice Hall, (1961), Page 256.

then the second order linear initial value problem

$$a_2(t)y'' + a_1(t)y' + a_0(t)y = f(t), \quad y(t_0) = y_0, \ y'(t_0) = y_1$$

has the solution $y(t)$ if and only if the first order linear system

$$y' = A(t)y + f(t), \quad y(t_0) = y_0$$

has the solution $y(t) = \begin{bmatrix} y(t) \\ y'(t) \end{bmatrix}$. You should convince yourself of the validity of this statement.

Linear Homogeneous Differential Equations

In Theorem 4 of Sect. 9.2, we showed that each solution to (1) takes the form $y = y_h + y_p$, where y_h is a solution to the *associated homogeneous system*

$$y'(t) = A(t)y(t) \tag{2}$$

and y_p is a fixed particular solution. We now focus on the solution set to (2).

Theorem 4. *If the $n \times n$ matrix $A(t)$ is continuous on an interval I, then the solution set to (2) is a linear space of dimension n. In other words,*

1. *There are n linearly independent solutions.*
2. *Given any set of n linear independent solutions $\{\phi_1, \phi_2, \ldots, \phi_n\}$, then any other solution ϕ can be written as*

$$\phi = c_1\phi_1 + \cdots + c_n\phi_n$$

for some scalars $c_1, \ldots, c_n \in \mathbb{R}$.

Proof. Let e_i be the $n \times 1$ matrix with 1 in the ith position and zeros elsewhere. By the existence and uniqueness theorem, there is a unique solution, $\psi_i(t)$, to (2) with initial condition $y(t_0) = e_i$. We claim $\{\psi_1, \psi_2, \ldots, \psi_n\}$ is linearly independent. To show this, let

$$\Psi(t) = \begin{bmatrix} \psi_1(t) & \psi_2(t) & \cdots & \psi_n(t) \end{bmatrix}. \tag{3}$$

Then Ψ is an $n \times n$ matrix of functions with $\Psi(t_0) = \begin{bmatrix} \psi_1(t_0) & \cdots & \psi_n(t_0) \end{bmatrix} = \begin{bmatrix} e_1 & \cdots & e_n \end{bmatrix} = I$, the $n \times n$ identity matrix. Now suppose there are scalars $c_1, \ldots, c_n$ such that $c_1\psi_1 + \cdots + c_n\psi_n = 0$, valid for all $t \in I$. We can reexpress this as $\Psi(t)c = 0$, where c is the column vector

$$c = \begin{bmatrix} c_1 \\ \vdots \\ c_n \end{bmatrix}. \tag{4}$$

Now evaluate at $t = t_0$ to get $c = Ic = \Psi(t_0)c = 0$. This implies $\{\psi_1, \psi_2, \ldots, \psi_n\}$ is linearly independent. Now suppose that ψ is any solution to (2). Let $\eta(t) = \Psi(t)\psi(t_0)$. Then η is a linear combination of $\psi_1, \ldots, \psi_n$, hence a solution, and $\eta(t_0) = \Psi(t_0)\psi(t_0) = \psi(t_0)$. Since η and ψ satisfy the same initial condition, they are equal by the existence and uniqueness theorem. It follows that every solution, $y(t)$, to (2) is a linear combination of $\psi_1, \ldots, \psi_n$ and may be expressed as

$$y(t) = \Psi(t)y(t_0), \tag{5}$$

where $y(t_0)$ is the initial condition, expressed as a column vector.

Now suppose that $\{\phi_1, \phi_2, \ldots, \phi_n\}$ is any set of n linearly independent solutions of (2). We wish now to show that any solution may be expressed as a linear combination of $\{\phi_1, \phi_2, \ldots, \phi_n\}$.[8] By (5), we have

$$\phi_i(t) = \Psi(t)\phi_i(t_0),$$

for each $i = 1, \ldots, n$. Let $\Phi(t)$ be the $n \times n$ matrix given by

$$\Phi(t) = \begin{bmatrix} \phi_1(t) & \phi_2(t) & \cdots & \phi_n(t) \end{bmatrix}. \tag{6}$$

Now we can write

$$\Phi(t) = \Psi(t)\Phi(t_0).$$

We claim $\Phi(t_0)$ is invertible. Suppose not. Then there would be a column matrix c as in (4), where $c_1, \ldots, c_n$ are not all zero, such that $\Phi(t_0)c = 0$. But this implies

$$c_1\phi_1(t) + \cdots + c_n\phi_n(t) = \Phi(t)c = \Psi(t)\Phi(t_0)c = 0.$$

This contradicts the linear independence of $\{\phi_1, \ldots, \phi_n\}$. It follows that $\Phi(t_0)$ must be invertible. We can thus write

$$\Psi(t) = \Phi(t)(\Phi(t_0))^{-1}. \tag{7}$$

Now suppose ϕ is a solution to (2). Then (5) and (7) give

$$\phi(t) = \Psi(t)\phi(t_0) = \Phi(t)(\Phi(t_0))^{-1}\phi(t_0),$$

which when multiplied out is a linear combination of the $\phi_1, \ldots, \phi_n$.   □

---

[8]This actually follows from a general result in linear algebra.

We will say that an $n \times n$ matrix function $\boldsymbol{\Phi}(t)$ is a **fundamental matrix** for a homogeneous system $\boldsymbol{y}'(t) = A(t)\boldsymbol{y}(t)$ if its columns form a linearly independent set of solutions, as in (6). A matrix function $\boldsymbol{\Psi}(t)$ is the **standard fundamental matrix at** $t = t_0$ if it is a fundamental matrix and $\boldsymbol{\Psi}(t_0) = I$, the $n \times n$ identity matrix, as in (3). Given a fundamental matrix $\boldsymbol{\Phi}(t)$ Theorem 4, shows that the solution set to the homogeneous system $\boldsymbol{y}'(t) = A(t)\boldsymbol{y}(t)$ is the span[9] of the columns of $\boldsymbol{\Phi}(t)$.

Theorem 5. *Suppose $A(t)$ is an $n \times n$ continuous matrix function on an interval I. A matrix function $\boldsymbol{\Phi}(t)$ is a fundamental matrix for $\boldsymbol{y}'(t) = A(t)\boldsymbol{y}(t)$ if and only if*

$$\boldsymbol{\Phi}'(t) = A(t)\boldsymbol{\Phi}(t) \quad and \quad \det \boldsymbol{\Phi}(t) \neq 0,$$

for at least one $t \in I$. If this is true for one $t \in I$, it is in fact true for all $t \in I$. The standard fundamental matrix $\boldsymbol{\Psi}(t)$ at $t = t_0$ is uniquely characterized by the equations

$$\boldsymbol{\Psi}'(t) = A(t)\boldsymbol{\Psi}(t) \quad and \quad \boldsymbol{\Psi}(t_0) = I.$$

Furthermore, given a fundamental matrix $\boldsymbol{\Phi}(t)$, the standard fundamental matrix $\boldsymbol{\Psi}(t)$ at $t = t_0$ is given by the formula

$$\boldsymbol{\Psi}(t) = \boldsymbol{\Phi}(t)\left(\boldsymbol{\Phi}(t_0)\right)^{-1}.$$

Proof. Suppose $\boldsymbol{\Phi}(t) = \begin{bmatrix} \boldsymbol{\phi}_1 & \cdots & \boldsymbol{\phi}_n \end{bmatrix}$ is a fundamental matrix for $\boldsymbol{y}'(t) = A(t)\boldsymbol{y}(t)$. Then

$$
\begin{aligned}
\boldsymbol{\Phi}'(t) &= \begin{bmatrix} \boldsymbol{\phi}_1'(t) & \cdots & \boldsymbol{\phi}_n'(t) \end{bmatrix} \\
&= \begin{bmatrix} A(t)\boldsymbol{\phi}_1(t) & \cdots & A(t)\boldsymbol{\phi}_n(t) \end{bmatrix} \\
&= A(t)\begin{bmatrix} \boldsymbol{\phi}_1(t) & \cdots & \boldsymbol{\phi}_n(t) \end{bmatrix} \\
&= A(t)\boldsymbol{\Phi}(t).
\end{aligned}
$$

As in the proof above, $\boldsymbol{\Phi}(t_0)$ is invertible which implies $\det \boldsymbol{\Phi}(t_0) \neq 0$. Since $t_0 \in I$ is arbitrary, it follows that $\boldsymbol{\Phi}(t)$ has nonzero determinant for all $t \in \mathcal{I}$. Now suppose that $\boldsymbol{\Phi}(t)$ is a matrix function satisfying $\boldsymbol{\Phi}'(t) = A(t)\boldsymbol{\Phi}(t)$ and $\det \boldsymbol{\Phi}(t) \neq 0$, for some point, $t = t_0$ say. Then the above calculation gives that each column $\boldsymbol{\phi}_i(t)$ of $\boldsymbol{\Phi}(t)$ satisfies $\boldsymbol{\phi}_i'(t) = A(t)\boldsymbol{\phi}(t)$. Suppose there are scalars $c_1, \ldots, c_n$ such that $c_1\boldsymbol{\phi}_1 + \cdots + c_n\boldsymbol{\phi}_n = 0$ as a function on I. If $\boldsymbol{c}$ is the column vector given as in (4), then

$$\boldsymbol{\Phi}(t_0)\boldsymbol{c} = c_1\boldsymbol{\phi}_1(t_0) + \cdots + c_n\boldsymbol{\phi}_n(t_0) = \boldsymbol{0}.$$

---

[9]Recall that "span" means the set of all linear combinations.

Since det $\boldsymbol{\Phi}(t_0) \neq 0$, it follows that $\boldsymbol{\Phi}(t_0)$ is invertible. Therefore, $c = 0$ and $\boldsymbol{\Phi}(t)$ is a fundamental matrix. Now suppose $\boldsymbol{\Psi}_1$ and $\boldsymbol{\Psi}_2$ are standard fundamental matrices at t_0. As fundamental matrices, their ith columns both are solutions to $y'(t) = A(t)y(t)$ and evaluate to e_i at t_0. By the existence and uniqueness theorem, they are equal. It follows that the standard fundamental matrix is unique. Finally, (7) gives the final statement $\boldsymbol{\Psi}(t) = \boldsymbol{\Phi}(t)(\boldsymbol{\Phi}(t_0))^{-1}$.   □

In the case $A(t) = A$ is a constant, the matrix exponential e^{At} is the standard fundamental matrix at $t = 0$ for the system $y' = Ay$. This follows from Lemma 9.5.1. More generally, $e^{A(t-t_0)}$ is the standard fundamental matrix at $t = t_0$.

Example 6. Show that

$$\boldsymbol{\Phi}(t) = \begin{bmatrix} e^{2t} & e^{-t} \\ 2e^{2t} & -e^{-t} \end{bmatrix}$$

is a fundamental matrix for the system $y' = \begin{bmatrix} 0 & 1 \\ 2 & 1 \end{bmatrix} y$. Find the standard fundamental matrix at $t = 0$.

▶ **Solution.** We first observe that

$$\boldsymbol{\Phi}'(t) = \begin{bmatrix} e^{2t} & e^{-t} \\ 2e^{2t} & -e^{-t} \end{bmatrix}' = \begin{bmatrix} 2e^{2t} & -e^{-t} \\ 4e^{2t} & e^{-t} \end{bmatrix}$$

and

$$A\boldsymbol{\Phi}(t) = \begin{bmatrix} 0 & 1 \\ 2 & 1 \end{bmatrix} \begin{bmatrix} e^{2t} & e^{-t} \\ 2e^{2t} & -e^{-t} \end{bmatrix} = \begin{bmatrix} 2e^{2t} & -e^{-t} \\ 4e^{2t} & e^{-t} \end{bmatrix}.$$

Thus, $\boldsymbol{\Phi}'(t) = A\boldsymbol{\Phi}(t)$. Observe also that

$$\boldsymbol{\Phi}(0) = \begin{bmatrix} 1 & 1 \\ 2 & -1 \end{bmatrix}$$

and this matrix has determinant -3. By Theorem 5, $\boldsymbol{\Phi}(t)$ is a fundamental matrix. The standard fundamental matrix at $t = 0$ is given by

$$\boldsymbol{\Psi}(t) = \boldsymbol{\Phi}(t)(\boldsymbol{\Phi}(0))^{-1}$$

$$= \begin{bmatrix} e^{2t} & e^{-t} \\ 2e^{2t} & -e^{-t} \end{bmatrix} \begin{bmatrix} 1 & 1 \\ 2 & -1 \end{bmatrix}^{-1}$$

$$= -\frac{1}{3} \begin{bmatrix} e^{2t} & e^{-t} \\ 2e^{2t} & -e^{-t} \end{bmatrix} \begin{bmatrix} -1 & -1 \\ -2 & 1 \end{bmatrix}$$

$$= \frac{1}{3} \begin{bmatrix} e^{2t} + 2e^{-t} & e^{2t} - e^{-t} \\ 2e^{2t} - 2e^{-t} & 2e^{2t} + e^{-t} \end{bmatrix}.$$

The reader is encouraged to compute e^{At} where $A = \begin{bmatrix} 0 & 1 \\ 2 & 1 \end{bmatrix}$ and verify that $\Psi(t) = e^{At}$.   ◄

Example 7. Show that

$$\Phi(t) = \begin{bmatrix} t^2 & t^3 \\ 2t & 3t^2 \end{bmatrix}$$

is a fundamental matrix for the system $y' = A(t)y$ where

$$A(t) = \begin{bmatrix} 0 & 1 \\ -6/t^2 & 4/t \end{bmatrix}.$$

Solve the system with initial condition $y(1) = \begin{bmatrix} 3 \\ 7 \end{bmatrix}$. Find the standard fundamental matrix at $t = 1$.

► **Solution.** Note that

$$\Phi'(t) = \begin{bmatrix} 2t & 3t^2 \\ 2 & 6t \end{bmatrix} = \begin{bmatrix} 0 & 1 \\ -6/t^2 & 4/t \end{bmatrix} \begin{bmatrix} t^2 & t^3 \\ 2t & 3t^2 \end{bmatrix} = A(t)\Phi(t),$$

while

$$\det \Phi(1) = \begin{bmatrix} 1 & 1 \\ 2 & 3 \end{bmatrix} = 1 \neq 0.$$

Hence, $\Phi(t)$ is a fundamental matrix. The general solution is of the form $y(t) = \Phi(t)c$, where $c = \begin{bmatrix} c_1 \\ c_2 \end{bmatrix}$. The initial condition implies

$$\begin{bmatrix} 3 \\ 7 \end{bmatrix} = \begin{bmatrix} 1 & 1 \\ 2 & 3 \end{bmatrix} \begin{bmatrix} c_1 \\ c_2 \end{bmatrix} = \begin{bmatrix} c_1 + c_2 \\ 2c_1 + 3c_2 \end{bmatrix}.$$

Solving for c gives $c_1 = 2$ and $c_2 = 1$. Thus,

$$y(t) = 2 \begin{bmatrix} t^2 \\ 2t \end{bmatrix} + \begin{bmatrix} t^3 \\ 3t^2 \end{bmatrix} = \begin{bmatrix} 2t^2 + t^3 \\ 4t + 3t^2 \end{bmatrix}.$$

The standard fundamental matrix is given by

$$\Psi(t) = \Phi(t)(\Phi(1))^{-1} = \begin{bmatrix} t^2 & t^3 \\ 2t & 3t^2 \end{bmatrix} \begin{bmatrix} 1 & 1 \\ 2 & 3 \end{bmatrix}^{-1}$$

$$= \begin{bmatrix} t^2 & t^3 \\ 2t & 3t^2 \end{bmatrix} \begin{bmatrix} 3 & -1 \\ -2 & 1 \end{bmatrix}$$

$$= \begin{bmatrix} 3t^2 - 2t^3 & -t^2 + t^3 \\ 6t - 6t^2 & -2t + 3t^2 \end{bmatrix}.$$

Observe that $\boldsymbol{\Phi}(0) = \begin{bmatrix} 0 & 0 \\ 0 & 0 \end{bmatrix}$ which has determinant 0. Why does this not prevent $\boldsymbol{\Phi}(t)$ from being a fundamental matrix?   ◀

You will recall that in Sect. 9.5, we used the inverse of the matrix exponential $\left(e^{tA}\right)^{-1} = e^{-tA}$ as an integrating factor for the constant coefficient system $\boldsymbol{y}' = A\boldsymbol{y} + \boldsymbol{f}$. As observed above, the matrix exponential e^{tA} is the standard fundamental matrix at $t = 0$. In the more general context, we will show that the inverse of any fundamental matrix $\boldsymbol{\Phi}(t)$ is an integrating factor for the system $\boldsymbol{y}'(t) = A(t)\boldsymbol{y}(t) + \boldsymbol{f}(t)$. To show this, however, particular care must be taken when calculating the derivative of the inverse of a matrix-valued function.

Lemma 8. *Suppose $\boldsymbol{\Phi}(t)$ is a differentiable and invertible $n \times n$ matrix-valued function. Then*

$$\frac{\mathrm{d}}{\mathrm{dt}}(\boldsymbol{\Phi}(t))^{-1} = -(\boldsymbol{\Phi}(t))^{-1} \boldsymbol{\Phi}'(t)(\boldsymbol{\Phi}(t))^{-1}.$$

Remark 9. Observe the order of the matrix multiplications. We are not assuming that $\boldsymbol{\Phi}(t)$ and its derivative $\boldsymbol{\Phi}'(t)$ commute.

Proof. We apply the definition of the derivative:

$$\frac{\mathrm{d}}{\mathrm{dt}}(\boldsymbol{\Phi}(t))^{-1} = \lim_{h \to 0} \frac{(\boldsymbol{\Phi}(t+h))^{-1} - (\boldsymbol{\Phi}(t))^{-1}}{h}$$

$$= \lim_{h \to 0} (\boldsymbol{\Phi}(t+h))^{-1} \frac{\boldsymbol{\Phi}(t) - \boldsymbol{\Phi}(t+h)}{h} (\boldsymbol{\Phi}(t))^{-1}$$

$$= -(\boldsymbol{\Phi}(t+h))^{-1} \lim_{h \to 0} \frac{\boldsymbol{\Phi}(t+h) - \boldsymbol{\Phi}(t)}{h} (\boldsymbol{\Phi}(t))^{-1}$$

$$= -(\boldsymbol{\Phi}(t))^{-1} \boldsymbol{\Phi}'(t)(\boldsymbol{\Phi}(t))^{-1}.$$

The second line is just a careful factoring of the first line. To verify this step, simply multiply out the second line.   □

If $\boldsymbol{\Phi}(t)$ is a scalar-valued function, then Lemma 8 reduces to the usual chain rule formula, $\frac{\mathrm{d}}{\mathrm{dt}}(\boldsymbol{\Phi}(t))^{-1} = -(\boldsymbol{\Phi}(t))^{-2}\boldsymbol{\Phi}'(t)$, since $\boldsymbol{\Phi}$ and $\boldsymbol{\Phi}'$ commute. For a matrix-valued function, we may not assume that $\boldsymbol{\Phi}$ and $\boldsymbol{\Phi}'$ commute.

Nonhomogeneous Linear Systems

We are now in a position to consider the solution method for the general linear system $y'(t) = A(t)y(t) + f(t)$ which we write in the form

$$y'(t) - A(t)y(t) = f(t). \tag{8}$$

Assume $A(t)$ and $f(t)$ are continuous on an interval I and an initial condition $y(t_0) = y_0$ is given. Suppose $\Phi(t)$ is a fundamental matrix for the associated homogeneous system $y'(t) = A(t)y(t)$. Then $(\Phi(t))^{-1}$ will play the role of an integrating factor and we will be able mimic the procedure found in Sect. 1.4. First observe from Lemma 8 and the product rule that we get

$$
\begin{aligned}
\left((\Phi(t))^{-1}y(t)\right)' &= (\Phi(t))^{-1}y'(t) - (\Phi(t))^{-1}\Phi'(t)(\Phi(t))^{-1}y(t) \\
&= (\Phi(t))^{-1}y'(t) - (\Phi(t))^{-1}A(t)\Phi(t)(\Phi(t))^{-1}y(t) \\
&= (\Phi(t))^{-1}y'(t) - (\Phi(t))^{-1}A(t)y(t) \\
&= (\Phi(t))^{-1}(y'(t) - A(t)y(t)).
\end{aligned}
$$

Thus, multiplying both sides of (8) by $(\Phi(t))^{-1}$ gives

$$\left((\Phi(t))^{-1}y(t)\right)' = (\Phi(t))^{-1}f(t).$$

Now change the variable from t to u and integrate both sides from t_0 to t where $t \in I$. We get

$$\int_{t_0}^t \left((\Phi(u))^{-1}y(u)\right)' du = \int_{t_0}^t (\Phi(u))^{-1}f(u)\,du.$$

The left side simplifies to $\left(\Phi(t)^{-1}y(t)\right) - \left(\Phi(t_0)^{-1}y(t_0)\right)$. Solving for $y(t)$, we get

$$y(t) = \Phi(t)(\Phi(t_0))^{-1}y_0 + \Phi(t)\int_{t_0}^t (\Phi(u))^{-1}f(u)\,du.$$

It is convenient to summarize this discussion in the following theorem.

Theorem 10. *Suppose $A(t)$ is an $n \times n$ matrix-valued function on an interval I and $f(t)$ is a $\mathbb{R}^n$-valued valued function, both continuous on an interval I. Let $\Phi(t)$ be any fundamental matrix for $y'(t) = A(t)y(t)$ and let $t_0 \in I$. Then*

$$y(t) = \Phi(t)(\Phi(t_0))^{-1}y_0 + \Phi(t)\int_{t_0}^t (\Phi(u))^{-1}f(u)\,du, \tag{9}$$

is the unique solution to

$$y'(t) = A(t)y(t) + f(t), \quad y(t_0) = y_0.$$

Analysis of the General Solution Set

The Particular Solution

When we set $y_0 = 0$, we obtain a single fixed solution which we denote by y_p. Specifically,

$$y_p(t) = \Phi(t) \int_{t_0}^{t} (\Phi(u))^{-1} f(u) \, du. \tag{10}$$

This is called a *particular solution*.

The Homogeneous Solution

When we set $f = 0$, we get the *homogeneous solution* y_h. Specifically,

$$y_h(t) = \Phi(t)(\Phi(t_0))^{-1} y_0. \tag{11}$$

Recall from Theorem 5 that the standard fundamental matrix at t_0 is $\Psi(t) = \Phi(t)(\Phi(t_0))^{-1}$. Thus, the homogeneous solution with initial value $y_h(t_0) = y_0$ is given by

$$y_h(t) = \Psi(t)y_0.$$

On the other hand, if we are interested in the set of homogeneous solutions, then we let y_0 vary. However, Theorem 5 states the $\Phi(t_0)$ is invertible so as y_0 varies over $\mathbb{R}^n$, so does $(\Phi(t_0))^{-1} y_0$. Thus, if we let $c = (\Phi(t_0))^{-1} y_0$, then we have that the set of homogeneous solution is

$$\{\Phi(t)c : c \in \mathbb{R}^n\}.$$

In other words, the set of homogeneous solution is the set of all linear combinations of the columns of $\Phi(t)$, as we observed earlier.

Example 11.   Solve the system $y'(t) = A(t)y(t) + f(t)$ where

$$A(t) = \begin{bmatrix} 0 & 1 \\ -6/t^2 & 4/t \end{bmatrix} \quad \text{and} \quad f(t) = \begin{bmatrix} t^2 \\ t \end{bmatrix}$$

▶ **Solution.** In Example 7, we verified that

$$\Phi(t) = \begin{bmatrix} t^2 & t^3 \\ 2t & 3t^2 \end{bmatrix}$$

is a fundamental matrix. It follows that the homogeneous solutions are of the form

$$y_h(t) = c_1 \begin{bmatrix} t^2 \\ 2t \end{bmatrix} + c_2 \begin{bmatrix} t^3 \\ 3t^2 \end{bmatrix},$$

where c_1 and c_2 are real scalars. To find the particular solution, it is convenient to set $t_0 = 1$ and use (10). We get

$$y_p(t) = \Phi(t) \int_1^t (\Phi(u))^{-1} f(u)\, du$$

$$= \Phi(t) \int_1^t \begin{bmatrix} 3/u^2 & -1/u \\ -2/u^3 & 1/u^2 \end{bmatrix} \begin{bmatrix} u^2 \\ u \end{bmatrix} du$$

$$= \Phi(t) \int_1^t \begin{bmatrix} 2 \\ -1/u \end{bmatrix} du$$

$$= \begin{bmatrix} t^2 & t^3 \\ 2t & 3t^2 \end{bmatrix} \begin{bmatrix} 2(t-1) \\ -\ln t \end{bmatrix}$$

$$= \begin{bmatrix} 2t^3 - 2t^2 - t^3 \ln t \\ 4t^2 - 4t - 3t^2 \ln t \end{bmatrix} = -2 \begin{bmatrix} t^2 \\ 2t \end{bmatrix} + 2 \begin{bmatrix} t^3 \\ 3t^2 \end{bmatrix} + \begin{bmatrix} -t^3 \ln t \\ -2t^2 - 3t^2 \ln t \end{bmatrix}.$$

It follows that the general solution may be written as

$$y(t) = y_h(t) + y_p(t)$$

$$= (c_1 - 2) \begin{bmatrix} t^2 \\ 2t \end{bmatrix} + (c_2 + 2) \begin{bmatrix} t^3 \\ 3t^2 \end{bmatrix} + \begin{bmatrix} -t^3 \ln t \\ -2t^2 - 3t^2 \ln t \end{bmatrix}$$

$$= C_1 \begin{bmatrix} t^2 \\ 2t \end{bmatrix} + C_2 \begin{bmatrix} t^3 \\ 3t^2 \end{bmatrix} + \begin{bmatrix} -t^3 \ln t \\ -2t^2 - 3t^2 \ln t \end{bmatrix},$$

where the last line is just a relabeling of the coefficients. In Example (7), we computed the standard fundamental matrix $\Psi(t)$. It could have been used in place of $\Phi(t)$ in the above computations. However, the simplicity of $\Phi(t)$ made it a better choice.   ◀

This example emphasized the need to have a fundamental matrix in order to go forward with the calculations of the particular and homogeneous solutions, (10) and (11), respectively. Computing a fundamental matrix is not always an easy task. In order to find a closed expression for it, we must place strong restrictions on the coefficient matrix $A(t)$. Let us consider one such restriction.

The Coefficient Matrix is a Functional Multiple of a Constant Matrix

Proposition 12. *Suppose $a(t)$ is a continuous function on an interval I. Let*

$$A(t) = a(t)A,$$

where A is a fixed $n \times n$ constant matrix. Then a fundamental matrix, $\Phi(t)$, for $y'(t) = A(t)y(t)$ is given by the formula

$$\Phi(t) = e^{b(t)A}, \tag{12}$$

where $b(t)$ is an antiderivative of $a(t)$. If $b(t) = \int_{t_0}^{t} a(u)\,du$, that is, $b(t)$ is chosen so that $b(t_0) = 0$, then (12) is the standard fundamental matrix at t_0.

Proof. Let

$$\Phi(t) = e^{b(t)A} = I + b(t)A + \frac{b^2(t)A^2}{2!} + \cdots.$$

Termwise differentiation gives

$$\begin{aligned}
\Phi'(t) &= b'(t)A + 2b(t)b'(t)\frac{A^2}{2!} + 3b^2(t)b'(t)\frac{A^3}{3!} + \cdots \\
&= a(t)A\left(I + b(t)A + \frac{b^2(t)A^2}{2!} + \cdots\right) \\
&= A(t)\Phi(t).
\end{aligned}$$

Since the matrix exponential is always invertible, it follows from Theorem 5 that $\Phi(t)$ is a fundamental matrix. If $b(t)$ is chosen so that $b(t_0) = 0$, then $\Phi(t_0) = e^{Au}|_{u=0} = I$, and hence, $\Phi(t)$ is the standard fundamental matrix at t_0.   □

Remark 13. We observe that $\Phi(t)$ may be computed by replacing u in e^{uA} with $b(t)$.

Example 14. Suppose $A(t) = (\tan t)A$, where

$$A = \begin{bmatrix} 2 & 3 \\ -1 & -2 \end{bmatrix}.$$

Find a fundamental matrix for $y'(t) = A(t)y(t)$. Assume $t \in (-\frac{\pi}{2}, \frac{\pi}{2})$.

▶ **Solution.** We first compute e^{Au}. The characteristic polynomial is

$$c_A(s) = \det(sI - A) = \det\begin{bmatrix} s-2 & -3 \\ 1 & s+2 \end{bmatrix} = s^2 - 1 = (s-1)(s+1).$$

It follows that $\mathcal{B}_{c_A} = \{e^u, e^{-u}\}$. Using Fulmer's method, we have $e^{Au} = e^u M_1 + e^{-u} M_2$. Differentiating and evaluating at $u = 0$ gives

$$I = M_1 + M_2,$$
$$A = M_1 - M_2.$$

It follows that $M_1 = \frac{1}{2}(A + I) = \frac{1}{2}\begin{bmatrix} 3 & 3 \\ -1 & -1 \end{bmatrix}$ and $M_2 = \frac{1}{2}(I - A) = \frac{1}{2}\begin{bmatrix} -1 & -3 \\ 1 & 3 \end{bmatrix}$. Thus,

$$e^{Au} = \frac{1}{2}e^u \begin{bmatrix} 3 & 3 \\ -1 & -1 \end{bmatrix} + e^{-u}\begin{bmatrix} -1 & -3 \\ 1 & 3 \end{bmatrix}.$$

Since $\ln \sec t$ is an antiderivative of $\tan t$, we have by Proposition 12

$$\Phi(t) = \frac{1}{2}e^{\ln \sec t}\begin{bmatrix} 3 & 3 \\ -1 & -1 \end{bmatrix} + \frac{1}{2}e^{-\ln \sec t}\begin{bmatrix} -1 & -3 \\ 1 & 3 \end{bmatrix}$$
$$= \frac{1}{2}\begin{bmatrix} 3\sec t - \cos t & 3\sec t - 3\cos t \\ -\sec t + \cos t & -\sec t + 3\cos t \end{bmatrix},$$

is the fundamental matrix for $y'(t) = A(t)y(t)$.   ◀

Example 15. Solve the mixing problem introduced at the beginning of this section. Specifically, solve

$$y' = A(t)y(t) + f(t), \quad y(0) = y_0,$$

where

$$A(t) = \begin{bmatrix} -\frac{2}{1+t} & \frac{1}{1+t} \\ \frac{1}{1+t} & -\frac{2}{1+t} \end{bmatrix}, \quad f(t) = \begin{bmatrix} 12 \\ 20 \end{bmatrix}, \quad \text{and} \quad y(0) = \begin{bmatrix} 8 \\ 0 \end{bmatrix}.$$

Determine the concentration of salt in each tank after 3 minutes. In the long term, what are the concentrations of salt in each tank?

▶ **Solution.** Let $a(t) = \frac{1}{t+1}$. Then $A(t) = a(t)A$ where $A = \begin{bmatrix} -2 & 1 \\ 1 & -2 \end{bmatrix}$. The characteristic polynomial of A is

$$c_A(s) = \det(sI - A) = \det \begin{bmatrix} s+2 & -1 \\ -1 & s+2 \end{bmatrix} = (s+1)(s+3).$$

A short calculation gives the resolvent matrix

$$(sI - A)^{-1} = \frac{1}{(s+1)(s+3)} \begin{bmatrix} s+2 & 1 \\ 1 & s+2 \end{bmatrix}$$

$$= \frac{1}{2(s+1)} \begin{bmatrix} 1 & 1 \\ 1 & 1 \end{bmatrix} + \frac{1}{2(s+3)} \begin{bmatrix} 1 & -1 \\ -1 & 1 \end{bmatrix}$$

and hence

$$e^{Au} = \frac{e^{-u}}{2} \begin{bmatrix} 1 & 1 \\ 1 & 1 \end{bmatrix} + \frac{e^{-3u}}{2} \begin{bmatrix} 1 & -1 \\ -1 & 1 \end{bmatrix}.$$

Since $\ln(t+1)$ is an antiderivative of $a(t) = \frac{1}{t+1}$, we have by Proposition 12

$$\Phi(t) = e^{\ln(t+1)A}$$

$$= \frac{e^{-\ln(t+1)}}{2} \begin{bmatrix} 1 & 1 \\ 1 & 1 \end{bmatrix} + \frac{e^{-3(\ln(t+1))}}{2} \begin{bmatrix} 1 & -1 \\ -1 & 1 \end{bmatrix}$$

$$= \frac{1}{2(t+1)} \begin{bmatrix} 1 & 1 \\ 1 & 1 \end{bmatrix} + \frac{1}{2(t+1)^3} \begin{bmatrix} 1 & -1 \\ -1 & 1 \end{bmatrix}$$

$$= \frac{1}{2(t+1)^3} \begin{bmatrix} (t+1)^2 + 1 & (t+1)^2 - 1 \\ (t+1)^2 - 1 & (t+1)^2 + 1 \end{bmatrix}$$

a fundamental matrix for $y'(t) = A(t)y(t)$. Observe that $\Phi(0) = I$ is the 2×2 identity matrix so that, in fact, $\Phi(t)$ is the standard fundamental matrix at $t = 0$. The homogeneous solution is now easily calculated:

$$y_h(t) = \Phi(t)y_0$$

$$= \frac{1}{2(t+1)^3} \begin{bmatrix} (t+1)^2 + 1 & (t+1)^2 - 1 \\ (t+1)^2 - 1 & (t+1)^2 + 1 \end{bmatrix} \begin{bmatrix} 8 \\ 0 \end{bmatrix}$$

$$= \frac{4}{(t+1)^3} \begin{bmatrix} (t+1)^2 + 1 \\ (t+1)^2 - 1 \end{bmatrix}.$$

A straightforward calculation gives

$$(\Phi(u))^{-1} = \frac{u+1}{2} \begin{bmatrix} (u+1)^2 + 1 & 1 - (u+1)^2 \\ 1 - (u+1)^2 & (u+1)^2 + 1 \end{bmatrix}$$

and

$$(\boldsymbol{\Phi}(u))^{-1}\boldsymbol{f}(u) = \frac{u+1}{2}\begin{bmatrix}(u+1)^2+1 & 1-(u+1)^2 \\ 1-(u+1)^2 & (u+1)^2+1\end{bmatrix}\begin{bmatrix}12 \\ 20\end{bmatrix}$$

$$= \begin{bmatrix}16(u+1)-4(u+1)^3 \\ 16(u+1)+4(u+1)^3\end{bmatrix}.$$

For the particular solution, we have

$$\boldsymbol{y}_{\mathrm{p}}(t) = \boldsymbol{\Phi}(t)\int_0^t (\boldsymbol{\Phi}(u))^{-1}\boldsymbol{f}(u)\,du$$

$$= \boldsymbol{\Phi}(t)\begin{bmatrix}8(u+1)^2-(u+1)^4 \\ 8(u+1)^2+(u+1)^4\end{bmatrix}\Bigg|_0^t$$

$$= \frac{1}{2(t+1)^3}\begin{bmatrix}(t+1)^2+1 & (t+1)^2-1 \\ (t+1)^2-1 & (t+1)^2+1\end{bmatrix}\begin{bmatrix}8(t+1)^2-(t+1)^4-7 \\ 8(t+1)^2+(t+1)^4-9\end{bmatrix}$$

$$= \frac{1}{(t+1)^3}\begin{bmatrix}7(t+1)^4-8(t+1)^2+1 \\ 9(t+1)^4-8(t+1)^2-1\end{bmatrix}.$$

Putting the homogeneous and particular solutions together gives

$$\boldsymbol{y}(t) = \boldsymbol{y}_{\mathrm{h}}(t) + \boldsymbol{y}_{\mathrm{p}}(t)$$

$$= \frac{4}{(t+1)^3}\begin{bmatrix}(t+1)^2+1 \\ (t+1)^2-1\end{bmatrix} + \frac{1}{(t+1)^3}\begin{bmatrix}7(t+1)^4-8(t+1)^2+1 \\ 9(t+1)^4-8(t+1)^2-1\end{bmatrix}$$

$$= \frac{1}{(t+1)^3}\begin{bmatrix}7(t+1)^4-4(t+1)^2+5 \\ 9(t+1)^4-4(t+1)^2-5\end{bmatrix}.$$

Finally, the amount of fluid in each tank is $v_1(t) = v_2(t) = t+1$. Thus, the concentration of salt in each tank is given by

$$\frac{1}{t+1}\boldsymbol{y}(t) = \frac{1}{(t+1)^4}\begin{bmatrix}7(t+1)^4-4(t+1)^2+5 \\ 9(t+1)^4-4(t+1)^2-5\end{bmatrix}.$$

Evaluating at $t=3$ gives concentrations

$$\begin{bmatrix}\frac{1733}{256} \\ \frac{2235}{256}\end{bmatrix} = \begin{bmatrix}6.77 \\ 8.73\end{bmatrix}.$$

In the long term, the concentrations are obtained by taking limits

$$\lim_{t \to \infty} \frac{1}{(t+1)^4} \begin{bmatrix} 7(t+1)^4 - 4(t+1)^2 + 5 \\ 9(t+1)^4 - 4(t+1)^2 - 5 \end{bmatrix} = \begin{bmatrix} 7 \\ 9 \end{bmatrix}.$$

Of course, the tank overflows in the long term, but for a sufficiently large tank, we can expect that the concentrations in each tank will be near 7g/L and 9g/L, respectively, for large values of t. ◄

Exercises

1–7. For each of the following pairs of matrix functions $\boldsymbol{\Phi}(t)$ and $A(t)$, verify that $\boldsymbol{\Phi}(t)$ is a fundamental matrix for the system $y' = A(t)y$. Use this fact to solve $y' = A(t)y$ with the given initial conditions $y(t_0) = y_0$. Next, determine the standard fundamental matrix at t_0.

1. $\boldsymbol{\Phi}(t) = \begin{bmatrix} e^{-t} & e^{2t} \\ e^{-t} & 4e^{2t} \end{bmatrix}$, $A(t) = \begin{bmatrix} -2 & 1 \\ -4 & 3 \end{bmatrix}$, $y(0) = \begin{bmatrix} 1 \\ -2 \end{bmatrix}$

2. $\boldsymbol{\Phi}(t) = \begin{bmatrix} e^{2t} & 3e^{3t} \\ e^{2t} & 2e^{3t} \end{bmatrix}$, $A(t) = \begin{bmatrix} 5 & -3 \\ 2 & 0 \end{bmatrix}$, $y(0) = \begin{bmatrix} 2 \\ 1 \end{bmatrix}$

3. $\boldsymbol{\Phi}(t) = \begin{bmatrix} \sin(t^2/2) & \cos(t^2/2) \\ \cos(t^2/2) & -\sin(t^2/2) \end{bmatrix}$, $A(t) = \begin{bmatrix} 0 & t \\ -t & 0 \end{bmatrix}$, $y(0) = \begin{bmatrix} 1 \\ 0 \end{bmatrix}$

4. $\boldsymbol{\Phi}(t) = \begin{bmatrix} 1+t^2 & 3+t^2 \\ 1-t^2 & -1-t^2 \end{bmatrix}$, $A(t) = \begin{bmatrix} t & t \\ -t & -t \end{bmatrix}$, $y(0) = \begin{bmatrix} 4 \\ 0 \end{bmatrix}$

5. $\boldsymbol{\Phi}(t) = \begin{bmatrix} -t\cos t & -t\sin t \\ t\sin t & -t\cos t \end{bmatrix}$, $A(t) = \begin{bmatrix} 1/t & 1 \\ -1 & 1/t \end{bmatrix}$, $y(\pi) = \begin{bmatrix} 1 \\ -1 \end{bmatrix}$

6. $\boldsymbol{\Phi}(t) = e^t \begin{bmatrix} 1 & t^2 \\ -1 & 1-t^2 \end{bmatrix}$, $A(t) = \begin{bmatrix} 1+2t & 2t \\ -2t & 1-2t \end{bmatrix}$, $y(0) = \begin{bmatrix} 1 \\ 0 \end{bmatrix}$

7. $\boldsymbol{\Phi}(t) = \begin{bmatrix} 1 & (t-1)e^t \\ -1 & e^t \end{bmatrix}$, $A(t) = \begin{bmatrix} 1 & 1 \\ 1/t & 1/t \end{bmatrix}$, $y(1) = \begin{bmatrix} -3 \\ 4 \end{bmatrix}$

8–12. For each problem below, compute the standard fundamental matrix for the system $y'(t) = A(t)y(t)$ at the point given in the initial value. Then solve the initial value problem $y'(t) = A(t)y(t) + f(t)$, $y(t_0) = y_0$.

8. $A(t) = \dfrac{1}{t}\begin{bmatrix} 1 & 0 \\ 0 & 1 \end{bmatrix}$, $f(t) = \begin{bmatrix} 1 \\ t \end{bmatrix}$, $y(1) = \begin{bmatrix} 1 \\ 2 \end{bmatrix}$

9. $A(t) = \dfrac{1}{t}\begin{bmatrix} 0 & -1 \\ 1 & 2 \end{bmatrix}$, $f(t) = \begin{bmatrix} 1 \\ -1 \end{bmatrix}$, $y(1) = \begin{bmatrix} 2 \\ 0 \end{bmatrix}$

10. $A(t) = \dfrac{2t}{t^2+1}\begin{bmatrix} 2 & 3 \\ -1 & -2 \end{bmatrix}$, $f(t) = \begin{bmatrix} -3t \\ t \end{bmatrix}$, $y(0) = \begin{bmatrix} 1 \\ -1 \end{bmatrix}$

11. $A(t) = \begin{bmatrix} 3\sec t & 5\sec t \\ -\sec t & -3\sec t \end{bmatrix}$, $f(t) = \begin{bmatrix} 0 \\ 0 \end{bmatrix}$, $y(0) = \begin{bmatrix} 2 \\ 1 \end{bmatrix}$

12. $A(t) = \begin{bmatrix} t & t \\ -t & -t \end{bmatrix}$, $f(t) = \begin{bmatrix} 4t \\ 4t \end{bmatrix}$, $y(0) = \begin{bmatrix} 4 \\ 0 \end{bmatrix}$

13–14. Solve the following mixing problems.

13. Two tanks are interconnected as illustrated below.

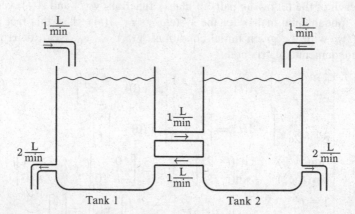

Assume that Tank 1 contains 2 Liters of pure water and Tank 2 contains 2 Liters of brine in which 20 grams of salt is initially dissolved. Moreover, the mixture is pumped from each tank to the other at a rate of 1 L/min. Assume that a brine mixture containing 6 grams salt/L enters Tank 1 at a rate of 1 L/min and pure water enters Tank 2 at a rate of 1 L/min. Assume the tanks are well stirred. Brine is removed from Tank 1 at the rate 2 L/min and from Tank 2 at a rate of 2 L/min. Let $y_1(t)$ be the amount of salt in Tank 1 at time t and let $y_2(t)$ be the amount of salt in Tank 2 at time t. Determine y_1 and y_2. What is the concentration of salt in each tank after 1 minute?

14. Two tanks are interconnected as illustrated below.

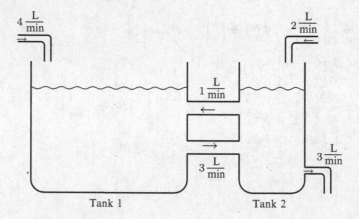

Assume initially that Tank 1 contains 2 Liters of pure water and Tank 2 contains 1 Liter of pure water. Moreover, the mixture from Tank 1 is pumped into Tank 2

at a rate of 3 L/min and the mixture from Tank 2 is pumped into Tank 1 at a rate of 1 L/min. Assume that a brine mixture containing 7 grams salt/L enters Tank 1 at a rate of 4 L/min and a brine mixture containing 14 grams salt/L enters Tank 2 at a rate of 2 L/min. Assume the tanks are well stirred. Brine is removed from Tank 2 at the rate 3 L/min. Let $y_1(t)$ be the amount of salt in Tank 1 at time t and let $y_2(t)$ be the amount of salt in Tank 2 at time t. Determine y_1 and y_2. Find the concentration of salt in each tank after 3 minutes. Assuming the tanks are large enough, what are the long-term concentrations of brine?

Appendix A
Supplements

A.1 The Laplace Transform is Injective

In this section, we prove Theorem 1 of Sect. 2.5 that states that the Laplace transform is injective on the set of Laplace transformable continuous functions.[1] Specifically, the statement is

Theorem 1. *Suppose f_1 and f_2 are continuous functions on $[0, \infty)$ and have Laplace transforms. Suppose*

$$\mathcal{L}\{f_1\} = \mathcal{L}\{f_2\}.$$

Then $f_1 = f_2$.

The proof of this statement is nontrivial. It requires a well-known result from advanced calculus which we will assume: the Weierstrass approximation theorem.

Theorem 2 (The Weierstrass Approximation Theorem). *Suppose h is a continuous function on $[0, 1]$. Then for any $\epsilon > 0$, there is a polynomial p such that*

$$|h(t) - p(t)| < \epsilon,$$

for all t in $[0, 1]$.

In essence, the Weierstrass approximation theorem states that a continuous function can be approximated by a polynomial to any degree of accuracy.

---

[1]The presentation here closely follows that found in *Advanced Calculus* by David Widder, published by Prentice Hall, 1961.

W.A. Adkins and M.G. Davidson, *Ordinary Differential Equations*,
Undergraduate Texts in Mathematics, DOI 10.1007/978-1-4614-3618-8,
© Springer Science+Business Media New York 2012

Lemma 3. *Suppose $h(t)$ is a continuous function so that*

$$\int_0^1 t^n h(t)\, dt = 0,$$

for each nonnegative integer n. Then $h(t) = 0$ for all $t \in [0, 1]$.

Proof. Let $\epsilon > 0$. By the Weierstrass approximation theorem, there is a polynomial p so that

$$|h(t) - p(t)| < \epsilon,$$

for all t in $[0, 1]$. Since a polynomial is a linear combination of powers of t, it follows by the linearity of the integral that $\int_0^1 p(t)h(t)\, dt = 0$. Now observe,

$$\int_0^1 (h(t))^2\, dt = \int_0^1 h(t)(h(t) - p(t))\, dt$$

$$\leq \int_0^1 |h(t)|\,|h(t) - p(t)|\, dt$$

$$\leq \epsilon \int_0^1 |h(t)|\, dt.$$

Since ϵ is arbitrary, it follows that $\int_0^1 (h(t))^2\, dt$ can be made as small as we like. This forces $\int_0^1 (h(t))^2\, dt = 0$. Since $(h(t))^2 \geq 0$, it follows that $(h(t))^2 = 0$, for all $t \in [0, 1]$. Therefore, $h(t) = 0$ for all $t \in [0, 1]$. $\square$

Theorem 4. *Suppose f is a continuous function on the interval $[0, \infty)$, $F(s) = \mathcal{L}\{f(t)\}(s)$ for $s \geq a$ and $F(a + nl) = 0$ for all $n = 0, 1, \ldots$, for some $l > 0$. Then $f \equiv 0$.*

Proof. Let $g(t) = \int_0^t e^{-au} f(u)\, du$. Since $F(a) = 0$, it follows that $\lim_{t \to \infty} g(t) = 0$. Write

$$F(a + nl) = \int_0^\infty e^{-(a+nl)t} f(t)\, dt = \int_0^\infty e^{-nlt} e^{-at} f(t)\, dt$$

and compute using integration by parts with $u = e^{-nlt}$ and $dv = e^{-at} f(t)$. Since $du = -nl e^{-nlt}$ and $v = \int_0^t e^{-au} f(u)\, du = g(t)$, we have

$$F(a + nl) = e^{-nlt} g(t)\Big|_0^\infty + nl \int_0^\infty e^{-nlt} g(t)\, dt.$$

$$= nl \int_0^\infty e^{-nlt} g(t)\, dt.$$

Since $F(a + nl) = 0$ we have

$$\int_0^\infty e^{-nlt} g(t) \, dt = 0,$$

for all $n = 1, 2, \ldots$. Now let $x = e^{-lt}$. Then $dx = -le^{-lt} dt = -lx \, dt$ and $t = -\frac{1}{l} \ln x = \frac{1}{l} \ln \frac{1}{x}$. Substituting and simplifying, we get

$$\int_0^1 x^{n-1} g\left(\frac{1}{l} \ln \frac{1}{x}\right) dx = 0,$$

for all $n = 1, 2, \ldots$. By Lemma 3, it follows that $g\left(\frac{1}{l} \ln \frac{1}{x}\right) = 0$ for all $x \in [0, 1]$, and hence, $g(t) = 0$ on $[0, \infty)$. Since $0 = g'(t) = e^{-at} f(t)$ it follows now that $f(t) = 0$ for all $t \in [0, \infty)$.

Proof (Proof of Theorem 1). Suppose $f(t) = f_1(t) - f_2(t)$. Then $\mathcal{L}\{f\}(s) = 0$ for all s. By Theorem, 4 it follows that f is zero and so $f_1 = f_2$.   □

A.2   Polynomials and Rational Functions

A *polynomial of degree* n is a function of the form

$$p(s) = a_n s^n + a_{n-1} s^{n-1} + \cdots + a_1 s + a_0,$$

where $a_n \neq 0$. The *coefficients* $a_0, \ldots, a_n$ may be real or complex. We refer to a_n as the *leading coefficient*. If the coefficients are all real, we say $p(s)$ is a *real polynomial*. The variable s may also be real or complex. A *root* of $p(s)$ is a scalar r such that $p(r) = 0$. Again r may be real or complex. If r is a root of $p(s)$, then there is another polynomial $p_1(s)$ of degree $n - 1$ such that

$$p(s) = (s - r) p_1(s).$$

The polynomial $p_1(s)$ may be obtained by the standard procedure of division of polynomials.

Even though the coefficients may be real, the polynomial $p(s)$ may only have nonreal complex roots. For example, $s^2 + 1$ only has i and $-$i as roots. Notice in this example that the roots are complex conjugates. This always happens with real polynomials.

Proposition 1. *Suppose $p(s)$ is a real polynomial and $r \in \mathbb{C}$ is a root. Then $\bar{r}$ is also a root.*

Proof. Suppose $p(s) = a_n s^n + \cdots + a_1 s + a_0$, with each coefficient in $\mathbb{R}$. We are given that $p(r) = 0$ from which follows

$$p(\bar{r}) = a_n \bar{r}^n + \cdots + a_1 \bar{r} + a_0$$
$$= \overline{a_n r^n} + \cdots + \overline{a_1 r} + \overline{a_0}$$
$$= \overline{a_n r^n + \cdots + a_1 r + a_0}$$
$$= \overline{p(r)} = \bar{0} = 0.$$

Thus $\bar{r}$ is also a root.   □

The fundamental theorem of algebra addresses the question of whether a polynomial has a root.

Theorem 2 (Fundamental Theorem of Algebra). *Let $p(s)$ be a polynomial of degree greater than 0. Then $p(s)$ has a root $r \in \mathbb{C}$.*

The following corollary follows immediately from the fundamental theorem of algebra.

Corollary 3. *Let $p(s)$ be a polynomial of degree n and $n \geq 1$. Then there are roots $r_1, \ldots, r_n \in \mathbb{C}$ such that*

$$p(s) = a_n(s - r_1) \cdots (s - r_n),$$

where a_n is the leading coefficient of $p(s)$.

Each term of the form $s - r$ is called a ***linear term***: if $r \in \mathbb{R}$ it is a real linear term, and if $r \in \mathbb{C}$, it is a complex linear term. An ***irreducible quadratic*** is a real polynomial $p(s)$ of degree 2 that has no *real* roots. In this case, we may write $p(s) = as^2 + bs + c$ as a sum of squares by a procedure called ***completing the square***:

$$p(s) = as^2 + bs + c$$
$$= a\left(s^2 + \frac{b}{a}s + \frac{c}{a}\right)$$
$$= a\left(s^2 + \frac{b}{a}s + \frac{b^2}{4a^2} + \frac{4ca - b^2}{4a^2}\right)$$
$$= a\left(\left(s + \frac{b}{2a}\right) + \left(\frac{\sqrt{4ca - b^2}}{2a}\right)^2\right)$$
$$= a\left((s - \alpha)^2 + \beta^2\right),$$

where we set $\alpha = -\frac{b}{2a}$ and $\beta = \frac{\sqrt{4ca-b^2}}{2a}$. From this form, we may read off the complex roots $r = \alpha + i\beta$ and $\bar{r} = \alpha - i\beta$. We further observe that $as^2 + bs + c = a(s - (\alpha + i\beta))(s - (\alpha - i\beta))$.

Corollary 4. *If $p(s)$ is a real polynomial, then $p(s)$ is a product of real linear terms or irreducible quadratics.*

Proof. By fundamental theorem of algebra, $p(s)$ is a product of linear factors of the form $s - r$. By Proposition 1, we have for each nonreal linear factor $s - r$ a corresponding nonreal factor $s - \bar{r}$ of $p(s)$. As observed above, $(s - r)(s - \bar{r})$ is an irreducible quadratic. It follows then that $p(s)$ is a product of real linear terms and irreducible quadratics.

Corollary 5. *Suppose $p(s)$ is a polynomial of degree n and has $m > n$ roots. Then $p(s) = 0$ for all $s \in \mathbb{R}$.*

Proof. This is an immediate consequence of Corollary 3.   □

Corollary 6. *Suppose $p_1(s)$ and $p_2(s)$ are polynomials and equal for all $s > A$, for some real number A. Then $p_1(s) = p_2(s)$, for all $s \in \mathbb{R}$.*

Proof. The polynomial $p_1(s) - p'_2(s)$ has infinitely many roots so must be zero, identically. Hence, $p_1(s) = p_2(s)$ for all $s \in \mathbb{R}$.   □

A **rational function** is a quotient of two polynomials, that is, it takes the form $\frac{p(s)}{q(s)}$. A rational function is **proper** if the degree of the numerator is less than the degree of the denominator.

Corollary 7. *Suppose $\frac{p_1(s)}{q_1(s)}$ and $\frac{p_2(s)}{q_2(s)}$ are rational functions that are equal for all $s > A$ for some real number A. Then they are equal for all s such that $q_1(s)q_2(s) \neq 0$.*

Proof. Suppose

$$\frac{p_1(s)}{q_1(s)} = \frac{p_2(s)}{q_2(s)}$$

for all $s > A$. Then

$$p_1(s)q_2(s) = p_2(s)q_1(s),$$

for all $s > A$. Since both sides are polynomials, this implies that $p_1(s)q_2(s) = p_2(s)q_1(s)$ for all $s \in \mathbb{R}$. Dividing by $q_1(s)q_2(s)$ gives the result.   □

A.3 $\mathcal{B}_q$ Is Linearly Independent and Spans $\mathcal{E}_q$

$\mathcal{B}_q$ Spans $\mathcal{E}_q$

This subsection is devoted to a detailed proof of Theorem 2 of Sect. 2.7. To begin, we will need a few helpful lemmas.

Lemma 1. *Suppose $q(s)$ is a polynomial which factors in the following way:*
$q(s) = q_1(s)q_2(s)$. *Then*

$$\mathcal{B}_{q_1} \subset \mathcal{B}_q,$$
$$\mathcal{R}_{q_1} \subset \mathcal{R}_q,$$
$$\mathcal{E}_{q_1} \subset \mathcal{E}_q.$$

(Of course, the same inclusions for q_2 are valid.)

Proof. Since any irreducible factor (linear or quadratic) of q_1 is a factor of q, it follows by the way $\mathcal{B}_q$ is defined using linear and irreducible quadratic factors of $q(s)$ that $\mathcal{B}_{q_1} \subset \mathcal{B}_q$. Suppose $p_1(s)/q_1(s) \in \mathcal{R}_{q_1}$. Then $p_1(s)/q_1(s) = p_1(s)q_2(s)/q_1(s)q_2(s) = p_1(s)q_2(s)/q(s) \in \mathcal{R}_q$. It follows that $\mathcal{R}_{q_1} \subset \mathcal{R}_q$. Finally, if $f \in \mathcal{E}_{q_1}$, then $\mathcal{L}\{f\} \in \mathcal{R}_{q_1} \subset \mathcal{R}_q$. Hence, $f \in \mathcal{E}_q$ and therefore $\mathcal{E}_{q_1} \subset \mathcal{E}_q$.   □

Lemma 2. *Let $q(s)$ be a polynomial of degree $n \geq 1$. Then*

$$\mathcal{B}_q \subset \mathcal{E}_q.$$

Proof. We proceed by induction on the degree of $q(s)$. If the degree of $q(s) = 1$, then we can write $q(s) = a(s - \lambda)$, and in this case, $\mathcal{B}_q = \{e^{\lambda t}\}$. Since

$$\mathcal{L}\{e^{\lambda t}\} = \frac{1}{s - \lambda} = \frac{a}{a(s - \lambda)} = \frac{a}{q(s)} \in \mathcal{R}_q$$

it follows that $e^{\lambda t} \in \mathcal{E}_q$. Hence, $\mathcal{B}_q \subset \mathcal{E}_q$. Now suppose $\deg q(s) > 1$. According to the fundamental theorem of algebra $q(s)$ must have a linear or irreducible quadratic factor. Thus, $q(s)$ factors in one of the following ways:

1. $q(s) = (s - \lambda)^k q_1(s)$, where $k \geq 1$ and $q_1(s)$ does not contain $s - \lambda$ as a factor.
2. $q(s) = ((s - \alpha)^2 + \beta^2)^k q_1(s)$, where $k \geq 1$ and $q_1(s)$ does not contain $(s - \alpha)^2 + \beta^2$ as a factor.

Since the degree of q_1 is less than the degree of q, we have in both cases by induction that $\mathcal{B}_{q_1} \subset \mathcal{E}_{q_1}$. Lemma 1 implies that $\mathcal{B}_{q_1} \subset \mathcal{E}_q$.

Case 1: $q(s) = (s - \lambda)^k q_1(s)$. Let $f \in \mathcal{B}_q$ be a simple exponential polynomial. Then either $f(t) = t^r e^{\lambda t}$, for some nonnegative integer r less than k, or $f \in \mathcal{B}_{q_1} \subset \mathcal{E}_q$. If $f(t) = t^r e^{\lambda t}$ then, $\mathcal{L}\{f\} = r!/(s - \lambda)^{r+1} \in \mathcal{R}_{(s-\lambda)^{r+1}} \subset \mathcal{R}_q$ by Lemma 1. Thus, $f \in \mathcal{E}_q$, and hence $\mathcal{B}_q \subset \mathcal{E}_q$.

Case 2: $q(s) = ((s - \alpha)^2 + \beta^2)^k q_1(s)$. Let $f \in \mathcal{B}_q$. Then either $f(t) = t^r e^{\alpha t}$ trig βt, where trig is sin or cos and r is a nonnegative integer less than k, or $f \in \mathcal{B}_{q_1} \subset \mathcal{E}_q$. If $f(t) = t^r e^{\alpha t}$ trig βt, then by Lemma 10 of Sect. 2.6, $\mathcal{L}\{f(t)\} \in \mathcal{R}_{((s-\alpha)^2+\beta^2)^k}$ and hence, by Lemma 1, $\mathcal{L}\{f(t)\} \in \mathcal{R}_q$. It follows that $f \in \mathcal{E}_q$. Hence $\mathcal{B}_q \subset \mathcal{E}_q$.   □

Proof (of Theorem 2 of Sect. 2.7). Since $\mathcal{E}_q$ is a linear space by Proposition 2 of Sect. 2.6 and $\mathcal{B}_q \subset \mathcal{E}_q$ by Lemma 2, we have

$$\text{Span } \mathcal{B}_q \subset \mathcal{E}_q.$$

To show $\mathcal{E}_q \subset \text{Span } \mathcal{B}_q$, we proceed by induction on the degree of q. Suppose $\deg q(s) = 1$. Then $q(s)$ may be written $q(s) = a(s - \lambda)$, for some constants $a \neq 0$ and $\lambda \in \mathbb{R}$ and $\mathcal{B}_q = \{e^{\lambda t}\}$. If $f \in \mathcal{E}_q$, then $\mathcal{L}\{f\} = c/(a(s - \lambda))$, for some constant c, and hence, $f = (c/a)e^{\lambda t}$. So $f \in \text{Span } \mathcal{B}_q$ and it follows that $\mathcal{E}_q = \text{Span } \mathcal{B}_q$.

Now suppose $\deg q > 1$. Then $q(s)$ factors in one of the following ways:

1. $q(s) = (s - \lambda)^k q_1(s)$, where $k \geq 1$ and $q_1(s)$ does not contain $s - \lambda$ as a factor.
2. $q(s) = ((s - \alpha)^2 + \beta^2)^k q_1(s)$, where $k \geq 1$ and $q_1(s)$ does not contain $(s - \alpha)^2 + \beta^2$ as a factor.

Since the degree of q_1, is less than the degree of q we have in both cases by induction that $\mathcal{E}_{q_1} = \text{Span } \mathcal{B}_{q_1}$.

Consider the first case: $q(s) = (s - \lambda)^k q_1(s)$. If $f \in \mathcal{E}_q$, then $\mathcal{L}\{f\}$ has a partial fraction decomposition of the form

$$\mathcal{L}\{f\}(s) = \frac{A_1}{s - \lambda} + \cdots + \frac{A_k}{(s - \lambda)^k} + \frac{p_1(s)}{q_1(s)},$$

for some constants $A_1, \ldots, A_k$ and polynomial $p_1(s)$. Taking the inverse Laplace transform, it follows that f is a linear combination of terms in $\{e^{\lambda t}, \ldots, t^{k-1}e^{\lambda t}\}$ and a function in $\mathcal{E}_{q_1}$. But since $\mathcal{E}_{q_1} = \text{Span } \mathcal{B}_{q_1}$ and $\mathcal{B}_q = \{e^{\lambda t}, \ldots, t^{k-1}e^{\lambda t}\} \cup \mathcal{B}_{q_1}$, it follows that $f \in \text{Span } \mathcal{B}_q$ and hence $\mathcal{E}_q = \text{Span } \mathcal{B}_q$.

Now consider the second case: $q(s) = ((s - \alpha)^2 + \beta^2)^k q_1(s)$. If $f \in \mathcal{E}_q$, then $\mathcal{L}\{f\}$ has a partial fraction decomposition of the form

$$\mathcal{L}\{f\}(s) = \frac{A_1 s + B_1}{(s - \alpha)^2 + \beta^2} + \cdots + \frac{A_k s + B_k}{((s - \alpha)^2 + \beta^2)^k} + \frac{p_1(s)}{q_1(s)},$$

for some constants $A_1, \ldots, A_k$, $B_1, \ldots, B_k$ and polynomial $p_1(s)$. Taking the inverse Laplace transform, it follows from Corollary 11 of Sect. 2.5 that f is a linear combination of terms in

$$\{e^{\alpha t} \cos \beta t, e^{\alpha t} \sin \beta t, \ldots, t^{k-1}e^{\alpha t} \cos \beta t, t^{k-1}e^{\alpha t} \sin \beta t\}$$

and a function in $\mathcal{E}_{q_1}$. But since $\mathcal{E}_{q_1} = \text{Span } \mathcal{B}_{q_1}$ and

$$\mathcal{B}_q = \{e^{\alpha t} \cos \beta t, e^{\alpha t} \sin \beta t, \ldots, t^{k-1}e^{\alpha t} \cos \beta t, t^{k-1}e^{\alpha t} \sin \beta t\} \cup \mathcal{B}_{q_1}$$

it follows that $f \in \text{Span } \mathcal{B}_q$ and hence $\mathcal{E}_q = \text{Span } \mathcal{B}_q$. $\qquad \square$

The Linear Independence of $\mathcal{B}_q$

Let $q(s)$ be a nonconstant polynomial. This subsection is devoted to showing that $\mathcal{B}_q$ is linearly independent. To that end, we derive a sequence of lemmas that will help us reach this conclusion.

Lemma 3. *Suppose* $q(s) = (s - \lambda)^m q_1(s)$ *where* $(s - \lambda)$ *is not a factor of* $q_1(s)$. *Suppose* $f_1 \in \mathcal{E}_{(s-\lambda)^m}$, $f_2 \in \mathcal{E}_{q_1}$, *and*

$$f_1 + f_2 = 0.$$

Then $f_1 = 0$ *and* $f_2 = 0$.

Proof. Let $F_1(s) = \mathcal{L}\{f_1\}(s)$ and $F_2(s) = \mathcal{L}\{f_2\}(s)$. Then $F_1(s) = \frac{p(s)}{(s-\lambda)^r}$, for some $r \le m$ and some polynomial $p(s)$ with no factor of $s - \lambda$. Similarly, $F_2(s) = \frac{p_1(s)}{q_1(s)}$, for some polynomial $p_1(s)$. Thus,

$$\frac{p(s)}{(s - \lambda)^r} + \frac{p_1(s)}{q_1(s)} = 0. \tag{1}$$

By Corollary 7 of Appendix A.2, equation (1) holds for all s different from λ and the roots of $q_1(s)$. Since $q_1(s)$ contains no factor of $s - \lambda$, it follows that $q_1(\lambda) \ne 0$. Consider the limit as s approaches λ in (1). The second term approaches the finite value $p_1(\lambda)/q_1(\lambda)$. Since the sum is 0, the limit of the left side is 0 and this implies that the limit of the first term is finite as well. But this can only happen if $p(s) = 0$. It follows that $F_1(s) = 0$ and hence $f_1 = 0$. This in turn implies that $f_2 = 0$. $\square$

Lemma 4. *Suppose* $q(s) = ((s - \alpha)^2 + \beta^2)^m q_1(s)$ *where* $(s - \alpha)^2 + \beta^2$ *is not a factor of* $q_1(s)$. *Suppose* $f_1 \in \mathcal{E}_{((s-\alpha)+\beta^2)^m}$, $f_2 \in \mathcal{E}_{q_1}$, *and*

$$f_1 + f_2 = 0.$$

Then $f_1 = 0$ *and* $f_2 = 0$.

Proof. Let $F_1(s) = \mathcal{L}\{f_1\}(s)$ and $F_2(s) = \mathcal{L}\{f_2\}(s)$. Then $F_1(s) = \frac{p(s)}{((s-\alpha)^2+\beta^2)^r}$, for some $r \le m$ and some polynomial $p(s)$ with no factor of $(s-\alpha)^2+\beta^2$. Similarly, $F_2(s) = \frac{p_1(s)}{q_1(s)}$, for some polynomial $p_1(s)$. Thus,

$$\frac{p(s)}{((s - \alpha)^2 + \beta^2)^m} + \frac{p_1(s)}{q_1(s)} = 0. \tag{2}$$

Again, by Corollary 7 of Appendix A.2, equation (2) holds for all s different from $\alpha \pm i\beta$ and the roots of $q_1(s)$. Since $q_1(s)$ contains no factor of $(s - \alpha)^2 + \beta^2$ it follows that $q_1(\alpha + i\beta) \ne 0$. Consider the limit as s approaches the complex number $\alpha + i\beta$ in (2). The second term approaches a finite value. Since the sum is 0, the limit of the left side is 0 and this implies that the limit of the first term is finite as well. But this can only happen if $p(s) = 0$. It follows that $F_1(s) = 0$ and hence $f_1 = 0$. This in turn implies that $f_2 = 0$. □

Lemma 5. *Let* $q(s) = (s - \lambda)^m$. *Then* $\mathcal{B}_q$ *is linearly independent.*

Proof. By definition, $\mathcal{B}_q = \{e^{\lambda t}, te^{\lambda t}, \ldots, t^{m-1}e^{\lambda t}\}$. Suppose

$$c_0 e^{\lambda t} + \cdots + c_{m-1} t^{m-1} e^{\lambda t} = 0.$$

Dividing both sides by $e^{\lambda t}$ gives

$$c_0 + c_1 t + \cdots + c_{m-1} t^{m-1} = 0.$$

Evaluating the at $t = 0$ gives $c_0 = 0$. Taking the derivative of both sides and evaluating at $t = 0$ gives $c_1 = 0$. In general, taking the rth derivative of both sides and evaluating at $t = 0$ gives $c_r = 0$, for $r = 1, 2, \ldots, m = 1$. It follows that all of the coefficients are necessarily 0, and this implies linear independence. □

Lemma 6. *Let* $q(s) = ((s - \alpha)^2 + \beta^2)^m$. *Then* $\mathcal{B}_q$ *is linearly independent.*

Proof. By definition, $\mathcal{B}_q = \{e^{\alpha t} \cos \beta t, \ e^{\alpha t} \sin \beta t, \ldots, \ t^{m-1} e^{\alpha t} \cos \beta t, \ t^{m-1} e^{\alpha t} \sin \beta t\}$. A linear combination of $\mathcal{B}_q$ set to 0 takes the form

$$p_1(t)e^{\alpha t} \cos \beta t + p_2(t)e^{\alpha t} \sin \beta t = 0, \tag{3}$$

where p_1 and p_2 are polynomials carrying the coefficients of the linear combination. Thus, it is enough to show p_1 and p_2 are 0. Dividing both sides of Equation (3) by $e^{\alpha t}$ gives

$$p_1(t) \cos \beta t + p_2(t) \sin \beta t = 0. \tag{4}$$

Let m be an integer. Evaluating at $t = \frac{2\pi m}{\beta}$ gives $p_1\left(\frac{2\pi m}{\beta}\right) = 0$, since $\sin 2\pi m = 0$ and $\cos 2\pi m = 1$. It follows that p_1 has infinitely many roots. By Corollary 5, of Sect. A.2 this implies $p_1 = 0$. In a similar way, $p_2 = 0$. □

Theorem 7. *Let* q *be a nonconstant polynomial. View* $\mathcal{B}_q$ *as a set of functions on* $I = [0, \infty)$. *Then* $\mathcal{B}_q$ *is linearly independent.*

Proof. Let k be the number of roots of q. Our proof is by induction on k. If q has but one root, then $q(s) = (s - \lambda)^n$ and this case is taken care of by Lemma 5. Suppose now that $k > 1$. We consider two cases. If $q(s) = (s - \lambda)^m q_1(s)$, where $q_1(s)$ has no factor of $s - \lambda$, then a linear combination of functions in $\mathcal{B}_q$ has the form $f_1(t) + f_2(t)$, where $f_1(t) = c_0 e^{\lambda t} + \cdots + c_{m-1} t^{m-1} e^{\lambda t}$ is a linear combination of functions in $\mathcal{B}_{(s-\lambda)^m}$ and $f_2(t)$ is a linear combination of functions in $\mathcal{B}_{q_1}$. Suppose

$$f_1(t) + f_2(t) = 0.$$

By Lemma 3, f_1 is identically zero and by Lemma 5, the coefficients $c_0, \ldots, c_{m-1}$ are all zero. It follows that $f_2 = 0$. Since $q_1(s)$ has one fewer root than q, it follows by induction the coefficients of the functions in $\mathcal{B}_{q_1}$ that make up f_2 are all zero. If follows that $\mathcal{B}_q$ is linearly independent.

Now suppose $q(s) = ((s - \alpha)^2 + \beta^2)^m q_1(s)$, where $q_1(s)$ has no factor of $(s - \alpha)^2 + \beta^2$. Then a linear combination of functions in $\mathcal{B}_q$ has the form $f_1(t) + f_2(t)$, where $f_1(t) = c_0 e^{\alpha t} \cos \beta t + \cdots + c_{m-1} t^{m-1} e^{\alpha t} \cos \beta t + d_0 e^{\alpha t} \sin \beta t + \cdots + d_{m-1} t^{m-1} e^{\alpha t} \sin \beta t$ is a linear combination of functions in $\mathcal{B}_{((s-\alpha)^2+\beta)^m}$ and $f_2(t)$ is a linear combination of functions in $\mathcal{B}_{q_1}$. Suppose

$$f_1(t) + f_2(t) = 0.$$

By Lemma 4 f_1 is identically zero, and by Lemma 6 the coefficients $c_0, \ldots, c_{m-1}$ and $d_0, \ldots, d_{m-1}$ are all zero. It follows that $f_2 = 0$. By induction, the coefficients of the functions in $\mathcal{B}_{q_1}$ that make up f_2 are all zero. If follows that $\mathcal{B}_q$ is linearly independent.   □

A.4   The Matrix Exponential

In this section, we verify some properties of the matrix exponential that have been used in Chap. 9.

First, we argue that the matrix exponential

$$e^A = I + A + \frac{A^2}{2!} + \frac{A^3}{3!} + \cdots + \frac{A^n}{n!} + \cdots$$

converges absolutely for all $n \times n$ matrices A. By this we mean that each entry of e^A converges absolutely. For convenience, let $E = e^A$. Let $a = \max\{|A_{i,j}| : 1 \le i, j \le n\}$. If $a = 0$, then $A = 0$ and the series that defines e^A reduces to I, the $n \times n$ identity. We will thus assume $a > 0$. Consider the (i, j) entry of A^2. We have

$$\left| (A^2)_{i,j} \right| = \left| \sum_{l=1}^{n} A_{i,l} A_{l,j} \right|$$

$$\le \sum_{l=1}^{n} |A_{i,l}| |A_{l,j}|$$

$$\le \sum_{l=1}^{n} a^2 = na^2.$$

In a similar way, the (i, j) entry of A^3 satisfies

$$\left|(A^3)_{i,j}\right| = \left|\sum_{l=1}^{n} A_{i,l}(A^2)_{l,j}\right|$$

$$\leq \sum_{l=1}^{n} |A_{i,l}|\left|(A^2)_{l,j}\right|$$

$$\leq \sum_{l=1}^{n} ana^2 = n^2 a^3.$$

By induction, we have

$$\left|(A^k)_{i,j}\right| \leq n^{k-1} a^k.$$

It follows from this estimate that

$$|E_{i,j}| = \left|(I)_{i,j} + (A)_{i,j} + \frac{(A^2)_{i,j}}{2!} + \frac{(A^3)_{i,j}}{3!} \cdots\right|$$

$$\leq \delta_{i,j} + |(A)_{i,j}| + \frac{|(A^2)_{i,j}|}{2!} + \frac{|(A^3)_{i,j}|}{3!} \cdots$$

$$\leq \delta_{i,j} + a + \frac{na^2}{2!} + \frac{n^2 a^3}{3!} + \frac{n^3 a^4}{4!} + \cdots$$

$$= \delta_{i,j} + \sum_{k=1}^{\infty} \frac{n^{k-1} a^k}{k!}$$

$$\leq 1 + \frac{1}{n}\sum_{k=0}^{\infty} \frac{(na)^k}{k!} = 1 + \frac{1}{n}e^{an}.$$

It follows now that the series for $E_{i,j}$ converges absolutely.

If we replace A by At, we get that each entry of e^{At} is an absolutely convergent power series in t with infinite radius of convergence. Furthermore, the last inequality above shows that the (i, j) entry of e^{At} is bounded by $1 + \frac{1}{n}e^{ant}$ and hence of exponential type.

A.5   The Cayley–Hamilton Theorem

Theorem 1 (Cayley–Hamilton). *Let A be an $n \times n$ matrix and $c_A(s)$ its characteristic polynomial. Then*

$$c_A(A) = 0.$$

The following lemma will be helpful for the proof.

Lemma 2. *Let j be a nonnegative integer and let $a \in \mathbb{C}$. Then*

$$\mathbf{D}^l\left(t^j e^{at}\right)\big|_{t=0} = \mathbf{D}^j t^l\big|_{t=a}.$$

Proof. The derivative formula $\mathbf{D}^l\left(y e^{at}\right) = \left((\mathbf{D}+a)^l y\right) e^{at}$ implies

$$\mathbf{D}^l(t^j e^{at})\big|_{t=0} = ((\mathbf{D}+a)^l t^j)\big|_{t=0}$$

$$= \sum_{k=0}^{l}\binom{l}{k} a^{l-k}(\mathbf{D}^k t^j)\big|_{t=0}$$

$$= \begin{cases} 0 & \text{if } l < j \\[2mm] \dfrac{a^{l-j} l!}{(l-j)!} & \text{if } l \geq j \end{cases}$$

$$= \mathbf{D}^j t^l\big|_{t=a}. \qquad\qquad \square$$

Proof (of Cayley-Hamilton Theorem). Let $c_A(s)$ be the characteristic polynomial of A. Suppose $\lambda_1, \cdots, \lambda_m$ are the roots of c_A with corresponding multiplicities $r_1, \cdots, r_m$. Then

$$\mathcal{B}_{c_A} = \left\{ t^j e^{\lambda_k t} : j = 0, \ldots, r_k - 1, k = 1, \ldots, m \right\}$$

is the standard basis for $\mathcal{E}_{c_A}$. As in Fulmer's method, Sect. 9.4, we may assume that there are $n \times n$ matrices $M_{j,k}$, $j = 0, \ldots, r_k - 1, k = 1, \ldots, m$, so that

$$e^{At} = \sum_{k=1}^{m}\sum_{j=0}^{r_k-1} t^j e^{\lambda_k t} M_{j,k}.$$

Differentiating both sides l times and evaluating at $t = 0$ gives

$$A^l = \sum_{k=1}^{m}\sum_{j=0}^{r_k-1}\mathbf{D}^l\left(t^j e^{\lambda_k t}\right)\big|_{t=0} M_{j,k} = \sum_{k=1}^{m}\sum_{j=0}^{r_k-1}\mathbf{D}^j t^l\big|_{t=\lambda_k} M_{j,k},$$

with the second equality coming from Lemma 2. Now let $p(s) = c_0 + c_1 s + \cdots + c_N s^N = \sum_{l=0}^{N} c_l s^l$ be any polynomial. Then

$$p(A) = \sum_{l=0}^{N} c_l A^l$$

$$= \sum_{l=0}^{N}\sum_{k=1}^{m}\sum_{j=0}^{r_k-1} c_l \mathbf{D}^j t^l\big|_{t=\lambda_k} M_{j,k}$$

$$= \sum_{k=1}^{m} \sum_{j=0}^{r_k-1} \mathbf{D}^j \left(\sum_{l=0}^{N} c_l t^l \right) \big|_{t=\lambda_k} M_{j,k}$$

$$= \sum_{k=1}^{m} \sum_{j=0}^{r_k-1} p^{(j)}(\lambda_k) M_{j,k}. \tag{1}$$

For the characteristic polynomial, we have $c_A(s) = \prod_{k=1}^{m}(s - \lambda_k)^{r_k}$, and hence, $c_A^{(j)}(\lambda_k) = 0$ for all $j = 0, \ldots, r_k - 1$ and $k = 1, \ldots m$. Now let $p(s) = c_A(s)$ in Equation (1) to get

$$c_A(A) = \sum_{k=1}^{m} \sum_{j=0}^{r_k-1} c_A^{(j)}(\lambda_k) M_{j,k} = 0.$$

$\square$

Appendix B
Selected Answers

Section 1.1

1. $P' = kP$

3. $h'(t) = \lambda \sqrt{h(t)}$

5. order: 2; Standard form: $y'' = t^3/y'$.

7. order: 2; Standard form: $y'' = -(3y + ty')/t^2$.

9. order: 4; standard form: $y^{(4)} = \sqrt[3]{(1 - (y''')^4)/t}$.

11. order: 3; standard form: $y''' = 2y'' - 3y' + y$.

13. y_1, y_2, and y_3

15. y_1, y_2, and y_4

17. y_1, y_2, and y_4

19.

$$y'(t) = 3ce^{3t}$$
$$3y + 12 = 3(ce^{3t} - 4) + 12 = 3ce^{3t} - 12 + 12 = 3ce^{3t}.$$

Note that $y(t)$ is defined for all $t \in \mathbb{R}$.

21.

$$y'(t) = \frac{ce^t}{(1 - ce^t)^2}$$

W.A. Adkins and M.G. Davidson, *Ordinary Differential Equations*,
Undergraduate Texts in Mathematics, DOI 10.1007/978-1-4614-3618-8,
© Springer Science+Business Media New York 2012

$$y^2(t) - y(t) = \frac{1}{(1 - ce^t)^2} - \frac{1}{1 - ce^t} = \frac{1 - (1 - ce^t)}{(1 - ce^t)^2} = \frac{ce^t}{(1 - ce^t)^2}.$$

If $c \leq 0$, then the denominator $1 - ce^t > 0$ and $y(t)$ has domain $\mathbb{R}$. If $c > 0$, then $1 - ce^t = 0$ if $t = \ln \frac{1}{c} = -\ln c$. Thus, $y(t)$ is defined either on the interval $(-\infty, -\ln c)$ or $(-\ln c, \infty)$.

23.

$$y'(t) = \frac{-ce^t}{ce^t - 1}$$

$$-e^y - 1 = -e^{-\ln(ce^t - 1)} - 1 = \frac{-1}{ce^t - 1} - 1 = \frac{-ce^t}{ce^t - 1}.$$

25.

$$y'(t) = -(c - t)^{-2}(-1) = \frac{1}{(c - t)^2}$$

$$y^2(t) = \frac{1}{(c - t)^2}.$$

The denominator of $y(t)$ is 0 when $t = c$. Thus, the two intervals where $y(t)$ is defined are $(-\infty, c)$ and (c, ∞).

27. $y(t) = \frac{e^{2t}}{2} - t + c$.

29. $y(t) = t + \ln|t| + c$

31. $y(t) = -\frac{2}{3}\sin 3t + c_1 t + c_2$

33. $y(t) = 3e^{-t} + 3t - 3$

35. $y(t) = -18(t + 1)^{-1}$

37. $y(t) = -te^{-t} - e^{-t}$.

Section 1.2

1. $y' = t$

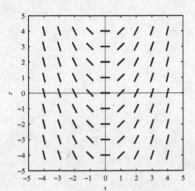

3. $y' = y(y + t)$

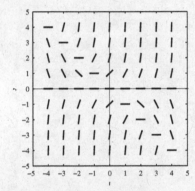

5.

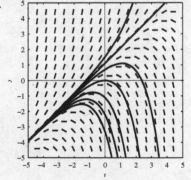

7.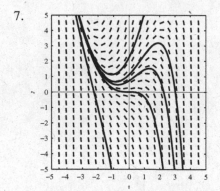

9.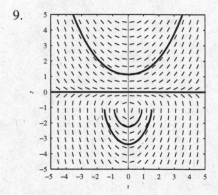

11. $y = 0$

13. $y = \pm 1$

15. $y = -t + (2n + 1)\pi, n \in \mathbb{Z}$

17. $2yy' - 2t - 3t^2 = 0$

19. $y' = \frac{3y}{t} - t$.

Section 1.3

1. Separable

3. Not separable

5. Separable

7. Not separable

9. Separable

11. $t^2(1 - y^2) = c$, c a real constant

13. $y^5 = \frac{5}{2}t^2 + 10t + c$, c a real constant

15. $y = 1 - c\cos t$, c any real constant

17. $y = \frac{4ce^{4t}}{1+ce^{4t}}$, c a real number, and $y = 4$

19. $y = \tan(t + c)$, c a real constant

21. $\ln(y + 1)^2 - y = \tan^{-1} t + c$, c a real constant, and $y = -1$

23. $y = \frac{1}{\ln|1-t|+c}$, c real constant

25. $y = 0$

27. $y = 4e^{-t^2}$

29. $y = 2\sqrt{u^2 + 1}$

31. $y(t) = \tan\left(-\frac{1}{t} + 1 + \frac{\pi}{3}\right)$, $(a, b) = (6/(6 + 5\pi),\ 6/(6 - \pi))$.

33. ≈ 212 million years old

35. ≈ 64 min

37. $t \approx 8.2$ min and $T(20) \approx 67.7°$

39. $\approx 205.5°$

41. 602

43. ≈ 3.15 years

45. 3857

47. $1,400$

Section 1.4

1. $y = \frac{1}{4}e^t - \frac{9}{4}e^{-3t}$.

3. $y = te^{2t} + 4e^{2t}$.

5. $y = \frac{e^t}{t} - \frac{e}{t}$

7. $y = \frac{\sin(t^2)}{2t} + \frac{c}{t}$.

9. $y = 4\sin 4t - 3\cos 4t + ce^{3t}$

11. $z = t^2 + 1 + ce^{t^2}$

13. $y = 1 - e^{-\sin t}$

15. $y = -t - \frac{1}{2} - \frac{3}{2}t^{-2}$.

17. $y = \frac{1}{a+b}e^{bt} + ce^{-at}$

19. $y = (t + c)\sec t$

21. $y = t^n e^t + ct^n$

23. $y = \frac{1}{5}t^2 - \frac{9}{5}t^{-3}$

25. $y(t) = \frac{1}{t} + (4a - 2)t^{-2}$

27. $y = (10 - t) - \frac{8}{10^4}(10 - t)^4$, for $0 \le t \le 10$. After 10 min, there is no salt in the tank.

29. (a) 10 min, (b) ≈ 533.33 g

31. (a) Differential equation: $P'(t) + (r/V)P(t) = rc$. If P_0 denotes the initial amount of pollutant in the lake, then $P(t) = Vc + (P_0 - Vc)e^{-(r/V)t}$. The limiting concentration is c.
(b) $t_{1/2} = (V/r)\ln 2; t_{1/10} = (V/r)\ln 10$
(c) Lake Erie: $t_{1/2} = 1.82$ years, $t_{1/10} = 6.05$ years, Lake Ontario: $t_{1/2} = 5.43$ years, $t_{1/10} = 18.06$ years

33. $y_2(t) = 10(5 + t) - \frac{500\ln(5+t)}{5+t} + \frac{500\ln 5 - 250}{5+t}$

Section 1.5

1. $y = t\tan(\ln|t| + \pi/4)$

3. $y = 2t$

5. $y = \pm t\sqrt{1 + kt}, k \in \mathbb{R}$

7. $y = t\sin(\ln|t| + c)$ and $y = \pm t$

9. $y = \dfrac{1}{1 - t}$

11. $y = \pm \dfrac{1}{\sqrt{1 + ce^{t^2}}}$ and $y = 0$

13. $y = \dfrac{1}{-5 + c\sqrt{1 - t^2}}$ and $y = 0$.

15. $y = -\sqrt{1 + 3e^{-t^2}}$

17. $y = \pm \dfrac{1}{\sqrt{t + \frac{1}{2} + ce^{2t}}}$ and $y = 0$

19. $-2y + \ln|2t - 2y| = c, c \in \mathbb{R}$ and $y = t$

21. $y - \tan^{-1}(t + y) = c, c \in \mathbb{R}$

23. $y = \pm\sqrt{ke^t - t}, k \in \mathbb{R}$

25. $y = e^{t-1+ce^{-t}}, c \in \mathbb{R}$

Section 1.6

1. $t^2 + ty^2 = c$

3. Not Exact

5. Not Exact

7. $(y - t^2)^2 - 2t^4 = c$

9. $y^4 = 4ty + c$

Section 1.7

1. $y(t) = 1 + \int_1^t uy(u)\, du$

3. $y(t) = 1 + \int_0^t \dfrac{u - y(u)}{u + y(u)}\, du$

5.

$$y_0(t) = 1$$

$$y_1(t) = \frac{1 + t^2}{2}$$

$$y_2(t) = \frac{5}{8} + \frac{t^2}{4} + \frac{t^4}{8}$$

$$y_3(t) = \frac{29}{48} + \frac{5t^2}{16} + \frac{t^4}{16} + \frac{t^6}{48}.$$

7. $y_1(t) = \frac{t^2}{2}$; $y_2(t) = \frac{t^2}{2} + \frac{t^5}{20}$; $y_3(t) = \frac{t^2}{2} + \frac{t^5}{20} + \frac{t^8}{160} + \frac{t^{11}}{4400}$.

9.

$$y_0(t) = 0$$

$$y_1(t) = t + \frac{t^3}{3}$$

$$y_2(t) = t + \frac{t^7}{7 \cdot 3^2}$$

$$y_3(t) = t + \frac{t^{15}}{15 \cdot 7^2 \cdot 3^4}$$

$$y_4(t) = t + \frac{t^{31}}{31 \cdot 15^2 \cdot 7^4 \cdot 3^8}$$

$$y_5(t) = t + \frac{t^{63}}{63 \cdot 31^2 \cdot 15^4 \cdot 7^8 \cdot 3^{16}}$$

11. Not guaranteed unique

13. Unique solution

15.

$$y_0(t) = 1$$

$$y_1(t) = 1 + at$$

$$y_2(t) = 1 + at + \frac{a^2 t^2}{2}$$

$$y_3(t) = 1 + at + \frac{a^2 t^2}{2} + \frac{a^3 t^3}{3!}$$

$$\vdots$$

$$y_n(t) = 1 + at + \frac{a^2 t^2}{2} + \cdots + \frac{a^n t^n}{n!}.$$

$y(t) = \lim_{n \to \infty} y_n(t) = e^{at}$; it is a solution; there are no other solutions.

17. Yes

19. 1. $y(t) = t + ct^2$.
    2. Every solution satisfies $y(0) = 0$. There is no contradiction to Theorem 5 since, in standard form, the equation is $y' = \frac{2}{t} y - 1 = F(t, y)$ and $F(t, y)$ is not continuous for $t = 0$.

21. No

Section 2.1

1. $Y(s) = \frac{2}{s-4}$ and $y(t) = 2e^{4t}$

3. $Y(s) = \frac{1}{(s-4)^2}$ and $y(t) = te^{4t}$

5. $Y(s) = \frac{1}{s+2} + \frac{1}{s-1}$ and $y(t) = e^{-2t} + e^{t}$

7. $Y(s) = \frac{3s+3}{s^2+3s+2} = \frac{3(s+1)}{(s+1)(s+2)} = \frac{3}{s+2}$ and $y(t) = 3e^{-2t}$

9. $Y(s) = \frac{s-1}{s^2+25}$ and $y(t) = \frac{-1}{5} \sin 5t + \cos 5t$

11. $Y(s) = \frac{1}{s+4}$ and $y(t) = e^{-4t}$

13. $Y(s) = \frac{1}{(s+2)^2} + \frac{1}{(s+2)^3} = \frac{1}{(s+2)^2} + \frac{1}{2} \frac{2}{(s+2)^3}$ and $y(t) = te^{-2t} + \frac{1}{2} t^2 e^{-2t}$

Section 2.2

Compute the Laplace transform of each function given below directly from the integral definition given in (1).

1. $\dfrac{3}{s^2} + \dfrac{1}{s}$

3. $\dfrac{1}{s-2} - \dfrac{3}{s+1}$

5. $\dfrac{5}{s-2}$

7. $\dfrac{2}{s^3} - \dfrac{5}{s^2} + \dfrac{4}{s}$

9. $\dfrac{1}{s+3} + \dfrac{7}{(s+4)^2}$

11. $\dfrac{s+2}{s^2+4}$

13. $\dfrac{2}{(s+4)^3}$

15. $\dfrac{2}{s^3} + \dfrac{2}{(s-2)^2} + \dfrac{1}{s-4}$

17. $\dfrac{24}{(s+4)^5}$

19. $\dfrac{1}{(s-3)^2}$

21. $\dfrac{12s^2 - 16}{(s^2+4)^3}$

23. $\dfrac{2s}{s^2+1} - \dfrac{2}{s}$

25. $\dfrac{\ln(s+6) - \ln 6}{s}$

27. $\dfrac{2b^2}{s(s^2+4b^2)}$

29. $\dfrac{1}{2}\left(\dfrac{a-b}{s^2+(a-b)^2} + \dfrac{a+b}{s^2+(a+b)^2}\right)$

31. $\dfrac{b}{s^2-b^2}$

33. $\dfrac{b}{s^2-b^2}$

(a) Show that $\Gamma(1) = 1$.
(b) Show that Γ satisfies the recursion formula $\Gamma(\beta+1) = \beta\Gamma(\beta)$.
(*Hint*: Integrate by parts.)
(c) Show that $\Gamma(n+1) = n!$ when n is a nonnegative integer.

Section 2.3

The $s-1$-chain	
$\dfrac{5s+10}{(s-1)(s+4)}$	$\dfrac{3}{s-1}$
$\dfrac{2}{s+4}$	

1.

The $s-5$-chain	
$\dfrac{1}{(s+2)(s-5)}$	$\dfrac{1/7}{(s-5)}$
$\dfrac{-1/7}{(s+2)}$	

3.

The $s-1$-chain	
$\dfrac{3s+1}{(s-1)(s^2+1)}$	$\dfrac{2}{s-1}$
$\dfrac{-2s+1}{s^2+1}$	

5.

The $s+3$-chain	
$\dfrac{s^2+s-3}{(s+3)^3}$	$\dfrac{3}{(s+3)^3}$
$\dfrac{s-2}{(s+3)^2}$	$\dfrac{-5}{(s+3)^2}$
$\dfrac{1}{s+3}$	$\dfrac{1}{s+3}$
0	

7.

	The $s + 1$ -chain	
9.	$\dfrac{s}{(s+2)^2(s+1)^2}$	$\dfrac{-1}{(s+1)^2}$
	$\dfrac{s+4}{(s+2)^2(s+1)}$	$\dfrac{3}{s+1}$
	$\dfrac{-3s-8}{(s+2)^2}$	

	The $s - 5$ -chain	
11.	$\dfrac{1}{(s-5)^5(s-6)}$	$\dfrac{-1}{(s-5)^5}$
	$\dfrac{1}{(s-5)^4(s-6)}$	$\dfrac{-1}{(s-5)^4}$
	$\dfrac{1}{(s-5)^3(s-6)}$	$\dfrac{-1}{(s-5)^3}$
	$\dfrac{1}{(s-5)^2(s-6)}$	$\dfrac{-1}{(s-5)^2}$
	$\dfrac{1}{(s-5)(s-6)}$	$\dfrac{-1}{s-5}$
	$\dfrac{1}{s-6}$	

13. $\dfrac{13/8}{s-5} - \dfrac{5/8}{s+3}$

15. $\dfrac{23}{12(s-5)} + \dfrac{37}{12(s+7)}$

17. $\dfrac{25}{8(s-7)} - \dfrac{9}{8(s+1)}$

19. $\dfrac{1}{2(s+5)} - \dfrac{1}{2(s-1)} + \dfrac{1}{s-2}$

21. $\dfrac{7}{(s+4)^4}$

23. $\dfrac{3}{(s+3)^3} - \dfrac{5}{(s+3)^2} + \dfrac{1}{s+3}$

25. $\dfrac{1}{54}\left(\dfrac{5}{s-5} + \dfrac{21}{(s+1)^2} + \dfrac{3}{(s-5)^2} - \dfrac{5}{s+1}\right)$

27. $\dfrac{-2}{(s+2)^2} - \dfrac{3}{s+2} - \dfrac{1}{(s+1)^2} + \dfrac{3}{s+1}$

29. $\dfrac{12}{(s-3)^3} + \dfrac{-14}{(s-3)^2} + \dfrac{15}{s-3} + \dfrac{-16}{s-2} + \dfrac{1}{s-1}$

31. $\dfrac{2}{(s-2)^2} + \dfrac{5}{s-2} + \dfrac{3}{(s-3)^2} - \dfrac{5}{s-3}$

33. $Y(s) = \frac{-3}{(s+1)^2} - \frac{1}{s+1} + \frac{1}{s-2}$ and $y(t) = -3te^{-t} - e^{-t} + e^{2t}$

35. $Y(s) = \frac{-30}{s^2} + \frac{24}{s} - \frac{26}{s+1} + \frac{1}{s-5}$ and $y(t) = -30t + 24 - 26e^{-t} + e^{5t}$

37. $Y(s) = \frac{2s-3}{(s-1)(s-2)} + \frac{4}{s(s-1)(s-2)} = \frac{2}{s} + \frac{3}{s-2} - \frac{3}{s-1}$ and $y(t) = 2 + 3e^{2t} - 3e^{t}$

Section 2.4

The $s^2 + 1$ -chain	
$\dfrac{1}{(s^2 + 1)^2(s^2 + 2)}$	$\dfrac{1}{(s^2 + 1)^2}$
$\dfrac{-1}{(s^2 + 1)(s^2 + 2)}$	$\dfrac{-1}{(s^2 + 1)}$
$\dfrac{1}{s^2 + 2}$	

1.

The $s^2 + 3$ -chain	
$\dfrac{8s + 8s^2}{(s+3)^3(s^2 + 1)}$	$\dfrac{12 - 4s}{(s^2 + 3)^3}$
$\dfrac{4(s - 1)}{(s^2 + 3)^2(s^2 + 1)}$	$\dfrac{2 - 2s}{(s^2 + 3)^2}$
$\dfrac{2(s - 1)}{(s^2 + 3)(s^2 + 1)}$	$\dfrac{1 - s}{s^2 + 3}$
$\dfrac{s - 1}{s^2 + 1}$	

3.

The $s^2 + 2s + 2$ -chain	
$\dfrac{1}{(s^2 + 2s + 2)^2(s^2 + 2s + 3)^2}$	$\dfrac{1}{(s^2 + 2s + 2)^2}$
$\dfrac{-(s^2 + 2s + 4)}{(s^2 + 2s + 2)(s^2 + 2s + 3)^2}$	$\dfrac{-2}{s^2 + 2s + 2}$
$\dfrac{2s^2 + 4s + 7}{(s^2 + 2s + 3)^2}$	

5.

7. $\dfrac{1}{10}\left(\dfrac{3}{s - 3} + \dfrac{1 - 3s}{s^2 + 1} \right)$

9. $\dfrac{9s^2}{(s^2 + 4)^2(s^2 + 1)} = \dfrac{12}{(s^2 + 4)^2} + \dfrac{1}{s^2 + 4} - \dfrac{1}{s + 1}$

11. $\dfrac{2}{s - 3} + \dfrac{6 - 2s}{(s - 3)^2 + 1}$

13. $\dfrac{-5s + 15}{(s^2 - 4s + 8)^2} + \dfrac{-s + 3}{s^2 - 4s + 8} + \dfrac{1}{s - 1}$

15. $\dfrac{s + 1}{(s^2 + 4s + 6)^2} + \dfrac{2s + 2}{s^2 + 4s + 6} + \dfrac{s + 1}{(s^2 + 4s + 5)^2} - \dfrac{2s + 2}{s^2 + 4s + 5}$

17. $Y(s) = \dfrac{1}{(s+2)^2} + \dfrac{4s}{(s^2+4)(s+2)^2} = \dfrac{1}{s^2+4}$ and $y(t) = \frac{1}{2}\sin 2t$

19. $Y(s) = \dfrac{1}{s^2+4} + \dfrac{3}{(s^2+9)(s^2+4)} = \frac{8}{5}\dfrac{1}{s^2+4} - \frac{3}{5}\dfrac{1}{s^2+9}$ and $y(t) = \frac{4}{5}\sin 2t - \frac{1}{5}\sin 3t$

Section 2.5

1. -5

3. $3t - 2t^2$

5. $3\cos 2t$

7. $-11te^{-3t} + 2e^{-3t}$

9. $e^{2t} - e^{-4t}$

11. $\dfrac{-1}{6}t^3e^{2t} + \dfrac{3}{2}t^2e^{2t} + 2te^{2t}$

13. $te^t + e^t + te^{-t} - e^{-t}$

15. $4te^{2t} - e^{2t} + \cos 2t - \sin 2t$

17. $3 - 6t + e^{2t} - 4e^{-t}$

19. $2e^{-t}\cos 2t - e^{-t}\sin 2t$

21. $e^{4t}\cos t + 3e^{4t}\sin t$

23. $e^t\cos 3t$

25. $2t\sin 2t$

27. $2te^{-2t}\cos t + (t-2)e^{-2t}\sin t$

29. $4te^{-4t}\cos t + (t-4)e^{-4t}\sin t$

31. $\dfrac{1}{256}\left((3-4t^2)e^t\sin 2t - 6te^t\cos 2t\right)$

33. $\dfrac{1}{48}\left((-t^3 + 3t)e^{4t}\sin t - 3t^2e^{4t}\cos t\right)$

35. $y(t) = \cos t - \sin t + 2(\sin t - t\cos t) = \cos t + \sin t - 2t\cos t$

37. $Y(s) = \dfrac{8s}{(s^2+1)^3}$ and $y(t) = t\sin t - t^2\cos t$

Section 2.6

1. $\mathcal{B}_q = \{e^{4t}\}$

3. $\mathcal{B}_q = \left\{1, e^{-5t}\right\}$

5. $\mathcal{B}_q = \left\{e^{3t}, te^{3t}\right\}$

7. $\mathcal{B}_q = \left\{e^{3t}, e^{-2t}\right\}$

9. $\mathcal{B}_q = \left\{e^{t/2}, e^{4t/3}\right\}$

11. $\mathcal{B}_q = \left\{e^{(2+\sqrt{3})t}, e^{(2-\sqrt{3})t}\right\}$

13. $\mathcal{B}_q = \left\{e^{-3t/2}, te^{-3t/2}\right\}$

15. $\mathcal{B}_q = \{\cos(5t/2), \sin(5t/2)\}$

17. $\mathcal{B}_q = \left\{e^t \cos 2t, e^t \sin 2t\right\}$

19. $\mathcal{B}_q = \left\{e^{-3t}, te^{-3t}, t^2 e^{-3t}, t^3 e^{-3t}\right\}$

21. $\mathcal{B}_q = \left\{e^t, te^t, t^2 e^t\right\}$

23. $\mathcal{B}_q = \left\{e^{-2t} \cos t, e^{-2t} \sin t, te^{-2t} \cos t, te^{-2t} \sin t\right\}$

25. $\mathcal{B}_q = \left\{\cos t, \sin t, t \cos t, t \sin t, t^2 \cos t, t^2 \sin t, t^3 \cos t, t^3 \sin t\right\}$

Section 2.7

1. Yes

3. Yes

5. Yes

7. No

9. No

11. No

13. $\mathcal{B}_q = \left\{e^t, e^{-t}, \cos t, \sin t\right\}$

15. $\mathcal{B}_q = \left\{e^t, te^t, t^2 e^t, e^{-7t}, te^{-7t}\right\}$

17. $\mathcal{B}_q = \left\{e^{-2t}, te^{-2t}, t^2 e^{-2t}, \cos 2t, \sin 2t, t \cos 2t, t \sin 2t\right\}$

19. $\mathcal{B}_q = \left\{e^{2t}, te^{2t}, e^{-3t}, te^{-3t}, t^2 e^{-3t}\right\}.$

21. $\mathcal{B}_q = \left\{e^{-4t}, te^{-4t}, e^{-3t} \cos 2t, e^{-3t} \sin 2t, te^{-3t} \cos 2t, te^{-3t} \sin 2t\right\}$

23. $\mathcal{B}_q = \left\{ e^{3t}, te^{3t}, t^2 e^{3t}, e^{-t} \cos 3t, e^{-t} \sin 3t, te^{-t} \cos 3t, te^{-t} \sin 3t \right\}$

25. $\mathcal{B}_q = \left\{ e^{t/2}, e^t, te^t \right\}$

27. $\mathcal{B}_q = \left\{ \cos \sqrt{3}t, \sin \sqrt{3}t, \cos \sqrt{2}t, \sin \sqrt{2}t \right\}$

Section 2.8

1. $\dfrac{t^3}{6}$

3. $3(1 - \cos t)$

5. $\frac{1}{13}(2e^{3t} - 2\cos 2t - 3\sin 2t)$

7. $\frac{1}{108}(18t^2 - 6t - 6 - e^{-6t})$

9. $\frac{1}{6}(e^{2t} - e^{-4t})$

11. $\frac{1}{a^2+b^2}(be^{at} - b\cos bt - a\sin bt)$

13. $\begin{cases} \dfrac{b\sin at - a\sin bt}{b^2 - a^2} & \text{if } b \neq a \\[2mm] \dfrac{\sin at - at\cos at}{2a} & \text{if } b = a \end{cases}$

15. $\begin{cases} \dfrac{a\sin at - b\sin bt}{a^2 - b^2} & \text{if } b \neq a \\[2mm] \dfrac{1}{2a}(at\cos at + \sin at) & \text{if } b = a \end{cases}$

17. $F(s) = \dfrac{4}{s^3(s^2 + 4)}$

19. $F(s) = \dfrac{6}{s^4(s + 3)}$

21. $F(s) = \dfrac{4}{(s^2 + 4)^2}$

23. $\dfrac{1}{4}(-e^t + e^{5t})$

25. $\dfrac{1}{2}t \sin t$

27. $\dfrac{1}{13}(2e^{3t} - 2\cos 2t - 3\sin 2t)$

29. $\dfrac{e^{at} - e^{bt}}{a - b}$

31. $\displaystyle\int_0^t g(x)\cos\sqrt{2}(t - x)\,dx$

33. $t - \sin t$

35. $\dfrac{1}{54}(2 - 6t + 9t^2 - 2e^{-3t})$

37. $\dfrac{1}{162}(2 - 3t\sin 3t - 2\cos 3t)$

Section 3.1

1. No

3. No

5. No

7. Yes; $(D^2 - 7D + 10)(y) = 0$, $q(s) = s^2 - 7s + 10$, homogeneous

9. Yes; $D^2(y) = -2 + \cos t$, $q(s) = s^2$, nonhomogeneous

11. (a) $6e^t$
    (b) 0
    (c) $\sin t - 3\cos t$

13. (a) 0
    (b) 0
    (c) 1

15. $y(t) = \cos 2t + c_1 e^t + c_2 e^{4t}$ where c_1, c_2 are arbitrary constants.

17. $y(t) = \cos 2t + e^t - e^{4t}$

19. $L(e^{rt}) = (ar^2 + br + c)e^{rt}$

Section 3.2

1. Linearly independent

3. Linearly dependent

5. Linearly dependent

7. Linearly independent

9. Linearly independent

11. Linearly independent

13. Linearly independent

15. t

17. $10t^{29}$

19. $e^{(r_1+r_2+r_3)t}(r_3 - r_1)(r_3 - r_2)(r_2 - r_1)$

21. 12

23. $c_1 = -2/5, c_2 = 1$

25. $a_1 = 3, a_2 = 3$

Section 3.3

1. $y(t) = c_1 e^{2t} + c_2 e^{-t}$

3. $y(t) = c_1 e^{-4t} + c_2 e^{-6t}$

5. $y(t) = c_1 e^{-4t} + c_2 t e^{-4t}$

7. $y(t) = c_1 e^{-t} \cos 2t + c_2 e^{-t} \sin 2t$

9. $y(t) = c_1 e^{-9t} + c_2 e^{-4t}$

11. $y(t) = c_1 e^{-5t} + c_2 t e^{-5t}$

13. $y = \frac{e^t - e^{-t}}{2}$

15. $y = t e^{5t}$

17. $q(s) = s^2 + 4s - 21, w(e^{3t}, e^{-7t}) = -10e^{-4t}, K = -10$.

19. $q(s) = s^2 - 6s + 9, w(e^{3t}, te^{3t}) = e^{6t}, K = 1$.

21. $q(s) = s^2 - 2s + 5, w(e^t \cos 2t, e^t \sin 2t) = 2e^{2t}, K = 2$.

Section 3.4

1. $y_p(t) = a_1 e^{3t}$

3. $y_p(t) = a_1 t e^{2t}$

5. $y_p(t) = a_1 \cos 5t + a_2 \sin 5t$

7. $y_p(t) = a_1 t \cos 2t + a_2 t \sin 2t$

9. $y_p(t) = a_1 e^{-2t} \cos t + a_2 e^{-2t} \sin t$

11. $y = -te^{-2t} + c_1 e^{-2t} + c_2 e^{5t}$

13. $y = \frac{1}{2} t^2 e^{-t} + c_1 e^{-t} + c_2 t e^{-t}$

15. $y = \frac{1}{2} e^{-3t} + c_1 e^{-2t} \cos t + c_2 e^{-2t} \sin t$

17. $y = -t^2 - 2 + c_1 e^t + c_2 e^{-t}$

19. $y = \frac{1}{2} t^2 e^{2t} + c_1 e^{2t} + c_2 t e^{2t}$.

21. $y = te^{2t} - \frac{2}{5} e^{2t} + c_1 e^{-3t} + c_2 t e^{-3t}$

23. $y = \frac{1}{4} t e^{-3t} \sin(2t) + c_1 e^{-3t} \cos(2t) + c_2 e^{-3t} \sin(2t)$

25. $y = \frac{-1}{12} e^{3t} + \frac{10}{21} e^{6t} + \frac{135}{84} e^{-t}$

27. $y = 2e^{2t} - 2 \cos t - 4 \sin t$

Section 3.5

1. $y = \frac{1}{32} e^{-6t} + c_1 e^{2t} + c_2 e^{-2t}$

3. $y = te^{-2t} + c_1 e^{-2t} + c_2 e^{-3t}$

5. $y = -te^{-4t} + c_1 e^{2t} + c_2 e^{-4t}$

7. $y = te^{2t} - \frac{2}{5}e^{2t} + c_1 e^{-3t} + c_2 t e^{-3t}$

9. $y = -3t^2 e^{4t} \cos 3t + t e^{4t} \sin 3t + c_1 e^{4t} \cos 3t + c_2 e^{4t} \sin 3t$

11. $y = \frac{1}{2}\sin t + c_1 e^{-t} + c_2 t e^{-t}$

Section 3.6

1. $k = 32$ lbs/ft

3. $k = 490$ N/m

5. $\mu = 8$ lbs s/ft

7. 400 lbs

9. $6y'' + 20y = 0$, $y(0) = .1$, $y'(0) = 0$; $y = \frac{1}{10}\cos\sqrt{\frac{10}{3}}\,t$, undamped free or simple harmonic motion; $A = 1/10$, $\beta = \sqrt{10/3}$, and $\phi = 0$

11. $\frac{1}{2}y'' + 2y' + 32y = 0$, $y(0) = 1$, $y'(0) = 1$; $y = e^{-2t}\cos\sqrt{60}t + \frac{3}{\sqrt{60}}e^{-2t}\sin\sqrt{60}t = \sqrt{\frac{23}{20}}e^{-2t}\cos(\sqrt{60}t + \phi)$, where $\phi = \arctan\sqrt{60}/20 \approx .3695$; underdamped free motion.

13. $y'' + 96y = 0$, $y(0) = 0$, $y'(0) = 2/3$; $y = \frac{\sqrt{6}}{36}\sin\sqrt{96}t = \frac{\sqrt{6}}{36}\cos\left(\sqrt{96}t - \frac{\pi}{2}\right)$; undamped free or simple harmonic motion and crosses equilibrium.

Section 3.7

1. $q(t) = -\frac{3}{100}e^{-20t}\cos 20t - \frac{3}{100}e^{-3t}\sin 20t + \frac{3}{100}$ and $I(t) = \frac{6}{5}e^{-20t}\cos(20t - \pi/2)$

3. $q(t) = \frac{4}{5}e^{-10t} + 7te^{-10t} - \frac{4}{5}\cos 5t + \frac{3}{5}\sin 5t$ and $I(t) = -e^{-10t} - 70te^{-10t} + 4\sin 5t + 3\cos 5t$

5. $q(t) = \frac{1}{75}(\cos 25t - \cos 100t)$; the capacitor will not overcharge.

Section 4.1

1. Yes; $(D^3 - 3D)y = e^t$, order 3, $q(s) = s^3 - 3s$, nonhomogeneous

3. No

5. (a) 0
   (b) 0
   (c) 0

7. (a) $10e^{-t}$
   (b) 0
   (c) 0

9. $y(t) = te^{2t} + c_1 e^{2t} + c_2 e^{-2t} + c_3$

11. $y(t) = te^{2t} + e^{2t} + 2e^{-2t} - 1$

Section 4.2

1. $y(t) = c_1 e^{-t} + c_2 e^{-\frac{1}{2}t} \cos \frac{\sqrt{3}}{2}t + c_3 e^{-\frac{1}{2}t} \sin \frac{\sqrt{3}}{2}t$

3. $y(t) = c_1 e^t + c_2 e^{-t} + c_3 \sin t + c_4 \cos t$

5. $y(t) = c_1 e^t + c_2 e^{-t} + c_3 e^{2t} + c_4 e^{-2t}$

7. $y(t) = c_1 e^{-2t} + c_2 \cos 5t + c_3 \sin 5t$

9. $y(t) = c_1 e^t + c_2 e^{-3t} + c_3 t e^{-3t} + c_4 t^2 e^{-3t}$

11. $y = e^t - 3e^{-t} + \cos t + 2 \sin t$

Section 4.3

1. $y = cte^{-t}$

3. $y = ce^{2t}$

5. $y = \frac{1}{2}te^t + c_1 e^{-t} + c_2 e^t + c_3$

7. $y = \frac{1}{12}te^{2t} + c_1 e^t + c_2 e^{-t} + c_3 e^{2t} + c_4 e^{-2t}$

9. $y = \frac{1}{2}te^t + c_1 e^{-t} + c_2 e^t + c_3$

11. $y = \frac{t^2}{8} + c_1 + c_2 \cos 2t + c_3 \sin 2t$

13. $y = -t(\sin t + \cos t) + c_1 e^t + c_2 \cos t + c_2 \sin t$

Section 4.4

1. $y_1(t) = -4e^{2t} + 6e^{4t}$ and $y_2(t) = -4e^{2t} + 3e^{4t}$

3. $y_1(t) = \cos 2t - \sin 2t$ and $y_2(t) = -\cos 2t + \sin 2t$

5. $y_1(t) = -20e^{-t} + 40e^t - 20e^{2t}$ and $y_2(t) = -6e^{-t} + 20e^t - 12e^{2t}$

7. $y_1(t) = 3e^{-t} - 3\cos 3t + \sin 3t$ and $y_2(t) = 3e^{-t} + 3\cos 3t + 3\sin 3t$

9. $y_1(t) = \cos t + \sin t + 2\cos 2t + \sin 2t$ and $y_2(t) = 2\cos t + 2\sin t - 2\cos 2t - \sin 2t$

13. $y_1(t) = 2e^t \cos 2t + 2e^t \sin 2t$ and $y_2(t) = -2e^t \cos 2t + 2e^t \sin 2t$

15. $y_1(t) = 20 - 19\cos t - 2\sin t$ and $y_2(t) = 8 - 8\cos t + 3\sin t$

Section 4.5

1. $y(t) = 10e^{-5t}$, asymptotically stable

3. $y(t) = e^t + e^{3t}$, unstable

5. $y(t) = e^{-2t} \sin(t)$, stable

7. $y(t) = e^{-3t} + 4te^{-3t}$, stable

9. $y(t) = e^t(\sin(t) + \cos(t))$, unstable

11. $y(t) = e^{-t}$, marginally stable

13. $y(t) = e^{-t}$

15. $y(t) = \frac{1}{4}(e^{2t} - e^{-2t})$

17. $y(t) = 1 - \cos(t)$

760   B   Selected Answers

Section 5.1

1. Not linear

3. Yes, nonhomogeneous, yes

5. Yes, nonhomogeneous, no

7. Yes, nonhomogeneous, no

9. Not linear

11. Yes, homogeneous, no

13. 1. $L(\frac{1}{t}) = 0$
    2. $L(1) = -1$
    3. $L(t) = 0$
    4. $L(t^r) = (r^2 - 1)t^r$

15. $C = \frac{-1}{2}$

17. (3) a. $y(t) = e^{-t} - e^t + 2t$
    (3) b. $y(t) = e^{-t} + (0)e^t + (1)t = e^{-t} + t$
    (3) c. $y(t) = e^{-t} + -e^t + 3t$
    (3) d. $y(t) = e^{-t} + (a-1)e^t + (b-a+2)t$

19. $(-\infty, 0)$

21. $(0, \pi)$

23. $(3, \infty)$

25. The initial condition occurs at $t = 0$ which is precisely where $a_2(t) = t^2$ has a zero. Theorem 6 does not apply.

27. The assumptions say that $y_1(t_0) = y_2(t_0)$ and $y_1'(t_0) = y_2'(t_0)$. Both y_1 and y_2 therefore satisfies the same initial conditions. By the uniqueness part of Theorem 6 $y_1 = y_2$.

Section 5.2

1. Dependent; $2t$ and $5t$ are multiples of each other.

3. Independent

5. Independent

11. 1. Suppose $at^3 + b |t^3| = 0$ on $(-\infty, \infty)$. Then for $t = 1$ and $t = -1$ we get

$$a + b = 0$$
$$-a + b = 0.$$

These equations imply $a = b = 0$. So y_1 and y_2 are linearly independent.

2. Observe that $y_1'(t) = 3t^2$ and $y_2'(t) = \begin{cases} -3t^2 & \text{if } t < 0 \\ 3t^2 & \text{if } t \geq 0. \end{cases}$ If $t < 0$,

then $w(y_1, y_2)(t) = \det \begin{pmatrix} t^3 & -t^3 \\ 3t^2 & -3t^2 \end{pmatrix} = 0$. If $t \geq 0$, then $w(y_1, y_2)(t) = \det \begin{pmatrix} t^3 & t^3 \\ 3t^2 & 3t^2 \end{pmatrix} = 0$. It follows that the Wronskian is zero for all $t \in (-\infty, \infty)$.

3. The condition that the coefficient function $a_2(t)$ be nonzero in Theorem 2 and Proposition 4 is essential. Here the coefficient function, t^2, of y'' is zero at $t = 0$, so Proposition 4 does not apply on $(-\infty, \infty)$. The largest open intervals on which t^2 is nonzero are $(-\infty, 0)$ and $(0, \infty)$. On each of these intervals, y_1 and y_2 are linearly dependent.

4. Consider the cases $t < 0$ and $t \geq 0$. The verification is then straightforward.

5. Again the condition that the coefficient function $a_2(t)$ be nonzero is essential. The uniqueness and existence theorem does not apply.

Section 5.3

1. The general solution is $y(t) = c_1 t + c_2 t^{-2}$.

3. The general solution is $y(t) = c_1 t^{\frac{1}{3}} + c_2 t^{\frac{1}{3}} \ln t$.

5. The general solution is $y(t) = c_1 t^{\frac{1}{2}} + c_2 t^{\frac{1}{2}} \ln t$.

7. $y(t) = c_1 t^{-3} + c_2 t^{-3} \ln t$

9. The general solution is $y(t) = c_1 t^2 + c_2 t^{-2}$.

11. The general solution is $y(t) = c_1 t^2 \cos(3 \ln t) + c_2 t^2 \sin(3 \ln t)$.

13. $y = 2t^{1/2} - t^{1/2} \ln t$

15. No solution is possible.

Section 5.4

1. $\ln\left(\dfrac{s-a}{s-b}\right)$

3. $s\ln\left(\dfrac{s^2+b^2}{s^2+a^2}\right) - 2b\tan^{-1}\left(\dfrac{b}{s}\right) + 2a\tan^{-1}\left(\dfrac{a}{s}\right)$

5. $y = c_1 e^{-t} + c_2(t-1)$

7. $y(t) = c_1 e^{-2t}$.

9. $y(t) = c_1\left((3-t^2)\sin t - 3t\cos t\right)$

11. $y(t) = c_1\dfrac{e^t - 1}{t}$.

13. $y(t) = c_1\left(\dfrac{e^{3t} - e^{2t}}{t}\right)$.

15. $y(t) = c_1\dfrac{\sin t}{t} + c_2\dfrac{1-\cos t}{t}$

Section 5.5

1. $y_2(t) = t^2\ln t$, and the general solution can be written $y(t) = c_1 t^2 + c_2 t^2\ln t$.

3. $y_2(t) = \sqrt{t}\ln t$, and the general solution can be written $y(t) = c_1\sqrt{t} + c + 2\sqrt{t}\ln t$.

5. $y_2(t) = te^t$. The general solution can be written $y(t) = c_1 t + c_2 te^t$.

7. $y_2(t) = \frac{-1}{2}\cos t^2$. The general solution can be written $y(t) = c_1\sin t^2 + c_2\cos t^2$.

9. $y_2(t) = -1 - t\tan t$. The general solution can be written $y(t) = c_1\tan t + c_2(1 + t\tan t)$.

11. $y_2(t) = -\sec t$. The general solution can be written $y(t) = c_1\tan t + c_2\sec t$.

13. $y_2 = -1 - \dfrac{t\sin 2t}{1+\cos 2t}$. The general solution can be written $y(t) = c_1\dfrac{\sin 2t}{1+\cos 2t} + c_2\left(1 + \dfrac{t\sin 2t}{1+\cos 2t}\right)$.

15. $y_2(t) = \frac{1}{2}t + \frac{1}{4}(1 - t^2)\ln\left(\frac{1+t}{1-t}\right)$, and the general solution can be written

$$y = c_1(1 - t^2) + c_2\left(\frac{1}{2}t + \frac{1}{4}(1 - t^2)\ln\left(\frac{1+t}{1-t}\right)\right).$$

Section 5.6

1. The general solution is $y(t) = \frac{-1}{2}t\cos t + c_1\cos t + c_2\sin t$.

3. The general solution is $y(t) = \frac{1}{4}e^t + c_1e^t\cos 2t + c_2e^t\sin 2t$.

5. The general solution is $y(t) = \frac{1}{2}e^{3t} + c_1e^t + c_2e^{2t}$.

7. The general solution is $y(t) = t\ln te^t + c_1e^t + c_2te^t$.

9. The general solution is $y(t) = \frac{t^4}{6} + c_1t + c_2t^2$.

11. The general solution is $y(t) = \frac{t}{2}\ln^2 t + c_1t + c_2t\ln t$.

13. The general solution is $y(t) = \frac{t^2}{2}\tan t + t + c_1\tan t + c_2\sec t$.

15. The general solution is $y(t) = t^2 + c_1\cos t^2 + c_2\sin t^2$.

19. $y_p(t) = \frac{1}{a}f(t) * \sinh at$

21. $y_p(t) = \frac{1}{a-b}f(t) * (e^{at} - e^{bt})$

Section 6.1

1. Graph (c)

3. Graph (e)

5. Graph (f)

7. Graph (h)

9. -22/3

11. 4

13. 11/2

15. 5

17. A, B true. C false.

19. A is true. B and C are false.

21. A and B are true. C and D are false.

23. A, B, C, D are all true.

25. $y(t) = \begin{cases} -1 + e^t & \text{if } 0 \le t < 2, \\ 1 - 2e^{t-2} + e^t & \text{if } 2 \le t < 4 \\ e^{t-4} - 2e^{t-2} + e^t & \text{if } 4 \le t < \infty. \end{cases}$

27. $y(t) = \begin{cases} 0 & \text{if } 0 \le t < 1, \\ -t + e^{t-1} & \text{if } 1 \le t < 2, \\ t - 2 - 2e^{t-2} + e^{t-1} & \text{if } 2 \le t < 3 \\ e^{t-3} - 2e^{t-2} + e^{t-1} & \text{if } 3 \le t < \infty. \end{cases}$

29. $y(t) = \begin{cases} -t + e^t - e^{-t} & \text{if } 0 \le t < 1, \\ e^t - e^{t-1} - e^{-1} & 1 \le t < \infty. \end{cases}$

Section 6.2

1.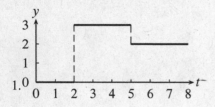

3.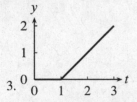

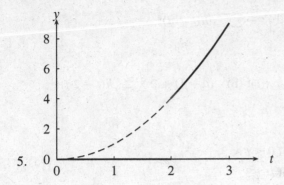

5.

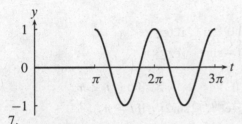

7.

9. (a) $(t-2)\chi_{[2,\infty)}(t)$; (b) $(t-2)h(t-2)$; (c) e^{-2s}/s^2.

11. (a) $(t+2)\chi_{[2,\infty)}(t)$; (b) $(t+2)h(t-2)$; (c) $e^{-2s}\left(\dfrac{1}{s^2}+\dfrac{4}{s}\right)$.

13. (a) $t^2\chi_{[4,\infty)}(t)$; (b) $t^2h(t-4)$; (c) $e^{-4s}\left(\dfrac{2}{s^3}+\dfrac{8}{s^2}+\dfrac{16}{s}\right)$.

15. (a) $(t-4)^2\chi_{[2,\infty)}(t)$; (b) $(t-4)^2h(t-2)$; (c) $e^{-2s}\left(\dfrac{2}{s^3}-\dfrac{4}{s^2}+\dfrac{4}{s}\right)$.

17. (a) $e^t\chi_{[4,\infty)}(t)$; (b) $e^t h(t-4)$; (c) $e^{-4(s-1)}\dfrac{1}{s-1}$.

19. (a) $te^t\chi_{[4,\infty)}(t)$; (b) $te^t h(t-4)$;
(c) $e^{-4(s-1)}\left(\dfrac{1}{(s-1)^2}+\dfrac{4}{s-1}\right)$.

21. (a) $t\chi_{[0,1)}(t)+(2-t)\chi_{[1,\infty)}(t)$; (b) $t+(2-2t)h(t-1)$;
(c) $\dfrac{1}{s^2}-\dfrac{2e^{-s}}{s^2}$.

23. (a) $t^2\chi_{[0,2)}(t)+4\chi_{[2\,3)}(t)+(7-t)\chi_{[3,\infty)}(t)$;
(b) $t^2+(4-t^2)h(t-2)+(3-t)h(t-3)$;
(c) $\dfrac{2}{s^3}-e^{-2s}\left(\dfrac{2}{s^3}+\dfrac{4}{s^2}\right)-\dfrac{e^{-3s}}{s^2}$.

25. (a) $\sum_{n=0}^{\infty}(t-n)\chi_{[n,n+1)}(t)$; (b) $t - \sum_{n=1}^{\infty}h(t-n)$;

(c) $\dfrac{1}{s^2} - \dfrac{e^{-s}}{s(1-e^{-s})}$.

27. (a) $\sum_{n=0}^{\infty}(2n+1-t)\chi_{[2n,2n+2)}(t)$; (b) $-(t+1) + 2\sum_{n=0}^{\infty}h(t-2n)$;

(c) $-\dfrac{1}{s^2} - \dfrac{1}{s} + \dfrac{2}{s(1-e^{-2s})}$.

29. $(t-3)h(t-3) = \begin{cases} 0 & \text{if } 0 \le t < 3, \\ t-3 & \text{if } t \ge 3. \end{cases}$

31. $h(t-\pi)\sin(t-\pi)$

$= \begin{cases} 0 & \text{if } 0 \le t < \pi, \\ \sin(t-\pi) & \text{if } t \ge \pi \end{cases} = \begin{cases} 0 & \text{if } 0 \le t < \pi, \\ -\sin t & \text{if } t \ge \pi. \end{cases}$

33. $\frac{1}{2}e^{-(t-\pi)}\sin 2(t-\pi)h(t-\pi) = \begin{cases} 0 & \text{if } 0 \le t < \pi, \\ \frac{1}{2}e^{-(t-\pi)}\sin 2t & \text{if } t \ge \pi. \end{cases}$

35. $\frac{1}{2}h(t-2)\sin 2(t-2) = \begin{cases} 0 & \text{if } 0 \le t < 2, \\ \frac{1}{2}\sin 2(t-2) & \text{if } t \ge 2. \end{cases}$

37. $h(t-4)\left(2e^{-2(t-4)} - e^{-(t-4)}\right)$

$= \begin{cases} 0 & \text{if } 0 \le t < 4, \\ 2e^{-2(t-4)} - e^{-(t-4)} & \text{if } t \ge 4. \end{cases}$

39. $t - (t-5)h(t-5) = \begin{cases} t & \text{if } 0 \le t < 5, \\ 5 & \text{if } t \ge 5. \end{cases}$

41. $h(t-\pi)e^{-3(t-\pi)}\left(2\cos 2(t-\pi) - \frac{5}{2}\sin 2(t-\pi)\right)$

$= \begin{cases} 0 & \text{if } 0 \le t < \pi, \\ e^{-3(t-\pi)}\left(2\cos 2t - \frac{5}{2}\sin 2t\right) & \text{if } t \ge \pi. \end{cases}$

Section 6.3

1. $y = \begin{cases} 0 & \text{if } 0 \le t < 1 \\ -\frac{3}{2}\left(1 - e^{-2(t-1)}\right) & \text{if } 1 \le t < \infty \end{cases}$

3. $y = \begin{cases} 0 & \text{if } 0 \le t < 2 \\ \frac{2}{3}\left(e^{3(t-2)} - 1\right) & \text{if } 2 \le t < 3 \\ \frac{2}{3}\left(e^{3(t-2)} - e^{3(t-3)}\right) & \text{if } 3 \le t < \infty \end{cases}$

5. $y = \begin{cases} 6e^{4t} - 4e^t & \text{if } 0 \le t < 1 \\ 6e^{4t} - e^{4t-3} - 3e & \text{if } 1 \le t < \infty \end{cases}$

7. $y = \begin{cases} 0 & \text{if } 0 \le t < 3 \\ \frac{1}{9}(1 - \cos 3(t - 3)) & \text{if } 3 \le t < \infty \end{cases}$.

9. $y = \begin{cases} 0 & \text{if } 0 \le t < 1 \\ 1 - 3e^{-2(t-1)} + 2e^{-3(t-1)} & \text{if } 1 \le t < 3 \\ 3e^{-2(t-3)} - 3e^{-2(t-1)} - 2e^{-3(t-3)} + 2e^{-3(t-1)} & \text{if } 3 \le t < \infty \end{cases}$

11. $y = \begin{cases} te^{-t} & \text{if } 0 \le t < 3 \\ 1 + te^{-t} - (t - 2)e^{-(t-3)} & \text{if } 3 \le t < \infty \end{cases}$

13. $y(t) = \begin{cases} 4 - 4e^{\frac{-t}{2}} & \text{if } 0 \le t < 3 \\ 20 - 4e^{\frac{-t}{2}} - 16e^{\frac{-(t-3)}{2}} & \text{if } t \ge 3. \end{cases}$

15. $y(t) = \begin{cases} 10 - 8e^{-3t/10} & \text{if } 0 \le t < 2 \\ 10e^{-3(t-2)/10} - 8e^{-3t/10} & \text{if } 2 \le t < 4 \\ 10 - 8e^{-3t/10} + 10e^{-3(t-2)/10} - 10e^{-3(t-4)/10} & \text{if } 4 \le t < \infty \end{cases}$

Section 6.4

1. $y = \begin{cases} 0 & \text{if } 0 \le t < 1 \\ e^{-2(t-1)} & \text{if } 1 \le t < \infty \end{cases}$

3. $y = \begin{cases} 2e^{4t} & \text{if } 0 \le t < 4 \\ 2e^{4t} + e^{4(t-4)} & \text{if } 4 \le t < \infty \end{cases}$

5. $y = \begin{cases} \frac{1}{2}\sin 2t & \text{if } 0 \le t < \pi, \\ \sin 2t & \text{if } t \ge \pi. \end{cases}$

7. $y = \begin{cases} e^{-t} & \text{if } 0 \le t < 2 \\ e^{-t} + e^{-(t-2)} - e^{-3(t-2)} & \text{if } 2 \le t < \infty \end{cases}$.

9. $y = \begin{cases} te^{-2t} - e^{-2t} & \text{if } 0 \le t < 1 \\ te^{-2t} - e^{-2t} + 3(t-1)e^{-2(t-1)} & \text{if } 1 \le t < \infty \end{cases}$

11. $y = \begin{cases} 24 - 24e^{-\frac{1}{4}t} & \text{if } 0 \le t < 3 \\ 24 - 24e^{-\frac{1}{4}t} + 4e^{-\frac{1}{4}(t-3)} & \text{if } 3 \le t < \infty \end{cases}$

13. $y = \begin{cases} e^{-t} & \text{if } 0 \le t < 2 \\ e^{-t} + e^{-(t-2)} & \text{if } 2 \le t < 4 \\ e^{-t} + e^{-(t-2)} + e^{-(t-4)} & \text{if } 4 \le t < 6 \\ e^{-t} + e^{-(t-2)} + e^{-(t-4)} + e^{-(t-6)} & \text{if } 6 \le t < \infty \end{cases}$ and $y(6) = 1.156\,\text{lbs.}$

15. $y = \begin{cases} \frac{1}{10}e^{-2t} & \text{if } 0 \le t < 4 \\ \frac{1}{10}e^{-2t} + (t-4)e^{-2(t-4)} & \text{if } 4 \le t < \infty \end{cases}$

17. $y = \begin{cases} \sin t & \text{if } 0 \le t < \pi \\ 0 & \text{if } \pi \le t < 2\pi \\ \sin t & \text{if } 2\pi \le t < 3\pi \\ 0 & \text{if } 3\pi \le t < 4\pi \\ \sin t & \text{if } 4\pi \le t < 5\pi \\ 0 & \text{if } 5\pi \le t < \infty. \end{cases}$

The graph is given below.

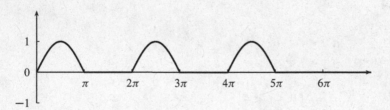

At $t = 0$, the hammer imparts a velocity to the system causing harmonic motion. At $t = \pi$, the hammer strikes in precisely the right way to stop the motion. Then at $t = 2\pi$, the process repeats.

19. $y = y_0 e^{-at} + k e^{-a(t-c)}$

Section 6.5

1. $f * g(t) = \begin{cases} e^t - 1 & \text{if } 0 \leq t < 1 \\ e^t - e^{t-1} & \text{if } 1 \leq t < \infty \end{cases}$

5. $f * g = \begin{cases} t & \text{if } 0 \leq t < 2 \\ -t + 4 & \text{if } 2 \leq t < 4 \\ 0 & \text{if } 4 \leq t < \infty \end{cases}$

7. $f * g = \begin{cases} \sin t & \text{if } 0 \leq t < \pi \\ 0 & \text{if } \pi \leq t < \infty \end{cases}$

9. $\zeta(t) = e^{3t}, y = \begin{cases} 2e^{3t} & \text{if } 0 \leq t < \infty \\ 2e^{3t} + \frac{1}{3}\left(e^{3(t-2)} - 1\right) & \text{if } 1 \leq t < \infty \end{cases}$

11. $\zeta(t) = e^{-8t}, y = \begin{cases} -2e^{-8t} & \text{if } 0 \leq t < 3 \\ -2e^{-8t} + \frac{1}{8}\left(1 - e^{-8(t-3)}\right) & \text{if } 3 \leq t < 5 \\ -2e^{-8t} + \frac{1}{8}\left(e^{-8(t-5)} - e^{-8(t-3)}\right) & \text{if } 5 \leq t < \infty \end{cases}$

13. $y = \frac{1}{9}\begin{cases} 1 - \cos 3t & \text{if } 0 \leq t < 2\pi \\ 0 & \text{if } 2\pi \leq t < \infty \end{cases}$

23. $y = 2e^t - 7te^t$

25. $y = 5 - 4\cos t$

Section 6.6

1. $(\langle t \rangle_1)^2$

3. $\frac{1}{9}([t]_3)^2$

5. $\langle t \rangle_2 + \frac{3}{2}[t]_2$

7. $\mathcal{L}\{f(\langle t \rangle_3)\} = \frac{1 - e^{-3(s-1)}}{1 - e^{-3s}}\frac{1}{s-1}$

9. $\mathcal{L}\{f(\langle t \rangle_{2p})\} = \frac{1 - e^{-ps}}{1 + e^{-ps}}\frac{1}{s}$

11. $\mathcal{L}\{[t]_p\} = \frac{p}{s(e^{ps} - 1)}$

13. $y = \frac{1-e^{-2s}}{1-e^{-2(s+1)}} \frac{1}{s}$

17. $\mathcal{L}^{-1}\left\{\frac{1-e^{-4(s-2)}}{(1-e^{-4s})(s-2)}\right\} = e^{2\langle t\rangle_4}$

Section 6.7

1.
$$y(t) = 10 - 10e^{\frac{-2t}{5}} + 10\,\text{sw}_2(t) - \frac{10e^{\frac{-2t}{5}}}{1+e^{\frac{4}{5}}}\left(1 + e^{\frac{4}{5}}(-1)^{[t/2]_1}e^{\frac{2}{5}[t]_2}\right)$$

The amount of salt fluctuates from 13.10 pounds to 16.90 pounds in the long term.

3. $y(t) = 5e^{-\frac{1}{2}t}\frac{e^{\frac{1}{2}[t]_2+1}-1}{e-1}$, and the salt fluctuation in the tank varies between 2.91 and 7.91 pounds for large values of t.

5. The mathematical model is

$$y' = ry - 40\delta_0(\langle t\rangle_1), \quad y(0) = 3000,$$

where $r = \frac{1}{12}\ln\frac{6}{5}$. The solution to the model is

$$y(t) = 3000e^{rt} - 40\frac{e^{rt} - e^{-r([t]_1-t+1)}}{1-e^{-r}}$$

and at the beginning of 60 months, there are $y(60) \approx 3477$ alligators.

Section 6.8

1.
$$y(t) = 2\left(2\,\text{sw}_1(t) - (-1)^{[t]_1}(\cos\langle t\rangle_1 - \alpha\sin\langle t\rangle_1)\right)$$
$$-2(\cos t + \alpha\sin t)\,,$$

where $\alpha = \frac{-\sin\sqrt{2}}{1+\cos\sqrt{2}}$. Motion is nonperiodic.

3. $y(t) = \frac{1}{\pi^2}\left(2\,\text{sw}_2(t) - \cos\pi t\left((-1)^{[t/2]_1} + 1\right)\right)$, the motion is periodic.

5. $y(t) = \frac{2}{\pi^2}\left(\text{sw}_1(t) - [t]_1\cos\pi t - \cos\pi t\right)$, with resonance.

7.

$$y(t) = \sin t + \sin\langle t\rangle_\pi$$
$$= (1 + (-1)^{[t/\pi]_1}) \sin t.$$

Motion is periodic.

9.

$$y(t) = \sin t + \gamma \cos t + \sin\langle t\rangle_1 - \gamma \cos\langle t\rangle_1,$$

where $\gamma = \frac{-\sin 1}{1-\cos 1}$. The motion is nonperiodic.

11.

$$y(t) = 2(\sin t)(1 + [t/2\pi]_1).$$

Resonance occurs.

Section 7.1

1. $R = 1$

3. $R = \infty$

5. $R = 0$

7. $R = \infty$

9. $R = 2$

11. $-\sum_{n=0}^{\infty} \frac{t^n}{a^{n+1}}$

13. $\sum_{n=0}^{\infty} \frac{(-1)^n t^{2n}}{(2n+1)!}$

15. $\sum_{n=0}^{\infty} (-1)^n \frac{t^{2n+1}}{2n+1}$

17. $\tan t = 1 + \frac{1}{3}t^3 + \frac{2}{15}t^5 + \frac{17}{315}t^7 + \cdots$

19. $e^t \sin t = t + t^2 + \frac{1}{3}t^3 - \frac{1}{30}t^5 + \cdots$

21. $(1-t)e^{-t}$

23. $f(t) = \frac{1}{(1-t)^2}$

25. $f(t) = -\frac{t}{2} + \frac{t^2-1}{4} \ln\left(\frac{1+t}{1-t}\right)$

27. The binomial theorem: $(a + b)^n = \sum_{k=0}^{n} \binom{n}{k} a^k b^{n-k}$

Section 7.2

1. $y(t) = c_0 \sum_{n=0}^{\infty} \frac{t^{2n}}{(2n)!} + c_1 \sum_{n=0}^{\infty} \frac{t^{2n+1}}{(2n+1)!} = c_0 \cosh t + c_1 \sinh t$

3. $y(t) = c_0 \sum_{n=0}^{\infty} (-1)^n \frac{k^{2n} t^{2n}}{(2n)!} + c_1 \sum_{n=0}^{\infty} (-1)^n \frac{k^{2n+1} t^{2n+1}}{(2n+1)!} = c_0 \cos kt + c_1 \sin kt$

5. $y(t) = c_0(1 - t^2) - c_1 \sum_{n=0}^{\infty} \frac{t^{2n+1}}{(2n+1)(2n-1)} = c_0(1 - t^2) - c_1 \left(\frac{t}{2} + \frac{t^2-1}{4} \ln\left(\frac{1-t}{1+t}\right)\right)$

7. $y(t) = c_0 \left(1 + \sum_{n=2}^{\infty} \frac{t^n}{n!}\right) + c_1 t = c_0(e^t - t) + c_1 t$

9. $y(t) = c_0(1 - 3t^2) + c_1\left(t - \frac{t^3}{3}\right)$

Section 7.3

1. -1 and 1 are regular singular points.

3. There are no singular points.

5. 0 is a regular singular point.

7. $q(s) = s(s + 1)$. The exponents of singularity are 0 and -1. Theorem 2 guarantees one Frobenius solution but there could be two.

9. $q(s) = s^2$. The exponent of singularity is 0 with multiplicity 2. Theorem 2 guarantees that there is one and only one Frobenius solution.

15. $y_1(t) = \sum_{m=0}^{\infty} \frac{(-1)^m (2m+2) t^{2m+3}}{(2m+3)!} = (\sin t - t \cos t)$ and $y_2(t) = \sum_{m=0}^{\infty} \frac{(-1)^{m+1}(2m-1) t^{2m}}{(2m)!} = (t \sin t + \cos t)$

17. $y_1(t) = \sum_{n=0}^{\infty} \frac{1}{n!} t^{n+1} = te^t$ and $y_2(t) = \left(te^t \ln t - t \sum_{n=1}^{\infty} \frac{s_n t^n}{n!}\right)$, where $s_n = 1 + \frac{1}{2} + \frac{1}{3} + \cdots \frac{1}{n}$.

19. $y_1(t) = \sum_{n=0}^{\infty} \frac{(-1)^n (n+1) t^n}{(n+3)!} = \left(\frac{(t+2)e^{-t}}{t} + \frac{t-2}{t}\right)$ and $y_2(t) = t^{-1}(1 - \frac{t}{2}) = \frac{1}{2}\left(\frac{2-t}{t}\right)$

21. $y_1(t) = t^2$ and $y_2(t) = t^2 \ln t + \left(-1 + 2t - \frac{t^3}{3} + \sum_{n=4}^{\infty} \frac{2(-1)^n t^n}{n!(n-2)}\right)$

23. The complex Frobenius series is $y(t) = (t^i + \left(\frac{1-2i}{1+2i}\right) t^{1+i})$; the real and imaginary parts are $y_1(t) = -3 \cos \ln t - 4 \sin \ln t + 5t \cos \ln t$ and $y_2(t) = -3 \sin \ln t + 4 \cos \ln t + 5t \sin \ln t$.

25. The complex Frobenius solution is $y(t) = \sum_{n=0}^{\infty} \frac{t^{n+i}}{n!} = t^i e^t$; the real and imaginary parts are $y_1(t) = e^t \cos \ln t$ and $y_2(t) = e^t \sin \ln t$.

Section 7.4

Section 8.1

1. $B + C = \begin{bmatrix} 1 & 1 \\ -1 & 7 \\ 0 & 3 \end{bmatrix}$, $B - C = \begin{bmatrix} 1 & -3 \\ 5 & -1 \\ -2 & 1 \end{bmatrix}$, and $2B - 3C = \begin{bmatrix} 2 & -8 \\ 13 & -6 \\ -5 & 1 \end{bmatrix}$

3. $A(B + C) = AB + AC = \begin{bmatrix} 3 & 4 \\ 1 & 13 \end{bmatrix}$, $(B + C)A = \begin{bmatrix} 3 & -1 & 7 \\ 3 & 1 & 25 \\ 5 & 0 & 12 \end{bmatrix}$

5. $AB = \begin{bmatrix} 6 & 4 & -1 & -8 \\ 0 & 2 & -8 & 2 \\ 2 & -1 & 9 & -5 \end{bmatrix}$

7. $CA = \begin{bmatrix} 8 & 0 \\ 4 & -5 \\ 8 & 14 \\ 10 & 11 \end{bmatrix}$

9. $ABC = \begin{bmatrix} 8 & 9 & -48 \\ 4 & 0 & -48 \\ -2 & 3 & 40 \end{bmatrix}$.

15. $\begin{bmatrix} 0 & 0 & 1 \\ 3 & -5 & -1 \\ 0 & 0 & 5 \end{bmatrix}$

17. (a) Choose, for example, $A = \begin{bmatrix} 0 & 1 \\ 0 & 0 \end{bmatrix}$ and $B = \begin{bmatrix} 0 & 0 \\ 1 & 0 \end{bmatrix}$.

(b) $(A + B)^2 = A^2 + 2AB + B^2$ precisely when $AB = BA$.

19. $B^n = \begin{bmatrix} 1 & n \\ 0 & 1 \end{bmatrix}$

21. (a) $\begin{bmatrix} 0 & 1 \\ 1 & 0 \end{bmatrix} A = \begin{bmatrix} v_2 \\ v_1 \end{bmatrix}$; the two rows of A are switched. (b) $\begin{bmatrix} 1 & c \\ 0 & 1 \end{bmatrix} A = \begin{bmatrix} v_1 + cv_2 \\ v_2 \end{bmatrix}$; to the first row is added c times the second row while the second row is unchanged. (c) To the second row is added c times the first row while the first row is unchanged. (d) The first row is multiplied by a while the second row is unchanged. (e) The second row is multiplied by a while the first row is unchanged.

Section 8.2

1. $A = \begin{bmatrix} 1 & 4 & 3 \\ 1 & 1 & -1 \\ 2 & 0 & 1 \\ 0 & 1 & -1 \end{bmatrix}$, $\mathbf{x} = \begin{bmatrix} x \\ y \\ z \end{bmatrix}$, $\mathbf{b} = \begin{bmatrix} 2 \\ 4 \\ 1 \\ 6 \end{bmatrix}$, and $[A|\mathbf{b}] = \begin{bmatrix} 1 & 4 & 3 & 2 \\ 1 & 1 & -1 & 4 \\ 2 & 0 & 1 & 1 \\ 0 & 1 & -1 & 6 \end{bmatrix}$.

3.
$$\begin{aligned} x_1 - \quad\quad x_3 + 4x_4 + 3x_5 &= 2 \\ 5x_1 + 3x_2 - 3x_3 - \quad x_4 - 3x_5 &= 1 \\ 3x_1 - 2x_2 + 8x_3 + 4x_4 - 3x_5 &= 3 \\ -8x_1 + 2x_2 \quad\quad + 2x_4 + \quad x_5 &= -4 \end{aligned}$$

5. RREF

7. $m_2(1/2)(A) = \begin{bmatrix} 0 & 1 & 0 & 3 \\ 0 & 0 & 1 & 3 \\ 0 & 0 & 0 & 0 \end{bmatrix}$

9. $t_{1,3}(-3)(A) = \begin{bmatrix} 1 & 0 & 1 & 0 & 3 \\ 0 & 1 & 3 & 4 & 1 \\ 0 & 0 & 0 & 0 & 0 \end{bmatrix}$

11. $\begin{bmatrix} 1 & 0 & 0 & -11 & -8 \\ 0 & 1 & 0 & -4 & -2 \\ 0 & 0 & 1 & 9 & 6 \end{bmatrix}$

13. $\begin{bmatrix} 1 & 2 & 0 & 0 & 3 \\ 0 & 0 & 1 & 0 & 2 \\ 0 & 0 & 0 & 1 & 0 \\ 0 & 0 & 0 & 0 & 0 \end{bmatrix}$

15. $\begin{bmatrix} 1 & 0 & 2 \\ 0 & 1 & 1 \\ 0 & 0 & 0 \\ 0 & 0 & 0 \\ 0 & 0 & 0 \end{bmatrix}$

17. $\begin{bmatrix} 1 & 4 & 0 & 0 & 3 \\ 0 & 0 & 1 & 0 & 1 \\ 0 & 0 & 0 & 1 & 3 \end{bmatrix}$

19. $\begin{bmatrix} x \\ y \\ z \end{bmatrix} = \begin{bmatrix} -1 \\ 1 \\ 0 \end{bmatrix} + \alpha \begin{bmatrix} -3 \\ 1 \\ 5 \end{bmatrix}, \alpha \in \mathbb{R}$

21. $\begin{bmatrix} x \\ y \end{bmatrix} = \alpha \begin{bmatrix} -2 \\ 1 \end{bmatrix}, \alpha \in \mathbb{R}$

23. $\begin{bmatrix} x \\ y \\ z \end{bmatrix} = \begin{bmatrix} 14/3 \\ 1/3 \\ -2/3 \end{bmatrix}$

25. $\begin{bmatrix} 0 \\ 3 \\ 4 \end{bmatrix} + \alpha \begin{bmatrix} 1 \\ 0 \\ 0 \end{bmatrix}, \alpha \in \mathbb{R}$

27. $\emptyset$

29. $\left\{ \begin{bmatrix} -1 \\ 0 \\ 1 \end{bmatrix} \right\}$

31. $\left\{ \begin{bmatrix} -34 \\ -40 \\ 39 \\ 1 \end{bmatrix} \right\}$

33. The equation $\begin{bmatrix} 5 \\ -1 \\ 4 \end{bmatrix} = a \begin{bmatrix} 1 \\ 1 \\ 2 \end{bmatrix} + b \begin{bmatrix} 1 \\ -1 \\ 0 \end{bmatrix}$ has solution $a = 2$ and $b = 3$. By

Proposition 7 $\begin{bmatrix} 5 \\ -1 \\ 4 \end{bmatrix}$, is a solution.

35. If $\mathbf{x}_i$ is the solution set for $A\mathbf{x} = \mathbf{b}_i$, then $\mathbf{x}_1 = \begin{bmatrix} -7/2 \\ 7/2 \\ -3/2 \end{bmatrix}$, $\mathbf{x}_2 = \begin{bmatrix} -3/2 \\ 3/2 \\ -1/2 \end{bmatrix}$, and

$\mathbf{x}_3 = \begin{bmatrix} 7 \\ -6 \\ 3 \end{bmatrix}$.

Section 8.3

1. $\begin{bmatrix} 4 & -1 \\ -3 & 1 \end{bmatrix}$

3. Not invertible

5. Not invertible

7. $\begin{bmatrix} -6 & 5 & 13 \\ 5 & -4 & -11 \\ -1 & 1 & 3 \end{bmatrix}$

9. $\begin{bmatrix} -29 & 39/2 & -22 & 13 \\ 7 & -9/2 & 5 & -3 \\ -22 & 29/2 & -17 & 10 \\ 9 & -6 & 7 & -4 \end{bmatrix}$

11. $\begin{bmatrix} 0 & 0 & -1 & 1 \\ 1 & 0 & 0 & 0 \\ 0 & 1 & 1 & -1 \\ -1 & -1 & 0 & 1 \end{bmatrix}$

13. $\mathbf{x} = \begin{bmatrix} 5 \\ -3 \end{bmatrix}$

15. $\mathbf{x} = \frac{1}{10} \begin{bmatrix} 16 \\ 11 \\ 18 \end{bmatrix}$

17. $\mathbf{x} = \begin{bmatrix} 19 \\ -4 \\ 15 \\ -6 \end{bmatrix}$

19. $(A^t)^{-1} = (A^{-1})^t$

21. $F(\theta)^{-1} = F(-\theta)$

Section 8.4

1. 1

3. 10

5. -21

7. 2

9. 0

11. $\frac{1}{s^2-6s+8} \begin{bmatrix} s-3 & 1 \\ 1 & s-3 \end{bmatrix}$ $s = 2, 4$

13. $\frac{1}{(s-1)^3} \begin{bmatrix} (s-1)^2 & 3 & s-1 \\ 0 & (s-1)^2 & 0 \\ 0 & 3(s-1) & (s-1)^2 \end{bmatrix}$ $s = 1$

15. $\frac{1}{s^3+s^2+4s+4} \begin{bmatrix} s^2+s & 4s+4 & 0 \\ -s-1 & s^2+s & 0 \\ s-4 & 4s+4 & s^2+4 \end{bmatrix}$ $s = -1, \pm 2i$

17. no inverse

19. $\frac{1}{8} \begin{bmatrix} 4 & -4 & 4 \\ -1 & 3 & -1 \\ -5 & -1 & 3 \end{bmatrix}$

21. $\frac{1}{6} \begin{bmatrix} 2 & -98 & 9502 \\ 0 & 3 & -297 \\ 0 & 0 & 6 \end{bmatrix}$

23. $\frac{1}{15} \begin{bmatrix} 55 & -95 & 44 & -171 \\ 50 & -85 & 40 & -150 \\ 70 & -125 & 59 & -216 \\ 65 & -115 & 52 & -198 \end{bmatrix}$

25. $\mathbf{x} = \begin{bmatrix} 5 \\ -3 \end{bmatrix}$

27. $\mathbf{x} = \frac{1}{10} \begin{bmatrix} 16 \\ 11 \\ 18 \end{bmatrix}$

Section 8.5

1. The characteristic polynomial is $c_A(s) = (s-1)(s-2)$. The eigenvalues are thus $s = 1, 2$. The eigenspaces are $E_1 = \mathrm{Span}\left\{\begin{bmatrix} 0 \\ 1 \end{bmatrix}\right\}$ and $E_2 = \mathrm{Span}\left\{\begin{bmatrix} 1 \\ -1 \end{bmatrix}\right\}$.

3. The characteristic polynomial is $c_A(s) = s^2 - 2s + 1 = (s-1)^2$. The only eigenvalue is $s = 1$. The eigenspace is $E_1 = \mathrm{Span}\left\{\begin{bmatrix} 1 \\ -1 \end{bmatrix}\right\}$.

5. The characteristic polynomial is $c_A(s) = s^2 + 2s - 3 = (s+3)(s-1)$. The eigenvalues are thus $s = -3, 1$. The eigenspaces are $E_{-3} = \mathrm{Span}\left\{\begin{bmatrix} 1 \\ -1 \end{bmatrix}\right\}$ and $E_1 = \mathrm{Span}\left\{\begin{bmatrix} -3 \\ 2 \end{bmatrix}\right\}$.

7. The characteristic polynomial is $c_A(s) = s^2 + 2s + 10 = (s+1)^2 + 3^2$. The eigenvalues are thus $s = -1 \pm 3i$. The eigenspaces are $E_{-1+3i} = \mathrm{Span}\left\{\begin{bmatrix} 7+i \\ 10 \end{bmatrix}\right\}$ and $E_{-1-3i} = \mathrm{Span}\left\{\begin{bmatrix} 7-i \\ 10 \end{bmatrix}\right\}$.

9. The eigenvalues are $s = -2, 3$. $E_{-2} = \mathrm{Span}\left\{\begin{bmatrix} 1 \\ 2 \\ 0 \end{bmatrix}, \begin{bmatrix} 0 \\ 1 \\ 1 \end{bmatrix}\right\}$, $E_3 = \mathrm{Span}\left\{\begin{bmatrix} 1 \\ 0 \\ -1 \end{bmatrix}\right\}$,

11. The eigenvalues are $s = 0, 2, 3$. $E_0 = \text{NS}(A) = \text{Span}\left\{\begin{bmatrix} 0 \\ 2 \\ 1 \end{bmatrix}\right\}$, $E_2 =$

Span $\left\{\begin{bmatrix} 2 \\ 2 \\ 1 \end{bmatrix}\right\}$, $E_3 = \text{Span}\left\{\begin{bmatrix} 0 \\ 1 \\ 1 \end{bmatrix}\right\}$.

13. We write $c_A(s) = (s - 2)((s - 2)^2 + 1)$ to see that the eigenvalues are

$s = 2, 2 \pm i$. $E_2 = \text{Span}\left\{\begin{bmatrix} 2 \\ 3 \\ 1 \end{bmatrix}\right\}$, $E_{2+i} = \text{Span}\left\{\begin{bmatrix} -4 + 3i \\ 4 + 2i \\ 5 \end{bmatrix}\right\}$, $E_{2-i} =$

Span $\left\{\begin{bmatrix} -4 - 3i \\ 4 - 2i \\ 5 \end{bmatrix}\right\}$.

Section 9.2

1. Nonlinear

3. $y' = \begin{bmatrix} \sin t & 0 \\ 1 & \cos t \end{bmatrix} y$; linear and homogeneous, but not constant coefficient.

5. $y' = \begin{bmatrix} 1 & 0 & 0 & 0 \\ 2 & 0 & 0 & 1 \\ 0 & 0 & 0 & 1 \\ 0 & 1 & 2 & 0 \end{bmatrix} y$; linear, constant coefficient, homogeneous.

11. $y' = \begin{bmatrix} 0 & 1 \\ -6 & -5 \end{bmatrix} y + \begin{bmatrix} 0 \\ e^{2t} \end{bmatrix}$, $\quad y(0) = \begin{bmatrix} 1 \\ -2 \end{bmatrix}$.

13. $y' = \begin{bmatrix} 0 & 1 \\ k^2 & 0 \end{bmatrix} y + \begin{bmatrix} 0 \\ \cos \omega t \end{bmatrix}$, $\quad y(0) = \begin{bmatrix} 0 \\ 0 \end{bmatrix}$.

15. $y' = \begin{bmatrix} 0 & 1 \\ -\frac{1}{t^2} & -\frac{2}{t} \end{bmatrix} y$, $\quad y(1) = \begin{bmatrix} -2 \\ 3 \end{bmatrix}$.

17. $A'(t) = \begin{bmatrix} -3e^{-3t} & 1 \\ & 2t & 2e^{2t} \end{bmatrix}$

19. $y'(t) = \begin{bmatrix} 1 \\ 2t \\ t^{-1} \end{bmatrix}$

21. $v'(t) = \left[-2e^{-2t} \; \frac{2t}{t^2+1} \; -3\sin 3t \right]$

23. $\frac{1}{4} \begin{bmatrix} e^2 - e^{-2} & e^2 + e^{-2} - 2 \\ 2 - e^2 - e^{-2} & e^2 - e^{-2} \end{bmatrix}$

25. $\begin{bmatrix} 4 & 8 \\ 12 & 16 \end{bmatrix}$

27. $\begin{bmatrix} \frac{1}{s} & \frac{1}{s^2} \\ \frac{2}{s^3} & \frac{1}{s-2} \end{bmatrix}$

29. $\begin{bmatrix} \frac{3!}{s^4} & \frac{2s}{(s^2+1)^2} & \frac{1}{(s+1)^2} \\ \frac{2-s}{s^3} & \frac{s-3}{s^2-6s+13} & \frac{3}{s} \end{bmatrix}$

31. $\frac{2}{s^2-1} \begin{bmatrix} 1 & -1 \\ -1 & 1 \end{bmatrix}$

33. $\begin{bmatrix} 1 & 2t & 3t^2 \end{bmatrix}$

35. $\begin{bmatrix} e^t + e^{-t} & e^t - e^{-t} \\ e^t - e^{-t} & e^t + e^{-t} \end{bmatrix}$

Section 9.3

1. $e^{At} = \begin{bmatrix} e^t & 0 \\ 0 & e^{-2t} \end{bmatrix}$

3. $e^{At} = \begin{bmatrix} \cosh t & \sinh t \\ \sinh t & \cosh t \end{bmatrix}$

5. $e^{At} = \begin{bmatrix} \frac{1}{2} + \frac{1}{2}e^{2t} & -\frac{1}{2} + \frac{1}{2}e^{2t} \\ -\frac{1}{2} + \frac{1}{2}e^{2t} & \frac{1}{2} + \frac{1}{2}e^{2t} \end{bmatrix}$

7. $e^{At} = \begin{bmatrix} \cos t & \sin t & 0 \\ -\sin t & \cos t & 0 \\ 0 & 0 & e^{2t} \end{bmatrix}$

9. $(sI - A)^{-1} = \begin{bmatrix} \frac{s-2}{s(s-3)} & \frac{-1}{s(s-3)} \\ \frac{-2}{s(s-3)} & \frac{s-1}{s(s-3)} \end{bmatrix}$ and $e^{At} = \begin{bmatrix} \frac{2}{3} + \frac{1}{3}e^{3t} & \frac{1}{3} - \frac{1}{3}e^{3t} \\ \frac{2}{3} - \frac{2}{3}e^{3t} & \frac{1}{3} + \frac{2}{3}e^{3t} \end{bmatrix}$

11. $(sI - A)^{-1} = \begin{bmatrix} \frac{s+1}{(s-1)^2+1} & \frac{5}{(s-1)^2+1} \\ \frac{-1}{(s-1)^2+1} & \frac{s-3}{(s-1)^2+1} \end{bmatrix}$

and $e^{At} = \begin{bmatrix} e^t \cos t + 2e^t \sin t & 5e^t \sin t \\ -e^t \sin t & e^t \cos t - 2e^t \sin t \end{bmatrix}$

13. $(sI - A)^{-1} = \begin{bmatrix} \frac{1}{s} & \frac{1}{s^2} & \frac{s+1}{s^3} \\ 0 & \frac{1}{s} & \frac{1}{s^2} \\ 0 & 0 & \frac{1}{s} \end{bmatrix}$ and $e^{At} = \begin{bmatrix} 1 & t & t+\frac{t^2}{2} \\ 0 & 1 & t \\ 0 & 0 & 1 \end{bmatrix}$

15. $e^{At} = \begin{bmatrix} \cos t & \sin t & 0 \\ -\sin t & \cos t & 0 \\ 0 & 0 & e^{2t} \end{bmatrix}$

Section 9.4

1. $e^{At} = \begin{bmatrix} e^t + te^t & -te^t \\ te^t & e^t - te^t \end{bmatrix}$

3. $e^{At} = \begin{bmatrix} 1 + 2t & t \\ -4t & 1 - 2t \end{bmatrix}$

5. $e^{At} = \begin{bmatrix} e^t \cos t + 3e^t \sin t & -10e^t \sin t \\ e^t \sin t & e^t \cos t - 3e^t \sin t \end{bmatrix}$

7. $e^{At} = \frac{1}{4}\begin{bmatrix} -7e^{2t} + 11e^{-2t} & 11e^{2t} - 11e^{-2t} \\ -7e^{2t} + 7e^{-2t} & 11e^{2t} - 7e^{-2t} \end{bmatrix}$

9. $e^{At} = \begin{bmatrix} e^{2t} \cos 3t + 8e^{2t} \sin 3t & 13e^{2t} \sin 3t \\ -5e^{2t} \sin 3t & e^{2t} \cos 3t - 8e^{2t} \sin 3t \end{bmatrix}$

11. $e^{At} = \begin{bmatrix} e^{-2t} - te^{-2t} & te^{-2t} \\ -te^{-2t} & e^{-2t} + te^{-2t} \end{bmatrix}$

13. $e^{At} = \begin{bmatrix} 2 - e^{-t} & 0 & -1 + e^{-t} \\ 0 & e^t & 0 \\ 2 - 2e^{-t} & 0 & -1 + 2e^{-t} \end{bmatrix}$

15. $e^{At} = \frac{1}{2}\begin{bmatrix} e^t + e^t \cos t & -e^t \sin t & e^t - e^t \cos t \\ 2e^t \sin t & 2e^t \cos t & -2e^t \sin t \\ e^t - e^t \cos t & e^t \sin t & e^t + e^t \cos t \end{bmatrix}$

17. $e^{At} = \begin{bmatrix} -e^t + 2\cos 2t - \sin 2t & e^t - \cos 2t \\ 2\sin 2t & \cos 2t & -\sin 2t \\ -2e^t + 2\cos 2t - \sin 2t & 2e^t - \cos 2t \end{bmatrix}$

19. $e^{At} = \begin{bmatrix} \cos t - t\cos t & \sin t - t\sin t & t\cos t & t\sin t \\ -\sin t + t\sin t & \cos t - t\cos t & -t\sin t & t\cos t \\ -t\cos t & -t\sin t & \cos t + t\cos t & \sin t + t\sin t \\ t\sin t & -t\cos t - \sin t - t\sin t & \cos t + t\cos t \end{bmatrix}$

Section 9.5

1. $y(t) = \begin{bmatrix} e^{-t} \\ -2e^{3t} \end{bmatrix}$

3. $y(t) = \begin{bmatrix} -e^{2t} + 2te^{2t} \\ 2e^{2t} \end{bmatrix}$

5. $y(t) = \begin{bmatrix} e^{-t} \\ 3e^{-t} \end{bmatrix}$

7. $y(t) = \begin{bmatrix} e^t - 2te^t \\ e^t - te^t \end{bmatrix}$

9. $y(t) = \begin{bmatrix} 2\cos 2t + 2\sin 2t \\ \cos 2t - \sin 2t \\ 2\cos 2t + 2\sin 2t \end{bmatrix}$

11. $y(t) = \begin{bmatrix} 1 + 2e^{-t}\sin 2t \\ -2 + 2e^{-t}\cos 2t \end{bmatrix}$

13. $y(t) = \begin{bmatrix} 2e^{-t}\cos 2t + 4e^{-t}\sin 2t \\ 1 + e^{-t}\sin 2t - 2e^{-t}\cos 2t \end{bmatrix}$

15. $y(t) = \begin{bmatrix} 2te^t + e^t - t - 1 \\ -4te^t - e^t + 2t + 2 \end{bmatrix}$

17. $y(t) = \begin{bmatrix} te^t \\ 2te^{2t} - e^{2t} + e^t \\ -2te^{2t} + te^t \end{bmatrix}$

19. $y_1(t) = 1 + 3e^{-2t}$, $y_2(t) = 2 + 4e^{-t} - 6e^{-2t}$, $t = 9.02$ seconds.

21. $y_1(t) = y_2(t) = 1 - e^{-2t}$

Section 9.6

1. $P = \begin{bmatrix} 1 & 3 \\ -1 & -1 \end{bmatrix}$, $J = P^{-1}AP = \begin{bmatrix} -1 & 0 \\ 0 & 1 \end{bmatrix}$, and the critical point is saddle.

3. $P = \begin{bmatrix} -3 & -1 \\ 5 & 0 \end{bmatrix}$, $J = \begin{bmatrix} -2 & -1 \\ 1 & -2 \end{bmatrix}$, and the origin is a stable spiral node.

5. A is of type J_3. The origin is an unstable star node.

7. $P = \begin{bmatrix} 1 & 3 \\ -1 & -1 \end{bmatrix}$, $J = P^{-1}AP = \begin{bmatrix} 2 & 0 \\ 0 & 4 \end{bmatrix}$, and the origin is an unstable node.

9. $P = \begin{bmatrix} -1 & 1 \\ 4 & 0 \end{bmatrix}$, $J = P^{-1}AP = \begin{bmatrix} 1 & -2 \\ 2 & 1 \end{bmatrix}$, and the origin is an unstable star node.

Section 9.7

1. $\boldsymbol{\Phi}(t)$ is a fundamental matrix, $y(t) = \begin{bmatrix} 2e^{-t} - e^{2t} \\ 2e^{-t} - 4e^{2t} \end{bmatrix}$, and the standard fundamental matrix at $t = 0$ is $\boldsymbol{\Psi}(t) = \dfrac{1}{3} \begin{bmatrix} 4e^{-t} - e^{2t} & -e^{-t} + e^{2t} \\ 4e^{-t} - 4e^{2t} & -e^{-t} + 4e^{2t} \end{bmatrix}$.

3. $\boldsymbol{\Phi}(t)$ is a fundamental matrix, $y(t) = \begin{bmatrix} \cos(t^2/2) \\ -\sin(t^2/2) \end{bmatrix}$, and the standard fundamental matrix at $t = 0$ is $\boldsymbol{\Psi}(t) = \begin{bmatrix} \cos(t^2/2) & \sin(t^2/2) \\ -\sin(t^2/2) & \cos(t^2/2) \end{bmatrix}$.

5. $\boldsymbol{\Phi}(t)$ is a fundamental matrix, $y(t) = \dfrac{t}{\pi} \begin{bmatrix} -\cos t + \sin t \\ \cos t + \sin t \end{bmatrix}$, and the standard fundamental matrix at $t = \pi$ is $\boldsymbol{\Psi}(t) = \dfrac{1}{\pi} \begin{bmatrix} -t \cos t & -t \sin t \\ t \sin t & -t \cos t \end{bmatrix}$.

7. $\boldsymbol{\Phi}(t)$ is a fundamental matrix, $y(t) = \begin{bmatrix} (t-1)e^{t-1} - 3 \\ e^{t-1} + 3 \end{bmatrix}$, and the standard

fundamental matrix at $t = 1$ is $\boldsymbol{\Psi}(t) = \begin{bmatrix} 1 + (t-1)e^{t-1} & (t-1)e^{t-1} \\ -1 + e^{t-1} & e^{t-1} \end{bmatrix}$.

9. $\boldsymbol{\Psi}(t) = \begin{bmatrix} t - t\ln t & -t\ln t \\ t\ln t & t + t\ln t \end{bmatrix}$ and $y(t) = \begin{bmatrix} 2t - t\ln t \\ t\ln t \end{bmatrix}$

11. $\boldsymbol{\Psi}(t) = \begin{bmatrix} \sec^2 t + 3\sec t\tan t + \tan^2 t & 5\sec t\tan t \\ -\sec t\tan t & \sec^2 t - 3\sec t\tan t + \tan^2 t \end{bmatrix}$ and

$y(t) = \begin{bmatrix} 2\sec^2 t + 11\sec t\tan t + 2\tan^2 t \\ \sec^2 t - 5\sec t\tan t + \tan^2 t \end{bmatrix}$.

13. $y_1(t) = 4(2-t) + (2-t)^2 - \dfrac{3}{4}(2-t)^4$ and $y_2(t) = 2(2-t) + (2-t)^2 + \dfrac{3}{4}(2-t)^3$.

The concentration (grams/L) of salt in Tank 1 after 1 min is $\dfrac{17}{4}$ and in Tank 2

is $\dfrac{15}{4}$.

Appendix C
Tables

C.1 Laplace Transforms

Table C.1 Laplace transform rules

$f(t)$	$F(s)$	Page
Definition of the Laplace transform		
1. $\quad f(t)$	$F(s) = \int_0^\infty e^{-st} f(t)\, dt$	111
Linearity		
2. $\quad a_1 f_1(t) + a_2 f_2(t)$	$a_1 F_1(s) + a_2 F_2(s)$	114
Dilation principle		
3. $\quad f(at)$	$\dfrac{1}{a} F\left(\dfrac{s}{a}\right)$	122
Translation principle		
4. $\quad e^{at} f(t)$	$F(s - a)$	120
Input derivative principle: first order		
5. $\quad f'(t)$	$s F(s) - f(0)$	115
Input derivative principle: second order		
6. $\quad f''(t)$	$s^2 F(s) - s f(0) - f'(0)$	115
Input derivative principle: nth order		
7. $\quad f^{(n)}(t)$	$s^n F(s) - s^{n-1} f(0) - s^{n-2} f'(0) - \cdots - s f^{(n-2)}(0) - f^{(n-1)}(0)$	116
Transform derivative principle: first order		
8. $\quad t f(t)$	$-F'(s)$	121
Transform derivative principle: second order		
9. $\quad t^2 f(t)$	$F''(s)$	
Transform derivative principle: nth order		
10. $\quad t^n f(t)$	$(-1)^n F^{(n)}(s)$	121
Convolution principle		
11. $\quad (f * g)(t) = \int_0^t f(\tau) g(t - \tau)\, d\tau$	$F(s) G(s)$	188
Input integral principle		
12. $\quad \int_0^t f(v)\, dv$	$\dfrac{F(s)}{s}$	190

(continued)

W.A. Adkins and M.G. Davidson, *Ordinary Differential Equations*,
Undergraduate Texts in Mathematics, DOI 10.1007/978-1-4614-3618-8,
© Springer Science+Business Media New York 2012

Table C.1   (continued)

$f(t)$	$F(s)$	Page
Transform integral formula		
13. $\dfrac{f(t)}{t}$	$\int_s^\infty F(\sigma)\,d\sigma$	357
$\dfrac{f(t)}{t}$ has a continuous extension to 0		
Second translation principle		
14. $f(t-c)h(t-c)$	$e^{-sc}F(s)$	405
Corollary to the second translation principle		
15. $g(t)h(t-c)$	$e^{-sc}\mathcal{L}\{g(t+c)\}$	405
Periodic functions		
16. $f(t)$, periodic with period p	$\dfrac{\int_0^p e^{-st}f(t)\,dt}{1-e^{-sp}}$	455
17. $f(\langle t\rangle_p)$	$\dfrac{\mathcal{L}\{f(t)-f(t)h(t-p)\}}{1-e^{-sp}}$	458
Staircase functions		
18. $f([t]_p)$	$\dfrac{1-e^{-ps}}{s}\sum_{n=0}^{\infty}f(np)e^{-nps}$	463
Transforms involving $\dfrac{1}{1\pm e^{-sp}}$		
19. $\sum_{N=0}^{\infty}\sum_{n=0}^{N}f(t-np)\chi_{[Np,(N+1)p)}$	$\dfrac{1}{1-e^{-sp}}F(s)$	461
20. $\sum_{N=0}^{\infty}\sum_{n=0}^{N}(-1)^n f(t-np)\chi_{[Np,(N+1)p)}$	$\dfrac{1}{1+e^{-sp}}F(s)$	461

Table C.2 Laplace transforms

$f(t)$	$F(s)$	Page
1. 1	$\dfrac{1}{s}$	116
2. t	$\dfrac{1}{s^2}$	
3. t^n $(n=0,2,3,\ldots)$	$\dfrac{n!}{s^{n+1}}$	116
4. t^α $(\alpha>0)$	$\dfrac{\Gamma(\alpha+1)}{s^{\alpha+1}}$	118
5. e^{at}	$\dfrac{1}{s-a}$	118
6. te^{at}	$\dfrac{1}{(s-a)^2}$	
7. $t^n e^{at}$ $(n=1,2,3,\ldots)$	$\dfrac{n!}{(s-a)^{n+1}}$	119
8. $\sin bt$	$\dfrac{b}{s^2+b^2}$	118
9. $\cos bt$	$\dfrac{s}{s^2+b^2}$	118

Table C.2 (continued)

	$f(t)$	$F(s)$	Page
10.	$e^{at}\sin bt$	$\dfrac{b}{(s-a)^2+b^2}$	120
11.	$e^{at}\cos bt$	$\dfrac{s-a}{(s-a)^2+b^2}$	120
12.	$\dfrac{\sin t}{t}$	$\tan^{-1}\dfrac{1}{s}$	358
13.	$\dfrac{\sin at}{t}$	$\tan^{-1}\left(\dfrac{a}{s}\right)$	365
14.	$\dfrac{e^{bt}-e^{at}}{t}$	$\ln\left(\dfrac{s-a}{s-b}\right)$	365
15.	$2\dfrac{\cos bt-\cos at}{t}$	$\ln\left(\dfrac{s^2+a^2}{s^2+b^2}\right)$	365
16.	$2\dfrac{\cos bt-\cos at}{t^2}$	$s\ln\left(\dfrac{s^2+b^2}{s^2+a^2}\right)-2b\tan^{-1}\left(\dfrac{b}{s}\right)$ $+2a\tan^{-1}\left(\dfrac{a}{s}\right)$	365

Laguerre polynomials

17.	$\ell_n(t)=\sum_{k=0}^{n}(-1)^k\binom{n}{k}\dfrac{t^k}{k!}$	$\dfrac{(s-1)^n}{s^{n+1}}$	361
18.	$\ell_n(at)$	$\dfrac{(s-a)^n}{s^{n+1}}$	366

The Heaviside function

19.	$h(t-c)$	$\dfrac{e^{-sc}}{s}$	404

The on-off switch

20.	$\chi_{[a,b)}$	$\dfrac{e^{-as}}{s}-\dfrac{e^{-bs}}{s}$	405

The Dirac delta function

21.	δ_c	e^{-cs}	428

The square-wave function

22.	sw_c	$\dfrac{1}{1+e^{-cs}}\dfrac{1}{s}$	456

The sawtooth function

23.	$\langle t\rangle_p$	$\dfrac{1}{s^2}\left(1-\dfrac{spe^{-sp}}{1-e^{-sp}}\right)$	457

Periodic dirac delta functions

24.	$\delta_0(\langle t\rangle_p)$	$\dfrac{1}{1-e^{-ps}}$	459

Alternating periodic dirac delta functions

25.	$(\delta_0-\delta_p)(\langle t\rangle_{2p})$	$\dfrac{1}{1+e^{-ps}}$	459

The matrix exponential

26.	e^{At}	$(sI-A)^{-1}$	459

Table C.3 Heaviside formulas

$f(t)$	$F(s)$
Heaviside formulas of the first kind	
1. $\dfrac{e^{at}}{a-b} + \dfrac{e^{bt}}{b-a}$	$\dfrac{1}{(s-a)(s-b)}$
2. $\dfrac{ae^{at}}{a-b} + \dfrac{be^{bt}}{b-a}$	$\dfrac{s}{(s-a)(s-b)}$
3. $\dfrac{e^{at}}{(a-b)(a-c)} + \dfrac{e^{bt}}{(b-a)(b-c)} + \dfrac{e^{ct}}{(c-a)(c-b)}$	$\dfrac{1}{(s-a)(s-b)(s-c)}$
4. $\dfrac{ae^{at}}{(a-b)(a-c)} + \dfrac{be^{bt}}{(b-a)(b-c)} + \dfrac{ce^{ct}}{(c-a)(c-b)}$	$\dfrac{s}{(s-a)(s-b)(s-c)}$
5. $\dfrac{a^2e^{at}}{(a-b)(a-c)} + \dfrac{b^2e^{bt}}{(b-a)(b-c)} + \dfrac{c^2e^{ct}}{(c-a)(c-b)}$	$\dfrac{s^2}{(s-a)(s-b)(s-c)}$
6. $\dfrac{r_1^k e^{r_1 t}}{q'(r_1)} + \cdots + \dfrac{r_n^k e^{r_n t}}{q'(r_n)},$ $\quad q(s) = (s-r_1)\cdots(s-r_n)$	$\dfrac{s^k}{(s-r_1)\cdots(s-r_n)},$ $\quad r_1,\dots,r_n,\text{ distinct}$
Heaviside formulas of the second kind	
7. te^{at}	$\dfrac{1}{(s-a)^2}$
8. $(1+at)e^{at}$	$\dfrac{s}{(s-a)^2}$
9. $\dfrac{t^2}{2}e^{at}$	$\dfrac{1}{(s-a)^3}$
10. $\left(t + \dfrac{at^2}{2}\right)e^{at}$	$\dfrac{s}{(s-a)^3}$
11. $\left(1 + 2at + \dfrac{a^2t^2}{2}\right)e^{at}$	$\dfrac{s^2}{(s-a)^3}$
12. $\left(\sum_{l=0}^{k}\binom{k}{l}a^{k-l}\dfrac{t^{n-l-1}}{(n-l-1)!}\right)e^{at}$	$\dfrac{s^k}{(s-a)^n}$

In each case, a, b, and c are distinct. See Page 165.

Table C.4 Laplace transforms involving irreducible quadratics

$f(t)$	$F(s)$
1. $\sin bt$	$\dfrac{b}{(s^2+b^2)}$
2. $\dfrac{1}{2b^2}(\sin bt - bt\cos bt)$	$\dfrac{b}{(s^2+b^2)^2}$
3. $\dfrac{1}{8b^4}\left((3-(bt)^2)\sin bt - 3bt\cos bt\right)$	$\dfrac{b}{(s^2+b^2)^3}$
4. $\dfrac{1}{48b^6}\left((15-6(bt)^2)\sin bt - (15bt-(bt)^3)\cos bt\right)$	$\dfrac{b}{(s^2+b^2)^4}$
5. $\cos bt$	$\dfrac{s}{(s^2+b^2)}$
6. $\dfrac{1}{2b^2}bt\sin bt$	$\dfrac{s}{(s^2+b^2)^2}$
7. $\dfrac{1}{8b^4}\left(bt\sin bt - (bt)^2\cos bt\right)$	$\dfrac{s}{(s^2+b^2)^3}$
8. $\dfrac{1}{48b^6}\left((3bt-(bt)^3)\sin bt - 3(bt)^2\cos bt\right)$	$\dfrac{s}{(s^2+b^2)^4}$

Table C.5 Reduction of order formulas

$$\mathcal{L}^{-1}\left\{\frac{1}{(s^2+b^2)^{k+1}}\right\} = \frac{-t}{2kb^2}\mathcal{L}^{-1}\left\{\frac{s}{(s^2+b^2)^k}\right\} + \frac{2k-1}{2kb^2}\mathcal{L}^{-1}\left\{\frac{1}{(s^2+b^2)^k}\right\}$$

$$\mathcal{L}^{-1}\left\{\frac{s}{(s^2+b^2)^{k+1}}\right\} = \frac{t}{2k}\mathcal{L}^{-1}\left\{\frac{1}{(s^2+b^2)^k}\right\}$$

See Page 155.

Table C.6 Laplace transforms involving quadratics

$f(t)$	$F(s)$	Page
Laplace transforms involving the quadratic $s^2 + b^2$		
1. $\dfrac{\sin bt}{(2b)^{2k}}\displaystyle\sum_{m=0}^{\lfloor\frac{k}{2}\rfloor}(-1)^m\binom{2k-2m}{k}\dfrac{(2bt)^{2m}}{(2m)!}$ $-\dfrac{\cos bt}{(2b)^{2k}}\displaystyle\sum_{m=0}^{\lfloor\frac{k-1}{2}\rfloor}(-1)^m\binom{2k-2m-1}{k}\dfrac{(2bt)^{2m+1}}{(2m+1)!}$	$\dfrac{b}{(s^2+b^2)^{k+1}}$	549
2. $\dfrac{2bt\sin bt}{k\cdot(2b)^{2k}}\displaystyle\sum_{m=0}^{\lfloor\frac{k-1}{2}\rfloor}(-1)^m\binom{2k-2m-2}{k-1}\dfrac{(2bt)^{2m}}{(2m)!}$ $-\dfrac{2bt\cos bt}{k\cdot(2b)^{2k}}\displaystyle\sum_{m=0}^{\lfloor\frac{k-2}{2}\rfloor}(-1)^m\binom{2k-2m-3}{k-1}\dfrac{(2bt)^{2m+1}}{(2m+1)!}$	$\dfrac{s}{(s^2+b^2)^{k+1}}$	549
Laplace transforms involving the quadratic $s^2 - b^2$		
3. $\dfrac{(-1)^k}{2^{2k+1}k!}\displaystyle\sum_{n=0}^{k}\dfrac{(2k-n)!}{n!(k-n)!}\left((-2t)^n\mathrm{e}^t - (2t)^n\mathrm{e}^{-t}\right)$	$\dfrac{1}{(s^2-1)^{k+1}}$	553
4. $\dfrac{(-1)^k}{2^{2k+1}k!}\displaystyle\sum_{n=1}^{k}\dfrac{(2k-n-1)!}{(n-1)!(k-n)!}\left((-2t)^n\mathrm{e}^t + (2t)^n\mathrm{e}^{-t}\right)$	$\dfrac{s}{(s^2-1)^{k+1}}$	553

C.2 Convolutions

Table C.7 Convolutions

	$f(t)$	$g(t)$	$(f*g)(t)$	Page
1.	$f(t)$	$g(t)$	$f*g(t) = \int_0^t f(u)g(t-u)\,\mathrm{d}u$	187
2.	1	$g(t)$	$\int_0^t g(\tau)\,\mathrm{d}\tau$	190
3.	t^m	t^n	$\dfrac{m!n!}{(m+n+1)!}t^{m+n+1}$	193
4.	t	$\sin at$	$\dfrac{at-\sin at}{a^2}$	
5.	t^2	$\sin at$	$\dfrac{2}{a^3}\left(\cos at - (1 - \frac{a^2t^2}{2})\right)$	
6.	t	$\cos at$	$\dfrac{1-\cos at}{a^2}$	
7.	t^2	$\cos at$	$\dfrac{2}{a^3}(at-\sin at)$	

(continued)

Table C.7 (continued)

	$f(t)$	$g(t)$	$(f * g)(t)$	Page
8.	t	e^{at}	$\dfrac{e^{at} - (1 + at)}{a^2}$	
9.	t^2	e^{at}	$\dfrac{2}{a^3}(e^{at} - (a + at + \frac{a^2 t^2}{2}))$	
10.	e^{at}	e^{bt}	$\dfrac{1}{b - a}(e^{bt} - e^{at}) \quad a \neq b$	192
11.	e^{at}	e^{at}	$t e^{at}$	192
12.	e^{at}	$\sin bt$	$\dfrac{1}{a^2 + b^2}(be^{at} - b\cos bt - a\sin bt)$	195
13.	e^{at}	$\cos bt$	$\dfrac{1}{a^2 + b^2}(ae^{at} - a\cos bt + b\sin bt)$	195
14.	$\sin at$	$\sin bt$	$\dfrac{1}{b^2 - a^2}(b\sin at - a\sin bt) \quad a \neq b$	195
15.	$\sin at$	$\sin at$	$\dfrac{1}{2a}(\sin at - at\cos at)$	195
16.	$\sin at$	$\cos bt$	$\dfrac{1}{b^2 - a^2}(a\cos at - a\cos bt) \quad a \neq b$	195
17.	$\sin at$	$\cos at$	$\dfrac{1}{2}t\sin at$	195
18.	$\cos at$	$\cos bt$	$\dfrac{1}{a^2 - b^2}(a\sin at - b\sin bt) \quad a \neq b$	195
19.	$\cos at$	$\cos at$	$\dfrac{1}{2a}(at\cos at + \sin at)$	195
20.	f	$\delta_c(t)$	$f(t - c)h(t - c)$	444
21.	f	$\delta_0(t)$	$f(t)$	445

Symbol Index

W.A. Adkins and M.G. Davidson, *Ordinary Differential Equations*,
Undergraduate Texts in Mathematics, DOI 10.1007/978-1-4614-3618-8,
© Springer Science+Business Media New York 2012

Index

W.A. Adkins and M.G. Davidson, *Ordinary Differential Equations*,
Undergraduate Texts in Mathematics, DOI 10.1007/978-1-4614-3618-8,
© Springer Science+Business Media New York 2012

Printed in the United States
By Bookmasters